Business Mathematics

FOURTEENTH EDITION

Gary Clendenen
Siena College

Stanley A. Salzman
American River College

330 Hudson Street, NY, NY 10013

Director, Portfolio Management: Michael Hirsch
Courseware Portfolio Manager: Matthew Summers
Content Producer: Lauren Morse
Managing Producer: Scott Disanno
Producer: Jonathan Wooding
Manager, Courseware QA: Mary Durnwald
Manager, Content Development: Eric Gregg
Product Marketing Manager: Fiona Murray
Marketing Assistant: Hanna Lafferty
Senior Author Support/Technology Specialist: Joe Vetere
Manager, Rights and Permissions: Gina Cheselka
Manufacturing Buyer: Carol Melville, LSC Communications
Associate Director of Design: Blair Brown
Program Design Lead: Barbara Atkinson
Text Design, Production Coordination, Composition, and Illustrations: iEnergizer Aptara®, Ltd.
Cover Image: Stuart Monk/Shutterstock

Copyright © 2018, 2015, 2012 by Pearson Education, Inc. All Rights Reserved. Printed in the United States of America. This publication is protected by copyright, and permission should be obtained from the publisher prior to any prohibited reproduction, storage in a retrieval system, or transmission in any form or by any means, electronic, mechanical, photocopying, recording, or otherwise. For information regarding permissions, request forms and the appropriate contacts within the Pearson Education Global Rights & Permissions department, please visit www.pearsoned.com/permissions/.

Attributions of third-party content appear on page xxx, which constitutes an extension of this copyright page.

PEARSON, ALWAYS LEARNING, and MyLab Math™ are exclusive trademarks owned by Pearson Education, Inc. or its affiliates in the U.S. and/or other countries.

Unless otherwise indicated herein, any third-party trademarks that may appear in this work are the property of their respective owners and any references to third-party trademarks, logos or other trade dress are for demonstrative or descriptive purposes only. Such references are not intended to imply any sponsorship, endorsement, authorization, or promotion of Pearson's products by the owners of such marks, or any relationship between the owner and Pearson Education, Inc. or its affiliates, authors, licensees or distributors.

Library of Congress Cataloging-in-Publication Data
Names: Clendenen, Gary, author. | Salzman, Stanley A., author.
Title: Business mathematics / Gary Clendenen, Siena College, Stanley A. Salzman, American River College.
Description: 14th edition. | Boston : Pearson, [2018] | Includes index.
Identifiers: LCCN 2017025606 | ISBN 9780134693323 (paperback) | ISBN 0134693329 (paperback)
Subjects: LCSH: Business mathematics. | Business mathematics—Programmed instruction.
Classification: LCC HF5691 .M465 2018 | DDC 650.01/513—dc23 LC record available at https://lccn.loc.gov/2017025606

2 18

ISBN 10:   0-13-469332-9
ISBN 13: 978-0-13-469332-3

Contents

Preface vi
The Business Mathematics, 14th Edition, Learning System vii
Learning Tips for Students xiv
Business Mathematics Pretest xv
Index of Applications xvi

Chapter 1
Whole Numbers and Decimals 1

1.1 Whole Numbers 2
1.2 Application Problems 14
1.3 Decimal Numbers 20
1.4 Addition and Subtraction of Decimals 24
1.5 Multiplication and Division of Decimals 28
   Chapter 1 Quick Review 35
   Chapter Terms 35
   Case Study: Cost of Getting Married 37
   Case in Point Summary Exercise: Subway 38
   Chapter 1 Test 39

Chapter 2
Fractions 41

2.1 Fractions 42
2.2 Addition and Subtraction of Fractions 48
2.3 Addition and Subtraction of Mixed Numbers 56
2.4 Multiplication and Division of Fractions 60
2.5 Converting Decimals to Fractions and Fractions to Decimals 68
   Chapter 2 Quick Review 72
   Chapter Terms 72
   Case Study: Operating Expenses at Woodline Moldings and Trim 74
   Case in Point Summary Exercise: The Home Depot 75
   Chapter 2 Test 76

Chapter 3
Percents 79

3.1 Writing Decimals and Fractions as Percents 80
3.2 Finding Part 87
3.3 Finding Base 94
   Supplementary Application Exercises on Base and Part 98
3.4 Finding Rate 100
   Supplementary Application Exercises on Base, Rate, and Part 104
3.5 Increase and Decrease Problems 108
   Chapter 3 Quick Review 115
   Chapter Terms 115
   Case Study: Self Employed Retirement Plan 117
   Case in Point Summary Exercise: Century 21 118
   Chapter 3 Test 119

Chapter 4
Equations and Formulas 121

4.1 Solving Equations 122
4.2 Applications of Equations 130
4.3 Business Formulas 139
4.4 Ratio and Proportion 148
   Chapter 4 Quick Review 157
   Chapter Terms 157
   Case Study: Forecasting Sales at Alcorn's Boutique 159
   Case in Point Summary Exercise: General Motors 160
   Chapter 4 Test 162
   Chapters 1–4 Cumulative Review 166

Chapter 5
Bank Services 171

5.1 Banking, Checking Accounts, and Check Registers 172
5.2 Checking Services and Credit-Card Transactions 182
5.3 Bank Statement Reconciliation 188
   Chapter 5 Quick Review 197
   Chapter Terms 197
   Case Study: Banking Activities of a Retailer 199
   Case in Point Summary Exercise: Jackson & Perkins 200
   Chapter 5 Test 202

Chapter 6
Payroll 204

6.1 Gross Earnings: Wages and Salaries 205
6.2 Gross Earnings: Piecework and Commissions 214
6.3 Social Security, Medicare, and Other Taxes 222
6.4 Income Tax Withholding 228
   Chapter 6 Quick Review 239
   Chapter Terms 239
   Case Study: Payroll: Finding Your Take-Home Pay 242
   Case in Point Summary Exercise: Payroll at Starbucks 243
   Chapter 6 Test 244

Chapter 7
Mathematics of Buying 246

7.1 Invoices and Trade Discounts 247
7.2 Series Discounts and Single Discount Equivalents 257

7.3 Cash Discounts: Ordinary Dating Methods 261
7.4 Cash Discounts: Other Dating Methods 267
    Chapter 7 Quick Review 274
    Chapter Terms 274
    Case Study: George Foreman 276
    Case in Point Summary Exercise: Discounts at Bed Bath & Beyond 277
    Chapter 7 Test 278

Chapter 8
Mathematics of Selling 280

8.1 Markup on Cost 281
8.2 Markup on Selling Price 288
    Supplementary Application Exercises on Markup 296
8.3 Markdown 298
8.4 Turnover and Valuation of Inventory 304
    Chapter 8 Quick Review 313
    Chapter Terms 313
    Case Study: Markdown: Reducing Prices to Move Merchandise 317
    Case in Point Summary Exercise: Recreational Equipment Inc. (REI) 318
    Chapter 8 Test 319
    Chapters 5–8 Cumulative Review 321

Chapter 9
Simple Interest 323

9.1 Basics of Simple Interest 324
9.2 Finding Principal, Rate, and Time 335
9.3 Simple Discount Notes 343
9.4 Discounting a Note Before Maturity 352
    Supplementary Application Exercises on Simple Interest and Simple Discount 360
    Chapter 9 Quick Review 364
    Chapter Terms 364
    Case Study: Banking in a Global World: How Do Large Banks Make Money? 368
    Case in Point Summary Exercise: Apple, Inc. 369
    Chapter 9 Test 370

Chapter 10
Compound Interest and Inflation 372

10.1 Compound Interest 373
10.2 Interest-Bearing Bank Accounts and Inflation 384
10.3 Present Value and Future Value 394
    Chapter 10 Quick Review 399
    Chapter Terms 399
    Case Study: Valuing a Chain of McDonald's Restaurants 401
    Case in Point Summary Exercise: Bank of America 402
    Chapter 10 Test 403
    Chapters 9–10 Cumulative Review 405

Chapter 11
Annuities, Stocks, and Bonds 407

11.1 Annuities and Retirement Accounts 408
11.2 Present Value of an Ordinary Annuity 416
11.3 Sinking Funds (Finding Annuity Payments) 424
    Supplementary Application Exercises on Annuities and Sinking Funds 431
11.4 Stocks and Mutual Funds 433
11.5 Bonds 443
    Chapter 11 Quick Review 449
    Chapter Terms 449
    Case Study: Financial Planning 452
    Case in Point Summary Exercise 453
    Chapter 11 Test 454

Chapter 12
Business and Consumer Loans 456

12.1 Open-End Credit and Charge Cards 457
12.2 Installment Loans 467
12.3 Early Payoffs of Loans 475
12.4 Personal Property Loans 482
12.5 Real Estate Loans 490
    Chapter 12 Quick Review 497
    Chapter Terms 497
    Case Study: Consolidating Loans 501
    Case in Point Summary Exercise: Underwater on a Home 503
    Chapter 12 Test 505
    Chapters 11–12 Cumulative Review 507

Chapter 13
Taxes and Insurance 510

13.1 Property Tax 511
13.2 Personal Income Tax 518
13.3 Fire Insurance 532
13.4 Motor-Vehicle Insurance 541
13.5 Life Insurance 549
    Chapter 13 Quick Review 556
    Chapter Terms 556
    Case Study: Financial Planning for Property Taxes and Insurance 559
    Case in Point Summary Exercise: Mattel Inc.—Taxes and Insurance 560
    Chapter 13 Test 562

Chapter 14
Depreciation 564

14.1 Straight-Line Method 565
14.2 Declining-Balance Method 573
14.3 Sum-of-the-Years'-Digits Method 580
    Supplementary Application Exercises on Depreciation 587

14.4 Units-of-Production Method 591
14.5 Modified Accelerated Cost Recovery System 596
    Chapter 14 Quick Review 604
    Chapter Terms 604
    Case Study: Comparing Depreciation Methods 606
    Case in Point Summary Exercise: Ford Motor Company 607
    Chapter 14 Test 608

Chapter 15
Financial Statements and Ratios 610

15.1 The Income Statement 611
15.2 Analyzing the Income Statement 616
15.3 The Balance Sheet 623
15.4 Analyzing the Balance Sheet 627
    Chapter 15 Quick Review 635
    Chapter Terms 635
    Case Study: Bicycle Shop 638
    Case in Point Summary Exercise: Apple, Inc. 640
    Chapter 15 Test 642

Chapter 16
Budgeting and Business Statistics 644

16.1 Planning and Budgeting 645
16.2 Frequency Distributions and Graphs 654
16.3 Mean, Median, and Mode 666
    Chapter 16 Quick Review 674
    Chapter Terms 674
    Case Study: Watching a Small Business Grow 677
    Case in Point Summary Exercise: Bobby Flay 678
    Chapter 16 Test 679

Appendix A
The Metric System A-1

Appendix B
Basic Calculators B-1

Appendix C
Financial Calculators C-1

Appendix D
Exponents and the Order of Operations D-1

Appendix E
Graphing Equations E-1

Answers to Selected Exercises AN-1
Glossary G-1
Index I-1
Photo Credits P-1

Preface

FROM THE AUTHORS

The fourteenth edition of Business Mathematics has been significantly revised to update the text, improve the discussions, and make the material more relevant to students. The focus on real-world applications has been sharpened. A different well-known company is highlighted at the beginning of each chapter and used throughout the chapter in examples, discussions, exercises, and a case at the end. Each chapter ends with two business application cases that will help students integrate concepts from the chapter. This edition is full of data, examples, graphs, photographs, and news clippings that will help students understand the relevance of the material as it teaches them to interpret data and information. A global perspective is emphasized through examples and exercises that highlight issues in other countries.

This book shows students how to use math to solve a wide variety of problems in business and also within families. Primary goals are to develop students' understanding of business, increase their ability to figure out how to work many different kinds of business problems, and motivate them using many actual business applications to which they can relate.

In this sense, we seek to develop a level of business "intuition" by having them work through the integrative cases, a wide-range of application exercises, writing and investigative questions, and discussions about current and relevant data. Additionally, we also seek to help students develop intuition related to business by discussing topics such as global supply chains, inventory, recessions, debt, etc. These topics are widely discussed in advanced courses in four year programs at colleges and universities throughout the world.

The new edition reflects the extensive business and teaching experience of the authors, college faculty who have previously worked in and owned businesses. It also incorporates ideas for improvement from reviewers nationwide as well as students who have taken the course. We focus on providing solid, practical, and up-to-date coverage of business mathematics topics beginning with a brief review of basic mathematics, and go on to introduce key business topics, such as bank services, payroll, business discounts and markups, simple and compound interest, stocks and bonds, consumer loans, taxes and insurance, depreciation, financial statements, and business statistics. A new section called Planning and Budgeting has been added as Section 16.1. It both emphasis the value of planning and budgeting in a business and in a family. Appendices expand material covered in the book to include the use of financial calculators, additional material on algebra related to exponents and order of operations, and the construction and use of graphs, a vitally important topic in today's world.

The traditional concept of learning has evolved based on knowledge that students learn in a variety of ways and that many classes are at least partly taught online or in labs. To support student learning in this multidimensional world, we have developed an outstanding supplemental learning package of print and digital products including the industry-leading MyLab Math. Numerous studies have shown that MyLab Math can greatly increase student learning and retention by presenting material in a variety of formats to suit all types of student learning styles.

Our state-of-the-art supplements package includes revised video lectures, new Case-in-Point videos, an enhanced PowerPoint package, student's solutions manual, an extensive instructor's manual, printed quick reference tables, and a wealth of online resources for instructors and students including MathXL online and MyLab Math. We hope this text and package satisfies all of your classroom needs. Please feel free to contact us with any questions or concerns. Use "Business Math" in the subject line.

Gary Clendenen
gclendenen@yahoo.com
Stanley Salzman
stan.salzman@comcast.net

The Business Mathematics, 14th Edition, Learning System

This textbook has evolved over the years as many thousands of students and hundreds of instructors have used the book and told us what works and what doesn't. *Business Mathematics*, 14th edition, Learning System is the result of this process of refinement that informs both the printed textbook and our MathXL and MyLab Math applications online. The goal of this textbook is for students to develop the computational skills they will need to be successful in the world of business along with a better understanding of business concepts and situations that require a mathematical solution. Each chapter is set up to teach a math concept and its applications in the following pattern:

1. A **"Case in Point" company profile** introduces the student to a company and a situation that requires math calculations.

   > A feature titled **Learning Catalytics** at the beginning of each chapter can be used to either introduce the topic quickly to students or test whether they have read the material.

2. A **clear explanation** of the math concept is presented, followed by **examples with detailed solutions**.
3. Students immediately apply the math concept to a similar problem in a **Quick-Check problem** to test their understanding.
4. **Solution steps**, detailing how to solve problems, are summarized in a shaded box.
5. **Quick Tips** provide students with helpful tips and cautions.
6. **Business applications** are found in examples, exercises, cases and discussion, and features such as Numbers in the News and newspaper clippings providing business and economic information.
7. An **Exercise Set** follows each section of the book providing a wealth of practice opportunities to develop computational skills. The exercises are paired, graded from simple to more complex, and conclude with numerous titled application word problems. Each type of exercise is preceded with a **Quick Start** worked example to help get students started.
8. **Additional Problem Sets** and **Supplementary Exercises** are embedded in select chapters for topics that students find difficult and typically require additional work.
9. A **Quick Review** section at the end of the chapter presents students with an overview of the math concepts covered in the chapter.
10. Two case studies require students to use math concepts to solve business problems in real companies. The first **Case Study** is a shorter case application, while the second **Case in Point Summary Exercise** revisits the chapter opening company with a more in-depth application. Both cases end with Discussion or Investigate questions that encourage further thinking.
11. Finally, a chapter concluding **Test** allows students to gauge their mastery of all chapter concepts and applications.
12. **Cumulative Review Problem Sets** appear every 2–4 chapters. These problems cover all math concepts covered in the preceding chapters and help students retain math concepts throughout the course.

Chapter	Case-in-Point Companies
1	Subway
2	Home Depot
3	Century 21
4	General Motors
5	Rose Gardens
6	Starbucks
7	Bed, Bath & Beyond
8	REI (sporting goods)
9	Apple, Inc.
10	Bank of America
11	Mayo Clinic
12	Citigroup
13	The Doll House (entrepreneur)
14	Capital Curb & Concrete
15	Apple, Inc.
16	Bev's Deli

BUILDING CALCULATOR SKILLS

This text provides the following resources to help students build calculator skills:

Calculator Solutions Calculator solutions, identified with the calculator symbol 🖩, appear after selected examples. These solutions show students the keystrokes needed to solve the Example.

Basic Calculator Instruction in Appendix B presents detailed coverage of basic calculators.

Financial Calculator Instruction in Appendix C reviews the basic functions of financial calculators using present value and future value. The financial calculator solutions are shown in shaded boxes along with the 🖩 for some examples.

NEW CONTENT HIGHLIGHTS

The fourteenth edition has far more changes than is possible to list, but here are many important changes listed by chapter(s).

- Chapters 1 through 4 have been revised and examples updated. If desired, the material in Chapter 4 on Equations and Formulas can be supplemented with additional material on Exponents and Order of Operations found in Appendix D and Graphing Equations found in Appendix E.

- Chapter 5 (**Bank Services**) has been completely revised to better align with today's reality. Students will begin to learn how banks operate in this chapter, which is then reinforced in Chapters 7, 8, 9, and 10. The ever-increasing role of Internet and mobile banking is emphasized.

- Chapter 6 (**Payroll**) has been extensively updated and includes the most recent information on Social Security, Medicare, and income tax withholding. Graphs and tables in the chapter show cost of living in different cities, average income and unemployment rate by level of education, and median income for a large number and wide range of careers.

- Chapter 8 (**Mathematics of Selling**) has been revised to better align with business practices, with slightly more focus on competition and the need to sometimes discount dated merchandise to move it out of inventory. Graphs included show annual sales at ten retail giants, percent of people who feel euphoric after making purchases of certain items (which helps drive sales), and cost comparisons across countries.

- Chapter 9 (**Simple Interest**) has been expanded and helps students understand the importance of interest rates. It includes a graph that shows how interest rates on consumer loans have greatly varied through the years and a discussion of how the government manipulates interest rates to help the economy grow or slow down depending on what it thinks the economy needs.

- Chapter 10 (**Compound Interest and Inflation**) discusses at length the benefits of compound interest over time including through the use of company-funded retirement plans. Inflation is defined and examples and exercises emphasize the effect of inflation on a family's income. Deflation is also described and discussed in terms of the Great Depression. The equation for finding compound interest is slightly more prominent in the chapter, along with a discussion of how to use the equation for those interested in a more algebraic approach.

- Chapter 11 (**Annuities, Stocks, and Bonds**) emphasizes the value of compound interest in long-term savings both for individuals using corporate-sponsored retirement plans and for businesses with a large expected expense coming up at some point. All of the material on stocks and bonds has been updated, and is discussed from the perspective of both corporations raising funds and investors.

- Chapter 12 (**Business and Consumer Loans**) has been extensively revised, and discusses the importance of loans for families, businesses, and the federal government. The chapter discusses many topics of interest to students: credit-card loans, student loans, FICO scores, consumer loans, business loans, and real estate loans. It highlights strategies for coping with debt by discussing refinancing and through a case highlighting a family that is "under water" or owes more on their home than it is worth.

- Chapter 13 (**Taxes and Insurance**) discusses taxes and insurance in terms of an entrepreneur. It gives students a sense of the tax and insurance complexity (property taxes, sales taxes, income taxes, payroll taxes, building insurance, and automobile insurance) that families and businesses face.

- Chapter 15 (**Financial Statements and Ratios**) discusses financial statements and ratios in terms of a company loved by students: Apple, Inc. The discussion on ratios has been expanded and examples are shown of other companies as well.

- Chapter 16 (**Budgeting and Business Statistics**) includes a NEW SECTION on planning and budgeting. First it discusses planning and budgeting for a family, with its known recurring expenses, and includes a discussion on how to plan for and deal with unexpected expenses. It then goes on to discuss planning and budgeting for a company called Bev's Deli, which is highlighted through the chapter. The inclusion of this new section will help students synthesize many topics from across the course, including: choosing a career and level of education to work toward, controlling costs, planning for expected and unexpected expenses, thinking long-term, managing debt, reflecting on the costs of insurance and taxes, and saving or investing. The remainder of the chapter discusses frequency distributions and graphs as well as measures of central tendency (mean, median, and mode). The material on graphs can be enhanced using Appendix E (Graphing Equations) for those interested.

The Business Mathematics, 14th Edition, Learning System ix

Get the Most out of MyLab Math for Business
Mathematics by Clendenen and Salzman

Used by over 2 million students a year, MyLab™ Math is the world's leading online program for teaching and learning mathematics. MyLab Math delivers assessment, tutorials, and multimedia resources that provide engaging and personalized experiences for each student, so learning can happen in any environment. Each course is developed to accompany Pearson's best-selling content, authored by thought leaders across the math curriculum, and can be easily customized to fit any course format. (Access code required.)

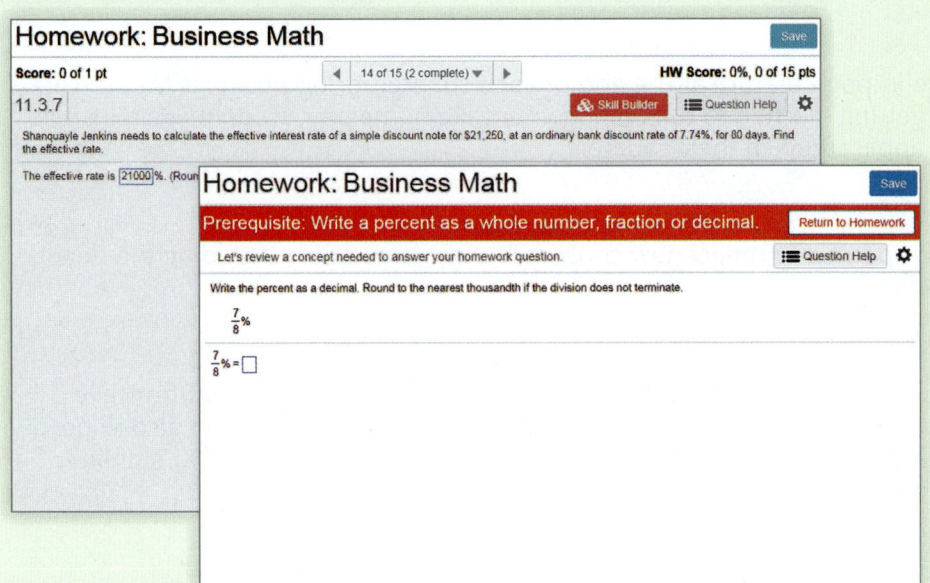

New! Skill Builder offers adaptive practice that is designed to increase students' ability to complete their assignments. By monitoring student performance on their homework, Skill Builder adapts to each student's needs and provides just-in-time, in-assignment practice to help them improve their proficiency of key learning objectives - including prerequisite skills if needed.

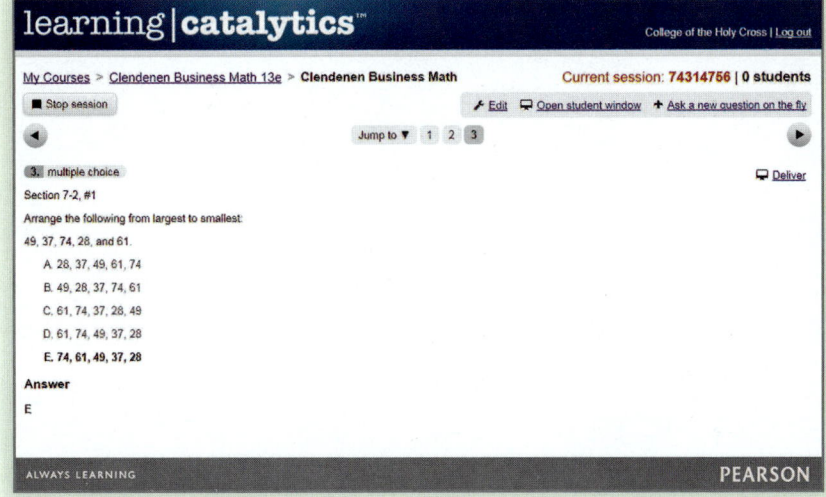

Learning Catalytics

Generate class discussion, guide your lecture, and promote peer-to-peer learning with real-time analytics. MyLab™ Math now provides Learning Catalytics—an interactive student response tool that uses students' smartphones, tablets, or laptops to engage them in more sophisticated tasks and thinking. MyLab™ Math access required.

pearson.com/mylab/math

The Business Mathematics, 14th Edition, Learning System

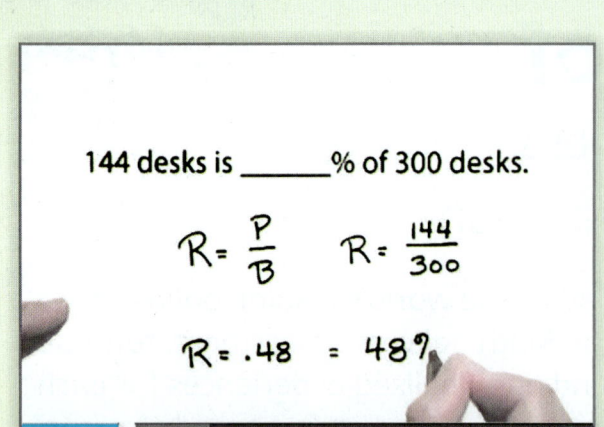

Updated Video Program
A variety of videos have been updated and added to the Clendenen Business Math course to walk students through concepts from every section of the text, giving them support when they need it - at home, in the lab, or on the go.

Case-in-point Videos
16 new videos, based off the Case-in-point feature at the end of each chapter, bring Business Math to life. From case studies on the cost of getting married to calculating your take home pay, students gain insight into the practical and day-to-day applications of their course.

Section Lecture Videos
Section Lecture Videos have been updated to reflect new content in the 14th edition, including the new section 16.1 on Planning & Budgeting.

MathXL® Online Course
With MathXL, instructors can create, edit, and assign online homework and tests using algorithmically generated exercises correlated at the objective level to *Business Mathematics*. Instructors can also import TestGen tests for added flexibility, and maintain records of all student work tracked in MathXL's online gradebook. (Access code required.)

Trade Application Library
Clendenen Business Math will be available with a library of MathXL applications focused on vocations and trades, allowing instructors to create assignments geared toward practical on-the-job applications.

pearson.com/mylab/math

Resources for Success

Instructor Resources

Instructor's Resource Manual
This manual contains suggestions for pacing the course and creating homework assignments. It discusses how to incorporate technology and how to structure project assignments. The manual also contains section-by-section suggestions for presenting lectures and for undertaking the explorations in the text.

PowerPoints
Available through **www.pearson.com** or inside your MyLab Math course, these fully editable lecture slides include definitions, key concepts, and examples for use in a lecture setting and are available for each section of the text.

Instructor's Solutions Manual
This free online manual includes complete solutions to the even-numbered exercises in the homework sections of the text.

TestGen
TestGen enables instructors to build, edit, print, and administer tests by using a computerized bank of questions developed to cover all the objectives of the text. TestGen is algorithmically based, allowing instructors to create multiple, but equivalent, versions of the same question or test with the click of a button. Instructors can also modify test-bank questions or add new questions. Tests can be printed or administered online. The software and test bank are available for free download from Pearson Education's online catalogue.

Student Resources

Student Solutions Manual
Fully worked solutions to odd-numbered exercises are available free online in MyLab© Math.

pearson.com/mylab/math

Acknowledgments

We would like to thank the many users of the thirteenth edition for their insightful observations and suggestions for improving this book. We also wish to express our appreciation and thanks to the following reviewers of this and previous editions for their contributions:

Joe Adamo, *Cazenovia College*
George Alexander, *Madison Area Technical College*
Cheryl Anderson, *Montana State University Billings*
Julia Angel, *North Arkansas College*
John Angeline, *Bucks County Community College*
Diana Baran, *Henry Ford College*
Yvonne Block, *College of Lake County*
Kathy Blondell, *St. Johns River Community College*
Scott Bryant, *Central Maine Community College*
Jesse Cecil, *College of the Siskiyous*
Carolyn Chapel, *Western Wisconsin Technical Institute*
Janet Ciccarelli, *Herkimer County Community College*
Vittoria Cosentino, *Metropolitan Community College*
Diallo Cummings, *Aiken Technical College*
Ron Deaton, *Grays Harbor College*
Elaine DiPerna, *Community College of Allegheny County—Allegheny*
Jacqueline Dlatt, *College of DuPage*
Dana Dye, *Gulf Coast State College*
James Grippe, *Central Maine Community College*
Perry Haan, *Tiffin University*
JoLynn Hightower, *Texas State Technical College*
Mysti Hobson, *Zane State College*
Chris Howell, *New Mexico Junior College*
Brandon Huff, *Lewis and Clark Community College*

Zafar D. Khan, *University of Virginia's College at Wise*
Amanda Kriesel, *Metropolitan Community College*
Jeanette Landin, *Empire College*
Young Jin Lee, *University of Wisconsin Green Bay*
Ping Lin, *California State University, Long Beach*
Chuck Lyons, *Hibbing Community College*
Krista Mahan, *Walla Walla Community College*
Carol Manigault, *Felician University*
Darin McGraw, *Washington County Community College*
Jennie Mitchell, *Saint Mary-of-the-Woods College*
Nikki Munden, *Lewis and Clark Community College*
Matt Njoku, *Montreat College*
Gary Rattray, *Central Maine Community College*
Frederick Reed, *Eastern New Mexico University–Ruidoso*
Bob Reese, *Illinois Valley Community College*
Randall Rinke, *Mercyhurst University*
Pam Rogers, *Spartanburg Community College*
Ellen Sawyer, *College of DuPage*
Farooq Sheikh, *SUNY Geneseo*
Robin Sirkis, *Delaware Technical and Community College—Owens*
Catherine Skura, *Sandhills Community College*
Sabrina Woodbery, *Guilford Technical Community College*

Our appreciation goes to Deana Richmond, and Leonda Clendenen, who checked all of the exercises and examples in the book for accuracy. We would also like to express our gratitude to our colleagues at American River College and Siena College who have helped us immeasurably with their support and encouragement.

The following individuals at Pearson Education had a large impact on this fourteenth edition of *Business Mathematics,* and we are grateful for their many efforts: Michael Hirsch, Editor in Chief; Matt Summers, Sponsoring Editor; and Lauren Morse, Content Producer. Thanks are due as well to iEnergizer Aptara®, Ltd., and Sherrill Redd in particular, for adeptly handling the production of this fourteenth edition.

As an author team, we are committed to providing the best possible text to help instructors teach and students succeed. As we continue to work toward this goal, we would welcome any comments or suggestions you might have via e-mail to gclendenen@yahoo.com. Please use "Business Math" in the subject line.

Gary Clendenen
Stanley A. Salzman

About the Authors

Gary Clendenen received bachelor's and master's degrees in mathematics before going into business for himself in the oil industry. He returned to academia and earned his Ph.D. in Business Management and has been a faculty member since then. His business experience includes working as an actuary for an insurance company and owning commercial real estate. He has published papers in numerous refereed journals and does volunteer work with several organizations. His hobbies include long bicycle rides, swimming, and reading on a wide variety of topics including history, economics, and natural resources. He has two sons and several grandchildren, and he and his wife "use miniature horses to encourage kids to read." Meet Hot Dog (brown) and Thor (gray).

Stanley A. Salzman has taught Business Math, Marketing, and Real Estate courses at American River College in Sacramento for 35 years. He says, "Some of my greatest moments in teaching have been seeing the look on the face of a student who understands a business math concept or idea for the first time." Stan and his wife have four children and eleven grandchildren. Stan likes outdoor activities, exercising, and collecting antique toy trains.

Learning Tips for Students

SUCCESS IN BUSINESS MATHEMATICS

This book focuses on using math to solve business problems. In the process, it will give you a better framework to understand business concepts related to payroll, supply chains, taxes, insurance, interest, debt, saving, financial statements, etc. It will teach you how to use math to solve a very wide variety of problems in business; yet it gives enough information to help you deal with personal financial issues. You will be using many of the concepts in this book throughout your life, so we encourage you to really understand the concepts well. Another goal of this book is to make you a better problem solver, which is what managers are looking for in the people they hire.

Studying business mathematics is different from studying subjects like English or history. The key to success is *regular practice*. This should not be surprising. After all, can you learn to ski or play a guitar without regular practice? The same is true for learning mathematics. Working problems nearly every day *is the key to becoming successful*. Here are some suggestions to help you succeed in business mathematics.

1. **Attend class regularly. Try to pay careful attention and take notes.**
2. **Ask questions in class.** It is not a sign of weakness, but of strength.
3. **Read the book carefully, maybe twice, and spend time using the online materials.** Studying each topic will help you solve the homework problems.
4. **Before doing your homework, look at the problems the teacher worked in class.** This will reinforce what you have learned.
5. **Read the section and review your notes before starting your homework.** Check your work against the answers in the back of the book. If you get a problem wrong and are unable to understand why, mark that problem and ask your instructor about it.
6. **Carefully organize your work.** This will help you think clearly and understand better.
7. **After you complete a homework assignment, quickly review the main concepts to reinforce what you have learned.**
8. **Use the chapter test at the end of each chapter as a practice test.** Carefully review any problem or concept you missed.
9. **Keep all quizzes and tests that are returned to you, and use them when you study for future tests and the final exam.** Correct any problems missed and look again at concepts related to that topic.
10. **Try not to worry if you do not understand a topic right away, and don't get stressed over tests.** No one understands all the concepts immediately! It takes time for every one of us to understand something new. If you understand the concepts well, have carefully looked at all examples both in the book and from your instructor, and done several exercises, you will probably do reasonably well on a test. Talk to your teacher if you have a lot of anxiety about tests.

Business Mathematics Pretest

This pretest will help you determine your areas of strength and weakness in the business mathematics presented in this book.

1. Round 5.46 to the nearest tenth.

2. Round $.064 to the nearest cent.

3. Round $399.49 to the nearest dollar.

4. Multiply: $\begin{array}{r} 7801 \\ \times\ 1758 \end{array}$

5. Divide: $35\overline{)11{,}032}$

6. Change $8\frac{7}{8}$ to an improper fraction.

7. Change $\frac{40}{26}$ to a mixed number.

8. Write $\frac{15}{21}$ in lowest terms.

9. Add: $\begin{array}{r} \frac{3}{4} \\ \frac{1}{2} \\ +\ \frac{7}{8} \end{array}$

10. Add: $\begin{array}{r} 2\frac{2}{3} \\ 7\frac{1}{4} \\ +\ 10\frac{1}{2} \end{array}$

11. Subtract: $\frac{3}{8} - \frac{7}{24}$

12. Subtract: $\begin{array}{r} 83\frac{3}{4} \\ -21\frac{2}{5} \end{array}$

13. Multiply: $\frac{3}{8} \times \frac{3}{5}$

14. Divide: $15\frac{1}{4} \div 5\frac{1}{8}$

15. Express .625 as a common fraction.

16. Express $\frac{3}{5}$ as a decimal.

17. Subtract: $\begin{array}{r} 598.316 \\ -79.839 \end{array}$

18. Multiply: $\begin{array}{r} 30.67 \\ \times\ 5.39 \end{array}$

19. Divide: $1.2\overline{)309.6}$

20. Express $\frac{7}{8}$ as a percent.

21. Intelnet spent 5.2% of its sales on advertising. If sales amounted to $864,250, what amount was spent on advertising?

22. What annual rate of return is needed to receive $930 in one year on an investment of $18,600?

23. Home Entertainment Systems offers an 80-inch LCD HDTV at a list price of $2459 less trade discounts of 20/10. What is the net cost?

24. A department head at Old Navy is paid $16.80 per hour with time and a half for all hours over 40 in a week. Find the employee's gross pay if she worked 43 hours in one week.

25. How long will it take an investment of $12,500 to earn $125 in interest at 4% per year? (*Hint:* Use Bankers Interest, i.e., assume 360 day year.)

26. An invoice from Collier Windows amounting to $20,250 is dated October 6 and offers terms of 3/10, n/30. If the invoice is paid on October 14, what amount is due?

27. Find the percent of markup based on selling price if some home exercise equipment costing $1584 is sold for $1980.

28. Find the single discount equivalent to a series discount of 30/20.

29. Using the straight-line method of depreciation, find the annual depreciation on a Bobcat loader that has a cost of $18,750, an estimated life of six years, and a scrap value of $750.

30. Whiting's Oak Furniture sells a dining room set for $1462.98 after deducting 26% from the original price. Find the original price.

xv

Index of Applications

A

Agriculture
Alligator hunting, 428
Christmas tree farm, 349
Commercial fertilizer, 66
Commercial fishing boats, 578–579, 603
Egg production, 13
Farmland prices, 112
Forestry operations, 487
Gardening, 53
Land area, 18
Land sale, 429
Landscaping, 479
Long-stemmed roses, 297
Pecan trees, 473
Race horses, 350
Trucks & sprayers, 349

Automotive/Transportation
Antilock brakes, 112
Automotive, 273
Auto parts startup, 146
Auto production, 155
Auto repair, 53, 138, 633
Auto sales, 113
Camaros and Mustangs, 98
Car emissions, 18
Car seats, 303
Chrome rims, 33
Company vehicles, 601
Diesel tractor, 589
Driving distractions, 91
Driving tests, 97
Fire truck, 430
Fuel consumption, 67
Garbage truck, 481
Harley Davidson, 138
Heavy-duty truck, 595
Hybrid Toyota, 33
Map reading, 154
Miles driven, 17
Motorcycles, 303, 516
Motorcycle safety, 104
Mustang, 145
New Toyota, 287
Petroleum transport, 53
Police cruisers, 265
Shuttle van, 588
Ski boat, 473
Tesla, 92
Toyota, 473
Tractor parts, 287
Tractor purchase, 358, 473, 487
Trailer load, 58
Transmission, 54
Transmission repair, 633
Truck accessories, 303
Vehicle depreciation, 571
Weighing freight, 18
Wheels with bling, 272
Yard Maintenance, 19

B

Banking
Accumulating $1,000,000, 428
Amortizing a loan, 488
Amount due, 273
Amount owed the IRS, 237
Appliance repair loan, 488
Bad credit history, 350
Bank balances, 180–181
Bank loan, 340
Benefit increase, 106
Budgeting, 106
Capital improvement, 333
Checking account records, 27
Check processing, 18
Commission with returns, 220
Completing check stubs, 179
Compound interest, 391
Corporate finance, 333, 382
Corporate savings, 392
Credit-card balance, 464–466
Credit-card deposits, 185–186
Credit union, 381
Emergency cash, 392
Financing college expenses, 397
Finding interest, 146
Finding time, 146
Home loan, 496
Inflation and retirement, 392
Inheritance, 147, 392
Interest earnings, 341
Interest rate, 341
International finance, 350, 381, 382
Loan amount, 147
Loan collateral, 391
Loan qualification, 97
Loans between banks, 333
Loans to minorities, 114
Loan to an uncle, 146
Maintaining bank records, 178–179
Maturity value, 146
Maximizing profit, 382
Partial invoice payment, 273
Partial payment, 479
Payment due, 273
Penalty on late payment, 341
Poor credit, 349
Promissory note, 341
Putting up collateral, 392
Reconciling checking accounts, 193–196
Retirement account, 98, 341
Retirement funds, 448
Retirement income, 392
Retirement planning, 415
Salary plus commission, 221
Saving for a home, 415
Saving for retirement, 340
Savings, 146, 381
Savings account, 391
Short-term savings, 340
Student loan, 146
Time of deposit, 341, 342, 391
Time or rate?, 382
United Kingdom, 381
Variable-commission payment, 220
Writing a will, 441

Business
Abbreviations on invoices, 253
Advertising expenses, 102, 103
Automotive supplies, 296
Barge depreciation, 572
Battery store, 358
Best Buy, 213
Book publishing, 664
Bridal shop, 145
Business expansion, 398
Business fixtures, 571
Business ownership, 93
Business safe, 588
Cadillac dealer, 632
Calculation gross earnings, 32–33
Call center, 19
Catering company, 265
Clothing shop, 154
Clothing store, 144
Coffee shop, 603, 620
Commercial carpeting, 66
Comparing discounts, 260
Convenience store, 517
Corporate profits, 91
Cost after markdown, 104
As of dating, 266
Dental-supply company, 615
Discount dates, 272
Distribution center, 587
Drilling equipment, 602
Electrical supplies, 265
Entrepreneur, E-12
Evaluating inventory, 312
Expanding manufacturing operations, 398
Finding discount dates, 266
Flower shop, 620
Food inflation, 112
Gift shop, 614
Global trade, 350
Grocery chain, 625
Grocery store, 423
Guitar shop, 620
Hardware store, 349
Hotel room costs, 18
Ice cream shop, 614
International business, 333
International shipments, 137
Inventory, 332, 480
Inventory purchase, 342
Juice company, 632
Lawsuit, 349
Luxury hotels, 18
Managerial earnings, 33
Natural-foods store, 54
Netflix, 112
New product failure, 92
New showroom, 429
Nike, 113
Oil profits, 155
Opening a restaurant, 487
Paint store, 334
Paper products manufacturing, 113
Partial invoice payment, 272
Partnership profits, 154
Print shop, 391
Product purchases, 10
Quality, E-11
Retail giants, 13
Ship building, 137
Shopping center, 603
Soft-drink bottling, 590
Spray-paint inventory, 311
Stock turnover at cost, 310
Stock turnover at retail, 310
Stock value, 114
Using invoices, 252–253
Value of a business, 398
Walmart Supercenter, 516
Women in business, 11

Index of Applications xvii

Business equipment
  Business signage, 588
  Canning machine, 473
  Car-wash machinery, 589
  Commercial fishing boats, 578–579, 603
  Commercial freezer, 585
  Commercial tile, 586
  Communication equipment, 297
  Company vehicles, 601
  Deep fryer, 595
  Dental office furniture, 603
  Depreciating equipment, 572
  Depreciating machinery, 572
  Depreciating office equipment, 584
  Double-pane windows, 297
  Drilling rig, 572
  Electronic equipment, 488
  English soccer equipment, 272
  Engraving, 587
  Factory, 584
  Forklift depreciation, 585
  George Foreman grill, 266
  Hospital equipment, 586
  Industrial forklift, 589
  Jewelry display cases, 588
  Kitchen equipment, 19
  Laboratory equipment, 571
  Machinery depreciation, 570
  Oak desk, 302
  Printer, 487
  Refrigerated display case, 590
  Scuba equipment, 489
  Storage tank, 601
  Surplus-equipment auction, 113
  Woodworking machinery, 587
  X-ray equipment, 428

C

Construction
  Airplane hangar, 538
  Asphalt crumb, 481
  Cabinet installation, 53, 54
  Commercial building, 137, 430, 538
  Concrete footings, 67
  Construction power tools, 579
  Conveyor system, 578
  Delivering concrete, 59
  Drilling equipment, 602
  Elderly housing, 496
  Financing construction, 358
  Finish carpentry, 54
  Forklift depreciation, 585
  Home construction, 342
  Landscape equipment, 586, 587
  New roof, 430
  Office complex, 516
  Parking lot fencing, 59
  Perimeter of fencing, 55
  Remodeling, 472, 480
  Restaurant kitchen, 430
  Road paving, 333
  Rock crusher, 357
  Security fencing, 59
  Stainless steel grill, 260
  Theater renovation, 19
  Triplex, 539
  Weather stripping, 67
  Window installation, 58
  Yacht construction, 631

D

Domestic
  Electricity rates, 65
  Fabric, 33
  Home beverage fountains, 255
  Household lubricant, 98
  Lights out, 104
  Material, 137
  Personal budgeting, 97
  Tailored clothing, 59

E

Education
  College bookstore, 185
  College enrollment, 96, 113
  College expenses, 113, 422, 423
  College textbooks, 18, 145
  Educational consultant, 213
  Exchange program, 137
  High school dropouts, 105
  Saving for college, 430
  School equipment, 579
  Student time management, 54
  Student union, 428
  Textbooks, 287
  University fees, 114
  Vocabulary, 103
Employment/Employee benefits
  Agricultural workers, 219
  Aiding disabled employees, 99
  Computer consultant, 479
  Earnings calculation, 65
  Educational consultant, 213
  Employee net pay, 236
  Employee population base, 96
  Female lawyers, 91
  Guaranteed hourly work, 220
  Heating-company representative, 238
  Hiring, 97
  Insurance office manager, 212
  Job cuts, 102
  Key employee insurance, 555
  Layoff alternative, 105
  Managerial earnings, 33
  Marketing representative, 237
  Nurses, 92
  Nursing, 103
  Office assistant, 212
  Part-time work, 58
  Payroll, 340
  Payroll deductions, 225, 226
  Piecework with overtime, 220
  Retail employment, 212
  Retirement, 415
  River raft manager, 238
  Self-employment deductions, 226, 397
  Starbucks district manager, 237
  Store manager, 213
  Women in the military, 102
  Women in the Navy, 91
  Working in China, 156
Entertainment/Sports
  Athletic shoes, 311
  Bowling equipment, 297
  Carnival, 144
  Casino, 349
  Competitive cyclist training, 17
  Dance shoes, 255
  Drums, 296
  Eating out, 112
  Elliptical trainer, 302
  Exercycle, 286
  Fishing boat, 18
  Fly-fishing, 296
  Gambling payback, 97
  Gaming, 145
  Golf clubs, 287, 296
  Home-workout equipment, 295
  Kayak, 302
  Lost overboard, 104
  Motorcyclists, 103
  Mountain bike, 297
  Movie projectors, 481
  Movies, 145
  Musical instruments, 144
  New auditorium, 428
  NY Yankees, 91
  Parachute jumps, 11
  Piano repair, 312
  Recreation equipment, 588
  Rock concert, 155
  Scuba diving, 428
  Scuba equipment, 489
  Scuba shoppe, 620
  Ski jackets, 286
  Snowboard, 296
  Sports complex, 429
  Sportswear, 297
  Sprint training, 54
  Super Bowl advertising, 92
  Swimming, 153
  Swimming pool pump, 295
  Table-tennis tables, 287
  Tanning salon, 103, 358
  Theater seating, 588
  T-shirts, 311
  Vacation, E-13
  Vacation mistakes, 104
  Virtual reality, 254
  Water skis, 137
  White water rafting, 12–13
  Wool socks, 311
  Yachts, 632
  Youth soccer, 19
Environment
  Alaska wilderness, 67
  Carbon dioxide, 18
  Car emissions, 18
  Climate change, 154
  Earthquake damage, 350
  Flooding, E-10
  Global warming, 104
  Hurricane Katrina, 12
  Recycling, 219
  Sea levels, 154
  Water scarcity, 107
  Winter-wheat planting, 113

F

Family
  Alimony, 26
  Child-care payments, 415
  Child support, 98
  Divorce settlement, 397
  Family budget, 92
  Family restaurant, 112
  Family size, 97
  Head of household, 531
  Married, 531
  Saving to buy a home, 106
Food service industry
  Bakery, 398
  Beef/turkey cost, 26
  Biscuits, 153
  Cake recipe, 54
  Campus vending machines, 12
  Chicken noodle soup, 106
  Coffee shop, 603, 620
  Fast-food restaurants, 586
  Food products, 271
  Frozen yogurt, 272
  Goat cheese, 474
  Health food, 333
  Hershey Kisses, 12
  Hershey mini chips, 12

xviii Index of Applications

Kitchen island, 255
McDonald's, E-10
Pizza, 145, 350
Restaurant tables, 590
Sales of health food, 154
Selling bananas, 295
Strawberry cheesecake, 66
Subway sandwiches, 17
Tiger food, 153
Wine, 296

G
General interest
Airport improvements, 429
Antiques, 147, 303
Apparel, 614
Bar soap, 91
Bed in a bag, 296
Beer consumption, 103
Blouses, 295
Christmas wreaths, 65
Crystal from Ireland, 272
Custom-made jewelry, 287
Dogs, 97
Engagement ring, 480
Fires, 114
Furniture, 663
Gold, 47
Iceberg volume, 155
Island area, 155
Japanese Yen, 156
Jewelry, 358
Lawsuit, 349
Liquid fertilizer, 255
Making jewelry, 67
Native-American jewelry, 55
Population forecasts, 105–106
Rare stamps, 429
Responder backpack, 286
Restaurant tips, 26
Sewer drain service, 586
Shampoo ingredients, 98
Social Security, 423
Songbird migration, 155
Government
American Chiropractic
  Association, 104
Biker helmet laws, 105
Criminal justice, 334
Disaster relief, 422
Gross national product, 11
Injury lawsuit, 422
Law enforcement, 340
Salvation Army, 12
Salvation army loss, 539
Total World War II veterans, 18
U.S. Paper money, 33
U.S. Patent recipients, 107
Voter registration, 97
War deaths, 664
World War II Veterans, 18

H
Healthcare
Alcohol level, 107
Blood cells, 153
Calories from fat, 98
Cholesterol levels, 98
Cone zone deaths, 114
Criminology lab, 212
Dental office furniture, 603
Diabetes, 91
Dispensing eye drops, 67

Flu pandemic of 1918, 98
Gambling with health, 71
Health food, 333
Health in a machine, 106
Medicine dose, 33
Motor Vehicle Accidents, 11
Overweight, 92
Oxygen supply, 632
Physically impaired, 18
Sick pet, 99
Side-impact collisions, 107
Smoking and cancer, 71
Smoking or nonsmoking, 97
Social Security and Medicare, 225–226
Weight-training equipment, 577

I
Insurance
20-pay life policy, 555
Adult auto insurance, 546–547
Bodily injury insurance, 547
Coinsured fire loss, 539
Fire insurance premium, 538
Fire loss, 539
Industrial building insurance, 538
Insurance company payment, 547–548
Key employee insurance, 555
Life insurance, 555
Major fire loss, 539
Medical expenses and property damage, 547
Multiple carriers, 539
Premium factors, 555
Property insurance, 98
Underinsured, 540
Universal life insurance, 555
Whole life insurance, 555
Youthful-operator auto insurance, 547
Youthful operator—no driver's training, 547

R
Real estate
Fourplex, 516
Home ownership, 96
Home prices, 105
Office building, 603
Pharmacy, 516
Radio station building, 516
Real estate commissions, 93
Real estate development, 153
Real estate fees, 33
Residential rental property, 602

S
Sales/Marketing
Auto sales, 113
Auto sales in China, 92
Boat purchase, 349
Cell phone, E-12
Computer sales, E-13
Condominium purchase, 496
Consumer internet sales, 93
Crayon sales, 12
Deli sales, 112
Department sales, 10
Furniture sales, 237
Hardware purchase, 255
Home purchase, 496
Hot tub purchase, 464
Inside sales, 212
iPad sales, 12
Jell-O sales, 12
Jetson aircraft sales, 626
Kitchen appliances, 271

Monthly sales, E-11
National home sales, 113
Nissan sales, 138
Nursing-care purchases, 254
Purchase of T-bills, 351, 359
Purchasing power, 392
Refrigerator, 473
Restaurant sales, 186–187
River-raft sales, 295
Sales of health food, 154
Selling a restaurant, 423
Selling bananas, 295
Smart phone discount, 104
Soda sales, 137
Subway sales, 26
Ticket sales, 153
Total sales, 106
Tractor purchase, 358, 473, 487
Travel-agency sales, 238
Tripod purchase, 255
Van purchase, 423
Vegetable sales, 138
Wholesale auto parts, 255
Science/Technology
Apple, Inc., E-11
Canning machine, 473
Carpet-cleaning equipment, 578
Ceiling fans, 296
Cell phones, E-12
Chip fabrication, 473
Communication equipment, 579
Computer chips, 145
Computer replacement, 422
Computer system, 473
Construction power tools, 579
Conveyor system, 578
Copy machines, 587
Device assembly, 53
Digital camera, 92
Digital thermometers, 296
Electric guitar, 472
Electronic analyzer, 578
Facebook users, 113
Flash drive, 297
Global Positioning Systems (GPS),
  272, 302
Graphic arts, 255
Laptops, 341, 603
Lawn mowing, 341, E-13
Measuring brass trim, 59
Metal lathe, 472
Notebook, 18
Outdoor lighting, 287
Printing, 481
Refrigerators, 144
Smart phones, E-14
Sound system, 137, 577
Surveillance cameras, 480
Video equipment, 589
Web design, 144, 333, 349, 481
Wind energy, 103
Wind turbine, 585
Stocks/Investments
Bond fund, 448
Bond purchase, 448
CDs or global stocks, 441
Fixed rate or stocks, 442
Investing, 382
Investing in bonds, 340
Investment, 381
Investment decision, 381
Mutual fund investing, 415
Stock price, E-12
Stock purchase, 441

T-bill and stock investing, 415
Trade discount, 256
Which investment?, 382

T

Taxes
  Comparing property tax rates, 517
  Federal withholding tax, 235–236
  Gasoline taxes, 91
  Income tax payment, 349
  Married—income tax, 530, 531
  Penalty on unpaid income tax, 334
  Penalty on unpaid property tax, 333
  Real estate taxes, 516
  Sales tax, 91, 112
  Sales-tax computation, 92
  Single—income tax, 530
  Social Security tax, 225
  State withholding tax, 236
  Taxes on home, 517
  Top tax rates, 665

Whole Numbers and Decimals

CHAPTER CONTENTS

1.1 Whole Numbers
1.2 Application Problems
1.3 Decimal Numbers
1.4 Addition and Subtraction of Decimals
1.5 Multiplication and Division of Decimals

CASE IN POINT

JESSICA FERNANDEZ worked part time for Subway when taking classes at a local community college, but she is now a manager who oversees 18 employees. She looks for employees who have a good work ethic, are honest and friendly, and can work with numbers. She uses numbers daily to schedule employees, compute sales, figure sales taxes, complete the payroll, and order inventory.

1.1 Whole Numbers

OBJECTIVES

1. Define whole numbers.
2. Round whole numbers.
3. Add whole numbers.
4. Round numbers to estimate an answer.
5. Subtract whole numbers.
6. Multiply whole numbers.
7. Multiply by omitting zeros.
8. Divide whole numbers.

CASE IN POINT To improve efficiency, Jessica Fernandez cross-trains each employee to do several tasks, including food preparation, cleanup, and operating the cash register. After watching an employee give a customer too much change for a second time, a frustrated Jessica Fernandez decided that any new hire had to pass a basic math test.

OBJECTIVE 1 Define whole numbers. The **decimal system** uses the ten one-place **digits**: 0, 1, 2, 3, 4, 5, 6, 7, 8, and 9. Combinations of these digits represent any number needed. The starting point of this system is the **decimal point (.)**. This section considers only the numbers made up of digits to the left of the decimal point—the **whole numbers**. The following diagram names the first fifteen places held by the digits to the left of the decimal point.

A researcher estimated that 581,075,900 smart phones will be sold in the fourth quarter of 2018. Use a **comma** to work with numbers this large. Starting at the decimal place and moving to the left, place a comma between each group of three digits. Although commas are always used when writing numbers, they are not shown on some calculators. Commas are not required for numbers with four digits such as 8475.

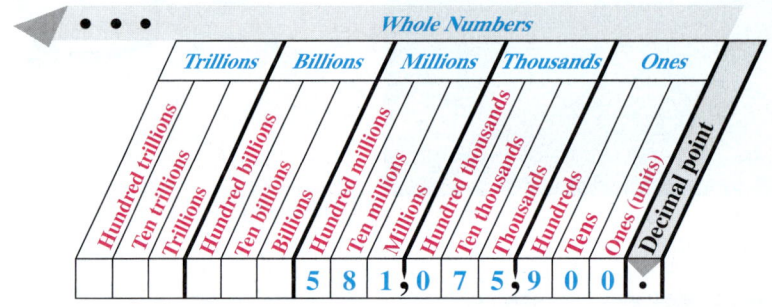

The number 581,075,900 is read as

five hundred eighty-one million, seventy-five thousand, nine hundred.

Notice that the word *and* is NOT USED with whole numbers. The word *and* is used for the decimal place, as discussed in Section 1.3.

Expressing Whole Numbers in Words **EXAMPLE 1** Write the following numbers in words.

(a) 7835 (b) 111,356,075 (c) 17,000,017,000

SOLUTION

(a) seven thousand, eight hundred thirty-five
(b) one hundred eleven million, three hundred fifty-six thousand, seventy-five
(c) seventeen billion, seventeen thousand

Quick TIP
Do not use the word *and* when reading or writing a whole number.

QUICK CHECK 1

At one point in 2017, the national debt of the United States was $20,750,361,119,450. Write the number in words.

OBJECTIVE 2 Round whole numbers. Large numbers usually have more detail (digits) than needed, as is the case for the national debt in Quick Check 1 above. So, these numbers are often rounded. For example, money amounts related to a large firm are often rounded to the nearest thousand or million dollars. Use these steps to **round whole numbers**.

1.1 Whole Numbers   **3**

Quick TIP

These steps are used to round numbers throughout this book.

> **Rounding Whole Numbers**
>
> **Step 1**   Locate the **place** to which the number is to be rounded. Draw a line under that place.
>
> **Step 2**   If the first digit to the *right* of the underlined place is **5 or more, increase** the digit in the place to which you are rounding by 1.
>
> If the digit is **4 or less, do not change**.
>
> **Step 3**   **Change** all digits to the right of the underlined digit to zeros.

Rounding Whole Numbers   **EXAMPLE 2**   Round each number as indicated.

(a) 579 to the nearest ten

(b) 34,127 to the nearest thousand

(c) 475,871 to the nearest ten thousand

(d) 79,625 to the nearest thousand

SOLUTION

(a) **Step 1**   Locate the tens place and underline.

$$5\underline{7}9$$
↑
└──── Round to this place.

     Step 2   The first digit to the right of the underlined digit is 9, which is greater than 5. Therefore, increase the digit in the tens place from 7 to 8.

     Step 3   Change all digits to the right of the tens place to zero. In other words, change the 9 in the ones place to a zero.

579 rounded to the nearest ten is 580.

Quick TIP

When rounding a number, look ONLY at the first digit to the right of the digit being rounded.

(b) **Step 1**   Locate the thousands place and underline. 3$\underline{4}$,127

     Step 2   Since the digit to the right of the thousands place is 1 (less than 5), do not change the 4 in the thousands place.

     Step 3   Change all digits to the right of the thousands place to zeros.

34,127 rounded to the nearest thousand is 34,000

(c) **Step 1**   Locate the ten thousands place and underline. 4$\underline{7}$5,871

     Step 2   Since the digit to the right of the ten thousands place is 5, which falls in the category of 5 or more, increase the 7 to an 8.

     Step 3   Change all digits to the right of the tens thousands place to zeros: **480,000**

(d) **Step 1**   Locate the thousands place and underline. 7$\underline{9}$,625

     Step 2   The number to the right of the underlined number 9 above is 5, which falls in the 5 or more category. Thus, increase the 9 by 1 to 10. Place a 0 in the thousands place and carry 1 to the ten thousands place changing the 7 to an 8.

     Step 3   Change all digits to the right of the thousands place to zeros: **80,000**

> **QUICK CHECK 2**
>
> Round each number.
>
> (a) 653,781 to the nearest ten thousand   (b) 6,578,321 to the nearest million
>
> (c) 499,100 to the nearest thousand   (d) 499,100 to the nearest hundred thousand

We will now review four basic **operations** with whole numbers: **addition, subtraction, multiplication,** and **division**.

OBJECTIVE 3 Add whole numbers. In **addition**, the numbers being added are **addends**, and the answer is the **sum**, or **total**, or **amount**.

$$\begin{array}{r} 8 \\ +\ 9 \\ \hline 17 \end{array}$$ addend
  addend
  sum (answer)

Add numbers by arranging them in a column with units above units, tens above tens, hundreds above hundreds, thousands above thousands, and so on. Use the decimal point as a reference for arranging the numbers. If a number does not include a decimal point, the decimal point is assumed to be at the far right. For example, 85 = 85. and 527 = 527.

Adding with Checking EXAMPLE 3 // To find total sales over the weekend at her Subway store, manager Jessica Fernandez needed to add the following amounts.

CASE IN POINT

> **Quick TIP**
> To minimize errors, check your work. You do not want to make a mistake and hand it to your boss.

$$\begin{array}{r}\$4028 \\ \$738 \\ 63 \\ 125 \\ 2617 \\ +\ 485 \\ \hline \$4028\end{array}$$

First, add down the columns → Then, check by adding up.

Adding from the top down results in an answer of $4028. Check for accuracy by adding again—this time from the bottom up. If the answers are the same, the sum is probably correct. If the answers are different, there is an error in either adding down or adding up, and the problem should be reworked. Both answers agree in this example, so the sum is correct.

QUICK CHECK 3
Find the total of the following expenses: $2805 + $871 + $28 + $169 + $1196

OBJECTIVE 4 Round numbers to estimate an answer. Answers can be quickly estimated using **front-end rounding**. This requires the first number to be rounded and all the following digits to be changed to zero. Only one nonzero digit remains.

Using Front-End Rounding to Estimate an Answer EXAMPLE 4 // The graphic shows the top oil producing areas in the U.S. Notice that a lot of oil is produced offshore in the Gulf of Mexico. Apply front-end rounding to estimate total oil production from these areas.

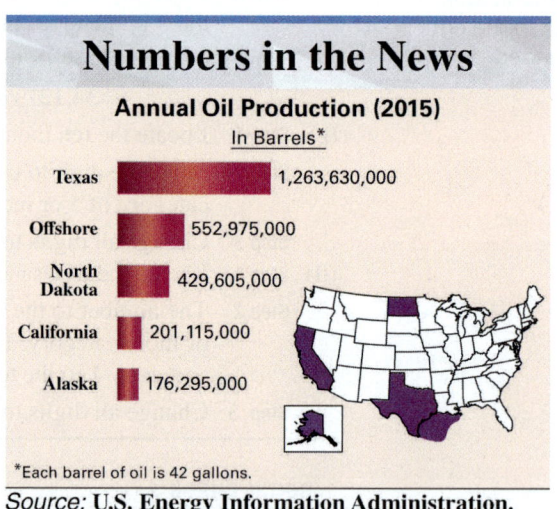

Numbers in the News
Annual Oil Production (2015)
In Barrels*

Texas	1,263,630,000
Offshore	552,975,000
North Dakota	429,605,000
California	201,115,000
Alaska	176,295,000

*Each barrel of oil is 42 gallons.
Source: U.S. Energy Information Administration.

SOLUTION

> **Quick TIP**
> In front-end rounding, only one nonzero digit (first digit) remains. All digits to the right are zeros.

	Actual		Front-End Rounded
Texas	1,263,630,000	→	1,000,000,000
Offshore	552,975,000	→	600,000,000
North Dakota	429,605,000	→	400,000,000
California	201,115,000	→	200,000,000
Alaska	176,295,000	→	200,000,000
		Estimated Total	2,400,000,000 barrels of oil

This rough estimate shows that total U.S. oil production was about 2.4 billion barrels in 2015. It is a rough estimate because some states that produce oil are not included in the list and we have used front-end rounding. For a more precise number, you need to get the data from all oil-producing areas and add.

1.1 Whole Numbers 5

QUICK CHECK 4

Use front-end rounding to estimate the total of the following numbers.

621,150; 38,400; 9682; 27,451; 435,620

OBJECTIVE 5 Subtract whole numbers. A subtraction problem is set up much like an addition problem. The top number is the **minuend**, the number being subtracted is the **subtrahend**, and the answer is the **difference**.

$$
\begin{array}{rl}
23 & \text{minuend} \\
-\ 7 & \text{subtrahend} \\
\hline
16 & \text{difference}
\end{array}
$$

Subtract one number from another by placing the subtrahend directly under the minuend with columns aligned. Begin the subtraction from the right-most column. When a digit in the subtrahend is *larger* than the corresponding digit in the minuend, **borrow** as shown in the next example.

Subtracting with Borrowing **EXAMPLE 5** Subtract 2894 Subway drink cups from 3783 Subway drink cups in inventory. First, write the problem as follows.

CASE IN POINT

$$
\begin{array}{r}
3783 \\
-2894
\end{array}
$$

In the ones (units) column, subtract 4 from 3 by borrowing a 1 from the tens column in the minuend to get 1 ten + 3, or 13, in the units column with 7 now in the tens column. Then subtract 4 from 13 for a result of 9. Complete the subtraction as follows.

$$
\begin{array}{r}
\overset{2}{\cancel{3}}\ \overset{16}{\cancel{7}}\ \overset{17}{\cancel{8}}\ \overset{13}{\cancel{3}} \\
-\ 2\ \ 8\ \ 9\ \ 4 \\
\hline
8\ \ 8\ \ 9
\end{array}
$$ drink cups

In this example, the tens are borrowed from the hundreds column, and the hundreds are borrowed from the thousands column.

QUICK CHECK 5

Subtract 7832 customers from 9511 customers.

Check the answer to a subtraction problem by adding the answer (difference) to the subtrahend. The result should equal the minuend.

Subtracting with Checking **EXAMPLE 6** Subtract 1635 from 5383 and check the answer.

Quick TIP

Do not change the order of the numbers when subtracting. For example, (9 − 5) is not the same thing as (5 − 9).

	Problem		**Check**	
Problem (subtract down) ↓	5383	minuend	5383 ↑	This result should equal the minuend.
	− 1635	subtrahend	+ 1635	
	3748	difference	3748	Check (add up)

QUICK CHECK 6

Subtract 2374 from 4165, and check the answer.

OBJECTIVE 6 Multiply whole numbers. Multiplication is actually a quick method of addition. For example, 3×4 means to add three fours: $4 + 4 + 4 = 12$. However, it is not practical to use addition for large numbers such as 103×92, which would require you to add 92 to itself 103 times. Instead, find this result with multiplication. The multiplication of 103 by 92 can be written in any of the following ways:

$$103 \times 92 = 103 \cdot 92 = 103 \cdot 92 = (103)(92)$$

It is okay to change the order when adding two numbers, e.g., $3 + 5 = 5 + 3$. It is also okay to change the order when multiplying two numbers, so $103 \times 92 = 92 \times 103$.

6 CHAPTER 1 Whole Numbers and Decimals

The number being multiplied is the **multiplicand**, the number doing the multiplying is the **multiplier**, and the answer is the **product**.

$$\begin{array}{rl} 3 & \text{multiplicand} \\ \times\ 4 & \text{multiplier} \\ \hline 12 & \text{product} \end{array}$$

When the multiplier contains more than one digit, **partial products** must be used, as in the next example, which shows the product of 25 and 34.

Multiplying Whole Numbers

EXAMPLE 7 Multiply 25×34 by first multiplying 25 by the 4 in the ones place as shown in Step 1. Then multiply 25 by 3 in the tens place as shown in Step 2, before adding to find the answer in Step 3.

Problem	Step 1	Step 2	Step 3	
25	25	25	25	multiplicand
× 34	× 34	× 34	× 34	multiplier
	100	100	100	partial product (25 × 4)
		75	+ 75	partial product (25 × 3)
			850	product

Step 1 Multiply 25 by 4 and write 100 aligning ones places.
Step 2 Multiply 25 by 3 and write 75 one position to the left since 3 is in the tenths place. The 5 in 75 will be in the ten's place.
Step 3 Add the two partial products to get the answer.

> **QUICK CHECK 7**
> Multiply 18 telemarketers by 36 phone calls per telemarketer per hour to estimate the number of calls made in one hour.

OBJECTIVE 7 Multiply by omitting zeros. If the multiplier or multiplicand end in zero, first omit any zeros at the right of the numbers and then replace omitted zeros at the right of the final answer. For example, find the product of 240 and 13 as follows.

$$\begin{array}{rl} 24\cancel{0} & \text{Omit the zero in the calculation.} \\ \times\ 13 & \\ \hline 72 & \\ 24\ \ & \text{Replace the omitted zero at the right of} \\ \hline 3120 & \text{312 for a final answer (product) of 3120.} \end{array}$$

Multiplying, Omitting Zeros

EXAMPLE 8 In the following multiplication problems, omit zeros in the calculation and then replace omitted zeros to obtain the product.

(a) 150   15 ← omit zeros
    × 70   × 7
            ─────
            105   ↙ attach 2 zeros
         10,500 ↙ answer

(b) 300   3 ← omit zeros
    × 90   × 9
            ─────
             27   ↙ attach 3 zeros
          27,000 ↙ answer

> **QUICK CHECK 8**
> Multiply 400 by 50. Omit zeros in the calculation and replace them in the product.

> **Quick TIP**
> A shortcut for multiplying by 10, 100, 1000, and so on is to just attach the number of zeros to the number being multiplied. For example,
>
> $33 \times 10 = 33$ and **1** zero $= 330$
> $56 \times 100 = 56$ and **2** zeros $= 5600$
> $732 \times 1000 = 732$ and **3** zeros $= 732{,}000$

1.1 Whole Numbers

OBJECTIVE 8 Divide whole numbers. The **dividend** is the number being divided, the **divisor** is the number doing the dividing, and the **quotient** is the answer. **Division** is indicated in any of the following ways.

$$15 \div 5 = 3 \qquad\qquad 5\overline{)15}$$

dividend divisor quotient $\qquad$ divisor dividend (quotient 3)

$$\frac{15}{5} = 3 \quad \text{quotient}$$

dividend / divisor

Dividing Whole Numbers **EXAMPLE 9** To divide 1095 baseball cards evenly among 73 collectors, divide 1095 by 73 as follows.

$$73\overline{)1095}$$

Since 73 is larger than 1 or 10, but smaller than 109, begin by dividing 73 into 109. There is one 73 in 109, so place 1 *over the digit 9* in the dividend as shown. Then multiply 1 and 73.

$$\begin{array}{r} 1 \\ 73\overline{)1095} \\ \underline{73} \\ 36 \end{array} \quad 1 \times 73 = 73$$

Subtract 73 from 109 to get 36. The next step is to bring down the 5 from the dividend, placing it next to the remainder 36. This gives the number 365. The divisor, 73, is then divided into 365 with a result of 5, which is placed to the right of the 1 in the quotient. Since 73 divides into 365 exactly 5 times, the final answer (quotient) is exactly 15.

$$\begin{array}{r} 15 \\ 73\overline{)1095} \\ \underline{73} \\ 365 \\ \underline{365} \\ 0 \end{array}$$

Check the answer by multiplying.

$$\begin{array}{r} 73 \\ \times\ 15 \\ \hline 365 \\ 73 \\ \hline 1095 \end{array}$$ Since this is the original number of cards, the answer checks.

QUICK CHECK 9
Divide $7506 evenly among 18 winners. How much will each receive?

Often, the divisor does not divide evenly into the dividend, leaving a remainder. The next example shows that remainders can be also be written using fractions or decimals. Fractions and decimals are covered in the next chapter. For now, write a remainder of 6 as R6.

Dividing with a Remainder in the Answer **EXAMPLE 10** Divide 126 by 24. Express the remainder in each of the three forms.

Remainder $\qquad$ **Fraction** $\qquad$ **Decimal**

$$\begin{array}{r} 5\ \text{R6} \\ 24\overline{)126} \\ \underline{120} \\ 6 \end{array} \qquad \begin{array}{r} 5\tfrac{6}{24} \\ 24\overline{)126} \\ \underline{120} \\ 6 \end{array} \qquad \begin{array}{r} 5.25 \\ 24\overline{)126.00} \\ \underline{120} \\ 6\ 0 \\ \underline{4\ 8} \\ 1\ 20 \\ \underline{1\ 20} \\ 0 \end{array}$$

QUICK CHECK 10
Divide 19 by 5.

8   **CHAPTER 1**   Whole Numbers and Decimals

If a divisor contains zeros at the far right, first drop the zeros in the divisor and then move the decimal point in the dividend the same number of places to the left as there were zeros dropped from the divisor.

$$900\overline{)108{,}000} \quad \text{becomes} \quad 9\overline{)1080}$$

Drop 2 zeros. — Move decimal point 2 places left.

Dropping Zeros to Divide   **EXAMPLE 11**   To divide 108,000 by 900, first drop two zeros from each number. Then divide.

$$\begin{array}{r} 120 \\ 9\overline{)1080} \\ \underline{9} \\ 18 \\ \underline{18} \\ 00 \\ \underline{00} \end{array}$$

Check Answer

$$\begin{array}{r} 120 \\ \times\;\;\;9 \\ \hline 1080 \end{array}$$ so the division is correct

You must change 9 back to 900 and multiply by 120 to get the original dividend of 108,000.

Quick TIP
After dropping zeros and dividing, do not add trailing zeros back to the answer.

Therefore, $108{,}000 \div 900 = 120$.

QUICK CHECK 11
First drop zeros, and then divide $19{,}200 \div 300$.

Checking Division Problems with Remainders   **EXAMPLE 12**   In a division problem, check the answer by multiplying the quotient (answer) and the divisor. Then add any remainder. If the result is not the same as the dividend, an error exists and the problem should be reworked. Check the following division problems.

(a)
$$\begin{array}{r} 37\text{ R}3 \\ 716\overline{)26{,}495} \\ \underline{21\;48} \\ 5\;015 \\ \underline{5\;012} \\ 3 \end{array}$$ remainder

(b)
$$\begin{array}{r} 85\text{ R}6 \\ 418\overline{)35{,}536} \\ \underline{33\;44} \\ 2\;096 \\ \underline{2\;090} \\ 6 \end{array}$$ remainder

SOLUTION

Quick TIP
Be sure to add the remainder to the product when checking a division problem with a remainder.

(a)
$$\begin{array}{r} 716 \\ \times\;\;\;37 \\ \hline 5012 \\ 2148 \\ \hline 26{,}492 \\ +\;\;\;\;\;3 \\ \hline 26{,}495 \end{array}$$
add remainder
correct

(b)
$$\begin{array}{r} 418 \\ \times\;\;\;85 \\ \hline 2090 \\ 3344 \\ \hline 35{,}530 \\ +\;\;\;\;\;6 \\ \hline 35{,}536 \end{array}$$
add remainder
correct

QUICK CHECK 12
Divide 9897 by 215. Check the answer by multiplying the quotient (answer) by the divisor.

1.1 Exercises   MyLab Math

The shaded sections below contain solutions to help you get a **QUICK START** *on the various types of exercises.*

Write the following numbers in words. (See Example 1.)

1. 7040   seven thousand, forty

2. 5310   five thousand, three hundred ten

C indicates an exercise that is related to the Case in Point feature.

3. 37,901 _____

4. 725,069 _____

5. 4,650,015 _____

6. 3,765,041,000 _____

*Round each of the following numbers first to the nearest ten, then to the nearest hundred, and finally to the nearest thousand. Go back to the **original number** each time before rounding to the next position. (See Example 2.)*

	Nearest Ten	Nearest Hundred	Nearest Thousand
7. 2065	2070	2100	2000
8. 8385	8390	8400	8000
9. 46,231	_____	_____	_____
10. 55,175	_____	_____	_____
11. 106,054	_____	_____	_____
12. 359,874	_____	_____	_____

13. Explain the three steps needed to round a number when the digit to the right of the place to which you are rounding is 5 or more. (See Objective 2.)

14. Explain the three steps needed to round a number when the digit to the right of the place to which you are rounding is 4 or less. (See Objective 2.)

Add each of the following. Check your answers. (See Example 3.)

15.	75 63 45 + 27 **210**	16.	57 26 43 + 18	17.	875 364 171 + 776	18.	135 594 415 + 276
19.	750 91 8 540 + 7	20.	371 45 839 3 + 47	21.	311,479 77,631 + 594,383	22.	803,526 759,991 + 36,024

Subtract each of the following. Check your answers. (See Examples 5 and 6.)

23.	896 − 228	24.	757 − 286	25.	3715 − 838	26.	6215 − 767
27.	65,198 − 43,652	28.	445,193 − 62,785	29.	7,025,389 − 936,490	30.	9,807,943 − 959,489

10 CHAPTER 1 Whole Numbers and Decimals

Solve the following problems. To serve as a check, the vertical and horizontal totals must be the same in the lower right-hand corner.

31. PRODUCT PURCHASES The following table shows monthly purchases at a Best Buy by product line for each of the first six months of the year. Complete the totals by adding horizontally and vertically.

Product	Jan.	Feb.	Mar.	Apr.	May	June	Totals
Software	$49,802	$36,911	$47,851	$54,732	$29,852	$74,119	**$293,267**
Computers	$86,154	$72,908	$31,552	$74,944	$85,532	$36,705	
Printers	$59,854	$85,119	$87,914	$45,812	$56,314	$91,856	
Smart phones	$73,951	$72,564	$39,615	$71,099	$72,918	$42,953	
Totals							

32. DEPARTMENT SALES The following table shows Jameson's Fashion expenses by department for the last six months of the year. Complete the totals by adding horizontally and vertically.

Department	July	Aug.	Sept.	Oct.	Nov.	Dec.	Totals
Office	$29,806	$31,712	$40,909	$32,514	$18,902	$23,514	
Production	$92,143	$86,599	$97,194	$72,815	$89,500	$63,754	
Sales	$31,802	$39,515	$58,192	$32,544	$41,920	$48,732	
Warehouse	$15,746	$12,986	$32,325	$41,983	$39,814	$20,605	
Totals							

Multiply each of the following. (See Example 7.)

33.
```
    218
  × 43
    654
   872
   9374
```

34.
```
    672
  ×  56
```

35.
```
   1896
  ×  62
```

36.
```
   7318
  ×  38
```

37.
```
   6452
  ×  263
```

38.
```
   7143
  ×  295
```

39.
```
   1109
  × 7311
```

40.
```
   9503
  × 3411
```

Estimate answers using front-end rounding. Then find the exact answers. (See Example 4.)

41. Estimate   rounds to   Exact
```
      8000   ←             8215
        60   ←               56
       700   ←              729
    + 4000   ←           + 3605
    12,760                 12,605
```

42. Estimate   Exact
```
            ←             2685
            ←               73
            ←              592
    +       ←           + 7183
```

43. Estimate   Exact
```
            ←              783
    −       ←            − 238
```

44. Estimate   Exact
```
            ←              942
    −       ←            − 286
```

45. Estimate   Exact
```
            ←              638
    ×       ←             × 47
```

46. Estimate   Exact
```
            ←              864
    ×       ←             × 74
```

Multiply, omitting zeros in the calculation and then replacing them at the right of the product to obtain the final answer. (See Example 8.)

47.
```
    370
  × 180
    37
  ×  18
   666   2 zeros
  66,600
```

48.
```
   520
 × 400
```

49.
```
  3760
 × 6000
```

50.
```
  7200
 × 1300
```

Divide each of the following. (See Examples 9 and 10.)

51.
```
     1241 R1
  4)4965
    4
    09
     8
    16
    16
     05
      4
      1
```

52. 7)13,214

53. 43)19,715

54. 93)81,452

55. Explain why checking the answer is a very important step in solving math problems.

56. In your personal and business life, when is it most important to check your math calculations? Why?

Divide each of the following, dropping zeros from the divisor. (See Examples 10 and 11.)

57.
```
   180)429,350
     2 385 R5
   18)42,935
      36
      6 9
      5 4
      1 53
      1 44
        95
        90
         5
```

58. 320)360,990

59. 1300)75,800

60. 1600)253,100

Rewrite the following numbers in words. (See Example 1.)

61. **MOTOR VEHICLE ACCIDENTS** According to the NHTSA, the number of people injured in accidents in 2015 was 2,443,000.

62. **WOMEN IN BUSINESS** A.G. Edwards reports that there are 8,534,350 businesses owned by women in the United States.

63. **PARACHUTE JUMPS** There are 3,200,000 parachute jumps in the United States each year according to the History Channel.

64. **GROSS NATIONAL PRODUCT** The market value of goods and services created in the U.S. increased last quarter to $18,036,650,000,000.

Rewrite the numbers from the following sentences using digits. (See Example 1.)

65. JELL-O SALES The average number of boxes of Jell-O gelatin sold every day is eight hundred fifty-four thousand, seven hundred ninety-five.

65. **854,795**

66. CRAYON SALES The Binney & Smith Company makes about two billion Crayola Crayons each year.

66. _____

67. SALVATION ARMY Last year, the Salvation Army served over fifty-six million, three hundred twelve thousand, seven hundred meals to hungry men, women, and children.

67. _____

68. HURRICANE KATRINA At a New Orleans pumping station, one of the pumps designed by Alexander Baldwin Wood pumped six hundred forty-eight million gallons of flood water (7500 gallons per second) in one day.

68. _____

Solve the following application problems.

69. HERSHEY MINI CHIPS A student estimated that there are approximately 5000 Mini Chips semisweet chocolate chips in 1 pound. How many chips are in 40 pounds?

$5 \times 4 = 20$ 4 zeros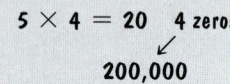
200,000

69. **200,000 chips**

70. HERSHEY KISSES Each day over 61,000,000 Hershey Kisses are produced. Estimate the number produced in 30 days.

70. _____

71. CAMPUS VENDING MACHINES On a normal weekday, the vending machines at American River College dispense 900 sodas, 400 candy bars, 500 snack items, and 200 cups of coffee. If it takes Jim Wilson four hours to restock the vending machines, how many items does he restock each hour?

71. _____

72. iPAD SALES The numbers of iPads sold weekly in one city were 1801, 927, 2088, 580, and 1049. Find the average number sold per week.

72. _____

WHITE WATER RAFTING *American River Raft Rentals lists the following daily raft rental fees. Notice that there is an additional $3 launch fee payable to the park system for each raft rented. Use this information to solve Exercises 73 and 74.*

American River Raft Rentals		
Size	Rental Fee	Launch Fee
4 persons	$70	$3
6 persons	$95	$3
10 persons	$165	$3
12 persons	$180	$3

73. On a recent Sunday, the following rafts were rented: 6 4-person rafts, 15 6-person rafts, 10 10-person rafts, and 5 12-person rafts. Find the total receipts, including the $3-per-raft launch fee.

73. _____

1.1 Exercises 13

74. During the July 4th weekend, the following rafts were rented: 38 4-person rafts, 73 6-person rafts, 58 10-person rafts, and 46 12-person rafts. Find the total receipts including the $3-per-raft launch fee.

74. _____

EGG PRODUCTION The following pictograph shows the states with the largest number of egg-laying chickens. Use this information to answer Exercises 75–78.

Numbers in the News

Top Egg-Laying States in 2016
(thousands of chickens)

State	Chickens
Iowa	51,062
Ohio	27,870
Pennsylvania	24,912
Indiana	24,353
California	20,024
Texas	18,769

Source: National Agricultural Statistics Service.

75. Find the number of egg-laying chickens in the top four states.
    $51,062 + 27,870 + 24,912 + 24,353 = 128,197$ thousand or 128,197,000

75. **128,197 thousand or 128,197,000**

76. Use front-end rounding to estimate the total number of egg-laying chickens from all states shown.

76. _____

77. How many more egg-laying chickens are there in Iowa than in Texas?

77. _____

78. How many more egg-laying chickens are there in Iowa and Ohio combined compared to California and Texas combined?

78. _____

RETAIL GIANTS The following pictograph shows annual sales and the number of stores for several large retailers. Use it to answer the questions that follow.

Numbers in the News

Giant Retailers
Annual Sales and Number of Stores (2016)

	Sales (billions)	Number of Stores
Dollar General	$20.3	🛒🛒🛒🛒🛒🛒🛒🛒🛒🛒🛒🛒
Walgreens	$76.6	🛒🛒🛒🛒🛒🛒🛒
Walmart	$340.0	🛒🛒🛒🛒
Target	$73.2	🛒🛒
Costco	$83.5	🛒
Amazon	$60.1	׀

🛒 = 1000 stores

Source: National Retail Federation.

79. Estimate the number of Dollar General stores.
    $12.5 \times 1000 = 12,500$ stores

79. **12,500 stores**

80. Estimate the number of Walmart stores.

80. _____

81. Which company has the fewest retail stores and how many does it have?

81. _____

82. Find the difference in annual sales between Walmart and Target.

82. _____

83. Does Costco or Target have more sales and if so how much more?

83. _____

84. How many more retail stores does Dollar General have than Walgreens?

84. _____

QUICK CHECK ANSWERS

1. Twenty trillion, seven hundred fifty billion, three hundred sixty-one million, one hundred nineteen thousand, four hundred fifty
2. (a) 650,000, (b) 7,000,000, (c) 499,000 (d) 500,000
3. $5069
4. 1,080,000
5. 1679 customers
6. 1791
7. 648 phone calls
8. 20 attach 3 zeros = 20,000
9. $417
10. 3 R4
11. 64
12. 46 R7

1.2 Application Problems

OBJECTIVES

1. Find indicator words in application problems.
2. Learn the four steps for solving application problems.
3. Learn to estimate answers.
4. Solve application problems.

CASE IN POINT When Jessica Fernandez became a manager at a Subway store, she brushed up on her math skills so she could train her employees.

Math has many applications in business. You need to read very carefully and critically to solve problems.

OBJECTIVE 1 Find indicator words in application problems. Sometimes there is an **indicator word** or phrase in an application problem that gives you an indication of which math operation to use. Here are some of these words.

Addition	Subtraction	Multiplication	Division	Equals
plus	less	product	divided by	is
more	subtract	double	divided into	the same as
more than	subtracted from	triple	quotient	equals
added to	difference	times	goes into	equal to
increased by	less than	of	divide	yields
sum	fewer	twice	divided equally	results in
total	decreased by	twice as much	per	are
sum of	loss of			
increase of	minus			
gain of	take away			
	reduced by			

The word *and* is not listed above since it can have many different meanings, including all the following:

1. sum of 3 **and** 4,
2. product of 6 **and** 8,
3. seventeen **and** one-half, or
4. six **and** seven tenths.

OBJECTIVE 2 Learn the four steps for solving application problems. When working problems, try to take your time and relax as if you were at the gym or a pool. Believe it or not, you are training your brain when working problems. It is okay if you do NOT know how to solve a problem when first reading it.

Solving Application Problems

Step 1   Read the problem carefully, twice if needed. Be sure that you understand what is being asked.

Step 2   Identify facts and look to see if there are any indicator words. Then work out a plan to solve the problem.

Step 3   Estimate a *reasonable answer* using rounding.

Step 4   Solve the problem. Does the answer make sense? If not, work it again.

OBJECTIVE 3 Learn to estimate answers. Each of the steps in solving an application problem is important, but special emphasis should be placed on step 3, estimating a reasonable answer. Many times an answer just *does not fit* the problem.

What is a *reasonable answer*? Read the problem and estimate the approximate size of the answer. Should the answer be part of a dollar, a few dollars, hundreds, thousands, or even millions of dollars? For example, if a problem asks for the retail price of a shirt, would an answer of $20 be reasonable? $1000? $.65? $65?

Always make an estimate of a reasonable answer. Always look at the answer and decide if it is reasonable. Doing this will help you become a better problem solver.

Using Word Indicators to Help Solve a Problem

EXAMPLE 1 Total sales at a neighborhood yard sale were $3584. The money was divided equally among the boys soccer club, the girls soccer club, the boys softball team, and the girls softball team. How much did each group receive?

SOLUTION

Sales of $3584 are to be *divided equally* among the four groups. Estimate the answer by first rounding $3584 to the nearest hundred ($3600); then divide by 4.

$3600 ÷ 4 = $900   **estimate of amount to each group**

Divide to find the exact answer.

```
      $896          Check Work
   4)3584
     32              $896
     ──              ×  4
     38              ─────
     36              $3584    The answer is correct.
     ──
     24
     24
     ──
      0
```

Each group will get $896.

QUICK CHECK 1

A budget surplus of $13,280 is divided evenly among four branch libraries. How much did each receive?

OBJECTIVE 4 Solve application problems. To improve your accuracy, use the four steps and estimate answers when solving application problems.

Solving an Application Problem

EXAMPLE 2 The daily sales figures at a busy Subway were $2358 on Monday, $3056 on Tuesday, $2515 on Wednesday, $1875 on Thursday, $3978 on Friday, $3219 on Saturday, and $3008 on Sunday. Find the total sales for the week.

SOLUTION

CASE IN POINT The phrase *total sales* indicates that you need to add daily sales to get the weekly total. Since the sales are about $3000 every day for a week of 7 days, a reasonable estimate would be around $21,000 (7 × $3000 = $21,000). Find the actual answer by adding the sales for each of the 7 days.

Quick TIP

Checking your work reduces errors. It is very important to check your work before presenting it to your boss!

```
              $20,009    Check
              $2358
              $3056
              $2515
   Add.       $1875
              $3978
              $3219
            + $3008
              ───────
              $20,009    $20,009 sales for the week
```

The estimate of $21,000 is close enough to actual sales of $20,009 to be reasonable. If you do not think your answer is reasonable, work the problem again to be sure.

QUICK CHECK 2

The numbers of visitors to a war veterans' memorial during one week are 5318, 2865, 4786, 1898, 3899, 2343, and 7221. First estimate the total attendance for the week. Then calculate exactly.

Solving an Application Problem

EXAMPLE 3 Many experts believe that some countries already have a water shortage, while others will be facing water shortages soon. The chart below shows an estimate of freshwater resources per person, by country.

(a) Find the difference in the water resources per person between China and India.

(b) Use division to compare water per person in the United States to that in India.

Droughts result in critical water shortages in India. Experts predict it will be even worse in 10 to 20 years.

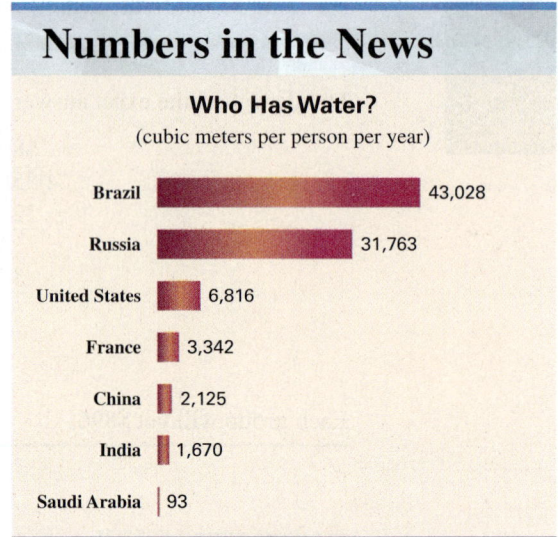

Numbers in the News
Who Has Water?
(cubic meters per person per year)

Country	Value
Brazil	43,028
Russia	31,763
United States	6,816
France	3,342
China	2,125
India	1,670
Saudi Arabia	93

DATA: CIA. The CIA maintains information on water resources by country because extreme droughts could cause mass migrations and result in political instability.

(c) The indicator word *difference* suggests a subtraction problem.

$$\begin{array}{r} 2125 \\ -\ 1670 \\ \hline 455 \end{array}$$ China
India
cubic meters per person per year

(d) $6816 \div 1670 = 4.08$ (rounded), or about 4 times as much water per person in the United States as in India.

The bar chart says nothing about the distribution of water within a country or whether that water is clean enough to drink. India does not have much water, and what water there is is heavily polluted.

QUICK CHECK 3
Find the difference in the water resources per person between Brazil and Saudi Arabia.

Solving a Two-Step Problem

EXAMPLE 4 In May, the landlord of an apartment building received $940 from each of eight tenants. After paying $2730 in expenses, how much money did the landlord have left?

SOLUTION

First, find the total rent before subtracting expenses. Since the rent is about $900 and there are eight tenants, a *reasonable estimate* would be around $7200 ($900 × 8 = $7200).

$$\begin{array}{r} \$940 \\ \times\quad 8 \\ \hline \$7520 \end{array}$$ total monthly rent is reasonable

Now subtract the expenses from the monthly income.

$$\begin{array}{r} \$7520 \\ -\ 2730 \\ \hline \$4790 \end{array}$$ after expenses

QUICK CHECK 4
A homeowner's association collected $385 from each of 62 homeowners. If the association paid $18,280 in expenses, how much remained?

Solving Application Problems **EXAMPLE 5** Use the data below that was taken from company websites to answer each question.

Subway	Calories	Total Fat (g)
Carved Turkey with Bacon	570	26
Oven Roasted Chicken	320	5
Classic Tuna	480	25
McDonalds		
Big Mac	540	28
Quarter Pounder with Cheese	530	27
Southwest Grilled Chicken Salad	350	12

(a) How many fewer calories and grams of fat are in an Oven Roasted Chicken sandwich from Subway than in a Big Mac?

(b) How many fewer calories and grams of fat are in a Southwest Grilled Chicken Salad from McDonalds than in a Carved Turkey with Bacon sandwich from Subway?

SOLUTION

(a) The word "fewer" suggests that subtraction should be used to answer the question.

$$\begin{array}{r}540 \text{ calories}\\-320 \text{ calories}\\\hline 220 \text{ fewer calories}\end{array} \qquad \begin{array}{r}28 \text{ grams of fat}\\-5 \text{ grams of fat}\\\hline 23 \text{ fewer grams of fat}\end{array}$$

(b) Again, fewer indicates subtraction.

$$\begin{array}{r}570 \text{ calories}\\-350 \text{ calories}\\\hline 220 \text{ calories}\end{array} \qquad \begin{array}{r}26 \text{ grams of fat}\\-12 \text{ grams of fat}\\\hline 14 \text{ grams of fat}\end{array}$$

QUICK CHECK 5

Find the total number of calories and grams of fat for a person eating both a Big Mac and a Southwest Grilled Chicken Salad.

1.2 Exercises // MyLab Math

The shaded sections below contain solutions to help you get a **QUICK START** *on the various types of exercises.*

Solve the following application problems.

 1. **SUBWAY SANDWICHES** Last week, Subway sold 602 Veggie Delite sandwiches, 935 ham sandwiches, 1328 turkey breast sandwiches, 757 roast beef sandwiches, and 1586 Subway Club sandwiches. Find the total number of sandwiches sold.
602 + 935 + 1328 + 757 + 1586 = 5208 sandwiches

1. **5208 sandwiches**

2. **COMPETITIVE CYCLIST TRAINING** During a week of training, Rob Andrews rode his road bicycle 80 miles on Monday, 75 miles on Tuesday, 135 miles on Wednesday, 40 miles on Thursday, and 52 miles on Friday. How many miles did he ride?
Total miles traveled = 80 + 75 + 135 + 40 + 52 = 382 miles

2. **382 miles**

3. **MILES DRIVEN** The Federal Highway Commission estimates that the total number of miles driven increased from about 1 trillion miles in 1970 to a little over 3 trillion miles this year. Estimate the increase.

3. _____

 indicates an exercise that is related to the Case in Point feature.

18 CHAPTER 1 Whole Numbers and Decimals

4. **CARBON DIOXIDE** The amount of carbon dioxide in the atmosphere has increased from 328 parts per million in 1970 to 405 parts per million today. Find the increase.

4. _____

5. **CAR EMISSIONS** A typical passenger vehicle emits about 4.7 metric tons of carbon dioxide per year. Assuming there are 253 million cars and trucks in the U.S., estimate the total emitted weight of carbon dioxide per year in the U.S.

5. _____

For fun, this is about the weight of 175,000,000 elephants emitted every year.

6. **TOTAL WORLD WAR II VETERANS** According to the Department of Veterans Affairs, there are about 850,000 U.S. World War II veterans alive today. If only 1 in 19 is still alive, estimate the total number who were World War II veterans.

6. _____

7. **FISHING BOAT** A fishing boat weighs 8375 pounds. If its 762-pound engine is removed and replaced with a 976-pound engine, find the weight of the boat after the engine change.

7. _____

8. **COLLEGE TEXTBOOKS** Tatum Palmer needs to buy three textbooks this semester and shops first at her college bookstore. The cost of new books are: $195, $180, and $205. The cost of used books are: $85, $62, and $92. Find the savings if she buys all three books used.

8. _____

9. **NOTEBOOK** The price of a mini notebook was lowered from $499 to $435. Find the decrease in price.

9. _____

10. **WEIGHING FREIGHT** A truck weighs 9250 pounds when empty. After being loaded with firewood, the truck weighs 21,375 pounds. What is the weight of the firewood?

10. _____

11. **LAND AREA** There are 43,560 square feet in 1 acre. How many square feet are there in 140 acres?

11. _____

12. **CHECK PROCESSING** One bank processes about 3000 checks every day. Find the number of checks processed in a year. (Use a 365-day year.)

12. _____

13. **HOTEL ROOM COSTS** In a recent study of hotel casinos, the cost per night at Harrah's Reno was $75, while the cost at Harrah's Lake Tahoe was $225 per night. Find the amount saved on a seven-night stay at Harrah's Reno instead of staying at Harrah's Lake Tahoe.

13. _____

14. **LUXURY HOTELS** A luxury hotel room at the Ritz-Carlton in Los Angeles costs $645 per night, while a nearby room at a Motel 6 costs $74 per night. What amount will be saved in a four-night stay at Motel 6 instead of staying at the Ritz-Carlton?

14. _____

15. **PHYSICALLY IMPAIRED** The Enabling Supply House purchased 6 wheelchairs at $1256 each and 15 speech compression disc players at $895 each. Find the total cost.

15. _____

16. **KITCHEN EQUIPMENT** Find the total cost if Subway buys 32 baking ovens at $1538 each and 28 warming ovens at $887 each.

16. _____

17. **YOUTH SOCCER** A youth soccer association raised $7588 through fund-raising projects. There were expenses of $838 that had to be paid first, and the remaining money was divided evenly among the 18 teams. How much did each team receive?

17. _____

18. **YARD MAINTENANCE** Doug Smith hires a crew of three men to mow yards. His total costs of wages, taxes, expenses, and the costs of equipment and truck is about $98 per hour. He charges an average of $60 per yard and his crew can mow and edge two yards per hour. Find his profit for a week in which his crews worked 35 hours.

18. _____

19. **THEATER RENOVATION** A theater owner is remodeling to provide enough seating for 1250 people. The main floor has 30 rows of 25 seats in each row. If the balcony has 25 rows, how many seats must be in each row of the balcony to satisfy the owner's seating requirements?

19. _____

20. **CALL CENTER** Beverly Stratton was hired to manage a call center for a market research firm that makes calls asking about consumer preferences. She anticipates about 82 outgoing calls per hour, 40 hours per week, 5 days a week, 50 weeks a year. Estimate the total number of calls per year. Also estimate the minimum number of call center operators needed if each can make about 17,000 calls a year. Round up to the next larger whole number.

20. _____

QUICK CHECK ANSWERS

1. $3320
2. 28,000; 28,330 visitors
3. 42,935 cubic meters per person per year
4. $5590
5. 890 calories; 40 grams of fat

1.3 Decimal Numbers

OBJECTIVES

1. Read and write decimal numbers.
2. Round decimal numbers.

OBJECTIVE 1 Read and write decimal numbers. A **decimal number** is any number written with a decimal point, such as 6.8, 5.375, or $7.50. Decimals, like fractions, can be used to represent parts of a whole. **Section 1.1** discussed how to read the digits to the *left* of the decimal point (whole numbers). Now we will see how to read the digits to the right of the decimal point, which always end in "th" or "ths."

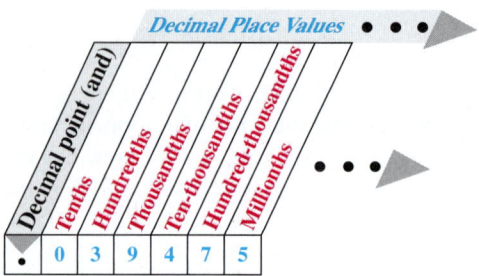

Read the decimal number .039475 as "39475 millionths," where millionths comes from the fact that the last digit of this number falls in the millionths place. Write this number using words as follows.

thirty-nine thousand, four hundred seventy-five millionths

When reading a number with digits both to the left and the right of the decimal place, use the word *and* to show the location of the decimal point. Here are some examples.

> **Quick TIP**
> Do not use commas to the right of a decimal place.

9.7	nine **and** seven tenths
11.59	eleven **and** fifty-nine hundredths
1045.658	one thousand, forty-five **and** six hundred fifty-eight thousandths
5,600,000.0072	five million, six hundred thousand **and** seventy-two ten thousandths

Reading Decimal Numbers **EXAMPLE 1** Write the following decimals in words.

(a) 19.08   (b) .097   (c) 7648.9713   (d) 3,068,001.7

SOLUTION

> **Quick TIP**
> Use the word "*and*" only to separate the whole number from the fractional (decimal) part.

(a) nineteen **and** eight hundredths

(b) ninety-seven thousandths

(c) seven thousand, six hundred forty-eight **and** nine thousand, seven hundred thirteen ten-thousandths

(d) three million, sixty-eight thousand, one **and** seven tenths

QUICK CHECK 1
Write (a) 0.068 and (b) 4,370.15 in words.

OBJECTIVE 2 Round decimal numbers. It is important to be able to round decimals. For example, Walgreens sells two candy bars for $.79, but you want to buy only one candy bar. The price of one bar is $.79 ÷ 2, which is $.395, but you cannot pay part of a cent. So the store rounds the price up to $.40 for one bar. The steps to round decimal numbers are the same as those used to round whole numbers.

> **Rounding Decimals**
> Step 1   Find the **place** to which the number is to be rounded. Draw a vertical line after that place to show that you are cutting off the rest of the digits.
> Step 2   Look at only the first digit to the right of your cut-off line. If the first digit is **5 or more, increase** the digit in the place to which you are rounding by 1. Otherwise, do not change the digit in the place to which you are rounding.
> Step 3   **Drop** all digits to the right of the place to which you have rounded.

To round 97.3892 to the nearest tenth, first draw a vertical line to the right of the 3 in the tenths place. The digit to the right of the vertical line is 5 or more, so round the 3 in the tenths position up to 4 and drop all digits to the right.

$$97.3\,|\,892$$

— digit to the right of tenths place
— tenths place

97.3892 rounded to the nearest tenth is 97.4.

Rounding Decimal Numbers

EXAMPLE 2 Round each as indicated.
(a) 87.562 to the nearest tenth
(b) 3678.5928 to the nearest hundredth

SOLUTION

(a) **Step 1** Locate the tenths place and put a vertical line immediately to the right.

$$87.5\,|\,62$$

— tenths place

Step 2 The number to the right of the line is 6, which is in the category of 5 or more. Therefore, round the 5 in the tenths place up to 6.

Step 3 Drop all digits to the right of the tenths place.

87.562 rounded to the nearest tenth is 87.6.

(b) **Step 1** Locate the hundredths place and put a vertical line to the right.

$$3678.59\,|\,28$$

Step 2 The number to the right of the line is 2, or less than 5. Therefore, leave the 9 as it is in the hundredths place.

Step 3 Drop all digits to the right of the hundredths place.

3678.5928 rounded to the nearest hundredth is 3678.59.

Rounding 3678.5928 to the nearest tenth would result in 3678.6, and rounding it to the nearest whole number would result in 3679.

QUICK CHECK 2

Round 72.8479 to the nearest thousandth.

Rounding the Same Decimal Number to Different Places

EXAMPLE 3 Round 24.918 to the nearest (a) hundredth, (b) tenth, and (c) whole number.

SOLUTION

(a) 24.918 rounded to the nearest hundredth is 24.92.
(b) 24.918 rounded to the nearest tenth is 24.9.
(c) 24.918 rounded to the nearest whole number is 25.

Quick TIP
Always refer back to the original number when rounding.

QUICK CHECK 3

Round 518.4464 to the nearest (a) thousandth, (b) hundredth, and (c) tenth.

1.3 Exercises  MyLab Math

*The shaded sections below contain solutions to help you get a **QUICK START** on the various types of exercises.*

Write the following decimals in words. (See Example 1.)

1. .38 thirty-eight hundredths
2. .91 ninety-one hundredths
3. 5.61

22 CHAPTER 1 Whole Numbers and Decimals

4. 6.53 _____
5. 7.408 _____
6. 1.254 _____
7. 37.593 _____
8. 20.903 _____
9. 4.0062 _____
10. 9.0201 _____

11. "My answer is right, but the decimal point is in the wrong place." Can this statement ever be correct? Explain. (See Objective 1.)

12. Explain the difference between thousands and thousandths.

Write the following decimals, using numbers.

13. four hundred thirty-eight and four tenths   **438.4**
14. six hundred five and seven tenths   **605.7**
15. ninety-seven and sixty-two hundredths   _____
16. seventy-one and thirty-three hundredths   _____
17. one and five hundred seventy-three ten-thousandths   _____
18. nine and three hundred eight ten-thousandths   _____
19. three and five thousand eight hundred twenty-seven ten-thousandths   _____
20. two thousand seventy-four ten-thousandths   _____

GROCERY SHOPPING *Alan Zagorin is grocery shopping. The store will round the amount he pays for each item to the nearest cent. Write the rounded amounts.* (See Examples 2–4.)

21. Apple pies are two for $11.99. So one pie is $5.995. Zagorin pays ____.

22. Four 12-packs of soda cost $69.94 or $17.485 per 12-pack. Zagorin pays ____.

23. Muffin mix is three packages for $1.75. So one package is $.58333. Zagorin pays ____.

24. Small candy bars are six for $3.94. So one bar is $.65666. Zagorin pays ____.

25. Barbeque sauce is three bottles for $11.98. So one bottle is $3.993. Zagorin pays ____.

26. Tony's Pizzas are five for $37.46. So one pizza is $7.492. Zagorin pays ____.

Round each of the decimals to the nearest tenth, the nearest hundredth, and the nearest thousandth. Remember to use the original number each time before rounding. (See Examples 2–4.)

	Nearest Tenth	Nearest Hundredth	Nearest Thousandth
27. 3.5218	3.5	3.52	3.522
28. 4.836	4.8	4.84	4.836
29. 2.54836	____	____	____
30. 7.44652	____	____	____
31. 27.32451	____	____	____
32. 89.53796	____	____	____

c// indicates an exercise that is related to the Case in Point feature.

	Nearest Tenth	Nearest Hundredth	Nearest Thousandth
33. 36.47249	_____	_____	_____
34. 58.95651	_____	_____	_____
35. .0562	_____	_____	_____
36. .0789	_____	_____	_____

Round each of the dollar amounts to the nearest cent.

37. $5.056 **$5.06**
38. $16.519 **$16.52**
39. $32.493 _____

40. $375.003 _____
41. $382.005 _____
42. $12,802.965 _____

43. $42.137 _____
44. $.846 _____
45. $.0015 _____

46. $.008 _____
47. $1.5002 _____
48. $7.6009 _____

49. $1.995 _____
50. $28.994 _____
51. $752.798 _____

Round each of the dollar amounts to the nearest dollar (nearest whole number).

52. $8.58 **$9**
53. $26.49 **$26**
54. $.57 _____

55. $.49 _____
56. $299.76 _____
57. $12,836.38 _____

58. $268.72 _____
59. $395.18 _____
60. $666.66 _____

61. $4699.62 _____
62. $11,285.13 _____
63. $378.59 _____

64. $233.86 _____
65. $722.38 _____
66. $8263.47 _____

67. Explain what happens when you round $.499 to the nearest dollar. (*See Objective 2.*)

68. Review Exercise 67. How else could you round $.499 to obtain a result that is more helpful? What kind of guideline does this suggest about rounding to the nearest dollar?

QUICK CHECK ANSWERS

1. (a) sixty-eight thousandths, **(b)** four thousand, three hundred seventy and fifteen hundredths

2. 72.848

3. (a) 518.446
   (b) 518.45
   (c) 518.4

1.4 Addition and Subtraction of Decimals

OBJECTIVES
1. Add decimals.
2. Estimate answers.
3. Subtract decimals.

CASE IN POINT As manager of a Subway, Jessica Fernandez is responsible for making bank deposits. These banking activities require the ability to accurately add and subtract decimal numbers.

OBJECTIVE 1 Add decimals. To add, write the numbers so that the decimal points are aligned, which causes the place values to be aligned. As with adding whole numbers, add decimal numbers in columns, beginning on the right and moving to the left.

You may wish to add **trailing zeros** to the right of the decimal point so that each number being added has the same number of digits to the right of the decimal point. This does not change the number and can make it easier to keep track of things.

$$45.93 = 45.930 = 45.9300$$

- two trailing zeros added
- one trailing zero added

Adding Decimal Numbers

EXAMPLE 1 Add $45.93 + 14.017 + 96.5432$.

SOLUTION

First, align place values by lining up the decimal points.

```
   45.93
   14.017
   96.5432
```
— decimals aligned

Although not required, you can add trailing zeros if you wish so that each number has the same number of digits to the right of the decimal point. Then, add from right to left.

```
  11  1
   45.9300
   14.0170
   96.5432
  ────────
  156.4902
```
— trailing zeros added

QUICK CHECK 1

Add 3.8, 14.604, 5.76, and 27.152.

OBJECTIVE 2 Estimate answers. Check that the numbers in Example 1 were correctly added by estimating the answer. Apply front-end rounding to the numbers as follows.

Problem		Estimate
45.93	→	50
14.017	→	10
+ 96.5432	→	+ 100
156.4902		160

The answer is relatively close to the actual value found of 156.4902 giving confidence that the original decimal numbers were added correctly.

Another option is to round each of the numbers to the nearest whole number and then compare as follows.

Problem		Estimate
45.93	→	46
14.017	→	14
+ 96.5432	→	+ 97
156.4902		157

The sum of 156.4902 is very close to the estimate of 157, so it appears the addition was done correctly.

Adding Dollars and Cents

EXAMPLE 2 During a recent week, a manager made the following bank deposits to a business account: $1783.38, $4341.15, $2175.94, $896.23, and $2562.53. Use front-end rounding to estimate the total deposits and then find the total deposits.

SOLUTION

Estimate		Problem
$ 2000	←	$ 1783.38
4000	←	4341.15
2000	←	2175.94
900	←	896.23
+ 3000	←	+ 2562.53
$11,900		$11,759.23

The total deposits for the week were $11,759.23, which is close to our estimate.

QUICK CHECK 2

The following bills were paid last week; $1268.72, $228.35, $2336.19, $176.68, and $1560.75. Use front-end rounding to estimate the total amount of the bills paid, and then find the total bills paid.

OBJECTIVE 3 Subtract decimals. Subtraction is done in much the same way as addition. Line up the decimal points and place as many zeros after each decimal as needed. For example, subtract 17.432 from 21.76 as follows.

$$\begin{array}{r} 21.76\underline{0} \\ -17.432 \\ \hline 4.328 \end{array}$$ Place one trailing zero after the top decimal.

Estimating and Then Subtracting Decimals

EXAMPLE 3 First estimate using front-end rounding and then subtract.

(a) $\begin{array}{r} 11.7 \\ -\ 4.923 \\ \hline \end{array}$ (b) $\begin{array}{r} 39.428 \\ -27.98 \\ \hline \end{array}$

SOLUTION

Attach zeros as needed and then subtract.

(a)
Estimate		Problem
10	←	11.700
− 5	←	− 4.923
5		6.777

(b)
Estimate		Problem
40	←	39.428
− 30	←	− 27.980
10		11.448

QUICK CHECK 3

First estimate using front-end rounding, and then subtract 5.32 from 26.952.

1.4 Exercises // MyLab Math

The shaded sections below contain solutions to help you get a QUICK START on the various types of exercises.

First use front-end rounding to estimate and then find the exact answer. (see Examples 1 and 2.)

1.
Estimate		Problem
40	←	43.36
20	←	15.8
+ 9	←	+ 9.3
69		68.46

2.
Estimate		Problem
600	←	623.15
700	←	734.29
+ 700	←	+ 686.26
2000		2043.70

3.
Estimate	Problem
	6.23
	3.6
	5.1
	7.2
	+ 1.69

26 CHAPTER 1 Whole Numbers and Decimals

4. Estimate Problem
12.79
2.15
16.28
4.39
+ 7.61

5. Estimate Problem
2156.38
5.26
2.791
+ 6.983

6. Estimate Problem
1889.76
21.42
19.35
+ 8.1

7. Estimate Problem
6133.78
506.124
18.63
+ 7.527

8. Estimate Problem
743.1
3817.65
2.908
4123.76
+ 21.98

9. Estimate Problem
1798.419
68.32
512.807
643.9
+ 428.

Place each of the following numbers in a column and then add. (See Example 1.)

10. 45.631 + 15.8 + 7.234 + 19.63 = **88.295**
11. 12.15 + 6.83 + 61.75 + 19.218 + 73.325 = **173.273**
12. 197.4 + 83.72 + 17.43 + 25.63 + 1.4 =
13. 27.653 + 18.7142 + 9.7496 + 3.21 =
14. 73.618 + 19.18 + 371.82 + 355.125 =

15. It is a good idea to estimate an answer before actually solving a problem. Why is this true? (See Objective 2.)

16. Explain why placing zeros after any digits to the right of the decimal point does not change the value of a number. (See Objective 1.)

Solve the following application problems.

17. **SUBWAY SALES** During a very busy week, daily sales at the Subway managed by Jessica Fernandez were $1815.79, $2367.34, $1976.22, $2155.81, $1698.14, $2885.26, and $2239.63. Find the total weekly sales.

17. _____

18. **RESTAURANT TIPS** Becky Waterton wants to know what her total tips were for the three days she worked as a waitress at a Spanish tapas restaurant. Her tips for Thursday, Friday, and Saturday were $85.25, $114.60, and $129.40. Find the total.

18. _____

19. **BEEF/TURKEY COST** At one store, the cost of T-bone steak is $12.99 per pound, while the cost of turkey is $1.89 per pound. How much more per pound is the price of T-bone steak than turkey?

19. _____

20. **ALIMONY** The average alimony paid to divorced spouses in one area is $1238.73 per month. However, the court ordered James Tuxon to pay his ex-spouse $1530 per month. Find the amount he is paying above the average.

20. _____

C// indicates an exercise that is related to the Case in Point feature.

First use front-end rounding to estimate the answer and then subtract. (See Example 3.)

	Estimate	Problem		Estimate	Problem		Estimate	Problem
21.	20 − 7 ――― 13	19.74 − 6.58 ――― 13.16	22.	40 − 8 ――― 32	35.86 − 7.91 ――― 27.95	23.		51.215 − 19.708
24.		27.613 − 18.942	25.		325.053 − 85.019	26.		3974.61 − 892.59
27.		7.8 − 2.952	28.		27.8 − 13.582	29.		5 − 1.9802

CHECKING-ACCOUNT RECORDS Jessica Fernandez, manager of Subway, had a bank balance of $5382.12 on March 1. During March, she deposited $60,375.82 received from sales, $3280.18 received as credits from suppliers, and $75.53 as a county tax refund. She paid out $27,282.75 to suppliers, $4280.83 for rent and utilities, and $12,252.23 for salaries and miscellaneous. Find each of the following.

30. How much did Fernandez deposit in March?
    $60,375.82 (sales) + $3280.18 (credits) + $75.53 (refund) = $63,731.53

30. _____

31. How much did she pay out?

31. _____

32. What was her final balance at the end of March?

32. _____

―――――― QUICK CHECK ANSWERS ――――――

1. 51.316
2. $5400; $5570.69
3. 25; 21.632

1.5 Multiplication and Division of Decimals

OBJECTIVES
1. Multiply decimals.
2. Divide a decimal by a whole number.
3. Divide a decimal by a decimal.

CASE IN POINT Managing a business requires the ability to multiply and divide decimal numbers. The manager at Subway applies these skills in many ways to find payroll, purchasing, and sales.

OBJECTIVE 1 Multiply decimals. Decimals are multiplied as if they were whole numbers. It is not necessary to line up the decimal points. The decimal point in the answer is then found using the following steps.

Positioning the Decimal Point

Step 1 Count the total number of digits to the *right* of the decimal point in each of the numbers being multiplied.

Step 2 In the answer, count from *right to left* the number of places found in step 1 and write the decimal point. It may be necessary to attach zeros to the left of the answer in order to correctly place the decimal point.

Multiplying Decimals **EXAMPLE 1** Multiply each of the following.
(a) 8.34×4.2 (b) $.032 \times .07$

SOLUTION

(a) First multiply the given numbers as if they were whole numbers.

$$
\begin{array}{r}
8.34 \\
\times \;\; 4.2 \\
\hline
1668 \\
3336 \\
\hline
35.028
\end{array}
$$

← 2 digits to right of decimal place
← 1 digit to right of decimal place

← 3 digits to right of decimal place

There are two digits to the right of the decimal place in 8.34 and one digit to the right of the decimal place in 4.2. Add the 2 and the 1 together to find that the answer should have 3 digits to the right of the decimal place.

35.028 3 places to the left

(b) Here, it is necessary to attach zeros at the left in the answer:

$$
\begin{array}{r}
.032 \\
\times \;\; .07 \\
\hline
.00224
\end{array}
$$

← 3 digits to the right of decimal place
← 2 digits to the right of decimal place
← total of 5 digits to right of decimal place

Two zeros are needed to have the correct number of digits to the right of the decimal place.

QUICK CHECK 1

Multiply each of the following
(a) 6.7×4.32 (b) $.086 \times .05$

The next example uses the formula for the gross pay or pay before deductions.

Gross pay = Number of hours worked × Pay per hour

1.5 Multiplication and Division of Decimals

Multiplying Two Decimal Numbers

EXAMPLE 2 Find the gross pay of a Subway employee working 31.5 hours at a rate of $8.65 per hour.

SOLUTION

CASE IN POINT Find gross pay by multiplying the number of hours worked by the pay per hour.

$$\begin{array}{r} 31.5 \\ \times\ 8.65 \\ \hline 1575 \\ 1890 \\ 2520 \\ \hline 272.475 \end{array}$$ ← 1 place
← 2 places

← 3 places in answer

This worker's gross pay, rounded to the nearest cent, is $272.48.

QUICK CHECK 2

A college student earns $15.75 per hour. How much is earned for 26.5 hours of work?

Applying Decimal Multiplication

EXAMPLE 3 Assume the cost of each 30-second television ad at the 2017 Super Bowl game was $5.9 million. If there were 60 advertising spots of 30 seconds during the game, find the total amount charged for advertising.

SOLUTION

Find the total amount charged for advertising by multiplying the charge for each advertisement by the number of ads.

$$\begin{array}{r} 5.9 \\ \times\ 60 \\ \hline 00 \\ 354 \\ \hline 354.0 \end{array}$$ ← 1 place
← 0 places

← 1 place

The total amount charged for advertising was $354,000,000.

QUICK CHECK 3

Find the total amount charged for Super Bowl ads in 2017 if two advertisers canceled, resulting in fifty-eight 30-second ads.

OBJECTIVE 2 Divide a decimal by a whole number. Divide the decimal number 21.93 by the whole number 3 by first writing the division problem as usual. Place the decimal point in the quotient directly above the decimal point in the dividend and perform the division.

Place a decimal point directly above dividend's decimal point.

$$3\overline{)21.93} \quad\text{with quotient } 7.31$$

Check by multiplying the divisor and the quotient. The answer should equal the dividend.

$$\begin{array}{r} 7.31 \\ \times\ \ \ 3 \\ \hline 21.93 \end{array}$$ ← matches dividend

Sometimes it is necessary to place zeros after the decimal point in the dividend. Do this if a remainder of 0 is not obtained. Attaching zeros *does not change* the value of the dividend. For example, divide 1.5 by 8 by dividing and placing zeros as needed.

30 CHAPTER 1 Whole Numbers and Decimals

```
    Problem           Check
    .1875             .1875
8)1.5000          ×       8      By the check,
    8                 1.5000      1.5 ÷ 8 = .1875
    70
    64
    60
    56    matches
    40
    40
     0
```

Sometimes, a remainder of 0 is never obtained when dividing. For example, dividing 4.7 by 3 results in repeating 6s that continue without end, as you can see from the following division. Continue the division to one decimal place more than needed so that you can **round the decimal** to the desired place. In this case, the division is carried out to the nearest ten-thousandth so that the final answer can be rounded to the nearest thousandth.

Quick TIP
Carry the division one place further than the position to which you wish to round.

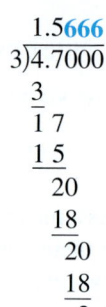

```
    1.5666
3)4.7000
  3
  1 7
  1 5
    20
    18
    20
    18
     2
```

The number 1.5666 rounded to the nearest thousandth is 1.567.

Dividing a Decimal by a Whole Number

EXAMPLE 4 Divide the following; then check using multiplication.
(a) 27.52 ÷ 32   (b) 153.4 ÷ 8

SOLUTION

```
          Problem      Check              Problem         Check
           .86          .86                19.175         19.175
(a) 32)27.52         ×  32      (b)  8)153.400         ×      8
        25 6          172                   8              153.400
        1 92          258                  73
        1 92          27.52                72
           0                                1 4
                                              8
                                             60
                                             56
                                             40
                                             40
                                              0
```

QUICK CHECK 4
Divide the following; then check using multiplication.
(a) 15)823.26   (b) 78.562 ÷ 4

1.5 Exercises

OBJECTIVE 3 Divide a decimal by a decimal. To divide by a decimal, first convert the divisor to a whole number. For example, to divide 27.69 by .3, convert .3 to a whole number by moving the decimal one place to the right. Then move the decimal point in the dividend, 27.69, one place to the right so that the value of the problem does not change. You may need to attach one or more trailing zeros to do this.

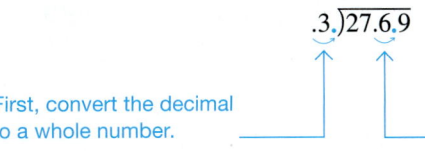

First, convert the decimal to a whole number.

Then move the decimal point in the dividend the same number of places to the right as done in the divisor.

Then divide as follows:

```
       92.3
    3)276.9
      27
      ──
       06
        6
       ──
       09
        9
       ──
        0
```

Dividing a Decimal by a Decimal

EXAMPLE 5 Divide and check the answers.

(a) 17.6 ÷ .25   (b) 5 ÷ .42

SOLUTION

Quick TIP
For an answer that has been rounded, the check answer will not be exactly equal to the original dividend.

		Check
	70.4	70.4
(a)	.25.)17.60.0 	× .25
	17 5	3520
	10	1408
	0	17.600
	10 0	
	10 0	
	0	

		Check	
	11.9047	Rounding the answer to the nearest thousandth gives 11.905.	11.905 (The check is off a little due to rounding.)
(b)	.42.)5.00.0000		× .42
	4 2		23810
	80		47620
	42		5.00010
	38 0		
	37 8		
	200		
	168		
	320		
	294		
	26		

QUICK CHECK 5
Divide; then check using multiplication.
(a) 22.5 ÷ .8   (b) 8.6 ÷ .32

1.5 Exercises // MyLab Math

The shaded sections below contain solutions to help you get a QUICK START on the various types of exercises.

First estimate using front-end rounding and then multiply. (See Example 1.)

1. Estimate		Problem
100	←	96.8
× 4	←	× 4.2
400		406.56

2. Estimate		Problem
20	←	16.6
× 4	←	× 4.2
80		69.72

C// indicates an exercise that is related to the Case in Point feature.

CHAPTER 1 Whole Numbers and Decimals

3. Estimate   Problem
               34.1
            × 6.8

4. Estimate   Problem
               70.35
            × 8.06

5. Estimate   Problem
               43.8
            × 2.04

6. Estimate   Problem
               69.3
            × 2.81

Multiply the following decimals.

7.   .532
    × 3.6
    .532 ← 3 decimals
   × 3.6 ← 1 decimal
   1.9152 ← 4 decimals

8.   .259
    × 6.2
    .259 ← 3 decimals
   × 6.2 ← 1 decimal
   1.6058 ← 4 decimals

9.   21.7
   × .431

10.   76.9
    × .903

11.   .0408
    × .06

12.   2481.9
    ×   .003

CALCULATING GROSS EARNINGS   *Find the gross pay for each employee at the given rate. Round to the nearest cent. (See Examples 2 and 3.)*

13. 18.5 hours at $14.50 per hour
    18.5 × $14.50 = $268.25
                                                  13. **$268.25**

14. 36.6 hours at $9.85 per hour   14. _____

15. 27.9 hours at $11.42 per hour, and 6.8 hours at $14.63 per hour   15. _____

16. 11.4 hours at $8.59 per hour, and 23.9 hours at $10.06 per hour   16. _____

Divide the following, and round your answer to the nearest thousandth. (See Examples 4 and 5.)

17. 6)48.45

18. 5)62.38

19. 411.63 ÷ 15

20. 2.43)9.6153

21. .65)37.6852

22. 15.62 ÷ .28

23. Explain how to place the decimal point in the answer of a decimal multiplication problem. (See Objective 1.)

24. Explain how to place the decimal point in the answer of a decimal division problem. Include the divisor, dividend, and quotient in your description. (See Objectives 2 and 3.)

Solve the following application problems:

25. **REAL ESTATE FEES** Robert Gonzalez recently sold his home for $246,500. He paid a commission of .06 times the price of the house. What was the amount of the commission?
$246,500 × .06 = $14,790

25. $14,790

26. **FABRIC** To make curtains for her new apartment, Janitha Williams needs 2.75 yards of material for each window. Find the total amount of material needed if she decides to do four windows.

26. _____

27. **HYBRID TOYOTA** To reduce his driving costs as a salesperson, Bill Chen bought a Toyota Prius. One week, he drove 519 miles in and around the city and used 10.2 gallons of gasoline. Find the miles per gallon to the nearest tenth.

27. _____

28. **MANAGERIAL EARNINGS** A Subway assistant manager earns $2528 each month for working a 48-hour week. Find (a) the number of hours worked each month and (b) the assistant manager's hourly earnings (1 month = 4.3 weeks). Round to the nearest cent.

(a) _____
(b) _____

29. **CHROME RIMS** Henry Barnes has a loan balance of $1170.18 on the new rims he bought for his SUV. If his payments are $106.38 per month, how many months will it take to pay off the balance?

29. _____

30. **MEDICINE DOSE** Each dose of a medication contains 1.62 milligrams of a certain ingredient. Find the number of doses that can be made from 57.13 milligrams of the ingredient. Round to the nearest whole number.

30. _____

31. **U.S. PAPER MONEY** The thickness of a $100 bill is .0043 inch.
    (a) If you had a pile of one hundred $100 bills, how high would it be?
    (b) How high would a pile of 1000 bills be?

(a) _____
(b) _____

32. (a) Use the information from Exercise 31 to find the number of $100 bills in a pile that is 43 inches high.
    (b) How much money would you have if the pile was all $20 bills, which have the same thickness as a $100 bill?

(a) _____
(b) _____

Use the following online information for a company going out of business to answer Exercises 33–36.

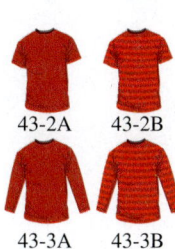

43-2A 43-2B
43-3A 43-3B

KNIT SHIRT ORDERING INFORMATION	
43–2A Short sleeve, solid colors	$14.75 each
43–2B Short sleeve, stripes	$16.75 each
43–3A Long sleeve, solid colors	$18.95 each
43–3B Long sleeve, stripes	$21.95 each
XXL size, add $2 per shirt.	
Monogram, $4.95 each. Gift box, $5 each.	

TOTAL PRICE OF ALL ITEMS (EXCLUDING MONOGRAMS AND GIFT BOXES)	SHIPPING, PACKING, AND HANDLING
$0–25.00	$3.50
$25.01–75.00	$5.95
$75.01–125.00	$7.95
$125.01+	$9.95
Shipping to each additional address add $4.25.	

CHAPTER 1 Whole Numbers and Decimals

33. Find the total cost of ordering four long-sleeve, solid-color shirts and two short-sleeve, striped shirts, all size XXL and all shipped to your home.

33. _____

34. What is the total cost of eight long-sleeve XXL shirts, five in solid colors and three striped? Include the cost of shipping the solid shirts to your home and the striped shirts to your brother's home.

34. _____

35. (a) What is the total cost, including shipping, of sending three short-sleeve, solid-color shirts, with monograms, in a gift box to your uncle for his birthday?

(a) _____

(b) _____

(b) How much did the monogram, gift box, and shipping add to the cost of your gift?

36. (a) Suppose you order one of each type of shirt for yourself, adding a monogram on each of the solid-color shirts. At the same time, you order three long-sleeve, striped, size-XXL shirts to be shipped to your father at a different address in a gift box. Find the total cost of your order.

(a) _____

(b) _____

(b) What is the difference in total cost (excluding shipping) between the shirts for yourself and the gift for your father?

QUICK CHECK ANSWERS

1. (a) 28.944 (b) .0043
2. $417.38
3. $342,200,000
4. (a) 54.884 (b) 19.6405
5. (a) 28.125 (b) 26.875

Chapter 1   Quick Review

Chapter Terms   Review the following terms to test your understanding of the chapter. For each term you do not know, refer to the page number found next to that term.

addends [p. 3]	difference [p. 5]	multiplicand [p. 6]	round whole numbers [p. 2]
addition [p. 3]	digits [p. 2]	multiplication [p. 3]	subtraction [p. 3]
amount [p. 3]	dividend [p. 7]	multiplier [p. 6]	subtrahend [p. 5]
borrow [p. 5]	division [p. 3]	operations [p. 3]	sum [p. 3]
comma [p. 2]	divisor [p. 7]	partial products [p. 6]	total [p. 3]
decimal number [p. 20]	front-end rounding [p. 4]	product [p. 6]	trailing zeros [p. 24]
decimal point [p. 2]	indicator word [p. 14]	quotient [p. 7]	whole numbers [p. 2]
decimal system [p. 2]	minuend [p. 5]	round [p. 2]	

CONCEPTS

1.1 Reading and writing whole numbers

The word *and* is not used. Commas help divide thousands, millions, and billions. A comma is not needed with a four-digit number.

1.1 Rounding whole numbers

Rules for rounding:
1. Identify the position to be rounded. Draw a line under that place.
2. If the digit to the right of the underlined place is 5 or more, increase by 1. If the digit is 4 or less, do not change.
3. Change all digits to the right of the underlined digit to zero.

1.1 Front-end rounding

Front-end rounding leaves only the first digit as a nonzero digit. All other digits are changed to zero.

1.1 Addition of whole numbers

Add from top to bottom, starting with the ones place and working to the left. To check, add from bottom to top.

1.1 Subtraction of whole numbers

Subtract the subtrahend from the minuend to get the difference, borrowing when necessary. To check, add the difference to the subtrahend to get the minuend.

EXAMPLES

Write 795 and 9,650,036 using words.

seven hundred ninety-five

nine million, six hundred fifty thousand, thirty-six

Round:

7 2̲ 6 to the nearest ten — tens position — 5 or more, so add 1 to tens position

So, 726 rounds to 730.

1,4̲98,586 to the nearest million — millions position — 4 or less, so do not change

So, 1,498,586 rounds to 1,000,000.

Round each of the following, using front-end rounding.

76 rounds to 80

348 rounds to 300

6512 rounds to 7000

23,751 rounds to 20,000

652,179 rounds to 700,000

Problem (add down)
```
  1140
   687
    26
     9
 + 418
  1140
```
Check (add up)

Problem
```
  6 2 1
  4 7̸ 3̸ 8̸
  −  6 4 9
    4 0 8 9
```

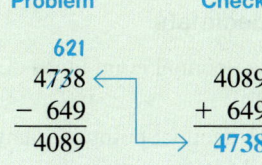

Check
```
   4089
 +  649
   4738
```

36 CHAPTER 1 Whole Numbers and Decimals

CONCEPTS	EXAMPLES

1.1 Multiplication of whole numbers

The multiplicand is multiplied by the multiplier, giving the product. When the multiplier has more than one digit, partial products must be used and then added.

```
      78    multiplicand
   ×  24    multiplier
     312    partial product
     156    partial product (one position left)
    1872    product
```

1.1 Division of whole numbers

÷ and ⟌ mean divide.
A —, as in $\frac{25}{5}$, means divide 25 by 5.
Also, the /, as in 25/5, means to divide 25 by 5.
Remainders are usually expressed as decimals.

```
                          44    quotient
           divisor    2⟌88      dividend
                      88
                       0
```

If answer is rounded, the check will not be perfect.

1.2 Application problems

Follow these steps.
1. Read the problem carefully.
2. Work out a plan using *indicator words* before starting.
3. Estimate a reasonable answer.
4. Solve the problem. If the answer is reasonable, check. If it is not, start over.

Shauna Gallegos earns $118 on Sunday, $87 on Monday, and $63 on Tuesday. Find her total earnings for the three days. Total means to add. An estimate using front-end rounding is $100 + $90 + $60 is $250.

```
   $268    Check
   $118
   $ 87
  +$ 63
   $268    total earnings are reasonable
```

1.3 Reading and rounding decimals

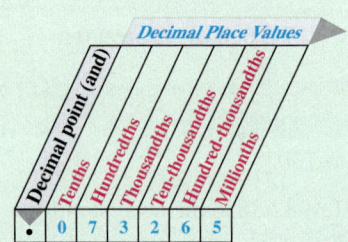

Read 1.35 as "one and thirty-five hundredths."
Round .073265 to the nearest ten-thousandth.

.0732|65
         ↑
         └── ten-thousandth position

Since the digit to the right is 6, increase the ten-thousandths digit by 1 and drop all digits to the right. So, .073265 rounds to .0733.

1.4 Addition and subtraction of decimals

Decimal points must be in a column. Attach zeros to keep digits in their correct columns.

Add: 5.68 + 785.3 + .007 + 10.1062
Line up the decimal points and add from right to left.

```
       5.6800   ⟵
     785.3000   ⟵  Attach zeros.
        .0070   ⟵
    + 10.1062
     801.0932
```

1.5 Multiplication of decimals

Multiply as if decimals are whole numbers. Place the decimal point as follows.
1. Count digits to the right of decimal points.
2. Count from right to left the same number of places as in step 1. Zeros must be attached on the left if necessary.

Multiply: .169 × .21

```
     .169     3 digits to the right of the decimal place
   × .21      2 digits to the right of the decimal place
     169
     338
   .03549    total of 5 digits to the right of the decimal
     ↑       place in answer
     └──── Attach one zero.
```

1.5 Division of decimals

1. Move the decimal point in the divisor all the way to the right.
2. Move the decimal point the same number of places to the right in the dividend.
3. Place a decimal point in the answer position directly above the dividend decimal point.
4. Divide as with whole numbers.

Divide 52.8 by .75

```
              70.4                Check
       .75.⟌52.80.0                70.4
            52 5                 ×  .75
               30                 3520
                0                 4928
               30 0              52.800
               30 0
                  0
```

Case Study

COST OF GETTING MARRIED

Interested in getting married? You might be surprised at the cost. A local newspaper said that the average costs of a wedding with 200 guests increased from $24,168 in 2010 to $28,540 in 2017. The bar graph gives the breakdown of average costs. Use the information to answer the questions that follow.

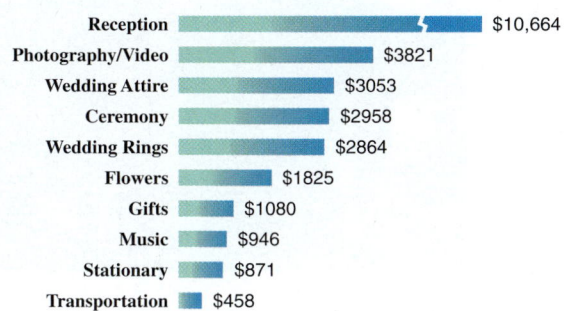

1. Find the combined cost of the reception, photography, wedding attire, and the ceremony.

    1. _____

2. Find the difference in the average costs of a wedding in 2010 versus 2017.

    2. _____

3. If you budget $7500 for the wedding reception and the cost per person is $42, find the number of guests you can invite and the amount remaining.

    3. _____

4. If costs force you to limit the number of guests to 115 and the cost of the reception to $5325, find the amount spent per person rounded to the nearest cent.

    4. _____

5. A couple plans to spend $22,000 for their wedding. Their parents have agreed to pay $8000 of the costs. If the couple has excess income of $650 per month, how many months will it require for them to pay their share of the costs?

    5. _____

INVESTIGATE

List five things that could be changed to bring the cost of a wedding down.

Case in Point Summary Exercise

SUBWAY

www.subway.com

Facts:

- 1965: First store opened in Connecticut
- 1974: First store franchised to an individual
- 1995: Supports cancer research
- 2018: More restaurants than any other restaurant chain

With the help of a family friend, seventeen-year-old Fred DeLuca opened the first Subway store in 1965. DeLuca believed in using only the freshest ingredients and drove many miles each week in his Volkswagen bug to buy fresh produce. Later, management decided to "give back" to the community and now supports research related to both heart disease and cancer.

Profit margins are surprisingly small in most businesses, so managers must work very hard to control costs. A primary responsibility of a store manager at Subway is to control costs. Of course, the store manager is also responsible for many issues related to employees, the quality of food and service, and also sales.

1. Jessica Fernandez is the manager of a Subway store. Find the total of an invoice with the following costs: produce, $486.12; meat and cheese, $1236.14; bread dough, $364.76; shipping charges, $103.75.

    1. _____

2. One employee worked the following hours during the week: Thursday, 3.5 hours; Friday, 4.5 hours; Saturday, 6 hours; and Sunday, 5.5 hours. He is paid $8.65 per hour. Find the total number of hours worked and the pay for the week rounded to the nearest cent.

    2. _____

3. Fernandez notes that meat and cheese costs went up from $1864.92 last week to $2065.48 this week and attributes it to additional customers. Find the difference between the two. If the average cost of meat and cheese per sub sandwich is $.94, estimate the number of additional customers this week compared to last week, rounded to the nearest whole number.

    3. _____

4. This month, Fernandez plans to spend four times as much on advertising as the $168.32 spent last week. Find the amount spent on advertising. If the increased advertising brings in 1.3 times last week's revenue of $10,984.76, estimate the revenue.

    4. _____

Discussion Question: *If sales increase by $1500 in a week, do you think profit for the week will also increase by about $1500? Why or why not?*

Source: Subway, www.subway.com

Chapter 1   Test

To help you review, the bracketed numbers indicate the section in which the topic is discussed.

Round as indicated.   [1.1]

1. 844 to the nearest ten
2. 21,958 to the nearest hundred
3. 671,529 to the nearest thousand

Round each of the following, using front-end rounding. [1.1]

4. 50,987
5. 851,004
6. One week, Katie Nopper earned the following commissions: Monday, $124; Tuesday, $88; Wednesday, $62; Thursday, $137; Friday, $195. Find her total amount of commissions for the week.   [1.2]
7. A rental business buys three airless sprayers at $1540 each, five rototillers at $695 each, and eight footstools at $38 each. Find the total cost of the equipment purchased.   [1.2]

Round as indicated.   [1.3]

8. $21.0568 to the nearest cent
9. $364.345 to the nearest cent
10. $7246.49 to the nearest dollar

Solve each problem.   [1.4 and 1.5]

11. $9.6 + 8.42 + 3.715 + 159.8 =$ _____

12. 2.715
    32.78
    426.3
    $+37$

13. 341.4
    -207.8

14. $3.8 - .0053$

15. 21.98
    $\times.72$

16. 218.6
    $\times.037$

17. $21.8\overline{)252.008}$

18. $24,500 \div 70 =$ ___

19. $57.358 \div 2.41 =$ ___

40 CHAPTER 1 Whole Numbers and Decimals

20. A manager at Subway wants to find the total cost of 24.8 pounds of sliced turkey at $1.89 a pound and 38.2 pounds of provolone cheese at $4.52 a pound. Find the final cost. [1.4 and 1.5]

20. _____

21. Roofing material costs $84.52 per square (10 ft × 10 ft). The roofer charges $55.75 per square for labor, plus $9.65 per square for supplies. Find the total cost for 26.3 squares of installed roof. Round to the nearest cent. [1.4 and 1.5]

21. _____

22. A federal law requires that all residential toilets sold in the United States use no more than 1.6 gallons of water per flush. Prior to this legislation, conventional toilets used 3.4 gallons of water per flush. Find the amount of water saved in one year by a family flushing the toilet 22 times each day (1 year = 365 days). [1.4 and 1.5]

22. _____

23. Steve Hamilton bought 135.5 meters of steel rod at $.86 per meter and 12 meters of brass rod at $2.18 per meter. How much change did he get from eight $20 bills? [1.4 and 1.5]

23. _____

24. A supermarket sells bananas for $1.74 per kilogram. Find the price per pound by dividing $1.74 by 2.2 pounds. Round to the nearest cent. [1.5]

24. _____

25. A concentrated fertilizer must be applied at the rate of .058 ounce per seedling. Find the number of seedlings that can be fertilized with 14.674 ounces of fertilizer. [1.5]

25. _____

26. A veterinarian decides that a cat with a kidney infection should get .65 milliliters of antibiotic twice a day for 7 days. Find the total amount of antibiotic needed rounded to the nearest milliliter.

26. _____

27. A family budgets $100 a month so that all four members can convert to the newest iPhone. At the store, they found that the iPhones will cost them $36.95 a month each, not including cell phone service. Find the monthly shortage in the budget related to just the cost of the phones.

27. _____

Fractions

2

CHAPTER CONTENTS

2.1 Fractions

2.2 Addition and Subtraction of Fractions

2.3 Addition and Subtraction of Mixed Numbers

2.4 Multiplication and Division of Fractions

2.5 Converting Decimals to Fractions and Fractions to Decimals

CASE IN POINT

KARA OAKS has been employed at The Home Depot for several years. During this time, she has worked in the hardware and plumbing departments, and she now manages the cabinetry department and helps people design replacement cabinets. Knowing and using fractions is a key part of Oaks's job. As shown in the Summary Case at the end of the chapter, she must be very accurate when determining the specifications for cabinets.

2.1 Fractions

OBJECTIVES
1. Recognize types of fractions.
2. Convert mixed numbers to improper fractions.
3. Convert improper fractions to mixed numbers.
4. Write a fraction in lowest terms.
5. Use the rules for divisibility.

CASE IN POINT The employees in the cabinetry department at Home Depot must be able to use decimals, fractions, and mixed numbers in order to work with customers. The measurements of cabinets, trim pieces, and room sizes never seem to be an even number of inches—they often involve fractions.

A **fraction** represents part of a whole. Fractions are written in the form of one number over another, with a line between the two numbers, as in the following.

$$\frac{5}{8} \quad \frac{1}{4} \quad \frac{9}{7} \quad \frac{13}{10} \quad \longleftarrow \text{numerator}$$
$$\longleftarrow \text{denominator}$$

The number above the line is the **numerator**, and the number below the line is the **denominator**. In the fraction $\frac{2}{3}$, the numerator is 2 and the denominator is 3. The denominator is the number of equal parts into which something is divided. The numerator tells how many of these parts are needed. For example, $\frac{2}{3}$ is "2 parts out of 3 equal parts," or $\frac{2}{3}$ of an apple.

$\frac{2}{3}$ means 2 parts out of 3 equal parts

OBJECTIVE 1 Recognize types of fractions. **Proper fractions**, also called common fractions have numerators that are smaller than their denominators. Proper fractions have a value of less than 1. **Improper fractions** have numerators that are equal to or greater than the denominators. Improper fractions have a value of 1 or more. Here are some examples.

Proper fractions: $\quad \frac{1}{2}, \frac{3}{4}, \frac{7}{13}, \frac{91}{100} \quad$ **numerator less than denominator**

Improper fractions: $\quad \frac{3}{3}, \frac{9}{7}, \frac{125}{4}, \frac{13}{12} \quad$ **numerator equal to or greater than denominator**

To write a whole number as a fraction, place the whole number as the numerator on top of a denominator of 1, as shown here.

$$8 = \frac{8}{1} \quad 25 = \frac{25}{1} \quad 4 = \frac{4}{1} \quad 135 = \frac{135}{1}$$

A **mixed number** is the sum of a whole number and a fraction. So, $5\frac{3}{8}$ is a mixed number meaning $5 + \frac{3}{8}$. It is even read as *five and three-eighths* or as the sum of a whole number and a fraction. Here are some other mixed numbers.

$17\frac{3}{5}$ seventeen and three-fifths

$92{,}407\frac{3}{4}$ ninety-two thousand, four hundred seven and three-fourths

It is important to realize that the number 0 is never used in the denominator of any fraction, since division by 0 is not defined.

2.1 Fractions

OBJECTIVE 2 Convert mixed numbers to improper fractions. Follow the steps shown to convert a mixed number to an improper fraction.

> **Converting a Mixed Number to an Improper Fraction**
> Step 1   Multiply the whole number by the denominator of the fraction.
> Step 2   Add the product to the numerator of the fraction to find the new numerator.
> Step 3   Keep the same denominator.

Here is an example.

Step 1 Multiply the whole number by the denominator.
Step 2 Add the product to the numerator.

$$4\frac{5}{8} = \frac{(4 \times 8) + 5}{8} = \frac{37}{8}$$

Step 3 Keep the same denominator.

Converting Mixed Numbers to Improper Fractions

EXAMPLE 1 Kara Oaks read the instructions that came with a new microwave oven. The opening in the kitchen cabinet needed to be $25\frac{3}{4}$ inches wide by $16\frac{1}{2}$ inches tall in order to install the microwave oven. Convert both of these mixed numbers to improper fractions.

(a) $25\frac{3}{4}$ (b) $16\frac{1}{2}$

SOLUTION

(a) Multiply the whole number 25 by the denominator 4 to find 100. Add the product to the numerator of the fraction 3 to find 103. Keep the same denominator of 4.

$$25\frac{3}{4} = \frac{(25 \times 4) + 3}{4} = \frac{103}{4}$$

(b) $16\frac{1}{2} = \frac{(16 \times 2) + 1}{2} = \frac{33}{2}$

> **QUICK CHECK 1**
> Convert to improper fractions.
> (a) $15\frac{1}{3}$ (b) $21\frac{3}{8}$

OBJECTIVE 3 Convert improper fractions to mixed numbers. Use the steps shown here to convert an improper fraction to a mixed number.

> **Converting an improper fraction to a mixed number**
> Step 1   Divide the numerator of the fraction by the denominator.
> Step 2   The quotient is the whole-number part of the mixed number.
> Step 3   The remainder is the numerator of the fraction.
> Step 4   Keep the same denominator in the fraction part of the mixed number.

For example, to convert $\frac{17}{5}$ to a mixed number, first divide 17 by 5. The **quotient** is the whole-number part and the remainder is the numerator of the fraction. Keep the same denominator.

$$\begin{array}{r} 3 \text{ R2} \\ 5\overline{)17} \\ \underline{15} \\ 2 \end{array} \qquad \text{So, } \frac{17}{5} = 3\frac{2}{5}.$$

44 CHAPTER 2 Fractions

Check the calculation by converting the mixed number $3\frac{2}{5}$ back to an improper fraction.

$$3\frac{2}{5} = \frac{(3 \times 5) + 2}{5} = \frac{17}{5}$$

The answer checks, so the fraction $\frac{17}{5}$ equals the mixed number $3\frac{2}{5}$. These are two different ways to write the same number.

Converting Improper Fractions to Mixed Numbers

EXAMPLE 2 Convert the following improper fractions to mixed numbers.

(a) $\frac{27}{4}$ (b) $\frac{29}{8}$ (c) $\frac{42}{7}$

SOLUTION

(a) Convert $\frac{27}{4}$ to a mixed number by dividing 27 by 4.

$$\begin{array}{r} 6 \\ 4\overline{)27} \\ \underline{24} \\ 3 \end{array} \qquad \frac{27}{4} = 6\frac{3}{4}$$

Therefore, $\frac{27}{4} = 6\frac{3}{4}$.

(b) Divide 29 by 8 to convert $\frac{29}{8}$ to a mixed number.

$$\begin{array}{r} 3 \\ 8\overline{)29} \\ \underline{24} \\ 5 \end{array} \qquad \frac{29}{8} = 3\frac{5}{8}$$

(c) Divide 42 by 7.

$$\begin{array}{r} 6 \\ 7\overline{)42} \\ \underline{42} \\ 0 \end{array} \qquad \frac{42}{7} = 6$$

QUICK CHECK 2

Convert to mixed numbers.

(a) $\frac{73}{4}$ (b) $\frac{32}{5}$

OBJECTIVE 4 Write a fraction in lowest terms. If both the numerator and denominator of a fraction cannot be divided without a remainder by any number other than 1, then the fraction is in **lowest terms**. For example, 2 and 3 cannot be divided without a remainder by any number other than 1, so the fraction $\frac{2}{3}$ is in lowest terms. In the same way, $\frac{1}{9}$, $\frac{4}{11}$, $\frac{12}{17}$, and $\frac{13}{15}$ are in lowest terms.

When both numerator and denominator *can* be divided without a remainder by a number other than 1, the fraction is *not* in lowest terms. For example, both 15 and 25 can be divided by 5 with no remainder, so the fraction $\frac{15}{25}$ is not in lowest terms. Write $\frac{15}{25}$ in lowest terms by dividing both numerator and denominator by 5, as follows.

$$\frac{15}{25} = \frac{15 \div 5}{25 \div 5} = \frac{3}{5} \quad \text{lowest terms}$$

Divide by 5.

2.1 Fractions

Writing Fractions in Lowest Terms

EXAMPLE 3 Write the following fractions in lowest terms.

(a) $\dfrac{15}{40}$ (b) $\dfrac{33}{39}$

SOLUTION

Look for a number that can be divided without a remainder into both the numerator and the denominator.

(a) Both 15 and 40 can be divided by **5**.

$$\dfrac{15}{40} = \dfrac{15 \div 5}{40 \div 5} = \dfrac{3}{8} \quad \text{lowest terms}$$

(b) Divide both numbers by 3.

$$\dfrac{33}{39} = \dfrac{33 \div 3}{39 \div 3} = \dfrac{11}{13} \quad \text{lowest terms}$$

QUICK CHECK 3

Write in lowest terms.

(a) $\dfrac{36}{45}$ (b) $\dfrac{48}{66}$

OBJECTIVE 5 Use the rules for divisibility. It is sometimes difficult to tell which numbers will divide evenly into another number. The following rules may help.

Rules for Divisibility

A whole number can be evenly divided by

2 if the last digit is an even number, such as 0, 2, 4, 6, or 8
3 if the sum of the digits is divisible by 3 with no remainder
4 if the last two digits are divisible by 4 with no remainder
5 if the last digit is 0 or 5
6 if the number is even and the sum of the digits is divisible by 3 with no remainder
8 if the last three digits are divisible by 8 with no remainder
9 if the sum of all the digits is divisible by 9 with no remainder
10 if the last digit is 0

Quick TIP
There is no easy rule to see if a number is evenly divisible by 7.

Using the Divisibility Rules

EXAMPLE 4 Determine whether the following statements are true.

(a) 3,746,892 is evenly divisible by 4.

(b) 15,974,802 is evenly divisible by 9.

SOLUTION

(a) The number 3,746,892 is evenly divisible by 4, since the last two digits form a number divisible by 4 without remainder.

$$3,746,\underline{892}$$

92 is divisible by 4.

Quick TIP
Testing for divisibility by adding the digits works only for 3 and 9.

(b) See if 15,974,802 is evenly divisible by 9 by adding the digits of the number.

$$1 + 5 + 9 + 7 + 4 + 8 + 0 + 2 = \mathbf{36}$$

36 is divisible by 9.

Since 36 is divisible by 9, the given number is divisible by 9.

The rules for divisibility help determine *only whether a number is evenly divisible by another number*. They cannot be used to find the result. The division must actually be done to find the quotient.

QUICK CHECK 4

Determine **(a)** whether 628,375,210 is evenly divisible by 5, and **(b)** whether 825,693,471 is evenly divisible by 3.

2.1 Exercises // MyLab Math

The shaded sections below contain solutions to help you get a QUICK START on the various types of exercises.

Convert the following mixed numbers to improper fractions. (See Example 1.)

1. $3\dfrac{5}{8} = \dfrac{29}{8}$

   $\dfrac{(8 \times 3) + 5}{8} = \dfrac{29}{8}$

2. $2\dfrac{4}{5} = \dfrac{14}{5}$

   $\dfrac{(5 \times 2) + 4}{5} = \dfrac{14}{5}$

3. $4\dfrac{1}{4} = $ _____

4. $3\dfrac{2}{3} = $ _____

5. $12\dfrac{2}{3} = $ _____

6. $2\dfrac{8}{11} = $ _____

7. $22\dfrac{7}{8} = $ _____

8. $17\dfrac{5}{8} = $ _____

9. $7\dfrac{6}{7} = $ _____

10. $21\dfrac{14}{15} = $ _____

11. $15\dfrac{19}{23} = $ _____

12. $7\dfrac{9}{16} = $ _____

Convert the following improper fractions to mixed or whole numbers and write in lowest terms. (See Examples 2 and 3.)

13. $\dfrac{13}{4} = 3\dfrac{1}{4}$

    $\begin{array}{r} 3 \\ 4\overline{)13} \\ \underline{12} \\ 1 \end{array}\quad 3\dfrac{1}{4}$

14. $\dfrac{9}{5} = 1\dfrac{4}{5}$

    $\begin{array}{r} 1 \\ 5\overline{)9} \\ \underline{5} \\ 4 \end{array}\quad 1\dfrac{4}{5}$

15. $\dfrac{8}{3} = $ _____

16. $\dfrac{23}{10} = $ _____

17. $\dfrac{38}{10} = $ _____

18. $\dfrac{56}{8} = $ _____

19. $\dfrac{40}{11} = $ _____

20. $\dfrac{78}{12} = $ _____

21. $\dfrac{125}{63} = $ _____

22. $\dfrac{195}{45} = $ _____

23. $\dfrac{183}{25} = $ _____

24. $\dfrac{720}{149} = $ _____

25. Explain how to change a mixed number to an improper fraction. (See Objective 2.)

26. Explain how to change an improper fraction to a mixed number. (See Objective 3.)

C// indicates an exercise that is related to the Case in Point feature.

2.1 Exercises 47

Write the following in lowest terms. (See Example 3.)

27. $\dfrac{8}{16} = \dfrac{1}{2}$

$\dfrac{8 \div 8}{16 \div 8} = \dfrac{1}{2}$

28. $\dfrac{15}{20} = \dfrac{3}{4}$

$\dfrac{15 \div 5}{20 \div 5} = \dfrac{3}{4}$

29. $\dfrac{25}{40} = $ ___

30. $\dfrac{36}{42} = $ ___

31. $\dfrac{27}{45} = $ ___

32. $\dfrac{112}{128} = $ ___

33. $\dfrac{165}{180} = $ ___

34. $\dfrac{12}{600} = $ ___

GOLD Pure gold is 24-karat gold. Due both to the high cost of pure gold and the fact that pure gold is soft and can be bent, gold jewelry is often mixed with an alloy and sold as 18-karat, 14-karat, or 10-karat. Write each fraction in lowest terms.

35. 24 karat = (24 parts gold, no alloy) $\dfrac{24}{24} = 1$ 35. 1 _____

36. 18 karat = (18 parts gold, 6 parts alloy)   36. _____

37. 14 karat = (14 parts gold, 10 parts alloy)   37. _____

38. 10 karat = (10 parts gold, 14 parts alloy)   38. _____

39. What does it mean when a fraction is expressed in lowest terms? (See Objective 4.)

40. Eight rules of divisibility were given. Write the three rules that are most useful to you. (See Objective 5.)

Put a check mark in the blank if the number at the left is evenly divisible by the number at the top. Put an X in the blank if the number is not divisible. (See Example 4.)

	2	3	4	5	6	8	9	10
41. 32	√	X	√	X	X	√	X	X
42. 45	X	√	X	√	X	X	√	X
43. 60	_	_	_	_	_	_	_	_
44. 72	_	_	_	_	_	_	_	_
45. 90	_	_	_	_	_	_	_	_
46. 105	_	_	_	_	_	_	_	_
47. 4172	_	_	_	_	_	_	_	_
48. 5688	_	_	_	_	_	_	_	_

QUICK CHECK ANSWERS

1. (a) $\dfrac{46}{3}$ (b) $\dfrac{171}{8}$
2. (a) $18\tfrac{1}{4}$ (b) $6\tfrac{2}{5}$
3. (a) $\tfrac{4}{5}$ (b) $\tfrac{8}{11}$

4. (a) 628,375,210 is evenly divisible by 5.
   (b) 825,693,471 is evenly divisible by 3.

2.2 Addition and Subtraction of Fractions

OBJECTIVES
1. Add and subtract like fractions.
2. Find the least common denominator.
3. Add and subtract unlike fractions.
4. Rewrite fractions with a common denominator.

OBJECTIVE 1 Add and subtract like fractions. Fractions with the same denominator are called **like fractions**. Such fractions have a **common denominator**. For example, $\frac{3}{4}$ and $\frac{5}{4}$ are *like* fractions with a common denominator of 4, while $\frac{4}{7}$ and $\frac{4}{9}$ are *not like* fractions.

Adding (or Subtracting) Like Fractions (Which Have Common Denominators)
Step 1   Add (or subtract) the numerators.
Step 2   Place the sum (or difference) over the common denominator.

Adding and Subtracting Like Fractions

EXAMPLE 1 Add or subtract.

(a) $\dfrac{3}{4} + \dfrac{1}{4} + \dfrac{5}{4}$ (b) $\dfrac{11}{15} - \dfrac{4}{15}$

SOLUTION

The fractions in both parts of this example are like fractions. Add or subtract the numerators and place the result over the common denominator.

(a) $\dfrac{3}{4} + \dfrac{1}{4} + \dfrac{5}{4} = \dfrac{3 + 1 + 5}{4}$ ← Add the numerators.
   ← Write the common denominator.

$= \dfrac{9}{4} = 2\dfrac{1}{4}$ ← Write the answer as a mixed number.

(b) $\dfrac{11}{15} - \dfrac{4}{15} = \dfrac{11 - 4}{15} = \dfrac{7}{15}$

Quick TIP
When adding or subtracting like fractions, keep the same denominator.

QUICK CHECK 1
Add or subtract.

(a) $\dfrac{5}{8} + \dfrac{7}{8} + \dfrac{1}{8}$ (b) $\dfrac{17}{21} - \dfrac{4}{21}$

Fractions are commonly used in the construction trades as shown in the drawing below. The fractions in the shelf-end base cross section on the left are *like fractions*, whereas the fractions in the shelf-end peninsula base on the right are *unlike fractions*, which do not have common denominators.

Shelf-End Base: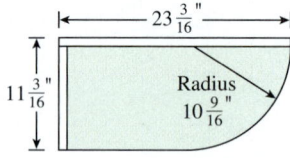
Cross Section

Shelf-End Peninsula Base: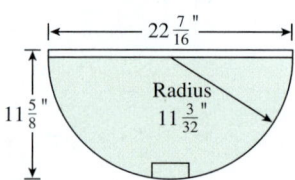
Cross Section

Quick TIP
The symbol " represents inches.

OBJECTIVE 2 Find the least common denominator. Fractions with different denominators, such as $\frac{3}{4}$ and $\frac{2}{3}$, are **unlike fractions**. Add or subtract unlike fractions by first writing the fractions with a common denominator. The **least common denominator (LCD)** for two or more fractions is the smallest whole number that can be divided, without a remainder, by all the denominators of the fractions. For example, the LCD of the fractions $\frac{3}{4}, \frac{5}{6}$, and $\frac{1}{2}$ is 12, since 12 is the smallest number that can be divided evenly by 4, 6, and 2.

There are two methods of finding the least common denominator:

Inspection. With small denominators, it may be possible to find the least common denominator by inspection. For example, the LCD for $\frac{1}{3}$ and $\frac{1}{5}$ is 15, the smallest number that can be divided evenly by both 3 and 5.

Method of prime numbers. If the LCD cannot be found by inspection, use the method of prime numbers, as shown in the next two examples. First, we define a prime number.

2.2 Addition and Subtraction of Fractions

> A **prime number** is a whole number that can be divided without a remainder by exactly two distinct numbers: itself and 1. Prime numbers are 2, 3, 5, 7, 11, 13, 17, and so on. The number 1 is *not* prime because it can be divided evenly by only *one* number: the number 1.

Finding the Least Common Denominator

EXAMPLE 2 Use the method of prime numbers to find the least common denominator for $\frac{5}{12}$, $\frac{7}{18}$, and $\frac{11}{20}$.

SOLUTION

First write the three denominators: 12   18   20

Begin by trying to divide the three denominators by the smallest prime number, 2. Write each quotient directly above the given denominator as follows.

$$\begin{array}{r} 6 \quad 9 \quad 10 \\ 2\overline{)12 \quad 18 \quad 20} \end{array}$$

This way of writing the division is just a handy way of writing the separate problems $2\overline{)12}$, $2\overline{)18}$, and $2\overline{)20}$. Two of the new quotients, 6 and 10, can still be divided by 2, so perform the division again. Since 9 cannot be divided evenly by 2, just bring up the 9.

$$\begin{array}{r} 3 \quad 9 \quad 5 \quad \text{Just bring 9 up.} \\ 2\overline{)6 \quad 9 \quad 10} \\ 2\overline{)12 \quad 18 \quad 20} \end{array}$$

Quick TIP
It does not matter which prime number you start with; the final list of prime numbers will be the same.

None of the new quotients in the top row can be divided by 2, so try the next prime number, 3. The numbers 3 and 9 can be divided by 3, and one of the new quotients can still be divided by 3, so the division is performed again.

$$\begin{array}{r} 1 \quad 1 \quad 5 \\ 3\overline{)1 \quad 3 \quad 5} \\ 3\overline{)3 \quad 9 \quad 5} \\ 2\overline{)6 \quad 9 \quad 10} \\ 2\overline{)12 \quad 18 \quad 20} \end{array}$$

Since none of the new quotients in the top row can be divided by 3, try the next prime number, 5. The number 5 can be used only once, as shown.

$$\begin{array}{r} 1 \quad 1 \quad 1 \\ 5\overline{)1 \quad 1 \quad 5} \\ 3\overline{)1 \quad 3 \quad 5} \\ 3\overline{)3 \quad 9 \quad 5} \\ 2\overline{)6 \quad 9 \quad 10} \\ 2\overline{)12 \quad 18 \quad 20} \end{array}$$

Now that the top row contains only 1s, find the least common denominator by multiplying the prime numbers in the left column: $2 \times 2 \times 3 \times 3 \times 5 = 180$.

QUICK CHECK 2

Use prime numbers to find the least common denominator for $\frac{3}{5}$, $\frac{5}{6}$, and $\frac{3}{20}$.

Finding the Least Common Denominator

EXAMPLE 3 Find the least common denominator for $\frac{3}{8}$, $\frac{5}{12}$, and $\frac{9}{10}$.

SOLUTION

Write the denominators in a row and use the method of prime numbers.

$$\begin{array}{r} 1 \quad 1 \quad 1 \\ 5\overline{)1 \quad 1 \quad 5} \\ 3\overline{)1 \quad 3 \quad 5} \\ 2\overline{)2 \quad 3 \quad 5} \\ 2\overline{)4 \quad 6 \quad 5} \\ \text{Start here} \longrightarrow 2\overline{)8 \quad 12 \quad 10} \end{array}$$

The least common denominator is $2 \times 2 \times 2 \times 3 \times 5 = 120$.

QUICK CHECK 3

Find the least common denominator for $\frac{4}{9}, \frac{5}{24}$, and $\frac{3}{4}$.

OBJECTIVE 3 Add and subtract unlike fractions. Unlike fractions may be added or subtracted using the following steps.

> **Adding or Subtracting Unlike Fractions**
> Step 1   Find the least common denominator (LCD).
> Step 2   Rewrite the unlike fractions as like fractions having the least common denominator.
> Step 3   Add or subtract numerators, placing answers over the LCD and reducing to lowest terms.

To add or subtract unlike fractions, rewrite the fractions with a common denominator. Since Example 2 shows that 180 is the least common denominator for $\frac{5}{12}, \frac{7}{18}$, and $\frac{11}{20}$, these three fractions can be added if each fraction is first written with a denominator of 180.

Step 1
$$\frac{5}{12} = \frac{}{180} \qquad \frac{7}{18} = \frac{}{180} \qquad \frac{11}{20} = \frac{}{180}$$

OBJECTIVE 4 Rewrite fractions with a common denominator. To rewrite the preceding fractions with a common denominator, first divide each denominator from the original fractions into the common denominator.

Step 2
$$12 \overline{)180}^{15} \qquad 18 \overline{)180}^{10} \qquad 20 \overline{)180}^{9}$$

Next multiply each quotient by the original numerator.

$$15 \times 5 = 75 \qquad 10 \times 7 = 70 \qquad 9 \times 11 = 99$$

Now, rewrite the fractions.

$$\frac{5}{12} = \frac{75}{180} \qquad \frac{7}{18} = \frac{70}{180} \qquad \frac{11}{20} = \frac{99}{180}$$

Finally, add.

Step 3
$$\frac{5}{12} + \frac{7}{18} + \frac{11}{20} = \frac{75}{180} + \frac{70}{180} + \frac{99}{180} = \frac{75 + 70 + 99}{180}$$

$$= \frac{244}{180} = 1\frac{64}{180} = 1\frac{16}{45} \qquad \text{Write the answer as a mixed number with the fraction in lowest terms.}$$

Adding and Subtracting Unlike Fractions

EXAMPLE 4 Add or subtract.

(a) $\frac{3}{4} + \frac{1}{2} + \frac{5}{8}$ (b) $\frac{9}{10} - \frac{3}{8}$

SOLUTION

(a) Inspection shows that the least common denominator is 8. Rewrite the fractions so each has a denominator of 8, then add.

$$\frac{3}{4} + \frac{1}{2} + \frac{5}{8} = \frac{6}{8} + \frac{4}{8} + \frac{5}{8} = \frac{6 + 4 + 5}{8} = \frac{15}{8} = 1\frac{7}{8}$$

(b) The least common denominator is 40. Rewrite the fractions so each has a denominator of 40, then subtract.

$$\frac{9}{10} - \frac{3}{8} = \frac{36}{40} - \frac{15}{40} = \frac{36 - 15}{40} = \frac{21}{40}$$

QUICK CHECK 4

Add or subtract.

(a) $\dfrac{2}{3} + \dfrac{5}{8} + \dfrac{3}{4}$ (b) $\dfrac{4}{5} - \dfrac{1}{3}$

Fractions can also be added or subtracted vertically, as shown in the next example.

Adding and Subtracting Unlike Fractions

EXAMPLE 5 Add or subtract.

(a) $\dfrac{2}{9} + \dfrac{3}{4}$ (b) $\dfrac{11}{16} + \dfrac{7}{12}$ (c) $\dfrac{7}{8} - \dfrac{5}{12}$

SOLUTION

First rewrite the fractions with a least common denominator.

(a) $\begin{aligned} \dfrac{2}{9} &= \dfrac{8}{36} \\ +\dfrac{3}{4} &= \dfrac{27}{36} \\ \hline &\dfrac{35}{36} \end{aligned}$

(b) $\begin{aligned} \dfrac{11}{16} &= \dfrac{33}{48} \\ +\dfrac{7}{12} &= \dfrac{28}{48} \\ \hline &\dfrac{61}{48} = 1\dfrac{13}{48} \end{aligned}$

(c) $\begin{aligned} \dfrac{7}{8} &= \dfrac{21}{24} \\ -\dfrac{5}{12} &= \dfrac{10}{24} \\ \hline &\dfrac{11}{24} \end{aligned}$

All calculator solutions are shown using a basic calculator. The calculator solution to part (b) uses the fraction key on the calculator.

11 [a^(b/c)] 16 [+] 7 [a^(b/c)] 12 [=] $1\dfrac{13}{48}$

Note: Refer to Appendix B for calculator basics.

QUICK CHECK 5

Add or subtract.

(a) $\dfrac{11}{12} + \dfrac{2}{3}$ (b) $\dfrac{9}{10} - \dfrac{7}{15}$

2.2 Exercises // MyLab Math

The shaded sections below contain solutions to help you get a **QUICK START** *on the various types of exercises.*

Write each fraction with the indicated denominator. (See Objective 3.)

1. $\dfrac{4}{5} = \dfrac{16}{20}$
   $20 \div 5 = 4$
   $4 \times 4 = 16$

2. $\dfrac{3}{4} = \dfrac{12}{16}$
   $16 \div 4 = 4$
   $4 \times 3 = 12$

3. $\dfrac{9}{10} = \dfrac{}{40}$

4. $\dfrac{7}{8} = \dfrac{}{56}$

5. $\dfrac{6}{5} = \dfrac{}{40}$

6. $\dfrac{7}{8} = \dfrac{}{64}$

7. $\dfrac{6}{7} = \dfrac{}{49}$

8. $\dfrac{11}{15} = \dfrac{}{120}$

C// indicates an exercise that is related to the Case in Point feature.

52 CHAPTER 2 Fractions

Find the least common denominator for each group of denominators using the method of prime numbers. (See Example 2.)

9. 3, 8, 24

	1	1
3)	3	1
2)	3	2
2)	3	4
2)	3	8

   $2 \times 2 \times 2 \times 3 = 24$

10. 18, 24, 72

	1	1	
3)	3	1	
3)	9	3	
2)	9	6	
2)	9	12	
2)	18	24	

    $2 \times 2 \times 2 \times 3 \times 3 = 72$

11. 12, 18, 20, _____

12. 18, 20, 24, _____

13. 15, 24, 32, _____

14. 6, 8, 10, 12, _____

15. 10, 35, 50, 60, _____

16. 5, 18, 25, 30, 36, _____

17. Prime numbers are used to find the least common denominator. Write the definition of a prime number in your own words. (See Objective 2.)

18. Explain the steps to change $\frac{3}{4}$ to a fraction having 24 as a denominator. (See Objective 4.)

Add or subtract. Write answers in lowest terms. (See Examples 4 and 5.)

19. $\frac{2}{5} + \frac{1}{5} = \frac{3}{5}$

    $\frac{2+1}{5} = \frac{3}{5}$

20. $\frac{2}{9} + \frac{4}{9} = \frac{2}{3}$

    $\frac{2+4}{9} = \frac{6}{9} = \frac{2}{3}$

21. $\frac{3}{10} + \frac{5}{10} =$ _____

22. $\frac{5}{8} + \frac{7}{8} =$ _____

23. $\frac{11}{12} - \frac{5}{12} =$ _____

24. $\frac{5}{7} - \frac{1}{7} =$ _____

25. $\frac{5}{12} - \frac{1}{16} =$ _____

26. $\frac{2}{3} - \frac{3}{8} =$ _____

27. $\frac{3}{4} + \frac{5}{9} + \frac{1}{3} =$ _____

28. $\frac{1}{4} + \frac{1}{8} + \frac{1}{12} =$ _____

29. $\frac{3}{7} + \frac{2}{5} + \frac{1}{10} =$ _____

30. $\dfrac{5}{6} + \dfrac{3}{4} + \dfrac{5}{8} =$ _____

31. $\dfrac{7}{10} + \dfrac{8}{15} + \dfrac{5}{6} =$ _____

32. $\dfrac{3}{10} + \dfrac{2}{5} + \dfrac{3}{20} =$ _____

33. $\begin{array}{r} \dfrac{3}{4} \\ \dfrac{2}{3} \\ +\dfrac{8}{9} \\ \hline \end{array}$

34. $\begin{array}{r} \dfrac{7}{12} \\ \dfrac{5}{8} \\ +\dfrac{7}{6} \\ \hline \end{array}$

35. $\begin{array}{r} \dfrac{8}{15} \\ \dfrac{3}{10} \\ +\dfrac{3}{5} \\ \hline \end{array}$

36. $\begin{array}{r} \dfrac{1}{6} \\ \dfrac{5}{9} \\ +\dfrac{13}{18} \\ \hline \end{array}$

37. $\begin{array}{r} \dfrac{7}{10} \\ -\dfrac{1}{4} \\ \hline \end{array}$

38. $\begin{array}{r} \dfrac{4}{5} \\ -\dfrac{2}{3} \\ \hline \end{array}$

39. $\begin{array}{r} \dfrac{5}{8} \\ -\dfrac{1}{3} \\ \hline \end{array}$

40. $\begin{array}{r} \dfrac{19}{24} \\ -\dfrac{5}{16} \\ \hline \end{array}$

41. Where are fractions used in everyday life? Think in terms of business applications, hobbies, and home life. Give three examples.

42. With the exception of the number 2, all prime numbers are odd numbers. However, not all odd numbers are prime numbers. Explain why these statements are true. (See Objective 2.)

Solve the following application problems.

43. GARDENING Zalia Todd is planting her flower bed and has ordered $\frac{1}{4}$ cubic yard of sand, $\frac{3}{8}$ cubic yard of mulch, and $\frac{1}{3}$ cubic yard of peat moss. Find the total cubic yards that she has ordered.

43. _____

44. AUTO REPAIR Chuck Manly has used his savings to repair his car. He spent $\frac{1}{4}$ of his savings for new tires, $\frac{1}{6}$ of his savings for brakes, $\frac{1}{10}$ of his savings for a tune-up, and $\frac{1}{12}$ of his savings for new belts and hoses. What fraction of his total savings has he spent?

44. _____

45. DEVICE ASSEMBLY When installing a cable to a computer, Ann Kuick must be certain that the proper type and size of mounting hardware is used. Find the total length of the bolt shown.

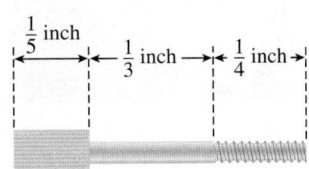

45. _____

46. CABINET INSTALLATION When installing cabinets for The Home Depot, Kara Oaks must be certain that the proper type and size of mounting screw are used. Find the total length of the screw.

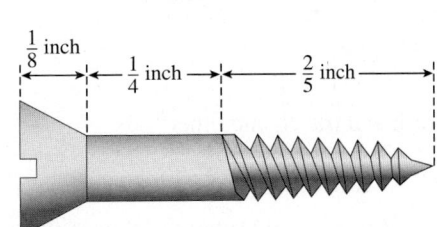

46. _____

47. PETROLEUM TRANSPORT Ken Faulk drives a tanker truck for Chemlake Transport. He leaves the refinery with his tanker filled to $\frac{7}{8}$ of capacity. If he delivers $\frac{1}{4}$ of the tank's contents at the first stop and $\frac{1}{3}$ of the tank's contents at the second stop, find the fraction of the tanker's contents remaining.

47. _____

54 CHAPTER 2 Fractions

48. **TRANSMISSION** A transmission contains $\frac{7}{8}$ gallon of hydraulic fluid. A cracked seal resulted in a loss of $\frac{1}{6}$ gallon of fluid in the morning and another $\frac{1}{3}$ gallon in the afternoon. Find the amount of fluid remaining.

48. _____

49. **CAKE RECIPE** Melissa Staple plans to triple the recipe for her favorite carrot cake in order to have enough for her grandmother's 88th birthday party. The recipe calls for $\frac{1}{4}$ cup of butter. Use addition to find the number of cups of butter needed.

49. _____

50. **NATURAL-FOODS STORE** Joan McKee wants to open a natural-foods store and has saved $\frac{2}{5}$ of the amount needed for startup costs. If she saves another $\frac{1}{8}$ of the amount needed and then $\frac{1}{6}$ more, find the total portion of the startup costs she has saved.

50. _____

51. **CABINET INSTALLATION** The mounting bracket shown was purchased at The Home Depot. Find the diameter of the hole, which is the distance across it.

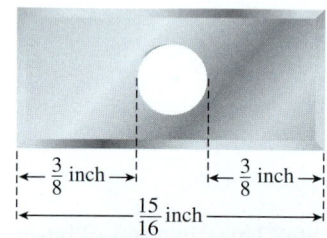

51. _____

52. **SPRINT TRAINING** Rona Martin competes in 100-meter sprints in high school. Today, she only has time to run $\frac{3}{4}$ mile. After stretching to warm up, her trainer has Martin sprint $\frac{1}{8}$ mile, $\frac{1}{4}$ mile, and $\frac{1}{4}$ mile. Find the additional distance she must run.

52. _____

STUDENT TIME MANAGEMENT *Refer to the circle graph to answer Exercises 53–56.*

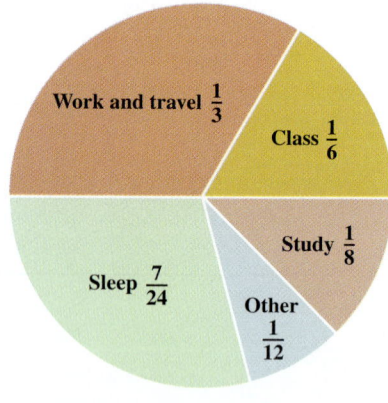

53. What fraction of the day was spent in class and study?

53. _____

54. What fraction of the day was spent in work and travel, and other?

54. _____

55. In which activity was the greatest amount of time spent? What fraction of the day was spent on this activity and class time?

55. _____

56. In which activity was the least amount of time spent? What fraction of the day was spent on this activity and study?

56. _____

Use the newspaper advertisement for this four-piece chisel set to answer Exercises 57 and 58.
(Source: Harbor Freight Tools.)

57. Find the difference in the cutting-edge width of the narrowest chisel and the second-to-widest chisel.

57. _____

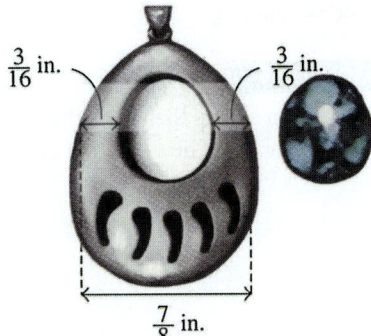

58. Find the difference in the cutting-edge width of the two chisels with the narrowest blades.

58. _____

59. **PERIMETER OF FENCING** A hazardous-waste site will require $\frac{7}{8}$ mile of security fencing. The site has four sides, three of which measure $\frac{1}{4}$ mile, $\frac{1}{6}$ mile, and $\frac{3}{8}$ mile. Find the length of the fourth side.

59. _____

60. **NATIVE-AMERICAN JEWELRY** Chakotay is fitting a turquoise stone into a bear-claw pendant. Find the diameter of the hole in the pendant. (The diameter is the distance across the center of the hole.)

60. _____

QUICK CHECK ANSWERS

1. (a) $\frac{13}{8} = 1\frac{5}{8}$ (b) $\frac{13}{21}$

2. 60

3. 72

4. (a) $\frac{49}{24} = 2\frac{1}{24}$ (b) $\frac{7}{15}$

5. (a) $1\frac{7}{12}$ (b) $\frac{13}{30}$

2.3 Addition and Subtraction of Mixed Numbers

OBJECTIVES
1. Add mixed numbers.
2. Add with carrying.
3. Subtract mixed numbers.
4. Subtract with borrowing.

OBJECTIVE 1 Add mixed numbers. To add mixed numbers, first add the fractions. Then add the whole numbers and combine the two answers. For example, add $16\frac{1}{8}$ and $5\frac{5}{8}$ as shown.

$$\text{First add the fractions: } \frac{1}{8} + \frac{5}{8} = \frac{1+5}{8} = \frac{6}{8} = \frac{3}{4} \quad \text{reduced}$$

Then add the whole numbers: $16 + 5 = 21$

Finally, write the sum of the fraction and whole-number parts as a mixed number.

$$21 + \frac{3}{4} = 21\frac{3}{4}$$

So, $16\frac{1}{8} + 5\frac{5}{8} = 21\frac{3}{4}$.

To add mixed numbers, change the mixed numbers, if necessary, so that the fraction parts have a common denominator.

Adding Mixed Numbers **EXAMPLE 1** Add $9\frac{2}{3}$ and $6\frac{1}{4}$.

SOLUTION

Inspection shows that 12 is the least common denominator. Write $9\frac{2}{3}$ as $9\frac{8}{12}$, and write $6\frac{1}{4}$ as $6\frac{3}{12}$. Then add. The work can be organized as follows.

$$9\frac{2}{3} = 9\frac{8}{12}$$
$$+6\frac{1}{4} = 6\frac{3}{12}$$
$$\overline{\phantom{+6\frac{1}{4} = }15\frac{11}{12}}$$

QUICK CHECK 1

Add $5\frac{1}{4}$ and $8\frac{3}{8}$.

OBJECTIVE 2 Add with carrying. If the sum of the fraction parts of mixed numbers is greater than 1, carry the excess from the fraction part to the whole-number part.

Adding with Carrying **EXAMPLE 2** A rubber gasket must extend around all four edges (perimeter) of the dishwasher door panel shown in the following picture before it is installed. Find the length of gasket material needed by adding $34\frac{1}{2}$ inches, $23\frac{3}{4}$ inches, $34\frac{1}{2}$ inches, and $23\frac{3}{4}$ inches.

Quick TIP

To add mixed numbers:
1. add fractions,
2. add whole numbers, and
3. combine the two to write as a mixed number.

SOLUTION

CASE IN POINT Write using the least common denominator and then add.

$$34\frac{1}{2} = 34\frac{2}{4}$$
$$23\frac{3}{4} = 23\frac{3}{4}$$
$$34\frac{1}{2} = 34\frac{2}{4}$$
$$+23\frac{3}{4} = 23\frac{3}{4}$$
$$\overline{114\frac{10}{4}} = 114 + \frac{10}{4} = 114 + 2\frac{2}{4} = 116\frac{2}{4} = 116\frac{1}{2} \text{ inches} \quad \text{length of gasket needed}$$

Dishwasher door panel $23\frac{3}{4}"\ h$

$34\frac{1}{2}"\ w$

QUICK CHECK 2

The four sides of a vegetable garden are $15\frac{1}{2}$ feet, $18\frac{3}{4}$ feet, $24\frac{1}{4}$ feet, and $30\frac{1}{2}$ feet. How many feet of fencing are needed to go around the garden?

OBJECTIVE 3 Subtract mixed numbers. To subtract two mixed numbers, change the mixed numbers, if necessary, so that the fraction parts have a common denominator. Then subtract the fraction parts and the whole-number parts separately. For example, subtract $3\frac{1}{12}$ from $8\frac{5}{8}$ by first finding that the least common denominator is 24. Then rewrite the problem as shown.

$$8\frac{5}{8} = 8\frac{15}{24} \quad \text{First subtract the fraction parts. Then subtract the whole-number parts.}$$
$$-3\frac{1}{12} = -3\frac{2}{24}$$
$$5\frac{13}{24} \quad \leftarrow \text{Subtract numerators.}$$
$$\uparrow \text{Subtract whole numbers.}$$

OBJECTIVE 4 Subtract with borrowing. The following example shows how to subtract when borrowing is needed.

EXAMPLE 3 Subtracting with Borrowing

(a) Subtract $6\frac{3}{4}$ from $10\frac{1}{8}$. (b) Subtract $15\frac{7}{12}$ from 41.

SOLUTION

Start by rewriting each problem with a common denominator.

(a) $10\frac{1}{8} = 10\frac{1}{8}$
$$-6\frac{3}{4} = -6\frac{6}{8}$$

Subtracting $\frac{6}{8}$ from $\frac{1}{8}$ requires borrowing from the whole number 10.
$$10\frac{1}{8} = 9 + 1 + \frac{1}{8}$$
$$= 9 + \frac{8}{8} + \frac{1}{8} = 9\frac{9}{8} \quad 1 = \frac{8}{8}$$

Rewrite the problem as shown. Check by adding $3\frac{3}{8}$ and $6\frac{3}{4}$. The answer should be $10\frac{1}{8}$.

$$10\frac{1}{8} = 9\frac{9}{8}$$
$$-6\frac{6}{8} = -6\frac{6}{8}$$
$$3\frac{3}{8}$$

(b) 41
$$-15\frac{7}{12}$$

To subtract the fraction $\frac{7}{12}$ requires borrowing 1 whole unit from 41.

$$41 = 40 + 1 = 40 + \frac{12}{12} = 40\frac{12}{12} \quad 1 = \frac{12}{12}$$

Rewrite the problem as shown. Check by adding $25\frac{5}{12}$ and $15\frac{7}{12}$. The answer should be 41.

$$41 = 40\frac{12}{12}$$
$$-15\frac{7}{12} = -15\frac{7}{12}$$
$$25\frac{5}{12}$$

🖩 The calculator solution to part (a) uses the fraction key.

10 $\boxed{a^{b/c}}$ 1 $\boxed{a^{b/c}}$ 8 $\boxed{-}$ 6 $\boxed{a^{b/c}}$ 3 $\boxed{a^{b/c}}$ 4 $\boxed{=}$ $3\frac{3}{8}$

Quick TIP
You do not have to write fractions using the least common denominator when adding or subtracting on a calculator.

QUICK CHECK 3
Subtract (a) $5\frac{2}{3}$ from $12\frac{3}{8}$ and (b) $17\frac{5}{9}$ from 73.

2.3 Exercises // MyLab Math

*The shaded sections below contain solutions to help you get a **QUICK START** on the various types of exercises.*

Add. Write each answer in lowest terms. (See Examples 1 and 2.)

1. $82\frac{3}{5}$ $82\frac{3}{5}$
   $+15\frac{1}{5}$ $+15\frac{1}{5}$
   $97\frac{4}{5}$ $97\frac{4}{5}$

2. $5\frac{1}{3}$ $5\frac{4}{12}$
   $+2\frac{1}{4}$ $+2\frac{3}{12}$
   $7\frac{7}{12}$ $7\frac{7}{12}$

3. $41\frac{1}{2}$
   $+39\frac{1}{4}$

4. $28\frac{1}{4}$
   $+23\frac{3}{5}$

58 CHAPTER 2 Fractions

5. $46\dfrac{3}{4}$
    $12\dfrac{5}{8}$
    $+\ 37\dfrac{4}{5}$
    _____

6. $26\dfrac{5}{8}$
    $17\dfrac{3}{14}$
    $+32\dfrac{2}{7}$
    _____

7. $32\dfrac{3}{4}$
    $6\dfrac{1}{3}$
    $+\ 14\dfrac{5}{8}$
    _____

8. $16\dfrac{7}{10}$
    $26\dfrac{1}{5}$
    $+\ 8\dfrac{3}{8}$
    _____

Subtract. Write each answer in lowest terms. (See Example 3.)

9. $16\dfrac{3}{4}\quad 16\dfrac{6}{8}$
    $-\ 12\dfrac{3}{8}\quad -\ 12\dfrac{3}{8}$
    _____
    $4\dfrac{3}{8}\quad4\dfrac{3}{8}$

10. $25\dfrac{13}{24}\quad 25\dfrac{13}{24}$
    $-\ 18\dfrac{5}{12}\quad -\ 18\dfrac{10}{24}$
    _____
    $7\dfrac{1}{8}\quad7\dfrac{3}{24}=7\dfrac{1}{8}$

11. $9\dfrac{7}{8}$
    $-6\dfrac{5}{12}$
    _____

12. 374
    $-211\dfrac{5}{6}$
    _____

13. 19
    $-\ 12\dfrac{3}{4}$
    _____

14. $71\dfrac{3}{8}$
    $-\ 62\dfrac{1}{3}$
    _____

15. $6\dfrac{1}{3}$
    $-\ 2\dfrac{5}{12}$
    _____

16. $72\dfrac{3}{10}$
    $-\ 25\dfrac{8}{15}$
    _____

17. In your own words, explain the steps you would take to add two large mixed numbers. (See Objective 1.)

18. When subtracting mixed numbers, explain when you need to borrow. Explain how to borrow using an example. (See Objective 4.)

Solve the following application problems.

19. **PART-TIME WORK** Loren Kabakov, a college student, works part time at the Cyber Coffeehouse. She worked $3\dfrac{3}{8}$ hours on Monday, $5\dfrac{1}{2}$ hours on Tuesday, $4\dfrac{3}{4}$ hours on Wednesday, $3\dfrac{1}{4}$ hours on Thursday, and 6 hours on Friday. How many hours did she work altogether?

    $3\dfrac{3}{8}+5\dfrac{1}{2}+4\dfrac{3}{4}+3\dfrac{1}{4}+6=3\dfrac{3}{8}+5\dfrac{4}{8}+4\dfrac{6}{8}+3\dfrac{2}{8}+6=21\dfrac{15}{8}=22\dfrac{7}{8}$ hours

    19. $22\dfrac{7}{8}$ hours

20. **TRAILER LOAD** A trailer is to be loaded with plasma televisions weighing $2\dfrac{5}{8}$ tons, DVD players weighing $6\dfrac{1}{2}$ tons, personal computers weighing $1\dfrac{5}{6}$ tons, and computer monitors weighing $3\dfrac{1}{4}$ tons. If the truck weighs $7\dfrac{3}{8}$ tons empty, find the total weight after it has been loaded.

    $2\dfrac{5}{8}+6\dfrac{1}{2}+1\dfrac{5}{6}+3\dfrac{1}{4}+7\dfrac{3}{8}=21\dfrac{7}{12}$ tons

    20. $21\dfrac{7}{12}$ tons

21. **WINDOW INSTALLATION** A contractor who installs windows for The Home Depot must attach a lead strip around all four sides of a custom-made stained glass window. If the window measures $34\dfrac{1}{2}$ by $23\dfrac{3}{4}$ inches, find the length of lead stripping needed.

    $34\dfrac{1}{2}+23\dfrac{3}{4}+34\dfrac{1}{2}+23\dfrac{3}{4}=$
    $34\dfrac{2}{4}+23\dfrac{3}{4}+34\dfrac{2}{4}+23\dfrac{3}{4}=114\dfrac{10}{4}=116\dfrac{1}{2}$ inches

    21. $116\dfrac{1}{2}$ inches

C// indicates an exercise that is related to the Case in Point feature.

22. **MEASURING BRASS TRIM** To complete a custom order, Kara Oaks of The Home Depot must find the number of inches of brass trim needed to go around the four sides of the lamp base plate shown. Find the length of brass trim needed.

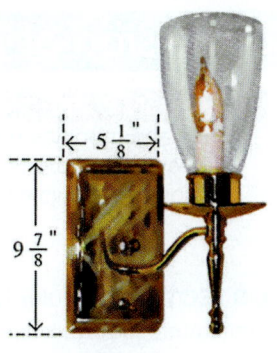

$5\frac{1}{8}"$

$9\frac{7}{8}"$

23. **SECURITY FENCING** The exercise yard at the corrections center has four sides and is enclosed with $527\frac{1}{24}$ feet of security fencing around it. If three sides of the yard measure $107\frac{2}{3}$ feet, $150\frac{3}{4}$ feet, and $138\frac{5}{8}$ feet, find the length of the fourth side.

24. **PARKING LOT FENCING** Three sides of a parking lot are $108\frac{1}{4}$ feet, $162\frac{3}{8}$ feet, and $143\frac{1}{2}$ feet. If the distance around the lot is $518\frac{3}{4}$ feet, find the length of the fourth side.

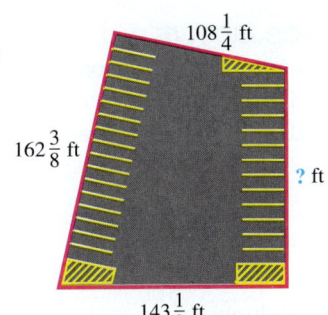

$108\frac{1}{4}$ ft

$162\frac{3}{8}$ ft

? ft

$143\frac{1}{2}$ ft

25. **DELIVERING CONCRETE** Chuck Stone has $8\frac{7}{8}$ cubic yards of concrete in a truck. If he unloads $2\frac{1}{2}$ cubic yards at the first stop, 3 cubic yards at the second stop, and $1\frac{3}{4}$ cubic yards at the third stop, how much concrete remains in the truck?

26. **TAILORED CLOTHING** Marv Levenson bought 15 yards of Italian silk fabric. He made two tops with $3\frac{3}{4}$ yards of the material, a suit for his wife with $4\frac{1}{8}$ yards, and a jacket with $3\frac{7}{8}$ yards. Find the number of yards of material remaining.

QUICK CHECK ANSWERS

1. $13\frac{5}{8} = 14\frac{1}{8}$
2. $87\frac{8}{4} = 89$ feet
3. (a) $6\frac{17}{24}$ (b) $55\frac{4}{9}$

2.4 Multiplication and Division of Fractions

OBJECTIVES
1. Multiply proper fractions.
2. Use cancellation.
3. Multiply mixed numbers.
4. Divide fractions.
5. Divide mixed numbers.
6. Multiply or divide by whole numbers.

CASE IN POINT Most of the cabinets sold by The Home Depot are standard size units and modules that can be combined to satisfy varied applications and room sizes. However, all too often, Kara Oaks finds that various components and trim pieces must be custom-sized, which requires her to work with fractions and mixed numbers.

OBJECTIVE 1 Multiply proper fractions. To multiply two fractions, first multiply the numerators to form a new numerator and then multiply the denominators to form a new denominator. Write the answer in lowest terms if necessary. For example, multiply $\frac{2}{3}$ and $\frac{5}{8}$ by first multiplying the numerators and then the denominators.

$$\frac{2}{3} \times \frac{5}{8} = \frac{2 \times 5}{3 \times 8} = \frac{10}{24} = \frac{5}{12} \quad \text{in lowest terms}$$

Multiply numerators.
Multiply denominators.

OBJECTIVE 2 Use cancellation. To **cancel**, find a number that divides evenly into both a numerator and denominator of fractions being multiplied. That number can be canceled, or divided out of the numerator and denominator before multiplying the fractions. For example, divide 2 into both 2 and 8 in the problem below. Once simplified, the fractions can be multiplied.

$$\frac{\overset{1}{2}}{3} \times \frac{5}{\underset{4}{8}} = \frac{1 \times 5}{3 \times 4} = \frac{5}{12}$$

Multiplying Common Fractions

EXAMPLE 1 Multiply.

(a) $\dfrac{8}{15} \times \dfrac{5}{12}$ (b) $\dfrac{35}{12} \times \dfrac{32}{25}$

SOLUTION

Use cancellation in both of these problems.

(a) $\dfrac{\overset{2}{\cancel{8}}}{\underset{3}{\cancel{15}}} \times \dfrac{\overset{1}{\cancel{5}}}{\underset{3}{\cancel{12}}} = \dfrac{2 \times 1}{3 \times 3} = \dfrac{2}{9}$

Divide 4 into both 8 and 12.
Divide 5 into both 5 and 15.

(b) $\dfrac{\overset{7}{\cancel{35}}}{\underset{3}{\cancel{12}}} \times \dfrac{\overset{8}{\cancel{32}}}{\underset{5}{\cancel{25}}} = \dfrac{7 \times 8}{3 \times 5} = \dfrac{56}{15} = 3\dfrac{11}{15}$

Divide 4 into both 12 and 32.
Divide 5 into both 35 and 25.

Quick TIP
When canceling, be certain that the numerator and the denominator are both divided by the same number.

QUICK CHECK 1
Multiply using cancellation.

(a) $\dfrac{7}{8} \times \dfrac{4}{21}$ (b) $\dfrac{36}{15} \times \dfrac{45}{24}$

OBJECTIVE 3 Multiply mixed numbers. To multiply mixed numbers, change the mixed numbers to improper fractions, cancel, and then multiply the fractions. For example, multiply $6\frac{1}{4}$ and $2\frac{2}{3}$ as follows.

$$6\dfrac{1}{4} \times 2\dfrac{2}{3} = \dfrac{25}{4} \times \dfrac{8}{3} = \dfrac{25}{\underset{1}{\cancel{4}}} \times \dfrac{\overset{2}{\cancel{8}}}{3} = \dfrac{25 \times 2}{1 \times 3} = \dfrac{50}{3} = 16\dfrac{2}{3}$$

Cancel.
Change to improper fractions.

60 CHAPTER 2 Fractions

2.4 Multiplication and Division of Fractions

Multiplying Mixed Numbers

EXAMPLE 2 Multiply.

(a) $3\dfrac{3}{4} \times 8\dfrac{2}{3}$

(b) $1\dfrac{3}{5} \times 3\dfrac{1}{3} \times 1\dfrac{3}{4}$

SOLUTION

Quick TIP
Be sure to change each mixed number to a fraction before multiplying.

(a) $3\dfrac{3}{4} = \dfrac{15}{4}$ and $8\dfrac{2}{3} = \dfrac{26}{3}$

$$\dfrac{\overset{5}{\cancel{15}}}{\underset{2}{\cancel{4}}} \times \dfrac{\overset{13}{\cancel{26}}}{\underset{1}{\cancel{3}}} = \dfrac{5 \times 13}{2 \times 1} = \dfrac{65}{2} = 32\dfrac{1}{2}$$

(b) $1\dfrac{3}{5} = \dfrac{8}{5}, 3\dfrac{1}{3} = \dfrac{10}{3}$, and $1\dfrac{3}{4} = \dfrac{7}{4}$

$$\dfrac{\overset{2}{\cancel{8}}}{\underset{1}{\cancel{5}}} \times \dfrac{\overset{2}{\cancel{10}}}{3} \times \dfrac{7}{\underset{1}{\cancel{4}}} = \dfrac{2 \times 2 \times 7}{1 \times 3 \times 1} = \dfrac{28}{3} = 9\dfrac{1}{3}$$

🖩 The calculator solution to part (b) uses the fraction key.

$1 \;\boxed{a^{b/c}}\; 3 \;\boxed{a^{b/c}}\; 5 \;\boxed{\times}\; 3 \;\boxed{a^{b/c}}\; 1 \;\boxed{a^{b/c}}\; 3 \;\boxed{\times}\; 1 \;\boxed{a^{b/c}}\; 3 \;\boxed{a^{b/c}}\; 4 \;\boxed{=}\; 9\dfrac{1}{3}$

QUICK CHECK 2

Multiply.

(a) $3\dfrac{3}{5} \times 1\dfrac{2}{3}$

(b) $2\dfrac{2}{3} \times 1\dfrac{5}{9} \times 3\dfrac{3}{4}$

You can double the following recipe by multiplying the amount of each item by 2.

Oatmeal Chocolate Chip Cookies

1 cup (2 sticks) margarine or butter, softened
$1\dfrac{1}{4}$ cups firmly packed brown sugar
$\dfrac{1}{2}$ cup granulated sugar
2 eggs
2 tablespoons milk
2 teaspoons vanilla
$1\dfrac{3}{4}$ cups all-purpose flour
1 teaspoon baking soda

$\dfrac{1}{2}$ teaspoon salt (optional)
$2\dfrac{1}{2}$ cups uncooked oats
One 12-ounce package (2 cups) semisweet chocolate morsels
1 cup coarsely chopped nuts (optional)

Heat oven to 375°F.
Beat margarine and sugars until creamy.
Add eggs, milk, and vanilla; beat well.
Add combined flour, baking soda, and salt; mix well. **Stir** in oats, chocolate morsels, and nuts; mix well.
Drop using rounded measuring tablespoonfuls onto ungreased cookie sheet.
Bake 9 to 10 minutes for a chewy cookie or 12 to 13 minutes for a crisp cookie.
Cool 1 minute on cookie sheet; remove to wire rack. Cool completely.

MAKES ABOUT 5 DOZEN

Multiplying a Mixed Number by a Whole Number

EXAMPLE 3 (a) Find the amount of uncooked oats needed if the preceding recipe for oatmeal chocolate chip cookies is doubled (multiplied by 2).

(b) How many cups of all-purpose flour are needed when the recipe is tripled (multiplied by 3)?

SOLUTION

(a) $2\dfrac{1}{2} \times 2 = \dfrac{5}{\underset{1}{\cancel{2}}} \times \dfrac{\overset{1}{\cancel{2}}}{1} = \dfrac{5 \times 1}{1 \times 1} = \dfrac{5}{1} = 5$ cups

(b) $1\dfrac{3}{4} \times 3 = \dfrac{7}{4} \times \dfrac{3}{1} = \dfrac{7 \times 3}{4} = \dfrac{21}{4} = 5\dfrac{1}{4}$ cups

QUICK CHECK 3

(a) Find the amount of brown sugar needed if the preceding recipe is tripled.

(b) How many cups of uncooked oats are needed if the recipe is multiplied by 15?

OBJECTIVE 4 Divide fractions. To divide fractions, first invert the divisor by exchanging the numerator and denominator. For example, when dividing $\frac{3}{8}$ by $\frac{7}{9}$, first invert $\frac{7}{9}$ to $\frac{9}{7}$. Then multiply the two remaining fractions.

$$\frac{3}{8} \div \frac{7}{9}$$

$$\frac{3}{8} \cdot \frac{9}{7} \quad \text{Invert second fraction and change to a multiplication problem.}$$

$$\frac{3 \cdot 9}{8 \cdot 7} \quad \text{Multiply numerators and denominators.}$$

$$\frac{27}{56}$$

So, $\frac{3}{8} \div \frac{7}{9} = \frac{27}{56}$.

Dividing Common Fractions

EXAMPLE 4 Divide.

(a) $\frac{7}{8} \div \frac{1}{4}$ (b) $\frac{25}{36} \div \frac{15}{18}$

Quick TIP

Only invert the second fraction, the divisor, when dividing fractions.

SOLUTION

First, invert the second fraction, and then multiply.

(a) $\frac{7}{8} \div \frac{1}{4} = \frac{7}{\underset{2}{\cancel{8}}} \times \frac{\overset{1}{\cancel{4}}}{1} = \frac{7 \times 1}{2 \times 1} = \frac{7}{2} = 3\frac{1}{2}$

(b) $\frac{25}{36} \div \frac{15}{18} = \frac{\overset{5}{\cancel{25}}}{\underset{2}{\cancel{36}}} \times \frac{\overset{1}{\cancel{18}}}{\underset{3}{\cancel{15}}} = \frac{5 \times 1}{2 \times 3} = \frac{5}{6}$

QUICK CHECK 4

Divide.

(a) $\frac{2}{3} \div \frac{1}{2}$ (b) $\frac{12}{21} \div \frac{18}{24}$

OBJECTIVE 5 Divide mixed numbers. To divide mixed numbers, first change all mixed numbers to improper fractions. Then divide as above by inverting the divisor and, finally, multiplying the fractions.

For example, work $3\frac{5}{9} \div 2\frac{2}{5}$ as follows.

$$3\frac{5}{9} \div 2\frac{2}{5}$$

$$\frac{32}{9} \div \frac{12}{5} \quad \text{Convert to improper fractions.}$$

$$\frac{32}{9} \cdot \frac{5}{12} \quad \text{Invert second fraction and change to a multiplication problem.}$$

$$\frac{32 \cdot 5}{9 \cdot 12} \quad \text{Multiply fractions.}$$

$$\frac{160}{108}$$

$$1\frac{13}{27} \quad \text{Change to a mixed number}$$

So, $3\frac{5}{9} \div 2\frac{2}{5} = 1\frac{13}{27}$.

2.4 Multiplication and Division of Fractions

OBJECTIVE 6 Multiply or divide by whole numbers. To multiply or divide a fraction by a whole number, write the whole number as a fraction over 1.

Multiply: $3\dfrac{3}{4} \times 16 = 3\dfrac{3}{4} \times \dfrac{16}{1} = \dfrac{15}{4} \times \dfrac{16}{1} = \dfrac{15}{\underset{1}{4}} \times \dfrac{\overset{4}{16}}{1} = 15 \times 4 = 60$

— Write as a whole number over 1.

Divide: $2\dfrac{2}{5} \div 3 = \dfrac{12}{5} \div \dfrac{3}{1} = \dfrac{\overset{4}{12}}{5} \times \dfrac{1}{\underset{1}{3}} = \dfrac{4 \times 1}{5 \times 1} = \dfrac{4}{5}$

— Write 3 as $\dfrac{3}{1}$.

Base-End Panel

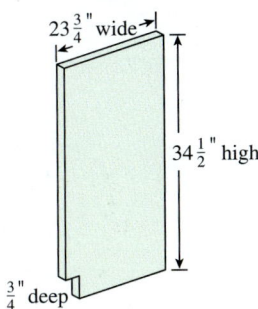

$23\tfrac{3}{4}"$ wide

$34\tfrac{1}{2}"$ high

$\tfrac{3}{4}"$ deep

Mills Pride manufactures cabinets for kitchens and baths. The specifications for base-end panels are shown in the diagram. The lumber used is $\tfrac{3}{4}$ inch deep and is cut down from 24 inches to a $23\tfrac{3}{4}$-inch width. The panel is then cut to a height of $34\tfrac{1}{2}$ inches. The materials used in the manufacture of cabinets, solid oak in this case, are very expensive. Every precaution is taken to minimize waste.

Multiplying a Whole Number by a Mixed Number

EXAMPLE 5 A cabinetmaker will need 80 base-end panels, shown above and to the left, to complete a job. If each panel is $34\tfrac{1}{2}$ inches in length, how many inches of oak material are needed, assuming no waste?

SOLUTION

Multiply the number of panels needed by the length of each panel: $34\tfrac{1}{2}$ (which is $\tfrac{69}{2}$).

$$80 \times \dfrac{69}{2} = \dfrac{\overset{40}{80}}{1} \times \dfrac{69}{\underset{1}{2}} = \dfrac{40 \times 69}{1 \times 1} = \dfrac{2760}{1} = 2760 \text{ inches}$$

The length of material needed by the cabinetmaker is 2760 inches.

QUICK CHECK 5

A plumber needs 68 pieces of 1-inch-diameter copper tubing. If each piece of tubing must be $28\tfrac{1}{2}$ inches long, how many total inches of tubing are needed?

Dividing a Whole Number by a Mixed Number

EXAMPLE 6 To complete a custom-designed cabinet, oak trim pieces must be cut exactly $2\tfrac{1}{4}$ inches long so that they can be used as dividers in a spice rack. Find the number of pieces that can be cut from a piece of oak that is 54 inches in length.

SOLUTION

To divide the length of the piece of oak by $2\tfrac{1}{4}$, first change the mixed number $2\tfrac{1}{4}$ to the fraction $\tfrac{9}{4}$. Then invert the divisor and multiply. The divisor is the number that follows the $\div$ sign.

$$54 \div 2\dfrac{1}{4} = 54 \div \dfrac{9}{4} = \dfrac{\overset{6}{54}}{1} \times \dfrac{4}{\underset{1}{9}} = \dfrac{6 \times 4}{1 \times 1} = \dfrac{24}{1} = 24$$

The number of trim pieces that can be cut from the oak stock is 24.

QUICK CHECK 6

A welder needs angle iron pieces that are $3\tfrac{1}{3}$ inches long. Find the number of pieces that can be cut from a piece of angle iron that is 70 inches in length, assuming no waste.

2.4 Exercises // MyLab Math

The shaded sections below contain solutions to help you get a QUICK START on the various types of exercises.

Multiply. Write each answer in lowest terms. (See Examples 1–3.)

1. $\dfrac{3}{4} \times \dfrac{2}{5} = \dfrac{3}{10}$

   $\dfrac{3}{\cancel{4}_{2}} \times \dfrac{\cancel{2}^{1}}{5} = \dfrac{3}{10}$

2. $\dfrac{2}{3} \times \dfrac{5}{8} = \dfrac{5}{12}$

   $\dfrac{\cancel{2}^{1}}{3} \times \dfrac{5}{\cancel{8}_{4}} = \dfrac{5}{12}$

3. $\dfrac{9}{10} \times \dfrac{11}{16} = $ _____

4. $\dfrac{2}{3} \times \dfrac{3}{8} = $ _____

5. $\dfrac{9}{22} \times \dfrac{11}{16} = $ _____

6. $\dfrac{5}{12} \times \dfrac{7}{10} = $ _____

7. $1\dfrac{1}{4} \times 3\dfrac{1}{2} = $ _____

8. $1\dfrac{2}{3} \times 2\dfrac{7}{10} = $ _____

9. $3\dfrac{1}{9} \times 3 = $ _____

10. $\dfrac{3}{4} \times \dfrac{8}{9} \times 2\dfrac{1}{2} = $ _____

11. $\dfrac{1}{4} \times 6\dfrac{2}{3} \times \dfrac{1}{5} = $ _____

12. $\dfrac{2}{3} \times \dfrac{9}{8} \times 3\dfrac{1}{4} = $ _____

13. $\dfrac{5}{9} \times 2\dfrac{1}{4} \times 3\dfrac{2}{3} = $ _____

14. $3 \times 1\dfrac{1}{2} \times 2\dfrac{2}{3} = $ _____

15. $5\dfrac{3}{5} \times 1\dfrac{5}{9} \times \dfrac{10}{49} = $ _____

Divide. Write each answer in lowest terms. (See Example 4.)

16. $\dfrac{1}{4} \div \dfrac{3}{4} = \dfrac{1}{3}$

    $\dfrac{1}{\cancel{4}_{1}} \times \dfrac{\cancel{4}^{1}}{3} = \dfrac{1}{3}$

17. $\dfrac{3}{8} \div \dfrac{5}{8} = \dfrac{3}{5}$

    $\dfrac{3}{\cancel{8}_{1}} \times \dfrac{\cancel{8}^{1}}{5} = \dfrac{3}{5}$

18. $\dfrac{13}{20} \div \dfrac{26}{30} = $ _____

19. $\dfrac{9}{10} \div \dfrac{3}{5} = $ _____

20. $\dfrac{7}{8} \div \dfrac{3}{4} = $ _____

21. $2\dfrac{1}{2} \div 3\dfrac{3}{4} = $ _____

22. $1\dfrac{1}{4} \div 4\dfrac{1}{6} = $ _____

23. $5 \div 1\dfrac{7}{8} = $ _____

24. $3 \div 1\dfrac{1}{4} = $ _____

C// indicates an exercise that is related to the Case in Point feature.

25. $\dfrac{3}{8} \div 2\dfrac{1}{2} =$ _____

26. $1\dfrac{7}{8} \div 6\dfrac{1}{4} =$ _____

27. $2\dfrac{5}{8} \div \dfrac{5}{16} =$ _____

28. $5\dfrac{2}{3} \div 6 =$ _____

29. In your own words, explain how to multiply fractions. Make up an example problem of your own showing how this works.

30. Describe how cancellation works in fractions and give an example. (See Objective 2.)

Find the time-and-a-half pay rate by multiplying the hourly wage given by $1\dfrac{1}{2}$ (See Example 5.)

31. $8 **$12**

   $\$8 \times 1\dfrac{1}{2} = \$12$

32. $17 _____

33. $12.50 _____

    (*Hint:* $\$12.50 = \$12\dfrac{1}{2}$)

34. $9.50 _____

35. Write the steps needed to divide a number by a fraction. (See Objective 4.)

36. When multiplying two proper fractions, the product is smaller than either of the fractions. For example, $\dfrac{1}{2} \cdot \dfrac{1}{3} = \dfrac{1}{6}$. Is the same thing true when multiplying a proper fraction by a mixed number? Explain and show some examples.

Solve the following application problems.

37. **ELECTRICITY RATES** The utility company says that the cost of operating a hair dryer is $\dfrac{3}{10}$ ¢ per minute. Find the cost of operating the hair dryer for 30 minutes.

37. _____

38. **ELECTRICITY RATES** The cost of electricity for brewing coffee is $\dfrac{2}{5}$ ¢ per minute. What is the cost of brewing coffee for 90 minutes?

38. _____

39. **CHRISTMAS WREATHS** Matthew Genaway wants to make 16 wreaths to sell at the craft fair. Each wreath needs $2\dfrac{1}{4}$ yards of ribbon. How many yards does he need?

39. _____

40. **EARNINGS CALCULATION** Jack Horner worked $38\dfrac{1}{4}$ hours at $10 per hour. How much money did he make?

40. _____

41. **FINISH CARPENTRY** Kara Oaks at The Home Depot estimates that a certain design for a kitchen and bathroom needs $109\dfrac{1}{2}$ feet of cabinet trim. How many homes can be fitted with cabinet trim if there are 1314 feet of cabinet trim available?

41. _____

42. COMMERCIAL FERTILIZER For 1 acre of a crop, $7\frac{1}{2}$ gallons of fertilizer must be applied. How many acres can be fertilized with 1200 gallons of fertilizer?

42. _____

43. A manufacturer of floor jacks is ordering steel tubing to make the handles for this jack. How much steel tubing is needed to make 135 of these jacks? (The symbol for inch is ".)

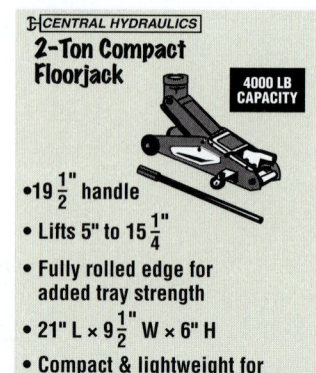

43. _____

44. A wheelbarrow manufacturer uses handles made of hardwood. Find the amount of wood that is needed to make 182 handles. The longest dimension shown is the handle length.

44. _____

45. STRAWBERRY CHEESECAKE The manager of a commercial kitchen has 40 cups of graham cracker crumbs on hand. Find the number of strawberry cheesecakes that can be made if each requires $1\frac{1}{4}$ cup of graham cracker crumbs.

45. _____

46. COMMERCIAL CARPETING The manager of the flooring department at The Home Depot determines that each apartment unit in a new complex requires $62\frac{1}{2}$ square yards of carpet. Find the number of apartment units that can be carpeted with 6750 square yards of carpet.

46. _____

47. FUEL CONSUMPTION A fishing boat uses $12\frac{3}{4}$ gallons of fuel on a full-day fishing trip and $7\frac{1}{8}$ gallons of fuel on a half-day trip. Find the total number of gallons of fuel used in 28 full-day trips and 16 half-day trips.

47. _____

48. MAKING JEWELRY One necklace can be completed in $6\frac{1}{2}$ minutes, while a bracelet takes $3\frac{1}{8}$ minutes. Find the total time that it takes to complete 36 necklaces and 22 bracelets.

48. _____

49. DISPENSING EYEDROPS How many $\frac{1}{8}$-ounce eyedrop dispensers can be filled with 11 ounces of eyedrops?

49. _____

50. CONCRETE FOOTINGS Each building footing requires $\frac{5}{16}$ cubic yard of concrete. How many building footings can be constructed from 10 cubic yards of concrete?

50. _____

51. ALASKA WILDERNESS "Grizzly" Hanson needs 40 crates of supplies to make it through the winter at his remote cabin. If he can carry only $8\frac{1}{2}$ crates with each load behind his snowmobile, find the number of round trips required. You may need to round up.

51. _____

G// 52. WEATHER STRIPPING Bill Rhodes, an employee at The Home Depot, sells a 200-yard roll of weather stripping material. Find the number of pieces of weather stripping $\frac{5}{8}$ yard in length that may be cut from the roll.

52. _____

QUICK CHECK ANSWERS

1. (a) $\frac{1}{6}$ (b) $\frac{9}{2} = 4\frac{1}{2}$
2. (a) 6 (b) $15\frac{5}{9}$
3. (a) $3\frac{3}{4}$ cups (b) 37.5 cups
4. (a) $\frac{4}{3} = 1\frac{1}{3}$ (b) $\frac{16}{21}$
5. 1938 inches
6. 21 pieces

2.5 Converting Decimals to Fractions and Fractions to Decimals

OBJECTIVES

1. Convert decimals to fractions.
2. Convert fractions to decimals.
3. Know common decimal equivalents.

Decimals and fractions are used to represent part of a whole. In this section, you will learn how to convert a decimal to a fraction, and vice versa.

OBJECTIVE 1 Convert decimals to fractions. Recall from the previous chapter that place values to the right of the decimal point are named as shown.

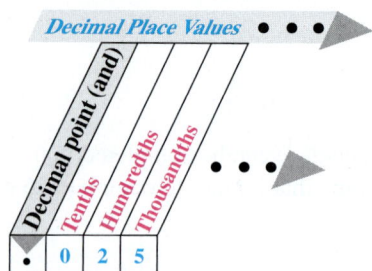

A decimal number can be written as a fraction just by reading it. For example, the number .025 is read as "twenty-five thousandths." Write it as a fraction by placing 25 in the numerator and 1000 in the denominator.

$$\frac{25}{1000} \text{ which reduces to } \frac{1}{40}$$

Decimal	Read as	Fraction Equivalent
0.6	six tenths	$\frac{6}{10} = \frac{3}{5}$
0.38	thirty-eight hundredths	$\frac{38}{100} = \frac{19}{50}$
0.875	eight hundred seventy-five thousandths	$\frac{875}{1000} = \frac{7}{8}$

Another method of converting a decimal to a fraction is by first removing the decimal point. The remaining number is the numerator of the fraction. The denominator of the fraction is 1 followed by as many zeros as there were digits to the right of the decimal point in the original number.

Converting Decimals to Fractions

EXAMPLE 1 Convert the following decimals to fractions.

(a) .3   (b) .98   (c) .654

SOLUTION

(a) After removing the decimal point in .3, you can see that the numerator of the fraction is 3. Since there is only one digit to the right of the decimal point, the denominator is 1 followed by one zero, or 10.

$$.3 = \frac{3}{10}$$

↑ 1 followed by 1 zero

(b) The numerator of the fraction is 98. Since there are two digits to the right of the decimal point, the denominator is 1 followed by two zeros, or 100.

$$.98 = \frac{98}{100} = \frac{49}{50} \text{ (lowest terms)}$$

↑ 1 followed by 2 zeros

(c) Since there are three digits to the right of the decimal point, the denominator is 1 followed by three zeros, or 1000.

$$.654 = \frac{654}{1000} = \frac{327}{500} \text{ (lowest terms)}$$

↑ 1 followed by 3 zeros

2.5 Converting Decimals to Fractions and Fractions to Decimals

QUICK CHECK 1

Convert the following decimals to fractions.

(a) .75   (b) .64   (c) .875

OBJECTIVE 2 Convert fractions to decimals. Convert a fraction to a decimal by dividing the numerator of the fraction by the denominator. Place a decimal point after the numerator and attach one zero at a time to the right of the decimal point as the division is performed. Keep going until the division produces a remainder of zero or until the desired degree of accuracy is reached.

Converting Fractions to Decimals

EXAMPLE 2 Convert the following fractions to decimals.

(a) $\frac{1}{8}$ (b) $\frac{2}{3}$

SOLUTION

(a) Convert $\frac{1}{8}$ to a decimal by dividing 1 by 8. Add trailing zeros after the 1 as needed.

$$\begin{array}{r} .125 \\ 8\overline{)1.000} \quad \text{Add zeros as needed.}\\ \underline{8} \\ 20 \\ \underline{16} \\ 40 \\ \underline{40} \\ 0 \quad \text{remainder of 0} \end{array}$$

Therefore, $\frac{1}{8} = .125$.

(b) Divide 2 by 3.

$$\begin{array}{r} .6666 \\ 3\overline{)2.0000} \quad \text{Keep attaching zeros.}\\ \underline{1\,8} \\ 20 \\ \underline{18} \\ 20 \\ \underline{18} \\ 20 \\ \underline{18} \\ 2 \end{array}$$

The division continues without ending and has a repeating decimal, which is often written as $.\overline{6}$, $.6\overline{6}$, or $.66\overline{6}$. It would be written as .67 when rounded to the nearest hundredth and as .667 when rounded to the nearest thousandth.

🖩 The calculator solution to this example is

$$2 \boxed{\div} 3 \boxed{=} 0.666666667$$

QUICK CHECK 2

Convert the following fractions to decimals.

(a) $\frac{4}{5}$ (b) $\frac{5}{8}$

Decimal Equivalents

$\frac{1}{16} = .0625$

$\frac{1}{10} = .1$

$\frac{1}{9} = .1111$ (rounded)

$\frac{1}{8} = .125$

$\frac{1}{7} = .1429$ (rounded)

$\frac{1}{6} = .1667$ (rounded)

$\frac{3}{16} = .1875$

$\frac{1}{5} = .2$

$\frac{1}{4} = .25$

$\frac{3}{10} = .3$

$\frac{1}{3} = .3333$ (rounded)

$\frac{3}{8} = .375$

$\frac{2}{5} = .4$

$\frac{1}{2} = .5$

$\frac{3}{5} = .6$

$\frac{5}{8} = .625$

$\frac{2}{3} = .6667$ (rounded)

$\frac{7}{10} = .7$

$\frac{3}{4} = .75$

$\frac{4}{5} = .8$

$\frac{5}{6} = .8333$ (rounded)

$\frac{7}{8} = .875$

$\frac{9}{10} = .9$

Quick TIP

Do not use both a decimal and a fraction in the same number.

OBJECTIVE 3 Know common decimal equivalents. Some of the more common **decimal equivalents** of fractions are listed in the margin. These decimals appear from least to greatest value and are rounded to the nearest ten-thousandth.

2.5 Exercises // MyLab Math

The shaded sections below contain solutions to help you get a QUICK START on the various types of exercises.

Convert the following decimals to fractions, and write each in lowest terms. (See Example 1.)

1. $.75 = \dfrac{3}{4}$

    $\dfrac{75}{100} = \dfrac{3}{4}$

2. $.55 = \dfrac{11}{20}$

    $\dfrac{55}{100} = \dfrac{11}{20}$

3. $.24 =$ _____

4. $.64 =$ _____

5. $.73 =$ _____

6. $.33 =$ _____

7. $.85 =$ _____

8. $.68 =$ _____

9. $.34 =$ _____

10. $.288 =$ _____

11. $.444 =$ _____

12. $.125 =$ _____

13. $.625 =$ _____

14. $.875 =$ _____

15. $.805 =$ _____

16. $.791 =$ _____

17. $.096 =$ _____

18. $.012 =$ _____

19. $.0375 =$ _____

20. $.0875 =$ _____

21. $.1875 =$ _____

22. $.9845 =$ _____

23. $.0016 =$ _____

24. $.0085 =$ _____

25. A classmate of yours is confused about how to convert a decimal to a fraction. Write an explanation of this for your classmate, including changing the fraction to lowest terms. (See Objective 1.)

26. Explain how to convert a fraction to a decimal. Be sure to mention rounding in your explanation. (See Objective 2.)

Convert the following fractions to decimals. If a division does not come out evenly, round the answer to the nearest thousandth. (See Example 2.)

27. $\dfrac{1}{4} = .25$

    $\begin{array}{r} .25 \\ 4\overline{)1.00} \\ \underline{8} \\ 20 \\ \underline{20} \\ 0 \end{array}$

28. $\dfrac{7}{8} =$ _____

29. $\dfrac{3}{8} =$ _____

30. $\dfrac{5}{8} =$ _____

31. $\dfrac{2}{3} =$ _____

32. $\dfrac{5}{6} =$ _____

33. $\dfrac{7}{9} =$ _____

34. $\dfrac{1}{9} =$ _____

35. $\dfrac{7}{11} =$ _____

36. $\dfrac{8}{25} =$ _____

37. $\dfrac{22}{25} =$ _____

38. $\dfrac{14}{25} =$ _____

39. $\dfrac{181}{205} =$ _____

40. $\dfrac{1}{99} =$ _____

41. GAMBLING WITH HEALTH A hospital study of 1521 heart-attack patients found that 1 out of 8 quit taking the life-saving drugs prescribed to them. **(a)** What fraction stopped taking their medicine? **(b)** Convert this fraction to a decimal. **(c)** How many patients in the study quit taking their medicine? Round to the nearest whole number.

(a) _____
(b) _____
(c) _____

42. SMOKING AND CANCER A study found that one out of every 16 smokers developed lung cancer. **(a)** Write the fraction as a decimal. **(b)** Out of a group of 8000 smokers, find the number expected to develop lung cancer.

(a) _____
(b) _____

QUICK CHECK ANSWERS

1. (a) $\dfrac{3}{4}$ **(b)** $\dfrac{16}{25}$ **(c)** $\dfrac{7}{8}$ **2. (a)** .8   **(b)** .625

Chapter 2 Quick Review

Chapter Terms Review the following terms to test your understanding of the chapter. For each term you do not know, refer to the page number found next to that term.

cancel [p. 60]
common denominator [p. 48]
common fraction [p. 42]
decimal equivalent [p. 69]
denominator [p. 42]

fraction [p. 42]
improper fraction [p. 42]
inspection [p. 48]
least common denominator (LCD) [p. 48]

like fractions [p. 48]
lowest terms [p. 44]
method of prime numbers [p. 48]
mixed number [p. 42]

numerator [p. 42]
prime number [p. 49]
proper fraction [p. 42]
quotient [p. 43]
unlike fractions [p. 48]

CONCEPTS

EXAMPLES

2.1 Types of fractions

Proper: Numerator smaller than denominator
Improper: Numerator equal to or greater than denominator
Mixed: Whole number plus proper fraction

$$4 + \frac{2}{3} = 4\frac{2}{3}$$

proper fractions: $\frac{2}{3}, \frac{3}{4}, \frac{15}{16}, \frac{1}{8}$

improper fractions: $\frac{17}{8}, \frac{19}{12}, \frac{11}{2}, \frac{5}{3}, \frac{7}{7}$

mixed numbers: $2\frac{2}{3}, 3\frac{5}{8}, 9\frac{5}{6}$

2.1 Converting fractions

Mixed to improper: Multiply whole number by denominator and add numerator.
Improper to mixed: Divide numerator by denominator and place remainder over denominator.

$$7\frac{2}{3} = \frac{(7 \times 3) + 2}{3} = \frac{23}{3}$$

$$\frac{17}{5} = 3\frac{2}{5} \qquad 5\overline{)17} \atop \underline{15} \atop 2$$

2.1 Writing fractions in lowest terms

Divide the numerator and denominator by the same number.

$$\frac{30}{42} = \frac{30 \div 6}{42 \div 6} = \frac{5}{7}$$

2.2 Adding like fractions

Keep the same denominator, add numerators, and reduce to lowest terms.

$$\frac{3}{4} + \frac{1}{4} + \frac{5}{4} = \frac{3 + 1 + 5}{4} = \frac{9}{4} = 2\frac{1}{4}$$

2.2 Finding a least common denominator (LCD)

Inspection method: Look to see if the LCD can be found.
Method of prime numbers: Use prime numbers to find the LCD.

$$\frac{1}{3} + \frac{1}{4} + \frac{1}{10}$$

$$5\overline{)1 \quad 1 \quad 5}$$
$$3\overline{)3 \quad 1 \quad 5}$$
$$2\overline{)3 \quad 2 \quad 5}$$
$$2\overline{)3 \quad 4 \quad 10}$$

Multiply the prime numbers.

$$2 \times 2 \times 3 \times 5 = 60 \text{ LCD}$$

2.2 Adding unlike fractions

1. Find the LCD.
2. Rewrite fractions using the LCD.
3. Add numerators, placing answers over the LCD, and reduce to lowest terms.

$$\frac{1}{3} + \frac{1}{4} + \frac{1}{10} \quad \text{LCD} = 60$$

$$\frac{1}{3} = \frac{20}{60}, \frac{1}{4} = \frac{15}{60}, \frac{1}{10} = \frac{6}{60}$$

$$\frac{20 + 15 + 6}{60} = \frac{41}{60}$$

CHAPTER 2 Quick Review

CONCEPTS	EXAMPLES
2.2 Subtracting fractions 1. Find the LCD. 2. Rewrite fractions using the LCD. 3. Keep the LCD as denominator and subtract numerators. 4. Reduce to lowest terms.	$\dfrac{5}{8} - \dfrac{1}{3} = \dfrac{15}{24} - \dfrac{8}{24} = \dfrac{15 - 8}{24} = \dfrac{7}{24}$
2.3 Adding mixed numbers 1. Find the LCD, then add fractions and reduce. 2. Add whole numbers. 3. Combine the sums of whole numbers and fractions.	$9\dfrac{2}{3} = 9\dfrac{8}{12}$ $+\ 6\dfrac{3}{4} = -6\dfrac{9}{12}$ LCD = 12 $15\dfrac{17}{12} = 16\dfrac{5}{12}$
2.3 Subtracting mixed numbers 1. Find the LCD and subtract fractions, borrowing if necessary. 2. Subtract whole numbers. 3. Combine the differences of whole numbers and fractions.	$8\dfrac{5}{8} = 8\dfrac{15}{24}$ $-\ 3\dfrac{1}{12} = -3\dfrac{2}{24}$ LCD = 24 $5\dfrac{13}{24}$
2.4 Multiplying proper fractions 1. Multiply numerators and multiply denominators. 2. Reduce the answer to lowest terms if cancelling was not done.	$\dfrac{6}{11} \times \dfrac{7}{8} = \dfrac{\overset{3}{\cancel{6}}}{11} \times \dfrac{7}{\underset{4}{\cancel{8}}} = \dfrac{21}{44}$
2.4 Multiplying mixed numbers 1. Change mixed numbers to improper fractions. 2. Cancel if possible. 3. Multiply fractions.	$1\dfrac{3}{5} \times 3\dfrac{1}{3} = \dfrac{8}{\underset{1}{\cancel{5}}} \times \dfrac{\overset{2}{\cancel{10}}}{3} = \dfrac{8}{1} \times \dfrac{2}{3}$ $= \dfrac{16}{3} = 5\dfrac{1}{3}$ Always reduce to lowest terms.
2.4 Dividing proper fractions Invert the divisor, multiply as proper fractions, and reduce the answer to lowest terms.	$\dfrac{25}{36} \div \dfrac{15}{18} = \dfrac{\overset{5}{\cancel{25}}}{\underset{2}{\cancel{36}}} \times \dfrac{\overset{1}{\cancel{18}}}{\underset{3}{\cancel{15}}} = \dfrac{5}{2} \times \dfrac{1}{3} = \dfrac{5}{6}$
2.4 Dividing mixed numbers Change mixed numbers to improper fractions. Invert the divisor, cancel if possible, and then multiply fractions.	$3\dfrac{5}{9} \div 2\dfrac{2}{5} = \dfrac{32}{9} \div \dfrac{12}{5} = \dfrac{\overset{8}{\cancel{32}}}{9} \times \dfrac{5}{\underset{3}{\cancel{12}}}$ $= \dfrac{40}{27} = 1\dfrac{13}{27}$
2.5 Converting decimals to fractions Think of the decimal as being written in words and write in fraction form. Reduce to lowest terms.	Convert .47 to a fraction. Think of .47 as **"forty-seven hundredths."** Then write as $\dfrac{47}{100}$.
2.5 Converting fractions to decimals Divide the numerator by the denominator. Round if necessary.	Convert $\dfrac{1}{8}$ to a decimal. $\begin{array}{r}.125\\8\overline{)1.000}\\\underline{8}\\20\\\underline{16}\\40\\\underline{40}\\0\end{array}$ $\dfrac{1}{8} = .125$

Case Study

OPERATING EXPENSES AT WOODLINE MOLDINGS AND TRIM

It is often said that a picture is worth a thousand words. Visual presentation of data is often used in business in the form of graphs. A commonly used graph that shows the relationships of various data is the circle graph, also called a pie chart. The circle, which contains 360 degrees, is divided into slices, or fractional parts. The size of each slice helps to show the relationship of the various slices to each other and to the whole.

The annual operating expenses for Woodline Moldings and Trim are shown below. Use this information to answer the questions that follow.

WOODLINE MOLDINGS AND TRIM OPERATING EXPENSES

Expense Item	Monthly Amount	Annual Amount	Fraction of Total Expenses
Salaries	$10,000	_____	_____
Rent	$6,000	_____	_____
Utilities	$2,000	_____	_____
Insurance	$1,500	_____	_____
Advertising	$1,500	_____	_____
Miscellaneous	$3,000	_____	_____
Total Expenses		_____	

1. Find the total annual operating expenses for Woodline Moldings and Trim.
2. What fraction should be used to represent each expense item as part of the total expenses?
3. Draw a circle (pie) graph using the fractions you found in part (b) to represent each expense item. Approximate the fractional part of the circle needed for each expense item. Label each segment of the circle graph with the fraction and the expense item.
4. Since there are 360 degrees in a circle, find the number of degrees that would be used to represent each expense item in the circle graph.

1. _____

2. _____

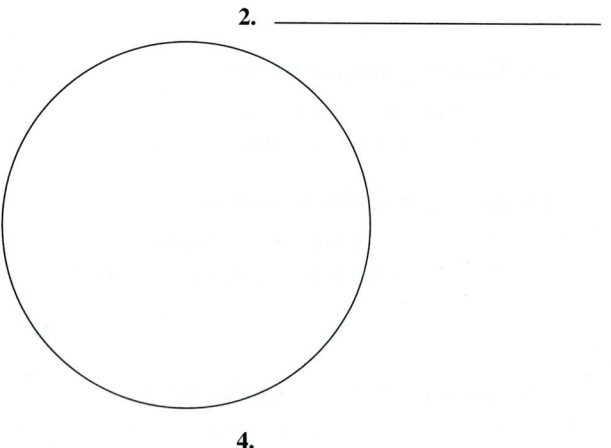

4. _____

INVESTIGATE

Business and economic data are often shown using pie charts. Find two pie charts with business or economic data and explain the contents. Look in newspapers such as *USA Today* or on the Internet.

Case in Point Summary Exercise

THE HOME DEPOT

www.homedepot.com

Facts:

- 1978: Founded in Atlanta, Georgia
- 2005: More than $80 billion in annual sales
- 2017: Nearly 2300 retail stores

The Home Depot is the world's largest home-improvement retailer, employing more than 300,000 employees called associates. Its goal is to provide a very high level of service, broad selection of products, and very competitive prices in a one-stop shopping environment. Management also believes in taking care of its employees by offering many benefits. It even has a program that matches employee charitable donations. Management also tries to encourage employees to donate their time to needy organizations.

1. Kara Oaks manages the cabinetry department. She has been asked to take measurements for a set of custom cabinets made of cherry wood. The side panel is $32\frac{1}{4}$ inches tall. If she needs 6 side panels, find the number of inches of cherry wood needed, assuming no waste when cut.

    1. _____

2. The width of each side panel is $14\frac{1}{2}$ inches. Find the area of each panel by multiplying the height of the panel given in 1. above by the width. The resulting area is in square inches.

    2. _____

3. Find the total area of cherry wood needed for all six side panels in square inches.

    3. _____

4. Ignoring waste, how many side panels can be made if the cabinetmaker has only 2250 square inches of cherry wood panels of the appropriate width?

    4. _____

Discussion Question: *Do you think that exact measurements are an important component of quality from the standpoint of a customer at The Home Depot? What are other important components of quality?*

Source: Home Depot, www.homedepot.com.

Chapter 2 Test

To help you review, the numbers in brackets show the section in which the topic was discussed.

Write the following fractions in lowest terms. **[2.1]**

1. $\dfrac{25}{30} = $ _____
2. $\dfrac{875}{1000} = $ _____
3. $\dfrac{84}{132} = $ _____

Convert the following improper fractions to mixed numbers, and write using lowest terms. **[2.1]**

4. $\dfrac{65}{8} = $ _____
5. $\dfrac{56}{12} = $ _____
6. $\dfrac{120}{45} = $ _____

Convert the following mixed numbers to improper fractions. **[2.1]**

7. $7\dfrac{3}{4} = $ _____
8. $18\dfrac{4}{5} = $ _____
9. $18\dfrac{3}{8} = $ _____

Find the LCD of each of the following groups of denominators. **[2.2]**

10. 2, 6, 5, _____
11. 6, 8, 15, _____
12. 6, 9, 12, 24, _____

Solve the following problems. **[2.2–2.4]**

13. $\dfrac{1}{5}$
    $\dfrac{3}{10}$
    $+ \dfrac{3}{8}$
    _____

14. $32\dfrac{5}{16}$
    $- 17\dfrac{1}{4}$
    _____

15. $126\dfrac{3}{16}$
    $- 89\dfrac{7}{8}$
    _____

16. $67\dfrac{1}{2} \times \dfrac{8}{15} = $

17. $33\dfrac{1}{3} \div \dfrac{200}{9} = $

Solve the following application problems.

18. Becky Finnerty, a pastry chef, used $23\frac{1}{2}$ pounds of powdered sugar for one recipe, $34\frac{3}{4}$ pounds powdered sugar for another recipe, and $17\frac{5}{8}$ pounds of powdered sugar for a third recipe. If Finnerty started with two 50-pound sacks of powdered sugar, find the amount of powdered sugar remaining. [2.3]

18. _____

19. Rhonda Goedeker received her Social Security check of $1275. After paying $\frac{1}{3}$ of this amount for rent, she paid $\frac{3}{5}$ of the remaining amount for food, utilities, and transportation. How much money does she have left?

19. _____

20. A painting contractor arrived at a 6-unit apartment complex with $147\frac{1}{2}$ gallons of paint. If his crew sprayed $68\frac{1}{2}$ gallons on the interior walls, rolled $37\frac{3}{8}$ gallons on the masonry exterior, and brushed $5\frac{3}{4}$ gallons on the window trim, find the number of gallons of paint remaining. [2.3]

20. _____

21. A pizza restaurant has 1000 ounces of mozzarella cheese in inventory. Find the number of regular pizzas that can be made if each uses $8\frac{1}{2}$ ounces of mozzarella cheese. Find the number of ounces of cheese remaining.

21. _____

Convert the following decimals to fractions. [2.5]

22. .625 =

23. .82 =

78 CHAPTER 2 Fractions

Use the advertisement for this four-piece chisel set to answer Exercises 24 and 25. [2.5]

24. Convert the cutting-edge width of the smallest chisel from a fraction to a decimal.

24. _____

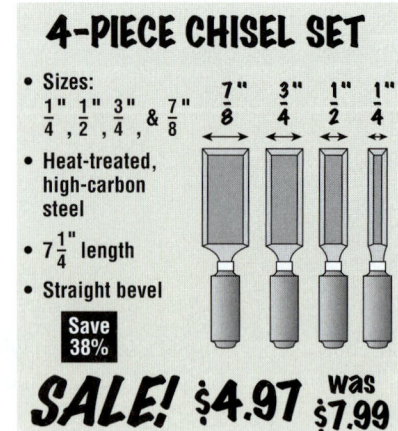

25. Convert the cutting-edge width of the largest chisel from a fraction to a decimal.

25. _____

Percents

3

CHAPTER CONTENTS

3.1 Writing Decimals and Fractions as Percents

3.2 Finding Part

3.3 Finding Base

3.4 Finding Rate

3.5 Increase and Decrease Problems

CASE IN POINT

TOM DUGALLY is a real estate agent with Century 21. He loves the business since it allows him to be his own boss. But he had to work for two years before he earned anything. Fortunately, his wife had a good job as a nurse, which gave them income as he was trying to build a business. Her job also provided them with medical insurance and paid into a retirement plan, so that they were building assets for the future.

Dugally uses percents on a daily basis to calculate monthly payments, determine the commission on a sale, determine whether someone will qualify for financing, and estimate real estate taxes. He simply *could not do his job* without knowing how to work with percents.

CHAPTER 3 Percents

Here are just a few examples where percents are used.

> Markdown on an older-model iPhone is 10% of the original price.
> Interest rate on a certificate of deposit is 2% of amount on deposit.
> Sales taxes in one city are 8.5% of the total.
> Real estate commission is 6% of the selling price of a house.
> Unemployment is 4.5% of the workforce.

Learn the topics in this chapter well—percents will be used throughout this course and you will see them throughout your life!

3.1 Writing Decimals and Fractions as Percents

OBJECTIVES

1. Write a decimal as a percent.
2. Write a fraction as a percent.
3. Write a percent as a decimal.
4. Write a percent as a fraction.
5. Write a fractional percent as a decimal.

Similar to fractions and decimals, **percents** represent parts of a whole. However, percents (**hundredths**) refer to parts out of 100. They are written using a percent sign (%). For example 1% means 1 of 100 equal parts. The number 12% is read "twelve percent."

	Fraction Form	Decimal Form
12% = 12 out of 100 equal parts	$= \dfrac{12}{100}$	$= .12$
25% = 25 out of 100 equal parts	$= \dfrac{25}{100}$	$= .25$
100% = 100 out of 100 equal parts	$= \dfrac{100}{100}$	$= 1.00$
150% = 150 out of 100 equal parts	$= \dfrac{150}{100}$	$= 1.50$

Since 100% is 1, it is the whole or entire amount. Any percent greater than 100% is more than the whole. You can make 100% on a test, but not 150%, unless there is extra credit of some type. However, a swimmer may eat $1\frac{1}{2}$ granola bars, thereby eating 150% of a granola bar.

OBJECTIVE 1 Write a decimal as a percent. From the table above, we see that $.12 = \frac{12}{100} = 12\%$. Effectively, the decimal point in .12 was moved two places to the right and then a percent sign (%) was attached to the end when converting to a percent.

> **Converting a decimal number to a percent**
> **Step 1** Move the decimal point two digits to the right attaching trailing zeros as needed. You are effectively multiplying by 100.
> **Step 2** Attach a percent sign.
>
> For example, to convert .75 to a percent:
>
> .75 ⇒ .75. ⇒ 75%
>
> (Move decimal point 2 places to the right. Attach percent sign.)

Changing Decimals to Percents

EXAMPLE 1 Change the following decimals to percents.
(a) .35   (b) .42   (c) 2.75

SOLUTION

Move the decimal point two places to the right and attach a percent sign.
(a) 35%   (b) 42%   (c) 275%

QUICK CHECK 1

Change the decimals to percents.
(a) .72   (b) .25   (c) 1.1

3.1 Writing Decimals and Fractions as Percents 81

If there is no digit in the hundredths position, place zeros to the right of the number to hold the hundredths position.

$$.5 = .50.\% = 50\% \qquad 1.2 = 1.20.\% = 120\%$$

attach zero ↑   attach zero ↑

Writing Decimals as Percents   **EXAMPLE 2** Write the following decimals as percents.
(a) .8   (b) 2.6   (c) .1   (d) 4

SOLUTION

It is necessary to attach zeros here.
(a) 80%   (b) 260%   (c) 10%   (d) 400%
  ↑ attach zero   ↑ attach zero   ↑ attach zero   ↑ attach 2 zeros

QUICK CHECK 2

Write the decimals as percents.
(a) .6   (b) 3.5   (c) 8

A percent may have digits to the right of a decimal point, as shown next.

Writing Decimals as Percents   **EXAMPLE 3** Write these decimals as percents.
(a) .625   (b) .0057   (c) 1.25

SOLUTION

(a) 62.5%   (b) .57%   (c) 125%

QUICK CHECK 3

Write the decimals as percents.
(a) .875   (b) .0038   (c) 3.5

Note that the percent .57% in the prior example is less than 1%, so .57% is less than 1 part out of 100 equal parts. In fact, .57% = .0057 is *a very small number*.

OBJECTIVE 2 Write a fraction as a percent. Write a fraction as a decimal by dividing the denominator into the numerator. Then, write the decimal number as a percent by moving the decimal point two places to the right and attaching a percent sign.

fraction   decimal   percent

$$\frac{2}{5} = .4 = 40\%$$

$$\begin{array}{r} .4 \\ 5\overline{)2.0} \\ \underline{2.0} \\ 0 \end{array}$$

Writing Fractions as Percents   **EXAMPLE 4** A marketing manager is given the following data in fraction form and must change the data to percents.

(a) $\frac{1}{4}$ (b) $\frac{3}{8}$ (c) $\frac{6}{5}$

SOLUTION

First write each fraction as a decimal, and then write the decimal as a percent.

(a) $\frac{1}{4} = .25 = 25\%$ (b) $\frac{3}{8} = .375 = 37.5\%$ (c) $\frac{6}{5} = 1.2 = 120\%$

82 CHAPTER 3 Percents

QUICK CHECK 4

Change the fractions to percents.

(a) $\frac{3}{4}$ (b) $\frac{2}{5}$ (c) $1\frac{5}{8}$

A second way to write a fraction as a percent is by multiplying the fraction by 100%. For example, write the fraction $\frac{4}{5}$ as a percent by multiplying $\frac{4}{5}$ by 100% (which is 1).

$$\frac{4}{5} = \frac{4}{5} \times 100\% = \frac{400\%}{5} = 80\%$$

OBJECTIVE 3 Write a percent as a decimal. Since percent means parts out of 100, change a percent to a decimal by dividing by 100.

Converting Percents to Decimals

Step 1 Drop the percent sign.
Step 2 Move the decimal point two places to the left. You are effectively dividing by 100.

For example, we show that 25% = .25.

$25\% \Rightarrow .25. \Rightarrow .25$

→ Drop the percent sign.
— Move the decimal point two places to the left.

Writing Percents as Decimals

EXAMPLE 5 To calculate insurance claims, a claims adjuster must change the following percents to decimals.

(a) 35% (b) 50% (c) 325% (d) $37\frac{1}{2}\%$ (Hint: $37\frac{1}{2}\% = 37.5\%$.)

Quick TIP

Change any fraction part of a percent to a decimal before converting from a percent to a decimal.

SOLUTION

Move the decimal point two places to the left and drop the percent sign.

(a) .35 (b) .5 (c) 3.25 (d) .375

QUICK CHECK 5

Change the percents to decimals.

(a) 75% (b) 40% (c) 280%

OBJECTIVE 4 Write a percent as a fraction. To write a percent as a fraction, first change the percent to a decimal, then write the decimal as a fraction in lowest terms.

Writing Percents as Fractions

EXAMPLE 6 Even though smoking is believed to account for 1 of every 5 deaths in the United States, 36 million Americans smoke. The bar chart shows the percent of people in each age group that smoke. Convert each percent to a fraction and reduce to lowest terms.

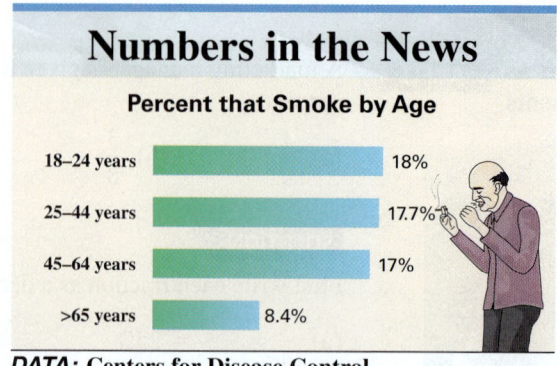

Numbers in the News

Percent that Smoke by Age

18–24 years   18%
25–44 years   17.7%
45–64 years   17%
>65 years   8.4%

DATA: Centers for Disease Control.

SOLUTION

Write each percent as a decimal and then as a fraction in lowest terms.

(a) $18\% = .18 = \dfrac{18}{100} = \dfrac{9}{50}$

(b) $17.7\% = .177 = \dfrac{177}{1000}$

(c) $17\% = .17 = \dfrac{17}{100}$

(d) $8.4\% = .084 = \dfrac{84}{1000} = \dfrac{21}{250}$

QUICK CHECK 6

Change the percents to fractions. Reduce to lowest terms.

(a) 35% (b) 22% (c) 88%

Quick TIP

The percent of student smokers in the U.S. has dropped from a peak of 36.4% in 1997 to 16.7% today.

We did not always know that smoking was harmful. During World War II, cigarettes were given to soldiers, who often traded them as currency. By the 1960s, a lot of scientific evidence indicated that smoking contributes to heart and lung disease. Today, smoking is considered to be *the* leading cause of preventable death in the world, killing more people than HIV/AIDS, tuberculosis, and malaria combined.

OBJECTIVE 5 Write a fractional percent as a decimal. A fractional percent such as $\frac{1}{2}\%$ is smaller than 1%. In fact, $\frac{1}{2}\%$ is equal to $\frac{1}{2}$ of 1%. Write a fractional percent as a decimal by first changing the fraction to a decimal, followed by the percent sign. Then, write that number as a decimal by moving the decimal point two places to the left and dropping the percent sign.

$$\frac{1}{2}\% = .5\% = .00.5 = .005$$

written as a decimal with percent sign remaining → move the decimal point two places to the left

Writing Fractional Percents as Decimals **EXAMPLE 7** The following percents appeared in an article on the use of robots in the workplace. Write each fractional percent as a decimal and drop the percent sign.

(a) $\dfrac{1}{5}\%$ (b) $\dfrac{3}{4}\%$ (c) $\dfrac{5}{8}\%$

SOLUTION

Begin by writing the fraction as a decimal percent.

(a) $\dfrac{1}{5}\% = .2\% = .002$

(b) $\dfrac{3}{4}\% = .75\% = .0075$

(c) $\dfrac{5}{8}\% = .625\% = .00625$

QUICK CHECK 7

Write the fractional percents as decimals.

(a) $\dfrac{1}{2}\%$ (b) $\dfrac{1}{4}\%$ (c) $\dfrac{7}{8}\%$

CHAPTER 3 Percents

The following chart shows the decimal and percent equivalents for several fractions.

Fraction, Decimal, and Percent Equivalents

$\frac{1}{100} = .01 = 1\%$	$\frac{9}{16} = .5625 = 56.25\%$ or $56\frac{1}{4}\%$
$\frac{1}{50} = .02 = 2\%$	$\frac{3}{5} = .6 = 60\%$
$\frac{1}{25} = .04 = 4\%$	$\frac{5}{8} = .625 = 62\frac{1}{2}\%$
$\frac{1}{20} = .05 = 5\%$	$\frac{2}{3} = .666\overline{6} = 66\frac{2}{3}\%$
$\frac{1}{16} = .0625 = 6.25\%$ or $6\frac{1}{4}\%$	$\frac{11}{16} = .6875 = 68.75\%$ or $68\frac{3}{4}\%$
$\frac{1}{12} = .083\overline{3} = 8\frac{1}{3}\%$	$\frac{7}{10} = .7 = 70\%$
$\frac{1}{10} = .1 = 10\%$	$\frac{3}{4} = .75 = 75\%$
$\frac{1}{9} = .111\overline{1} = 11\frac{1}{9}\%$	$\frac{4}{5} = .8 = 80\%$
$\frac{1}{8} = .125 = 12.5\%$ or $12\frac{1}{2}\%$	$\frac{13}{16} = .8125 = 81.25\%$ or $81\frac{1}{4}\%$
$\frac{1}{7} = .1428 = 14\frac{2}{7}\%$	$\frac{5}{6} = .833\overline{3} = 83\frac{1}{3}\%$
$\frac{1}{6} = .166\overline{6} = 16\frac{2}{3}\%$	$\frac{7}{8} = .875 = 87\frac{1}{2}\%$
$\frac{3}{16} = .1875 = 18\frac{3}{4}\%$	$\frac{9}{10} = .9 = 90\%$
$\frac{1}{5} = .2 = 20\%$	$\frac{15}{16} = .9375 = 93.75\%$ or $93\frac{3}{4}\%$
$\frac{1}{4} = .25 = 25\%$	$1 = 1.00 = 100\%$
$\frac{3}{10} = .3 = 30\%$	$1\frac{1}{10} = 1.1 = 110\%$
$\frac{5}{16} = .3125 = 31.25\%$	$1\frac{1}{4} = 1.25 = 125\%$
$\frac{1}{3} = .333\overline{3} = 33\frac{1}{3}\%$	$1\frac{1}{3} = 1.1333\overline{3} = 133\frac{1}{3}\%$
$\frac{3}{8} = .375 = 37\frac{1}{2}\%$	$1\frac{1}{2} = 1.5 = 150\%$
$\frac{2}{5} = .4 = 40\%$	$1\frac{2}{3} = 1.666\overline{6} = 166\frac{2}{3}\%$
$\frac{7}{16} = .4375 = 43.75\%$ or $43\frac{3}{4}\%$	$1\frac{3}{4} = 1.75 = 175\%$
$\frac{1}{2} = .5 = 50\%$	$2 = 2.00 = 200\%$

3.1 Exercises // MyLab Math

The shaded sections below contain solutions to help you get a **QUICK START** *on the various types of exercises.*

Write the following decimals as percents. (See Examples 1–3.)

1. .25 = **25%**
2. .4 = **40%**
3. .72 = _____
4. 1.3 = _____
5. 2.034 = _____
6. .625 = _____
7. 3.625 = _____
8. 4.6 = _____
9. .875 = _____
10. .005 = _____
11. .0005 = _____
12. .0012 = _____
13. 3.45 = _____
14. .2108 = _____
15. .0308 = _____

Write the following as decimals. (See Examples 4–6.)

16. $\frac{1}{5}$ = **.2**
17. $\frac{5}{8}$ = **.625**
18. 64% = _____
19. 65% = _____
20. $\frac{1}{100}$ = _____
21. $\frac{1}{8}$ = _____
22. $8\frac{1}{2}\%$ = _____
23. $12\frac{1}{2}\%$ = _____
24. $\frac{1}{200}$ = _____
25. $\frac{1}{400}$ = _____
26. $50\frac{3}{4}\%$ = _____
27. $84\frac{3}{4}\%$ = _____
28. $3\frac{3}{8}$ = _____
29. $1\frac{3}{4}$ = _____
30. 350% = _____

Determine the fraction, decimal, or percent equivalents for each of the following, as necessary. Write fractions in lowest terms and convert to a mixed number if appropriate.

	Fraction	Decimal	Percent
31.	$\frac{1}{2}$	.5	50%
32.	$\frac{3}{50}$	.06	6%
33.	_____	.875	_____
34.	$\frac{4}{5}$	_____	_____
35.	_____	_____	.8%
36.	_____	.00625	_____
37.	$10\frac{1}{2}$	_____	_____
38.	_____	_____	675%
39.	_____	.65	_____
40.	$4\frac{3}{8}$	_____	_____
41.	_____	.005	_____
42.	_____	_____	$\frac{1}{8}\%$
43.	$\frac{1}{3}$	_____	_____
44.	_____	_____	12.5%
45.	_____	2.5	_____
46.	$\frac{7}{20}$	_____	_____
47.	_____	_____	$4\frac{1}{4}\%$
48.	_____	1.25	_____
49.	$2\frac{1}{2}$	_____	_____
50.	_____	5.125	_____
51.	_____	_____	1037.5%
52.	_____	_____	$\frac{3}{4}\%$

CHAPTER 3 Percents

	Fraction	Decimal	Percent
53.	_____	.0025	_____
54.	$\dfrac{5}{8}$	_____	_____
55.	_____	_____	$37\dfrac{1}{2}\%$
56.	_____	_____	$6\dfrac{3}{4}\%$

57. Fractions, decimals, and percents are all used to describe a part of something. The use of percents is much more common than fractions and decimals. Why do you suppose this is true?

58. List five uses of percent that are or will be part of your life. Consider the activities of working, shopping, saving, and planning for the future.

59. Choose a decimal number and write it as a percent and as a fraction. Explain the steps used to do each. (See Objectives 2 and 3.)

60. The fractional percent $\dfrac{1}{2}\%$ is equal to .005. Is this percent smaller or larger than 1 (the whole)? (See Objective 4.)

QUICK CHECK ANSWERS

1. (a) 72%   (b) 25%   (c) 110%
2. (a) 60%   (b) 350%   (c) 350%
3. (a) 87.5%   (b) .38%   (c) 162.5%
4. (a) 75%   (b) 40%   (c) 62.5%
5. (a) .75   (b) .4   (c) 2.8
6. (a) $\dfrac{7}{20}$ (b) $\dfrac{11}{50}$ (c) $\dfrac{22}{25}$
7. (a) $\dfrac{1}{2}\% = .5\% = .005$
   (b) $\dfrac{1}{4}\% = .25\% = .0025$
   (c) $\dfrac{7}{8}\% = .875\% = .00875$

3.2 Finding Part

OBJECTIVES

1. Know the three components of a percent problem.
2. Learn the basic percent formula.
3. Solve for part.
4. Recognize the terms associated with base, rate, and part.
5. Calculate sales tax.
6. Learn the standard format of percent problems.

CASE IN POINT As a real estate agent, Tom Dugally is paid a commission when he produces income as a result of a sale, of a sale of his listing by someone else, or of a completed rental agreement. Currently, Dugally is looking for a home in the $210,000 price range for Scott and Andrea Abriani, a couple he met at an open house.

OBJECTIVE 1 Know the three components of a percent problem. Problems involving percents have three main components. Usually, two of these components are given, and the third component must be found.

> **Three components of a percent problem**
> 1. **Base:** The whole or total, starting point, or that to which something is being compared.
> 2. **Rate:** A number followed by %, or **percent**.
> 3. **Part:** The result of multiplying the base and the rate. The part is a *part* of the base. For example, sales tax is a part of total sales.

OBJECTIVE 2 Learn the basic percent formula. The base, rate, and part are related by the basic **percent formula**.

$$P = R \times B$$
$$\text{Part} = \text{Rate} \times \text{Base}$$

OBJECTIVE 3 Solve for part. Tom Dugally finds a house that the Abrianis purchase for $210,000. A 6% real estate commission must be paid. The $210,000 price of the house is the whole (or base), and the 6% commission is the rate. The unknown is the actual commission, which is the part. Find the commission as follows.

$$P = R \times B$$
$$P = 6\% \times \$210{,}000$$
$$= .06 \times \$210{,}000 \quad \text{Change percent to decimal.}$$
$$= \$12{,}600$$

The real estate commission of $12,600 will be split between the listing agent, the selling agent, and their two brokers. So only part of it goes to Dugally.

Solving for Part   EXAMPLE 1 Solve for part, using $P = R \times B$.
(a) 4% of 50 (b) 1.2% of 180 (c) 140% of 225 (d) $\frac{1}{4}$% of 560

(Hint: $\frac{1}{4}\% = .25\%$.)

SOLUTION

(a) $\begin{array}{r} 50 \\ \times\ .04 \\ \hline 2.00 \end{array}$
(b) $\begin{array}{r} 180 \\ \times\ .012 \\ \hline 2.160 \end{array}$
(c) $\begin{array}{r} 225 \\ \times\ 1.4 \\ \hline 315.0 \end{array}$
(d) $\begin{array}{r} 560 \\ \times\ .0025 \\ \hline 1.4000 \end{array}$

QUICK CHECK 1

Solve for part.
(a) 6% of 200 (b) 1.8% of 150 (c) 210% of 310 (d) $\frac{1}{2}$% of 1300

EXAMPLE 2 The bar graph shows the unemployment rate by age. Use the data provided to estimate the number of unemployed teenagers out of a total of 32,000 working-age teenagers in one city.

Quick TIP
Clearly, unemployment is much higher amoung younger people.

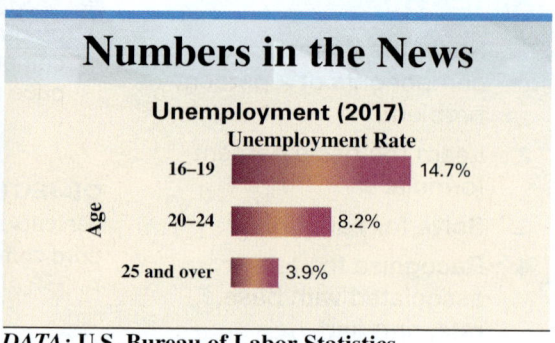

Numbers in the News
Unemployment (2017)

Age	Unemployment Rate
16–19	14.7%
20–24	8.2%
25 and over	3.9%

DATA: U.S. Bureau of Labor Statistics.

SOLUTION
The base is 32,000. The rate for unemployed teenagers is 14.7%. The number of unemployed teenagers is part of the whole, so part (P) is the unknown.

$$P = R \times B$$
$$P = 14.7\% \times 32,000$$
$$= .147 \times 32,000$$
$$= 4704 \quad \text{unemployed youth}$$

The calculator solution to this example is

14.7 [%] [×] 32,000 [=] 4704

Note: Refer to Appendix B for calculator basics.

QUICK CHECK 2
How many of the 83,200 men 25 and over would you expect to be unemployed?

OBJECTIVE 4 Recognize the terms associated with base, rate, and part. Percent problems have certain similarities. For example, some phrases are associated with the base in the problem. Other phrases lead to the part, while % or *percent* following a number identifies the rate. The following chart helps distinguish between the base and the part.

Words and Phrases Associated with Base and Part

Usually indicates the base (B)	Usually indicates the part (P)
Sales	Sales tax
Investment	Return on investment
Savings	Interest
Retail price	Discount
Last year's figure	Increase or decrease
Old salary	Raise
Earnings	Expenditures

OBJECTIVE 5 Calculate sales tax. Calculating **sales tax** is a good example of finding part. States, counties, and cities often collect taxes on retail sales to the consumer. The sales tax is a percent of the sale. This percent varies from as low as 3% in some states to over 9% in others.

Sales Tax Formula

$$P = R \times B$$
$$\text{Sales tax} = \text{Sales tax rate} \times \text{Sales}$$

Calculating Sales Tax **EXAMPLE 3** Becky Smith finally saved enough to buy the guitar she had dreamed about. Her goal was to start a band with her two sisters as backup and a friend as a drummer. The list price on the guitar was $1199.99 and the sales tax was 8.5%. Find the sales tax and total cost.

SOLUTION

The whole (B) is $1199.99, and the rate ($R$) is 8.5%.

$$P = R \times B$$
$$P = 8.5\% \times \$1199.99$$
$$= .085 \times \$1199.99$$
$$= \$101.99915, \text{ or } \$102.00 \text{ rounded to the nearest cent}$$

Now, add the sales tax cost to the cost of the guitar to find the total.

$$\text{Total} = \$1199.99 + \$102 = \$1301.99$$

QUICK CHECK 3

One day, the Lock Shoppe had sales of $1485 and charged a sales tax of 6%. Find the sales tax and the total sales including the tax.

Identify the rate, base, and part with the following hints.

> **Base** tends to be preceded by the word *of* or *on*; tends to be the *whole*.
> **Rate** is followed by a percent sign or the word *percent*.
> **Part** is in the same units as the base and is usually a portion of the base.

OBJECTIVE 6 Learn the standard format of percent problems. Percent problems can be written in the form "% of whole is/are part," as shown by these common examples.

Rate	Whole		Part
7.5% of the	total	is	the sales tax
8.5% of the	workers	are	unemployed
74% of the	students	are	full-time
18% of the	children	are	obese

The following data shows the percent of 1582 people surveyed that purchased each type of ice cream. To find the number out of 1582 surveyed that bought Breyers, multiply the rate (R) of 14.7% by the whole (B) of 1582.

$$14.7\% \times 1582 = 232.554, \text{ or 233 people (rounded)}$$

Quick TIP
Since B × R = R × B, it makes no difference which term is used first when finding the part in the percent equation, i.e.,

14.7% × 1582 = 1582 × 14.7%

Numbers in the News

Get the Scoop

Last year, Americans spent a total of $4.5 billion on ice cream. These are the brands they purchased.

- 37.7% Other
- 22.3% Private label
- 14.7% Breyers
- 10.6% Dreyer's/Edy's Grand
- 5.4% Blue Bell
- 4.8% Häagen Dazs
- 4.5% Ben & Jerry's

DATA: Information Resources Inc., NPD Group.

CHAPTER 3 Percents

Identifying the Pieces in Percent Problems

EXAMPLE 4 Identify the whole and rate in the following; then find the part.
(a) A refrigerator with an original price of $949 was marked down 10%.
(b) Expenses for the weekend were 92% of total sales of $1850.
(c) Corporate income taxes were 30% of total profit of $18,240,000.

SOLUTION

	Rate	×	Base	=	Part
(a)	10%	×	$949	=	$94.90 discount
(b)	92%	×	$1850	=	$1702 expenses
(c)	30%	×	$18,240,000	=	$5,472,000 corporate income taxes

QUICK CHECK 4

A video game set is priced at $280 less a 15% discount. First, identify the whole and rate. Then calculate the part (the discount).

3.2 Exercises // MyLab Math

The shaded sections below contain solutions to help you get a QUICK START on the various types of exercises.

Solve for part in each of the following. Round to the nearest hundredth. (See Example 1.)

1. 10% of 620 homes = __62 homes__
2. 25% of 3500 websites = __875 websites__
3. 75.5% of $800 = _____
4. 20.5% of $1500 = _____
5. 4% of 120 feet = _____
6. 125% of 2000 products = _____
7. 175% of 5820 miles = _____
8. 15% of 75 crates = _____
9. 17.5% of 1040 cell phones = _____
10. 52.5% of 1560 trucks = _____
11. 118% of 125.8 yards = _____
12. 110% of 150 apartments = _____
13. $90\frac{1}{2}$% of $5930 = _____
14. $7\frac{1}{2}$% of $150 = _____

15. Identify the three components in a percent problem. Use your own words to describe how to identify each of these components. (See Objective 1.)

16. There are words and phrases that are usually associated with base and part. Give three examples of words that usually identify the base and the accompanying word for the part. (See Objective 4.)

Solve for part in each of the following application problems. Round to the nearest cent unless otherwise indicated. (See Examples 2 and 3.)

17. **SUMMER VACATION** Of 350 people surveyed recently in New York, 68% said they prefer to vacation in Florida. Find the number preferring Florida.
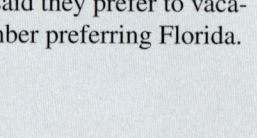

17. __238 people__

 indicates an exercise that is related to the Case in Point feature.

18. **CORPORATE PROFITS** A survey of 30 large retailers showed that expenses averaged 92.4% of total sales. Estimate the expenses for a firm with $138 million in quarterly sales to the nearest tenth of a million dollars.

18. _____

19. **SALES TAX** A real estate broker wants to purchase a new personal digital assistant priced at $249. Find the total cost if the sales tax rate is 7.75%.

19. _____

20. Thomas Dugally of Century 21 Real Estate is working with a mortgage company that charges borrowers $350 plus 2% of the loan amount. What is the total charge to get a home loan of $190,000?

20. _____

21. **WOMEN IN THE NAVY** The navy guided-missile destroyer USS *Sullivans* has a 335-person crew of which 19.1% are female. Find the number of female crew members. Round to the nearest whole number.

21. _____

22. **NY YANKEES** The average price of a ticket for the better box seats at Yankee Stadium was $90 in 2005. Estimate the average cost today if it has increased by 250%.

22. _____

23. **BAR SOAP** A bar of Ivory Soap is $99\frac{44}{100}$% pure. If the bar of soap weighs 9 ounces, how many ounces are pure? Round to the nearest hundredth.

23. _____

24. **GASOLINE TAXES** An economics class estimated that about 22% of the price of gasoline is used to pay the various taxes along the supply chain including income taxes, property taxes, and excise taxes. If the price of gasoline is $2.68 per gallon, find the amount going to pay taxes rounded to the nearest cent.

24. _____

25. **DRIVING DISTRACTIONS** It is estimated that 29.5% of automobile crashes are caused by driver distractions, such as texting. If there are 16,450 automobile crashes in a study, what number would you expect to be caused by driver distractions?

25. _____

26. **DIABETES** A research team trained and encouraged 786 people with diabetes to exercise 4 hours a week, reduce meal sizes, and substitute vegetables for carbohydrates, snacks, and desserts. They found that 16.8% of the group lost more than 10 pounds and were able to get off of their diabetes medications within one year. Find the number getting off of their medication.

26. _____

27. **FEMALE LAWYERS** There are 1,300,705 lawyers in the U.S. and 65% are male. Find **(a)** the percent of the lawyers who are female and **(b)** the number of lawyers who are female.

(a) _____

(b) _____

CHAPTER 3 Percents

28. OVERWEIGHT The World Health Organization states that 25.3% of the 7.5 billion people in the world are overweight. Find **(a)** the percent who are not overweight and **(b)** the number who are not overweight.

(a) _____
(b) _____

29. DIGITAL CAMERA A Sony 18-megapixel digital camera priced at $319 is marked down 25%. Find the price of the camera after the markdown.

29. _____

30. NURSES In a survey of 146 hospital managers, 30% said they have difficulty recruiting enough nurses willing to work the hours needed to do the job. Find the number of managers making this claim rounded to the nearest whole number.

30. _____

31. NEW PRODUCT FAILURE Marketing Intelligence Service says that there were 15,401 new products introduced last year. If 86% of the products introduced last year failed to reach their business objectives, find the number of products that did reach their objectives. Round to the nearest whole number.

31. _____

32. FAMILY BUDGET A family of four with a monthly after-tax income of $5150 spends 90% of its earnings and saves the balance for the down payment on a house. Find **(a)** the monthly savings and **(b)** the annual savings of this family.

(a) _____
(b) _____

33. AUTO SALES IN CHINA General Motors' sales in China were up 5.2% compared to the 3,424,778 vehicles sold there last year. Find this years' sales rounded to the nearest whole number.

33. _____

34. SUPER BOWL ADVERTISING The average cost of 30 seconds of advertising during the Super Bowl 6 years ago was $3.5 million. If the increase in cost over the last 6 years was 71%, find the average cost of 30 seconds of advertising during the Super Bowl this year. Round to the nearest tenth of a million.

34. _____

35. SALES-TAX COMPUTATION As the owner of a copy and print shop, you must collect $6\frac{1}{2}\%$ of the amount of each sale for sales tax. If sales for the month are $48,680, find the combined amount of sales and tax.

35. _____

36. TESLA An owner decides to trade in his three-year-old Tesla (electric car) for a new one. The dealer allows her a credit of $22,800 off the $42,000 price of the new car for her trade-in. Find her total cost assuming she also has to pay the 8.25% sales tax on the difference between the price of the new car and the trade-in credit.

36. _____

37. REAL ESTATE COMMISSIONS Thomas Dugally of Century 21 Real Estate sold a small home for $174,900. The commission was 6% of the sale price. Since Dugally both listed and sold the home, he receives 60% of the commission, and his broker receives the remainder. Find the amount received by Dugally.

37. _____

38. BUSINESS OWNERSHIP Jenny Ruiz has an 82% ownership in a company called Jenny's Cell Phones. If the company has a value of $98,400 and Ruiz receives an income of 45% of the value of her ownership, find her income.

38. _____

CONSUMER INTERNET SALES *Country Store has a unique selection of merchandise that it sells by catalog and over the Internet. Use the shipping and insurance delivery chart below and a sales tax rate of 5% to solve Exercises 39–42. There is no sales tax on shipping and insurance.*

Shipping and Insurance Delivery Chart
Up to $15.00..........................add $3.95
$15.01 to $25.00....................add $5.95
$25.01 to $35.00....................add $6.95
$35.01 to $50.00....................add $7.95
$50.01 to $70.00....................add $8.95
$70.01 to $99.99....................add $9.95
$100.00 or more...................add $10.95

39. Find the total cost of 6 Small Fry Handi-Pan electric skillets at a cost of $29.99 each.

39. _____

40. A customer ordered 5 sets of flour-sack towels at a cost of $12.99 each. What is the total cost?

40. _____

41. Find the total cost of 3 pop-up hampers at a cost of $9.99 each and 4 nonstick minidonut pans at $10.99 each.

41. _____

42. What is the total cost of 5 coach lamp bird feeders at a cost of $19.99 each and 6 garden weather centers at $14.99 each?

42. _____

QUICK CHECK ANSWERS

1. (a) 12 (b) 2.7 (c) 651 (d) 6.5
2. 3245
3. $89.10 sales tax; $1574.10 total
4. $B = \$280; R = .15;$
$P =$ discount amount
$= \$42$ discount

3.3 Finding Base

OBJECTIVES

1. Use the basic percent formula to solve for base.
2. Find sales when tax amount and tax rate are known.
3. Find the investment when interest payment and rate of interest are known.

CASE IN POINT Thomas Dugally of Century 21 Real Estate helps buyers select properties they can afford. Real estate lenders have strict guidelines that determine the maximum loan that they will give a buyer. Usually, the lender will limit the borrowers' monthly house payment to no more than about 28% of their monthly income.

OBJECTIVE 1 Use the basic percent formula to solve for base. In some problems, the rate and part are given, but the base, or whole, must be found. The formula $P = R \times B$ can be used to get the **formula for base**. The following circle diagram may help. To find the formula for base, cover B with your finger. This leaves P divided by R, so $B = \dfrac{P}{R}$.

$$\text{Base} = \dfrac{\text{Part}}{\text{Rate}} \quad \text{or} \quad B = \dfrac{P}{R}$$

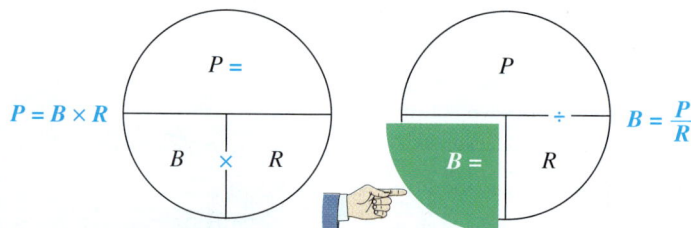

A lender qualifies the Abriana family for a home loan with a monthly payment of $1386 or 28% of their income. Substitute these values into the following form of the percent equation to find monthly income.

$$B = \dfrac{P}{R}$$

$$B = \dfrac{1386}{.28} = 4950 \quad \text{Monthly income is \$4950.}$$

Solving for Base **EXAMPLE 1** Solve for base, using the formula $B = \dfrac{P}{R}$.
(a) 8 is 4% of _____.
(b) 135 is 15% of _____.
(c) 1.25 is 25% of _____.

SOLUTION

(a) $\dfrac{8}{.04} = 200$
(b) $\dfrac{135}{.15} = 900$
(c) $\dfrac{1.25}{.25} = 5$

QUICK CHECK 1

Solve for base, using $B = \dfrac{P}{R}$.
(a) 15 is 10% of _____.
(b) 62 is 1% of _____.
(c) 1.6 is 40% of _____.

OBJECTIVE 2 Find sales when tax amount and tax rate are known. In business problems involving sales tax, the amount of sales is always the base.

Finding Sales When Sales Tax Is Given  One week, an athletic shoe store collected sales taxes of $780. If the sales tax rate is 5%, find total sales for the week.

SOLUTION

Sales tax equals 5% of total sales. So the rate is 5%, part is $780, and total sales are unknown. Arrange the problem in standard form.

$$R \times B = P$$
$$5\% \text{ of total sales is } \$780 \text{ (tax)}$$

Dividing both sides by R (5%) results in $B = \dfrac{P}{R}$.

$$B = \frac{780}{.05} = \$15{,}600 \text{ total sales}$$

The calculator solution to this example is

$$780 \;\boxed{\div}\; .05 \;\boxed{=}\; 15600$$

QUICK CHECK 2

The number of people who passed the real estate license exam was 832. If this was a 65% pass rate, how many took the exam?

It is very important to check the reasonableness of an answer. Intuitively, total sales should be much higher than the sales tax. This is true in the last example because the sales tax is $780 and total sales are $15,600. If $780 had mistakenly been used as the base, the resulting answer would be $780 × 5% = $39, which would not be a reasonable figure for total sales.

OBJECTIVE 3 Find the investment when interest payment and rate of interest are known.

Finding the Amount of an Investment  Roberta Gonzales received $162.50 in interest from an investment that paid 1.25% interest for the year. Find the amount of money invested in the account.

SOLUTION

The part (P) is the $162.50 in interest and the rate (R) is 1.25%. The whole, or base (B), is unknown.

$$B = \frac{P}{R}$$
$$B = \frac{\$162.50}{1.25\%}$$
$$= \frac{\$162.50}{.0125}$$
$$= \$13{,}000$$

The original investment was $13,000.

QUICK CHECK 3

One quarter, the administrator of a school district's retirement funds received $37,500 interest from an investment in bonds that paid 1.5% for the quarter. Find the amount invested.

3.3 Exercises // MyLab Math

The shaded sections below contain solutions to help you get a QUICK START on the various types of exercises.

Solve for base in each of the following. Round to the nearest hundredth. (See Example 1.)

1. 530 firms is 25% of __2120__ firms.
2. 240 letters is 80% of __300__ letters.
3. 130 salads is 40% of _____ salads.
4. 32 shipments is 8% of _____ shipments.
5. 110 lab tests is 5.5% of _____ lab tests.
6. $850 is $4\frac{1}{4}$% of _____.
7. 36 students is .75% of _____ students.
8. 23 workers is .5% of _____ workers.
9. 66 files is .15% of _____ files.
10. 54,600 boxes is 60% of _____ boxes.
11. 50 doors is .25% of _____ doors.
12. 39 bottles is .78% of _____ bottles.
13. $33,870 is $37\frac{1}{2}$% of _____.
14. $8500 is $27\frac{1}{2}$% of _____.
15. $12\frac{1}{2}$% of _____ people is 135 people.
16. $18\frac{1}{2}$% of _____ circuits is 370 circuits.
17. 375 crates is .12% of _____ crates.
18. 3.5 quarts is .07% of _____ quarts.
19. .5% of _____ homes is 327 homes.
20. 6.5 barrels is .05% of _____ barrels.
21. 12 audits is .03% of _____ audits.
22. 8 banks is .04% of _____ banks.

23. The basic percent formula is $P = R \times B$. Show how to find the formula to solve for B (base). (See Objective 1.)

24. Explain how to identify the base, rate, and part in a problem involving sales tax. (See Objective 2.)

Solve the following application problems. Round where appropriate. (See Examples 2 and 3.)

25. **HOME OWNERSHIP** The number of households in the U.S. that own homes is 79.4 million, or 63.5% of all households. Find the total number of households to the nearest tenth of a million.

    $$B = \frac{P}{R} = \frac{79.4 \text{ million}}{.635} = 125.0 \text{ million}$$

    25. __125.0 million__

26. **EMPLOYEE POPULATION BASE** In a large metropolitan area, 81% of the employed population is enrolled in a health maintenance organization (HMO). If 700,650 employees are enrolled, find the total number of people in the employed population.

    26. _____

27. **COLLEGE ENROLLMENT** This semester there are 1785 married students on campus. If this figure represents 23% of the total enrollment, what is the total enrollment?

    27. _____

C// indicates an exercise that is related to the Case in Point feature.

28. VOTER REGISTRATION Registered Republican voters make up 13.8% of the population of one county. If there are 345,000 registered Republican voters in the county, find the total population in the county.

28. _____

29. LOAN QUALIFICATION Thomas Dugally found a home for Scott and Andrea Abriani that will require a monthly loan payment of $1350. If the lender insists that the buyer's monthly payment not exceed 30% of the buyer's after-tax monthly income, find the minimum after-tax monthly income required by the lender.

29. _____

30. DRIVING TESTS In analyzing the success of driver's license applicants, the state finds that 58.3% of those examined received a passing mark. If the records show that 8370 new driver's licenses were issued, what was the number of applicants? Round to the nearest whole number.

30. _____

31. PERSONAL BUDGETING Jim Lawler spends 28% of his income on housing, 15% on food, 11% on clothing, 15% on transportation, 11% on education, 7% on recreation, 7% on miscellaneous, and saves the balance. If his savings amount to $200 per month, what are his monthly earnings?

31. _____

32. FAMILY SIZE One survey among college students found that 28.4% grew up in a home with two or more siblings and 44.7% grew up in a home with one sibling. If 194 students grew up in a home with no siblings, find the size of the survey group.

32. _____

33. DOGS In a recent survey of dog owners, it was found that 901, or 34%, of the owners take their dogs on vacation with them. Find the number of dog owners in the survey who do not take their dogs on vacation.

33. _____

34. HIRING Apple, Inc. recently announced that it will be hiring 5000 employees. If this is about 4.3% of the current workforce, estimate the number of people who work for Apple.

34. _____

35. GAMBLING PAYBACK A casino advertises that it gives a 97.4% payback on slot machines, and the balance is retained by the casino. If the amount retained by the casino one day is $4823, find the total amount played on the slot machines.

35. _____

36. SMOKING OR NONSMOKING A casino hotel in Barbados states that 70% of its rooms are for nonsmokers. If the resort allows smoking in 264 rooms, find the total number of rooms.

36. _____

Supplementary Application Exercises on Base and Part

Solve the following application problems. Read each problem carefully to determine whether base or part is being asked for.

1. **SHAMPOO INGREDIENTS** Most shampoos contain 75% to 90% water. If there are 12.5 ounces of water in a bottle of shampoo that contains 78% water, what is the size of the bottle of shampoo? Round to the nearest whole number.

   1. _____

2. **HOUSEHOLD LUBRICANT** The lubricant WD-40 is used in 82.3 million U.S. homes, which is 65.8% of all homes in the United States. Find the total number of homes in the U.S. rounded to the nearest tenth of a million.

   2. _____

3. **PROPERTY INSURANCE** Thomas Dugally of Century 21 Real Estate sold a commercial building valued at $423,750. If the building is insured for 68% of its value, find the amount of insurance coverage.

   3. _____

4. **FLU PANDEMIC OF 1918** The Spanish Flu epidemic of 1918 killed more than 50 million people worldwide. Among others, Eskimos had little resistance to this flu strain. Ninety percent of the 80 villagers of Brevig Mission, Alaska died of the flu within five days, killing entire families. Find the number of villagers that died.

   4. _____

5. **CAMAROS AND MUSTANGS** The Chevrolet Camaro was introduced in 1967. Camaro sales that year were 220,917, which was 46.2% of the number of Ford Mustangs sold in the same year. Find the number of Mustangs sold in 1967. Round to the nearest whole number.

   5. _____

6. **CHILD SUPPORT** Sean Eden has 12.4% of his earnings withheld for child support. If this amounts to $396.80 per month, find his annual earnings.

   6. _____

7. **CALORIES FROM FAT** Häagen-Dazs vanilla ice cream has 270 calories per serving. If 60% of these calories come from fat, find the number of calories coming from fat.

   7. _____

8. **CHOLESTEROL LEVELS** At a recent health fair, 32% of the people tested were found to have high cholesterol levels. If 350 people were tested, find the number having a high cholesterol level.

   8. _____

9. **RETIREMENT ACCOUNT** Erin Joyce has 9.5% of her earnings deposited into a retirement account. If this amounts to $308.75 per month, find her annual earnings.

   9. _____

10. **SICK PET** Mary Johnson was very upset when her golden Labrador Retriever lost 6 pounds due to a kidney infection. If that was 6.9% of the Lab's original weight, find the original weight to the nearest pound.

10. _____

AIDING DISABLED EMPLOYEES The bar graph below shows how companies have accommodated their employees with disabilities. The data were collected from personnel directors, human resources directors, and executives responsible for hiring at 501 companies. Use this information to solve Exercises 11–14.

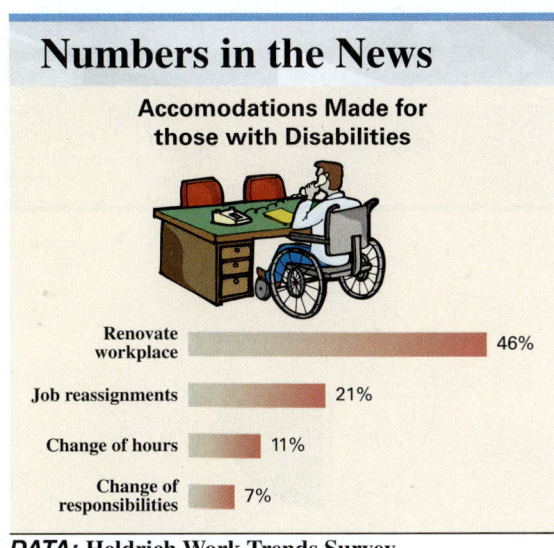

Numbers in the News

Accomodations Made for those with Disabilities

- Renovate workplace: 46%
- Job reassignments: 21%
- Change of hours: 11%
- Change of responsibilities: 7%

DATA: Heldrich Work Trends Survey.

11. How many companies have renovated the workplace to aid employees with disabilities?

11. _____

12. How many companies changed worker responsibilities to aid employees with disabilities?

12. _____

13. Find the number of companies that changed worker hours to aid employees with disabilities.

13. _____

14. Find the number of companies that made job reassignments to aid employees with disabilities.

14. _____

QUICK CHECK ANSWERS

1. (a) 150 (b) 6200 (c) 4   3. $2,500,000
2. 1280 took the exam

3.4 Finding Rate

OBJECTIVES
1. Use the basic percent formula to solve for rate.
2. Solve for the rate in application problems.

CASE IN POINT As managing broker at a Century 21 Real Estate office, Tom Dugally must prepare monthly, quarterly, and annual reports. He often calculates each expense item as a percent of income. He watches these numbers carefully to better control costs.

OBJECTIVE 1 Use the basic percent formula to solve for rate. In this type of percent problem, the part and base are given, and the rate must be found. The **formula for rate** is found from the formula $P = B \times R$. To find the rate using the diagram, cover R to get $\frac{P}{B}$, or part ÷ base.

$$\text{Rate} = \frac{\text{Part}}{\text{Base}} \quad \text{or} \quad R = \frac{P}{B}$$

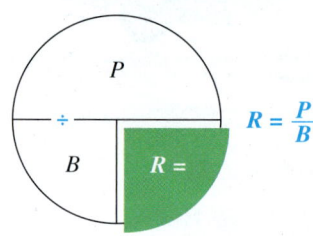

Actually, the formula $P = B \times R$ can be used to find either P, B, or R as long as the values of two of the three components are known.

To find P: Use $P = B \times R$.

To find R: $P = B \times R$

$\dfrac{P}{B} = \dfrac{B \times R}{B}$ Divide both sides by B.

$\dfrac{P}{B} = R,$ or $R = \dfrac{P}{B}$

To find B: $P = B \times R$

$\dfrac{P}{R} = \dfrac{B \times R}{R}$ Divide both sides by R.

$\dfrac{P}{R} = B,$ or $B = \dfrac{P}{R}$

Solving for Rate EXAMPLE 1 Solve for rate.
(a) 26 is _____ % of 104.   (b) _____ % of 300 is 60.   (c) 54 is _____ % of 12.

SOLUTION

(a) $\dfrac{26}{104} = .25 = 25\%$ (b) $\dfrac{60}{300} = .2 = 20\%$ (c) $\dfrac{54}{12} = 4.5 = 450\%$

Quick TIP
When finding rate, be sure to change your decimal answer to a percent.

QUICK CHECK 1

Solve for rate.
(a) 63 is ___ % of 84.   (b) ___ % of 128 is 544.

OBJECTIVE 2 Solve for the rate in application problems. Total revenue at Dugally's busy Century 21 Real Estate office one month was $113,000. The $9884 cost of maintaining the computer system that month was unusually high. Find the percent of revenue that went to maintain the computer system. Note first that the base is $113,000 and the part is $9884. The unknown is rate.

$$R = \frac{P}{B}$$

$$R = \frac{\$9884}{\$113,000}$$

$$= .0875 \quad \text{(rounded)}$$

$$= 8.75\%$$

Investors constantly try to invest funds where they can get the best rate of return on investment. The rate of return is found using the basic percent formula.

Finding the Rate of Return **EXAMPLE 2** An investment officer at Graham Bank placed $3,000,000 in a government security. It remained there for one year and earned $48,000 in interest. Find the rate of return.

SOLUTION

The whole is $3,000,000 and the part is $48,000. Find the rate as follows.

$$R = \frac{P}{B}$$

$$R = \frac{\$48,000}{\$3,000,000}$$

$$= .016, \quad \text{or} \quad 1.6\%$$

The investment earned 1.6% for the year.

QUICK CHECK 2

An investment of $40,500 resulted in a profit of $1012.50. What was the rate of return?

Solving for the Percent Remaining **EXAMPLE 3** Sheila Jones is running a marathon (26.2 miles) but is exhausted after 22 miles. Find the percent of the race that she must still run.

SOLUTION

Distance yet to be run = 26.2 − 22 = 4.2 miles

$$R = \frac{P}{B}$$

$$R = \frac{4.2}{26.2}$$

$$= .160 \quad \text{(rounded)}$$

$$= 16\%$$

Jones still has to run 16% of the total distance.

QUICK CHECK 3

As long as it is maintained, an inexpensive hot-water heater should last 12 years before it needs to be replaced. If a water heater is 10 years old, find the percent of the water heater's life that remains, to the nearest percent.

3.4 Exercises — MyLab Math

The shaded sections below contain solutions to help you get a QUICK START on the various types of exercises.

Solve for rate in each of the following. Round to the nearest tenth of a percent. (See Example 1.)

1. __10__ % of 2760 listings is 276 listings.
2. __40__ % of 850 showings is 340 showings.
3. 35 rail cars is _____ % of 70 rail cars.
4. 144 desks is _____ % of 300 desks.
5. _____ % of 78.57 ounces is 22.2 ounces.
6. _____ % of 728 miles is 509.6 miles.
7. 114 tuxedos is _____ % of 150 tuxedos.
8. $310.75 is _____ % of $124.30.
9. _____ % of $53.75 is $2.20.
10. _____ % of 850 liters is 3.4 liters.
11. 46 shirts is _____ % of 780 shirts.
12. 5.2 smart phones is _____ % of 28.4 smart phones.
13. _____ % of 600 acres is 7.5 acres.
14. _____ % of $8 is $.06.
15. 170 cartons is _____ % of 68 cartons.
16. _____ % of 425 orders is 612 orders.
17. _____ % of $330 is $91.74.
18. _____ % of 752 employees is 470 employees.

19. The basic percent formula is $P = B \times R$. Show how to find the formula to solve for R (rate). (See Objective 1.)

20. A problem includes last year's sales and this year's sales and asks for the percent of increase. Explain how you would identify the base, rate, and part in this problem. (See Objective 4.)

Solve for rate in the following application problems. Round to the nearest tenth of a percent. (See Examples 2–5.)

21. **ADVERTISING EXPENSES** Thomas Dugally of Century 21 Real Estate reports that office income last month was $315,600, while office expenses were $19,567.20. What percent of last month's income was spent on office expenses?

    $$R = \frac{P}{B} = \frac{19{,}567.20}{315{,}600} = .062 = 6.2\%$$

    21. 6.2%

22. **JOB CUTS** One Carrier Corporation factory builds air conditioning systems for commercial buildings. Sales are traditionally low as summer draws to an end and the manager decides he must lay off 80 of the 760 employees working on the assembly lines. Find the percent of workers that will be laid off.

    22. _____

23. **WOMEN IN THE MILITARY** A recent study by Rand's National Defense Research Institute examined 48,000 military jobs, such as Army attack-helicopter pilot and Navy gunner's mate. It was found that 8640 of the jobs were filled by women. What percent of these jobs are filled by women?

    23. _____

C// indicates an exercise that is related to the Case in Point feature.

24. **VOCABULARY** There are 55,000-plus words in *Webster's Dictionary*, but most educated people can identify only 20,000 of these words. What percent of the words in the dictionary can these people identify?

24. _____

25. **ADVERTISING EXPENSES** Advertising expenditures for Bailey's Roofers are as follows.

Newspaper	$2250	Television	$1425
Radio	$954	Yellow Pages	$1605
Outdoor	$1950	Miscellaneous	$2775

What percent of the total advertising expenditures is spent on radio advertising?

25. _____

26. **TANNING SALON** Only 250 of Thai Beach's 1100 customers tan regularly. Find the percent that tan regularly.

26. _____

27. **BEER CONSUMPTION** Cricket's is a popular club for students from the nearby college. The club had 840 customers during Labor Day weekend. If 487 of those customers were under 26 years of age, find the percent under 26.

27. _____

28. **MOTORCYCLISTS** Of the 2380 motorcycle riders that participated in a recent bike rally through the California mountains, 2041 were men. Find the percent who were men.

28. _____

29. **NURSING** In a recent survey of 230 nurses, 159 were unhappy with their jobs due to the difficult working hours and work-related stresses. Find the percent unhappy with their jobs.

29. _____

30. **WIND ENERGY** In the early 1980s, wind-generated electricity cost about 30 cents per kilowatt hour. Today, the best wind systems can produce electricity for 2 cents per kilowatt hour. Find today's cost as a percent of the cost in the 1980s.

30. _____

QUICK CHECK ANSWERS

1. (a) 75% (b) 425%   3. 17%
2. 2.5%

Supplementary Application Exercises on Base, Rate, and Part

In the equation $P = B \times R$, P = part, B = base, or whole, and R = rate, or percent. Use the following forms of the equation to solve the application problems.

$$P = B \times R \qquad B = \frac{P}{R} \qquad R = \frac{P}{B}$$

You will first need to read the problem carefully to determine whether base, part, or rate is unknown. Round rates to the nearest tenth of a percent.

1. **VACATION MISTAKES** Out of 571 employees interviewed, 17% said they thought about work while on vacation. How many people thought about work?

2. **AMERICAN CHIROPRACTIC ASSOCIATION** There are 60,000 licensed chiropractors in the United States. If 25% of these chiropractors belong to the American Chiropractic Association (ACA), find the number of chiropractors in the ACA.

3. **MOTORCYCLE SAFETY** Only 20 of the 50 states require all motorcycle riders to wear helmets. What percent of the states require motorcycle riders to wear helmets?

4. **GLOBAL WARMING** Periodically, 50% or more of all species of animals have gone extinct in a relatively short period of time. Many scientists believe we are entering such a period now as Earth warms. In fact, one scientist predicts that as many as 5 million out of the estimated 8.7 million species on Earth will disappear by 2100. Find the percent that may go extinct.

5. **LOST OVERBOARD** In a recent insurance company study of boaters who had lost items overboard, 88 boaters or 8% said that they lost their cell phones. Find the total number of boaters in the survey.

6. **LIGHTS OUT** There are still 88,000 households in the United States that do not have electricity. If this is .08% of the homes, find the total number of households.

7. **COST AFTER MARKDOWN** A commercial copier and fax machine priced at $398 is marked down 7% to promote the new model. Find the reduced price.

8. **SMART PHONE DISCOUNT** An older model smart phone was marked down by 15% from its original price of $199. Find the reduced price of the old model.

A monthly sales report for the top four salespeople at Active Sports is shown below. Use this information to answer Exercises 9–12.

Employee	Sales	Commission Rate	Commission
Strong, A.	$18,960	3%	$568.80
Ferns, K.	$21,460	3%	$643.80
Keyes, B.	$17,680	4%	$707.20
Vargas, K.	$23,104	5%	$1152.20

9. Find the commission for Strong.   9. _____

10. Find the commission for Ferns.   10. _____

11. What is the rate of commission for Keyes?   11. _____

12. What is the rate of commission for Vargas?   12. _____

13. **BIKER HELMET LAWS** An estimated 2.48 million motorcycle riders in the country support biker helmet laws. If this was 62% of the total motorcycle riders in the country, what is the total number of motorcycle riders?   13. _____

14. **HOME PRICES** In one city, Century 21 estimates that the average selling price of a home increased 2.4%, or by $5088, compared to last year. Find the average selling price of a house last year.   14. _____

15. **HIGH SCHOOL DROPOUTS** In one inner city school district in a large city, 257 out of 414 entering students dropped out before graduating. Find the percent of dropouts.   15. _____

16. **LAYOFF ALTERNATIVE** Instead of laying off workers, a company cut all employee hours from 40 hours a week to 30 hours a week. By what percent were employee hours cut?   16. _____

POPULATION FORECASTS *The figure shows current population with a forecast of population for 2030 for the six most populated countries of the world. Use the data to answer Exercises 17–20.*

Numbers in the News
Most Populous Countries

	2015 Population	Forecast for 2030
China	1376 million	1415 million
India	1311 million	1539 million
U.S.	322 million	356 million
Indonesia	258 million	295 million
Brazil	208 million	229 million
Pakistan	189 million	245 million
Nigeria	182 million	263 million

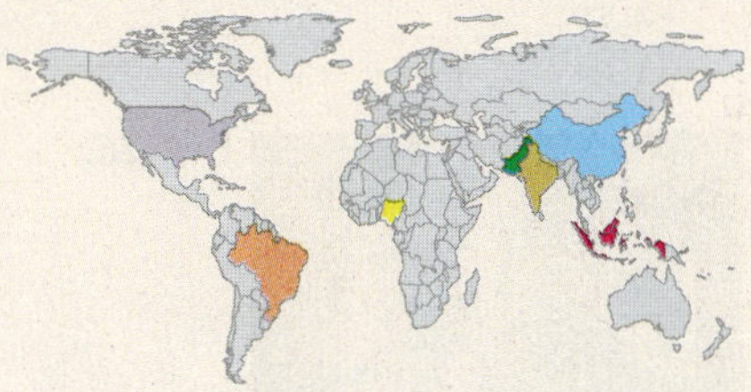

Source: **United Nations.**

17. Determine which two countries are projected to grow most rapidly by first finding the percent increase in population projected for 2030 for each country.

18. Use the percents from the prior exercise to list the countries from those growing most rapidly to those growing the slowest. Explain why Nigeria is higher up on this list than India when India is expected to add a lot more people by 2030.

19. Asia is the more populous of all the continents. Use the 2015 data in the table for countries in Asia (China, India, Indonesia, and Pakistan) to estimate the number of Asians as a percent of the world population of 7.5 billion. Explain why this estimate is too low.

20. China has been the most populous country in the world for decades. What country is forecast to be the most populous by 2030? Find the difference in the projected populations of the two countries.

Work the following exercises.

21. **HEALTH IN A MACHINE** Vending machines on campus must include healthy food choices such as fruits, fruit juices, and healthy snacks. Of the total items sold in the machines this past month, 1440 or 25% have been in the healthy foods group. Find the total number of items sold in the vending machines.

21. _____

22. **TOTAL SALES** If the sales tax rate is $7\frac{1}{2}\%$ and the sales tax collected is $942.30, find the total sales.

22. _____

23. **SAVING TO BUY A HOME** Bob and Ginny McCurdy wish to save to buy a home. They set up the following budget based on percent of take-home income: 25% for rent and utilities, 22% for food and pharmacy, 12% for health care, and 32% for other, with the remainder to savings. Their take-home pay after taxes is $5450 per month. Find the annual savings.

23. _____

24. **CHICKEN NOODLE SOUP** In one year, there were 350 million cans of chicken noodle soup sold (all brands). If 60% of this soup is sold in the cold-and-flu season (October through March), find the number of cans sold in the cold-and-flu season.

24. _____

25. **BUDGETING** The owner of Omni Web Designs decides to save in order to purchase the small building he currently rents. Currently, after tax income averages $6800 per month. The owner budgets the following: 35% for rent, 17% for utilities and insurance, and 40% for salaries and supplies, with the remainder to savings. Find the annual savings.

25. _____

26. **BENEFIT INCREASE** One 80-year-old couple gets $1582 in monthly Social Security benefits. They receive a letter saying that their benefits will go up by 2.4% per month. Find the amount of the increase and the new total monthly benefit, both to the nearest dollar.

26. _____

27. SIDE-IMPACT COLLISIONS Automobile accidents involving side-impact collision resulted in 9000 deaths last year. If automobiles were manufactured to meet an improved side-impact standard, it is estimated that 63.8% of these deaths would have been prevented. How many deaths would have been prevented?

27. _____

28. WATER SCARCITY A soda bottling company in India is installing new equipment that will reduce its current water consumption of 6.2 million gallons per year by 4.2%. Find the amount of water used after the installation of the new equipment and round to the nearest tenth of a million gallons.

28. _____

29. U.S. PATENT RECIPIENTS Among the 50 companies receiving the greatest number of U.S. patents last year, 18 were Japanese companies. **(a)** What percent of the top 50 companies were Japanese companies? **(b)** What percent of the top 50 companies were not Japanese companies?

(a) _____

(b) _____

30. ALCOHOL LEVEL Of the roughly 231 million drivers in the country, 3.5 million have a commercial license. The blood alcohol limit for driving under the influence of alcohol for non-commercial drivers is .08% but it is .04% for drivers with a commercial license.
(a) Find the percent of drivers who have a blood-alcohol limit of .08%.
(b) Find the percent who have a limit of .04%.

(a) _____

(b) _____

3.5 Increase and Decrease Problems

OBJECTIVES

1. Learn to identify an increase or decrease problem.
2. Apply the basic diagram for increase word problems.
3. Use the basic percent formula to solve for base in increase problems.
4. Apply the basic diagram for decrease word problems.
5. Use the basic percent formula to solve for base in decrease problems.

CASE IN POINT Tom Dugally of Century 21 Real Estate knows that real estate values are always changing, usually going up, occasionally going down, but never staying the same. Dugally needs to keep track of the market and must be able to calculate increases and decreases in value on a regular basis.

OBJECTIVE 1 Learn to identify an increase or decrease problem. Managers commonly look at how amounts change, either up or down. For example, a manager might need to know the percent by which sales have **increased** or costs have **decreased**. A consumer might want to know the percent by which the price of an item has changed. Identify these **increase** and **decrease** problems as follows.

Identifying Increase and Decrease Problems

Increase problem. The base (100%) *plus* some portion of the base gives a new value, which is the part. Phrases such as *after an increase of*, *more than*, or *greater than* often indicate an increase problem. The basic formula for an increase problem is

$$\text{Original} + \text{Increase} = \text{New value}$$
$$\text{(base)} \qquad\qquad\qquad \text{(part)}$$

Decrease problem. The part equals the base (100%) *minus* some portion of the base, giving a new value. Phrases such as *after a decrease of*, *less than*, or *after a reduction of* often indicate a decrease problem. The basic formula for a decrease problem is

$$\text{Original} - \text{Decrease} = \text{New value}$$
$$\text{(base)} \qquad\qquad\qquad \text{(part)}$$

OBJECTIVE 2 Apply the basic diagram for increase word problems.

EXAMPLE 1 Using a Diagram to Understand an Increase Problem

The value of a home sold by Tom Dugally this year is $203,500, which is 10% more than last year's value. Find the value of the home last year.

SOLUTION

CASE IN POINT The phrase "10% more than last year's value," tells us that last year's value is the base and that it is the unknown. The figure below shows last year's value as 100% and this year's value which is:

(the base + 10% more than last year's value) = 110% of the base

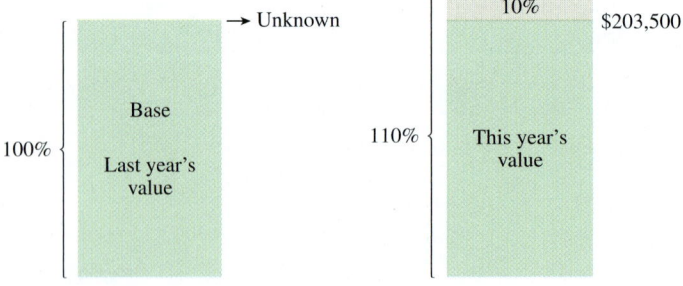

The diagram shows that the 10% increase is based on last year's value which is not known. It is not based on this year's value of $203,500. This year's value is last year's value plus 10% of last year's value.

OBJECTIVE 3 Use the basic percent formula to solve for base in increase problems.

$$\begin{array}{ccccc} \text{Original} & + & \text{Increase} & = & \text{New value} \\ 100\% & + & 10\% & = & 110\% \end{array}$$

3.5 Increase and Decrease Problems

Substituting in the percent formula with part = $203,500 and rate = 110% results in the following.

$$\$203,500 = 110\% \times B$$

Solving for B results in the following equation, which is solved for B.

$$B = \frac{\$203,500}{110\%}$$

$$B = \frac{\$203,500}{1.1}$$

$$B = \$185,000$$

So the value of the house last year was $185,000. Now check the answer.

$185,000 ← last year's value
\+ 18,500 ← 10 % of $185,000
$203,500 ← this year's value

$203,500 is 110% of $185,000.

> **Quick TIP**
> It is very important to take the time to understand which is the part and which is the whole before working percent problems.

QUICK CHECK 1

The total sales at Office Products this year are $713,340, which is 35% more than last year's sales. What is the amount of last year's sales?

Finding Base after Two Increases  Due to increased demand for a patented process that will help analyze the DNA of mice, Biotics Genome has increased production of testing kits by 20% per year for each of the two past years. This year's production is 93,600 kits. Find the number of kits produced two years ago.

SOLUTION

The two 20% increases cannot be added together since the two increases act on different bases. Solve this problem in two steps, one for each of the 20% increases. First, use a diagram to find last year's production of DNA kits.

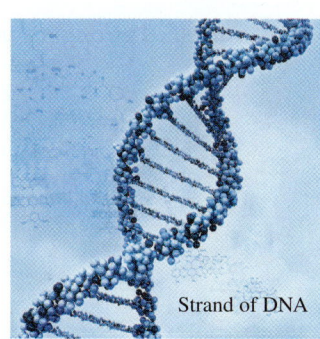
Strand of DNA

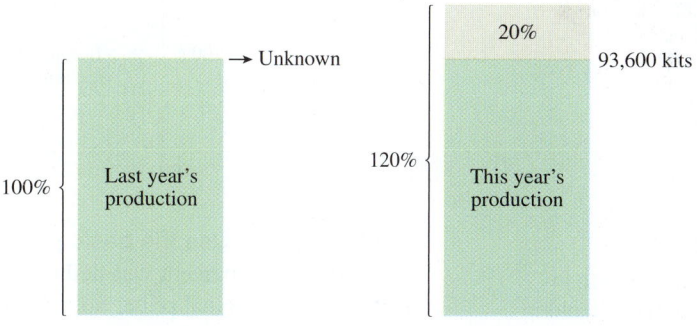

Last year's production plus 20% of last year's production equals this year's production. So, part (P) = 93,600 and rate (R) = 120%.

$$B = \frac{P}{R} = \frac{93,600}{120\%} = \frac{93,600}{1.2} = 78,000 \quad \text{last year's production}$$

The number of DNA testing kits produced last year was 78,000. Use this number and knowledge that the sales volume increased 20% from 2 years ago to last year to find production 2 years ago.

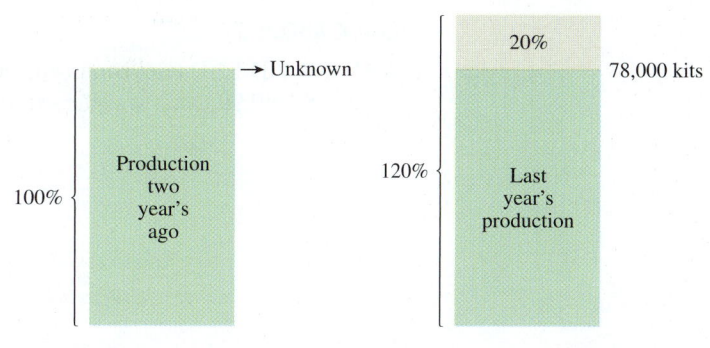

> **Quick TIP**
> It is important to realize that the two 20% increases cannot be added together to equal one increase of 40%. Each 20% increase is calculated on a different base.

Production two years ago added to 20% of that production equals last year's production. So, part $(P) = 78{,}000$ and rate $(R) = 120\%$ once again.

$$B = \frac{P}{R} = \frac{78{,}000}{120\%} = \frac{78{,}000}{1.2} = 65{,}000 \quad \text{production two years ago}$$

 The calculator solution to this example is done by dividing in a series.

$$93600 \div 1.2 \div 1.2 = 65000$$

Check the answer.

65,000	production 2 years ago
+ 13,000	20% increase
78,000	production last year
+ 15,600	20% increase
93,600	production this year

QUICK CHECK 2

Due to a large gift from a donor, Blalock College has increased the number of scholarships by 10% per year for each of the past two years. If 1815 scholarships will be offered this year, find the number offered two years ago.

OBJECTIVE 4 Apply the basic diagram for decrease word problems.

Using a Diagram to Understand a Decrease Problem

EXAMPLE 3 After Nike deducted 10% from the price of a pair of competition running shoes, Katie Small paid $135. What was the original price of the shoes?

SOLUTION

The phrase "deducted 10% from the price" indicates it is a decrease problem. The original price is the unknown.

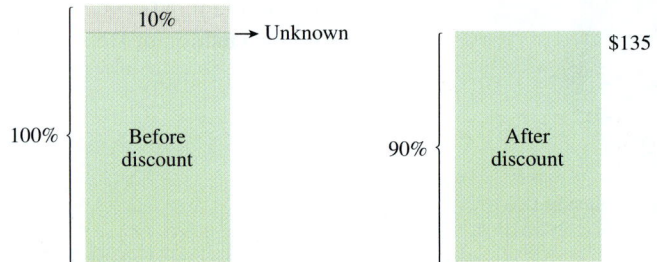

OBJECTIVE 5 Use the basic percent formula to solve for base in decrease problems. Ten percent was deducted from the original price. The resulting price paid of $135 is 90% of the original price. So, part $(P) = \$135$ and rate $(R) = 90\%$.

$$B = \frac{P}{R} = \frac{135}{90\%} = \frac{135}{.9} = \$150 \quad \text{Original price.}$$

Quick TIP

The common error made in decrease problems is thinking that the reduced price is the base. The original price is actually the base.

Check the answer.

$150	original price
− 15	10% discount
$135	price paid

QUICK CHECK 3

A 65-inch HD LCD television is discounted 25% from the original price. If the discounted price is $1799.99, what was the original price of the television?

3.5 Exercises // MyLab Math

The shaded sections below contain solutions to help you get a QUICK START on the various types of exercises.

Solve for base in each of the following.
(Hint: Original + Increase = New Value.) (See Examples 1 and 2.)

New Value (after increase)	Rate of Increase	Base
1. $450	20%	$375
2. $800	25%	_____
3. $30.70	10%	_____
4. $10.09	5%	_____

Solve for base in each of the following.
(Hint: Original − Decrease = New Value.) (See Example 3.)

New Value (after decrease)	Rate of Decrease	Base
5. $20	20%	$25
6. $1530	15%	_____
7. $598.15	30%	_____
8. $98.38	15%	_____

9. Certain words or word phrases help to identify an increase problem. Discuss how you identify an increase problem. (See Objective 1.)

10. Certain words or word phrases help to identify a decrease problem. Discuss how you identify a decrease problem. (See Objective 1.)

Solve the following application problems. Read each problem carefully to decide which are increase or decrease problems, and work accordingly. (See Examples 1–3.)

C// 11. HOME-VALUE APPRECIATION Thomas Dugally of Century 21 Real Estate just listed a home for $205,275. If this is 5% more than what the home sold for last year, find last year's selling price.

$$B = \frac{P}{R} = \frac{205{,}275}{1.05} = \$195{,}500$$

11. $195,500

12. DEALER'S COST Cruz Electronics sold an Xbox system for $345, a loss of 8% of the dealer's original cost. Find the original cost.

$$B = \frac{P}{R} = \frac{345}{.92} = \$375$$

12. $375

C// indicates an exercise that is related to the Case in Point feature.

112 CHAPTER 3 Percents

13. FAMILY RESTAURANT Santiago Rowland owns a small restaurant and charges 8% sales tax on all orders. At the end of the day he has a total of $1026 including the sales and sales tax in his cash register. **(a)** What were his sales not including sales tax? **(b)** Find the amount that is sales tax.

(a) _____

(b) _____

14. SALES TAX Tom Dugally of Century 21 Real Estate purchased a new GPS for $317.37, including $6\frac{1}{2}$% sales tax. Find **(a)** the price of the GPS and **(b)** the amount of sales tax.

(a) _____

(b) _____

15. EATING OUT There are roughly 550,000 fast-food places in the world. If fast food represents 21% of the total restaurants, find the total number of restaurants.

15. _____

16. ANTILOCK BRAKES In a recent test of an automobile antilock braking system (ABS) on wet pavement, the stopping distance was 114 feet. If this was 28.75% less than the distance needed to stop the same automobile without the ABS, find the distance needed to stop without the ABS.

16. _____

17. FOOD INFLATION Food inflation refers to the annual rate of increase in food prices. Food inflation in one South American country was 10% for each of the past two years. Estimate the cost, two years ago, of a basket of food that costs $80 today. Round each calculation to the nearest cent before proceeding.

17. _____

18. DELI SALES Sara Rasic, owner of Sara's Deli, says that her sales have increased exactly 20% per year for the last two years. Her sales this year are $170,035.20. Find her sales two years ago.

18. _____

19. NETFLIX Netflix has 91 million subscribers streaming movies for an increase of 225% over five years ago. Find the number of subscribers five years ago.

19. _____

20. FARMLAND PRICES The value of Iowa farmland increased 4.3% this year to a statewide average value of $7003 per acre. How much per acre did Iowa farmland increase this year? Round to the nearest dollar.

20. _____

3.5 Exercises

21. **AUTO SALES** Automobile sales in Asia were 41.2 million this year, a 4.3% increase over last year. Find the number of auto sales in Asia last year, rounded to the nearest tenth of a million.

21. _____

22. **FACEBOOK USERS** Amazingly, 1.8 billion people use Facebook today, an increase of 85% from 2012. Estimate the number of users in 2012.

22. _____

23. **SURPLUS-EQUIPMENT AUCTION** In a three-day public auction of Jackson County's surplus equipment, the first day brought $5750 in sales and the second day brought $4186 in sales, with 28% of the original equipment left to be sold on the third day. Find the value of the remaining surplus equipment.

23. _____

24. **COLLEGE EXPENSES** After spending $3450 for tuition and $4350 for dormitory fees, Edgar Espina finds that 35% of his original savings remains. Find the amount of his savings that remains.

24. _____

25. **PAPER PRODUCTS MANUFACTURING** The world's largest paper-manufacturing company reported a 16% drop in third-quarter earnings. If earnings had dropped to $122 million, find the earnings before the drop. Round to the nearest hundredth of a million.

25. _____

26. **NATIONAL HOME SALES** During the recession, sales of existing homes decreased 2.3% to an annual number of 4.87 million units. Find the annual number of homes sold before the decrease. Round to the nearest hundredth of a million.

26. _____

27. **WINTER-WHEAT PLANTING** Even though wheat prices rose during the planting season, farmers planted only 50.2 million acres of winter wheat varieties. If this is 2% fewer acres than last year, find the number of acres planted last year. Round to the nearest tenth of a million.

27. _____

28. **NIKE** Last year, shares of Nike stock fell 17.5% to $53.77. Find the value of the stock before the decrease.

28. _____

29. **COLLEGE ENROLLMENT** The enrollment at one college has grown 6% per year for the past two years. If the current enrollment is 33,708 students, find the number of students enrolled two years ago.

29. _____

114 CHAPTER 3 Percents

30. UNIVERSITY FEES Students at one state university are outraged. The annual university fees were 30% more last year than they were the year before. If the fees are $3380 per year this year, which is 30% more than they were last year, find the annual student fees two years ago.

30. _____

31. CONE ZONE DEATHS This year there were 1181 deaths related to road construction zones in the United States. If this is an increase of 70% in the last five years, what was the number of deaths five years ago? Round to the nearest whole number.

31. _____

32. LOANS TO MINORITIES A mortgage lender made 52% more loans to minorities this month than last month. If the number of loans made to minorities this year is 2660, estimate the number of loans made to minorities last year.

32. _____

33. FIRES One city reported that the number of fires caused by arson dropped 17% in one year to 182. Find the number of fires caused by arson in the previous year.

33. _____

34. STOCK VALUE The stock value of drugstore operator CVS dropped 7.4% to close at $79.51 per share. Find the value of each share of stock before the drop. Round to the nearest cent.

34. _____

QUICK CHECK ANSWERS

1. $528,400
2. 1500 scholarships
3. $2399.99

Chapter 3 Quick Review

Chapter Terms Review the following terms to test your understanding of the chapter. For each term you do not know, refer to the page number found next to that term.

base [p. 89]
decrease problem [p. 108]
formula for base [p. 94]
formula for rate [p. 100]

hundredths [p. 80]
increase problem [p. 108]
part [p. 89]
percent [p. 80]

percent formula [p. 87]
percents [p. 80]
rate [p. 89]

sales tax [p. 88]
whole [p. 89]

CONCEPTS

EXAMPLES

3.1 Writing a decimal as a percent

Move the decimal point two places to the right and attach a percent sign (%).

.75 (.75.) = 75%

3.1 Writing a fraction as a percent

First change the fraction to a decimal. Then move the decimal point two places to the right and attach a percent sign (%).

$$\frac{2}{5} = .4$$
.4 (.40.) = 40%

3.1 Writing a mixed number as a percent

First change the mixed number to a decimal. Then move the decimal point two places to the right and attach a percent sign.

$$\frac{13}{8} = 1.375 \,(1.37.5\%) = 137.5\%$$

3.1 Writing a percent as a decimal

Move the decimal point two places to the left and drop the percent sign (%).

50% (.50.%) = .5

3.1 Writing a percent as a fraction

First change the percent to a decimal. Then write the decimal as a fraction in lowest terms.

$$15\% \,(.15.\%) = .15 = \frac{15}{100} = \frac{3}{20}$$

3.1 Writing a fractional percent as a decimal

First change the fraction to a decimal, keeping the percent sign. Then move the decimal point two places to the left and drop the percent sign (%).

$$\frac{1}{2}\% = .5\%$$
.5% = .005

3.2 Solving for part, using the percent formula

Part = Base × Rate
$P = B \times R$
$P = BR$

A company offered a 15% discount on all sales. Find the discount on sales of $1850.

$P = B \times R$
$P = \$1850 \times 15\%$
$P = \$1850 \times .15 = \$277.50$ discount

3.2 Using the standard format to solve percent problems

Express the problem in the format

$R \quad \times \quad B \quad = \quad P$
_____ % of _____ is _____

Notice that *of* means × and *is* means =.

A shop gives a 10% discount on all repairs. Find the discount on a $175 repair.

Rate = 10%, Base = $175, and discount is unknown.
$P = R \times B$
Discount = 10% × $175 = $17.50

CONCEPTS

3.3 Using the percent formula to solve for base
Use $P = B \times R$.

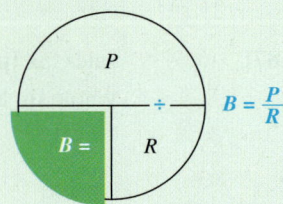

3.4 Using the percent formula to solve for rate
Use $P = R \times B$.

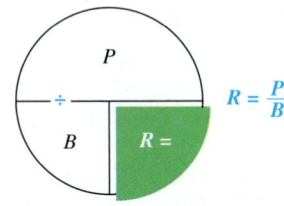

3.5 Drawing a diagram and using the percent formula to solve increase problems
Solve for the base when given the rate (110%) and the part (after increase).

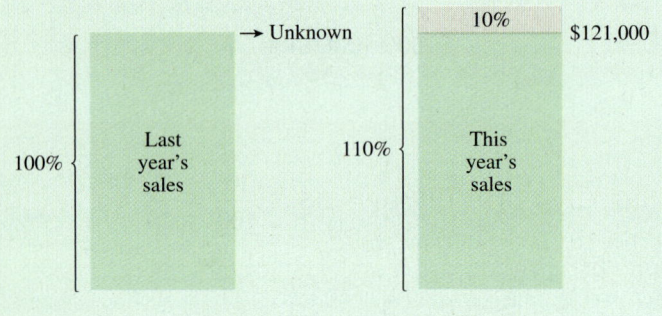

3.5 Drawing a diagram and using the percent formula to solve decrease problems
Solve for the base when given the rate (90%) and the part (after decrease).

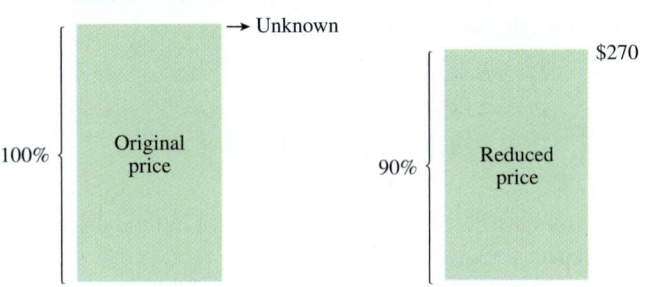

EXAMPLES

If the sales tax rate is 4%, find amount of sales when the sales tax is $18.

$$R \times B = P, \quad \text{or} \quad B = \frac{P}{R}$$

$$B = \frac{18}{.04} = \$450 \text{ sales}$$

The return is $307.80 on an investment of $3420. Find the rate of return.

$$R \times B = P, \quad \text{or} \quad R = \frac{P}{B}$$

$$R = \frac{307.8}{3420} = .09 = 9\%$$

This year's sales are $121,000, which is 10% more than last year's sales. Find last year's sales.

Original + Increase = New value
100% + 10% = 110%

Use $B = \dfrac{P}{R}$.

$$B = \frac{\$121,000}{110\%} = \frac{\$121,000}{1.1}$$

$$= 110,000 \text{ last year's sales}$$

Check: $110,000   last year's sales
       + 11,000   10% of $110,000
       $121,000   this year's sales

After a deduction of 10% from the price, a customer paid $270. Find the original price.

Original − Decrease = New value
100% − 10% = 90%

Use $B = \dfrac{P}{R}$.

$$B = \frac{270}{.9} = \$300 \text{ original price}$$

Check: $300   original price
       − 30   10% discount
       $270   price paid

Case Study

SELF EMPLOYED RETIREMENT PLAN

Betty and Juan Martinez are self-employed and have a retirement plan. They are thinking about buying individual stocks and are considering the well-known companies in the list. Find the stock price last year, the percent of change from last year, or the stock price this year as indicated. Round dollar amounts to the nearest cent and percents to the nearest tenth.

Numbers in the News

Stock Prices of Well-Known Companies

Company	Stock Symbol	Stock Price Last Year	Stock Price This Year	% Change from Last Year
Apple	AAPL	$94.09	$121.89	29.5%
ExxonMobil	XOM	$76.29	$85.29	11.8%
McDonald's	MCD	$123.79	$122.30	−1.2%
Coca-Cola Co.	KO	$43.00	$41.88	−2.6%
Wal-Mart Stores	WMT	$66.36	$66.96	.9%

After looking at the wide range of annual returns and also considering risk, the Martinezes decide to use a financial advisor rather than choosing their own stocks for investments.

INVESTIGATE

List five large publicly held companies that you know or whose products and services you enjoy. Then use the Web to find the most recent closing stock price. Ignoring commissions, estimate the cost if you purchased ten shares of each of the five companies.

Case in Point Summary Exercise

CENTURY 21

www.century21.com

Facts:

- 1971: Founded in California
- 1977: Went public
- 2015: 7000 independently owned offices
- 2018: Over 100,000 sales professionals

Century 21 Real Estate Corporation franchises real estate offices. It has locations in all 50 states and more than 70 countries.

Tom Dugally worked as an agent for Century 21 for several years before he received his broker's license. A few years later, he contacted Century 21 and they helped him set up his own real estate office under a Century 21 franchise. He uses percents on a daily basis to find real estate commissions, monitor costs, follow interest rates, estimate property taxes, and keep up with home prices.

1. Dugally owned a small apartment complex that was worth $865,000 last year. He sold it for $892,680 this year. Find the percent increase in value.

    1. _____

2. Mr. Makin wants Dugally's agency to sell a similar property. Last year, this property had a value of $1,145,000. Use the percent increase from the exercise above to estimate the current value of this property.

    2. _____

3. A homeowner has a home that is currently valued at $435,000. If his home price has increased by the amount indicated in exercise 1 above, find the amount by which it increased in value compared to one year ago. Round all dollar values to the nearest dollar.

    3. _____

4. The total real estate commissions paid to Dugally's real estate agency for the quarter were $237,075. If 65% of those funds are paid directly to the self-employed real estate agents that work in the office under Dugally, find the amount the real estate office keeps. He uses these funds to pay for office, computer, advertising expenses, unilities, etc.

    4. _____

Discussion Question: It is very difficult for a new real estate agent to earn enough to live on while learning the business. What can Dugally do to train and help a new agent?

Source: CENTURY 21, www.century21.com.

Chapter 3 Test

To help you review, the numbers in brackets show the section in which the topic was discussed.

Determine the fraction, decimal, or percent equivalent for each of the following as necessary. Write fractions in lowest terms. **[3.1]**

	FRACTION	DECIMAL	PERCENT
1.	$\frac{3}{8}$	.375	_____
2.	_____	.35	35%
3.	$\frac{3}{125}$	_____	2.4%
4.	_____	.14	14%
5.	$5\frac{7}{8}$	5.875	_____
6.	$1\frac{3}{4}$	_____	175%

Solve the following problems. **[3.1–3.4]**

7. 36 home sales is 12% of what number of home sales?

7. _____

8. What is $\frac{1}{4}$% of $1260?

8. _____

9. Find the fractional equivalent of 24%.

9. _____

10. 48 purchase orders is $2\frac{1}{2}$% of how many purchase orders?

10. _____

11. Change 87.5% to its fractional equivalent.

11. _____

12. Change $1\frac{3}{8}$ to a percent.

12. _____

13. The Los Angeles Lakers made 46.7% of their 92 field goal attempts in a recent game. Find the number of field goals made.

13. _____

120 CHAPTER 3 Percents

14. An underwriter at an insurance company rejected 32 applications for car insurance in one month due to the poor driving record of the applicants. If this was 2.3% of the total number of applications, find the total number of applications. [3.3]

14. _____

15. The Honda Civic GX runs on natural gas so that it uses no gasoline. It gets 38 miles per gallon on the highway and has one of the cleanest engines available. Although it is listed at one dealer at $28,350, it is offered at a discount of 6.3% based partly on a rebate from the state. Find the sale price. [3.5]

15. _____

16. There are 42.2 million Americans who are 65 or older which is 13.1% of the population. Estimate the population of the U.S. to the nearest tenth of a million. [3.3]

16. _____

17. A retail store with a monthly advertising budget of $3400 decides to set up a media budget. It plans to spend 22% for television advertising, 38% for newspaper advertising, 14% for outdoor signs, 15% for radio advertising, and the remainder for bumper stickers.
(a) What percent of the total budget do they plan to spend on bumper stickers?
(b) How much do they plan to spend on bumper stickers for the entire year? [3.2]

(a) _____

(b) _____

18. Americans lose about 300 million golf balls each year, and about 225 million of these are recovered and resold in what has become a $200 million annual business. What percent of the lost golf balls are recovered and resold? [3.4]

18. _____

19. A digital media player is marked "Reduced 20%, Now Only $149." Find the original price of the media player. [3.5]

19. _____

20. Last year's backpack sales were 10% more than they were the year before. This year's sales are 1452 units, which is 10% more than last year. Find the number of backpacks sold two years ago. [3.5]

20. _____

21. The local real estate board reports that the number of condominium listings last month was 379. If 357 condominiums were listed in the same month last year, find the percent of increase. Round to the nearest tenth of a percent. [3.4]

21. _____

22. Procter & Gamble, the maker of Tide detergent, Pampers diapers, and Clairol hair-care products, had quarterly earnings that rose 20% to $1.76 billion. Find the earnings in the previous quarter. Round to the nearest hundredth of a billion. [3.4]

22. _____

Equations and Formulas

4

CHAPTER CONTENTS

4.1 Solving Equations
4.2 Applications of Equations
4.3 Business Formulas
4.4 Ratio and Proportion

CASE IN POINT

BEN JAMES works in the research department of General Motors (GM) as a statistician. His duties include making forecasts of demand for individual vehicle models produced by the company. He uses a variety of tools, including equations, to make the forecasts.

It is very important that management have an accurate forecast of demand. A forecast of demand that is too low results in lost sales, since customers often will not wait to buy. A forecast that is too high results in too much inventory, which is expensive to carry in stock. Sometimes, excess inventory must be marked down in order to sell it. Either way, the firm loses money, potentially a *lot* of money. So, James has a very important job at GM.

122 CHAPTER 4 Equations and Formulas

Equations and formulas are often used in business. For example, they are used to find markup, interest, depreciation, the future value of annuities, and many other things. In this chapter, we show how to solve and work with basic equations and formulas.

4.1 Solving Equations

OBJECTIVES

1. Learn the basic terminology of equations.
2. Use basic rules to solve equations.
3. Solve equations requiring more than one operation.
4. Combine like terms in equations.
5. Use the distributive property to simplify equations.

OBJECTIVE 1 Learn the basic terminology of equations. Some of the basic terms related to equations are given in the table.

Definitions	Examples
A **variable** is a letter used to represent an unknown value. Any letter can be used for a variable.	x, t, s, z, R, y, or A
A **term** is a number, a variable, or the product or quotient of a number and a variable.	$53, y, 6x, 9.5z, \frac{4}{5}y, \frac{s}{3}$
An **expression** can be a single term, but it is often the sum or difference of two or more terms. An expression does not contain an *equal* sign.	$7b, 6x + 9, z - \frac{1}{2}y, 19.4t - 5$
An **equation** is two expressions that are equal to one another.	$x + 5 = 9$
Each equation has a **left side** and a **right side**.	$x + 5 = 9$
A **solution** to an equation is the number that can be substituted in place of a variable that makes the equation true.	$x = 4$ is a solution to the equation $x + 5 = 9$.

To see that $x = 4$ is a solution to the equation $x + 5 = 9$, **substitute** 4 in place of x and determine if the resulting equation is true.

$$x + 5 = 9$$
$$4 + 5 = 9 \quad \text{Substitute 4 in place of } x \text{ in the equation.}$$
$$9 = 9 \quad \text{Replace } 4 + 5 \text{ with 9 on the left side.}$$

Since the final equation is true, we know that $x = 4$ is a solution to the equation.

OBJECTIVE 2 Use basic rules to solve equations. To solve a basic equation for the unknown, change it so that

1. all terms with a variable are on one side of the equation and
2. all terms with only numbers are on the other side of the equation.

It makes no difference whether the variables are on the left side or the right side of the equation.

To move a term to the other side of the equation, do the opposite operation, or "undo," as shown in Example 1, using the following two rules. It is important that what you do to one side of the equation is done to the other side.

> **Rules for Solving Equations**
>
> **Addition Rule** The same number or term may be added or subtracted on both sides of an equation.
>
> **Multiplication Rule** Both sides of an equation may be multiplied or divided by the same nonzero number or term. Actually it is okay to multiply both sides of an equation by 0, but it is not okay to divide by 0 since division by 0 is not defined in mathematics.

An intermediate goal of solving basic equations is for all terms with variables to be on one side of the equation and all terms without variables on the other side.

Solving a Linear Equation Using Addition

EXAMPLE 1 Solve $x - 9 = 15$ for the unknown.

SOLUTION

The number 9 is being subtracted from the x on the left side. Undo this operation and move 9 to the other side of the equation by adding 9 to both sides.

$$x - 9 = 15$$
$$x - 9 + 9 = 15 + 9 \quad \text{Add 9 to both sides.}$$
$$x + 0 = 24$$
$$x = 24$$

To confirm that the value of the unknown is 24, substitute 24 for x in the original equation.

$$x - 9 = 15 \quad \text{Original equation}$$
$$24 - 9 = 15 \quad \text{Let } x = 24.$$
$$15 = 15 \quad \text{True}$$

Therefore, x is equal to 24. This is the solution to the equation.

QUICK CHECK 1

Solve $y - 22 = 45$.

Solving a Linear Equation Using Subtraction

EXAMPLE 2 Solve $k + 7 = 18$.

SOLUTION

To isolate k on the left side, do the opposite of adding 7, which is *subtracting* 7 from both sides.

$$k + 7 = 18$$
$$k + 7 - 7 = 18 - 7 \quad \text{Subtract 7 from both sides.}$$
$$k = 11$$

QUICK CHECK 2

Solve $m + 19 = 32$.

Solving a Linear Equation Using Division

EXAMPLE 3 Solve $5p = 60$.

SOLUTION

The term $5p$ indicates the multiplication of 5 and p. Since division is the inverse of multiplication, solve the equation by *dividing* both sides by 5.

$$5p = 60$$
$$\frac{5p}{5} = \frac{60}{5} \quad \text{Divide both sides by 5.}$$
$$p = 12$$

Check by substituting 12 for p in the original equation.

QUICK CHECK 3

Solve $17g = 153$.

124 CHAPTER 4 Equations and Formulas

Solving a Linear Equation Using Multiplication

EXAMPLE 4 Solve $\dfrac{y}{3} = 9$.

SOLUTION

The bar in $\dfrac{y}{3}$ means to divide. Since multiplication is the opposite of division, multiply both sides by 3.

$$\dfrac{y}{3} = 9$$

$$\dfrac{y}{\cancel{3}} \cdot \cancel{3} = 9 \cdot 3 \quad \text{Multiply both sides of the equation by 3.}$$

$$y = 27$$

QUICK CHECK 4

Solve $\dfrac{1}{4}t = 28$.

Example 5 shows how to solve an equation using a reciprocal. To get the **reciprocal** of a nonzero fraction, exchange the numerator and the denominator. For example, the reciprocal of $\dfrac{7}{9}$ is $\dfrac{9}{7}$. The product of two reciprocals is 1.

$$\dfrac{\overset{1}{\cancel{7}}}{\underset{1}{\cancel{9}}} \cdot \dfrac{\overset{1}{\cancel{9}}}{\underset{1}{\cancel{7}}} = 1$$

So $\dfrac{9}{7}$ is the reciprocal of $\dfrac{7}{9}$.

Solving a Linear Equation Using Reciprocals

EXAMPLE 5 Solve $\dfrac{3}{4}z = 9$.

SOLUTION

Solve this equation by multiplying both sides by $\dfrac{4}{3}$, the reciprocal of $\dfrac{3}{4}$. This process will give $1z$, or just z, on the left.

$$\dfrac{3}{4}z = 9$$

$$\dfrac{3}{4}z \cdot \dfrac{4}{3} = 9 \cdot \dfrac{4}{3} \quad \text{Multiply both sides by } \dfrac{4}{3}.$$

$$\dfrac{\cancel{3}}{\cancel{4}}z \cdot \dfrac{\cancel{4}}{\cancel{3}} = 9 \cdot \dfrac{4}{3} \quad \text{Cancel.}$$

$$z = 12$$

QUICK CHECK 5

Solve $\dfrac{1}{2}z = 99$.

OBJECTIVE 3 Solve equations requiring more than one operation. To solve equations that require more than one step, first isolate the terms involving the unknown (or variable) on one side of the equation and the constants (or numbers) on the other side by using addition and subtraction.

Solving a Linear Equation Using Several Steps

EXAMPLE 6 Solve $2m + 5 = 17$.

SOLUTION

$$2m + 5 = 17$$

$$2m + 5 - 5 = 17 - 5 \quad \text{Subtract 5 from both sides.}$$

$$2m = 12 \quad \text{All terms with variables are on the left side of the equation and only a number is on the other side.}$$

Quick TIP
The unknown can be on either side of the equal sign, so $6 = m$ is the same as $m = 6$.

Now divide both sides by 2.

$$\frac{2m}{2} = \frac{12}{2} \quad \text{Divide both sides by 2.}$$
$$m = 6$$

As before, check by substituting 6 for m in the original equation.

QUICK CHECK 6

Solve $3y + 9 = 45$.

Solving a Linear Equation Using Several Steps

EXAMPLE 7 Solve $3y - 12 = 52$.

SOLUTION

$$3y - 12 = 52$$
$$3y - 12 + 12 = 52 + 12 \quad \text{Add 12 to both sides.}$$
$$3y = 64$$
$$\frac{3y}{3} = \frac{64}{3} \quad \text{Divide both sides by 3.}$$
$$y = 21\frac{1}{3}$$

QUICK CHECK 7

Solve $9q - 14 = 58$.

OBJECTIVE 4 Combine like terms in equations. Some equations have more than one term with the same variable. Terms with the same variables are called **like terms**. They can be *combined* by adding or subtracting the coefficients just as 5 apples + 2 apples can be combined to be 7 apples.

$$5y + 2y = (5 + 2)y = 7y$$
$$11k - 8k = (11 - 8)k = 3k$$
$$12p - 5p + 2p = (12 - 5 + 2)p = 9p$$
$$2z + z = 2z + 1z = (2 + 1)z = 3z$$

Terms with different variables in them *cannot* be combined into a single term just as 5 apples + 2 bananas cannot be combined to be either 7 apples or 7 bananas. These are called **unlike terms**. For example, $12y + 5x$ cannot be combined to make one term, since y and x are different variables.

Solving a Linear Equation Using Several Steps

EXAMPLE 8 Solve $8y - 6y + 4y = 24$.

SOLUTION

Start by combining terms on the left: $8y - 6y + 4y = 2y + 4y = 6y$.

$$8y - 6y + 4y = 24$$
$$(8 - 6 + 4)y = 24 \quad \text{Combine like terms.}$$
$$6y = 24$$
$$\frac{6y}{6} = \frac{24}{6} \quad \text{Divide both sides by 6.}$$
$$y = 4$$

QUICK CHECK 8

Solve $17z - 12z = 42$.

126 CHAPTER 4 Equations and Formulas

Solving a Linear Equation

EXAMPLE 9 Solve $7z - 3z + .5z = 15$.

SOLUTION

$$7z - 3z + .5z = 15$$
$$(7 - 3 + .5)z = 15 \quad \text{Combine like terms.}$$
$$4.5z = 15$$
$$\frac{4.5z}{4.5} = \frac{15}{4.5} \quad \text{Divide both sides by 4.5.}$$
$$z = 3\frac{1}{3}$$

QUICK CHECK 9

Solve $3t - .5t = 17$.

OBJECTIVE 5 Use the distributive property to simplify equations. An equation may include a number being multiplied by the sum of two or more terms in parentheses. Use the **distributive property** to multiply the number by each term in the parentheses.

$$a(b + c) = ab + ac$$

The following diagram may help in remembering the distributive property.

Multiply *a* by *b* and by *c*.

$$a(b + c) = ab + ac$$

More examples:

Multiply 2 by *m* and 7.

$$2(m + 7) = 2m + 2 \cdot 7 = 2m + 14$$
$$8(k - 5) = 8k - 8 \cdot 5 = 8k - 40$$

Solving a Linear Equation Using the Distributive Property

EXAMPLE 10 Solve $8(t - 5) = 16$.

SOLUTION

First use the distributive property on the left to remove the parentheses.

$$8(t - 5) = 16$$
$$8t - 40 = 16$$
$$8t - 40 + 40 = 16 + 40 \quad \text{Add 40 to both sides.}$$
$$8t = 56$$
$$\frac{8t}{8} = \frac{56}{8} \quad \text{Divide both sides by 8.}$$
$$t = 7$$

QUICK CHECK 10

Solve $9(s - 16) = 135$.

Now that you have seen how to solve a few equations, look at the steps to keep in mind when solving equations.

Solving an Equation

Step 1 Remove all parentheses on both sides of the equation using the distributive property.

Step 2 Combine all like terms on both sides of the equation.

Step 3 Place all terms containing a variable on the same side of the equation and all terms not containing a variable on the other side of the equation.

Step 4 Multiply or divide the term with a variable by numbers as needed to produce a term with a coefficient of 1 in front of the variable.

Solving a Linear Equation Using the Distributive Property

EXAMPLE 11 Solve $5r - 2 = 2(r + 5)$.

SOLUTION

$$5r - 2 = 2(r + 5)$$
$$5r - 2 = 2r + 10 \quad \text{Use the distributive property on the right side.}$$
$$5r - 2 + 2 = 2r + 10 + 2 \quad \text{Add 2 to both sides to get all terms with only numbers on the right side.}$$
$$5r = 2r + 12$$
$$5r - 2r = 2r + 12 - 2r \quad \text{Subtract } 2r \text{ from both sides to get all terms with only variables on the left side.}$$
$$5r - 2r = 12$$
$$3r = 12 \quad \text{Combine like terms on the left side.}$$
$$\frac{3r}{3} = \frac{12}{3} \quad \text{Divide both sides by 3.}$$
$$r = 4$$

Quick TIP
Be sure to check the answer in the *original* equation.

Check by substituting 4 for r in the original equation.

QUICK CHECK 11

Solve $5(s - 4) = 3(s + 9)$.

Solving a Linear Equation Using the Distributive Property

EXAMPLE 12 Solve $3(t - .8) = 14 + t$.

SOLUTION

$$3(t - .8) = 14 + t$$
$$3t - 2.4 = 14 + t \quad \text{Use the distributive property.}$$
$$3t - 2.4 - t = 14 + t - t \quad \text{Subtract } t \text{ from both sides.}$$
$$2t - 2.4 = 14 \quad \text{Combine like terms on the left side.}$$
$$2t - 2.4 + 2.4 = 14 + 2.4 \quad \text{Add 2.4 to both sides.}$$
$$2t = 16.4$$
$$\frac{2t}{2} = \frac{16.4}{2} \quad \text{Divide both sides by 2.}$$
$$t = 8.2$$

QUICK CHECK 12

Solve $12u - 8 = 8(u + 4)$.

4.1 Exercises // MyLab Math

The shaded sections below contain solutions to help you get a QUICK START on the various types of exercises.

Solve each equation for the unknown.

1. $s + 12 = 15$ **3**
$$s + 12 = 15$$
$$s + 12 - 12 = 15 - 12$$
$$s = 3$$

2. $k + 15 = 22$ **7**
$$k + 15 = 22$$
$$k + 15 - 15 = 22 - 15$$
$$k = 7$$

3. $b - 7 = 24$ _____

4. $P - 13 = 52$ _____

5. $12 = b + 9$ _____

6. $7 = m - 3$ _____

7. $8k = 56$ _____

8. $3q = 120$ _____

9. $60 = 30m$ _____

10. $94 = 2z$ _____

11. $\dfrac{m}{5} = 6$ _____

12. $\dfrac{r}{7} = 1$ _____

13. $\dfrac{2}{3} a = 5$ _____

14. $\dfrac{3}{4} m = 18$ _____

15. $\dfrac{9}{5} r = 18$ _____

16. $2x = \dfrac{5}{3}$ _____

17. $3m + 5 = 17$ _____

18. $2y - 5 = 39$ _____

19. $4r + 3 = 9$ _____

20. $2p + \dfrac{1}{2} = \dfrac{3}{2}$ _____

21. $11r - 5r + 6r = 84$ _____

C// indicates an exercise that is related to the Case in Point feature.

22. $5m + 6m - 2m = 72$ _____ **23.** $3(2x + 3) = 3x + 12$ _____ **24.** $4z + 2 = 2(z + 3)$ _____

25. Define *variable*, *term*, *expression*, and *equation*.

26. Explain the difference between an expression and an equation. Then explain what is meant by the *solution to an equation*.

QUICK CHECK ANSWERS

1. $y = 67$
2. $m = 13$
3. $g = 9$
4. $t = 112$
5. $z = 198$
6. $y = 12$
7. $q = 8$
8. $z = 8.4$
9. $t = 6.8$
10. $s = 31$
11. $s = 23.5$
12. $u = 10$

4.2 Applications of Equations

OBJECTIVES
1. Translate phrases into mathematical expressions.
2. Write equations from given information.
3. Solve application problems.

CASE IN POINT At General Motors, Ben James looks for patterns using past sales data. He also studies economic trends that affect buyers, such as unemployment or age of existing vehicles on the road. He uses the knowledge gained to develop equations that help him make forecasts of demand. Without a good forecast, managers would not know how many cars to build and suppliers would not know how many parts to produce.

OBJECTIVE 1 Translate phrases into mathematical expressions. Most problems in business are expressed in words. Before these problems can be solved, they must be converted into mathematical language.

Word problems often have certain phrases that occur again and again. The key to solving word problems is to correctly translate these expressions into mathematical expressions. The next few examples illustrate this process.

Translating Verbal Expressions Involving Addition

 Write the following verbal expressions as mathematical expressions. Use x to represent the unknown quantity. Note that other letters can also be used to represent this unknown quantity.

SOLUTION

Verbal Expression	Mathematical Expression	Comments
(a) 5 plus a number	$5 + x$	x represents the number, and *plus* indicates **addition**.
(b) Add 20 to a number	$x + 20$	x represents the number, and *add* indicates **addition**.
(c) The sum of a number and 12	$x + 12$	x represents the number, and *sum* indicates **addition**.
(d) 6 more than a number	$x + 6$	x represents the number, and *more than* indicates **addition**.

QUICK CHECK 1
Write a number plus 7 as a mathematical expression using y for the unknown.

Translating Verbal Expressions Involving Subtraction

 Write each of the following verbal expressions as a mathematical expression. Use p as the variable.

SOLUTION

Verbal Expression	Mathematical Expression	Comments
(a) 3 less than a number	$p - 3$	p represents the number, and *less than* indicates **subtraction**.
(b) A number decreased by 14	$p - 14$	p represents the number, and *decreased by* indicates **subtraction**.
(c) 10 fewer than p	$p - 10$	p represents the number, and *fewer than* indicates **subtraction**.

QUICK CHECK 2
Use s for the unknown, and write a mathematical expression for 19 minus an unknown.

Translating Verbal Expressions Involving Multiplication and Division

EXAMPLE 3 Write the following verbal expressions as mathematical expressions. Use y as the variable.

SOLUTION

Verbal Expression	Mathematical Expression	Comments
(a) The product of a number and 3	$3y$	The word *product* indicates **multiplication**.
(b) Four times a number	$4y$	The word *times* indicates **multiplication**.
(c) Two-thirds of a number	$\frac{2}{3}y$	The word *of* indicates **multiplication**.
(d) A number divided by 2	$\frac{y}{2}$	The word *quotient* indicates **division**.
(e) The sum of 3 and a number is multiplied by 5	$5(3 + y)$	This requires **parentheses**.
(f) 7 is multiplied by the difference of an unknown number and 14	$7(y - 14)$	This requires **parentheses**.

Quick TIP
The order of operations (see Appendix D) indicates that the math operations inside the parentheses must be done first.

QUICK CHECK 3
Use t for the variable, and write an expression for the product of 7 and the sum of 9 plus an unknown.

When multiplying or adding, it does not make any difference which variable is written first. For example:

$$15 \times y = y \times 15 \qquad x + 3 = 3 + x$$

Quick TIP
The symbol $\neq$ means "is not equal to."

However, the order *does* make a difference when dividing or subtracting, as you can see.

$$\frac{10}{x} \neq \frac{x}{10} \quad (\text{unless } x = 10) \qquad z - 150 \neq 150 - z \quad (\text{unless } z = 150)$$

OBJECTIVE 2 Write equations from given information. Read the problem carefully as you develop the equation(s) needed to solve it.

Writing an Equation from Words

EXAMPLE 4 Translate "the product of 5 and a number decreased by 8 is 100" into an equation. Then, solve the equation.

SOLUTION
Use the letter y as the variable and translate as follows.

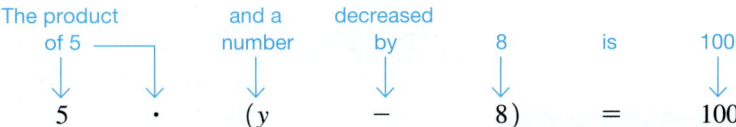

Quick TIP
Translating the word problem to $5 \cdot y - 8 = 100$ would not give the same result. The 5 must be multiplied by the difference: $(y - 8)$.

Simplify and complete the solution of the equation.

$$5 \cdot (y - 8) = 100$$
$$5y - 40 = 100 \quad \text{Apply the distributive property.}$$
$$5y = 140 \quad \text{Add 40 to both sides.}$$
$$y = 28 \quad \text{Divide by 5.}$$

QUICK CHECK 4
Write "ninety-four is the product of 12 and the sum of a number and 2.5" as an equation, using p for the variable.

Writing an Equation from Words

EXAMPLE 5 Write "the sum of an unknown and 6, when divided by 15, is equal to 7" as an equation, using r as the variable.

SOLUTION

The sum of an unknown and 6	when divided by 15	is equal to	7.
↓	↓	↓	↓
$(r + 6)$	$\div\ 15$	$=$	7

Simplify and solve the equation.

$$(r + 6) \div 15 = 7$$
$$(r + 6) \div 15 \cdot 15 = 7 \cdot 15 \quad \text{Multiply both sides by 15.}$$
$$r + 6 = 105$$
$$r + 6 - 6 = 105 - 6 \quad \text{Subtract 6 from both sides.}$$
$$r = 99$$

QUICK CHECK 5

Write "the quantity 17 minus an unknown, divided by 12, is equal to 14" as an equation, using r as the variable.

OBJECTIVE 3 Solve application problems. Now that statements have been translated into mathematical expressions, you can use this knowledge to solve problems.

> **Quick TIP**
> Be patient. It takes years of practice to become a good problem solver.

Solving Application Problems

- Step 1 Read the problem carefully. A sketch may help.
- Step 2 Identify the unknown and choose a variable to represent it. If possible, write any other unknowns in terms of the same variable.
- Step 3 Translate the problem into an equation.
- Step 4 Solve the equation.
- Step 5 Answer the question(s) asked in the problem.
- Step 6 Check your solution using the original words of the problem.
- Step 7 Be sure your answer is reasonable.

Step 3 can be the most difficult. To write an equation from the information given in the problem, convert the facts stated in words into mathematical expressions.

Solving a Business Problem

EXAMPLE 6 Ben James is studying monthly demand for Chevrolet Impalas in two different regions of the country. He has 14,200 new Impalas to ship to the two regions, and he decides to ship 3350 more to the Southeast region than to the mid–Atlantic Coast region. Find the number to be shipped to the mid–Atlantic Coast region.

CASE IN POINT Let m = number of Impalas to be shipped to the mid–Atlantic Coast region. Then $(m + 3350)$ = number to be shipped to the Southeast region. The sum of the two shipments is 14,200.

SOLUTION

Number to mid–Atlantic Coast region	+	Number to Southeast region	=	Total
m	+	$(m + 3350)$	=	$14{,}200$

$$m + m + 3350 = 14{,}200 \quad \text{Remove parentheses}$$
$$2m + 3350 = 14{,}200 \quad \text{Combine like terms.}$$
$$2m + 3350 - 3350 = 14{,}200 - 3350 \quad \text{Subtract 3350 from both sides.}$$
$$2m = 10{,}850 \quad \text{Simplify.}$$
$$\frac{2m}{2} = \frac{10{,}850}{2} \quad \text{Divide both sides by 2.}$$
$$m = 5425$$

4.2 Applications of Equations

So the numbers to be shipped are

$$m = 5425 \text{ to the mid-Atlantic Coast region}$$
$$(m + 3350) = 8775 \text{ to the Southeast region}$$
$$= 14{,}200 \text{ to be shipped}$$

Since this agrees with the total to be shipped, the answer checks.

QUICK CHECK 6

A professor was taking 28 students to Guadalajara, Mexico, on a two-week, travel-study program. If there were 14 more women than men, find the number of men and the number of women.

Applying Equation Solving

EXAMPLE 7 A mattress for a single bed is on sale for $200, which is $\frac{4}{5}$ of its original price. Find the original price.

SOLUTION

Let p represent the original price, $200 is the sale price, and the sale price is $\frac{4}{5}$ of the original price. Use this information to write the equation.

Sale price	is	$\frac{4}{5}$	of	original price.
↓	↓	↓	↓	↓
$200	=	$\frac{4}{5}$	×	p

Solve the equation.

$$200 = \frac{4}{5} \cdot p$$

$$\frac{5}{4} \cdot 200 = \frac{5}{4} \cdot \frac{4}{5} \cdot p \quad \text{Multiply by reciprocal.}$$

$$\frac{1000}{4} = 1 \cdot p$$

$$250 = p$$

The original price is $250.

QUICK CHECK 7

Fearful of bankruptcy, the top manager of a company laid off enough people so that the workforce was only $\frac{2}{3}$ of the original workforce. Find the original number of employees if there were 34 employees after the layoff.

Solving Investment Problems

EXAMPLE 8 Laurie Zimmerman has $15,000 to invest. She places a portion of the funds in a savings account and $3000 more than twice this amount in a retirement account. How much is put into the savings account? How much is placed in the retirement account?

SOLUTION

Let z represent the amount invested in the savings account. To find the amount invested in the retirement account, translate as follows.

3000	more than	2 times the amount
↓	↓	↓
3000	+	$2z$

Since the sum of the two investments must be $15,000, an equation can be formed as follows.

Amount invested in savings		Amount invested in retirement account		Total amount invested
z	+	$(3000 + 2z)$	=	$15{,}000

Now solve the equation.

$$z + (3000 + 2z) = 15{,}000$$
$$3z + 3000 = 15{,}000$$
$$3z = 12{,}000 \quad \text{Subtract 3000.}$$
$$z = 4000 \quad \text{Divide by 3.}$$

The amount invested in the passbook account is z, or $4000. The amount invested in the retirement account is $3000 + 2z$ or $3000 + 2(4000) = \$11{,}000$.

> **QUICK CHECK 8**
>
> A mutual fund company has $24 million to invest. They place part of the funds in the bonds of a utility company and $4 million more than three times that amount in New York City bonds. Find the amount placed in utility bonds and the amount in New York City bonds.

Solving a Business Problem

EXAMPLE 9 The Eastside Nursery ordered 27 tree seedlings. Some of the seedlings were elms, costing $17 each. The remainder of the seedlings were maples at $11 each. The total cost of the seedlings is $375. Find the number of elms and the number of maples.

SOLUTION

Let x represent the number of elm seedlings in the shipment. Since the shipment contained 27 seedlings, the number of maples is found by subtracting the number of elms from 27.

$$27 - x = \text{number of maples}$$

If each elm seedling costs $17, then x elm seedlings will cost $17x$ dollars. Also, the cost of $(27 - x)$ maple seedlings at $11 each is $11(27 - x)$. The total cost of the shipment was $375.

A table can be very helpful in identifying the knowns and unknowns.

	COST PER SEEDLING	NUMBER OF SEEDLINGS	TOTAL COST
Elms	$17	x	$17x$
Maples	$11	$(27 - x)$	$11(27 - x)$
Totals		27	375

The information in the table is used to develop the following equation.

$$\text{Cost of elms} + \text{Cost of maples} = \text{Total cost}$$
$$17x + 11(27 - x) = 375$$

Now solve this equation.

$$17x + 297 - 11x = 375 \quad \text{Use distributive property.}$$
$$6x + 297 = 375 \quad \text{Combine terms.}$$
$$6x = 78 \quad \text{Subtract 297 from each side.}$$
$$x = 13 \quad \text{Divide each side by 6.}$$

There were $x = 13$ elm seedlings and $(27 - 13) = 14$ maple seedlings.

> **QUICK CHECK 9**
>
> A beer-bottling company ordered 2100 more 12-ounce bottles than 1-quart bottles. The cost of the 12-ounce bottles was $.08 each, and the cost of the 1-quart bottles was $.11 each. The total cost was $310.50 not including shipping. Find the number of each type of bottle ordered.

4.2 Exercises // MyLab Math

The shaded sections below contain solutions to help you get a QUICK START on the various types of exercises.

Write the following as mathematical expressions. Use x as the variable. Be sure to use parentheses as needed.

1. 27 plus a number $27 + x$
2. the sum of a number and $16\frac{1}{2}$ $x + 16\frac{1}{2}$
3. a number added to 22   _____
4. 6.8 added to a number   _____
5. 4 less than a number   _____
6. 12 fewer than a number   _____
7. subtract $3\frac{1}{2}$ from a number   _____
8. subtract a number from 5.4   _____
9. triple a number   _____
10. the product of a number and 9   _____
11. three-fifths of a number   _____
12. four-thirds of a number   _____
13. the quotient of 9 and a number   _____
14. the quotient of a number and 11   _____
15. 16 divided by a number   _____
16. a number divided by 4   _____
17. the product of 2.1 and the sum of 4 and a number   _____
18. the quantity of a number plus 4, divided by 9   _____
19. 7 times the difference of a number and 3   _____
20. the difference of a number and 2, multiplied by 7   _____

Write mathematical expressions for each of the following.

21. **WINTER GLOVES**   A ski shop purchases 15 pairs of ski gloves at y dollars each.   **15y**
22. **TUITION FEES**   Find the cost of x students paying their one-semester tuition of $2800 each.   **2800x**
23. **LIVESTOCK FEED**   The demand forecast for next month is 472 tons of livestock feed. Find the amount that should be ordered if inventory is x tons.   _____
24. **LAPTOPS**   Eighty-three of the x employees use laptops. How many do not?   _____
25. **UNION MEMBERSHIP**   A company has 73 employees of whom x are members of a union. How many employees are not union members?   _____
26. **CARD SALES**   The inventory of a small card shop is valued at $73,000. The value of the greeting cards is x. Find the value of the rest of the inventory.   _____
27. **TEXTBOOK PURCHASES**   A college bookstore paid $20,210 to purchase x textbooks for a biology class. Find the cost of one textbook.   _____
28. **ADMISSION FEES**   A lodge paid $1853 for tickets to a concert for its x employees. Find the cost of one ticket.   _____
29. **CHARITABLE DONATIONS**   Robin has 21 books on computers. She donates x of them to the school library. How many does she have left?   _____
30. **VIDEO STORE**   A video rental store is x years old. How old will it be in 8 years?   _____

C// indicates an exercise that is related to the Case in Point feature.

136 CHAPTER 4 Equations and Formulas

Solve the following application problems. The steps to solve application problems are repeated here for your convenience—use them.

Solving Applied Problems

Step 1   Read the problem carefully. A sketch may help.
Step 2   Identify the unknown and choose a variable to represent it. If possible, write any other unknowns in terms of the same variable.
Step 3   Translate the problem into an equation.
Step 4   Solve the equation.
Step 5   Answer the question(s) asked in the problem.
Step 6   Check your solution using the original words of the problem.
Step 7   Be sure your answer is reasonable.

31. Four times a number, plus 6 equals 58. Find the number.

    $4x + 6 = 58$
    $4x = 52$
    $x = 13$

    31. 13

32. Seventeen times a number, plus 5 equals 107. Find the number.

    $17x + 5 = 107$
    $17x = 102$
    $x = 6$

    32. 6

33. Six times the quantity of 4 minus a number is 15. Find the number.

    33. _____

34. Twelve times the quantity of a number less 1 is 72. Find the number.

    34. _____

35. When 6 is added to a number, the result is 7 times the number. Find the number.

    35. _____

36. If 6 is subtracted from three times a number, the result is 4 more than the number. Find the number.

    36. _____

37. When 5 times a number is added to twice the number, the result is 10. Find the number.

    37. _____

38. If 7 times a number is subtracted from 11 times the number, the result is 9. Find the number.

    38. _____

39. SOUND SYSTEM Last month, Ben Jamison sold 17 more sound systems than did the other salesperson at the store. If the two salespeople sold a total of 101 systems, find the number sold by Jamison.

39. _____

40. SODA SALES A grocery store sold 19 more cases of Coke than Sprite. Given that 43 cases were sold, find the number of cases of Coke sold.

40. _____

41. SHIP BUILDING One hundred eighty-five more people work in building the ships than are in management, accounting, finance, and marketing. Given a total of 229 employees, find the number working in building ships.

41. _____

42. EXCHANGE PROGRAM Twenty-one students went on a student exchange program to Hong Kong. There were 11 more women than men. Find the number of women.

42. _____

43. WATER SKIS A new set of slalom water skis are priced at $314.10 which is 9/10ths of the original price. Find the original price to the nearest dollar.

43. _____

44. INTERNATIONAL SHIPMENTS Because of unusually high handling and freight charges, Western Oil Equipment charges $\frac{5}{4}$ of the list price for an item shipped to Indonesia. Find the list price of an item that was charged at $725.

44. _____

45. COMMERCIAL BUILDING Jane Anderson purchased a commercial building and plans to fix it up. The building can be used for both retail stores and office space. She thinks she can get a total annual rent of $135,000, with $3\frac{1}{2}$ times as much rent coming from retail stores as from office space. Find the rent she expects from office space and also the rent she expects from retail stores.

45. _____

46. MATERIAL Karen Cherie has a piece of fabric that is 106 inches long. She wishes to cut it into two pieces so that one piece is 12 inches longer than the other. What should be the length of each piece?

46. _____

138 CHAPTER 4 Equations and Formulas

47. HARLEY DAVIDSON Excluding managers, there are 22 full-time employees at a large Harley Davidson shop. New employees receive $10.75 per hour, while more experienced and higher-skilled workers average $19.45 per hour. The company spends $401.80 per hour in wages, not counting benefits. Find the number of each type of worker.

47. _____

48. VEGETABLE SALES Florida Vegetables, Inc. makes $0.10 on a head of lettuce and $0.08 on a bunch of carrots. Last week, a total of 12,900 heads of lettuce and bunches of carrots were sold, with a total revenue of $1174. How many heads of lettuce and how many bunches of carrots were sold?

48. _____

49. NISSAN SALES At one dealership the profits on Nissan Altimas and Sentras average $1200 and $850, respectively. The total profit in a month in which they sold 120 of these models was $130,350. Find the number of each sold.

49. _____

50. AUTO REPAIR One week, revenue from the 95 repairs at an auto repair shop totaled $20,040. The average charges for repairs on personal and commercial vehicles were $250 and $180, respectively. Find the number of each type of vehicle repaired.

50. _____

51. Are the problems in this section difficult for you? Explain why or why not. How can you become better at solving application problems?

52. Write out the steps necessary to solve an application problem. (See Objective 3.)

QUICK CHECK ANSWERS

1. $y + 7$
2. $19 - s$
3. $7(9 + t)$
4. $94 = 12(p + 2.5)$
5. $\dfrac{17 - r}{12} = 14$
6. 7 men, 21 women
7. 51 employees
8. utility bonds, $5 million; New York City bonds, $19 million
9. 2850 12-ounce bottles; 750 1-quart bottles

4.3 Business Formulas

OBJECTIVES

1. Evaluate formulas for given values of the variables.
2. Solve formulas for a specific variable.
3. Use standard business formulas to solve word problems.
4. Evaluate formulas containing exponents.

OBJECTIVE 1 Evaluate formulas for given values of the variables. Many of the most useful rules and procedures in business are given as **formulas,** which are equations showing how one number is found from other numbers. One useful formula in business is the one for simple interest.

$$\text{Interest} = \text{Principal} \times \text{Rate} \times \text{Time}$$

When written out in words, as shown, a formula can take up too much space and be hard to remember. For this reason, it is common to *use letters as variables for the words* in a formula. Many times the first letter in each word of a formula is used, to make it easier to remember the formula. By this method, the formula for simple interest is written as follows.

$$\text{Interest} = \text{Principal} \times \text{Rate} \times \text{Time}$$
$$I = P \times R \times T$$

By using letters to express the relationship among interest, principal, rate, and time, we have generalized the relationship so that any value can be substituted into the formula. When three values are substituted into the formula, we can find the value of the remaining variable.

Evaluating a Formula **EXAMPLE 1** Use the formula $I = PRT$ to find I if $P = \$7000$, $R = .09$, and $T = 2$.

SOLUTION

Substitute $7000 for P, .09 for R, and 2 for T in the formula $I = PRT$. Remember that writing P, R, and T together as PRT indicates the product of the three letters.

$$I = PRT$$
$$I = \$7000(.09)(2)$$

Multiply on the right to get the solution.

$$I = \$1260$$

QUICK CHECK 1

Use $F = ma$ to find F if $m = 1700$ and $a = 9.8$.

Evaluating a Formula **EXAMPLE 2** Use $I = PRT$ to find P if $I = \$5760$, $R = .06$, and $T = 3$.

SOLUTION

Substitute the given numbers for the letters of the formula.

$$I = PRT$$
$$\$5760 = P(.06)(3)$$
$$5760 = .18P$$
$$\frac{5760}{.18} = \frac{.18P}{.18} \qquad \text{Divide both sides by .18.}$$
$$\$32,000 = P$$

QUICK CHECK 2

Use the formula $I = PRT$ to find I if $P = \$40,000$, $T = 1$, and $R = .0125$.

Evaluating a Formula **EXAMPLE 3** Solve for rate (R) given $M = \$12{,}270$, $P = \$12{,}000$, and $T = .5$ in the equation $M = P(1 + RT)$.

SOLUTION

$$M = P(1 + RT)$$
$$\$12{,}270 = \$12{,}000(1 + R \cdot .5)$$
$$12{,}270 = 12{,}000 + 12{,}000 \cdot R \cdot .5 \quad \text{Use the distributive property.}$$
$$12{,}270 = 12{,}000 + 6000 \cdot R$$
$$12{,}270 - 12{,}000 = 12{,}000 + 6000 \cdot R - 12{,}000 \quad \text{Subtract 12,000 from each side.}$$
$$270 = 6000 \cdot R$$
$$\frac{270}{6000} = \frac{6000 \cdot R}{6000} \quad \text{Divide by 6000.}$$
$$.045 = R$$

Rate is .045, or 4.5%.

QUICK CHECK 3

Solve for rate (R) if $M = \$8300$, $P = \$8000$, and $T = .5$ in the equation $M = P(1 + RT)$.

OBJECTIVE 2 Solve formulas for a specific variable. Sometimes, you may need to rearrange a formula to solve for a particular variable. For example, rearrange the formula $I = PRT$ to solve for the variable P as follows.

$$I = PRT$$
$$\frac{I}{RT} = \frac{P\cancel{RT}}{\cancel{RT}} \quad \text{Divide both sides by } RT.$$
$$\frac{I}{RT} = P, \quad \text{or} \quad P = \frac{I}{RT}$$

Solving a Formula for a Specific Variable **EXAMPLE 4** Solve for T in the formula $M = P(1 + RT)$. This formula gives the maturity value (M) of an initial amount of money (P) invested at a specific rate (R) for a certain period of time (T).

SOLUTION

Start by using the distributive property on the right side.

$$M = P(1 + RT)$$
$$M = P + PRT$$

Now subtract P from both sides.

$$M - P = P + PRT - P$$
$$M - P = PRT$$

Divide each side by PR.

$$\frac{M - P}{PR} = \frac{\cancel{P}R\cancel{T}}{\cancel{P}R}$$
$$\frac{M - P}{PR} = T, \quad \text{or} \quad T = \frac{M - P}{PR}$$

QUICK CHECK 4

Solve $S = (kT - 12)$ for T.

Solving a Formula for a Specific Variable

EXAMPLE 5 Solve for T in the formula $D = \dfrac{B}{MT}$.

SOLUTION

This formula gives the discount rate (D) of a note in terms of the face value (B), the time of a note (T), and the maturity value (M). Solve for T.

$$D = \dfrac{B}{MT}$$

$$DMT = \dfrac{B}{MT}MT \quad \text{Multiply both sides by } MT.$$

$$DMT = B$$

$$\dfrac{\cancel{D}MT}{\cancel{D}M} = \dfrac{B}{DM} \quad \text{Divide both sides by } DM.$$

$$T = \dfrac{B}{DM}$$

QUICK CHECK 5

Solve $T = \dfrac{R}{JK}$ for J.

OBJECTIVE 3 Use standard business formulas to solve word problems. In the following examples, application problems that use some common business formulas are solved.

Finding Gross Sales  **EXAMPLE 6** Find the gross sales amount from selling 481 fishing lures at $2.65 each.

SOLUTION

The formula for gross sales is $G = NP$, where N is the number of items sold and P is the price per item. To find the gross sales from selling 481 fishing lures at $2.65 each, use the formula.

$$G = NP$$
$$G = 481(\$2.65)$$
$$G = \$1274.65$$

The gross sales are $1274.65.

QUICK CHECK 6

Find the gross revenue from selling 27 cases of wine at $156 each.

Finding Selling Price  **EXAMPLE 7** A retailer purchased a boat for $2170. It then adds a markup of $650 before placing it on the lot to sell. Find the selling price.

SOLUTION

The selling price is found by adding the cost of the item and the markup.

$$S = C + M$$

The variable C is the cost and M is the markup, which is the amount added to the cost to cover expenses and profit. The selling price is found as shown.

$$S = \$2170 + \$650$$
$$S = \$2820$$

The selling price is $2820.

QUICK CHECK 7

A discount Web retailer buys software for $26.92 each and marks it up by $8.07 each. Find the selling price.

142 CHAPTER 4 Equations and Formulas

OBJECTIVE 4 Evaluate formulas containing exponents. Exponents are used to show repeated multiplication of a quantity called the *base*. For example,

Exponent: the number of times the base is multiplied by itself

$$x^2 = x \cdot x$$

Base: the quantity being multiplied.

Similarly,

$$z^3 = z \cdot z \cdot z \quad \text{and} \quad 5^4 = 5 \cdot 5 \cdot 5 \cdot 5 = 625$$

Forecasting Monthly Sales **EXAMPLE 8** After significant research, Ben James believes that the dollar amount of Chevrolet vehicles that can be sold in one state can be approximated using the following equation.

$$S = \$40 + 16.2 \times A^2$$

CASE IN POINT Both advertising (A) and sales (S) are in millions of dollars. Estimate sales for July if advertising is $1.4 million.

SOLUTION

Substitute $1.4 into the equation for advertising and solve for S to find the forecast.

$S = \$40 + 16.2 \times A^2$ Multiply.
$S = \$40 + 16.2 \times (\$1.4)^2$
$S = 40 + 16.2 \times 1.96$ Evaluate exponent first.
$S = 40 + 31.752$
$S = 71.752$

Quick TIP
Since the $1.4 substituted in place of A was in millions, the final answer of 71.752 is in millions.

Based on the model, July sales are forecast to be $71,752,000.

QUICK CHECK 8

Stress on a suspension cable on a bridge can be estimated using $S = 7.2(\text{weight})^2 + 3480$. Find the stress if the weight is 29 tons.

4.3 Exercises // MyLab Math

*The shaded sections below contain solutions to help you get a **QUICK START** on the various types of exercises.*

In the following exercises a formula is given, along with the values of all but one of the variables in the formula. Find the value of the variable that is not given. Round to the nearest hundredth, if applicable.

1. $I = PRT$; $P = \$4600$, $R = .085$, $T = 1\frac{1}{2}$ $\underline{\$586.50}$
2. $F = ma$; $m = 820$, $a = 12$ $\underline{9840}$

3. $P = B \times R$; $B = \$168{,}000$, $R = .06$ _____
4. $B = \dfrac{P}{R}$; $P = \$1200$, $R = .08$ _____

5. $s = c + m$; $c = \$14$, $m = \$2.50$ _____
6. $m = s - c$; $s = \$24{,}200$, $c = \$2800$ _____

7. $P = 2L + 2W$; $P = 40$, $W = 6$ _____
8. $P = 2L + 2W$; $P = 340$, $L = 70$ _____

9. $P = \dfrac{I}{RT}$; $T = 3$, $I = 540$, $R = .08$ _____
10. $M = P(1 + RT)$; $R = .15$, $T = 2$, $M = 481$ _____

11. $y = mx^2 + c$; $m = 3$, $x = 7$, $c = 4.2$ _____
12. $C = \$5 + \$.10N$; $N = 38$ _____

13. $M = P(1 + i)^n$; $P = \$640$, $i = .02$, $n = 8$ _____
14. $M = P(1 + i)^n$; $M = \$2400$, $i = .05$, $n = 4$ _____

15. $E = mc^2$; $m = 7.5$, $c = 1$ _____
16. $x = \dfrac{1}{2}at^2$; $t = 5$, $x = 150$ _____

17. $A = \frac{1}{2}(b + B)h; A = 105, b = 19, B = 11$ _____

18. $A = \frac{1}{2}(b + B)h; A = 70, b = 15, B = 20$ _____

19. $P = \frac{S}{1 + RT}; S = 24{,}600, R = .06, T = \frac{5}{12}$ _____

20. $P = \frac{S}{1 + RT}; S = 23{,}815, R = .09, T = \frac{11}{12}$ _____

Solve each formula for the indicated variable.

21. $A = LW$; for L
$A = LW$
$\frac{A}{W} = \frac{LW}{W}$
$L = \frac{A}{W}$

22. $d = rt$; for t
$d = rt$
$\frac{d}{r} = \frac{rt}{r}$
$t = \frac{d}{r}$

21. $L = \frac{A}{W}$

22. $t = \frac{d}{r}$

23. $PV = nRT$; for V

24. $I = PRT$; for R

23. _____

24. _____

25. $M = P(1 + i)^n$; for P

26. $R(1 - DT) = D$; for R

25. _____

26. _____

27. $P = \frac{A}{1 + i}$; for i

28. $M = P(1 + RT)$; for R

27. _____

28. _____

29. $P = M(1 - DT)$; for D

30. $P = \frac{M}{1 + RT}$; for R

29. _____

30. _____

C// indicates an exercise that is related to the Case in Point feature.

31. $A = \dfrac{1}{2}(b + B)h$; for h

32. $P = 2L + 2W$; for L

31. _____

32. _____

Solve the following application problems.

33. CARNIVAL A carnival purchases 1800 stuffed animals to give to people who win at carnival games. Given a total cost of $4320, find the cost per stuffed animal.

$y = $ cost per stuffed animal
$1800 \cdot y = \$4320$
$\dfrac{1800y}{1800} = \dfrac{4320}{1800}$
$y = \$2.40$

33. $2.40

34. WEB PAGES A retailer paid $1305 to a college student who built and added 15 pages to the firm's website. Find the cost per web page.

$c = $ cost per web page
$15 \cdot c = \$1305$
$\dfrac{15c}{15} = \dfrac{1305}{15}$
$c = \$87$

34. $87

35. MUSICAL INSTRUMENTS The Guitar Shoppe bought 6 sets of bongo drums and 7 Alvarez guitars for $2445.80. The guitars cost $269 apiece. Find the cost for a set of bongo drums.

35. _____

36. REFRIGERATORS An appliance store bought 8 large refrigerators and 10 washer/dryer combination sets. The store's cost for a washer/dryer was $462 and the total paid for all of the appliances was $10,860. Find the cost of a refrigerator.

36. _____

37. CLOTHING STORE The weekly salary for each salesperson at a clothing store is found using the formula $S = \$280 + .05x$, where $x = $ employee's total sales for the week. Find the salary of two part-time employees with the following weekly sales: **(a)** $2940 and **(b)** $4450.

37. _____

38. MUSTANG Adam Nguyen needs $5000 for a down payment on a new Mustang. He has $4650, which he invests in an investment called a master limited partnership that pays 6% simple interest per year. Estimate the number of months he must wait to have the required down payment.

38. _____

39. COMPUTER CHIPS One quarter, a computer chip manufacturer had net sales of $230 million with returns equal to $\frac{1}{40}$ of gross sales. Find gross sales rounded to the nearest million given that net sales equals gross sales less returns.

39. _____

40. BRIDAL SHOP One month, the Bridal Shop had net sales of $33,000 and return(s) of $\frac{1}{12}$ of gross sales. If net sales equal gross sales minus returns, find gross sales.

40. _____

41. MOVIES The manager of one theater marks up the cost of a box of chocolate-covered raisins by $\frac{3}{4}$ of the cost of the raisins. Find the cost to the nearest cent if the theatre sells a box of raisins for $6.50. (*Hint:* Selling price is cost plus markup.)

41. _____

42. COLLEGE TEXTBOOKS One college bookstore chain marks up the books it sells by an average of 33%. Find the cost to the bookstore of a book they sell for $195. (*Hint:* Selling price is cost plus markup.)

42. _____

43. PIZZA Last year, the expenses at Mack's Pizzeria were $\frac{5}{6}$ of the total revenue and the remainder was profit. The profit, or the total revenue less expenses, was $107,400. Find the total revenue.

43. _____

44. GAMING Best Gameworks has expenses that run $\frac{15}{16}$ of revenue. One month's profit (the difference between revenue and expenses) was $18,000. Find the revenue.

44. _____

146 CHAPTER 4 Equations and Formulas

In Exercises 45–49, use the formula I = PRT, where I = interest in dollars, P = principal or loan amount, R = interest rate written as a decimal, and T = time in years.

45. FINDING INTEREST Find the interest if principal of $5200 is invested at $7\frac{1}{2}$% (or .075) for one year.

45. _____

46. LOAN TO AN UNCLE Ben Cross loaned $8000 to his uncle for 4 years and received $1920 in interest. Find the interest rate.

46. _____

47. AUTO PARTS STARTUP A loan company made a $22,000 loan to Jennifer Fleur so that she can start an auto parts business. The loan was for 2 years, and interest was $5720. Find the rate of interest.

47. _____

48. FINDING TIME Fred Tausz loaned $39,000 to his sister. The loan was at 7% (or 0.07), with interest of $13,650. Find the time for the loan.

48. _____

49. STUDENT LOAN Joan Summers borrowed $5850 (P) from her brother to help pay for her last year in college. They agreed to an interest rate (R) of 3% and a final interest amount (I) of $702. Find the time.

49. _____

In Exercises 50–52, use M = P(1 + RT) where M = maturity value, P = principal or loan amount, R = annual interest rate written as a decimal, and T = time in years.

50. MATURITY VALUE Mary Scott invests $1000 in a bond fund that she hopes will yield 7% per year (or .07) for 5 years. What maturity value (M) will she have in her account at the end of 5 years?

50. _____

51. SAVING John Wood had $4560 in his retirement account after 2 years. The account paid 7% (or .07) per year. How much did John initially deposit in his account? $[M = P(1 + RT)]$

51. _____

52. ANTIQUES Jan Reus borrowed $12,500 from her uncle to help start an antique shop. She repaid $14,750 exactly 2 years later. Find the interest rate using $[M = P(1 + RT)]$.

52. _____

Solve using the equation given. Note that M = maturity value, P = principal or loan amount, i = annual interest rate, and n is the number of years.

53. LOAN AMOUNT Ignacia Jaquez paid a maturity value of $5989.50 on a 3-year note with annual interest of 10% (or .10). Use $M = P(1 + i)^n$ to solve for the amount borrowed.

53. _____

54. INHERITANCE Maybelle Jackson inherited $8500 when her grandfather died. She placed the money in a dividend paying bond fund that pays 3.5% per year and plans to leave it there for 20 years until she retires. Use $M = P(1 + i)^n$ to solve for the maturity value.

54. _____

55. Write a step-by-step explanation of the procedure you would use to solve the equation $A = P + PRT$ for R. (See Objective 2.)

56. Formulas are used in business, physics, biology, chemistry, engineering, economics, and many other places. Why are formulas so commonly used? (See Objective 1.)

QUICK CHECK ANSWERS

1. $F = 16,660$
2. $I = \$500$
3. $R = 7.5\%$
4. $T = \dfrac{(S + 12)}{k}$
5. $J = \dfrac{R}{KT}$
6. $4212
7. $34.99

4.4 Ratio and Proportion

OBJECTIVES
1. Define a ratio.
2. Set up a proportion.
3. Solve a proportion for unknown values.
4. Use proportions to solve problems.

OBJECTIVE 1 Define a ratio. A **ratio** is a quotient of two quantities that is used to *compare* the quantities. The ratio of the number a to the number b is written in any of the following ways.

$$a \text{ to } b, \qquad a{:}b, \qquad \frac{a}{b}$$

All are pronounced "a to b" or "a is to b." The third way shown of writing a ratio is most common in mathematics, while $a{:}b$ is perhaps more common in business.

Writing Ratios EXAMPLE 1 Write a ratio in the form $\frac{a}{b}$ for each word phrase. Notice in each example that the number mentioned first is always the numerator.

SOLUTION

(a) The ratio of 5 hours to 3 hours is $\frac{5}{3}$.

(b) To find the ratio of 5 hours to 3 days, *first convert* 3 *days* to *hours*. Since there are 24 hours in 1 day, **3 days** $= 3 \cdot 24 =$ **72 hours**. Then the ratio of 5 hours to 3 days is the quotient of 5 and 72.

$$\frac{5}{72}$$

(c) The ratio of $700,000 in sales to $950,000 in sales is written this way:

$$\frac{\$700{,}000}{\$950{,}000}$$

Reduce to write this ratio in lowest terms:

$$\frac{\$700{,}000}{\$950{,}000} = \frac{14}{19}$$

As in part (b) in Example 1, you may need to convert both numbers in a ratio to the same unit (of time, weight, etc.) before writing the ratio.

> **QUICK CHECK 1**
> Write 40 sheep to 15 goats as a ratio.

Writing Ratios EXAMPLE 2 Burger King sold the following items in a one-hour period last Friday afternoon.
- 70 bacon cheeseburgers
- 15 plain hamburgers
- 30 salad combos
- 45 chicken sandwiches
- 40 fish sandwiches

Write ratios for the following items sold:
(a) bacon cheeseburgers to fish sandwiches
(b) salad combos to chicken sandwiches
(c) plain hamburgers to salad combos
(d) fish sandwiches to total items sold

SOLUTION

(a) $\dfrac{\text{bacon cheeseburgers}}{\text{fish sandwiches}} = \dfrac{70}{40} = \dfrac{7}{4}$

(b) $\dfrac{\text{salad combos}}{\text{chicken sandwiches}} = \dfrac{30}{45} = \dfrac{2}{3}$

(c) $\dfrac{\text{plain hamburgers}}{\text{salad combos}} = \dfrac{15}{30} = \dfrac{1}{2}$

(d) $\dfrac{\text{fish sandwiches}}{\text{total items sold}} = \dfrac{40}{200} = \dfrac{1}{5}$

QUICK CHECK 2
Use the data from Example 2 to write a ratio for the sum of bacon cheeseburgers and plain hamburgers, to salad combos.

OBJECTIVE 2 Set up a proportion. A ratio is used to compare two numbers or amounts. A **proportion** says that two ratios are equal, as shown here.

$$\frac{3}{4} = \frac{15}{20}$$

A proportion can be simplified by multiplying both sides of the equation by the product of the two denominators.

$$\frac{a}{b} = \frac{c}{d}$$

$$\frac{a}{\cancel{b}} \cdot \cancel{b} d = \frac{c}{\cancel{d}} \cdot b\cancel{d} \quad \text{Multiply both sides by the product of two denominators.}$$

$$ad = bc$$

Thus, $\frac{a}{b} = \frac{c}{d}$ only if the cross products $a \cdot d$ and $b \cdot c$ are equal to one another. This **method of cross products** can be used to determine if a proportion is true.

> **Method of Cross Products**
> The proportion
> $$\frac{a}{b} = \frac{c}{d}$$
> is true if the cross products $a \cdot d$ and $b \cdot c$ are equal (i.e., if $ad = bc$).

Determining If a Proportion Is True

EXAMPLE 3 Decide whether the following proportions are true.

(a) $\frac{3}{5} = \frac{12}{20}$ (b) $\frac{2}{3} = \frac{9}{16}$

SOLUTION

(a) Find each cross product.

$$\frac{3}{5} = \frac{12}{20}$$
$$3 \times 20 \stackrel{?}{=} 5 \times 12$$
$$60 = 60$$

Since the cross products are equal, the proportion is true.

(b) Find the cross products.

$$\frac{2}{3} = \frac{9}{16}$$
$$2 \times 16 \stackrel{?}{=} 3 \times 9$$
$$32 \neq 27 \quad \text{Not equal.}$$

This proportion is false, so $\frac{2}{3} \neq \frac{9}{16}$.

Quick TIP
The symbol $\neq$ means is "not equal to."

QUICK CHECK 3
Decide whether the following proportion is true or false: $\frac{2}{9} = \frac{8}{18}$.

OBJECTIVE 3 Solve a proportion for unknown values. Four numbers and/or variables are used in a proportion. If any three of the values are known, the fourth can be found. The two methods to solve a proportion are to:

1. multiply both sides by the product of the two denominators or
2. use the method of cross products.

Solving a Proportion   EXAMPLE 4 Find the value of the unknowns. Use the first method for **(a)** and the second method for **(b)**.

(a) $\dfrac{3}{5} = \dfrac{x}{40}$ (b) $\dfrac{3}{10} = \dfrac{5}{k}$

SOLUTION

(a) Solve for the unknown by multiplying both sides of the equation by the product of the denominators.

$$\dfrac{3}{5} = \dfrac{x}{40}$$

$$\dfrac{3}{\cancel{5}} \cdot (\cancel{5} \cdot 40) = \dfrac{x}{\cancel{40}} \cdot (5 \cdot \cancel{40})$$ Multiply both sides by product of denominators.

$$120 = 5x$$

$$\dfrac{120}{5} = \dfrac{\cancel{5}x}{\cancel{5}}$$ Divide both sides by 5.

$$24 = x, \quad \text{or} \quad x = 24$$

(b) Solve for the unknown using the method of cross products.

$$\dfrac{3}{10} = \dfrac{5}{k}$$

$$3k = 50 \quad \text{Use method of cross products.}$$

$$\dfrac{\cancel{3}k}{\cancel{3}} = \dfrac{50}{3}$$

$$k = \dfrac{50}{3} = 16\dfrac{2}{3}$$

QUICK CHECK 4

Find y in the following proportion: $\dfrac{7}{y} = \dfrac{84}{228}$.

Solving Proportions   EXAMPLE 5 A food wholesaler charges a restaurant chain $83 for 3 crates of lettuce. How much should it charge for 5 crates of lettuce?

SOLUTION

Let x be the cost of 5 crates of lettuce. Set up a proportion with one ratio the number of crates and the other ratio the costs. Use this pattern.

$$\dfrac{\text{Crates}}{\text{Crates}} = \dfrac{\text{Cost}}{\text{Cost}}$$

Now substitute the given information.

$$\dfrac{3}{5} = \dfrac{83}{x}$$

→ 3 crates cost $83
→ 5 crates cost $x

Cross multiply and use the cross products to solve the proportion.

$$3x = 5(83)$$

$$3x = \$415$$

$$x = \$138.33 \quad \text{(rounded to the nearest cent)}$$

The 5 crates cost $138.33.

QUICK CHECK 5

A fish wholesaler buys 40 pounds of fresh salmon from fishermen for $232. Estimate the cost of 300 pounds of salmon.

OBJECTIVE 4 Use proportions to solve problems. Proportions are used in many practical applications, as shown in the next two examples.

Solving Applications **EXAMPLE 6** A firm in Hong Kong and one in Thailand agree to jointly develop an engine-control microchip to be sold to North American auto manufacturers. They agree to split the development costs in a ratio of 8:3 (Hong Kong firm to Thailand firm), resulting in a cost of $9,400,000 to the Hong Kong firm. Find the cost to the Thailand firm.

SOLUTION

Let x represent the cost to the Thailand firm. Then

$$\frac{8}{3} = \frac{9{,}400{,}000}{x}$$
$$8x = 3 \cdot 9{,}400{,}000 \quad \text{Cross multiply.}$$
$$8x = 28{,}200{,}000$$
$$x = \mathbf{3{,}525{,}000} \quad \text{Divide by 8.}$$

The Thailand firm's share of the costs is $3,525,000.

QUICK CHECK 6

Two real estate developers agree to build a project estimated to cost $17.6 million. They decide to split costs in a ratio of 7:4. Find the cost to each.

Solving Applications **EXAMPLE 7** James McIntosh wants to cut down a tree in his front yard, but he is worried about it falling into his garage. He needs to know the height of the tree. One morning, he notices that his own 6-foot body casts an 8-foot shadow at the same time that a typical tree casts a 34-foot shadow. Find the height of the tree.

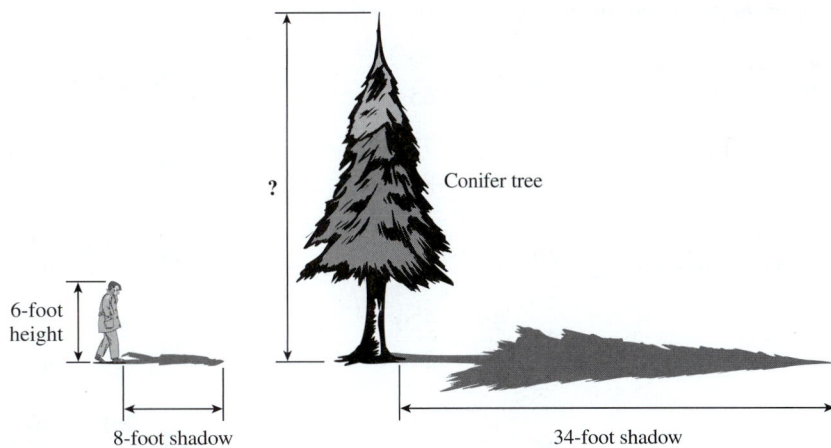

SOLUTION

Set up a proportion in which the height of the tree is given the variable name x.

$$\frac{6}{8} = \frac{x}{34}$$
$$6 \cdot 34 = 8 \cdot x \quad \text{Cross multiply.}$$
$$\frac{204}{8} = \frac{\cancel{8} \cdot x}{\cancel{8}} \quad \text{Divide by 8.}$$
$$x = 25.5 \text{ feet}$$

The height of the tree is 25.5 feet.

QUICK CHECK 7

Late afternoon one day, a 20-foot flagpole casts a 32-foot shadow. Find the shadow cast by a boy 4 feet tall.

152 CHAPTER 4 Equations and Formulas

4.4 Exercises // MyLab Math

The shaded sections below contain solutions to help you get a **QUICK START** *on the various types of exercises.*

Write the following ratios. Write each ratio in lowest terms.

1. 18 kilometers to 64 kilometers $\frac{9}{32}$

2. 18 defects out of 580 items $\frac{9}{290}$

3. 216 students to 8 faculty _____

4. $80 in returns to $8360 in sales _____

5. 8 men to 6 women _____

6. 12 feet to 1 inch _____

7. 30 kilometers (30,000 meters) to 8 meters _____

8. 30 inches to 5 yards _____

9. 90 dollars to 40 cents _____

10. 148 minutes to 4 hours _____

11. 4 dollars to 10 quarters _____

12. 35 dimes to 6 dollars _____

13. 20 hours to 5 days _____

14. 6 days to 9 hours _____

15. $.80 to $3 _____

16. $1.20 to $.75 _____

17. $3.24 to $.72 _____

18. $3.57 to $.42 _____

Decide whether the following proportions are true or false.

19. $\frac{3}{5} = \frac{21}{35}$ T _____

20. $\frac{6}{13} = \frac{30}{65}$ T _____

21. $\frac{9}{7} = \frac{720}{480}$ _____

22. $\frac{54}{14} = \frac{270}{70}$ _____

23. $\frac{69}{320} = \frac{7}{102}$ _____

24. $\frac{17}{19} = \frac{72}{84}$ _____

25. $\frac{19}{32} = \frac{33}{77}$ _____

26. $\frac{19}{30} = \frac{57}{90}$ _____

27. $\frac{110}{18} = \frac{160}{27}$ _____

28. $\frac{46}{17} = \frac{212}{95}$ _____

29. $\frac{32}{75} = \frac{61}{108}$ _____

30. $\frac{28}{75} = \frac{224}{600}$ _____

31. $\frac{7.6}{10} = \frac{76}{100}$ _____

32. $\frac{95}{64} = \frac{320}{217}$ _____

33. $\frac{2\frac{1}{4}}{5} = \frac{9}{20}$ _____

34. $\frac{\frac{3}{4}}{80} = \frac{\frac{9}{8}}{120}$ _____

35. $\frac{4\frac{1}{5}}{6\frac{1}{8}} = \frac{27}{41}$ _____

36. $\frac{1\frac{1}{2}}{12} = \frac{5\frac{1}{4}}{42}$ _____

37. $\frac{8.15}{2.03} = \frac{61.125}{15.225}$ _____

38. $\frac{423.88}{17.119} = \frac{330.6264}{13.35282}$ _____

Solve the following proportions.

39. $\frac{x}{15} = \frac{49}{105}$ 7 _____

40. $\frac{y}{35} = \frac{27}{315}$ 3 _____

41. $\frac{6}{9} = \frac{r}{108}$ _____

42. $\frac{16}{41} = \frac{112}{I}$ _____

43. $\frac{63}{s} = \frac{3}{5}$ _____

44. $\frac{260}{390} = \frac{x}{3}$ _____

45. $\frac{1}{2} = \frac{r}{7}$ _____

46. $\frac{2}{3} = \frac{5}{s}$ _____

47. $\frac{\frac{3}{4}}{6} = \frac{3}{x}$ _____

48. $\frac{3}{x} = \frac{11}{9}$ _____

49. $\frac{12}{P} = \frac{23.571}{15.714}$ _____

50. $\frac{86.112}{57.408} = \frac{k}{15}$ _____

C// indicates an exercise that is related to the Case in Point feature.

51. Explain the difference between ratio and proportion. (See Objective 2.)

52. Explain cross products using the rules of algebra. (See Objective 2.)

Solve the following application problems.

53. TICKET SALES One Ticketmaster outlet sold 350 rock-concert tickets in 2 days. At that rate, find the number of tickets it can expect to sell in 9 days.

$$\frac{350}{2} = \frac{x}{9}$$
$$350 \cdot 9 = 2x$$
$$2x = 3150$$
$$x = 1575 \text{ tickets}$$

53. 1575 tickets

54. BLOOD CELLS A 170-pound person has about 30 trillion blood cells. Estimate the number of blood cells in a 140-pound person to the nearest tenth of a trillion.

$$\frac{170}{30} \text{ trillion} = \frac{140}{x} \text{ trillion}$$
$$170x = 30 \cdot 140$$
$$170x = 4200$$
$$x = 24.7 \text{ trillion} \quad \text{(rounded)}$$

54. 24.7 trillion

55. REAL ESTATE DEVELOPMENT Mike George paid $310,000 for a new 5-unit apartment house. Find the cost for a 12-unit apartment house.

55. _____

56. TIGER FOOD A 450-pound circus tiger eats 15 pounds of meat per day. How many pounds of meat would you expect a 360-pound tiger to eat per day?

56. _____

57. SWIMMING Penny Thorpe can swim 400 meters in a nearby lake in 7.1 minutes. At this same rate, estimate the amount of time it will take her to swim one mile (1609 meters). Round to the nearest tenth of a minute.

57. _____

58. BISCUITS A biscuit recipe that feeds 7 requires 2 cups of flour. How much flour is needed for enough biscuits to feed a church group of 125? Round to the nearest whole number.

58. _____

154 CHAPTER 4 Equations and Formulas

59. CLIMATE CHANGE The concentration of carbon dioxide in the atmosphere has increased from 315 parts per million to 405 parts per million in the past 50 years. During the same period of time, the global average temperature increased by 1.5°F. At that rate, estimate the amount of further increase in global average temperature to the nearest tenth of a degree, if the concentration of carbon dioxide (CO_2) in the atmosphere increases from 405 parts per million up to 550 parts per million, as predicted by some.

59. _____

60. SEA LEVELS Many scientists believe that climate change contributes to the melting of ice in glaciers, on Greenland, and at both poles and thus contributes to rising sea levels. Assume sea levels rise at a constant rate of 1.3 inches per decade (10 years) and estimate the rise in sea level between 2020 and 2100.

60. _____

61. MAP READING The distance between two cities on a road map is 2 inches. Actually, the cities are 120 miles apart. The distance between two other cities is 17 inches. How far apart are these cities?

61. _____

62. CLOTHING SHOP Martha Vinn opened a small consignment store and had sales of $3720 during the first 3 weeks. At that rate, estimate sales for the first 4 weeks.

62. _____

63. SALES OF HEALTH FOOD Natural Harvest had sales of $274,312 for the first 20 weeks of the year. Estimate sales for the entire 52-week year.

63. _____

64. PARTNERSHIP PROFITS Chester and Gaines have a partnership agreement that calls for profits to be paid in the ratio of 2:5, respectively. Find the amount that goes to Chester if Gaines receives $45,000.

64. _____

65. OIL PROFITS The two partners in Alamo Energy agreed to split profits in a ratio of 3:8. If the first partner received $48,000 in profits one year, find the profit earned by the second partner.

65. _____

66. ROCK CONCERT A blues group held a concert in a city of 650,000 people and 8200 people attended. At the same rate, find the number they can expect to attend in a city with 1,000,000 people, assuming a big enough building is available.

66. _____

67. SONGBIRD MIGRATION Small songbirds in one area migrate at about 20 miles per hour whereas eider ducks migrate at 35 miles per hour. How far would the eider ducks migrate in the same amount of time it would take songbirds to migrate 200 miles?

67. _____

68. ISLAND AREA Indonesia has an area of 741,101 square miles and is made up of 13,677 islands. Assume the United States, with an area of 3,618,770 square miles, were similarly broken up into islands. How many islands would there be (to the nearest whole number)?

68. _____

69. ICEBERG VOLUME Seven-eighths of an iceberg is below the water since icebergs are made up of freshwater, which is not as dense as seawater. Find the amount of an iceberg that is under water if the amount above water has a volume of 500,000 cubic meters. (*Hint:* $1 - \frac{7}{8} = \frac{1}{8}$ above water.)

69. _____

70. AUTO PRODUCTION An auto plant produces 3 red sports models for every 7 blue family models. Find the number of red sports models produced if the plant produces 868 blue family models.

70. _____

156 CHAPTER 4 Equations and Formulas

71. JAPANESE YEN Benjamin Lopez was in Japan for 2 months on a business trip. Find the number of U.S. dollars he will receive for 20,355 Japanese yen if $1 U.S. can be exchanged for 115 yen.

71. _____

72. WORKING IN CHINA Gina Harden was offered a job as a teacher at an elite private high school in Beijing at an annual salary of 471,200 yuan. Find the salary in U.S. dollars if 6.70 yuan can be exchanged for $1 U.S.

72. _____

QUICK CHECK ANSWERS

1. 8:3   2. 17:6   3. false   4. $y = 19$ 5. $1740
6. $11.2 million and $6.4 million   7. 6.4 fee

Chapter 4 Quick Review

Chapter Terms Review the following terms to test your understanding of the chapter. For each term you do not know, refer to the page number found next to that term.

addition rule [p. 122]	expression [p. 122]	multiplication rule [p. 122]	solution [p. 122]
cross multiply [p. 150]	left side [p. 122]	proportion [p. 149]	substitute [p. 122]
cross product [p. 149]	like terms [p. 125]	ratio [p. 148]	term [p. 122]
distributive property [p. 126]	method of cross product [p. 149]	reciprocal [p. 124]	unlike terms [p. 125]
equation [p. 122]		right side [p. 122]	variable [p. 122]

CONCEPTS	EXAMPLES
4.1 Use the addition rule to solve basic equations. Add or subtract the same number from both sides of the equation.	Solve $x - 17.5 = 50$. $x - 17.5 = 50$ $x - 17.5 + 17.5 = 50 + 17.5$ **Add 17.5** $x = 67.5$
4.1 Use the multiplication rule to solve basic equations. Multiply or divide both sides of the equation by the same number.	Solve $6x = 72$. $6x = 72$ $\dfrac{6x}{6} = \dfrac{72}{6}$ **Divide by 6** $x = 12$
4.1 Solve more complicated equations. Use the preceding addition and multiplication rules. Move all terms with only numbers to one side of the equation and all terms with variables to the other side.	Solve $10y = 8y + 42$. $10y = 8y + 42$ $10y - 8y = 8y + 42 - 8y$ **Subtract 8y** $2y = 42$ $\dfrac{2y}{2} = \dfrac{42}{2}$ **Divide by 2** $y = 21$
4.1 Solve equations with parentheses. First use the distributive property to remove the parentheses; then solve.	Solve $15(p - 1) = 3p + 2$. $15(p - 1) = 3p + 2$ $15p - 15 = 3p + 2$ **By distributive property** $15p - 15 - 3p = 3p + 2 - 3p$ **Subtract 3p** $12p - 15 = 2$ $12p - 15 + 15 = 2 + 15$ **Add 15** $12p = 17$ $\dfrac{12p}{12} = \dfrac{17}{12}$ **Divide by 12** $p = 1\dfrac{5}{12}$

4.2 Translate phrases into expressions.
  7 plus a number
  sum of a number and 12
  a number minus $\frac{1}{2}$
  product of 2 and a number
  a number multiplied by 19
  a number divided by .68
  8 times the sum of a number and 1
	Any letter can be used as a variable.
	$7 + x$
	$y + 12$
	$z - \frac{1}{2}$
	$2A$
	$19t$
	$\dfrac{T}{.68}$
	$8(P + 1)$

CHAPTER 4 Equations and Formulas

CONCEPTS	EXAMPLES
4.2 Solve a basic application problem. Read problem carefully. Define variables. Write an equation and solve for the unknown. Check to make sure the answer is reasonable.	The sum of two consecutive odd numbers is 96. Find both numbers. Let x be the smaller number. Then $(x + 2)$ is the larger number. $$x + (x + 2) = 96$$ $$2x + 2 = 96$$ $$2x = 94$$ $$x = 47 \quad \text{and} \quad (x + 2) = 49$$
4.3 Evaluate a formula given values. Substitute known values into the formula and solve for the unknown.	Use $I = PRT$ to find I when $P = \$10{,}000$, $R = 4.5\%$, and $T = 1$. $$I = PRT$$ $$I = \$10{,}000 \cdot .045 \cdot 1$$ $$I = \$450$$
4.3 Solve a formula for a variable. Isolate the variable of interest on one side of the equation.	Solve $I = PRT$ for P. $$I = PRT$$ $$\frac{I}{RT} = \frac{PRT}{RT}$$ $$\frac{I}{RT} = P, \quad \text{or} \quad P = \frac{I}{RT}$$
4.3 Evaluate formulas with exponents. Substitute known values into the formula and solve for the unknown.	Use $M = P(1 + i)^n$ with $P = \$5000$, $i = 5\%$, and $n = 3$ to find M. $$M = P(1 + i)^n$$ $$M = \$5000(1 + .05)^3$$ $$M = 5000(1.05)^3$$ $$M = 5000 \cdot 1.157625$$ $$M = \$5788.13$$
4.4 Determine if a proportion is true. Check to see that cross products are equal to one another.	Is $\frac{4}{5} = \frac{28}{35}$ true? $$4 \cdot 35 \stackrel{?}{=} 5 \cdot 28 \quad \text{Use cross products}$$ $$140 = 140$$ The proportion is true.
4.4 Solve a proportion for an unknown. Use the method of cross products and solve the resulting equation.	Find x: $\frac{35}{17} = \frac{x}{153}$. $$35 \cdot 153 = 17x \quad \text{Use cross products}$$ $$5355 = 17x$$ $$\frac{5355}{17} = \frac{17x}{17}$$ $$x = 315$$
4.4 Use a proportion to solve a problem. 1. Define the unknown variable. 2. Set up the proportion. 3. Use the method of cross products. 4. Solve the equation.	A hiker walked 5.8 miles in 2 hours. At this rate, estimate the number of hours needed to hike 10 miles. Let x = number of hours needed. $$\frac{5.8 \text{ miles}}{2 \text{ hours}} = \frac{10 \text{ miles}}{x \text{ hours}}$$ $$5.8x = 2 \cdot 10 \quad \text{Use cross products}$$ $$5.8x = 20$$ $$\frac{5.8x}{5.8} = \frac{20}{5.8}$$ $$x = 3.4 \text{ hours} \quad \text{(rounded)}$$

Case Study

FORECASTING SALES AT ALCORN'S BOUTIQUE

Jane Alcorn had sold items on eBay for several years but felt it was time to open a retail store. After thinking about the options and competition, she decided to open a shop specializing in women's clothing. Sales were slow for her first few months and she began to wonder how much she should be spending on advertising. So, she decided to keep track of the monthly amount spent on advertising and the monthly sales as shown.

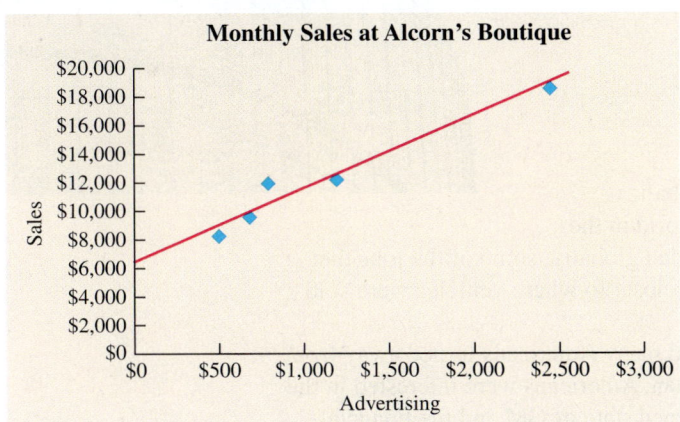

Jane's son Benton used statistics to help find the equation that was a best fit for the data. The equation is shown as the straight line in the graph. He told his mother it might not help much since it did not take into account factors such as seasonality or competition. And of course, Alcorn still had to decide where to advertise and exactly what the advertisements should look like. But she thought the equation might give her some idea of how much she should spend on advertising each month.

1. Estimate sales for advertising of $800 and $2000 if Sales = $6500 + $4.95 · Advertising.

    1. _____

2. Ah, Alcorn thought, this is easy. If you increase advertising from $800 up to $2000, then sales go up from $10,500 to $16,500. So an increase in advertising expenses of $1200 results in an increase in sales of $6000, at least during the first months her business was open. Her gross profit margin on the fashion goods she carried in her tiny boutique was 50%. Find the increased gross profit by multiplying the 50% by the increase in sales.

    2. _____

3. Therefore, an increase in advertising of $1200 results in $3000 more gross profit. Given this information, Alcorn's first thought was to borrow $5000 from a bank and advertise a lot. Then she began to wonder. What suggestions do you have for her? Should she borrow $5000 and advertise? Why or why not (based on your intuition)?

    3. _____

INVESTIGATE

List all of the factors you can think of that can influence sales at a small local store. You may wish to talk to the manager or owner of store and compare to your list.

Case in Point Summary Exercise

GENERAL MOTORS

www.GM.com

Facts:

- 1908: Founded
- 1982: Built huge factory in Spain
- 2009: Filed for bankruptcy
- 2010: Emerged from bankruptcy
- 2017: 60% of sales were outside U.S.

In some respects, General Motors (GM) *is* the story of globalization. It was one of the most successful companies in the world in the 1960s under the gifted manager Alfred Sloan. As it expanded globally, some of the jobs that had gone to Americans were moved overseas, partly to be closer to where vehicles were sold but also to cut costs.

Gasoline prices increased sharply during the 1970s and many Americans turned away from GM products and bought cars from Toyota, Honda, or Nissan. Americans were interested in the smaller, high-quality cars built by the Japanese. The weakened state of GM and the financial crisis of 2008–2010 forced the company to file for bankruptcy in 2009, when the U.S. government stepped in to help. But the company has since recovered.

Ben James received a bachelor's degree in statistics and a master's degree in business administration. He has worked at GM for several years. Among other things, he studies data such as that shown in the following two figures and develops equations to forecast demand.

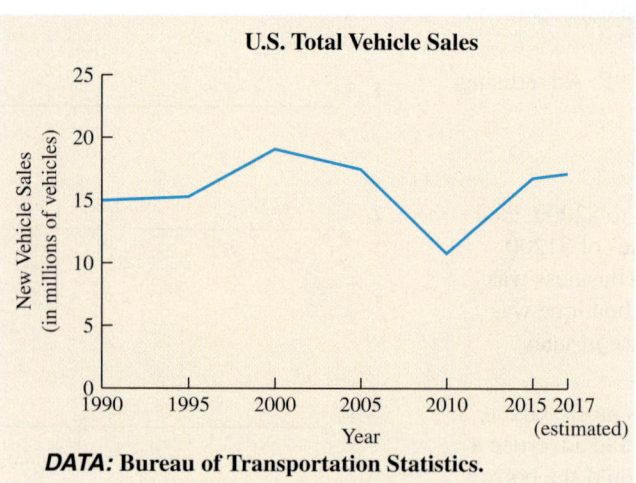

DATA: Bureau of Transportation Statistics.

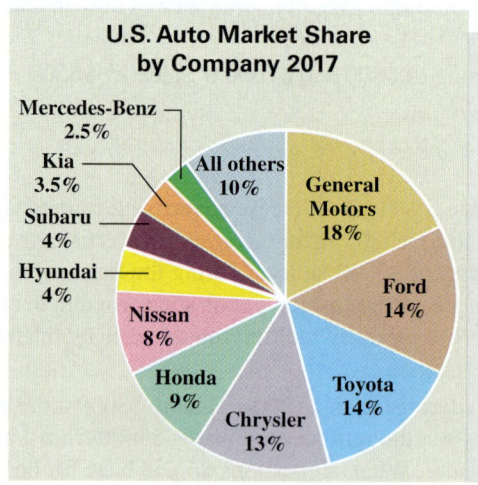

1. What can you learn from the two figures?

2. Many factors are involved in making a forecast. To illustrate, use the following equation to make a monthly forecast of car sales (in thousands) in one area given Advertising = $6.2 (millions), Economic Growth = 2.8% (.028), and Level of Competition = .6. (See Section 4.2, Examples 1–8.)

   Sales = $34.8 + 5.3 \cdot$ Advertising $+ 485 \cdot$ Economic Growth $- 34.2 \cdot$ Level of Competition

2. _____

Source: GENERAL MOTORS, www.GM.com.

3. GM has responded to increasing competition, and rapid economic growth in China by building and selling cars in China. The strategy has worked for GM, since China is now its largest market. Forecast monthly GM sales in one region of China using the following equation (in thousands) when Advertising = $5.3 (millions), Economic Growth = 6.5%, and Level of Competition = .7. (See Section 4.3, Examples 1–8.)

   Sales = 50.9 + 9.6 · Advertising + 720 · Economic Growth − 28.7 · Level of Competition

3. _____

4. In one European country, the sales of GM products increased from 23,850 to 36,400 in the third year that GM was in that market. Use a proportion to estimate the number of sales in the third year in a neighboring country with a very similar economy, if year two sales were 14,910. (See Section 4.4, Examples 5–7.)

4. _____

Discussion Question: Many factors other than those listed here affect sales of new automobiles. Think about your own family or a family you know and list things that influence the purchase of an automobile, new or used.

Chapter 4 Test

To help you review, the numbers in brackets show the section in which the topic was discussed.

Solve each equation for the unknown. [4.1]

1. $x + 45 = 96$ _____

2. $r - 36 = 14.7$ _____

3. $8t + 45 = 175.4$ _____

4. $4x - 6 = 15$ _____

5. $\dfrac{s}{6} = 43$ _____

6. $\dfrac{5z}{8} = 85$ _____

7. $\dfrac{m}{4} - 5 = 9$ _____

8. $5(x - 3) = 3(x + 4)$ _____

9. $6y = 2y + 28$ _____

10. $3r - 7 = 2(4 - 3r)$ _____

11. $.15(2x - 3) = 5.85$ _____

12. $.6(y - 3) = .1y$ _____

In Exercises 13–22, a formula is given, along with the values of all but one of the variables in the formula. Find the value of the variable that is not given. [4.3]

13. $I = PRT$; $P = 2800$, $R = .09$, $T = 2$ _____

14. $S = C + M$; $C = 275$, $M = 49$ _____

15. $G = NP$; $N = 840$, $P = 3.79$ _____

16. $M = P(1 + RT)$; $P = 420$, $R = .07$, $T = 2\dfrac{1}{2}$ _____

17. $R = \dfrac{D}{1 - DT}$; $D = .04$, $T = 5$ _____

18. $A = \dfrac{S}{1 + RT}$; $S = 12{,}600$, $R = .12$, $T = \dfrac{5}{12}$ _____

CHAPTER 4 Test   163

19. $T = \dfrac{D}{S}$; $T = 100$, $S = 2$ _____

20. $\dfrac{I}{PR} = T$; $P = 1000$, $R = .05$, $T = 1\dfrac{1}{2}$ _____

21. $d = rt$; $r = .07$, $t = 12$ _____

22. $I = PRT$; $P = 500$, $R = .08$, $T = 3$ _____

In Exercises 23–28, solve for the indicated variables. **[4.3]**

23. $A = LW$; for W _____

24. $d = rt$; for r _____

25. $I = PRT$; for T _____

26. $P = 1 + RT$; for R _____

27. $A = P + PRT$; for T _____

28. $R(1 - DT) = D$; for R _____

In Exercises 29–34, write the ratio in lowest terms. **[4.4]**

29. 250 pesos to 1250 pesos _____

30. 45 women to 110 men _____

31. $1.20 to 75¢ _____

32. 20 hours to 5 days _____

33. 35 dimes to 6 dollars _____

34. 30 inches to five yards _____

In Exercises 35–40, decide whether the proportions are true or false. **[4.4]**

35. $\dfrac{2}{3} = \dfrac{42}{63}$ _____

36. $\dfrac{6}{9} = \dfrac{36}{52}$ _____

37. $\dfrac{18}{20} = \dfrac{56}{60}$ _____

38. $\dfrac{12}{18} = \dfrac{8}{12}$ _____

39. $\dfrac{420}{600} = \dfrac{14}{20}$ _____

40. $\dfrac{7.6}{10} = \dfrac{76}{100}$ _____

In Exercises 41–48, solve the proportions for the unknown. **[4.4]**

41. $\dfrac{y}{35} = \dfrac{25}{5}$ _____

42. $\dfrac{15}{s} = \dfrac{45}{117}$ _____

43. $\dfrac{a}{25} = \dfrac{4}{20}$ _____

44. $\dfrac{6}{x} = \dfrac{4}{18}$ _____

45. $\dfrac{z}{20} = \dfrac{80}{200}$ _____

46. $\dfrac{25}{100} = \dfrac{8}{m}$ _____

47. $\dfrac{1}{2} = \dfrac{r}{7}$ _____

48. $\dfrac{2}{3} = \dfrac{5}{s}$ _____

Solve the following application problems. [4.2–4.4]

49. The sum of an unknown and eight is twenty. Find the unknown.

    49. _____

50. The sum of four plus an unknown equals fifty-one. Find the unknown.

    50. _____

51. An unknown times thirty equals one thousand eight hundred. Find the unknown.

    51. _____

52. Twenty-four equals three times an unknown. Find the unknown.

    52. _____

53. Three times an unknown plus five is equal to fifty. Find the unknown.

    53. _____

54. Four plus seven times an unknown is eighteen. Find the unknown.

    54. _____

55. The sum of two consecutive whole numbers is equal to ninety-one. Find the numbers.

    55. _____

56. The sum of two consecutive odd whole numbers is equal to two hundred forty. Find both numbers.

    56. _____

57. Tom throws some coins onto a table. His twin brother Joe throws coins worth twice as much onto the table. The total value of the coins on the table is $2.61. How much money did Joe place on the table?

    57. _____

58. A business math class has 47 students, with 9 more women than men. Find the number of men and women in the class.

    58. _____

59. Cajun Boatin' Inc. bought 5 small boats and 3 skiffs for $14,878. A small boat costs $1742. Find the cost of a skiff.

    59. _____

60. Mike Anderson bought a 4-unit apartment house for $172,000. Use a proportion to find the cost of a 10-unit apartment house.

    60. _____

61. The tax on a $40 item is $3. Find the tax on a $160 item.

    61. _____

62. Sam's Phones bought 17 smart phones for $1942.25. Find the cost of one phone.

    62. _____

CHAPTER 4 Test   165

63. The bookstore at Hudson Community College has a markup that is $\frac{1}{4}$ its cost on a book. Find the cost to the bookstore of a novel selling for $25.

63. _____

64. An unknown principal (P) loaned out at 8% for $1\frac{3}{4}$ years yields a maturity value (M) of $1368. Use $M = P(1 + RT)$ to find the principal.

64. _____

65. Explain why all terms with a variable should be placed on one side of the equation, and all terms without a variable should be placed on the other side, when solving an equation. [4.1]

66. In your own words, explain the terms *formula*, *ratio*, and *proportion* [4.3–4.4]

Chapters 1–4 Cumulative Review

CHAPTERS 1–4

To help you review, the numbers in brackets show the section in which the topic was introduced.

Round each of the following numbers as indicated. [1.1, 1.3]

1. 65,462 to the nearest hundred
2. 4,732,489 to the nearest thousand
3. 78.35 to the nearest tenth
4. 328.2849 to the nearest hundredth

Solve the following problems. [1.1–1.5]

5. 351
    763
    2478
    + 17

6. 45,867
    − 37,985

7. 634
    × 38

8. 2450
    × 320

9. $6290 \div 74 =$ _____

10. $22,899 \div 102 =$ _____

11. $.46 + 9.2 + 8 + 17.514 =$ _____

12. 45.36
    − 23.7

13. 29.8
    × .41

14. $21.8\overline{)396.76}$

Solve the following application problems.

15. Felix Schmid decides to establish a budget. He will spend $700 for rent, $325 for food, $420 for child care, $182 for transportation, $300 for other expenses, and he will put the remainder in savings. If his monthly take-home pay is $2025, find his savings. [1.1]

16. Clancy Strock wrote a feature article called "I know … I was there" for each issue of *Reminisce* magazine. He has written an article for each monthly issue of the magazine for 15 years. How many of these monthly articles has he written? [1.1]

17. Software Depot had a bank balance of $29,742.18 at the beginning of April. During the month, the firm made deposits of $14,096.18 and $6529.42. A total of $18,709.51 in checks was paid by the bank during the month. Find the firm's checking account balance at the end of April. [1.4]

18. Cara Groff pays $128.11 each month to the Bank of Brazil. How many months will it take her to pay off $4099.52? [1.5]

Solve the following problems. [2.1–2.4]

19. Write $\dfrac{48}{54}$ in lowest terms.

20. Write $8\dfrac{1}{8}$ as an improper fraction.

21. Write $\dfrac{107}{15}$ as a mixed number.

22. $1\dfrac{2}{3} + 2\dfrac{3}{4} =$

23. $5\dfrac{7}{8} + 7\dfrac{2}{3} =$

24. $6\dfrac{1}{3} - 4\dfrac{7}{12} =$

25. $8\dfrac{1}{2} \times \dfrac{9}{17} \times \dfrac{2}{3} =$ _____

26. $3\dfrac{3}{4} \div \dfrac{27}{16} =$ _____

Solve the following application problems.

27. The size of a prison cell at Alcatraz Prison in the San Francisco Bay is 5 feet by 9 feet. The average size of a shark cage is 5 feet by $6\dfrac{1}{2}$ feet. How many more square feet are there in the prison cell than in the shark cage?

27. _____

28. To prepare for the state real estate exam, Mia Dawson studied $5\dfrac{1}{2}$ hours on the first day, $6\dfrac{1}{4}$ hours on the second day, $3\dfrac{3}{4}$ hours on the third day, and 7 hours on the fourth day. How many hours did she study altogether? [2.3]

28. _____

29. The storage area at American River Raft Rental has four sides and is enclosed with $527\dfrac{1}{24}$ feet of security fencing around it. If three sides of the yard measure $107\dfrac{2}{3}$ feet, $150\dfrac{3}{4}$ feet, and $138\dfrac{5}{8}$ feet, find the length of the fourth side. [2.3]

29. _____

30. Play-It-Now Sports Center has decided to divide $\dfrac{2}{3}$ of the company's profit-sharing funds evenly among the eight store managers. What fraction of the total amount will each receive? [2.4]

30. _____

Solve the following problems. [2.5]

31. Change .65 to a fraction.

31. _____

32. Change $\dfrac{2}{3}$ to a decimal. Round to the nearest thousandth.

32. _____

Solve the following problems. [3.1–3.4]

33. Change $\dfrac{7}{8}$ to a percent. _____

34. Change .25% to a decimal. _____

35. Find 35% of 6200 home loans. _____

36. Find 134% of $80. _____

37. 275 sales is what percent of 1100 sales? _____

38. 375 patients is what percent of 250 patients? _____

168 CHAPTER 4 Equations and Formulas

Solve the following application problems.

39. Currently there are 38,990 movie screens in the United States. It is expected that only 400 screens will be added in the next five years. Find the percent to be added.

39. _____

40. A Bose Home Theater System normally priced at $2499.99 is on sale for 15% off. Find the amount of discount and the sales price. Round to the nearest cent.   [3.2]

40. _____

41. Bookstore sales of the *Physicians' Desk Reference*, which contains prescription drug information, rose 13.7% this year. If sales this year were 111,150 copies, find last year's sales. Round to the nearest whole number.   [3.5]

41. _____

42. After deducting 11.8% of total sales as her commission, George-Ann Hornor, a salesperson for Marx Toy Company, deposited $35,138.88 to the company account. Find the total amount of her sales.   [3.5]

42. _____

43. The number of Facebook users increased by 225 times in one of its earlier years in business. Find the percent increase.   [3.1]

43. _____

44. The value of a stock used to be 6 times what it is worth today. The value today is what percent of the past value? Round to the nearest tenth of a percent. Show your work, explaining how you arrived at your answer.   [3.1]

44. _____

WHERE'S THE BEEF?   The United States exported 1.067 million metric tons of beef. Use the circle graph below to solve Exercises 45–48.   [3.2]

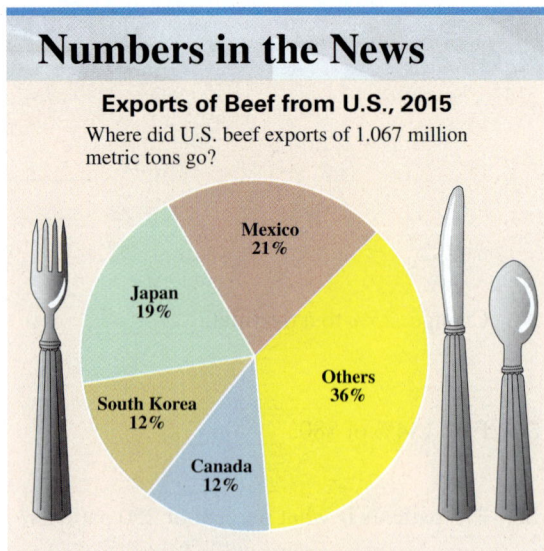

Source: U.S. Department of Agriculture.

45. (a) What percent of the exported beef was shipped to Japan, Mexico, and South Korea combined? **(b)** Find the total number of metric tons exported to these three countries.

(a) _____

(b) _____

46. (a) What percent of the exported beef was shipped to countries other than Japan, Mexico, and South Korea combined? **(b)** Find the total number of metric tons exported to other countries.

(a) _____

(b) _____

47. How much more beef was exported to Japan than to South Korea?

47. _____

48. How much more beef was exported to Mexico than to Canada?

48. _____

Solve each equation for the unknown. **[4.1]**

49. $9r - 23 = 31$ _____

50. $\dfrac{3}{4}t = 120$ _____

51. $5y - 10 = 26$ _____

52. $20 + 5x = 83 - x$ _____

Find the value of the variable that is not given. **[4.3]**

53. $I = PRT; P = \$45{,}000, R = .015, T = .5$ _____

54. $d = rt; r = 378, t = 8.5$ _____

55. $A = P + PRT; A = 1368, P = 1200, T = 2$ _____

56. $M = P(1 + i)^n; n = 6, i = 0.02, M = \$42{,}231.09$ _____

Solve the equation for the variable indicated. **[4.3]**

57. Solve $I = PRT$ for T. _____

58. Solve $PV = nRT$ for P. _____

59. Solve $M = P(1 + i)^n$ for P. _____

60. $A = \dfrac{S}{1 + RT}$ for R. _____

Write the ratio in lowest terms. **[4.4]**

61. 500 euros to 725 dollars _____

62. 24 ounces to 15 gallons _____

63. 14 lockers to 40 minutes _____

64. 630 apples to 168 cans _____

Solve the following application problems.

65. The sum of an unknown and 47 is 93. Find the unknown. **[4.2]**

65. _____

66. The sum of two consecutive odd numbers is 228. Find the numbers. **[4.2]**

66. _____

67. A shoe retailer marks up the price of running shoes manufactured in China by 68%. Much of the markup goes to pay the costs of the retailer. Find the cost to the retailer if the selling price is $89.

67. _____

68. Forty-seven students signed up for a two-week travel/study trip to Greece. There were 9 more women than men. Find the number of men and women. **[4.2]**

68. _____

69. The cost of replacing a roof on a 1450-square-foot house is $6600. Find the cost to replace the roof on a 2400-square-foot house. Round to the nearest dollar. **[4.4]**

69. _____

70. A 32-unit apartment complex generates total monthly rents of $21,920. Assuming comparable rents, estimate the total monthly rent generated by a 70-unit complex. **[4.4]**

70. _____

Bank Services

5

CHAPTER CONTENTS

5.1 Banking, Checking Accounts, and Check Registers

5.2 Checking Services and Credit-Card Transactions

5.3 Bank Statement Reconciliation

CASE IN POINT

BARBARA WIFFY owns a small nursery named Rose Gardens. Although she sells all types of outdoor trees and shrubs, her specialty is roses. For the best quality roses, she has always bought from Jackson & Perkins Wholesale, Inc. This firm has research scientists who experiment with hybridizing roses with the goal of creating new, beautiful flowers.

As her business began to grow, Barbara Wiffy carefully looked at several different banks and talked to bank managers before choosing a bank. The relationship her firm has with a bank is very important!

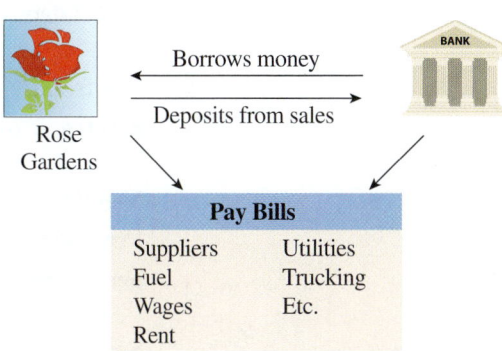

CHAPTER 5 Bank Services

Historically, banks offered few services other than checking accounts, savings accounts, and making loans. Today, they also offer automated teller machines, electronic banking, credit and debit cards, note collection, investment services, mobile banking, guaranteed checks, and even payroll services for firms. In this chapter, you will learn about banks. You will also learn how to write checks, make deposits, and reconcile account balances.

5.1 Banking, Checking Accounts, and Check Registers

OBJECTIVES

1. Understand banks and recent trends.
2. Understand checking accounts and checks.
3. Calculate the monthly service charges.
4. Identify the parts of a deposit slip.
5. Identify the parts of a check stub.
6. Complete the parts of a check register.

CASE IN POINT As a business owner, Barbara Wiffy receives payments in all these ways: cash, checks, credit cards, debit cards, electronic payments, and even traveler's checks. In turn, she pays bills using checks, credit cards, debit cards, and electronic payments. She chose a bank that will support her growing company by efficiently handling all these types of transactions.

OBJECTIVE 1 Understand banks and recent trends. Banks are VERY important for our economy. They hold money from depositors in checking and savings accounts as well as certificates of deposit, and then lend money to individuals and firms. Banks earn revenue by charging **interest** on loans and charging fees such as monthly account fees, low-balance fees, returned-check fees, foreign transaction fees, ATM fees, and so on. Without banks it would be difficult or impossible for individuals to buy a car or a house and for businesses to meet payroll or buy supplies.

Banks face a lot of competition from other banks and also from **Internet service providers** such as Lending Club or Prosper.com, which make loans to individuals or firms through a website. Rather than going to banks, people with good business ideas and work experience can sometimes get startup funding through **crowdsourcing** funding firms such as gofundme.com and kickstarter.com, where multiple individuals help finance the startup costs for a new business or, say, recording a new music disk. Some college students with particular needs have even gotten help with bills through crowdsourcing firms or borrowed money for college from Internet service providers.

Banks have responded to the pressures of increased competition by finding ways to reduce costs, such as buying smaller banks and merging with large banks. They have also done so by developing **electronic** or **online banking**, and **mobile banking**, both of which reduce costs per **transaction**. This way of doing business is also convenient for customers, so it's a win-win situation. Customers can use smart phones, other mobile devices, or computers to check balances, transfer funds, receive alerts, pay bills, and even make deposits. Roughly half of mobile phone owners now use mobile banking.

Feature	Description
Automated teller machine (ATM)	A machine connected to a worldwide network and used to make deposits, transfer money, pay bills, or get cash. Its use may result in a charge.
Debit card	A card that looks like a charge card but instantly transfers funds from the purchaser's bank account to the merchant's. Its use requires a **personal identification number (PIN)**.
Direct deposit or **payment**	Allows for direct deposits to be made into an account or direct payments to be made from an account.
Electronic commerce (EC)	Buying or selling products or services over the Internet.
Electronic (online) banking	Electronically transferring funds or paying bills. Used extensively by business and individuals.
Mobile ticketing	Process of purchasing tickets using a mobile device.
Prepaid card	A card with a prepaid balance.
Service fees	Fees charged by a bank for low account balance, overdrafts, returned checks, etc.
Smart card	A card with an embedded microchip used to make payments, e.g. a Visa card.
Point of sale (POS)	The time and place where a sale is made at a retail store.

Large retailers use computer networks to manage the flow of money, keep costs low, and help manage inventory. For example, Target uses a **point of sale (POS) system** at cash registers to record transactions. Debit-card transaction amounts are routed to the customer's bank for immediate payment. Credit-card transactions are routed to the company issuing the credit card and from there to the customer's bank for payment. **Uniform Product Codes (UPC)** identifying each item sold during a day are forwarded nightly to the distribution center for that particular store, so that only items needed are put onto trucks for delivery to that store. Checks are used for fewer payments than ever before, and are bundled together before being submitted to the merchant's bank for collection. Merchants prefer payments with debit cards, since they have the lowest cost and funds are almost immediately moved from the customer to the Target account, reducing the chances of nonpayment.

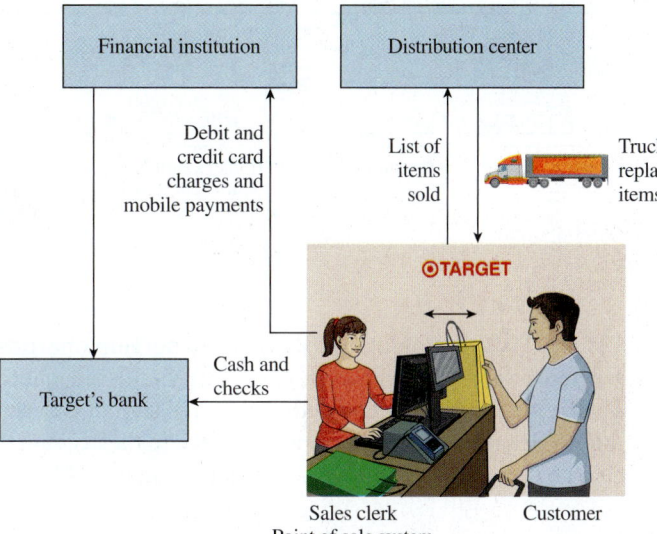

Thief from the Past

Thief of Today

Quick TIP

Maintain a good credit history by making all payments when due.

Electronic banking is not without problems. For example, it is easy to authorize a payment for which you do not have available funds in your account, resulting in a service charge. Clearly, account balances should be carefully watched to avoid service charges—which can add up. Another problem is unauthorized use of your account by someone else. For example, a growing problem is **identity theft**, in which someone gathers enough information about you to fraudulently establish credit cards or borrow money using your name and personal information. It can be very difficult to stop the thief since it is difficult to find him or her. It is extremely important to protect your financial information as well as your log-on ID and password to all financial accounts. Experts recommend using complex passwords that cannot easily be discovered by criminals. They also recommend destroying documents that include credit-card numbers, account numbers, Social Security numbers, and all bank records.

Individuals and firms create their own credit history based on how well they make payments. **Credit history** is an indication of how likely the individual or firm is to repay debt on time. The companies Equifax, Experian, and TransUnion track your credit history and sell that information to lenders. If you make late payments or try to get away with not paying a debt, that information is recorded and then sold to lenders when you try to borrow funds in the future. At minimum, a bad credit history will result in your incurring higher interest rates and therefore higher monthly payments on future loans. If your credit history is really bad, you may not be able to find any company that will lend you money.

OBJECTIVE 2 Understand checking accounts and checks. Checking accounts are used by individuals and businesses for daily transactions. The two basic types of checking accounts are discussed here. There are dozens of variations of these account types.

Personal checking accounts are used by individuals. The bank usually charges for the printing of checks and often charges a monthly fee for any checking account whose balance falls below a minimum amount at any time during the month. Individuals usually have the option of receiving a debit and/or credit card, as well as the ability to use ATMs and pay bills or receive deposits either electronically or using mobile banking. Some accounts pay a small amount of **interest**.

Business checking accounts often receive more services and have greater activity than do personal accounts. For example, banks often collect debts on behalf of businesses. The bank automatically credits the amount to the business account.

174 CHAPTER 5 Bank Services

Although most transactions are done electronically today, there are still hundreds of millions of checks written each year. So it is important to understand how to write a check.

SAMPLE CHECK

- Name and address of account holder.
- Bank and Federal Reserve district number
- Check number
- Write clearly in ink the name of the person or place receiving the check (**payee**).
- Correctly date each check.
- Write the amount of the check close to the dollar sign so additional digits cannot be added by others.
- Start at the far left and write the amount of the check in words using *and* to represent the decimal point. Draw a heavy, wavy line from the end of the written amount to the word *dollars*.
- Always sign checks with the same signature (**payor**).
- Numbers along the bottom row are printed in magnetic ink.
- Indicate reason for writing check.
- Bank number
- Account number
- The check number appears here *and* in the upper-right corner.
- When the check is processed, the amount of the check is imprinted here. It should match the check amount ($159.90).

Check details:
- Betty Squier, 583 Woodglen Circle, Atlanta, GA 30214
- 12-045 / 6789
- 358
- December 10, 20 18
- Pay to the order of Green Giant Nursery $ 159 90/100
- One hundred fifty-nine and 90/100 DOLLARS
- American National Bank
- For Outdoor Plants
- Betty Squier
- ⑈678904512⑈ ⑈149735505⑈ ⑈0058 ⑈00000015990⑈

OBJECTIVE 3 Calculate the monthly service charges. Banks often charge monthly **service fees** based on the average balance in the account for the month covered by the statement. The amount charged is often the sum of two charges: a **maintenance charge per month** determined by the average balance in the account during the month plus a **per-debit charge** (per-check charge). Use the data from the table below in the next example.

Average Balance	Maintenance Charge per Month	Per-Check Charge
Less than $500	$12.00	$.20
$500–$1999	$7.50	$.20
$2000–$4999	$5.00	$.10
$5000 or more	0	0

Finding the Checking-Account Service Charge

EXAMPLE 1 Find the monthly service charge for the following business accounts.

(a) Omni Computer, 38 checks written, average balance $983

Based on the average balance of $983, the maintenance charge is $7.50 and the per-check charge is $.20.

$$\text{Charge} = \text{Maintenance charge} + \text{Per-check charge}$$
$$= \$7.50 + 38 \text{ checks} \times \$.20$$
$$= \$7.50 + \$7.60$$
$$= \$15.10$$

(b) Jamison Auto Repair, 62 checks written, average balance $4632.25

Based on the average balance, the maintenance charge is $5.00 and the per-check charge is $.10.

$$\text{Charge} = \text{Maintenance charge} + \text{Per-check charge}$$
$$= \$5.00 + 62 \text{ checks} \times \$.10$$
$$= \$5.00 + \$6.20$$
$$= \$11.20$$

The calculator solutions to this example use chain calculations, with the calculator observing the order of operations.

(a) 7.5 [+] 38 [×] .2 [=] 15.1   **(b)** 5 [+] 62 [×] .1 [=] 11.2

Note: Refer to Appendix B for calculator basics.

> **QUICK CHECK 1**
>
> Find the monthly service charge for the following business accounts.
>
> (a) Towne Florist, 76 checks written, average balance $2180
> (b) Stacy's Coffee Shop, 58 checks written, average balance $1850

OBJECTIVE 4 Identify the parts of a deposit slip. Cash and checks are placed into an account using a **deposit slip** or **deposit ticket**. The account number is printed at the bottom in magnetic ink. The slip contains blanks for entering any currency (bills) or coins (change) as well as any checks that are to be deposited.

SAMPLE DEPOSIT SLIP

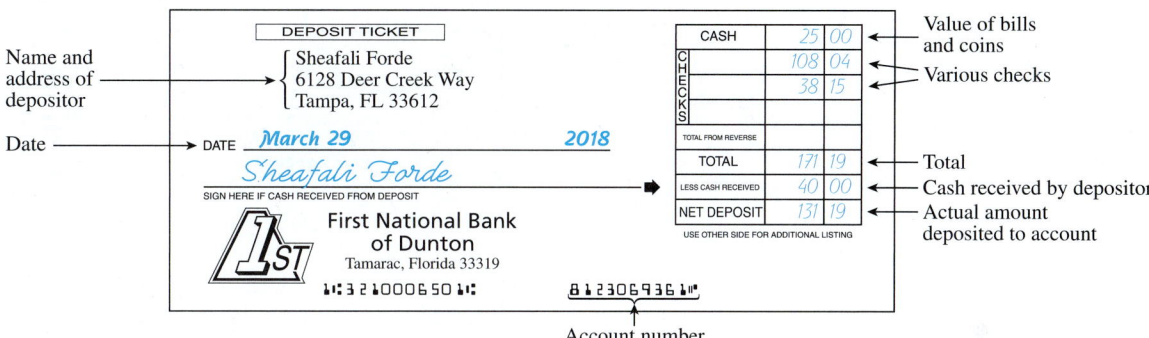

Before depositing a check in your bank, you must **endorse** (sign) it on the back within 1.5 inches of the edge as shown in the following figure. A **blank endorsement** is simply the signature of the person being paid. A lost check with a blank endorsement can be cashed by anyone finding the check. It is better to use a **restricted endorsement** that includes "For Deposit Only" followed by other information such as the name of a person or company and checking account number. A restricted check that is lost is worthless to anyone finding the check. A **special endorsement** is used to pass the check on to someone else.

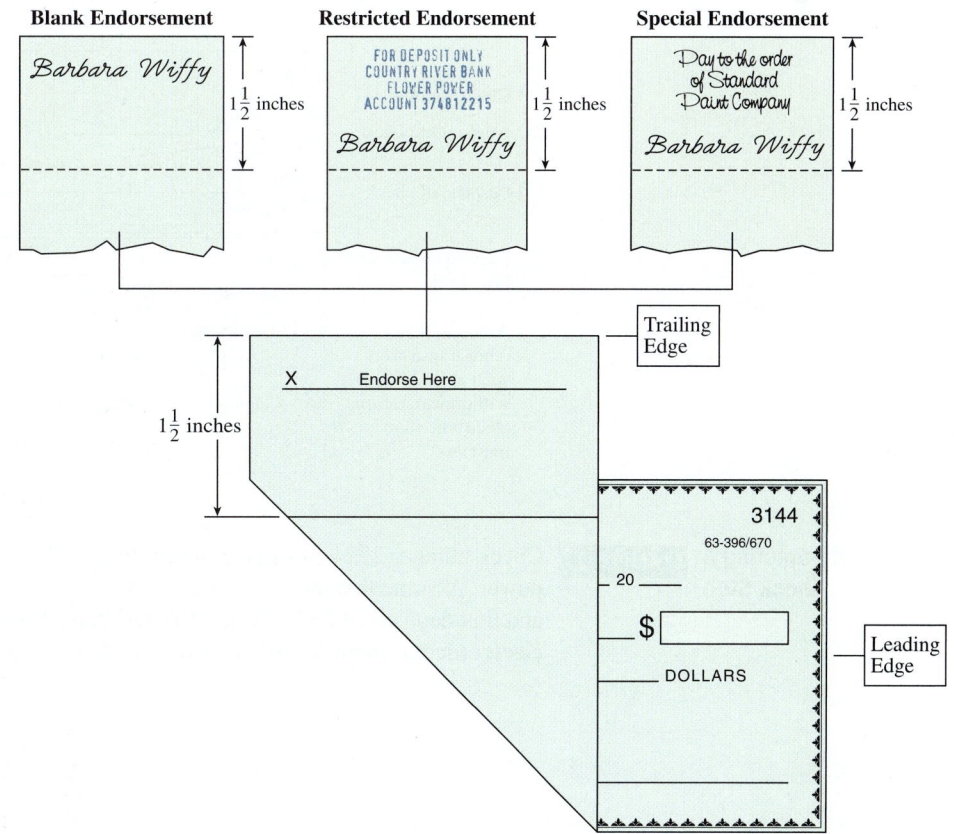

176 CHAPTER 5 Bank Services

Quick TIP
Mobile phones can be used to deposit checks. A photo of each side of the check is required.

After being endorsed, a check may be deposited along with the deposit slip listing the check and showing the account number to which the check is to be deposited. Once the funds are transferred, the check is **canceled** and no longer valid. Photos are taken of both sides of the original check and kept by the bank, which destroys the original check. The information on the back of the check shown below shows the processing history of the check. Banks use photos of checks to save the time and costs of mailing the checks back. This reduces **transaction costs.** Processing checks in this fashion reduces the time between writing a check and the time that funds are deposited into an account.

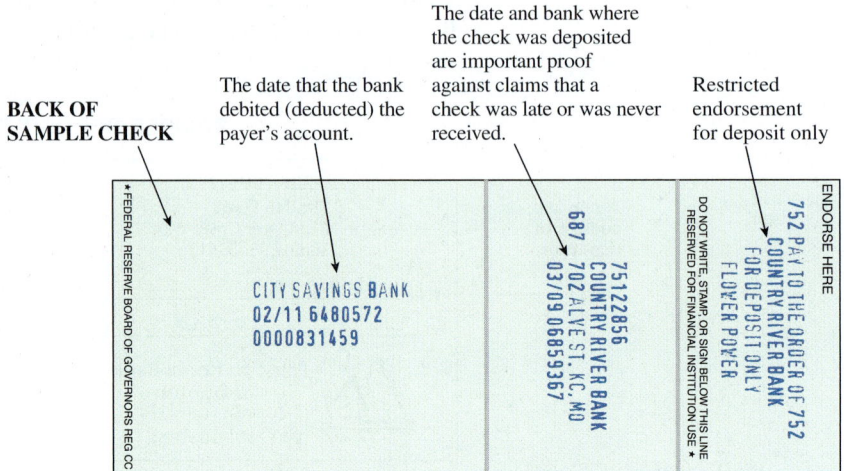

OBJECTIVE 5 Identify the parts of a check stub. It is important to keep a careful record of all deposits, checks, and electronic transfers. This is the only way to know exactly how much is in the account and to maintain the accurate records needed for tax returns and financial statements.

Although most businesses keep records on computers today, we will show the ideas using **check stubs** and **registers** since the ideas are transferable to the computer. **Total** refers to the balance from the bottom of the previous check stub. Add deposits and subtract the amount of this check and add or subtract other charges/credits to find the **balance** in the account. Although paying the bills is tedious work, accurate and quick payment of bills forms the backbone of business and personal finance.

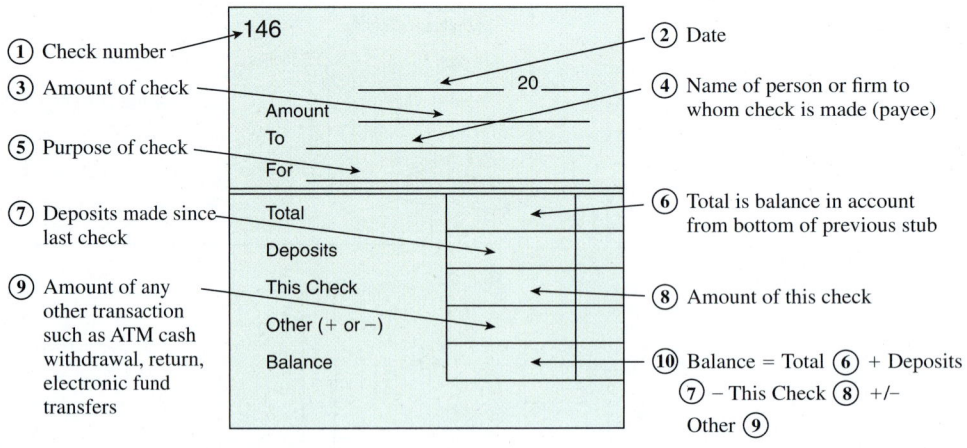

Completing a Check Stub

EXAMPLE 2 Check number 2724 was made out on June 8, 2018 to Lillburn Utilities as payment for water and power. Assume that the check was for $182.15, that the balance brought forward is $4245.36, and that deposits of $337.71 and $193.17 have been made since the last check was written. An **electronic payment** for $820.36 was made to James Carpenter. Complete the check stub.

5.1 Banking, Checking Accounts, and Check Registers

SOLUTION

2724		
June 8 20 18		
Amount $182.15		
To Lillburn Utilities		
For Water and Power		
Total	$4245 36	← Balance from end of previous stub
Deposits	+ 530 88	← Sum of deposits made since last check
This Check	– 182 15	← Amount of this check
Other (+ or –)	– 820 36	← Other credits to, or debits from, the account
Balance	$3773 73	← Current balance goes to next stub

QUICK CHECK 2

Use the check stub to find the balance forward. Check number 3502 is made out on November 15, 2018 to City Hall for $273.28 for garbage collection. The balance brought forward was $6750.17, and deposits of $1876.22 and $879.65 have been made since the last check was written. An ATM withdrawal of $200 was also made.

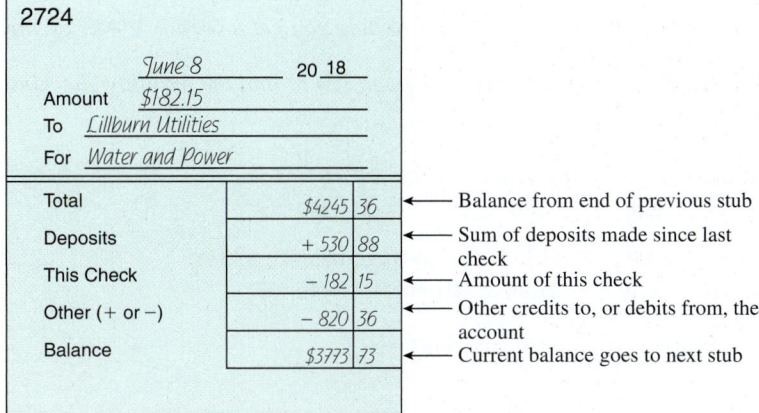

OBJECTIVE 6 Complete the parts of a check register. Although a few people use check stubs to keep track of funds, most use a **check register**, also called a **transaction register**, which can either be on paper or created using computer software. The check register is used to keep track of all transactions including deposits, checks, ATM withdrawals, refunds, and electronic deposits or payments. Update the balance by subtracting the amount of withdrawal or adding the amount of deposit or credit as you go from one line to the next.

Sample Check Register

CHECK NO.	DATE	TRANSACTION DESCRIPTION	AMOUNT OF WITHDRAWAL	✓	AMOUNT OF DEPOSIT/CREDIT	BALANCE
		BALANCE BROUGHT FORWARD →				3518 72
electronic	5/8	Swan Brothers	378 93			3139 79
1436	5/8	Class Acts	25 14			3114 65
1437	5/9	Mirror Lighting	519 65			2595 00
	5/10	Deposit			3821 17	6416 17
electronic	5/10	Woodlake Auditorium	750 00			5666 17
	5/12	Deposit			500 00	6166 17
1438	5/12	Rick's Clowns	170 80			5995 37
1439	5/14	Y.M.C.A.	219 17			5776 20
	5/14	ATM	120 00			5656 20
	5/15	Deposit			326 15	5982 35
electronic	5/16	Stage Door Playhouse	825 00			5157 35
electronic	5/17	Gilbert Eckern	1785 00			3372 35
	5/19	Deposit			1580 25	4952 60

5.1 Exercises // MyLab Math

The shaded sections below contain solutions to help you get a QUICK START on the various types of exercises.

CHECKING CHARGES Use the table on page 174 to find the monthly checking-account service charge for the following accounts. (See Example 1.)

1. Rose Gardens, 92 checks, average balance $4618
   $5.00 + (92 × $.10) = $5.00 + $9.20 = $14.20   1. $14.20

2. Fresh Choice Restaurant, 114 checks, average balance $3318
   $5.00 + (114 × $.10) = $5.00 + $11.40 = $16.40   2. $16.40

3. Pest-X, 40 checks, average balance $491   3. _____

4. Kitchen Supplies, Inc., 76 checks, average balance $468 4. _____

5. Budget Dry Cleaning, 48 checks, average balance $1763   5. _____

6. Medical Supplies, 272 checks, average balance $8205   6. _____

7. Software and More, 72 checks, average balance $516   7. _____

8. Pet Supplies, 85 checks, average balance $3241   8. _____

MAINTAINING BANK RECORDS Use the following information to complete each check stub. (See Example 2.)

	Date	To	For	Amount of This Check	Total from Previous Stub	Deposits	Other
9.	Mar. 8, 2018	Wayne Plumbing	Septic	$1835.42	$6215.13	$135; $1459.12	$100 cash withdrawal
10.	Nov. 4, 2018	Target	Misc.	$85.45	$1239.56	$122	$35 credit
11.	Oct. 7, 2018	PetSmart	Pet Food	$138.94	$2105.09	$832.47	$280.65 electronic payment

9. 857
   Date _____
   Amount _____
   To _____
   For _____
   Total _____
   Deposits _____
   This Check _____
   Other (+ or −) _____
   Balance _____

10. 1248
    Date _____
    Amount _____
    To _____
    For _____
    Total _____
    Deposits _____
    This Check _____
    Other (+ or −) _____
    Balance _____

11. 735
    Date _____
    Amount _____
    To _____
    For _____
    Total _____
    Deposits _____
    This Check _____
    Other (+ or −) _____
    Balance _____

C// indicates an exercise that is related to the Case in Point feature.

12. List and explain at least six parts of a check. Draw a sketch showing where these parts appear on a check. (See Objective 2.)

13. Discuss the advantages and disadvantages of mobile banking from your phone. (See Objective 1.)

14. Write an explanation of two types of check endorsements. Describe where these endorsements must be placed. (See Objective 4.)

15. Explain in your own words the factors that determine the service charges on a business checking account. (See Objective 3.)

COMPLETING CHECK STUBS *Using the information provided for 2018, complete the following check stubs for Rose Gardens.* **The balance brought forward for check stub 5311 is $7223.69.** *(See Example 2.)*

Checks Written

Number	Date	To	For	Amount
5311	Oct. 7	Julie Davis	Seeds	$1250.80
5312	Oct. 10	Post Office	Postage	$39.12
5313	Oct. 15	United Parcel	Shipping	$356.28

Deposits Made

Date	Amount
Oct. 8	$732.18
Oct. 9	$23.32
Oct. 13	$1025.45

Electronic Payments

Date	Amount
Oct. 8	$650.07
Oct. 12	$147.32

16. Stub 5311 — Rose Gardens
- Date: Oct. 7
- Amount: $1250.80
- To: Julie Davis
- For: Seeds
- Total: $7223.69
- Deposits: $0.00
- This Check: $1250.80
- Other (+ or −): $0.00
- Balance: $5972.89

17. Stub 5312 — Rose Gardens
- Date: Oct. 10
- Amount: $39.12
- To: Post Office
- For: Postage
- Total: $5972.89
- Deposits: $755.50
- This Check: $39.12
- Other (+ or −): −$650.07
- Balance: $6039.20

18. Stub 5313 — Rose Gardens
- Date: Oct. 15
- Amount: $356.28
- To: United Parcel
- For: Shipping
- Total: $6039.20
- Deposits: $1025.45
- This Check: $356.28
- Other (+ or −): −$147.32
- Balance: $6561.05

180 **CHAPTER 5** Bank Services

BANK BALANCES *In Exercises 19–22, complete the balance column in the following company check registers after each check or deposit transaction. Electronic payments are indicated by "elec." (See Objective 6.)*

19. Rose Gardens

CHECK NO.	DATE	TRANSACTION DESCRIPTION	AMOUNT OF WITHDRAWAL	✓	AMOUNT OF DEPOSIT/CREDIT	BALANCE
		BALANCE BROUGHT FORWARD →				9628 35
1221	10/4	Delta Contractors	215 71			
elec.	10/5	Hand Fabricating	573 78			
1222	10/5	Photo Specialties	112 15			
	10/6	Deposit			753 28	
	10/8	Deposit			1475 69	
elec.	10/9	Young Marketing	426 55			
elec.	10/11	Jackson & Perkins (hybrid roses)	637 93			
	10/11	ATM (cash withdrawal)	65 62			
elec.	10/14	Light and Power Utilities	248 17			
	10/16	Deposit			335 85	
1223	10/16	License Board	450 50			

20. Jessica Tate Consulting

CHECK NO.	DATE	TRANSACTION DESCRIPTION	AMOUNT OF WITHDRAWAL	✓	AMOUNT OF DEPOSIT/CREDIT	BALANCE
		BALANCE BROUGHT FORWARD →				1629 86
elec.	7/3	Ahwahnee Hotel	250 45			
862	7/5	Willow Creek	149 00			
863	7/5	Void				
	7/7	Deposit			117 73	
elec.	7/9	Del Campo High School	69 80			
	7/10	Deposit			329 86	
	7/12	Deposit			418 30	
864	7/14	Big 5 Sporting Goods	109 76			
865	7/14	Dr. Yates	614 12			
866	7/16	Office Supplies	32 18			
	7/16	Payment from Totem School District			1374 25	

21. Stencils by Loree

CHECK NO.	DATE	TRANSACTION DESCRIPTION	AMOUNT OF WITHDRAWAL	✓	AMOUNT OF DEPOSIT/CREDIT	BALANCE
		BALANCE BROUGHT FORWARD →				832 15
elec.	3/17	AirTouch Cellular	257 29			
elec.	3/18	Curry Village	190 50			
	3/19	Deposit			78 29	
	3/21	Deposit			157 42	
1123	3/22	San Juan District	38 76			
elec.	3/23	Office Supplies	175 88			
	3/23	Deposit			379 28	
1124	3/24	Class Video	197 20			
1125	3/24	Water World	25 10			
1126	3/25	Bel Air Market	75 00			
	3/29	Deposit			722 35	

22. Beverly's Event Planning

CHECK NO.	DATE	TRANSACTION DESCRIPTION	AMOUNT OF WITHDRAWAL	✓	AMOUNT OF DEPOSIT/CREDIT	BALANCE
		BALANCE BROUGHT FORWARD →				3852 48
elec.	12/6	Web Masters	143 16			
elec.	12/7	Water and Power	118 40			
	12/8	Deposit			286 32	
	12/10	ATM (cash)	80 00			
2312	12/11	Ann Kuick	986 22			
elec.	12/11	Account Temps	375 50			
	12/13	Deposit			1201 82	
elec.	12/14	Patterson's Foods	735 68			
2313	12/15	ABC Cleaners	223 94			
	12/13	Deposit			498 01	
2314	12/18	Federal Parcel	78 24			

QUICK CHECK ANSWERS

1. (a) $12.60
   (b) $19.10

2. November 15, 2018; $273.28; City Hall; Garbage Collection; $6750.17; $2755.87; $273.28; $200; $9032.76

5.2 Checking Services and Credit-Card Transactions

OBJECTIVES

1. Identify bank services available to customers.
2. Understand interest-paying checking plans.
3. Understand credit-card, debit-card, and mobile payment fees.

Percent of Students Having a Type of Debt (2017)

Student loans	52%
Credit card	23%
Automobile loan	13%
Personal loan	9%
Mortgage	8%

Source: "How American College Students Manage Their Finances," news.salliemae.com.

OBJECTIVE 1 Identify bank services available to customers. Some services provided by banks, along with the *typical charges*, are listed here.

- An **ATM card** is a payment card issued by a bank that allows a customer to use an ATM. There is usually no fee to use an ATM at any branch of your bank. However, the use of ATMs belonging to any other bank may result in a charge of up to $2.50 and use of an ATM in another country may cost $5 per transaction.
- An **overdraft** occurs when a check is written for which there are **nonsufficient funds (NSF)** in the checking account and the customer has no overdraft protection, also referred to as bouncing a check. The typical charge to the writer of the "bad" check is $25 to $35 per bad check. The same charge occurs when a check is returned because it was improperly completed.
- **Overdraft protection** is given when an account balance is too small to cover the amount of a check and an overdraft occurs. Charges for overdraft protection vary among banks.
- A **returned-deposit item** is a check that was deposited and then returned to the bank, usually because of lack of funds in the account of the person or firm writing the check. A common charge to the depositor of the check is $25.
- A **stop-payment order** is a request by an account owner that the bank not honor a check already written ($30 per request).
- A **cashier's check** is a check written by the bank and it is viewed as being as good as cash ($8 per check).
- A **money order** is a purchased instrument that is often used in place of cash and is sometimes required instead of a personal or business check ($4 each). The payee should be written on the money order by the purchaser immediately after purchase to avoid it being cashed by someone who finds or steals the money order.
- A **notary service** is an official certification of a signature on a document that is required on certain business documents. Occasionally this service is free to customers, but there is usually a charge ($10).

OBJECTIVE 2 Understand interest-paying checking plans. Federal regulations allow banks to offer checking and savings accounts that pay interest. Interest rates are currently very low due to the lingering effects of the financial crisis that began in 2008, but those rates are likely to increase in the future. **Money market accounts** are interest-bearing accounts that require a higher balance and can often be used like a checking account. Before the financial crisis, money market accounts paid higher interest rates and yet still gave the depositor the flexibility of withdrawing funds. Today, the rates paid on money market accounts are also very low. Banks also offer longer-term investments called **certificates of deposit (CDs),** which offer higher interest rates. They are discussed in Chapter 10.

OBJECTIVE 3 Understand credit-card, debit-card, and mobile payment fees. Businesses are charged fees whether customers pay using **credit cards**, **debit cards**, or a device such as a smartphone to make a **mobile payment**. Actually, businesses also have costs when paid in cash. They have to protect the cash against theft, which requires safes, armored car shipments, guards, and other processes to prevent theft. So businesses have costs no matter how they are paid. Managers continually work to keep transaction costs low, including costs based on payments.

After a customer has made a *credit-card* charge at a merchant, information on the transaction is transferred over the Internet to the bank through the bank's processor, as shown in the following figure. The processor then shares the information with the merchant's bank as well as the issuer of the credit card. The issuer of the credit card accumulates charges for the month and sends a bill to the card owner on a monthly basis. The statement may include returned items with credits to the card owner.

5.2 Checking Services and Credit-Card Transactions

Financial Payments System

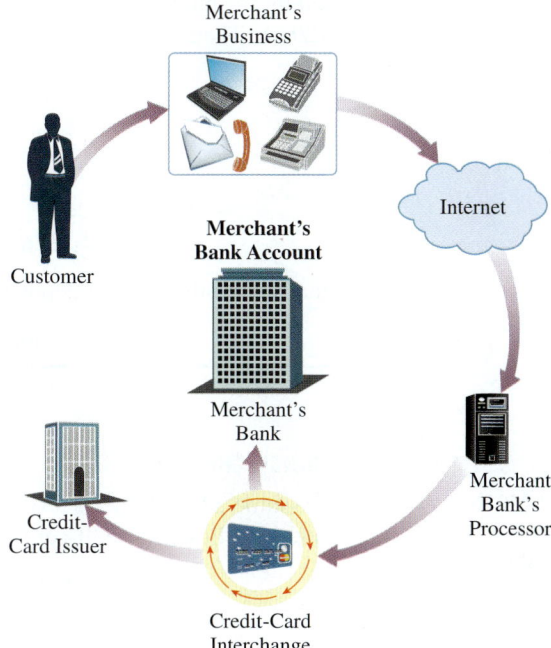

In contrast, debit cards result in an immediate charge to the card holder's bank account. The use of debit cards has become very popular, as evidenced by the $2 trillion per year in debit charges and a continuing growth rate of over 15% per year.

Mobile payments can be made by tapping or waving smartphones above a device owned by the merchant. A disadvantage of mobile payments is that there is no paper record, so it can be difficult to correct an improper charge. Mobile payment systems continue to change rapidly.

Method of Payment	Charges Paid by Merchants
Credit card	2% or more of the amount of the charge
Debit card	Average of 21 cents per swipe
Mobile payment	Supports credit-card, debit-card, and prepaid card payment options. Costs vary widely.

You may have noticed that the price of gasoline sometimes differs depending on which payment method is used. Now you know why—the merchant has higher costs when a customer pays with a credit card and many merchants try to pass the extra costs on to customers.

Now you will learn how to find the net deposit for a business after deducting fees.

> **Finding Net Deposit for Credit-Card Sales:**
> 1. Net credit-card sales = Total credit-card sales − Total credit-card refunds
> 2. Total fee = Appropriate percent × Net credit-card sales
> 3. Net deposit = Net credit-card sales − Total fee

Find the Fees and Net Deposit on Credit-Card Sales

EXAMPLE 1  Here are credit-card sales and credit-card refunds made early one morning at Rose Gardens. Find the total credit-card sales, the total credit-card refunds, and the net credit-card sales. Then find the bank fee charged, assuming a fee of 2% of the net credit-card sales. Finally, find the amount deposited to Rose Gardens's checking account.

CASE IN POINT

Credit-Card Sales		Credit-Card Refunds
$82.31	$146.50	$43.83
$38.18	$78.80	$85.95
$249.33	$470.15	
$46.80	$320.90	

SOLUTION

Adding the numbers above results in credit-card sales of $1432.97 and credit-card refunds of $129.78. Then find the net credit-card sales as follows.

$$\begin{aligned} \text{Total sales} &= \$1432.97 \\ \text{Total refunds} &= -129.78 \\ \text{Net sales} &\quad \$1303.19 \end{aligned}$$

$$\begin{aligned} \text{Bank fee} &= 2\% \text{ of } \$1303.19 \\ &= \$26.06 \quad (\text{rounded}) \end{aligned}$$

Finally, find the amount deposited to Rose Gardens's checking account.

$$\text{Amount} = \$1303.19 - \$26.06 = \$1277.13$$

Quick TIP

Cash rewards on prior Master-Card and Visa charges are really a rebate on previous purchases.

QUICK CHECK 1

Assume a fee of 2.25% and find the amount deposited to a business owner's account based on the following credit-card transactions.

Sales: $532.15, $650.19, $127.45, $314.98
Refunds: $120.15

Find the Difference in Service Fees between Credit Cards and Debit Cards

EXAMPLE 2 Mershon Printing had the following sales late one afternoon: $58.20, $41.17, $18.67, and $137.10. Find the extra amount of service charges that must be paid if all transactions were made using a credit card with a bank fee of 2.25% compared to all transactions made using a debit card with a charge of $.21 per swipe.

SOLUTION

Service charge if all transactions were done with a credit card:

$$(\$58.20 + \$41.17 + \$18.67 + \$137.10) \times .0225 = \$5.74$$

Service charge if all transactions were done with a debit card:

$$4 \text{ transactions} \times \$.21 \text{ per swipe} = \$.84$$

Extra service charges if all transactions were done using a credit card:

$$\$5.74 - \$.84 = \$4.90$$

QUICK CHECK 2

Last hour, Monroe Muffler had sales of $185.40, $431.25, and $654.13. Find the amount of extra service charges that must be paid if all transactions were made using a credit card with a bank fee of 2% compared to all transactions made using a debit card with a charge of $.20 per swipe.

Example 2 shows that credit-card fees result in meaningfully higher service charges than debit-card fees. You may be thinking these fees are small and meaningless for a large bank, especially debit-card charges, which are clearly much smaller than credit-card fees. However, consider that there are tens of billions of credit- and debit-card transactions every year and it becomes apparent that these service charge fees add up quickly. In fact, merchants believe the fees are far too high and they are working through the National Retail Federation to get them reduced.

5.2 Exercises // MyLab Math

The shaded sections below contain solutions to help you get a QUICK START on the various types of exercises.

COLLEGE BOOKSTORE *A college bookstore accepts cash, checks, debit cards, credit cards, cashier's checks, and mobile payments. The following credit-card transactions occurred in the first hour of business one morning.*

Sales		Refunds
$66.68	$18.95	$62.16
$119.63	$496.28	$106.62
$53.86	$21.85	$38.91
$178.62	$242.78	
$219.78	$176.93	

1. What is the total amount of the credit-card sales?
   $66.68 + $119.63 + $53.86 + $178.62 + $219.78 + $18.95 + $496.28 + $21.85 + $242.78 + $176.93 = $1595.36

   1. $1595.36

2. Find the total of the credit-card refunds.
   $62.16 + $106.62 + $38.91 = $207.69

   2. $207.69

3. Find credit-card sales less refunds.

   3. _____

4. If the fee paid by the business is 2%, find the amount of the charge at the statement date.

   4. _____

5. Find the amount of the credit given to College Bookstore after the fee is subtracted.

   5. _____

C// CREDIT-CARD DEPOSITS *Rose Gardens accepts credit cards. In a recent period, the business had the following credit-card charges and credits. (See Examples 1 and 2.)*

Sales		Refunds
$78.56	$38.15	$29.76
$875.29	$18.46	$102.15
$330.82	$22.13	$71.95
$55.24	$707.37	
$47.83	$245.91	

6. What is the total amount of the credit-card sales?

   6. _____

7. Find the total of the credit-card refunds.

   7. _____

8. Find credit-card sales less refunds.

   8. _____

9. If the bank charges Rose Gardens a 3% fee, find the amount of the charge at the statement date.

   9. _____

10. Find the amount of the credit given to Rose Gardens after the fee is subtracted.

    10. _____

C// indicates an exercise that is related to the Case in Point feature.

186 CHAPTER 5 Bank Services

CREDIT-CARD DEPOSITS *A local motorcycle repair shop had the following credit-card charges and credits during a recent period.*

Sales		Refunds
$7.84	$98.56	$13.86
$33.18	$318.72	$58.97
$50.76	$116.35	
$12.72	$23.78	
$9.36	$38.95	
$118.68	$235.82	

11. Find the total amount of the credit-card sales.   11. _____

12. Find the total of the credit-card refunds.   12. _____

13. Find credit-card sales less refunds.   13. _____

14. If the fee paid by the shop is 2%, find the amount of the charge at the statement date.   14. _____

15. Find the amount of the credit given to Campus Bicycle Shop after the fee is subtracted. 15. _____

RESTAURANT SALES *Entrepreneur Christina Thompson noted her credit-card transactions for the evening before her first radio ads. She wants to know if the ads will be effective and needs some data for comparison.*

Sales		Refunds
$14.86	$76.15	$43.15
$49.70	$226.17	$17.06
$183.60	$63.95	
$238.75	$111.10	
$18.36	$77.86	
$52.08	$132.62	

16. Find the total amount of the credit-card sales.   16. _____

17. Find the total of the credit-card refunds.   17. _____

18. Find credit-card sales less refunds.   18. _____

19. If the fee paid by the shop is 2.5%, find the amount of the charge at the statement date.   19. _____

20. Find the amount of credit given to Christina Thompson's restaurant after the fee is subtracted.   20. _____

21. List and describe in your own words four services offered to business checking-account customers. (See Objective 1.)

22. The merchant accepting a credit card from a customer must pay a fee of 2% or more of the transaction amount. Why is the merchant willing to do this? Who really pays this fee? (See Objective 3.)

23. Kitchen Supplies, Inc. had the following sales: $58.60, $95.32, $283.40, and $204.50. Find the extra amount of service charges that must be paid if all transactions were made using a credit card with a bank fee of 2.5% compared to all transactions made using a debit card with a charge of $.21 per swipe. (See Example 2.)

24. During the lunch hour, Fashion Jewelry had sales of $145.34, $455.60, $92.50, $137.80, and $243.90. Find the extra amount of service charges that must be paid if all transactions were made using a credit card with a bank fee of 2% compared to all transactions made using a debit card with a charge of $.21 per swipe. (See Example 2.)

QUICK CHECK ANSWER

1. $1470.77   2. $24.82

5.3 Bank Statement Reconciliation

OBJECTIVES

1. Know the importance of reconciling a checking account.
2. Reconcile a bank statement.
3. List the outstanding checks.
4. Find the *adjusted bank balance* or *current balance*.

CASE IN POINT The owner of Rose Gardens knows the importance of keeping accurate checking-account records. She has received customers' checks drawn on accounts with nonsufficient funds (NSF) and both lost the revenue of the sale and been charged a returned check fee by her bank. Yet she has never written a bad check herself, as doing so could hurt her business reputation.

OBJECTIVE 1 Know the importance of reconciling a checking account. Each month, a bank creates a **bank statement** that shows all activity in the account for the month, including checks, debits, ATM transactions, electronic funds transfers, service charges, interest, and fees such as the fees on credit-card sales found in the previous section. The bank statement is available electronically on the bank's website or on paper. Here are a few reasons business owners look carefully at their bank statements every month:

> **Reasons to look at Bank Statements**
> 1. The fees charged by the bank vary from month to month.
> 2. A deposited check may be returned due to insufficient funds. The amount of the check and any fees must be deducted from the account balance.
> 3. A scheduled deposit may not have been made.
> 4. The account holder may have made an error in the check register or when entering data into the firm's computer system.
> 5. **Electronic payments** may be for an incorrect amount, or two payments may have been made when only one should have been.
> 6. Bank employees may make an error.

Writing a check without sufficient funds in an account is also called **bouncing a check**. The penalty for this can be both financial and legal. Banks charge high fees for bouncing a check and repeatedly doing it can get you in legal trouble.

Employee theft is a constant worry for managers, who often separate duties to prevent it. For example, a manager may set up a system in which one employee examines the bank statements for any errors while a second authorizes payments and a third audits bank statements at the end of every month. This makes it more difficult for any one person to steal from the firm without getting caught. Employees caught stealing are often fired and face criminal charges.

Quick TIP
It is NOT a good idea to steal anything from your employer!

OBJECTIVE 2 Reconcile a bank statement. The process of checking the bank statement against the check register is called **reconciliation**. Reconciling the bank statement every month for every account is an important step in preventing employee theft.

Notice the variety of items shown in the bank statement including electronic deposits when customers pay Rose Gardens electronically, other deposits, checks, and electronic withdrawals, which include electronic payments, ATM withdrawals, debits, and bank fees.

Electronic transfers are easy to track since they show up very quickly on the bank's computer system. Checks are different. For example, you may send your aunt a check for her birthday that she does not deposit at her bank for five weeks. A firm often has written and mailed checks that do not yet show up on the bank statement. These unpaid checks are called **checks outstanding**.

5.3 Bank Statement Reconciliation

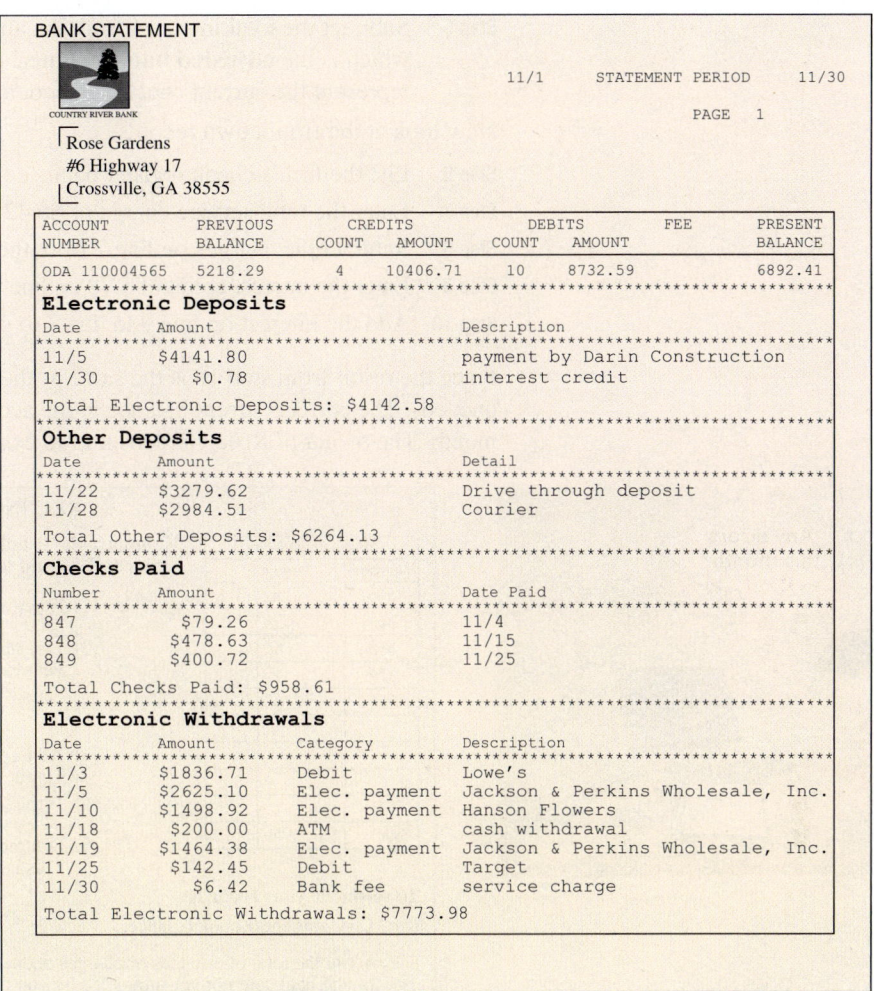

Reconciling a Checking Account

EXAMPLE 1 The bank statement for Rose Gardens shown above shows a balance of $6892.41 after a bank service charge of $6.42 and an interest credit of $.78. The check register maintained by the owner of Rose Gardens shows a current balance of $7580.38. Reconcile the account as follows.

CASE IN POINT

OBJECTIVE 3 List the outstanding checks. Find the checks outstanding by comparing the list of checks on the bank statement against the list of checks written by the firm. You may wish to put a check mark on the bank statement and/or check register as you compare them. In particular, note any difference between the amounts shown in the check register and the bank statement. Here are the outstanding checks.

Number	Amount
846	$42.73
852	$598.71
853	$68.12
857	$79.80
858	$160.30

After listing the outstanding checks in the space provided on the form, total them. The total is $949.66.

OBJECTIVE 4 Find the *adjusted bank balance* or *current balance*. Follow the steps listed here and shown on the reconciliation form on the next page.

Step 1 Enter the new balance of $6892.41 in the space at the top of the reconciliation form.

Step 2 List any deposits that have not yet been recorded by the bank. These are called **deposits in transit (DIT)**. Suppose that Rose Gardens has deposits of $892.41 and $739.58 that are not yet recorded. These numbers are written at step 2 on the form.

Step 3 Add the numbers from steps 1 and 2. At this point, the total is $8524.40.

Step 4 Write the total of outstanding checks. The total is $949.66.

Step 5 Subtract the total in step 4 from the amount from step 3. The result here is $7574.74, which is the **adjusted bank balance**, or the **current balance**. This number should represent the current checking-account balance.

Now look at the firm's own records.

Step 6 List the firm's check register balance of $7580.38 on line 6 on the reconciliation form.

Step 7 Enter the total service charge of $6.42 on line 7.

Step 8 Subtract the charges on line 7 from the checkbook balance on line 6 to get $7573.96.

Step 9 Enter the interest credit of $.78 on line 9. This is interest paid on money in the account.

Step 10 Add the interest on line 9 to line 8 to get $7574.74, the same result as in step 5.

Since the result from step 10 is the same as the result from step 5, the account is **balanced** (reconciled). The correct current balance in the account is $7574.74, and there were no errors this month. The owner of Rose Gardens knows exactly how much is in her checking account.

RECONCILIATION

Checks Outstanding	
Number	Amount
846	$ 42.73
852	598.71
853	68.12
857	79.80
858	160.30
Total	$ 949.66

Compare the list of checks paid by the bank with your records. List and total the checks not yet paid.

(1) Enter new balance from bank statement: $ 6892.41

(2) List any deposits made by you and not yet recorded by the bank: + 892.41
+ 739.58
+
+

(3) Add all numbers from lines above. Total: 8524.40

(4) Write total of checks outstanding: − 949.66

(5) Subtract (4) from (3). This is adjusted bank balance: $ 7574.74

To reconcile your records:

(6) List your checkbook balance: $ 7580.38

(7) Write the total of any fees or charges deducted by the bank and not yet subtracted by you from your checkbook: − 6.42

(8) Subtract line (7) from line (6). 7573.96

(9) Enter interest credit: (Add to your checkbook) + .78

(10) Add line (9) to line (8). Adjusted checkbook balance. $ 7574.74

New balance of your account; this number should be same as (5).

QUICK CHECK 1

Find the current balance for a checking account given the following: balance from bank statement, $7622.18; checks outstanding of $318.36, $1752.44, $738.35, $66.78; and deposits in transit of $1197.22 and $578.91.

There are several typical reasons a checking account might not balance.

Quick TIP
It is difficult to balance the checking accounts of a large business such as Target, which often has over 100,000 transactions every day.

Why a Checking Account Might Not Balance

- Forgetting to enter a check, a debit, an electronic payment, or an ATM withdrawal.
- Transposing numbers (entering $961.20 as $916.20, for example).
- Making addition or subtraction errors.
- Forgetting to subtract one of the bank service fees, such as those charged for a **returned check** or for ATM use at a different bank.
- Bounced checks, deposits not made on time, service charges, etc.

It is necessary to carefully compare the check register (or computerized account records) to the bank statement when an account does not reconcile. It can be difficult to find any error(s) in an active account, but it is important that managers keep track of funds so that they know how much they have in an account. Usually, errors are made by the account owner, although banks make mistakes occasionally.

Bank fees vary widely and individuals with higher bank balances may have no bank fees charged whatsoever. Active firms may see charges of hundreds of dollars in bank fees every month. Before choosing a bank, look carefully at the fees they charge.

5.3 Bank Statement Reconciliation

Reconciling a Checking Account

EXAMPLE 2 Using the information in the following check register and the bank statement, reconcile the checking account. Compare the items appearing on the check register to the bank statement. A check ✓ indicates that the transaction was on the prior month's bank statement, so that it does not need to be added or subtracted again.

Check Register ← Maintained by Business Employees

CHECK NO.	DATE	TRANSACTION DESCRIPTION	AMOUNT OF WITHDRAWAL	✓	AMOUNT OF DEPOSIT/CREDIT	BALANCE
		BALANCE BROUGHT FORWARD →				2782 95
elec.	7/11	Miller's Outpost	138 50	✓		2644 45
elec.	7/12	Barber Advertising	73 08			2571 37
723	7/18	Wayside Lumber	318 62	✓		2252 75
	7/24	Deposit			983 37	3236 12
724	7/25	I.R.S.	836 15			2399 97
elec.	7/26	Supplies	450 00			1949 97
725	7/28	Sacramento Bee	67 80			1882 17
726	8/2	T.V.A.	59 25			1822 92
727	8/3	Carmichael Office	97 37			1725 55
	8/6	Deposit			875 45	2601 00
ATM	8/5	ATM Cash	80 00			2521 00

Bank Statement ← Produced by Bank

```
ACCOUNT      PREVIOUS      CREDITS         DEBIT         PRESENT
NUMBER       BALANCE   COUNT  AMOUNT   COUNT   AMOUNT    BALANCE
7009875433   $2325.83    2    $983.59    7    $1555.15  $1754.27
```

Electronic Deposits
```
Date      Amount                    Description
7/24      $983.37                   payment from JP Morgan
8/4       $.22                      interest credit
Total Electronic Deposits: $983.59
```

Other Deposits
```
Date      Amount                    Detail
Total Other Deposits: $0
```

Checks Paid
```
Number    Amount                    Date Paid
724       $836.15                   7/28
726       $59.25                    8/4
Total Checks Paid: $895.40
```

Electronic Withdrawals
```
Date      Amount      Category         Description
7/22      $73.08      elec. payment    Barber Advertising
7/30      $450.00     debit            supplies
7/30      $49.07      debit            returned check
8/4       $80.00      ATM              cash withdrawal
8/5       $7.60       Bank fee         service charge
Total Electronic Withdrawals: $659.75
```

RECONCILIATION

Checks Outstanding

Number	Amount
725	$ 67 80
727	97 37
Total	$ 165 17

Compare the list of checks paid by the bank with your records. List and total the checks not yet paid.

(1) Enter new balance from bank statement: **$ 1754.27**

(2) List any deposits made by you and not yet recorded by the bank: **+ 875.45**
 +
 +
 +

(3) Add all numbers from lines above.
 Total: **2629.72**

(4) Write total of checks outstanding: **− 165.17**

(5) Subtract (4) from (3).
 This is adjusted bank balance: **$ 2464.55**

To reconcile your records:

(6) List your checkbook balance: **$ 2521.00**

(7) Write the total of any fees or charges deducted by the bank and not yet subtracted by you from your checkbook: **− 56.67** (Returned check and service charge)

(8) Subtract line (7) from line (6). **2464.33**

(9) Enter interest credit: (Add to your checkbook) **+ .22**

(10) Add line (9) to line (8). Adjusted checkbook balance. **$ 2464.55** ← New balance of your account; this number should be same as (5).

Quick TIP
Errors are often found when reconciling accounts. Careful investigation is needed to find the errors.

Since the adjusted bank balance from step 5 is the same as the adjusted checkbook balance from step 10, the account is reconciled (balanced). The correct current balance in the account is $2464.55.

QUICK CHECK 2

Use the form in Example 2 to reconcile the following account. The checkbook of Dottie Fogel Furnishings shows a balance of $7779. When the bank statement was received, it showed a balance of $6237.44, a returned check amounting to $246.70, a service charge of $15.60, and a check-printing charge of $18.50. There were unrecorded deposits of $1442.44 and $479.50, and checks outstanding of $146.36, $91.52, $43.78, and $379.52.

5.3 Exercises — MyLab Math

The shaded sections below contain solutions to help you get a QUICK START on the various types of exercises.

CURRENT CHECKING BALANCE Find the current balance for each of the following accounts. (See Example 1.)

	Balance from Bank Statement	Checks Outstanding	Deposits Not Yet Recorded	Balance
1.	$4572.15	$225.23 $97.68	$418.25 $348.17	$816.14 $571.28	$4870.24

$4572.15 − $225.23 − $97.68 − $418.25 − $348.17 + $816.14 + $571.28 = $4870.24

2.	$6274.76	$381.40 $875.14	$681.10 $83.15	$346.65 $198.96	$4799.58

$6274.76 − $381.40 − $875.14 − $681.10 − $83.15 + $346.65 + $198.96 = $4799.58

3.	$7911.42	$52.38 $95.42	$528.02 $76.50	$492.80 $38.72	_____
4.	$9343.65	$840.71 $78.68	$665.73 $87.00	$971.64 $3382.71	_____
5.	$19,523.20	$6853.60 $795.77	$340.00 $22.85	$6724.93 $78.81	_____
6.	$32,489.50	$3589.70 $263.15	$18,702.15 $7269.78	$7110.65 $2218.63	_____

7. Explain in your own words the significance of writing a bad check. What might the cost be in dollars? What are the other consequences? (See Objective 1.)

8. What are the financial costs to a business owner who receives a bad check? What would the business owner likely do regarding this customer? (See Objective 1.)

9. Briefly describe the importance of reconciling a checking account. What benefits are derived from keeping good checking records? (See Objective 2.)

10. List the types of errors you (or a parent or friend) have had when balancing a checking account. (See Objective 4.)

C// indicates an exercise that is related to the Case in Point feature.

5.3 Exercises

RECONCILING CHECKING ACCOUNTS *For Exercises 11 and 12, use the following table to reconcile each account and find the current balance. (See Example 1.)*

		Exercise 11.		Exercise 12.
Balance from bank statement		$6875.09		$14,928.42
Checks outstanding	421	$371.52	512	$84.76
(check number is given first)	424	$429.07	515	$109.38
	427	$883.69	517	$42.03
	429	$35.62	519	$1429.12
Deposits not yet recorded		$701.56		$54.21
		$421.78		$394.76
		$689.35		$1002.04
Bank charge		$8.75		$7.00
Interest credit		$10.71		$22.86
Checkbook balance		$6965.92		$14,698.28
Current balance				

11.

RECONCILIATION

Compare the list of checks paid by the bank with your records. List and total the checks not yet paid.

Checks Outstanding
Number	Amount

Total

(1) Enter new balance from bank statement: _____

(2) List any deposits made by you and not yet recorded by the bank: + _____ + _____ + _____ + _____

(3) Add all numbers from lines above. Total:

(4) Write total of checks outstanding: − _____

(5) Subtract (4) from (3). This is adjusted bank balance: _____

To reconcile your records:

(6) List your checkbook balance: _____

(7) Write the total of any fees or charges deducted by the bank and not yet subtracted by you from your checkbook: − _____

(8) Subtract line (7) from line (6). _____

(9) Enter interest credit: (Add to your checkbook) + _____

(10) Add line (9) to line (8). Adjusted checkbook balance. _____

New balance of your account; this number should be same as (5).

12.

RECONCILIATION

Compare the list of checks paid by the bank with your records. List and total the checks not yet paid.

Checks Outstanding
Number	Amount

Total

(1) Enter new balance from bank statement: _____

(2) List any deposits made by you and not yet recorded by the bank: + _____ + _____ + _____ + _____

(3) Add all numbers from lines above. Total:

(4) Write total of checks outstanding: − _____

(5) Subtract (4) from (3). This is adjusted bank balance: _____

To reconcile your records:

(6) List your checkbook balance: _____

(7) Write the total of any fees or charges deducted by the bank and not yet subtracted by you from your checkbook: − _____

(8) Subtract line (7) from line (6). _____

(9) Enter interest credit: (Add to your checkbook) + _____

(10) Add line (9) to line (8). Adjusted checkbook balance. _____

New balance of your account; this number should be same as (5).

CHECKING-ACCOUNT RECONCILIATION *Reconcile the checking accounts for the following companies. Compare the items appearing on the bank statement with the check register. A ✓ indicates that the transaction appeared on the previous month's statement. (Codes indicate the following: RC means returned check; SC means service charge; IC means interest credit; CP means check printing charge; ATM means automated teller machine.) (See Example 2.)*

13. Rose Gardens

CHECK NO.	DATE	TRANSACTION DESCRIPTION	AMOUNT OF WITHDRAWAL	✓	AMOUNT OF DEPOSIT/CREDIT	BALANCE
		BALANCE BROUGHT FORWARD →				6669 34
elec.	2/8	Floors to Go	248 96			6420 38
762	2/9	Healthways Dist.	125 63			6294 75
	2/11	Deposit			626 34	6921 09
763	2/12	Franchise Tax	770 41	✓		6150 68
764	2/14	Foothill Repair	22 86	✓		6127 82
elec.	2/15	Yellow Pages	91 24			6036 58
	2/17	Deposit			826 03	6862 61
elec.	2/17	Morning Herald	71 59			6791 02
765	2/18	San Juan Electric	63 24			6727 78
ATM	2/22	ATM Gas	15 26			6712 52
766	2/23	West Construction	405 07			6307 45
767	2/24	Heater Repairs	525 00			5782 45
	2/26	Deposit			220 16	6002 61
768	2/28	Capital Alarm	135 76			5866 85

```
Bank Statement
****************************************************
ACCOUNT      PREVIOUS    CREDITS          DEBIT          PRESENT
NUMBER       BALANCE   COUNT  AMOUNT  COUNT  AMOUNT     BALANCE
****************************************************
2352276478   $5876.07    3   $1452.49    9   $1692.05   $5636.51
****************************************************
Electronic Deposits
Date     Amount          Description
****************************************************
2/16     $626.34         payment from Jayne Inc.
2/24       $.12          interest credit

Total Electronic Deposits: $626.46
****************************************************
Other Deposits
Date     Amount          Detail
****************************************************
2/17     $826.03         Regions Bank

Total Other Deposits: $826.03
****************************************************
Checks Paid
Number   Amount          Date Paid
****************************************************
762      $125.63         2/21
766      $405.07         2/26
767      $525.00         2/28
Total Checks Paid: $1055.70
****************************************************
Electronic Withdrawals
Date     Amount     Category    Description
****************************************************
2/16      $91.24    bill pay    Central Electric
2/17     $248.96    debit       Wal-Mart Stores
2/19      $71.59    bill pay    Michaels(crafts)
2/22     $198.17    debit       check returned due to
                                nonsufficient funds
2/22      $15.26    debit       lunch
2/27      $11.13    bank fees   service charge
Total Electronic Withdrawals: $636.35
```

RECONCILIATION

Compare the list of checks paid by the bank with your records. List and total the checks not yet paid.

Checks Outstanding	
Number	Amount
Total	

(1) Enter new balance from bank statement: _____

(2) List any deposits made by you and not yet recorded by the bank:
  + _____
  + _____
  + _____
  + _____

(3) Add all numbers from lines above.
Total: _____

(4) Write total of checks outstanding: − _____

(5) Subtract (4) from (3).
This is adjusted bank balance: _____

To reconcile your records:

(6) List your checkbook balance: _____

(7) Write the total of any fees or charges deducted by the bank and not yet subtracted by you from your checkbook: − _____

(8) Subtract line (7) from line (6). _____

(9) Enter interest credit: (Add to your checkbook) + _____

(10) Add line (9) to line (8).
Adjusted checkbook balance. _____

New balance of your account; this number should be same as (5).

14. **Capital Video & Digital**

CHECK NO.	DATE	TRANSACTION DESCRIPTION	AMOUNT OF WITHDRAWAL	✓	AMOUNT OF DEPOSIT/CREDIT	BALANCE	
		BALANCE BROUGHT FORWARD →				7682	07
elec.	3/3	Action Packing Supplies	451 16			7230	91
663	3/3	Crown Paper	954 29	✓		6276	62
664	3/5	ATM Cash	80 00	✓		6196	62
	3/7	Deposit			913 28	7109	90
elec.	3/10	Fairless Water District	72 37			7037	53
665	3/12	Audia Temporary	340 88			6696	65
elec.	3/13	Lionel Toys	618 65			6078	00
666	3/14	Fairless Hills Power	100 50			5977	50
	3/16	Deposit			450 18	6427	68
	3/18	Deposit			163 55	6591	23
667	3/20	Hunt Roofing	238 50			6352	73
668	3/22	Standard Brands	315 62			6037	11
669	3/23	Penny-Saver Products	67 29			5969	82
	3/24	Deposit			830 75	6800	57

```
Bank Statement
**************************************************************
ACCOUNT      PREVIOUS    CREDITS           DEBIT            PRESENT
NUMBER       BALANCE     COUNT   AMOUNT    COUNT   AMOUNT   BALANCE
**************************************************************
1278992431   $6647.78    4       $1549.49  7       $1816.41 $6380.86
**************************************************************
Electronic Deposits
Date      Amount              Description
**************************************************************
3/7       $913.28             Credit from Nike
3/20      $22.48              interest credit from CD
Total Electronic Deposits: $935.76
**************************************************************
Other Deposits
Date      Amount              Detail
**************************************************************
3/16      $450.18             Invoice A65912
3/22      $163.55             John Tyler High School
Total Other Deposits: $613.73
**************************************************************
Checks Paid
Number    Amount              Date Paid
**************************************************************
665       $340.88             3/16
667       $238.50             3/26
Total Checks Paid: $579.38
**************************************************************
Electronic Withdrawals
Date      Amount    Category    Description
**************************************************************
3/11      $451.16   bill pay    Baseball Equipment, Inc.
3/20      $72.37    debit       Office Max
3/22      $618.65   bill pay    Wells Fargo-mortgage
3/16      $82.15    debit       returned check
3/26      $12.70    bank fee    service charge
Total Electronic Withdrawals: $1237.03
```

RECONCILIATION

Checks Outstanding	
Number	Amount
Total	

Compare the list of checks paid by the bank with your records. List and total the checks not yet paid.

(1) Enter new balance from bank statement: _____

(2) List any deposits made by you and not yet recorded by the bank:
+ _____
+ _____
+ _____
+ _____

(3) Add all numbers from lines above.
Total: _____

(4) Write total of checks outstanding: − _____

(5) Subtract (4) from (3).
This is adjusted bank balance: _____

To reconcile your records:

(6) List your checkbook balance: _____

(7) Write the total of any fees or charges deducted by the bank and not yet subtracted by you from your checkbook: − _____

(8) Subtract line (7) from line (6). _____

(9) Enter interest credit: (Add to your checkbook) + _____

(10) Add line (9) to line (8).
Adjusted checkbook balance. _____

New balance of your account; this number should be same as (5).

QUICK CHECK ANSWERS

1. $6522.38   2. $7498.20

Chapter 5 Quick Review

Chapter Terms *Review the following terms to test your understanding of the chapter. For each term you do not know, refer to the page number found next to that term.*

adjusted bank balance [p. 190]
ATM card [p. 182]
automated teller machine (ATM) [p. 172]
balance [p. 176]
balanced [p. 190]
bank statement [p. 188]
blank endorsement [p. 175]
bouncing a check [p. 188]
business checking account [p. 174]
canceled (check) [p. 176]
cashier's check [p. 182]
certificates of deposit (CDs) [p. 182]
check register [p. 178]
check stubs [p. 176]
checks outstanding [p. 188]
credit card [p. 182]
credit history [p. 173]
crowdsourcing [p. 172]
current balance [p. 190]
debit card [p. 172]
deposit slip [p. 175]
deposit ticket [p. 175]
deposits in transit (DIT) [p. 189]
direct deposit [p. 172]
direct payment [p. 173]
electronic banking [p. 172]
electronic commerce (EC) [p. 172]
electronic payment [p. 177]
endorse [p. 175]
identity theft [p. 173]
interest [p. 174]
Internet service providers [p. 172]
maintenance charge per month [p. 174]
mobile banking [p. 172]
mobile payment [p. 182]
mobile ticketing [p. 172]
money market account [p. 182]
money order [p. 182]
nonsufficient funds (NSF) [p. 182]
notary service [p. 182]
online banking [p. 172]
overdraft [p. 182]
overdraft fees [p. 000]
overdraft protection [p. 182]
per-debit charge [p. 174]
personal checking account [p. 174]
personal identification number (PIN) [p. 172]
point of sale (POS) [p. 173]
prepaid card [p. 172]
reconciliation [p. 188]
restricted endorsement [p. 175]
returned check [p. 190]
returned-deposit item [p. 182]
service fees [p. 172]
smart card [p. 172]
special endorsement [p. 175]
stop-payment order [p. 182]
transaction [p. 172]
transaction costs [p. 176]
transaction register [p. 177]
Uniform Product Code (UPC) [p. 173]

CONCEPTS

5.1 Checking-account service charges
A checking-account maintenance fee is usually charged and there is often a per-check charge.

5.2 Bank services offered
The checking account customer must be aware of various banking services that are offered.

5.2 Deposits with credit-card transactions
All credit-card refunds must be subtracted from total credit-card sales to find the net deposit. The fee is then subtracted from this total.

EXAMPLES

Find the monthly checking-account service charge for a business with 36 checks, given a monthly maintenance charge of $7.50 and a $.20 per check charge.

$$\$7.50 + 36(\$.20) = \$7.50 + \$7.20 =$$
$$\$14.70 \text{ monthly service charge}$$

Overdraft protection Offered to protect the customer from bouncing a check (NSF)

Debit card Used at an automated teller machine to get cash or at a merchant to make a purchase

Stop-payment order Stops payment on a check written in error

Cashier's check A check written by a financial institution, such as a bank, and that is viewed to be as good as cash

Money order An instrument used in place of cash

Notary service An official certification of a signature or document

Online banking Allows customers to perform many banking functions on the Internet

The following are credit-card charges and credits.

Charges		Credits
$28.15	$78.59	$21.86
$36.92	$63.82	$19.62

(a) Find total charges.

$$\$28.15 + \$36.92 + \$78.59 + \$63.82 = \$207.48$$

CONCEPTS	EXAMPLES
	(b) Find total credits. $21.86 + $19.62 = $41.48 **(c)** Find gross deposit. $207.48 − $41.48 = $166 **(d)** Given a 3% fee, find the amount of the charge. $166 × .03 = $4.98 **(e)** Find the amount of credit given to the business. $166 − $4.98 = $161.02
5.2 Comparing credit-card charges to debit charges Businesses pay an average fee of 2% to banks for any credit charges but only about $.21 per debit-card swipe.	**Charges** $35.79 $323.42 $164.12 $12.64 **(a)** Find total credit-card charges assuming a 2% fee. ($35.79 + $164.12 + $323.42 + $12.64) × .02 = $10.72 (rounded) **(b)** Find the total debit-card charges assuming $.21 per swipe. 4 swipes × $.21 per swipe = $.84 **(c)** Find the extra bank charge if all four charges were made using a credit card compared to if all four charges were made using a debit card. $10.72 − $.84 = $9.88
5.3 Reconciliation of a checking account A checking-account customer must periodically reconcile checking-account records with those of the bank or financial institution. Use a bank statement and reconciliation form to do this.	The accuracy of all checks written, deposits made, service charges incurred, and interest paid is checked and verified. The customer's checkbook balance and bank balance must be the same for the account to reconcile, or balance.

Case Study

BANKING ACTIVITIES OF A RETAILER

Jessica Lange owns her own high-end hair salon and employs several hair stylists. Customers usually pay using either a debit card or a credit card, but they sometimes write checks or pay with cash. Her credit-card sales last week were $8752.40. During the same period, she had $573.94 in credit refunds and she pays fees of $2\frac{1}{2}\%$ on the net credit-card sales.

The balance at the beginning of the week was $4228.34. Checks outstanding were $758.14, $38.37, $1671.88, $120.13, $2264.75, $78.11, $3662.73, $816.25, and $400. Lange had credit-card and bank deposits of $458.23, $771.18, $235.71, $1278.55, $663.52, and $1475.39 that were not recorded.

1. Find the net deposit given credit-card sales and refunds.   1. _____

2. Find the fee on the net credit-card sales.   2. _____

3. What is the total of the checks outstanding?   3. _____

4. Find the total of the deposits that were not recorded.   4. _____

5. Find the current balance in the account.   5. _____

INVESTIGATE

Look at your bank statement and identify any fees you pay and interest you earn. Name three factors other than cost that are important to you when selecting a bank. Does your bank meet your needs, or can you find one better suited to your situation?

Case in Point Summary Exercise

JACKSON & PERKINS

www.jacksonandperkins.com

Facts:

- 1872: Founded
- 1901: Sold first hybridized rose
- 1939: Began selling using mail order
- 1960s: Moved growing operations to California
- 2018: Bud, grow, and harvest more than 10 million plants

The Jackson & Perkins Company was founded in 1872 to wholesale strawberry and grape plants. But it quickly found its specialty in roses. It takes 7 to 10 years of hybridizing work to find a handful of the best and hardiest new varieties of roses. In order to help sell flowers, marketing at Jackson & Perkins has given fun names to specific hybridized roses, including Sweet Intoxication and Pope John Paul II.

Barbara Wiffy at Rose Gardens buys roses from Jackson & Perkins Wholesale, Inc., marks up the prices, and then sells them to her customers. Rose Gardens is an established customer that often places several orders with the company during a month, but usually only makes one or two electronic payments to them each month. Following are copies of Rose Gardens's check register for the month and also the bank statement. Help Wiffy reconcile the business account to see what other charges should be added and subtracted from her balance and see if there are any errors in her check register.

Check Register—Rose Gardens

Check Num.	Date	Amount of Withdrawal	Amount of Deposit	Comment	Balance
				Balance Brought Forward	$9,277.59
	6/24	$1,000.00	-	Online Transfer to Savings	$8,277.59
761	6/25	$195.00	-	Wal-Mart	$8,082.59
	6/26	$2,446.82	-	Electronic—Mortgage Payment	$5,635.77
	6/26	$4,080.97	-	Electronic—Office Max	$1,554.80
	6/27	$395.07	-	Electronic—Utilities	$1,159.73
	6/28	-	$4,500.00	Online Transfer from ***5332	$5,659.73
762	6/29	$32.34	-	CVS—Pharmacy	$5,627.39
	6/29	-	$7,550.80	Direct deposit—debit	$13,178.19
	6/30	$2,153.93	-	Electronic—Dice Plants	$11,024.26
	7/1	-	$5,781.48	Regular Deposit	$16,805.74
763	7/3	$723.35	-	Max's Remodeling	$16,082.39
	7/14	$6,230.19	-	Electronic—Jackson & Perkins	$9,852.20
764	7/15	$3,290.41	-	Bacon Farms	$6,561.79
	7/15	-	$2,705.00	Regular Deposit	$9,266.79

Source: JACKSON & PERKINS.

GLEN MEADOWS BANK
Bank Statement

Period: June 17 through July 16
Year: 2018

Checking Summary
Balance Calculation

Previous Balance	9,277.59
Checks	227.34
Withdrawals	16,315.93
Deposits	20,537.28
Interest Paid	0.57
Current Balance	**13,272.17**

Checks
#761	195.00
#762	32.34

Interest
7/16	0.57

Date	Description	Withdrawals	Deposits	Balance
6/24	Online Transfer—to savings	1,000.00		
6/26	Online Pmt.—Mtge. Pymt.	2,446.82		
6/26	Debit Pmt.—Office Max	4,080.97		
6/27	Online Pmt.—Utilities	395.07		
6/28	Online Transfer—from checking		4,500.00	
6/29	Direct Deposit—Numerous Debit Purchases		7,550.80	
6/30	Online Pmt.—Dice Plants	2,153.93		
7/01	Regular Deposit		5,781.48	
7/14	Online Pmt.—Jackson & Per.	6,230.19		
7/15	Regular Deposit		2,705.00	
7/16	Fee to Bank	8.95		

RECONCILIATION

Checks Outstanding

Number	Amount
Total	

(1) Enter new balance from bank statement: _____

(2) List any deposits made by you and not yet recorded by the bank:
 + _____
 + _____
 + _____
 + _____

(3) Add all numbers from lines above. _____
(4) Write total of checks outstanding: − _____
(5) Subtract (4) from (3). _____

To reconcile your records:

(6) List your checkbook balance: _____
(7) Write the total of any fees. − _____
(8) Subtract line (7) from line (6). _____
(9) Enter interest credit: + _____
(10) Add line (9) to line (8). Adjusted checkbook balance. _____

New balance

Discussion Question: Many people no longer reconcile their checking account, effectively trusting the bank. Is it reasonable for a business with a large volume of banking activity not to reconcile its checking account? Why or why not?

Chapter 5 Test

To help you review, the numbers in brackets show the section in which the topic was discussed.

Use the table on page 174 to find the monthly checking-account service charge for the following accounts. **[5.1]**

1. Tino's Italian Grocery, 62 checks, average balance $1834

2. Gifts Galore, 44 checks, average balance $2398

3. Batista Tile Works, 27 checks, average balance $418

Complete the following three check stubs for Mad Men Advertising in 2018. Find the balance at the bottom of each stub. **[5.1]**

Checks Written

Number	Date	To	For	Amount
2261	Aug. 6	WBC Broadcasting	Airtime	$6892.12
2262	Aug. 8	Lakeland Weekly	Ad	$1258.36
2263	Aug. 14	W. Wilson	Freelance Art	$416.14

Deposits made: $1572 on Aug. 7, $10,000 on Aug. 10.

4. 2261 — Date ___ Amount ___ To ___ For ___
   Total $16,409.82
   Deposits ___
   This Check ___
   Other (+ or −) ___
   Balance ___

5. 2262 — Date ___ Amount ___ To ___ For ___
   Total $9517.70
   Deposits ___
   This Check ___
   Other (+ or −) ___
   Balance ___

6. 2263 — Date ___ Amount ___ To ___ For ___
   Total $9831.34
   Deposits ___
   This Check ___
   Other (+ or −) ___
   Balance ___

EMILIO GIFTS FOR EVERYONE, INC. *The gift shop has gifts for all occasions. Use the following information about credit-card transactions to answer the next five questions.* **[5.2]**

Sales		Refunds
$218.68	$135.82	$45.63
$37.84	$67.45	$36.36
$33.18	$461.82	
$20.76	$116.35	
$12.72	$23.78	
$8.97	$572.18	

7. Find the total amount of credit-card sales.

8. What is the total amount of the refunds?

9. Find credit-card sales less refunds.

C// indicates an exercise that is related to the Case in Point feature.

10. Assuming that the bank charges the retailer a $3\frac{1}{2}\%$ fee, find the amount of the fee at the statement date.

10. _____

11. Find the amount of the credit given to Emilio Gifts For Everyone, Inc. after the fee is subtracted.

11. _____

C// 12. Weddings by Bobbi is a regular customer of Rose Gardens. Use the information in the following table to reconcile her checking account on the form that follows. [5.3]

12. _____

Balance from bank statement		$4721.30
Checks outstanding	3221	$82.74
(check number is given first)	3229	$69.08
	3230	$124.73
	3232	$51.20
Deposits not yet recorded		$758.06
		$32.51
		$298.06
Bank charge		$2.00
Interest credit		$9.58
Checkbook balance		$5474.60
Current balance		_____

RECONCILIATION

Checks Outstanding

Number	Amount
Total	

Compare the list of checks paid by the bank with your records. List and total the checks not yet paid.

(1) Enter new balance from bank statement: _____

(2) List any deposits made by you and not yet recorded by the bank:
+ _____
+ _____
+ _____
+ _____

(3) Add all numbers from lines above. Total: _____

(4) Write total of checks outstanding: − _____

(5) Subtract (4) from (3). This is adjusted bank balance: _____

To reconcile your records:

(6) List your checkbook balance: _____

(7) Write the total of any fees or charges deducted by the bank and not yet subtracted by you from your checkbook: − _____

(8) Subtract line (7) from line (6). _____

(9) Enter interest credit: (Add to your checkbook) + _____

(10) Add line (9) to line (8). Adjusted checkbook balance. _____

New balance of your account; this number should be same as (5).

Payroll 6

CHAPTER CONTENTS

- 6.1 Gross Earnings: Wages and Salaries
- 6.2 Gross Earnings: Piecework and Commissions
- 6.3 Social Security, Medicare, and Other Taxes
- 6.4 Income Tax Withholding

CASE IN POINT

SARAH BRYNSKI manages a Starbucks store and knows that payroll is complicated. She looked into the history of two programs relating to deductions: Social Security and Medicare.

Brynski found that many elderly people suffered greatly during the Great Depression of the 1930s. The Social Security Act was passed in 1935 to help supplement the retirement income of the elderly as well as provide income for the disabled. Medicare was established in 1965 to help people 65 and over pay for hospital care, medical care, and, later, prescriptions. Experts believe that without these two programs, poverty among the elderly would greatly increase.

Sarah also discovered that both Social Security and Medicare are projected to have financial problems by 2030. Sarah realizes that these two programs are far too important to disappear, so she expects there will be some changes to both in the next few years.

It is important that payroll is done accurately and completed on a timely basis. Making mistakes with payroll means that individuals, the state, or the federal government are not paid the correct amount(s), which can result in penalties.

6.1 Gross Earnings: Wages and Salaries

OBJECTIVES

1. Calculate gross earnings for salaries and wages.
2. Find overtime earnings for more than 40 hours of work per week.
3. Use the overtime premium method of calculating gross earnings.
4. Find overtime earnings for over 8 hours of work per day.
5. Understand double time, shift differentials, and split-shift premiums.
6. Find equivalent earnings for different pay periods.
7. Find overtime for salaried employees.

CASE IN POINT As a manager at Starbucks, Sarah Brynski hires, schedules, and oversees employees. She must also make sure each is paid the correct amount. Occasionally, she authorizes overtime when a scheduled employee does not show up. However, she tries to use as little overtime as possible to help control costs.

The first step in preparing payroll is to determine **gross earnings** or total amount earned. Several **deductions** must be subtracted from gross earnings to arrive at **net pay**, the amount the employee actually receives. The employer must keep records to satisfy legal requirements. Many businesses use the services of a company to professionally prepare their payroll. Other businesses use computer software such as QuickBooks to help them complete this task.

$$\text{Net Pay} = \text{Gross Earnings} - \text{Deductions}$$

OBJECTIVE 1 Calculate gross earnings for salaries and wages. Several methods are used to find an employee's pay. Salaries and wages are discussed in this section, and piecework and commission are discussed in the next section. Regular payroll checks are essential for most people since they do not have enough savings to live on for more than a few weeks or months.

Payroll is often based on hours worked. The hours worked can be recorded on a computer or on **time cards**. The records include: name and other personal information; the days, times, and hours worked; and the total number of hours worked. The information from the records is transferred to a **payroll ledger** showing all payroll information, as shown in Example 1.

Katie Nopper, whose payroll card is shown, is a shift manager at a local coffee shop and is paid an **hourly wage** of $14.80. Her gross earnings can be calculated with the following formula.

$$\text{Gross earnings} = \text{Number of hours worked} \times \text{Rate per hour}$$

For example, if Nopper works 8 hours at $14.80 per hour, her gross earnings are

$$\text{Gross earnings} = 8 \text{ hours} \times \$14.80 \text{ per hour} = \$118.40$$

PAYROLL CARD
NO TIME CLOCK REQUIRED

EMPL. NO. 1375 CARD NO. _____

FULL NAME: Katie Nopper AGE (IF UNDER 18): _____
ADDRESS: 412 Fawndale Drive
SOCIAL SECURITY NO. 123-45-6789
DATE EMPLOYED: February 17, 2018 POSITION: Shift MGR RATE: $14.80
PAY PERIOD STARTING: 7/23/2018 ENDING: 7/27/2018

DATE	REGULAR TIME					OVER TIME		
	IN	OUT	IN	OUT	DAILY TOTALS	IN	OUT	DAILY TOTALS
7/23	8:00	11:50	12:20	4:30	8	4:30	6:30	2
7/24	7:58	12:00	12:30	4:30	8	5:00	7:30	2.5
7/25	8:00	12:00	12:30	4:32	8			
7/26	7:58	12:05	12:35	4:30	8	4:30	5:00	.5
7/27	8:01	12:00	1:00	5:00	8			

APPROVED BY: PD
TOTAL REGULAR TIME: 40   5

REGULAR DAYS WORKED: 5 @ 8 HRS. @ 14.80   EARNINGS $ 592.00
ADDITIONAL COMPENSATION:
VALUE OF MEALS, LODGING, GIFTS, ETC.   AMOUNT $ _____
COMMISSIONS, FEES, BONUSES, GOODS, ETC. OT 5 @ 22.20   AMOUNT $ 111.00
OTHER REMUNERATIONS (KIND) $ _____
DEDUCTIONS: TOTAL EARNINGS $ 703.00

I CERTIFY THE FOREGOING TO BE A CORRECT ACCOUNT OF THE TIME WORKED AND WAGES RECEIVED.
SIGNATURE: Katie Nopper   DATE PAID: 8/1/2018

Completing a Payroll Ledger

EXAMPLE 1 Sarah Brynski is doing the payroll for two employees at her Starbucks store. She must first complete a payroll ledger.

Employee	\multicolumn{7}{c}{Hours Worked}	Total Hours	Rate	Gross Earnings
	S	M	T	W	TH	F	S			
Nelson, L.	—	2	4	8	6	3	—		$10.75	
Orr, T.	—	3.5	3	7	6.75	7	—		$9.80	

SOLUTION

CASE IN POINT First, find the total number of hours worked by each person.

Nelson: $2 + 4 + 8 + 6 + 3 =$ **23 hours**

Orr: $3.5 + 3 + 7 + 6.75 + 7 =$ **27.25 hours**

Then multiply the number of hours worked by the rate per hour to find the gross earnings.

$$
\begin{array}{cc}
\text{Nelson:} & \text{Orr:} \\
23 & 27.25 \\
\times\ \$10.75 & \times\ \$9.80 \\
\hline
\$247.25 & \$267.05
\end{array}
$$

The payroll ledger can now be completed.

Employee	S	M	T	W	TH	F	S	Total Hours	Rate	Gross Earnings
Nelson, L.	—	2	4	8	6	3	—	23	$10.75	$247.25
Orr, T.	—	3.5	3	7	6.75	7	—	27.25	$9.80	$267.05

QUICK CHECK 1

Ellie Kugler worked 6 hours on Monday, 4.5 hours on Tuesday, 4 hours on Wednesday, 5.75 hours on Thursday, and 3 hours on Friday. If she is paid $10.18 per hour, find her gross earnings for the week.

The pay for workers varies by country and even by region within a country. The Big Mac Index shown below shows how long it takes an average worker in different countries to earn enough to buy a Big Mac at a McDonalds.

Numbers in the News

Big Mac Index

Measures how long an individual must work on average to earn enough to buy a Big Mac.

Tokyo	10 minutes
Los Angeles	11 minutes
Paris	15 minutes
Buenos Aires	29 minutes
Bejing	42 minutes
Mexico City	87 minutes

Source: www.statistia.com (OECD).

OBJECTIVE 2 Find overtime earnings for more than 40 hours of work per week. The **Fair Labor Standards Act** covers most full-time employees in the United States. It establishes a workweek of 40 hours and sets the minimum hourly wage. The law states that an **overtime** wage (a higher-than-normal wage) must be paid for all hours worked over 40 hours per workweek. Many companies not covered by the Fair Labor Standards Act pay a **time-and-a-half rate** ($1\frac{1}{2}$, or 1.5, times the normal rate) for any work over 40 hours per week. Gross earnings before deductions are found as follows.

Gross earnings = Earnings at regular rate + Earnings at time-and-a-half rate

6.1 Gross Earnings: Wages and Salaries

Completing a Payroll Ledger with Overtime

EXAMPLE 2 Complete the following payroll ledger.

Employee	\multicolumn{7}{c	}{Hours Worked}	\multicolumn{2}{c	}{Total Hours}	Reg. Rate	\multicolumn{3}{c	}{Gross Earnings}
	S	M	T	W	TH	F	S	Reg.	O.T.		Reg.	O.T.	Total
Lanier, D.	6	9	8.25	8	9	4.5	—			$10.30			
Morse, T.	—	10	6.75	9	6.25	10	4.25			$9.48			

SOLUTION

Total hours worked:

Lanier: 6 + 9 + 8.25 + 8 + 9 + 4.5 = **44.75 hours**
Morse: 10 + 6.75 + 9 + 6.25 + 10 + 4.25 = **46.25 hours**

Amount of overtime worked by each:

Lanier: 44.75 hours worked − 40 hours = **4.75 hours of overtime**
Morse: 46.25 hours worked − 40 hours = **6.25 hours of overtime**

Regular time pay:

Lanier: 40 hours × $10.30 = $412.00
Morse: 40 hours × $9.48 = $379.20

Overtime pay at time and a half:

Lanier: 4.75 hours × $10.30 × 1.5 = $73.39
Morse: 6.25 hours × $9.48 × 1.5 = $88.88

Total pay = Regular time pay + Overtime pay:

Lanier: $412.00 + $73.39 = $485.39
Morse: $379.20 + $88.88 = $468.08

The ledger can now be completed.

Employee	\multicolumn{7}{c	}{Hours Worked}	\multicolumn{2}{c	}{Total Hours}	Reg. Rate	\multicolumn{3}{c	}{Gross Earnings}
	S	M	T	W	TH	F	S	Reg.	O.T.		Reg.	O.T.	Total
Lanier, D.	6	9	8.25	8	9	4.5	—	40	4.75	$10.30	$412.00	$73.39	$485.39
Morse, T.	—	10	6.75	9	6.25	10	4.25	40	6.25	$9.48	$379.20	$88.88	$468.08

QUICK CHECK 2

Joseph Monti works 5 hours on Monday, 10 hours on Tuesday, 9.5 hours on Wednesday, 8.25 hours on Thursday, 6 hours on Friday, and 7.5 hours on Saturday. Monti is paid $9.85 per hour and time and a half for all hours over 40 per week. Find his gross earnings.

OBJECTIVE 3 Use the overtime premium method of calculating gross earnings. Another method of finding the gross earnings with overtime is the **overtime premium method**, which produces the same result. Add the total hours at the regular rate to the overtime hours at one-half the regular rate to arrive at gross earnings.

Overtime Premium Method

Straight-time earnings ← total hours worked × regular rate
+ Overtime premium ← overtime hours worked × $\frac{1}{2}$ regular rate
Gross earnings

Using the Overtime Premium Method

EXAMPLE 3 This week, Holly Kelly worked 40 regular hours and 12 overtime hours. Her regular rate of pay is $17.40 per hour. Find her total gross pay, using the overtime premium method.

SOLUTION

Kelly's total hours are 40 + 12 = 52, and her overtime premium rate is .5 × $17.40 = $8.70.

$$
\begin{array}{rl}
52 \text{ hours} \times \$17.40 = \$904.80 & \text{regular rate earnings} \\
12 \text{ overtime hours} \times \$8.70 = \$104.40 & \text{overtime premium} \\
\hline
\$1009.20 & \text{gross earnings}
\end{array}
$$

The calculator solution uses the order of operations to find the regular earnings and the overtime earnings, and then adds these together.

52 ⊠ 17.4 ⊞ 12 ⊠ 17.4 ⊠ .5 ⊟ 1009.2

Note: Refer to Appendix B for calculator basis.

QUICK CHECK 3

Janice Garinger worked 40 regular hours and 10 overtime hours this week. If her regular pay is $12.10 per hour, find her gross earning using the overtime premium method.

OBJECTIVE 4 Find overtime earnings for over 8 hours of work per day. Some companies pay the time-and-a-half rate for any time worked over 8 hours in any one day no matter how many hours are worked in a week. This **daily overtime** is shown in the next example.

Finding Overtime Each Day

EXAMPLE 4 One week, Jason Polanski worked the hours shown. Given that his regular pay is $10.10 per hour, find gross earnings.

Quick TIP
Although the individual in Example 4 did not work 40 hours during the week, he still receives overtime pay, since he worked over 8 hours on two different days.

	S	M	T	W	TH	F	S	Total Hours
Reg.	—	8	5	7	8	—	—	28
O.T.	—	2	—	—	4	—	—	6

SOLUTION

Polanski worked fewer than 40 hours this week. Still, his company pays time and a half for any hours over 8 in a day, so he will earn some overtime. The table shows that he worked 28 regular hours and 6 overtime hours.

Overtime hourly pay = $10.10 × $1\frac{1}{2}$ = **$15.15 per hour**
Wages at regular rate: 28 hours × $10.10 = $282.80
Wages at overtime rate: 6 hours × **$15.15** = $90.90
Total gross earnings $373.70

QUICK CHECK 4

Luke Stansbury is paid $9.60 per hour and time and a half for all hours over 8 worked per day. He worked 9 hours on Monday, 6 hours on Tuesday, 12 hours on Wednesday, and 3 hours on Friday. Find his gross earnings for the week.

Some careers require unusual hours and do not pay overtime for over 40 hours worked in a week or 8 hours in a day. An example is a firefighter, who may work 24 hours and then get 48 hours off.

OBJECTIVE 5 Understand double time, shift differentials, and split-shift premiums. Some companies pay **double time** for working on holidays or on Saturdays and Sundays. A **shift differential** may be given to compensate for working less-desirable hours such as the swing shift (4:00 P.M. to midnight) or graveyard (midnight to 8:00 A.M.) shifts.

Restaurant employees often receive a **split-shift premium** for working, say, 4 hours followed by a few hours off before working another 4 hours. Yet other employers offer **compensatory time (comp time)** for overtime hours worked. Workers get time off for overtime hours worked rather than additional money. For example, an employee might be given 12 hours off for working 8 hours of overtime.

OBJECTIVE 6 Find equivalent earnings for different pay periods. Some employees are paid by the hour; others, such as most managers, earn a **salary** or a fixed amount per **pay period**. The following table lists common pay periods.

Common Pay Periods	
Monthly	12 paychecks each year
Semimonthly	Twice each month; 24 paychecks each year
Biweekly	Every 2 weeks; 26 paychecks each year
Weekly	52 paychecks each year

Determining Equivalent Earnings

EXAMPLE 5 A human resources manager needs to compare the earnings of employees:

Scott Perrine—weekly salary of $546   Julie Circle—semimonthly salary of $1850
Tonya McCarley—biweekly salary of $1686   Bill Leonard—a monthly salary of $2890

For each, find **(a)** earnings per year, **(b)** earnings per month, and **(c)** earnings per week.

SOLUTION

Scott Perrine:
(a) $546 × 52 = $28,392 per year
(b) $28,392 ÷ 12 = $2366 per month
(c) $546 per week

Tonya McCarley:
(a) $1686 × 26 = $43,836 per year
(b) $43,836 ÷ 12 = $3653 per month
(c) $1686 ÷ 2 = $843 per week

Julie Circle:
(a) $1850 × 24 = $44,400 per year
(b) $1850 × 2 = $3700 per month
(c) $44,400 ÷ 52 = $853.85 per week

Bill Leonard:
(a) $2890 × 12 = $34,680 per year
(b) $2890 per month
(c) $34,680 ÷ 52 = $666.92 per week

QUICK CHECK 5

Glen Lewis receives a weekly salary of $852. Find his **(a)** earnings per year, **(b)** earnings per month, and **(c)** semimonthly earnings.

Hmm, what level of education should I work toward?

The figure below shows that more education tends to result in higher incomes and a lower chance of being unemployed. For example, getting an associate's degree raises the average high school graduate's weekly earnings from $678 to $798. It also decreases the chances of being unemployed from 5.4% to 3.8% and usually results in additional benefits such as health insurance, retirement plan, etc. Earning a bachelor's degree really increases income. Education usually pays!

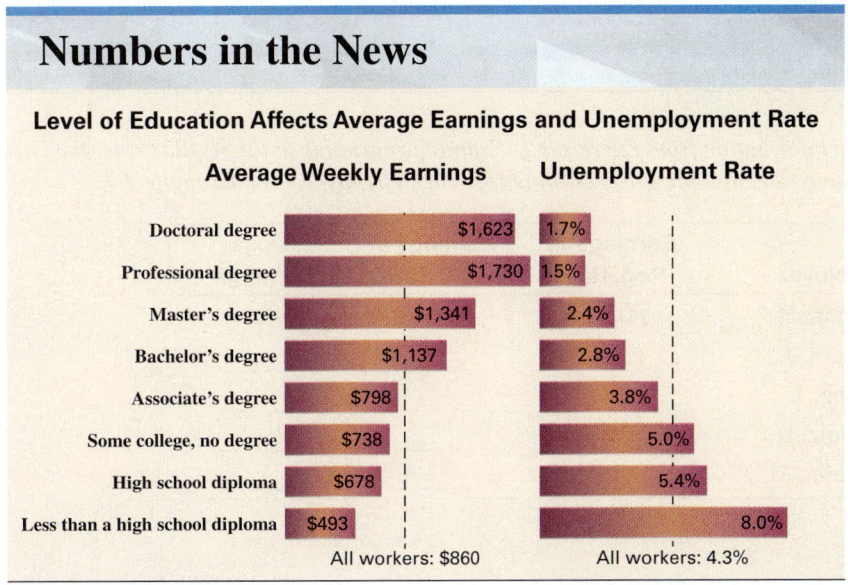

Numbers in the News

Level of Education Affects Average Earnings and Unemployment Rate

	Average Weekly Earnings	Unemployment Rate
Doctoral degree	$1,623	1.7%
Professional degree	$1,730	1.5%
Master's degree	$1,341	2.4%
Bachelor's degree	$1,137	2.8%
Associate's degree	$798	3.8%
Some college, no degree	$738	5.0%
High school diploma	$678	5.4%
Less than a high school diploma	$493	8.0%
	All workers: $860	All workers: 4.3%

Source: U.S. Census Bureau.

OBJECTIVE 7 Find overtime for salaried employees.
Managers and some other employees often receive a salary to do a job no matter how many hours were worked. However, some firms pay these employees time and a half for all work over 40 hours per week.

Finding Overtime for Salaried Employees

EXAMPLE 6 Monica Gonzales earns $936 a week as an assistant to the vice president of a community bank. A normal workweek for her is 40 hours. Find her gross earnings if she works 45 hours one week and is paid time and a half for everything over 40 hours.

SOLUTION

$$\text{Hourly equivalent to salary} = \frac{\$936}{40 \text{ hours}} = \$23.40$$

$$\text{Overtime rate} = 1\tfrac{1}{2} \times \$23.40 = \mathbf{\$35.10}$$

Find gross earnings as follows.

Regular wages		$936.00
Overtime wages	5 hours × $35.10 =	$175.50
		$1111.50 gross earnings

The calculator solution to this example is

936 [+] 936 [÷] 40 [×] 1.5 [×] 5 [=] 1111.5

QUICK CHECK 6

Elinor Nunes has a normal workweek of 35 hours and is paid $675.50 each week. Find her gross earnings if she works 42 hours one week and is paid time and a half for all hours worked above 40.

6.1 Exercises // MyLab Math

*The shaded sections below contain solutions to help you get a **QUICK START** on the various types of exercises.*

Find the number of regular hours and the overtime hours (any hours over 40) for each employee. Then calculate the overtime rate (time and a half) for each employee. (See Examples 1 and 2.)

Employee	S	M	T	W	TH	F	S	Reg. Hrs.	O.T. Hours	Reg. Rate	O.T. Rate
1. Simon, D.	—	7	4	7	10	8	4	40	0	$10.80	**$16.20**
2. Elbern, J.	—	6.5	9	7.5	8	9.5	7			$11.20	
3. Kling, J.	3	5	8.25	9	8.5	5	—			$18.70	
4. Scholz, K.	8.5	9	7.5	8	10	8.25	—			$9.50	
5. Tomlin, M.	—	9.5	7	9	9.25	10.5	—			$11.48	

Using the information from Exercises 1–5, find the earnings at the regular rate, the earnings at the overtime rate, and the gross earnings for each employee. (See Example 2.)

Employee	Earnings at Reg. Rate	Earnings at O.T. Rate	Gross Earnings
6. Simon, D.	$432	$0	$432
7. Elbern, J.			
8. Kling, J.			
9. Scholz, K.			
10. Tomlin, M.			

C// indicates an exercise that is related to the Case in Point feature.

Find the overtime rate, the amount of earnings at regular pay, the amount at overtime pay, and the total gross wages for each employee. (See Example 2.)

	Total Hours		Reg.	O.T.	Gross Earnings		
Employee	Reg.	O.T.	Rate	Rate	Regular	Overtime	Total
11. Triolo, J.	30	—	$11.30	$16.95	$339	$0	$339
12. Demaree, D.	36.25	—	$10.20	____	____	____	____
13. Snow, P.	40	4.5	$14.40	____	____	____	____
14. Taylor, O.	40	6.75	$12.08	____	____	____	____
15. Weyers, C.	40	4.25	$9.18	____	____	____	____

Some companies use the overtime premium method to determine gross earnings. Use this method to complete the following payroll ledger. Overtime is paid at the time-and-a-half rate for all hours over 40. (See Example 3.)

	Hours Worked							Total	Reg.	O.T.	O.T. Premium	Gross Earnings		
Employee	S	M	T	W	TH	F	S	Hours	Rate	Hours	Rate	Reg.	O.T.	Total
16. Averell, B.	10	9	8	5	12	7	—	51	$11.40	11	$5.70	$581.40	$62.70	$644.10
17. Brownlee, K.	7.75	10	5	9.75	8	10	—	___	$9.50	___	___	___	___	___
18. Carter, M.	—	12	11	8	8.25	11	—	___	$8.60	___	___	___	___	___
19. Parks, K.	—	8.5	5.5	10	12	10.5	7	___	$12.50	___	___	___	___	___
20. Parr, J.	—	10	9.75	9	11.5	10	—	___	$10.20	___	___	___	___	___

Some companies pay overtime for all time worked over 8 hours in a given day. Use this method to complete the following payroll ledger. Overtime is paid at the time-and-a-half rate. (See Example 4.)

	Hours Worked							Total Hours		Reg.	O.T.	Gross Earnings		
Employee	S	M	T	W	TH	F	S	Reg.	O.T.	Rate	Rate	Reg.	O.T.	Total
21. Bailey, M.	—	10	9	11	6	5	—	35	6	$9.40	$14.10	$329.00	$84.60	$413.60
22. Campbell, C.	—	9	8.75	7	8.5	10	—	___	___	$9.85	___	___	___	___
23. Ruhkala, B.	—	7.5	8	9	10.75	8	—	___	___	$10.80	___	___	___	___
24. Salorin, B.	—	9	10	8	6	9.75	—	___	___	$17.20	___	___	___	___
25. Warren, L.	—	9.5	8.5	7.75	8	9.5	—	___	___	$21.50	___	___	___	___

26. Describe two payroll payment plans choosing from double time, shift differential, split-shift premium, and compensatory time. (See Objective 5.)

27. If you were given a choice of overtime pay or compensatory time, which would you choose? Why? (See Objective 5.)

Find the equivalent earnings for each of the following salaries as indicated. (See Example 5.)

	Earnings				
	Weekly	Biweekly	Semimonthly	Monthly	Annual
28.	$496	$992	$1074.67	$2149.33	$25,792
29.	$443.08	$886.15	$960	$1920	$23,040
30.	___	$852	___	___	___
31.	___	___	___	$5200	___
32.	___	___	___	___	$26,100
33.	$830	___	___	___	___
34.	___	___	___	___	$32,000

Find the weekly gross earnings for the following people who are on salary and normally work a 40-hour week. Overtime is paid at the time-and-a-half rate. (See Example 6.)

Employee	Weekly Salary	Hours Worked	Weekly Gross Earnings
35. Burckle, E.	$398	42	_____
36. de Bouchel, V.	$468	45	_____
37. Feist-Milker, R.	$420	43	_____
38. Johnson, J.	$520	56	_____
39. Mader, C.	$640	48	_____
40. Logan, L.	$890	50	_____

Solve the following application problems.

41. **RETAIL EMPLOYMENT** Last week, Roberto Gonzales worked 48 hours at Starbucks. Find his gross earnings for the week if he is paid $9.80 per hour and earns time and a half for all hours over 40 worked in a week.
    48 − 40 = 8 overtime hours
    1.5 × $9.80 = $14.70 overtime rate
    40 × $9.80 = $392 regular wages
    8 × $14.70 = $117.60 overtime wages
    $392 + $117.60 = $509.60 gross earnings

    41. $509.60

42. **INSIDE SALES** Jessica Parker is a sales associate at JC Penney and is paid $13.60 per hour for straight time and time and a half for all hours over 40 worked in a week. Find her gross earnings for a week in which she worked 52 hours.

    42. _____

43. **CRIMINOLOGY LAB** Benito Zamora earns $12.80 per hour at his job as an assistant in a criminology lab. He is paid time and a half for everything worked over 8 hours in one day. Find his gross earnings for a week in which he worked the following hours: Monday, 9.5; Tuesday, 7; Wednesday, 10.75; Thursday, 4.5; and Friday, 8.75.

    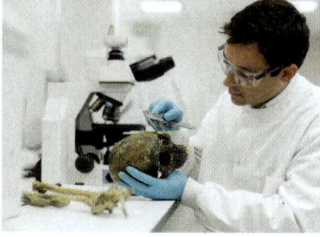

    43. _____

44. **OFFICE ASSISTANT** Sheila Spinnay is an office assistant and worked 10 hours on Monday, 9.75 hours on Tuesday, 5.5 hours on Wednesday, 12 hours on Thursday, and 7.25 hours on Friday. Her regular rate of pay is $11.50 an hour, with time and a half paid for all hours over 8 worked in a given day. Find her gross earnings for the week.

    44. _____

45. **INSURANCE OFFICE MANAGER** Alicia Klein is paid $728 a week as an office manager. Her normal workweek is 40 hours. She gets paid time and a half for overtime. Find her gross earnings for a week in which she works 46 hours.

    45. _____

46. **BEST BUY** The office assistant to the manager of a Best Buy earns $630 weekly. Find the equivalent earnings if the employee is paid (a) biweekly, (b) semimonthly, (c) monthly, and (d) annually.

(a) _____
(b) _____
(c) _____
(d) _____

47. **STORE MANAGER** Michelle Renda is assistant manager at Barnes and Noble and is paid $42,900 annually. Find the equivalent earnings if this amount is paid (a) weekly, (b) biweekly, (c) semimonthly, and (d) monthly.

(a) _____
(b) _____
(c) _____
(d) _____

48. **EDUCATIONAL CONSULTANT** P. K. Dickenson is an educational consultant who works full time for a large school district and earns $68,000 per year. Find the equivalent earnings if his salary is paid (a) weekly, (b) biweekly, (c) semimonthly, and (d) monthly.

(a) _____
(b) _____
(c) _____
(d) _____

49. Would you prefer a weekly paycheck or a monthly paycheck? Why? (See Objective 6.)

50. Managers usually make more than their employees. Why? Think in terms of duties managers have that employees do not.

51. It is unusual for a new college graduate to be hired immediately as a manager. Why do you think this is the case?

QUICK CHECK ANSWERS

1. $236.69
2. $486.34
3. $665.50
4. $312
5. (a) $44,304 (b) $3692 (c) $1846
6. $878.15

6.2 Gross Earnings: Piecework and Commissions

OBJECTIVES

1. Find the gross earnings for piecework.
2. Determine the gross earnings for differential piecework.
3. Find the gross pay for piecework with a guaranteed hourly wage.
4. Calculate the overtime earnings for piecework.
5. Find the gross earnings using commission rate times sales.
6. Determine a commission using the variable commission rate.
7. Find the gross earnings with a salary plus commission.

OBJECTIVE 1 Find the gross earnings for piecework. The wages of the previous section are **time rates**, since they depend on the time an employee is on the job, although managers are often paid a fixed amount no matter how many hours they work. The methods described in this section are **incentive rates** that pay an employee for actual production. The help-wanted ads below are for jobs offering incentive rates of pay.

The truck driver job pays 43 cents per mile (cpm). The Salesperson's Dream, Automotive Sales Person, Brink's Home Security, and Collector jobs pay a commission based on sales. The Construction job for lathers-stucco work offers piecework and hourly compensation.

A **piecework rate** pays an employee a given amount per item as follows:

> Gross earnings = Pay per item × Number of items produced

For example, a truck driver who drives 520 miles in one day and is paid a piecework rate of $.43 per mile has the following earnings:

$$\text{Gross earnings} = \$.43 \times 520 = \$223.60$$

EXAMPLE 1 Finding Gross Earnings for Piecework

Stacy Arrington is paid $.73 for sewing a jacket collar, $.86 for a sleeve with cuffs, and $.94 for a lapel. One week she sewed 318 jacket collars, 112 sleeves with cuffs, and 37 lapels. Find her gross earnings.

SOLUTION

Multiply the rate per item by the number of that type of item.

Item	Rate		Number		Total
Jacket collars	$.73	×	318	=	$232.14
Sleeves with cuffs	$.86	×	112	=	$96.32
Lapels	$.94	×	37	=	$34.78

Find the gross earnings by adding the three totals from the table.

$$\$232.14 + \$96.32 + \$34.78 = \mathbf{\$363.24}$$

QUICK CHECK 1

A production worker is paid $.58 for assembling a ceiling lamp, $1.23 for assembling a ceiling fan, and $.86 for assembling a box fan. One week a worker assembled 220 ceiling lamps, 318 ceiling fans, and 174 box fans. Find the worker's gross earnings.

6.2 Gross Earnings: Piecework and Commissions

OBJECTIVE 2 Determine the gross earnings for differential piecework. There are many variations to the straight piecework rate just described. For example, some rates have **quotas** that must be met, with a premium for each item produced beyond the quota. A typical **differential piece rate** plan is one where the rate paid per item depends on the number of items produced.

EXAMPLE 2 Using Differential Piecework

Adecco is looking for individuals who enjoy working with their hands and are willing to bend, cut, and insert discrete electronic components such as integrated circuits into standard printed circuit boards and solder as required. The pay scale follows.

1–100 units	$2.10 each
101–150 units	$2.25 each
151 or more units	$2.40 each

Find the gross earnings of a worker producing **214 units**.

SOLUTION

```
 214   ← total units
-100   ← first 100 units →     100 units at $2.10 each = $210.00
 114
- 50   ← next 50 units →        50 units at $2.25 each = $112.50
  64   ← number over 150 →      64 units at $2.40 each = $153.60
                               214 total units          = $476.10
```

The gross earnings are $476.10.

Quick TIP
With differential piecework, the highest amount paid applies to only the last units produced.

QUICK CHECK 2

Scooter Frame Company pays welders as follows: 1–150 frames, $2.85 each; 151–250, $3.20 each; and 251 or more, $3.45 each. Find the gross earnings of a worker who welds 282 frames.

OBJECTIVE 3 Find the gross pay for piecework with a guaranteed hourly wage. The piecework and differential piecework rates are frequently modified to include a guaranteed hourly pay rate. With this method, the employer must pay either the minimum wage or the piecework earnings, whichever is higher.

EXAMPLE 3 Finding Earnings with a Guaranteed Hourly Wage

A tire installer at the Tire Center is paid $10.50 per hour for an 8-hour day or $3.20 per tire installed, whichever is higher. Find the weekly earnings for an employee having the following rate of production.

Monday	32 tires
Tuesday	24 tires
Wednesday	34 tires
Thursday	22 tires
Friday	33 tires

Quick TIP
The company guarantees that a tire installer will not earn less than $10.50 per hour or $84 for an 8-hour day.

SOLUTION

The hourly earnings for an 8-hour day are **$84 (8 × $10.50)**. The larger of hourly earnings or the piecework earnings is paid each day.

Monday	32 × $3.20 =	$102.40
Tuesday	24 × $3.20 =	$84.00 **since it is greater than $76.80 (24 × $3.20)**
Wednesday	34 × $3.20 =	$108.80
Thursday	22 × $3.20 =	$84.00 **since it is greater than $70.40 (22 × $3.20)**
Friday	33 × $3.20 =	$105.60
	Total	$484.80 weekly earnings

216 CHAPTER 6 Payroll

> **QUICK CHECK 3**
>
> A cabinet door finisher is paid $14.70 per hour for an 8-hour day or $.95 per cabinet door finished, whichever is higher. Find the gross earnings for a worker who finished 106 doors on Monday, 127 doors on Tuesday, 152 doors on Wednesday, 120 doors on Thursday, and 138 doors on Friday.

OBJECTIVE 4 Calculate the overtime earnings for piecework. Similar to other employees, piecework employees are paid time and a half for overtime.

Calculating Earnings with Overtime Piecework

EXAMPLE 4 Jon Taylor fulfills order shipments at a large distribution center. He pulls items from stacks of inventory before handing orders to an operator who shrink-wraps the order and puts an address label on it. Jon is paid $1.85 per order fulfilled during regular time and time and a half for all orders filled during overtime. Find his gross earnings for a week in which he fulfilled 293 orders during regular time and 36 orders during overtime.

SOLUTION

$$\begin{aligned}
\text{Gross earnings} &= \text{Earnings at regular piece rate} + \text{Earnings at overtime piece rate} \\
&= \$1.85 \cdot 293 \quad\quad + \quad\quad \$1.85 \cdot 1\tfrac{1}{2} \cdot 36 \\
&= \$542.05 \quad\quad\quad + \quad\quad \$99.90 \\
&= \$641.95
\end{aligned}$$

> **QUICK CHECK 4**
>
> An assembler is paid $.84 for each child car seat assembled. During a recent week, she assembled 400 car seats during regular time and 138 car seats during overtime hours. If time and a half is paid for each overtime assembly, find the gross earnings for the week.

Salespeople are often paid a **commission rate** that is either a percent of sales or an amount per item sold. The idea is to motivate salespeople by paying them on the basis of actual results (sales).

OBJECTIVE 5 Find the gross earnings using commission rate times sales. A salesperson on **straight commission** is paid a fixed percent of sales, calculated as follows:

> Gross earnings = Commission rate × Amount of sales

The following data shows average incomes for some sales positions. However, the average does not mean a lot since some salespeople make far more than others. You can imagine that a mother of two young children who just started working as a real estate agent probably does not make anything near the income shown in the table. However, a real estate agent who has worked in the business for years may make quite a bit more than that shown.

Numbers in the News

Average Salaries for Sales Positions

- Sales Engineer: $87,390
- Insurance Agent: $46,330
- Advertising Sales: $45,350
- Real Estate Agent: $42,680
- Travel Agent: $31,870
- Retail Sales: $20,990

DATA: Bureau of Labor Statistics, U.S. Department of Labor.

Quick TIP
Some of the sales positions in the table require a Bachelor's Degree.

Determining Earnings Using Commission EXAMPLE 5  A real estate broker charges a 6% commission. Find the commission on a house selling for $385,000.

SOLUTION

$$\text{Commission} = 6\% \text{ of } \$385{,}000 = \$23{,}100$$

QUICK CHECK 5

An advertising sales representative is paid a 6% commission. Find the commission earned on sales of $14,680.

Before the commission is calculated, any **returns** from customers, or any **allowances**, such as discounts, must be subtracted from sales.

Subtracting Returns When Using Commission EXAMPLE 6  Amanda Roach, a food-supplements sales representative, had sales of $10,230 one month, with returns and allowances of $1120. Find her gross income if the commission rate is 12%.

SOLUTION

The returns and allowances must first be subtracted from gross sales. Then multiply the difference, net sales, by the commission rate.

$$\begin{aligned}
\text{Gross earnings} &= (\$10{,}230 \text{ gross sales} - \$1120 \text{ returns and allowances}) \times 12\% \\
&= \$9110 \text{ net sales} \times .12 \\
&= \$1093.20 \text{ gross earnings}
\end{aligned}$$

QUICK CHECK 6

Chesiree Jones had sales of $38,290 one month with returns of $864. Find her gross earnings if the commission rate is 8.5%.

OBJECTIVE 6 Determine a commission using the variable commission rate. The **sliding scale**, or **variable commission**, is a method of pay designed to retain top-producing salespeople. Under such a plan, a higher rate of commission is paid as sales get larger and larger.

Finding Earnings Using a Variable Commission EXAMPLE 7  Maureen O'Connor sells food and bakery products to businesses such as Starbucks and is paid as follows based on monthly sales.

Sales	Rate
Up to $10,000	6%
$10,001–$20,000	8%
$20,001 and up	9%

Find O'Connor's earnings if she has sales of $32,768 during a two-week interval.

SOLUTION

CASE IN POINT

```
 $32,768   ← total sales
-  10,000  ← first $10,000 →   $10,000 at 6%   =  $600.00
 $22,768
-  10,000  ← next $10,000  →   $10,000 at 8%   =  $800.00
 $12,768   ← over $10,000  →   $12,768 at 9%   = $1149.12
           $32,768 total sales    $2549.12 total commission
```

QUICK CHECK 7

Timmy Heslin sells office copiers to businesses and is paid a variable commission rate. His commission rate on sales up to $20,000 is 2%; sales of $20,001 to $30,000, 2.5%; and sales of $30,001 and up, 3%. If he has sales of $38,400, find the commission earned.

OBJECTIVE 7 Find the gross earnings with a salary plus commission. With a **salary plus commission**, the salesperson is paid a fixed sum per pay period, plus a commission on all sales. This method of payment is commonly used by large retail stores.

> Gross earnings = Fixed amount per pay period + Amount earned on commission

This method is particularly attractive to new salespeople who lack experience. It offers the security of a guaranteed income to cover basic living costs while providing an incentive to sell more.

Adding Commission to a Salary

EXAMPLE 8 Jaime Bailey is paid $450 per week by Beverly's Creations, plus 3% on all sales over $500. Find his gross earnings for a week with sales of $8255.

SOLUTION

$$\begin{align} \text{Gross earnings} &= \text{Weekly salary} + 3\% \text{ on sales above } \$500 \\ &= \$450 \quad\quad + .03(\$8255 - \$500) \\ &= \$450 + (.03 \times \$7755) \\ &= \$450 + \$232.65 \\ &= \$682.65 \end{align}$$

QUICK CHECK 8

Sanford Howard is paid a salary of $500 a week, plus a 6% commission on all sales over $2500. Find his gross earnings for a week in which his total sales were $14,862.

Commission-based earning plans may be a strong deterrent to hiring new salespeople, so many companies offer a salary plus commission instead of a straight commission.

Subtracting a Draw to Find Earnings

EXAMPLE 9 Craig Johnson has office product sales of $36,850 for the month and is paid an 8% commission on all sales. Find his gross earnings after subtracting an $850 draw he took earlier in the month.

SOLUTION

$$\begin{align} \text{Gross earnings} &= \text{Commissions} - \text{Draw} \\ &= (.08 \times \$36,850) - \$850 \\ &= \$2948 - \$850 \\ &= \$2098 \end{align}$$

QUICK CHECK 9

Bob Smith has toy train sales of $18,540 for the month and is paid a 10% commission on all sales. If he had draws of $975 for the month, find his gross earnings after repaying the drawing account.

Quick TIP
It should not be just about the money. It is also important to find a career you will enjoy.

Wages vary significantly, depending on career. Look carefully at the following table, which lists the median weekly wage for different careers. Recall from the definition of **median** that half of the people make more than the median and the other half make less than the median. Those careers on the left usually require a college degree, whereas those on the right may not require a college degree but may require specialized training.

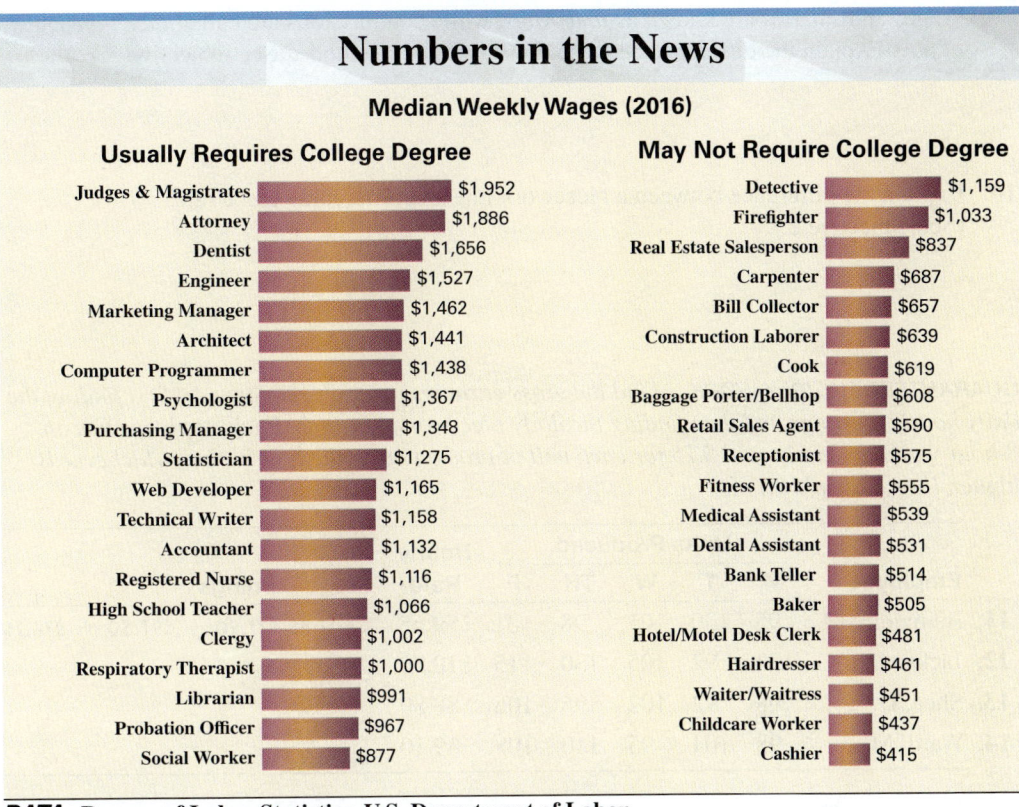

DATA: Bureau of Labor Statistics, U.S. Department of Labor.

6.2 Exercises // MyLab Math

The shaded sections below contain solutions to help you get a **QUICK START** *on the various types of exercises.*

RECYCLING Earth Plus pays workers $.48 per container for sorting recyclable materials. Find the daily gross earnings for each worker. (See Example 1.)

Employee	Number of Containers	Gross Earnings	Employee	Number of Containers	Gross Earnings
1. Crossett, J. 194 × $.48 = $93.12	194	$93.12	2. Biron, C.	292	_____
3. Campbell, K.	320	_____	4. Birch, J.	285	_____

AGRICULTURAL WORKERS Find the daily gross earnings for each employee. (See Example 2.) Suppose that avocado pickers are paid as follows.

1–500 avocados	$.10 each
501–700 avocados	$.14 each
Over 700 avocados	$.18 each

Employee	Number of Avocados	Gross Earnings	Employee	Number of Avocados	Gross Earnings
5. Hoch, R. (500 × $.10) + (200 × $.14) + (120 × $.18) = $99.60	820	$99.60	6. Leonard, M. K.	907	_____
7. Matysek, J.	852	_____	8. Panunzio, K.	1108	_____

C// indicates an exercise that is related to the Case in Point feature.

9. Wages and salaries are known as *time rates*, while commissions are called *incentive rates of pay*. Explain the difference between these payment methods. (See Objective 5.)

10. Explain the difference between a piecework rate and differential piece rate. (See Objective 2.)

GUARANTEED HOURLY WORK Find the gross earnings for each employee by first finding the daily hourly earnings and then finding the daily piecework earnings. Each employee has an 8-hour workday and is paid $.75 for each unit of production or the hourly rate, whichever is higher. (See Example 3.)

Employee	\multicolumn{5}{c	}{Units Produced}	Hourly Rate	Gross Earnings
	M	T	W	TH	F		
11. Benjamin, M.	95	130	105	98	101	$9.65	**$407.85** $77.20 + $97.50 + $78.75 + $77.20 + $77.20 = $407.85
12. Jackson, S.	123	92	105	130	115	$10.20	
13. Shea, D.	98	92	102	96	106	$8.80	
14. Ward, M.	98	104	95	110	108	$9.10	

PIECEWORK WITH OVERTIME Find the gross earnings for each employee. Overtime is 1.5 times the normal rate per piece. (See Example 4.)

Employee	Units Produced Reg.	O.T.	Rate per Unit	Gross Earnings
15. Wood, J.	430	82	$.76	**$420.28**
16. McDonald, M.	530	58	$.74	
17. Jennings, A.	470	70	$.82	
18. O'Brien, K.	504	52	$.85	

COMMISSION WITH RETURNS Find the gross earnings for each of the following salespeople. (See Examples 5 and 6.)

Employee	Total Sales	Returns and Allowances	Rate of Commission	Gross Earnings
19. Mares, E.	$6210	$129	10%	**$608.10**
($6210 − $129) × 10% = $6081 × .1 = $608.10				
20. Peterson, J.	$12,240	$415	7%	
21. Sanchez, L.	$26,480	$1250	7%	
22. Kling, J.	$3806	$108	20%	

VARIABLE-COMMISSION PAYMENT Find the gross earnings for each of the following employees. (See Example 7.) Livingston's Concrete pays its salespeople the following commissions.

6% on first $7500 in sales
8% on next $7500 in sales
10% on any sales over $15,000

Employee	Total Sales	Gross Earnings	Employee	Total Sales	Gross Earnings
23. Christensen, C.	$18,550	$1405	24. Hubbard, P.	$11,225	_____
($7500 × .06) + ($7500 × .08) + ($3550 × .10) = $1405					
25. Steed, F.	$10,480	_____	26. Goldstein, S.	$25,860	_____

SALARY PLUS COMMISSION *Fire Fighting Equipment pays salespeople as follows: $452 per week plus a commission of .9% on sales between $15,000 and $25,000, with 1.1% paid on sales in excess of $25,000. Find the gross earnings for each of the following salespeople. (No commission is paid on the first $15,000 of sales.) (See Example 8.)*

Employee	Total Sales	Gross Earnings	Employee	Total Sales	Gross Earnings
27. Wilson, F.	$52,620	$845.82	28. Barnes, A.	$36,300	_____
$452 + $10,000 × .009 + $27,620 × .011 = $845.82					
29. Feathers, C.	$32,874	_____	30. Maguire, J.	$28,400	_____

QUICK CHECK ANSWERS

1. $668.38 **2.** $857.90 **3.** $631.35 **4.** $509.88 **5.** $880.80 **6.** $3181.21 **7.** $902
8. $1241.72 **9.** $879

6.3 Social Security, Medicare, and Other Taxes

OBJECTIVES
1. Understand FICA.
2. Find FICA tax and Medicare tax.
3. Determine the FICA tax and the Medicare tax paid by a self-employed person.
4. Find the state disability insurance deduction.

CASE IN POINT Sarah Brynski needs to show a new employee how her pay is calculated. It begins with total hours and hourly rate to find total income. However, there are various deductions from total income, including FICA (Social Security) and Medicare.

Finding gross earnings is only the first step in preparing a payroll. The employer must then subtract all required deductions from gross earnings. Deductions usually include Social Security tax, Medicare tax, federal income tax withholding, and state income tax withholding. Other deductions may include state disability insurance, union dues, retirement, vacation pay, credit union savings or loan payments, purchase of bonds, uniform expenses, group insurance plans, and charitable contributions. Subtracting these deductions from gross earnings results in **net pay**, the amount the employee receives.

OBJECTIVE 1 Understand FICA. The **Great Depression** of the 1930s was very difficult for many elderly people. President Franklin D. Roosevelt and Congress responded by passing the **Federal Insurance Contributions Act (FICA)**, which created a plan called **Social Security**. The plan required all employees, employers, and self-employed individuals to make regular contributions from wages and self-employment income. In turn, it generated monthly benefits for retired workers, spouses of retired workers, and disabled workers. The table below shows that the maximum annual contribution for high-earning individuals has increased from $45 (1.5% of $3000) in 1950 to $7886.40 (6.2% of $127,200) in 2017.

Medicare was established in the 1960s to help those 65 and over, as well as younger people with disabilities, pay medical expenses. This program spreads the risk related to the costs of illness across a large group of people. It too requires contributions from employees, employers, and self-employed.

Looking for Work During the Great Depression

	Employee Contributions			
	Social Security Tax		Medicare Tax	
Year	Social Security Tax Rate	Employee Earnings Subject to the Tax	Medicare Tax Rate	Employee Earnings Subject to the Tax
1950	1.5%	$3,000	—	—
1970	4.2%	$7,800	1.05%	$7,800
1990	6.2%	$51,300	1.45%	$51,300
2010	6.2%	$106,800	1.45%	all wages
2017	6.2%	$127,200	1.45%	all wages*

*The Health Care Act of 2010 resulted in extra Medicare taxes for high-income individuals and couples than those shown in the table. We will not discuss these in this textbook.

As you can see by looking at any payroll statement, the contributions listed in the table above are paid by all employees. It is important to know that the employer must match the payments made by each employee. Individuals who pay into Social Security and Medicare receive retirement and healthcare benefits when older. You might think of these as savings accounts that will eventually benefit you.

Whether a citizen or not, every employee must have a Social Security card. Applications for a Social Security card are made through Social Security offices or at post offices. Each year the Social Security Administration sends out a statement to every worker showing past contributions and providing an estimate of future benefits. It is important to check the statement for errors and to correct them with the Social Security Administration. Remember, your contributions will define your benefits later in life.

The maximum earnings subject to Social Security as well as the tax rates for both Social Security and Medicare are determined by Congress. Since these numbers change almost every year, we will use the following to find Social Security tax and Medicare tax throughout the book.

6.3 Social Security, Medicare, and Other Taxes

Finding Social Security and Medicare Taxes for Employees

Social Security tax = 6.2% of earnings up to $130,000

Medicare tax = 1.45% of total earnings

Note: No Social Security tax is paid on earnings above $130,000 per year.

Over 65 million people receive regular Social Security checks, and more than 55 million receive Medicare benefits. Chances are someone in your family either receives or will soon receive these benefits.

OBJECTIVE 2 Find FICA tax and Medicare tax. Remember to look up the most current rates and maximum income before calculating the amount to be withheld for Social Security and Medicare.

Finding FICA Tax and Medicare Tax

EXAMPLE 1 Help Jessica Ramirez, the manager at a local Panera Bread, find the Social Security and Medicare tax that must be withheld from the weekly earnings of two employees.

(a) Kelleher: $362.40 gross earnings (b) Kimbrel: $468.02 gross earnings

SOLUTION

In each case, the tax is rounded to the nearest cent.

(a) Wage × Tax rate = Tax

$362.40 × .062 = $22.47 **Social Security tax**

$362.40 × .0145 = $5.25 **Medicare tax**

(b) Wage × Tax rate = Tax

$468.02 × .062 = $29.02 **Social Security tax**

$468.02 × .0145 = $6.79 **Medicare tax**

Quick TIP
Calculations are the same for part-time and full-time employees.

QUICK CHECK 1

Find **(a)** the Social Security tax and **(b)** the Medicare tax that must be withheld from gross earnings of $418.50.

Finding FICA Tax

EXAMPLE 2 Shannon Woolums has earned $127,634.05 so far this year. Her gross earnings for the current pay period are $5224.03. Find her Social Security tax.

SOLUTION

The Social Security tax applies only to the first $130,000 in earnings. Since the current pay period amount would increase Woolums's income to over $130,000, first subtract earnings from $130,000.

Maximum earnings subject to Social Security	$130,000.00
Prior earnings for the year	− 127,634.05
Amount subject to Social Security tax	$2,365.95

Social Security tax = 0.62 × $2365.95 = **$146.69**

No additional Social Security tax will be withheld for the remainder of the year.

QUICK CHECK 2

Mary Single has earned $128,750.10 so far this year. If her gross earnings are $3080 this week, find the Social Security tax to be withheld.

Numbers in the News

Social Insurance Contributions as a Percent of Total Gross Earnings

Country	Employer	Employee
China	up to 20%	8%
Germany	9.35%	9.35%
Italy	23.81%	9.19%
Japan	9.15%	9.15%
Sweden	18.79%	7.41%
U.S.	6.2%	6.2%

Source: Social Security Administration.

The table at the left shows that contributions to social programs vary greatly between countries. In general, higher contributions result in higher benefits later in life.

OBJECTIVE 3 Determine the FICA tax and the Medicare tax paid by a self-employed person. Self-employed individuals have no employer to match the employee contribution. So they are required pay a rate that is double that of an employee:

12.4% of adjusted earnings for Social Security tax and

2.9% of adjusted earnings for Medicare tax.

Thus, self-employed individuals essentially pay both the employee and the employer taxes.

Finding FICA and Medicare Tax for the Self-Employed

EXAMPLE 3 Find the Social Security tax and the Medicare tax paid by Ta Shon Williams, a self-employed web designer who had adjusted earnings of $53,820 this year.

SOLUTION

Social Security tax = $53,820 × **12.4%** = $53,820 × .124 = $6673.68
Medicare tax = $53,820 × **2.9%** = $53,820 × .029 = + $1560.78
Total $8234.46

QUICK CHECK 3

Find **(a)** the Social Security tax and **(b)** the Medicare tax for a self-employed accountant who had adjusted earnings of $73,875 this year.

OBJECTIVE 4 Find the state disability insurance deduction. States that have a state disability insurance (SDI) program require employers to withhold a portion of their earnings for disability insurance. This program pays the employee if she is injured and unable to work. The tax rates vary widely, so in our example we will use an **SDI deduction** of 1% of gross earnings on the first $31,800 earned in a year. There are no SDI deductions on earnings above this amount.

Finding the State Disability Insurance Deduction

EXAMPLE 4 Find the state disability insurance deduction for an employee at Comet Auto Parts with gross earnings of $418 this pay period. The SDI rate is 1%, and the employee has not earned $31,800 this year.

SOLUTION

The state disability insurance deduction is $4.18 ($418 × **.01**).

QUICK CHECK 4

An employee has gross earnings of $525.80 this week. The employee has not earned $31,800 this year and the SDI rate is 1%. What is the state disability insurance deduction?

Knowing SDI Maximum Deductions

EXAMPLE 5 Jenoa Perkins has earned $29,960 so far this year. Find the SDI deduction if gross earnings this pay period are $2872. Use an SDI rate of 1% on the first $31,800.

SOLUTION

The SDI deduction will be taken on $1840 of the current gross earnings.

$31,800   maximum earnings subject to SDI
−$29,960   earnings this year
$1,840   earnings subject to SDI

The SDI deduction is $18.40 ($1840 × .01).

QUICK CHECK 5

Linda Shirley has earned $30,780 so far this year. Her earnings this pay period are $3289. Find the SDI deduction using an SDI rate of 1% on the first $31,800.

6.3 Exercises // MyLab Math

The shaded sections below contain solutions to help you get a QUICK START on the various types of exercises.

Find the Social Security tax and the Medicare tax for each of the following amounts of gross earnings. Assume a 6.2% FICA rate and a 1.45% Medicare tax rate. (See Example 1.)

1. $420 $26.04 $6.09
2. $942.50 $58.44 $13.67
3. $463.24 _____ _____
4. $606.35 _____ _____
5. $854.71 _____ _____
6. $683.65 _____ _____

SOCIAL SECURITY TAX *Find the Social Security tax for each employee for the current pay period. Assume a 6.2% FICA rate up to a maximum of $130,000. (See Example 2.)*

Employee	Gross Earnings This Year (So Far)	Earnings Current Pay Period	Social Security Tax
7. Barnby, Z.	$126,945.32	$6218.48	$189.39
8. Hale, R.	$127,438.75	$5200.00	$158.80
9. Hall, T.	$125,016.22	$7260.00	_____
10. Maurin, J.	$128,971.95	$4487.52	_____
11. Saraniti, S.	$129,329.75	$3053.73	_____
12. De Bouchel, V.	$128,974.08	$6160.86	_____

PAYROLL DEDUCTIONS *Find the regular earnings, overtime earnings, gross earnings, Social Security tax (6.2%), Medicare tax (1.45%), and state disability insurance deduction (1%) for each employee. Assume that no employee will have earned more than the FICA or SDI maximum at the end of the current pay period. Assume that time and a half is paid for any overtime in a 40-hour week. (See Examples 1–4.)*

Employee	Hours Worked	Regular Rate	Regular Earnings	Overtime Earnings	Gross Earnings	Social Security Tax	Medicare Tax	SDI Deduction
13. Garrett, R.	45.5	$9.22	$368.80	$76.07	$444.87	$27.58	$6.45	$4.45
14. Harcos, W.	47.75	$8.50	___	___	___	___	___	___
15. Plescia, P.	45	$14.20	___	___	___	___	___	___
16. Eckern, G.	45	$10.20	___	___	___	___	___	___
17. McIntosh, R.	47	$11.68	___	___	___	___	___	___
18. Cox, Z.	48	$11.50	___	___	___	___	___	___

Solve the following application problems. Round to the nearest cent.

19. **SOCIAL SECURITY AND MEDICARE** Maria Ortega worked 43.5 hours last week at Starbucks. She is paid $10.20 per hour, plus time and a half for all hours over 40 per week. Find her **(a)** Social Security tax and **(b)** Medicare tax for the week.
    Gross income is ($40 × $10.20) + (3.5 × $1\frac{1}{2}$) × $10.20 = $461.55
    (a) Social Security tax = $461.55 × .062 = $28.62
    (b) Medicare tax = $461.55 × .0145 = $6.69

    (a) $28.62
    (b) $6.69

indicates an exercise that is related to the Case in Point feature.

CHAPTER 6 Payroll

20. SOCIAL SECURITY AND MEDICARE Chriscelle Merquillo receives 8% commission on all sales. Her sales on Monday of last week were $1412.20, with $1928.42 on Tuesday, $598.14 on Wednesday, $1051.12 on Thursday, and $3640.12 on Friday. Find her **(a)** Social Security tax and **(b)** Medicare tax for the week.

(a) _____
(b) _____

21. PAYROLL DEDUCTIONS Donna Laughman is paid an 8% commission on sales. This week she had sales of $19,482 and returns and allowances of $193. Find the amount of **(a)** her Social Security tax, **(b)** her Medicare tax, and **(c)** her state disability insurance deduction for this pay period. (Assume the SDI rate is 1% and that earnings will not exceed $31,800.)

(a) _____
(b) _____
(c) _____

22. PAYROLL DEDUCTIONS Martin Rodriguez works in sales for FedEx and is paid $675 per week plus a commission of 2% on sales. His sales last week were $17,240. Find the amount of **(a)** his Social Security tax, **(b)** his Medicare tax, and **(c)** his state disability insurance deduction for the pay period. (Assume the SDI rate is 1% and that earnings will not exceed $31,800.)

(a) _____
(b) _____
(c) _____

SELF-EMPLOYMENT DEDUCTIONS *The following problems refer to self-employed individuals. These people pay a Social Security tax of 12.4% and Medicare tax of 2.9%. Find both of the taxes on the following annual adjusted earnings. (See Example 3.)*

23. Tony Romano, owner of The Cutlery, earned $58,238.74.

23. _____

24. Rachel Leach, an interior designer, earned $36,724.72.

24. _____

25. Krystal McClellan, cosmetics consultant, earned $29,104.80.

25. _____

26. Ron Morris, a Chic-Filet franchise owner, earned $78,007.14.

26. _____

27. Ben Walker, clothing sales, earned $26,843.60.

27. _____

28. Sadie Chambers, senior account executive, earned $92,748.32.

28. _____

29. Calculate your FICA and Medicare deductions from a recent paycheck and make sure the correct amounts were withheld. (See Objectives 1–2.)

30. Describe the difference between the FICA paid by an employee and FICA paid by a self-employed person. (See Objective 3.)

QUICK CHECK ANSWERS

1. (a) $25.95 (b) $6.07
2. $77.49
3. (a) $9160.50 (b) $2142.38
4. $5.26
5. $10.20

6.4 Income Tax Withholding

OBJECTIVES

1. Understand the Employee's Withholding Allowance Certificate.
2. Find the federal withholding tax using the wage bracket method.
3. Find the federal withholding tax using the percentage method.
4. Find the state withholding tax using the state income tax rate.
5. Find net pay when given gross wages, taxes, and other deductions.
6. Find the quarterly amount owed to the Internal Revenue Service.
7. Understand additional employer responsibilities and employee benefits.

CASE IN POINT After completing the payroll at Starbucks, all FICA taxes, Medicare taxes, and federal withholding taxes withheld from employees are sent to the Internal Revenue Service. Managers must stay current on all changes that affect withholding.

In this section, we will talk about withholding for the **personal income tax** that must be paid by individuals. The **income tax withheld** is periodically sent to the **Internal Revenue Service (IRS)**.

OBJECTIVE 1 Understand the Employee's Withholding Allowance Certificate. The income taxes withheld from a paycheck are based on the marital status and the number of **withholding allowances** claimed on the W-4 form filed with the employer. **Marital status** is based on whether an individual is married on the last day of the year. A married person with three children normally claims five allowances, one for each person in the family. However, people who want a refund when filing taxes can choose to have more withheld each pay period by decreasing the number of withholding allowances. The allowances claimed on the W-4 form are used to estimate the amount to withhold. The exact number of allowances must be used by the employee on his tax return at the end of the year so that the correct income tax can be calculated.

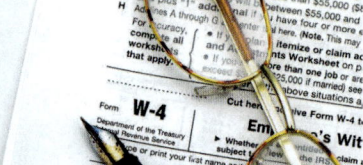

Source: Department of the Treasury Internal Revenue Service.

Employers may use either of two methods to determine the amount of federal withholding tax to deduct from paychecks: the **wage bracket method** and the **percentage method**. The process at this step simply results in an estimate of the amount of income tax that must be paid by the employee. The exact amount of income tax that must be paid for the year is not known until the income tax return is completed.

OBJECTIVE 2 Find the federal withholding tax using the wage bracket method. The Internal Revenue Service supplies many different withholding tax tables to be used with the wage bracket method. We show only a portion of a few of the tables here.

WAGE BRACKET METHOD

SINGLE Persons—WEEKLY Payroll Period
(For Wages Paid through December 31, 2016)

And the wages are —		And the number of withholding allowances claimed is —										
At least	But less than	0	1	2	3	4	5	6	7	8	9	10
		The amount of income tax to be withheld is —										
350	360	38	26	16	8	0	0	0	0	0	0	0
360	370	39	28	17	9	1	0	0	0	0	0	0
370	380	41	29	18	10	2	0	0	0	0	0	0
380	390	42	31	19	11	3	0	0	0	0	0	0
390	400	44	32	20	12	4	0	0	0	0	0	0
400	410	45	34	22	13	5	0	0	0	0	0	0
410	420	47	35	23	14	6	0	0	0	0	0	0
420	430	48	37	25	15	7	0	0	0	0	0	0
430	440	50	38	26	16	8	0	0	0	0	0	0
440	450	51	40	28	17	9	1	0	0	0	0	0
450	460	53	41	29	18	10	2	0	0	0	0	0
460	470	54	43	31	19	11	3	0	0	0	0	0
470	480	56	44	32	21	12	4	0	0	0	0	0
480	490	57	46	34	22	13	5	0	0	0	0	0
490	500	59	47	35	24	14	6	0	0	0	0	0

Source: Department of the Treasury Internal Revenue Service.

MARRIED Persons—WEEKLY Payroll Period
(For Wages Paid through December 31, 2016)

And the wages are —		And the number of withholding allowances claimed is —										
At least	But less than	0	1	2	3	4	5	6	7	8	9	10
		The amount of income tax to be withheld is —										
500	510	34	26	18	11	3	0	0	0	0	0	0
510	520	35	27	19	12	4	0	0	0	0	0	0
520	530	36	28	20	13	5	0	0	0	0	0	0
530	540	38	29	21	14	6	0	0	0	0	0	0
540	550	39	30	22	15	7	0	0	0	0	0	0
550	560	41	31	23	16	8	0	0	0	0	0	0
560	570	42	32	24	17	9	1	0	0	0	0	0
570	580	44	33	25	18	10	2	0	0	0	0	0
580	590	45	34	26	19	11	3	0	0	0	0	0
590	600	47	35	27	20	12	4	0	0	0	0	0
600	610	48	37	28	21	13	5	0	0	0	0	0
610	620	50	38	29	22	14	6	0	0	0	0	0
620	630	51	40	30	23	15	7	0	0	0	0	0
630	640	53	41	31	24	16	8	0	0	0	0	0
640	650	54	43	32	25	17	9	1	0	0	0	0

Source: Department of the Treasury Internal Revenue Service.

SINGLE Persons—MONTHLY Payroll Period
(For Wages Paid through December 31, 2016)

And the wages are—		And the number of withholding allowances claimed is—										
At least	But less than	0	1	2	3	4	5	6	7	8	9	10
		The amount of income tax to be withheld is—										
1,600	1,640	176	126	76	42	8	0	0	0	0	0	0
1,640	1,680	182	132	81	46	12	0	0	0	0	0	0
1,680	1,720	188	138	87	50	16	0	0	0	0	0	0
1,720	1,760	194	144	93	54	20	0	0	0	0	0	0
1,760	1,800	200	150	99	58	24	0	0	0	0	0	0
1,800	1,840	206	156	105	62	28	0	0	0	0	0	0
1,840	1,880	212	162	111	66	32	0	0	0	0	0	0
1,880	1,920	218	168	117	70	36	3	0	0	0	0	0
1,920	1,960	224	174	123	74	40	7	0	0	0	0	0
1,960	2,000	230	180	129	78	44	11	0	0	0	0	0
2,000	2,040	236	186	135	84	48	15	0	0	0	0	0
2,040	2,080	242	192	141	90	52	19	0	0	0	0	0
2,080	2,120	248	198	147	96	56	23	0	0	0	0	0
2,120	2,160	254	204	153	102	60	27	0	0	0	0	0
2,160	2,200	260	210	159	108	64	31	0	0	0	0	0

Source: Department of the Treasury Internal Revenue Service.

MARRIED Persons—MONTHLY Payroll Period
(For Wages Paid through December 31, 2016)

And the wages are—		And the number of withholding allowances claimed is—										
At least	But less than	0	1	2	3	4	5	6	7	8	9	10
		The amount of income tax to be withheld is—										
2,200	2,240	151	117	83	50	16	0	0	0	0	0	0
2,240	2,280	155	121	87	54	20	0	0	0	0	0	0
2,280	2,320	161	125	91	58	24	0	0	0	0	0	0
2,320	2,360	167	129	95	62	28	0	0	0	0	0	0
2,360	2,400	173	133	99	66	32	0	0	0	0	0	0
2,400	2,440	179	137	103	70	36	2	0	0	0	0	0
2,440	2,480	185	141	107	74	40	6	0	0	0	0	0
2,480	2,520	191	145	111	78	44	10	0	0	0	0	0
2,520	2,560	197	149	115	82	48	14	0	0	0	0	0
2,560	2,600	203	153	119	86	52	18	0	0	0	0	0
2,600	2,640	209	158	123	90	56	22	0	0	0	0	0
2,640	2,680	215	164	127	94	60	26	0	0	0	0	0
2,680	2,720	221	170	131	98	64	30	0	0	0	0	0
2,720	2,760	227	176	135	102	68	34	0	0	0	0	0
2,760	2,800	233	182	139	106	72	38	4	0	0	0	0

Source: Department of the Treasury Internal Revenue Service.

Finding Federal Withholding Using the Wage Bracket Method

EXAMPLE 1 Benito Flores works for Apple. He is single and claims no withholding allowances since he prefers to get money back from the government when filing his income tax return. This choice will increase the amount withheld from his paycheck every week. Use the wage bracket method to find his withholding tax for a 25-hour workweek in which his gross earnings were $368.

Quick TIP
It is important to use the correct table when finding the amount to withhold.

SOLUTION
Use the table for single person, weekly payroll. The earnings of $368 are found in the row of "at least 360 but less than 370." Look in that row and the column for 0 withholding allowances to find $39, which is the amount that should be withheld.

QUICK CHECK 1
Rachel Leach is single and claims two withholding allowances. Use the wage bracket method to find her withholding tax if her weekly gross earnings are $358.

Using the Wage Bracket Method for Federal Withholding

EXAMPLE 2 Pat Rowell is married, claims three withholding allowances, and has monthly gross earnings of $2780.50. Find her withholding tax using the wage bracket method.

SOLUTION
Use the table for married persons—monthly payroll period. Look down the two left columns, and find the range that includes Rowell's gross earnings: "at least 2,760 but less than 2,800." Read across the table to the column headed "3" (for the three withholding allowances). The withholding tax is $106. Had Rowell claimed five withholding allowances, her withholding tax would have been only $38.

> **QUICK CHECK 2**
>
> Bob Martinez has monthly earnings of $2700. He is married and claims four withholding allowances. Find his withholding tax using the wage bracket method.

OBJECTIVE 3 Find the federal withholding tax using the percentage method. Many companies today prefer to use the *percentage method* to determine federal withholding tax. The percentage method does not require the several pages of tables needed with the wage bracket method and is more easily adapted to computer applications in the processing of payrolls.

Percentage Method: Amount for One Withholding Allowance

Payroll Period	One Withholding Allowance
Weekly	$77.90
Biweekly	$155.80
Semimonthly	$168.80
Monthly	$337.50

Percentage Method Tables for Income Tax Withholding (For Wages Paid in 2016)

TABLE 1—WEEKLY Payroll Period

(a) SINGLE person (including head of household)—

If the amount of wages (after subtracting withholding allowances) is:

Not over $43 . . . . $0

Over—	But not over—		of excess over—
$43	—$222	. . . $0.00 plus 10%	—$43
$222	—$767	. . . $17.90 plus 15%	—$222
$767	—$1,796	. . . $99.65 plus 25%	—$767
$1,796	—$3,700	. . . $356.90 plus 28%	—$1,796
$3,700	—$7,992	. . . $890.02 plus 33%	—$3,700
$7,992	—$8,025	. . . $2,306.38 plus 35%	—$7,992
$8,025	. . . .	. . . $2,317.93 plus 39.6%	—$8,025

(b) MARRIED person—

If the amount of wages (after subtracting withholding allowances) is:

Not over $164 . . . . $0

Over—	But not over—		of excess over—
$164	—$521	. . . $0.00 plus 10%	—$164
$521	—$1,613	. . . $35.70 plus 15%	—$521
$1,613	—$3,086	. . . $199.50 plus 25%	—$1,613
$3,086	—$4,615	. . . $567.75 plus 28%	—$3,086
$4,615	—$8,113	. . . $995.87 plus 33%	—$4,615
$8,113	—$9,144	. . . $2,150.21 plus 35%	—$8,113
$9,144	. . . .	. . . $2,511.06 plus 39.6%	—$9,144

TABLE 2—BIWEEKLY Payroll Period

(a) SINGLE person (including head of household)—

If the amount of wages (after subtracting withholding allowances) is:

Not over $87 . . . . $0

Over—	But not over—		of excess over—
$87	—$443	. . . $0.00 plus 10%	—$87
$443	—$1,535	. . . $35.60 plus 15%	—$443
$1,535	—$3,592	. . . $199.40 plus 25%	—$1,535
$3,592	—$7,400	. . . $713.65 plus 28%	—$3,592
$7,400	—$15,985	. . . $1,779.89 plus 33%	—$7,400
$15,985	—$16,050	. . . $4,612.94 plus 35%	—$15,985
$16,050	. . . .	. . . $4,635.69 plus 39.6%	—$16,050

(b) MARRIED person—

If the amount of wages (after subtracting withholding allowances) is:

Not over $329 . . . . $0

Over—	But not over—		of excess over—
$329	—$1,042	. . . $0.00 plus 10%	—$329
$1,042	—$3,225	. . . $71.30 plus 15%	—$1,042
$3,225	—$6,171	. . . $398.75 plus 25%	—$3,225
$6,171	—$9,231	. . . $1,135.25 plus 28%	—$6,171
$9,231	—$16,227	. . . $1,992.05 plus 33%	—$9,231
$16,227	—$18,288	. . . $4,300.73 plus 35%	—$16,227
$18,288	. . . .	. . . $5,022.08 plus 39.6%	—$18,288

TABLE 3—SEMIMONTHLY Payroll Period

(a) SINGLE person (including head of household)—

If the amount of wages (after subtracting withholding allowances) is:

Not over $94 . . . . $0

Over—	But not over—		of excess over—
$94	—$480	. . . $0.00 plus 10%	—$94
$480	—$1,663	. . . $38.60 plus 15%	—$480
$1,663	—$3,892	. . . $216.05 plus 25%	—$1,663
$3,892	—$8,017	. . . $773.30 plus 28%	—$3,892
$8,017	—$17,317	. . . $1,928.30 plus 33%	—$8,017
$17,317	—$17,388	. . . $4,997.30 plus 35%	—$17,317
$17,388	. . . .	. . . $5,022.15 plus 39.6%	—$17,388

(b) MARRIED person—

If the amount of wages (after subtracting withholding allowances) is:

Not over $356 . . . . $0

Over—	But not over—		of excess over—
$356	—$1,129	. . . $0.00 plus 10%	—$356
$1,129	—$3,494	. . . $77.30 plus 15%	—$1,129
$3,494	—$6,685	. . . $432.05 plus 25%	—$3,494
$6,685	—$10,000	. . . $1,229.80 plus 28%	—$6,685
$10,000	—$17,579	. . . $2,158.00 plus 33%	—$10,000
$17,579	—$19,813	. . . $4,659.07 plus 35%	—$17,579
$19,813	. . . .	. . . $5,440.97 plus 39.6%	—$19,813

TABLE 4—MONTHLY Payroll Period

(a) SINGLE person (including head of household)—

If the amount of wages (after subtracting withholding allowances) is:

Not over $188 . . . . $0

Over—	But not over—		of excess over—
$188	—$960	. . . $0.00 plus 10%	—$188
$960	—$3,325	. . . $77.20 plus 15%	—$960
$3,325	—$7,783	. . . $431.95 plus 25%	—$3,325
$7,783	—$16,033	. . . $1,546.45 plus 28%	—$7,783
$16,033	—$34,633	. . . $3,856.45 plus 33%	—$16,033
$34,633	—$34,775	. . . $9,994.45 plus 35%	—$34,633
$34,775	. . . .	. . . $10,044.15 plus 39.6%	—$34,775

(b) MARRIED person—

If the amount of wages (after subtracting withholding allowances) is:

Not over $713 . . . . $0

Over—	But not over—		of excess over—
$713	—$2,258	. . . $0.00 plus 10%	—$713
$2,258	—$6,988	. . . $154.50 plus 15%	—$2,258
$6,988	—$13,371	. . . $864.00 plus 25%	—$6,988
$13,371	—$20,000	. . . $2,459.75 plus 28%	—$13,371
$20,000	—$35,158	. . . $4,315.87 plus 33%	—$20,000
$35,158	—$39,625	. . . $9,318.01 plus 35%	—$35,158
$39,625	. . . .	. . . $10,881.46 plus 39.6%	—$39,625

> **Finding the amount of withholding using the percentage method**
>
> **Step 1** Multiply the number of withholding allowances by the amount for one withholding allowance from the table titled "Percentage Method: Amount for One Withholding Allowance."
>
> **Step 2** Subtract the amount found in step 1 from gross earnings.
>
> **Step 3** Find the appropriate table for Percentage Method of Withholding and then find the correct row.
>
> **Step 4** Follow the directions in the table to find the tax to be withheld.

Finding Federal Withholding Using the Percentage Method

EXAMPLE 3 Sarah Brynski at Starbucks is married, claims two withholding allowances, and had weekly gross earnings of $690 one week. Use the percentage method to find the withholding tax.

SOLUTION

CASE IN POINT

Step 1 From the table titled "Percentage Method: Amount for One Withholding Allowance," the amount for one withholding allowance, weekly payroll is $77.90.

Number of allowances × Amount for one allowance = 2 × $77.90 = **$155.80**

Step 2 Gross earnings − Amount from step 1 = $690 − **$155.80** = **$534.20**

Step 3 Use the Percentage Method Tables for weekly payroll period, married person.

Step 4 Go to the row of **over $521 but not over $1,613.**

$$\begin{aligned}\text{Withholding} &= \$35.70 + \textbf{15\% of excess over \$521} \\ &= \$35.70 + .15 \cdot (\$534.20 - \$521) \\ &= \$35.70 + \$1.98 \\ &= \$37.68\end{aligned}$$

The calculator solution to this example is

690 $-$ 2 $\times$ $77.90 $=$ $534.20 STO

$35.70 $+$ .15 $\times$ (RCL $-$ $521) $=$ $37.68

QUICK CHECK 3

Sadie Simms is married and claims three withholding allowances. Use the percentage method to find her withholding tax in a week when she has earnings of $1263.

Finding Federal Withholding Using the Percentage Method

EXAMPLE 4 Dac Kien is married, claims three withholding allowances, and receives $7850 a month working as a chemical engineer for Dupont. Use the percentage method to find his withholding tax.

SOLUTION

Step 1 The amount of one withholding allowance monthly is $337.50.

Number of allowances × Amount for one allowance = 3 × $337.50 = **$1012.50**

Step 2 Gross earnings − Amount from step 1 = $7850 − **$1012.50** = $6837.50

Step 3 Use the table on the prior page for monthly payroll and married person.

Step 4 Go to the row of over $2,258 but not over $6,988.

$$\begin{aligned}\text{Withholding} &= \$154.50 + \textbf{15\% of excess over \$2,258} \\ &= \$154.50 + .15 \times (\$6837.50 - \$2,258) \\ &= \$841.43\end{aligned}$$

QUICK CHECK 4

Howard Martin has earnings of $5735 for the month. He is married and claims four withholding allowances. Use the percentage method to find his withholding tax.

Finding Federal Withholding Using the Percentage Method

EXAMPLE 5 Ben Fisher is a single architect who claims two exemptions. He has semimonthly (twice a month) gross earnings of $4430. Use the percentage method to find the amount of withholding for each paycheck.

SOLUTION

One withholding allowance for semimonthly is $168.80.

Number of allowances × Amount for one allowance = 2 × $168.80 = **$337.60**

Gross earnings − Amount from above = $4430 − **$337.60** = $4092.40

Use the table for semimonthly payroll and single. Go to the row of over $3,892 but not over $8,017.

$$\begin{aligned} \text{Withholding} &= \$773.30 + 28\% \text{ of excess over } \$3{,}892 \\ &= \$773.30 + .28 \times (\$4092.40 - \$3{,}892) \\ &= \$829.41 \end{aligned}$$

QUICK CHECK 5

Carol Dixon has biweekly earnings of $4350, is single, and claims two withholding allowances. Use the percentage methods to find her withholding tax.

The amount of withholding tax found using the wage bracket method will vary slightly from the amount of withholding tax found using the percentage method. Remember, both of these methods give an estimate of how much tax should be withheld. The exact tax is calculated at the end of the year on the federal income tax return.

OBJECTIVE 4 Find the state withholding tax using the state income tax rate. Many states and cities also have an income tax collected by withholding. **State income taxes** vary, with no state income tax in Alaska, Florida, Nevada, South Dakota, Texas, Washington, and Wyoming. Employees and the self-employed must pay any state income taxes in addition to all federal taxes, including FICA and Medicare.

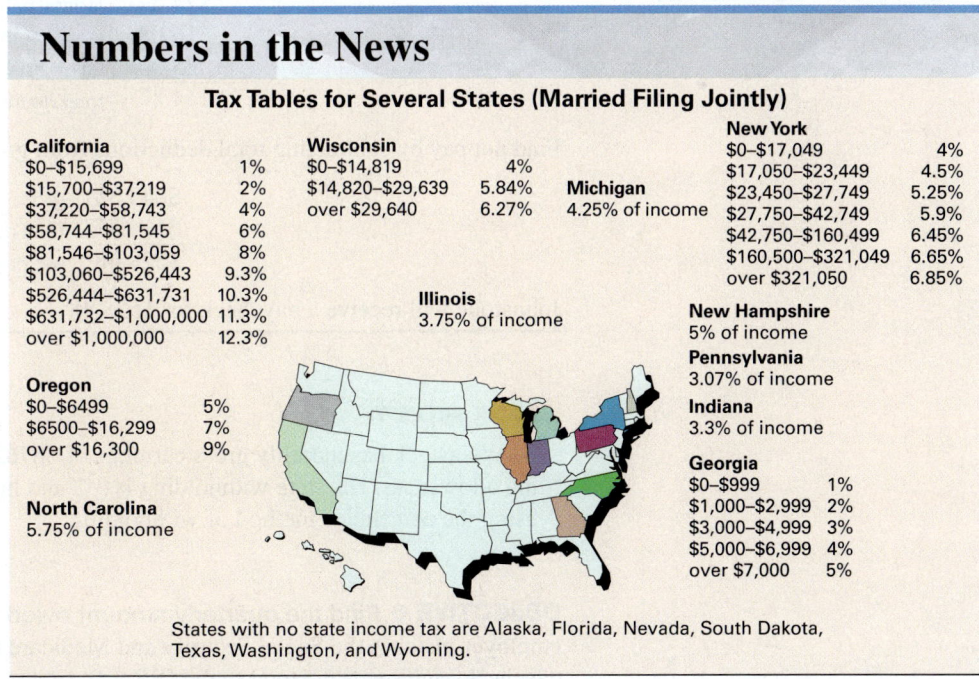

Numbers in the News

Tax Tables for Several States (Married Filing Jointly)

States with no state income tax are Alaska, Florida, Nevada, South Dakota, Texas, Washington, and Wyoming.

DATA: State tax websites, Tax-Brackets.org.

Finding the State Withholding Tax

EXAMPLE 6 Hilda Worthington works as a nurse in Michigan and earns $4250 for the month. Find the tax rate from the chart and calculate the state withholding tax.

SOLUTION

Michigan has a flat tax rate of 4.25%.

$$\text{Tax} = \$4250 \times .0425 = \$180.63$$

234 CHAPTER 6 Payroll

> **QUICK CHECK 6**
>
> An optometrist assistant lives in Illinois and earns $3225 one month. Find the tax rate from the table and calculate the state withholding tax.

OBJECTIVE 5 Find net pay when given gross wages, taxes, and other deductions. Net pay is found after subtracting all deductions, including all taxes and any union dues, payments, insurance premiums, retirement contributions, etc.

Finding Net Pay

Gross earnings
− FICA tax (Social Security)
− Medicare tax
− Federal withholding tax
− State withholding tax
− Other deductions
Net pay

Determining Net Pay after Deductions

EXAMPLE 7 Kizzy Johnstone is married and claims three withholding allowances. Her weekly gross earnings are $643.35. Her state withholding is 2.5% of her gross income and her union dues are $25. Find her net pay using the percentage method of withholding.

SOLUTION

First find FICA (Social Security) tax, which is $39.89; then Medicare, which is $9.33. Federal withholding tax is $24.57 and state withholding is $16.08. Total deductions are

$39.89	FICA tax (6.2%)
9.33	Medicare tax (1.45%)
$24.57	federal withholding
16.08	state withholding (2.5%)
+ 25.00	union dues
$114.87	total deductions

Find net pay by subtracting total deductions from gross earnings.

$643.35	gross earnings
−$114.87	total deductions
$528.48	net pay

Johnstone will receive a paycheck for $528.48.

> **QUICK CHECK 7**
>
> Al Weinstock has monthly gross earnings of $8762, is married, and claims three withholding allowances. His state withholding is 3% and his union dues are $58. Find his *net pay* using the percentage method of withholding.

OBJECTIVE 6 Find the quarterly amount owed to the Internal Revenue Service. The employer matches the Social Security and Medicare amounts withheld from all employees' paychecks and sends the total to the IRS.

Finding the Amount of FICA and Medicare Tax Due

EXAMPLE 8 If the employees at Fair Oaks Automotive Repair pay a total of $789.11 in Social Security tax and $184.55 in Medicare tax, how much must the employer send to the Internal Revenue Service?

SOLUTION

Withheld from employees: $789.11 + $184.55 = $973.66
Employer match: + 973.66
Amount employer must send to IRS: $1947.32

QUICK CHECK 8

Payless Dry Cleaners withheld $461.90 in Social Security tax and $108.03 in Medicare tax from employees. How much must the employer send to the Internal Revenue Service?

In addition to withholding the employee's Social Security tax and a matching amount paid by the employer, the employer must also send the amount withheld from employees' paychecks for income tax to the Internal Revenue Service.

Finding the Employer's Amount Due the IRS

EXAMPLE 9 The manager at a movie theater collected $2765.42 from its employees for FICA tax, $646.75 for Medicare tax, and $3572.86 in federal withholding tax. Compute the total amount that must be sent to the government.

SOLUTION

$2765.42	collected from *employees* for FICA tax
2765.42	equal amount paid by *employer* for FICA tax
646.75	collected from *employees* for Medicare tax
646.75	equal amount paid by *employer* for Medicare tax
3572.86	federal withholding tax
$10,397.20	total due to government

The movie theater must send $10,397.20 to the Internal Revenue Service.

QUICK CHECK 9

In one quarter, SmartPhones in the Mall collected $1953.82 in Social Security tax, $456.94 in Medicare tax, and $3780.47 in federal withholding tax from its employees. Find the total amount due to the government.

OBJECTIVE 7 Understand additional employer responsibilities and employee benefits. Each quarter, employers must file **Form 941**, the **Employer's Quarterly Federal Tax Return**. The form itemizes total employee wages and earnings, the income taxes withheld from employees, the FICA taxes withheld from employees, and the FICA taxes paid by the employer.

The **Federal Unemployment Tax Act (FUTA)** requires employers to pay an additional tax. This **unemployment insurance tax**, paid entirely by employers, is used to pay unemployment benefits to an individual who has become unemployed and is unable to find work. In general, all employers who paid wages of $1500 or more in a calendar quarter must file an employer's annual Federal Unemployment Tax (FUTA) return.

Although rates vary, the employer must pay 6% of the first $7000 in earnings for that year for each employee. As soon as the employee reaches earnings of $7000, no additional unemployment tax must be paid.

Fringe benefits are extra benefits such as health care, dental care, retirement plan contributions, day care, life insurance, tuition reimbursement, sick leave, vacation, etc. Not every employer offers all of these.

6.4 Exercises   MyLab Math

The shaded sections below contain solutions to help you get a **QUICK START** *on the various types of exercises.*

FEDERAL WITHHOLDING TAX *Find the federal withholding tax for the following employees. Use the wage bracket method. (See Examples 1 and 2.)*

	Employee	Gross Earnings	Married?	Withholding Allowances	Federal Withholding Tax
1.	Safron, I.	$2784.30 monthly	yes	3	$106
2.	Stanger, J.	$425.76 weekly	no	2	$25
3.	Costa, P.	$507.52 weekly	yes	0	_____

C// indicates an exercise that is related to the Case in Point feature.

Employee	Gross Earnings	Married?	Withholding Allowances	Federal Withholding Tax
4. Elle, M.	$2416.88 monthly	yes	1	_____
5. Alexander, L.	$1868.29 monthly	no	2	_____
6. Schaller, M.	$1953.35 monthly	no	1	_____
7. Moss, L.	$2641.20 monthly	yes	5	_____
8. Nguyen, B.	$458.20 weekly	no	1	_____
9. Mehta, S.	$1622.41 monthly	no	3	_____
10. Sommerfield, K.	$645.30 weekly	yes	0	_____
11. Weisner, W.	$2134.86 monthly	no	5	_____
12. O'Sullivan, S.	$585.87 weekly	yes	2	_____

STATE WITHHOLDING TAX Use the state income tax rate given to find the state withholding tax for the following employees. (See Example 6.)

Employee	Gross Weekly Earnings	State Income Tax Rate	State Withholding
13. Romano, J.	$690.12	2.8%	**$19.32**
14. Burner, P.	$368.53	3.5%	**$12.90**
15. Davis, L.	$466.71	6%	_____
16. Lundborg, J.	$541.45	5%	_____
17. Fox, D.	$1607.23	4.3%	_____
18. Ticarro, C.	$2802.58	4.95%	_____

EMPLOYEE NET PAY Use the percentage method of withholding to find federal withholding tax, a 6.2% FICA rate to find FICA tax, and 1.45% to find Medicare tax for the following employees. Then find the net pay for each employee. The number of withholding allowances and the marital status are listed after each employee's name. Assume that no employee has earned over $130,000 so far this year. (See Examples 3 and 5.)

Employee	Gross Earnings	FICA	Medicare Tax	Federal Withholding Tax	Net Pay
19. Hamel; 4, M	$576.28 weekly	**$35.73**	**$8.36**	**$10.07**	**$522.12**
20. Guardino; 3, S	$2878.12 monthly	**$178.44**	**$41.73**	**$213.04**	**$2444.91**
21. Zhang; 1, S	$4503.16 monthly	_____	_____	_____	_____
22. Erb; 2, M	$625 weekly	_____	_____	_____	_____
23. Terry; 3, M	$2276.83 semimonthly	_____	_____	_____	_____
24. Galluccio; 1, S	$420.17 weekly	_____	_____	_____	_____
25. Derma; 6, M	$2971.06 semimonthly	_____	_____	_____	_____
26. Eddy; 2, M	$1020 weekly	_____	_____	_____	_____
27. Wann; 3, S	$3753.18 biweekly	_____	_____	_____	_____
28. Zamost; 2, S	$6625.24 monthly	_____	_____	_____	_____
29. Balik; 1, S	$1786.44 biweekly	_____	_____	_____	_____
30. Gertz; 4, M	$2618.52 biweekly	_____	_____	_____	_____

31. Explain how to find the federal withholding tax using the wage bracket (tax tables) method. (See Objective 2.)

32. Explain how to find the federal withholding tax using the percentage method. (See Objective 3.)

33. In your job, which deductions were subtracted from gross earnings to arrive at net pay? Which was the largest deduction?

34. Use the tables in this section and confirm that individuals with higher incomes pay a higher percent of their income in taxes. Should people with higher salaries pay a higher percent of their income in taxes? Explain. (See Objective 3.)

AMOUNT OWED THE IRS Calculate the total amount owed to the Internal Revenue Service from each of the following firms. (See Example 9.)

	Firm	FICA Tax Collected from Employees	Medicare Tax Collected from Employees	Total Federal Withholding Tax	Amount Due IRS
35.	Starbucks	$1483.59	$342.37	$5096.13	$8748.05
36.	Atlasta Ranch	$265.36	$61.24	$4111.68	_____
37.	Tony Balony's	$8212.18	$1895.37	$33,117.42	_____
38.	Plescia Produce	$212.78	$49.10	$958.68	_____
39.	Todd Consultants	$7271.39	$1678.24	$26,423.84	_____
40.	Hartmann Shoes	$6538.42	$1508.87	$22,738.57	_____

Use the percentage method of withholding, a FICA rate of 6.2%, a Medicare rate of 1.45%, an SDI rate of 1%, and a state withholding tax of 3.4% in the following problems.

41. MARKETING REPRESENTATIVE Daniel Baker manages a small warehouse and has weekly earnings of $975. He is married and claims four withholding allowances. His deductions include FICA, Medicare, federal withholding, state disability insurance, state withholding, union dues of $7, and credit union savings of $100. Find his net pay for a week in February.
Net pay = $975 − $60.45 − $14.14 − $57.06 − $9.75 − $33.15 − $7 − $100 = $693.45

41. $693.45

42. STARBUCKS DISTRICT MANAGER Awanata Jackson, district manager for Starbucks, has earnings of $1147 in one week of March. She is single and claims four withholding allowances. Her deductions include FICA, Medicare, federal withholding, state disability insurance, state withholding, a United Way contribution of $15, and a savings bond of $100. Find her net pay for the week.

42. _____

43. FURNITURE SALES Kathy Bates, a salesperson for American Home, is paid a salary of $800 per week plus 7% of all sales over $5000. She is single and claims two withholding allowances. Her deductions include FICA, Medicare, federal withholding, state disability insurance, state withholding, credit union savings of $50, a Salvation Army contribution of $10, and dues of $15 to the National Association of Professional Saleswomen. Find her net pay for a week in April during which she had sales of $11,284 with returns and allowances of $424.48.

43. _____

238 CHAPTER 6 Payroll

44. **TRAVEL-AGENCY SALES** Scott Salman, a travel agent, is paid on a variable commission, is married, and claims four withholding allowances. He receives 3% of the first $20,000 in sales, 4% of the next $10,000 in sales, and 6% of all sales over $30,000. This week he has sales of $45,550 and the following deductions: FICA, Medicare, federal withholding, state disability insurance, state withholding, a retirement contribution of $45, a savings bond of $50, and charitable contributions of $20. Find his net pay after subtracting all of his deductions.

44. _____

45. **HEATING-COMPANY REPRESENTATIVE** Evelyn Beaton, a commission sales representative for Alternative Heating Company, is paid a monthly salary of $4200 plus a bonus of 1.5% on monthly sales. She is married and claims three withholding allowances. Her deductions include FICA, Medicare, federal withholding, state disability insurance, no state withholding, car payment of $285 to a credit union, charitable contributions of $25, and a savings bond of $50. Find her net pay for a month in which her sales were $42,618. The state in which Beaton works has no state income tax.

45. _____

46. **RIVER RAFT MANAGER** River Raft Adventures pays its manager, Kathryn Speers, a monthly salary of $2880 plus a commission of .8% based on total monthly sales volume. In May, River Raft Adventures has total sales of $86,280. Speers is married and claims five withholding allowances. Her deductions include FICA, Medicare, federal withholding, state disability insurance, state withholding of $159.30, credit union payment of $300, and a retirement plan contribution of $150. Find her net pay.

46. _____

QUICK CHECK ANSWERS

1. $16
2. $64
3. $111.95
4. $473.55
5. $838.64
6. 3.75%; $120.94
7. $6716.47
8. $1139.86
9. $8601.99

Chapter 6 Quick Review

Chapter Terms Review the following terms to test your understanding of the chapter. For each term you do not know, refer to the page number found next to that term.

allowances [p. 217]
commission rate [p. 216]
compensatory time (comp time) [p. 208]
daily overtime [p. 208]
deductions [p. 205]
differential piece rate [p. 215]
double time [p. 208]
Employer's Quarterly Federal Tax Return [p. 235]
Fair Labor Standards Act [p. 206]
Federal Insurance Contributions Act (FICA) [p. 222]
Federal Unemployment Tax Act (FUTA) [p. 235]

Form 941 [p. 235]
fringe benefits [p. 235]
Great Depression [p. 222]
gross earnings [p. 205]
hourly wage [p. 205]
incentive rates [p. 214]
income tax withheld [p. 228]
Internal Revenue Service (IRS) [p. 228]
marital status [p. 228]
median [p. 218]
Medicare [p. 222]
net pay [p. 205]
overtime [p. 206]

overtime premium method [p. 207]
pay period [p. 209]
payroll ledger [p. 205]
percentage method [p. 228]
personal income tax [p. 228]
piecework rate [p. 214]
quotas [p. 215]
returns [p. 217]
salary [p. 209]
salary plus commission [p. 218]
SDI deduction [p. 224]
shift differential [p. 208]

sliding scale [p. 217]
Social Security [p. 222]
split-shift premium [p. 208]
state income taxes [p. 233]
straight commission [p. 216]
time-and-a-half rate [p. 206]
time card [p. 205]
time rates [p. 214]
unemployment insurance tax [p. 235]
variable commission [p. 217]
wage bracket method [p. 228]
withholding allowances [p. 228]

CONCEPTS

EXAMPLES

6.1 Gross earnings

Gross earnings = Hours worked × Rate per hour

40 hours worked at $11.60 per hour
Gross earnings = 40 × $11.60 = $464

6.1 Gross earnings with overtime

First, find the regular earnings. Then, determine overtime pay at overtime rate. Finally, add regular and overtime earnings.

Gross earnings =
Earnings at regular rate + Earnings at time-and-half rate

40 regular hours at $11.60 per hour
10 overtime hours at time and a half

Gross earnings = (40 × $11.60) + (10 × $11.60 × 1.5)
= $464 + $174
= $638

6.1 Common pay periods

Pay Period	Paychecks Per Year
Monthly	12
Semimonthly	24
Biweekly	26
Weekly	52

Find the earnings equivalent of $2800 per month for other pay periods.

Annually = $2800 × 12 = $33,600

Semimonthly = $\dfrac{\$2800}{2}$ = $1400

Biweekly = $\dfrac{\$2800 \times 12}{26}$ = $1292.31

Weekly = $\dfrac{\$2800 \times 12}{52}$ = $646.15

6.1 Overtime for salaried employees

First, find the hourly equivalent. Next, multiply the hourly equivalent rate by the overtime hours by 1.5. Finally, add overtime earnings to the salary.

Salary is $648 per week for 40 hours. Find the earnings for 46 hours.

$648 ÷ 40 = $16.20 per hour
$16.20 × 6 × 1.5 = $145.80 overtime
$648 + $145.80 = $793.80

6.2 Gross earnings for piecework

Gross earnings = Pay per item × Number of items

Items produced, 175; pay per item, $.65; find the gross earnings for the day.

$.65 × 175 = $113.75

CONCEPTS	EXAMPLES
6.2 Gross earnings for differential piecework The rate paid per item produced varies with level of production.	1–100 items, $1.75 each 101–150 items, $2.00 each 151 or more items, $2.20 each Find the gross earnings for producing 223 items. $100 \times \$1.75 = \$175$ first 100 units $50 \times \$2.00 = \$100$ next 50 units $73 \times \$2.20 = \$160.60$ number over 150 223 total items $435.60 total earnings
6.2 Overtime earnings on piecework Gross earnings = Earnings at regular rate + Earnings at overtime rate	Items produced on regular time, 530; items produced on overtime, 110; piece rate $.60; find the gross earnings. Gross earnings = $(530 \times \$.60) + 110(1.5 \times \$.60) = \$318 + \$99$ $= \$417$
6.2 Straight commission Gross earnings = Commission rate × Amount of sales	Sales of $25,800; commission rate is 5%. $.05 \times \$25{,}800 = \$1290$
6.2 Variable commission Commission rate varies at different sales levels.	Up to $10,000, 6% $10,001–$20,000, 8% $20,001 and up, 9% Find the commission on sales of $32,768. $.06 \times \$10{,}000 = \$600.00$ first $10,000 $.08 \times \$10{,}000 = \$800.00$ next $10,000 $.09 \times \$12{,}768 = \$1149.12$ amount over $20,000 $32,768 $2549.12 total commission
6.2 Salary and commission Gross earnings = Fixed earnings + Commission	Salary, $450 per week; commission rate, 3%; find the gross earnings on sales of $6848. Gross earnings = $\$450 + (.03 \times \$6848) = \$450 + \$205.44$ $= \$655.44$
6.2 Commission with a drawing account Gross earnings = Commission − Draw	Sales for month, $38,560; commission rate, 7%; draw, $750 for month; find the gross earnings. Gross earnings = $(.07 \times \$38{,}560) - \$750 = \$2699.20 - \$750$ $= \$1949.20$
6.3 FICA; Social Security tax The gross earnings are multiplied by the tax rate. When the maximum earnings are reached, no additional FICA is withheld that year.	Gross earnings, $458; Social Security tax rate, 6.2%; find the Social Security tax. $\$458 \times .062 = \$28.40$
6.3 Medicare tax The gross earnings are multiplied by the Medicare tax rate. Medicare tax is paid on all earnings.	Gross earnings, $458; Medicare tax rate, 1.45%; find the Medicare tax. $\$458 \times .0145 = \$6.64$
6.3 State disability insurance deductions Multiply the gross earnings by the SDI tax rate. When the maximum earnings are reached, no additional taxes are paid in that year.	Gross earnings, $2880; SDI tax rate, 1%; find SDI tax. $\$2880 \times .01 = \$28.80$

CHAPTER 6 Quick Review

CONCEPTS	EXAMPLES
6.4 Federal withholding tax—wage bracket A tax must be paid on the total earnings. Look up the amount in the wage bracket table.	Single employee, 3 allowances, weekly earnings of $468. Find the amount to withhold. Wage bracket table for single, weekly, row with at least $460 but less than $470, 3 allowances. Withholding = **$19**
6.4 Federal withholding tax—percentage A tax must be paid on the total earnings. 1. Multiply number of withholding allowances by the amount for one allowance from table. 2. Find gross earnings − amount in step 1. 3. Use the appropriate table and row. 4. Follow the directions in the table.	Single employee, 2 withholding allowances, weekly earnings of $535. Find the amount to withhold. 1. 2 allowances × $77.90 = **$155.80** 2. $535 − **$155.80** = **$379.20** 3. Table for single, weekly, row for over $222 but not over $767 4. $17.90 + .15 × (**$379.20** − $222) = **$41.48**
6.4 State withholding tax Tax is paid on total earnings. No maximum as with FICA.	Married employee with weekly earnings of $692; find the state withholding tax given a state withholding tax rate of 4.5%. **4.5%** × $692 = .045 × $692 = $31.14

6.4 Quarterly report, Form 941
Filed each quarter and calculated as follows:
    FICA paid by employees
    FICA paid by employer
    Medicare paid by employees
    Medicare paid by employer
 + Federal withholding tax paid by employees
    Total to be sent to the IRS
	If quarterly FICA withheld from employees is $5269, Medicare tax is $1581, and federal withholding tax is $14,780, find the total owed to the IRS by the employer.
	(**$5269** + **$1581**) × 2 + $14,780 = **$28,480**

Case Study

PAYROLL: FINDING YOUR TAKE-HOME PAY

Janice Wong receives an annual salary of $42,536, which is paid weekly. Her normal workweek is 40 hours, and she is paid time and a half for all overtime. She is single and claims two withholding allowances. Her deductions include FICA, Medicare, federal withholding, state disability insurance, state withholding at 4.4%, credit union payments of $125, retirement deductions of $75, association dues of $12, and a Diabetes Association contribution of $25. Find each of the following for a week in which she works 52 hours.

1. Regular weekly earnings

2. Overtime earnings

3. Total gross earnings

4. FICA

5. Medicare

6. Federal withholding using the percentage method

7. State disability insurance deduction

8. State withholding

9. Pay

10. Find the percent of her total gross earnings that she actually receives, rounded to the nearest tenth of a percent. It is amazing how much is withheld, isn't it?

INVESTIGATE

Look at the statement that you (or a parent or friend) received with your last paycheck. Be certain that your gross earnings are correct. Understand and check all of the deductions made by your employer. Subtract all deductions from your gross earnings to be certain that your net pay is accurate.

Case in Point Summary Exercise

PAYROLL AT STARBUCKS

www.ssa.gov

Facts:

- 1935: Social Security Act passed
- 1937: First Social Security taxes collected
- 1965: Medicare established
- 2018: Medicare and Social Security benefits are helping over 60 million people

Sarah Brynski is busy running a Starbucks and thinking about inventory, making sure the store looks great, ensuring good service, hiring good people, motivating employees, maintaining inventories, calculating wages, deducting correct amounts for the various taxes, and getting the payroll done on time every week. She is also interested in finding ways to increase sales since part of her income is based on sales. Overtime is paid for more than 40 hours a week, the state tax rate is 3.07%, and the state disability income rate (SDI) is 1%.

1. Use the percentage method and help her calculate the weekly payroll for the employees listed.

Employee	Hourly Wage	Hours Worked	Total Pay	FICA	Medicare	Federal Tax	State Tax	SDI	Other	Net Pay
Chavez, S., 1	$11.50	30	___	___	___	___	___	___	$0	___
Parton, M., 2	$12.25	45	___	___	___	___	___	___	$25	___
Dickens, S., 0	$9.35	30	___	___	___	___	___	___	$10	___

2. Assume Brynski receives a base salary of $450 per week plus a commission of 6% of sales over $5000. Do her weekly payroll for a week with $13,290 in sales.

Employee	Base Salary	Comm.	Total Pay	FICA	Medicare	Federal Tax	State Tax	SDI	Other	Net Pay
Brynski, M., 2	___	___	___	___	___	___	___	___	$65	___

3. Find the amount Brynski must send to the Internal Revenue Service for all of the above employees including herself.

4. Find the amount Brynski must send to the state, including income tax and disability.

Discussion Question: Actually, payroll calculations in large firms are usually done at a corporate or regional office using specialized software. Describe the advantages of such a system for local managers as well as for the corporation.

Chapter 6 Test

To help you review, the numbers in brackets show the section in which the topic was discussed.

Complete the following payroll ledger. Find the total gross earnings for each employee. Time and a half is paid on all hours over 40 in one week. **[6.1]**

Employee	Hours Worked	Reg. Hrs.	O.T. Hrs.	Reg. Rate	Gross Earnings
1. Bianchi	46.5	___	___	$10.80	___
2. Hanna	47.5	___	___	$12.48	___

Solve the following application problems.

3. Judy Martinez is paid $34,060 annually. Find the equivalent earnings if this amount is paid (a) weekly, (b) biweekly, (c) semimonthly, and (d) monthly. **[6.2]**

   (a) _____
   (b) _____
   (c) _____
   (d) _____

4. At Jalisco Electronics, assemblers are paid according to the following differential piece rate scale: 1–20 units in a day, $4.50 each; 21–30 units, $5.50 each; and $7 each for every unit over 30. Adrian Ortega assembled 70 units in one week. Find his gross pay. **[6.2]**

   4. _____

5. Rheonna Winston receives a commission of 6% for selling a $235,500 house. One-half of the commission goes to the broker and one-half of the remainder to another salesperson. Winston gets the rest. Find the amount she receives. **[6.2]**

   5. _____

An employee is paid a salary of $11,050 per month. If the current FICA rate is 6.2% on the first $130,000 of earnings, and the Medicare tax rate is 1.45% of all earnings, how much should be withheld for (a) FICA tax and (b) Medicare tax during the following months? **[6.3]**

6. March: (a) _____ (b) _____   7. December: (a) _____ (b) _____

Find the federal withholding tax using the wage bracket method for each of the following employees. **[6.4]**

8. Ahearn: 2 withholding allowances, single, $415.82 weekly earnings   8. _____
9. Zanotti: 2 withholding allowances, married, $642.30 weekly earnings   9. _____
10. Allgier: 3 withholding allowances, married, $2743.30 monthly earnings   10. _____
11. Yeoman: 4 withholding allowances, single, $2046.75 monthly earnings   11. _____
12. Benner: 5 withholding allowances, married, $2390.00 monthly earnings   12. _____

Find the net pay for each of the following employees after FICA, Medicare, federal withholding tax, state disability insurance, and other deductions have been taken out. Assume that none has earned over $130,000 so far this year. Assume a FICA rate of 6.2%, Medicare rate of 1.45%, and a state disability insurance rate of 1%. Use the percentage method of withholding. **[6.3 and 6.4]**

13. Murphy: $1852.75 monthly earnings, 1 withholding allowance, single, $37.80 in other deductions   13. _____

14. Pupek: $1028 weekly earnings, 3 withholding allowances, married, state withholding of $50.50, credit union savings of $50, contribution of $20   14. _____

15. Bunson, $984.20 weekly earnings, 6 withholding allowances, married, state withholding of $22.18, credit union payment of $75

15. _____

Solve the following application problems.

16. Joseph Flores is paid $1000 per week plus a commission of 5% on all personal sales. Flores sold $712 worth of goods on Monday, $523 on Tuesday, $1002 on Wednesday, $391 on Thursday, and $609 on Friday. Returns and allowances for the week were $114. Find the employee's **(a)** Social Security tax (6.2%), **(b)** Medicare tax (1.45%), and **(c)** state disability insurance deduction (1%) for the week. **[6.3 and 6.4]**

(a) _____
(b) _____
(c) _____

17. Neta Fitzgerald earned $127,375.60 so far this year. This week she earned $2649.78. Find her **(a)** FICA tax and **(b)** Medicare tax for this week's earnings. **[6.3]**

(a) _____
(b) _____

For Exercises 18 and 19, find (a) the Social Security tax and (b) the Medicare tax that each self-employed person must pay. Since they are self-employed, use a FICA tax rate of 12.4% and a Medicare tax rate of 2.9%. **[6.3]**

18. Kirby: $36,714.12

(a) _____
(b) _____

19. Biondi: $42,380.62

(a) _____
(b) _____

20. The employees of Quick-Lube paid a total of $837.12 in Social Security tax last month, $195.78 in Medicare tax, and $1217.34 in federal withholding tax. Find the total amount that the employer must send to the Internal Revenue Service.

20. _____

Mathematics of Buying

7

CHAPTER CONTENTS

7.1 Invoices and Trade Discounts

7.2 Series Discounts and Single Discount Equivalents

7.3 Cash Discounts: Ordinary Dating Methods

7.4 Cash Discounts: Other Dating Methods

CASE IN POINT

BED BATH & BEYOND INC. operates a chain of home-furnishing specialty stores selling items such as bed linens, bath accessories, cookware, and small electric appliances. Customer service in a family-style atmosphere is an essential part of the firm's marketing strategy.

Jack Williams works in the main offices of Bed Bath & Beyond in the purchasing department. He works with firms that supply Bed Bath & Beyond with products. His job requires that he remain in close contact with marketing to make sure he knows what is selling and what is not. He also needs to understand consumer trends since products are often ordered months before they are available for sale in the stores. The last thing he wants to do is order products that no one will buy.

Since controlling costs is so important, Williams works every day to get the largest discounts possible from each. He also wants to help the company prosper, which may work to his benefit in terms of an eventual job promotion.

Managing inventories is at the heart of retailing. Look at the generalized **supply chain**, or **distribution chain**, below showing **suppliers, manufacturer, wholesaler (distributor), retailer**, and customer. Products move forward through the supply chain toward customers and money moves backward from customer to retailer to wholesaler, etc. Although some retailers sell products only on the Web, most larger retailers sell using both the Internet and actual stores.

Supply Chain

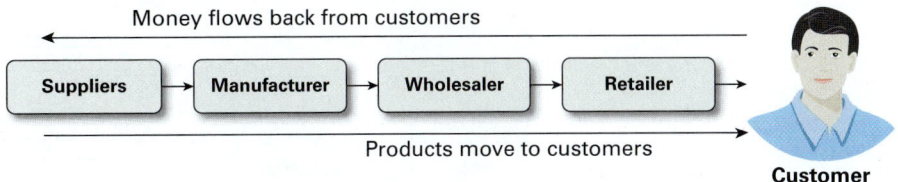

Manufacturers have a **suggested retail price** or **list price** for products. However, they often discount the list price due to price changes, quantity purchased, geographic location, seasonal fluctuations, levels of inventory in stock, method and time of payment, and competition. Several common types of discounts are discussed in this chapter.

7.1 Invoices and Trade Discounts

OBJECTIVES
1. Complete an invoice.
2. Understand common shipping terms.
3. Identify invoice abbreviations.
4. Calculate trade discounts and understand why they are given.
5. Differentiate between single and series discounts.
6. Calculate each series discount separately.
7. Use complements to calculate series discounts.
8. Use a table to find the net cost equivalent of series discounts.

CASE IN POINT Since he works with invoices every day in his job in purchasing at Bed Bath & Beyond, Jack Williams must have a good understanding of trade discounts, cash discounts, and shipping terms.

OBJECTIVE 1 Complete an invoice. An **invoice**, such as the one from J.B. Sherr Co. on the next page, is a document issued by a seller to a buyer. It is used to record the details of a transaction. It is a **sales invoice** for the seller, but a **purchase invoice** for the buyer. It shows the quantity ordered, number shipped, the **unit price**, and the **extension total**, which is the number of items shipped multiplied by the unit price for the item. The **invoice total** is the sum of the extension totals. Most firms send and receive invoices electronically.

OBJECTIVE 2 Understand common shipping terms. Since goods can be damaged, lost, or stolen during shipment, it is important for everyone to know which firms own products that are in shipment. That is the only way it will be clear which firm must pay for goods lost or damaged while in shipment. Here are some common terms related to shipping, so these terms are found on invoices.

Free on board (FOB) shipping point	Buyer pays for shipping costs. Ownership of goods passes to the purchaser prior to the shipment.
FOB destination	Seller pays the shipping charge and retains ownership until goods reach the destination.
Cash on delivery (COD)	Payment for goods is made at the time of delivery.
Free alongside ship (FAS)	Goods are delivered to the dock with all freight charges to that point paid by the seller.

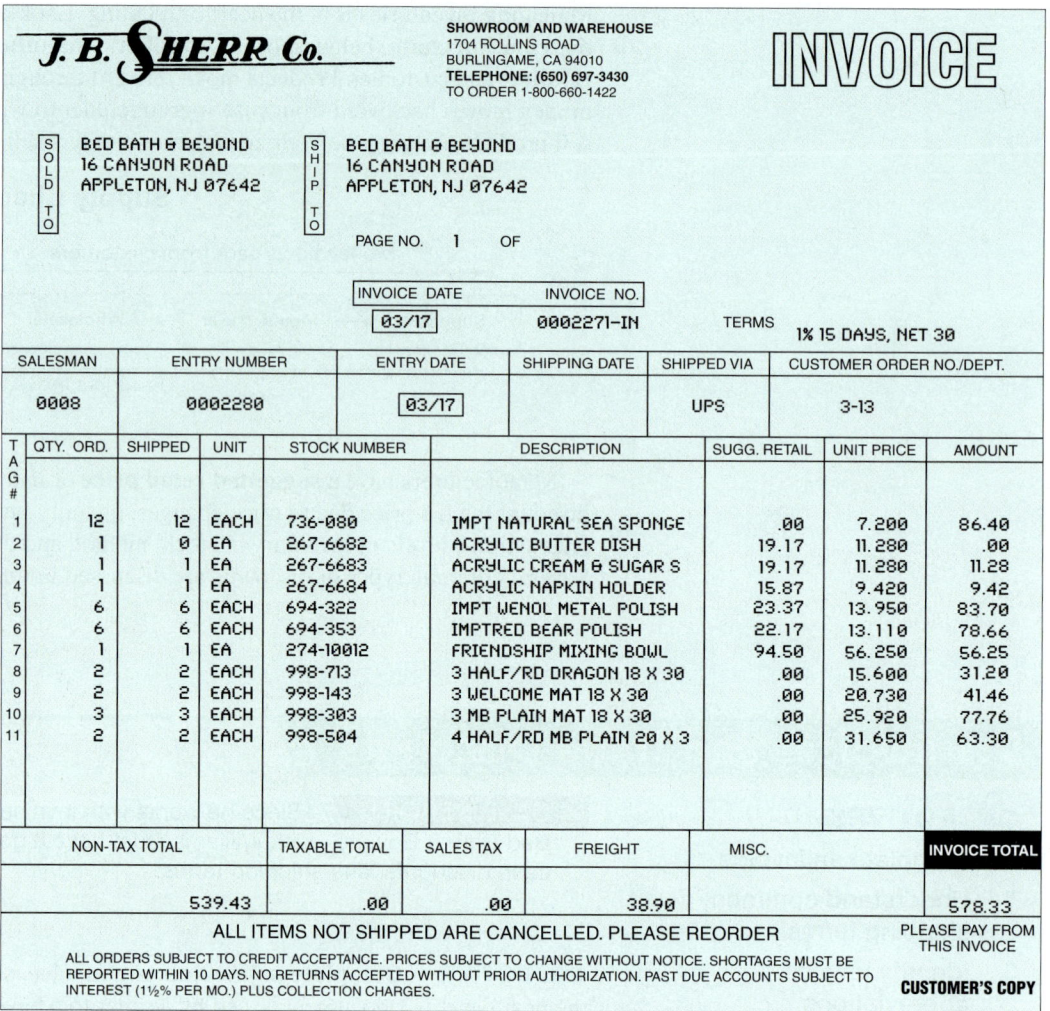

OBJECTIVE 3 Identify invoice abbreviations. The table below includes some commonly used abbreviations.

Invoice Abbreviations

ea.	each	drm.	drum
doz.	dozen	cs.	case
gro.	gross (144 items)	pr.	pair
gr gro.	great gross (12 gross)	C	Roman numeral for 100
qt.	quart	M	Roman numeral for 1000
gal.	gallon (4 quarts)	cwt.	per hundredweight
bbl.	barrel	cpm.	cost per thousand
cL	centiliter	lb.	pound
L	liter	oz.	ounce
in.	inch	g	gram
ft.	foot	kg	kilogram
yd.	yard	ROG	receipt of goods
mm	millimeter	ex. or x	extra dating
cm	centimeter	FOB	free on board
m	meter	EOM	end of month
km	kilometer	COD	cash on delivery
ct.	crate	FAS	free alongside ship

7.1 Invoices and Trade Discounts   249

OBJECTIVE 4 Calculate trade discounts and understand why they are given. Firms often give discounts to businesses that buy products. The seller provides a list price but then often offers a **trade discount** that is subtracted from the list price, resulting in a **net price** or **net cost** paid by the purchaser. Insurance and shipping charges are usually NOT discounted.

> **Finding the Net Cost**
>
> Net cost = List price − Trade Discount

Calculating a Single Trade Discount — **EXAMPLE 1**

The list price of a commercial-grade Cuisinart food processor is $189.20 with a trade discount of 25% given to Bed Bath & Beyond due to the volume of purchases. Find the net cost.

CASE IN POINT

SOLUTION

Trade discount = 25% of $189.20 = **$47.30**

Net cost = List price − Trade discount
= $189.20 − **$47.30**
= $141.90

QUICK CHECK 1

A wine refrigerator has a list price of $589.99 and a trade discount of 35%. Find the net cost.

The advertisements suggest either that the retail store may have received a large discount or that it is trying to move inventory quickly.

OBJECTIVE 5 Differentiate between single and series discounts. In Example 1, a **single discount** of 25% was offered. Sometimes two or more discounts are combined into a **series discount**, also called **chain discount**. The three common methods used to calculate a series discount are shown in the next three objectives.

OBJECTIVE 6 Calculate each series discount separately. The first of the three methods of finding the discount requires the discounts to be calculated separately.

Calculating Series Trade Discounts — **EXAMPLE 2**

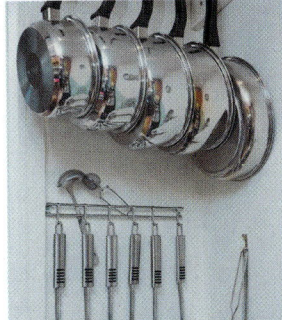

Bed Bath & Beyond is offered a series discount of 20/10 on a Cooking with Calphalon stainless steel cookware set with a list price of $150. Find the net cost after the series discount.

CASE IN POINT

SOLUTION

Apply the first discount in 20/10, which is 20%.

Discount = **20% of list price of $150** = **$30**

Amount after first discount = $150 − **$30** = $120

Now apply the second discount in 20/10, which is 10%. Importantly, apply the discount to the amount after the first discount, or $120.

Discount = 10% of discounted price of $120 = **$12**

Net cost = $120 − **$12** = $108

The net cost is $108.

QUICK CHECK 2

A 12-place setting of stainless steel flatware is list priced at $249.99. If a series discount of 10/15 is offered, what is the net cost after the series discount is taken?

It is important to *never add series discounts together*. For example, a 20/10 discount is **NOT** the same as a 30% discount. Apply the discounts sequentially, or one after the other.

OBJECTIVE 7 Use complements to calculate series discounts. Complements can also be used to find series discounts. The **complement** of a discount is found by subtracting the decimal form of the discount from 1 as shown.

Discount	Decimal Equivalent	Finding the Complement		Complement
10%	.1	1 − .1	=	.90
20%	.20	1 − .2	=	.80
25%	.25	1 − .25	=	.75
30%	.30	1 − .3	=	.70
62%	.62	1 − .62	=	.38

The complement of the discount is the portion actually paid. For example, a 10% discount means that 90% must be paid, and a 25% discount means that 75% must be paid. The complements can be used to find the **net cost equivalent**, which is then used to find the net cost after the discount.

> **Finding the Discount Using Net Cost Equivalent**
>
> 1. Find the complement of each discount in the series discount.
> 2. Find the net cost equivalent by multiplying the complements from step 1.
> 3. The net cost after the discount is the list price times the net cost equivalent found in step 2.

Using Complements to Find the Net Cost

EXAMPLE 3 Kitchen Crafters is offered a series discount of 20/10 on a George Foreman Grilling Machine with a list price of $139.99. Find the net cost after the series discount.

SOLUTION

Apply the preceding steps to find the net cost after the discount.

	Discount	Decimal Equivalent	Finding the Complement		Complement
Step 1	20%	.2	1 − .2	=	.8
	10%	.1	1 − .1	=	.9

Step 2 Net cost equivalent = .8 × .9 = .72
Step 3 Net cost after discount = List price × Net cost equivalent

= $139.99 × .72
= $100.79

To find the actual discount, subtract the net cost after discount from the original price.

Discount = Original price − Net cost after discount
= $139.99 − $100.79
= $39.20

On many calculators, you can subtract the discount percents from the list price in a series calculation.

139.99 − 20 % − 10 % = 100.79 (rounded)

Note: Refer to Appendix B for calculator basics.

QUICK CHECK 3

A supplier offers a series discount of 25/15 on a waffle iron. If the list price of the waffle iron is $135, use the complement with respect to 1 of each of the single discounts to find (a) the net cost and (b) the amount of the discount.

Using Complements to Solve Series Discounts

EXAMPLE 4 The list price of a Heartland 30-inch combination gas and electric stove is $3095. Find the net cost after a series discount of 20/10/10.

SOLUTION

Start by finding the complements with respect to 1 of each discount.

series discount ⟶ 20/10/10
                    ↓ ↓ ↓
Find complements
with respect to 1. ⟶ .8 .9 .9
                     ↓ ↓ ↓
Multiply
complements. ⟶ .8 × .9 × .9 = .648   net cost equivalent

$3095   list price
× .648   net cost equivalent
$2005.56   net cost

QUICK CHECK 4

A 31-cubic-foot bottom-freezer refrigerator is list priced at $2249.99. What is the net cost after a series discount of 15/20/5?

OBJECTIVE 8 Use a table to find the net cost equivalent of series discounts. The following table can be used to find the net cost equivalents for common series discounts. Look in the left column for the first one or two discounts in the series discount. Then, look across the top of the table for the last discount in the series discount. For example, a 15/10/25 series discount is found by looking at the 15/10 row and the 25% column: .57375.

Quick TIP
Do not round the net cost equivalent, as this may result in a wrong answer.

Net Cost Equivalents of Series Discounts

	5%	10%	15%	25%	30%	35%	40%
5	.9025	.855	.8075	.7125	.665	.6175	.57
10	.855	.81	.765	.675	.63	.585	.54
10/5	.81225	.7695	.72675	.64125	.5985	.55575	.513
15	.8075	.765	.7225	.6375	.595	.5525	.51
15/10	.72675	.6885	.65025	.57375	.5355	.49725	.459
20	.76	.72	.68	.6	.56	.52	.48
20/15	.646	.612	.578	.51	.476	.442	.408
25	.7125	.675	.6375	.5625	.525	.4875	.45
25/20	.57	.54	.51	.45	.42	.39	.36
25/25	.534375	.50625	.478125	.421875	.39375	.365625	.3375
30	.665	.63	.595	.525	.49	.455	.42
40	.57	.54	.51	.45	.42	.39	.36

CHAPTER 7 Mathematics of Buying

Since the commutative property of math applies to multiplication, it does not matter which discount is listed first in a series discount. For example, a 15/20/5 discount is the same as a 20/15/5 discount or a 5/15/20 discount.

Using a Table of Net Cost Equivalents

EXAMPLE 5 Use the table to find the net cost equivalents of the following series discounts.
(a) 10/10   (b) 20/10   (c) 25/25/5   (d) 35/20/15

SOLUTION
First find the row and column needed to read the net cost equivalent from the table, then get the net cost equivalent.
(a) Row 10, Column 10%:   Net cost equivalent = .81
(b) Row 20, Column 10%:   Net cost equivalent = .72
(c) Row 25/25, Column 5%:   Net cost equivalent = .534375
(d) There is no row 35/20 corresponding to the first two numbers in 35/20/15. But 35/20/15 is the same thing as 15/20/35 or as 20/15/35. Using the last of these, go to Row 20/15 and Column 35%:   Net cost equivalent = .442

> **QUICK CHECK 5**
> Use the table of net cost equivalents to find the net cost equivalents of the following series discounts: **(a)** 30/5, **(b)** 10/40, **(c)** 20/15/10, **(d)** 10/5/5.

7.1 Exercises // MyLab Math

The shaded sections below contain solutions to help you get a **QUICK START** *on the various types of exercises.*

USING INVOICES   Compute the extension totals and the invoice total for the following invoices.

HOME ACCESSORIES WHOLESALERS

Sold to: Kitchen Crafters
10100 Fair Oaks Blvd.
Fair Oaks, CA 95628

Date: June 10
Order. No.: 796152
Shipped by: UPS
Terms: Net

#	Quantity	Order No./Description	Unit Price	Extension Total
1.	6 doz.	pastry brush, wide	$37.80 doz.	$226.80
2.	3 gro.	napkins, cotton	$12.60 gro.	$37.80
3.	9 doz.	cherry pitters	$14.04 doz.	
4.	8	food processors (3 qt.)	$106.12 ea.	
5.	53 pr.	stainless tongs	$68.12 pr.	
6.			Invoice Total	
			Shipping and Insurance	$85.60
7.			Total Amount Due	

	J & K'S MUSTANG PARTS **New and Used**		Date: July 17 Order No.: 100603 Shipped by: Emery Terms: Net
	Sold to: Dave's Auto Body & Paint 4443-B Auburn Blvd. York, PA 17402		

	Quantity	Order No./Description	Unit Price	Extension Total
8.	24	filler tube gaskets	$2.25 ea.	
9.	12 pr.	taillight lens gaskets	$4.75 pr.	
10.	6 pr.	taillight bezels to body	$10.80 pr.	
11.	2 gr.	door panel fasteners	$14.20 gr.	
12.	18	bumper bolt kits	$16.50 ea.	
13.			Invoice Total	
			Shipping and Insurance	$139.40
14.			Total Amount Due	

ABBREVIATIONS ON INVOICES *What does each of the following abbreviations represent?*

15. ft. **foot**
16. sk. **sack**
17. pr. _____
18. gr. gro. _____
19. kg _____
20. qt. _____
21. cs. _____
22. gro. _____
23. drm. _____
24. yd. _____
25. L _____
26. cpm. _____
27. gal. _____
28. cwt. _____
29. COD _____
30. FOB _____

31. Name six items that appear on an invoice. (See Objective 1.)

32. Explain in your own words the difference between *FOB shipping point* and *FOB destination*. In each case, who pays for shipping and when does ownership of the merchandise transfer? (See Objective 2.)

Using complements (with respect to 1) of the single discounts, find the net cost equivalents for each of the following discounts. Do not round. (See Examples 3 and 4.)

33. 10/20 $.9 \times .8 = .72$
34. 20/20 $.8 \times .8 = .64$
35. 10/10/10 $.9 \times .9 \times .9 = .729$
36. 15/20/25 $.85 \times .8 \times .75 = .51$
37. 25/5 _____
38. 5/15 _____
39. 40/30/20 _____
40. 20/20/10 _____
41. 5/10/15 _____
42. 25/10/20/10 _____

indicates an exercise that is related to the Case in Point feature.

254 CHAPTER 7 Mathematics of Buying

Find the net cost of each of the following list prices. Round to the nearest cent. (See Examples 1–4.)

43. $418 less 20/20 **$267.52**
44. $148 less 25/10 **$99.90**
45. $16.40 less 5/10 _____
46. $860 less 20/40 _____
47. $1260 less 15/25/10 _____
48. $69.20 less 10/10/20 _____
49. $380 less 20/10/20 _____
50. $2008 less 10/5/20 _____
51. $22 less 10/15 _____
52. $25 less 30/20 _____
53. $980 less 10/10/10 _____
54. $8220 less 30/5/10 _____
55. $2000 less 10/40/10 _____
56. $1630 less 10/5/10 _____
57. $1250 less 20/20/20 _____
58. $1410 less 10/20/5 _____

59. Identify five possible reasons that trade discounts are used.

60. Explain the difference between a single trade discount and a series or chain trade discount. (See Objectives 5 and 6.)

61. Explain complement with respect to 1 and give an example. (See Objective 7.)

62. Using complements, explain how to find the net cost equivalent of a 25/20 series discount. Also explain why a 25/10/10 series discount is not the same as a 25/20 discount. (See Objective 7.)

Solve the following application problems in trade discount. Round to the nearest cent.

63. **VIRTUAL REALITY** An Oculus Rift virtual reality headset lists for $649. Find the net cost after trade discounts if the series discount is 10/10.

63. _____

64. **NURSING-CARE PURCHASES** Roger Wheatley, a restorative nursing assistant (RNA), finds that the list price of one dozen adjustable walkers is $1680. Find the cost per walker if a series discount of 40/25 is offered.

64. _____

65. **KITCHEN ISLAND** Kitchen Crafters purchases a tile-topped wooden kitchen island list priced at $480. It is available at either a 10/15/10 discount or a 20/15 discount. **(a)** Which discount gives the lower price? **(b)** Find the difference in net cost.

(a) _____

(b) _____

66. **HARDWARE PURCHASE** Oaks Hardware purchases an extension ladder list priced at $120. It is available at either a 10/10/10 discount or a 15/15 discount. **(a)** Which discount gives the lower price? **(b)** Find the difference.

(a) _____

(b) _____

67. **TRIPOD PURCHASE** The list price of an aluminum tripod is $65. It is available at either a 15/10/10 discount or a 15/20 discount. **(a)** Which discount gives the lower price? **(b)** Find the difference.

(a) _____

(b) _____

68. **LIQUID FERTILIZER** Continental Fertilizer Supply offers a series discount of 10/20/20 on major purchases. If a 58,000-gallon tank (bulk) of liquid fertilizer is list priced at $27,200, what is the net cost after trade discounts?

68. _____

69. **HOME BEVERAGE FOUNTAINS** Bed Bath & Beyond receives a 10/5/20 series trade discount from a supplier. If they purchase 4 dozen Bella Home beverage fountains list priced at $468 per dozen, find the net cost.

69. _____

70. **WHOLESALE AUTO PARTS** Kimara Swenson, an automotive mechanics instructor, is offered mechanics' net prices on all purchases at Foothill Auto Supply. If mechanics' net prices mean a 10/20 discount, how much will Swenson spend on a dozen sets of metallic brake pads that are list priced at $648 per dozen?

70. _____

71. **GRAPHIC ARTS** Hewlett Packard introduces a new high-speed, Indigo Digital Press that can be used to print brochures, business cards, posters, and photos at a list price of $126,300. Find the net cost if a trade discount of 5/20/5 is given on the first ten sold.

71. _____

72. **DANCE SHOES** How much will Giselle, a dance instructor, pay for three dozen pairs of dance shoes if the list price is $144 per dozen and a series discount of 10/25/30 is offered?

72. _____

CHAPTER 7 Mathematics of Buying

73. TRADE DISCOUNT The Door Store offers a series trade discount of 30/20 to its builder customers. Robert Gonzalez, a new employee in the billing department, understood the 30/20 terms to mean 50% and computed this trade discount on a list price of $5440. How much difference did this error make in the amount of the invoice?

73. _____

74. TRADE DISCOUNT Carrier corporation offers a series trade discount of 15/10/5 for certain heat pumps to its largest customer. A new employee believes that the 15/10/5 series discount is the same thing as a 30% discount (15% + 10% + 5%). Find the error on a commercial heat pump with a list price of $4850.

74. _____

QUICK CHECK ANSWERS

1. $383.49
2. $191.24
3. (a) $86.06   (b) $48.94
4. $1453.49
5. (a) .665   (b) .54   (c) .612   (d) .81225

7.2 Series Discounts and Single Discount Equivalents

OBJECTIVES

1. Express a series discount as an equivalent single discount.
2. Find the list price given the series discount and the net cost.

CASE IN POINT At Bed Bath & Beyond, Jack Williams pays close attention to all invoices received from suppliers. Invoices sometimes have mistakes, and he always wants to use the maximum trade discounts allowed, so he needs to look at the due dates, amounts, and discount terms.

OBJECTIVE 1 Express a series discount as an equivalent single discount. Series or chain discounts are often expressed as a single discount rate. Find a **single discount equivalent** to a series discount by subtracting the net cost equivalent from 1. This discount is equivalent to the series discount and is always written as a percent.

Finding the Single Discount Equivalent

$$\text{Single discount equivalent} = 1 - \text{Net cost equivalent}$$

Finding a Single Discount Equivalent

EXAMPLE 1 Find the single discount equivalent if Spectral Heating offers a 20/10 discount on all heating systems.

SOLUTION

series discount → 20/10

Find complements with respect to 1. → .8   .9

Multiply complements. → .8 × .9 = .72   net cost equivalent

```
1.00   base (100%)
- .72   net cost equivalent (remains)
 .28   discount is 28%
```

The single discount equivalent of a 20/10 series discount is 28%.

QUICK CHECK 1

Baltimore Wholesale Electric offers a 30/20 discount on wholesale purchases of all small appliances. What is the single discount equivalent?

OBJECTIVE 2 Find the list price given the series discount and the net cost. Sometimes the net cost after trade discounts is given along with the series discount, and the list price must be found.

Solving for the List Price

EXAMPLE 2 Find the list price of a Kohler kitchen sink that has a net cost of $243.20 after trade discounts of 20/20.

SOLUTION

Use a net cost equivalent. Start by finding the percent paid, using complements.

series discount → 20/20

complements with respect to 1 → .8 × .8 = .64   remains (net cost equivalent)

Thus **64%** of the list price was paid. Find the list price with the standard percent formula.

$$R \times B = P$$
$$64\% \text{ of list price} = \$243.20$$

or

$$B = \frac{P}{R} = \frac{243.2}{.64} = \$380 \text{ list price}$$

The list price of the sink is $380.

258 CHAPTER 7 Mathematics of Buying

QUICK CHECK 2

After trade discounts of 10/30, the net cost of a 10-quart crock pot is $62.36. Find the list price.

Solving for the List Price

EXAMPLE 3 Find the list price of a Bunn 10-cup Generation home brewer having a series discount of 10/30/20 and a net cost of $60.48.

CASE IN POINT

SOLUTION

Use complements to find the percent paid.

series discount ⟶ 10/30/20

complements with respect to 1: $.9 \times .7 \times .8 = .504$ remains (percent paid)

Therefore, .504 of the list price is $60.48. Use the formula for base.

$$B = \frac{P}{R} = \frac{\$60.48}{.504} = \$120 \text{ list price}$$

The list price of the Bunn home brewer is $120. Check this answer as in Example 3.

QUICK CHECK 3

Find the list price of an Oster Toaster Oven having a series discount of 10/10/10 and a net cost of $43.73.

Examples 2 and 3 are percent decrease problems similar to those shown in **Section 3.5**, although they may look different because the discount is a series of two or more discounts.

7.2 Exercises // MyLab Math

The shaded sections below contain solutions to help you get a **QUICK START** *on the various types of exercises.*

Find the net cost equivalent and the single discount equivalent of each of the following series discounts. Do not round net cost equivalents or single discount equivalents. (See Example 1.)

	Series Discount	Net Cost Equivalent	Single Discount Equivalent
1.	10/20	.72	28%
	$.9 \times .8 = .72; 1.00 - .72 = 28\%$		
2.	10/10	.81	19%
	$.9 \times .9 = .81; 1.00 - .81 = 19\%$		
3.	20/15	_____	_____
4.	25/25	_____	_____
5.	10/30/20	_____	_____
6.	5/10/15	_____	_____
7.	20/10/10/20	_____	_____
8.	5/5/10/10	_____	_____

9. Using complements, show that the single discount equivalent of a 25/20/10 series discount is 46%. (See Objective 1.)

10. Suppose that you own a business and are offered a choice of a 10/20 trade discount or a 20/10 trade discount. Decide which is better and explain why. (See Objective 1.)

Find the list price, given the net cost and the series discount. (See Examples 3 and 4.)

11. Net cost $518.40; trade discount 20/10
    .8 × .9 = .72; $518.40 ÷ .72 = $720   11. **$720**

12. Net cost $813.75; trade discount 30/25   12. _____

13. Net cost $1559.52; trade discount 5/10/20   13. _____

14. Net cost $2697.30; trade discount 10/10/10   14. _____

15. Net cost $265.39; trade discount 10/20/5   15. _____

16. Net cost $613.60; trade discount 25/25   16. _____

17. Net cost $4312.40; trade discount 5/10/15   17. _____

18. Net cost $43.17; trade discount 5/40   18. _____

Solve the following application problems. Round to the nearest cent.

19. **JEWELRY** A diamond ring costs Zales Jewelry $1338.93 after a series trade discount   19. _____
    of 10/10/5. Find the list price.

20. **APPLE iPAD** Apple offers a 20/10/10 discount to a   20. _____
    school district on bulk purchases of new iPads,
    resulting in a cost of $453.60. Find the list price.

21. **GPS** After a series trade discount of 20/20, a GPS had a net cost of $132.54. Find the list   21. _____
    price.

22. **NINTENDO** After a series trade discount of 10/5, a new Nintendo game had a net cost of   22. _____
    $18.69. Find the list price.

C// indicates an exercise that is related to the Case in Point feature.

CHAPTER 7 Mathematics of Buying

Solve the following two review application exercises on trade discounts.

23. **COMPARING DISCOUNTS** A S'mores Maker Kit has a list price of $39.95 and is offered to wholesalers with a series discount of 20/10/10. The same appliance is offered to Kitchen Crafters (a retailer) with a series discount of 20/10. **(a)** Find the wholesaler's price. **(b)** Find Kitchen Crafters' price. **(c)** Find the difference between the two prices.

(a) _____
(b) _____
(c) _____

24. **STAINLESS STEEL GRILL** A stainless steel gas grill is list priced at $495. The manufacturer offers a series discount of 25/20/10 to wholesalers and a 25/20 series discount to retailers. **(a)** What is the wholesaler's price? **(b)** What is the retailer's price? **(c)** What is the difference between the prices?

(a) _____
(b) _____
(c) _____

QUICK CHECK ANSWERS

1. 44%
2. $98.98
3. $59.99

7.3 Cash Discounts: Ordinary Dating Methods

OBJECTIVES

1. Calculate net cost after discounts.
2. Use the ordinary dating method.
3. Determine whether cash discounts are earned.
4. Use postdating when calculating cash discounts.

OBJECTIVE 1 Calculate net cost after discounts. Cash discounts are offered by sellers to encourage prompt payment by customers. In effect, the seller is saying, "Pay me quickly and receive a discount." Businesses often borrow money for their day-to-day operation. Immediate cash payments from customers *reduce* the need for the seller to borrow money and therefore reduce the interest cost that must be paid on borrowed funds.

> **Finding the Net Cost**
>
> Net cost = (List price − Trade discount) − Cash discount

OBJECTIVE 2 Use the ordinary dating method. There are many methods for finding cash discounts, but nearly all of these are based on the **ordinary dating method**. The methods discussed here and in the next section are the most common in use today. The ordinary dating method of cash discount is expressed on an invoice as

$$2/10, n/30 \quad \text{or} \quad 2/10, \text{net } 30$$

which is read as "two ten, net thirty." This means that a 2% discount applies if the invoice is paid within 10 days of the invoice date. If no payment is made within 10 days, then the full invoice amount must be paid within 30 days of the invoice date. If no payment is made by the end of 30 days from the invoice date, then a penalty may apply.

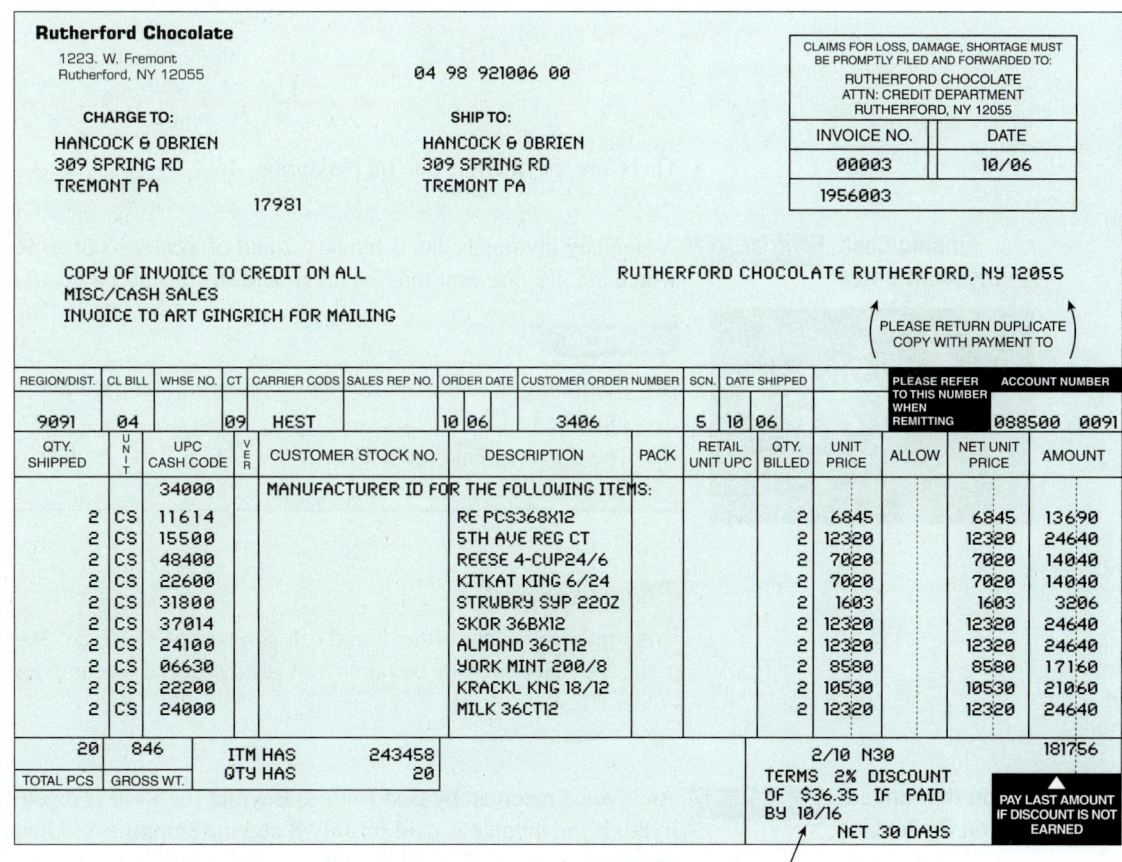

Discount Terms

To find the due date of an invoice, use the number of days in each month as shown below.

The Number of Days in Each Month

30-Day Months	31-Day Months		Exception
April	January	August	February
June	March	October	(28 days normally;
September	May	December	29 days in leap year)
November	July		

The number of days in each month of the year can also be remembered using the following rhyme and "knuckle" methods:

Rhyme Method:
30 days hath September,
April, June, and November.
All the rest have 31, except
February, which has 28 and
in a leap year 29.

Knuckle Method:

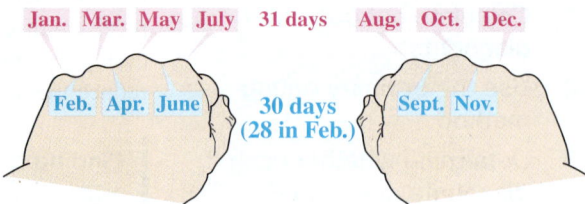

Leap years occur every four years: 2020, 2024, 2028, etc.

OBJECTIVE 3 Determine whether cash discounts are earned. Determine the due date for an invoice by adding the given number of days to the starting date. For example, to determine 10 days from April 7, add the number of days to the date ($7 + 10 = 17$). The due date, or 10 days from April 7, is April 17.

If the discount date or payment date is in the following month, first subtract the number of days remaining from the invoice date to the end of the current month. Then find the number of days in the next month needed to equal the discount period or net payment period. For example, find the payment due date if it is 15 days from the invoice date of October 20.

$$\begin{array}{rl} 31 & \text{days in October} \\ -20 & \text{invoice date} \\ \hline 11 & \text{days remaining in October} \end{array}$$

$$\begin{array}{rl} 15 & \text{total number of days} \\ -11 & \text{days remaining in October} \\ \hline 4 & \text{November (future date)} \end{array}$$

Therefore, payment is due on November 4.

Finding Cash Discount Dates

EXAMPLE 1 A Hershey invoice is dated January 2 and offers terms of 2/10, net 30. Find **(a)** the last date on which the 2% discount may be taken and **(b)** the net payment date.

SOLUTION

(a) Beginning with the invoice date, January 2, the last date for taking the discount is January 12 ($2 + 10$).

(b) The net payment date is February 1 ($31 - 2 = 29$ days remaining in January plus 1 day in February.)

QUICK CHECK 1

An invoice is dated June 8 and offers terms of 3/15, net 30. Find **(a)** the last date on which the 3% discount may be taken and **(b)** the net payment date.

Finding the Amount Due on the Invoice

EXAMPLE 2 An invoice received by Bed Bath & Beyond for $840 is dated July 1 and offers terms of 2/10, n/30. If the invoice is paid on July 8 and the shipping and insurance charges, which were FOB shipping point, are $35.60, find the total amount due.

CASE IN POINT

SOLUTION

Since the invoice was paid in 7 days ($8 - 1 = 7$), which is less than the 10-day requirement identified by 2/10, the 2% cash discount applies.

$$\text{Cash discount} = 2\% \text{ of } \$840 = \$16.80$$

$$\begin{aligned} \text{Amount due} &= \text{Invoice amount} - \text{Cash discount} \\ &= \quad \$840 \quad - \quad \$16.80 \\ &= \mathbf{\$823.20} \end{aligned}$$

However, the shipping and insurance charges must be added to the amount due.

$$\text{Total to be paid} = \text{Amount due} + \text{Shipping and insurance charges}$$
$$= \$823.20 + \$35.60$$
$$= \$858.80$$

The total amount due is $858.80.

QUICK CHECK 2

An invoice received for $2830.15, is dated March 21, and offers 3/15, n/30. If the invoice is paid on April 4 and the shipping and insurance charges are $124.96, find the amount due.

Generally the cash discount is not applied to any shipping and insurance charges. To find the total amount due, first deduct applicable trade and cash discounts, and then add shipping and insurance charges.

OBJECTIVE 4 Use postdating when calculating cash discounts. In the ordinary dating method, the cash discount date and net payment date are both counted from the date of the invoice. Occasionally, an invoice is **postdated**, or dated in the future. This may be done to give the purchaser more time to take advantage of the cash discount. The seller places a date that is after the actual invoice date, sometimes labeling it **AS OF**.

For example, the following Bakersfield Clothing invoice is dated 07/25/18 AS OF 08/01/18. Both the cash discount period and the net payment date are counted from 08/01/18 (August 1). The result is to give the purchaser additional time to pay the invoice and receive the discount.

Quick TIP
The retailer pays a much lower price for items than a consumer does, and then marks up the price to the consumer.

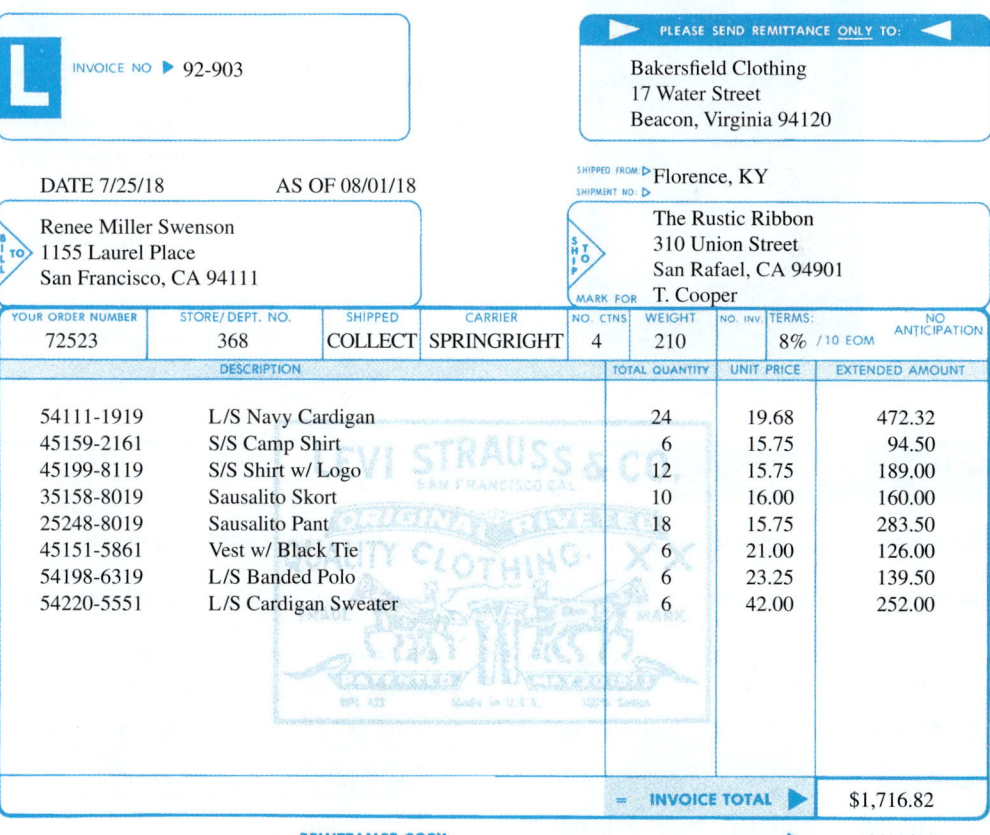

264 CHAPTER 7 Mathematics of Buying

Using Postdating AS OF with Invoices

EXAMPLE 3 An invoice for a shipment of Henkels cutlery from Germany is dated October 21 AS OF November 1 with terms of 3/15, n/30. Find **(a)** the last date on which the cash discount may be taken and **(b)** the net payment date.

ONLY 199.99
HENKELS PREMIER CLASSIC 15-PIECE SET
Includes five kitchen knives, six steak knives, fork, shears, steel, and block

SOLUTION

(a) Beginning with the postdate (AS OF) of November 1, the last date for taking the discount is November 16 (**1 + 15**).

(b) The net payment date is December 1 **(29 days remaining in November and 1 day in December)**.

QUICK CHECK 3

An invoice for Waterford Crystal from Ireland is dated August 24 AS OF September 1 with terms of 1/10, n/30. Find **(a)** the last date on which the cash discount may be taken and **(b)** the net payment date.

Determining Cash Discount Due Dates for a Sliding Scale

EXAMPLE 4 An invoice from Cellular Products is dated May 18 and offers terms of 4/10, 3/25, 1/40, n/60. Find **(a)** the three final dates for each cash discount and **(b)** the net payment date.

SOLUTION

(a) The three final cash discount dates are

4% if paid by May 28	10 days from May 18
3% if paid by June 12	25 days from May 18
1% if paid by June 27	40 days from May 18

(b) The net payment date is July 17 (20 days beyond the last cash discount date of June 27).

QUICK CHECK 4

An invoice is dated April 14 and offers terms of 3/10, 2/30, 1/40, n/60. Find **(a)** the three final dates for each cash discount and **(b)** the net payment date.

Never take more than one of the cash discounts. *With all methods of giving cash discounts, if the net payment period is not given, the net payment due date is assumed to be 20 days beyond the cash discount period.* After that date, the invoice is considered overdue. If either the final discount date or the net payment date is on a Sunday or holiday, the next business day is used.

7.3 Exercises // MyLab Math

The shaded sections below contain solutions to help you get a **QUICK START** *on the various types of exercises.*

Find the final discount date and the net payment date for each of the following. (See Examples 1 and 3.)

	Invoice Date	AS OF	Terms	Final Discount Date	Net Payment Date
1.	May 4		2/10, n/30	May 14	June 3
2.	Apr. 12		3/10, net 30	April 22	May 12
3.	June 30	July 10	3/15, n/60	_____	_____
4.	Nov. 7	Nov. 18	3/10, n/40	_____	_____
5.	Sept. 11		4/20, n/30	_____	_____
6.	July 31		2/15, net 20	_____	_____

Solve for the amount of discount and the total amount due on each of the following invoices. Add shipping and insurance charges if given. (See Examples 2 and 4.)

	Invoice Amount	Invoice Date	Terms	Date Invoice Paid	Shipping and Insurance	Amount of Discount	Total Amount Due
7.	$85.18	Nov. 2	2/10, net 30	Nov. 11	$8.72	$1.70	$92.20
	$85.18 − $1.70 + $8.72 = $92.20						
8.	$842	Mar. 8	3/10, n/30	Mar. 14	$54.80	_____	_____
9.	$78.07	May 5	net 30	June 1	$3.18	_____	_____
10.	$294	Apr. 5	net 30	May 2	$16.20	_____	_____
11.	$1080	July 8	5/10, 2/20, n/30	July 26	$62.15	_____	_____
12.	$1282	July 1	4/15, net 40	July 7	$21.40	_____	_____

13. Describe the difference between a trade discount and a cash discount. Why are cash discounts offered? (See Objective 1.)

14. Using 2/10, n/30 as an example, explain what an ordinary dating cash discount means. (See Objective 2.)

Solve the following application problems. Assume no insurance or shipping charges.

G// 15. CATERING COMPANY Bed Bath & Beyond offers cash discounts of 4/10, 2/20, net 30 to all catering companies. An invoice is dated June 18 amounting to $4635.40 and is paid on July 7. Find the amount needed to pay the invoice.

June 18 to July 7 = (30 − 18) + 7 = 19 days
Discount is 2%; $4635.40 × .98 = $4542.69

15. **$4542.69**

16. **ELECTRICAL SUPPLIES** A shipment of electrical supplies is received from Lyskovo Elec-trotechnical Works. The invoice is dated March 8, amounts to $6824.58, and has terms of 2/15, 1/20 AS OF March 20. Find the amount needed to pay the invoice on April 2.

16. _____

17. **POLICE CRUISERS** Based on a purchase of 100 police cars, a city is offered a discount of 5/5/2 with terms of net 30. Find the net price if the list price of the 100 cars was $2,604,000 and the invoice was paid within 30 days.

17. _____

G// indicates an exercise that is related to the Case in Point feature.

18. GEORGE FOREMAN GRILL A George Foreman Rotisserie Grill is list priced at $59.99 with a trade discount of 20/5/10 and terms of 4/10, n/30. Find the cost to Kitchen Crafters, assuming that both discounts are earned.

18. _____

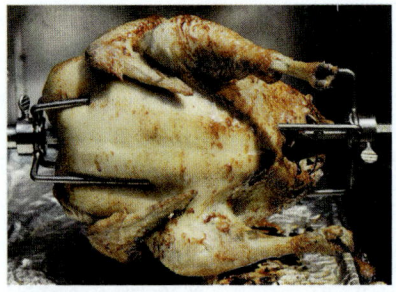

19. FINDING DISCOUNT DATES An invoice is dated January 18 and offers terms of 6/10, 4/20, 1/30, n/50. Find **(a)** the three final discount dates and **(b)** the net payment date.

(a) _____
(b) _____

20. FINDING DISCOUNT DATES An invoice with terms of 4/15, 3/20, 1/30, n/60 is dated September 4. Find **(a)** the three final discount dates and **(b)** the net payment date.

(a) _____
(b) _____

21. AS OF DATING An invoice is dated March 28 AS OF April 5 with terms of 4/20, n/30. Find **(a)** the final discount date and **(b)** the net payment date.

(a) _____
(b) _____

22. AS OF DATING An invoice is dated May 20 AS OF June 5 with terms of 2/10, n/30. Find **(a)** the final discount date and **(b)** the net payment date.

(a) _____
(b) _____

23. How do you remember the number of days in each month of the year? List the months and the number of days in each. (See Objective 2.)

24. Explain in your own words how AS OF dating (postdating) works. Why is it used? (See Objective 4.)

QUICK CHECK ANSWERS

1. (a) June 23 **(b)** July 8
2. $2870.21
3. (a) September 11 **(b)** October 1
4. (a) April 24; May 14; May 24 **(b)** June 13

7.4 Cash Discounts: Other Dating Methods

OBJECTIVES

1. Solve cash discount problems with end-of-month dating.
2. Use receipt-of-goods dating to solve cash discount problems.
3. Use extra dating to solve cash discount problems.
4. Determine credit given for partial payment of an invoice.

CASE IN POINT In addition to the ordinary dating method of cash discounts, several other cash discount methods are in common use. Jack Williams works with many invoices every day and must be able to use each discount method.

OBJECTIVE 1 Solve cash discount problems with end-of-month dating. End-of-month and **proximo** dating, abbreviated **EOM** and **prox.**, are treated the same. For example, both

3/10 EOM   and   3/10 prox.

mean that 3% may be taken as a cash discount if payment is made by the 10th of the month following the sale. The 10 days are counted from the *end of the month* in which the invoice is dated. For example, an invoice dated July 14 with terms of 3/10 EOM has a discount date 10 days from the end of the month, or the 10th of August (August 10).

Since this method increases the length of time during which a discount may be taken, it is common to add an extra month when the date of an invoice is the 26th of the month or later. For example, if an invoice is dated March 25 and the discount offered is 3/10 EOM, the last date on which the discount may be taken is April 10. *However, if the invoice is dated March 26 (or any later date in March) and the cash discount offered is 3/10 EOM, then the last date on which the discount may be taken is May 10.*

The practice of adding an extra month when the invoice is dated the 26th of a month or after is used *only* with the end-of-month (proximo) dating cash discount. It does *not* apply to any of the other cash discount methods.

Using End-of-Month Dating

EXAMPLE 1 If an invoice from Oster is dated June 10 with terms of 3/20 EOM, find **(a)** the final date on which the cash discount may be taken and **(b)** the net payment date.

SOLUTION

(a) The discount date is July 20 (20 days after the end of June).

(b) When no net payment due date is given, common business practice is to allow 20 days after the last discount date. The net payment date is August 9, which is **20 days** after the last discount date (July 20), since no net payment date is given.

QUICK CHECK 1

An invoice from American Hotel, is dated February 15 with terms of 2/10 EOM. Find **(a)** the final date on which the discount may be taken and **(b)** the net payment date.

Using Proximo Dating

EXAMPLE 2 Find the amount due on an invoice of $782 for some Black and Decker Belgian waffle makers dated August 3, if terms are 1/10 prox. and the invoice is paid on September 4.

SOLUTION

The discount may be taken up to ten days after the end of August, or until September 10. The discount is earned, since the payment date of September 4 falls before September 10.

$$\begin{aligned}\text{Amount due} &= \text{Invoice amount} - \text{Cash discount}\\ &= \$782 \quad\quad\quad\quad - .01 \times \$782 \\ &= \$774.18\end{aligned}$$

QUICK CHECK 2

An invoice for firefighting supplies is for $1475 and dated October 17. Find the amount due if the invoice terms are 3/20 proximo and the invoice is paid on November 18.

OBJECTIVE 2 Use receipt-of-goods dating to solve cash discount problems. Receipt-of-goods dating, abbreviated **ROG**, offers cash discounts determined from the date goods are actually received. This method is used in situations with longer shipping times. Electronic

268 CHAPTER 7 Mathematics of Buying

Quick TIP
If the net payment period is not given, the payment due date is assumed to be 20 days *after the cash discount date.*

invoices arrive immediately, but the buyer is given time to inspect the shipment before making payment.

For example, the discount 3/15 ROG allows a 3% cash discount if the invoice is paid within 15 days from receipt of goods. The date that goods are received is determined by the delivery date. If the invoice was dated March 5 and goods were received on April 7, the last date to take the 3% cash discount would be April 22 (April 7 plus 15 days). Since it is not given, the net payment date is 20 days after the last discount date, or May 12 (April 22 plus 20 days).

Using Receipt-of-Goods Dating

EXAMPLE 3 Best Buy received an invoice dated December 12, with terms of 2/10 ROG. The goods were received on January 2. Find **(a)** the final date on which the cash discount may be taken and **(b)** the net payment date.

SOLUTION

(a) The discount date is January 12 (**10 days** after receipt of goods, January 2 plus **10 days**).
(b) The net payment date is February 1 (**20 days** after the last discount date).

QUICK CHECK 3

An invoice from The Home Depot is dated June 4, with terms of 4/15 ROG. The merchandise was received on July 25. Find **(a)** the final date on which the cash discount may be taken and **(b)** the net payment date.

Working with ROG Dating

EXAMPLE 4 Find the amount due on a FedEx invoice for $285 with terms of 3/10 ROG. The invoice is dated June 8, goods are received on June 18, and the invoice is paid on June 30.

SOLUTION

The last date to take the 3% cash discount is June 28, 10 days after June 18. Since the invoice is paid on June 30, 2 days after the last discount date, **no cash discount may be taken**. The entire amount of the invoice must be paid.

$285   invoice amount
−   0   no cash discount
$285   amount due

QUICK CHECK 4

Find the amount due on an invoice of $896, with terms of 4/20 ROG, if the invoice is dated March 10, the goods are received April 28, and the invoice is paid on May 15.

OBJECTIVE 3 Use extra dating to solve cash discount problems. Extra dating (**extra, ex., or x**) gives the buyer additional time to take advantage of a cash discount. For example, the discount

2/10–50 extra   or   2/10–50 ex.   or   2/10–50 x

allows a 2% cash discount if the invoice is paid within 10 + 50 = 60 days from the date of the invoice. The discount is expressed as 2/10–50 ex. (rather than 2/60) to show that the 50 days are *extra*, or in addition to the normal 10 days offered.

There are several reasons for using extra dating. A supplier might extend the discount period during a slow sales season to generate more sales or gain a competitive advantage. For example, the seller might offer Christmas merchandise with extra dating to allow the buyer to take the cash discount after the holiday selling period.

Using Extra Dating

EXAMPLE 5 An invoice for several drones is dated November 23 with terms of 2/10–50 ex. Find **(a)** the final date on which the cash discount may be taken and **(b)** the net payment date.

SOLUTION

(a) The discount date is January 22 (**7 days** remaining in November + 31 days in December = 38; thus, **22 more days** are needed in January to total 60).

(b) The net payment date is February 11 (**20 days** after the last discount date).

QUICK CHECK 5

An invoice is dated July 7 with terms of 1/10–60 ex. Find **(a)** the final date on which the cash discount may be taken and **(b)** the net payment date.

Understanding Extra Dating

EXAMPLE 6 An invoice from Wind Turbines & Solar, Inc., is dated August 5, amounts to $8180, offers terms of 3/10–30 x, and is paid on September 12. Find the net payment.

SOLUTION

Step 1 The last day to take the 3% cash discount is September 14 (**August 5 + 40 days = September 14**). Since the invoice is paid on September 12, the **3%** discount may be taken.

Step 2 The **3%** cash discount is computed on $8180, the amount of the invoice. The discount to be taken is $245.40.

Step 3 Subtract the cash discount from the invoice amount to determine the amount of payment.

$$
\begin{array}{rl}
\$8180.00 & \text{invoice amount} \\
- \$245.40 & \text{3\% cash discount} \\
\hline
\$7934.60 & \text{amount of payment}
\end{array}
$$

> **Quick TIP**
> The amount of payment in Example 6 can be found by multiplying the invoice amount by (100 − 3)% or 97%.

QUICK CHECK 6

An invoice for $5412 is dated May 22 and offers terms of 2/20–40 x. If the invoice is paid on July 15, what is the amount of payment due?

OBJECTIVE 4 Determine credit given for partial payment of an invoice. Occasionally, a customer pays only a portion of the total amount due on an invoice. If this **partial payment** is made within a discount period, the customer is entitled to a discount on the portion of the invoice that is paid.

If the terms of an invoice are 3%, 10 days, then only 97% (100% − 3%) of the invoice amount must be paid during the first 10 days. So, for each $.97 paid, the customer is entitled to $1.00 of credit.

> **Finding the Balance Due After a Partial Payment Is Made in Time to Earn a Discount**
>
> 1. Find the complement of the discount percent.
> 2. Divide the partial payment by the complement of the cash discount percent.
> 3. Subtract the credit earned from the invoice amount to find the balance due.

Finding Credit for Partial Payment

EXAMPLE 7 Dave's Body and Paint receives an invoice for $1140 dated March 8 that offers terms of 2/10 prox. A partial payment of $450 is made on April 5. Find **(a)** the amount credited for the partial payment, **(b)** the balance due on the invoice, and **(c)** the cash discount earned.

SOLUTION

Step 1 Since the payment of $450 was made on April 5 or before the last discount date of April 10, it earns a cash discount.

$$\text{Complement of the discount percent} = 100\% - 2\% = \mathbf{98\%}$$

$$\text{Amount credited for partial payment} = \frac{\text{Part}}{\text{Base}} = \frac{\$450}{98\%} = \$459.18$$

Step 2 Balance due = Invoice amount − Credit given
Balance due = $1140 − **$459.18** = $680.82

Step 3 Cash discount = Credit given − Partial payment
Cash discount = **$459.18** − $450 = $9.18

A calculator solution to this example includes these three steps. First, find the amount of credit given.

$$450 \; \boxed{\div} \; .98 \; \boxed{=} \; 459.18 \quad \text{(rounded)} \; \boxed{\text{STO}}$$

Then, store the amount of credit and subtract this amount from the invoice amount to find the balance due.

$$1140 \; \boxed{-} \; \boxed{\text{RCL}} \; \boxed{=} \; 680.82$$

Finally, subtract the partial payment from the amount of credit given to find the cash discount.

$$\boxed{\text{RCL}} \; \boxed{-} \; 450 \; \boxed{=} \; 9.18$$

QUICK CHECK 7

Central Parts receives an invoice for $1082 dated September 14 that offers terms of 2/20 prox. A partial payment of $590 is paid on October 15. Find **(a)** the amount credited for the partial payment, **(b)** the balance due on the invoice, and **(c)** the cash discount earned.

Discounts are important, and managers often work to pay invoices in time to earn discounts. However, firms struggling with cash flow cannot always manage this.

7.4 Exercises // MyLab Math

The shaded sections below contain solutions to help you get a QUICK START on the various types of exercises.

Find the discount date and net payment date for each of the following. (The net payment date is 20 days after the final discount date. See Examples 1, 3, and 5.)

	Invoice Date	Terms	Date Goods Received	Final Discount Date	Net Payment Date
1.	Feb. 8	3/10 EOM		Mar. 10	Mar. 30
2.	July 6	2/10 EOM		_____	_____
3.	July 14	2/15 ROG	Sept. 3	_____	_____
4.	Jan. 15	3/15 ROG	Feb. 5	_____	_____
5.	Nov. 22	1/10–20 x		_____	_____
6.	Apr. 12	3/15–50 ex.		_____	_____

Solve for the amount of discount and the amount due on each of the following invoices. Round to the nearest cent. (See Examples 2, 4, and 6.)

	Invoice Amount	Invoice Date	Terms	Date Goods Received	Date Invoice Paid	Amount of Discount	Amount Due
7.	$682.28	June 4	3/20 ROG	July 25	Aug. 10	$20.47	$661.81
	$20.47 discount; $682.28 − $20.47 = $661.81 due						
8.	$2847.90	Mar. 29	2/15 ROG	Apr. 15	Apr. 29	_____	_____
9.	$356.20	May 17	3/15 prox.		June 12	_____	_____
10.	$11,480	Apr. 6	2/15 prox.		Apr. 30	_____	_____
11.	$2935.40	Jul. 24	4/20 EOM		Aug. 10	_____	_____
12.	$1380	May 28	1/15 EOM		June 10	_____	_____
13.	$3540	Aug. 2	3/10–20 extra		Sept. 1	_____	_____
14.	$3250.60	Oct. 17	3/15–20 ex.		Oct. 20	_____	_____

15. Explain the common business practice when no net payment date is given on an invoice. (See Objective 1.)

16. Describe why ROG dating is offered to customers. Use an example in your description. (See Objective 2.)

Solve the following application problems.

17. **KITCHEN APPLIANCES** Nostalgia Kitchen Appliances offers terms of 2/10–30 ex. to stimulate slow sales in the winter months. Bed Bath & Beyond purchased $2382.58 worth of appliances and was offered the above terms. If the invoice was dated November 3, find (a) the final date on which the cash discount may be taken and (b) the amount paid if the discount was earned.

 (a)   27   days remain in November
     + 13   days in December
      40   days is December 13

 (b) $2382.58 × .02 = 47.651 = $47.65 discount
     $2382.58 − $47.65 = $2334.93 paid

 (a) Dec. 13
 (b) $2334.93

18. **FOOD PRODUCTS** A wholesaler of food products offers terms of 4/15–40 ex. to encourage sales. A retailer was offered these terms on an order of $9864.18. If the invoice was dated March 10, find (a) the final date on which the cash discount may be taken and (b) the amount paid if the discount was earned.

 (a) _____
 (b) _____

272 CHAPTER 7 Mathematics of Buying

19. GLOBAL POSITIONING SYSTEMS A recent invoice for 24 Garmin GPS systems amounting to $6720.50 was dated February 20 and offered terms of 2/20 ROG. If the equipment was received on March 20 and the invoice was paid on April 8, find the amount due.

19. _____

20. WHEELS WITH BLING Scott Ryder purchased some 21-inch and 23-inch custom wheels for his performance auto parts store and was offered a cash discount of 2/10 EOM. The invoice amounted to $7218.80 and was dated June 2. The wheels were received 7 days later, and the invoice was paid on July 7. Find the amount necessary to pay the invoice in full.

20. _____

21. ENGLISH SOCCER EQUIPMENT An invoice dated December 8 is received with a shipment of soccer equipment from England on April 18 of the following year. The list price of the equipment is $2538, with allowed series discounts of 25/10/10. If cash terms of sale are 3/15 ROG, find the amount necessary to pay in full on April 21.

21. _____

22. CRYSTAL FROM IRELAND Bed Bath & Beyond receives an invoice for Waterford crystal from Ireland amounting to $5382.40 and dated May 17. The terms of the invoice are 5/20–90 x and the invoice is paid on September 2. Find the amount necessary to pay the invoice in full.

22. _____

23. PARTIAL INVOICE PAYMENT PetSmart receives an invoice amounting to $2016.90 with terms of 8/10, net 30 and dated August 20 AS OF September 1. If a partial payment of $1350 is made on September 8, find **(a)** the credit given for the partial payment and **(b)** the balance due on the invoice.

(a) _____

(b) _____

24. FROZEN YOGURT Yogurt for You receives an invoice amounting to $263.40 with terms of 2/20 EOM and dated September 6. If a partial payment of $150 is made on October 15, find **(a)** the credit given for the partial payment and **(b)** the balance due on the invoice.

(a) _____

(b) _____

25. DISCOUNT DATES An invoice received by Auto Zone is dated May 12 with terms of 2/10 prox. Find **(a)** the final date on which the discount may be taken and **(b)** the net payment date.

(a) _____

(b) _____

26. DISCOUNT DATES An invoice received by Family Dollar is dated November 11 with terms of 3/20 ROG, and the goods are received on December 3. Find **(a)** the final date on which the cash discount may be taken and **(b)** the net payment date.

(a) _____

(b) _____

27. AMOUNT DUE Find the amount due on an invoice of $1525 with terms of 1/20 ROG. The invoice is dated October 20, goods are received December 1, and the invoice is paid on December 20.

27. _____

28. PAYMENT DUE Find the payment that should be made on an invoice dated September 28, amounting to $4680, offering terms of 2/10–50 x and paid on November 25.

28. _____

29. PARTIAL INVOICE PAYMENT An invoice received for computer network equipment has terms of 3/15–30 x and is dated May 20. The amount of the invoice is $4402.58, and a partial payment of $3250 is made on July 1. Find **(a)** the credit given for the partial payment and **(b)** the balance due on the invoice.

(a) _____

(b) _____

30. AUTOMOTIVE Orange County Choppers makes a partial payment of $8726 on an invoice of $17,680.38. If the invoice is dated April 14 with terms of 4/20 prox. and the partial payment is made on May 13, find **(a)** the credit given for the partial payment and **(b)** the balance due on the invoice.

(a) _____

(b) _____

31. Write a short explanation of partial payment. Why would a company accept a partial payment? Why would a customer make a partial payment? (See Objective 4.)

32. Of all the different types of cash discounts presented in this section, which type seemed most interesting to you? Explain your reasons.

QUICK CHECK ANSWERS

1. (a) March 10 (b) March 30
2. $1430.75
3. (a) August 9 (b) August 29
4. $860.16
5. (a) September 15 (b) October 5
6. $5303.76
7. (a) $602.04 (b) $479.96
   (c) $12.04

274 CHAPTER 7 Mathematics of Buying

Chapter 7 Quick Review

Chapter Terms Review the following terms to test your understanding of the chapter. For each term you do not know, refer to the page number found next to that term.

AS OF [p. 263]
cash discount [p. 261]
cash on delivery (COD) [p. 247]
chain discount [p. 249]
complement [p. 250]
distribution chain [p. 247]
end of month (EOM) [p. 267]
extension total [p. 247]
extra dating (extra, ex., x) [p. 268]

free alongside ship (FAS) [p. 247]
free on board (FOB) [p. 247]
FOB destination [p. 247]
invoice [p. 247]
invoice total [p. 247]
list price [p. 247]
manufacturer [p. 247]
net cost [p. 249]
net cost equivalent [p. 250]

net price [p. 249]
ordinary dating method [p. 261]
partial payment [p. 269]
postdated "AS OF" [p. 263]
prox. [p. 267]
proximo [p. 267]
purchase invoice [p. 247]
receipt-of-goods dating (ROG) [p. 267]
retailer [p. 247]

sales invoice [p. 247]
series discount [p. 249]
single discount [p. 249]
single discount equivalent [p. 257]
suggested retail price [p. 247]
suppliers [p. 247]
supply chain [p. 247]
trade discounts [p. 249]
unit price [p. 247]
wholesaler [p. 247]

CONCEPTS

7.1 Trade discount and net cost

First find the amount of the trade discount. Then find net cost using:

Net cost = List price − Trade discount

EXAMPLES

List price, $340; discount, 15%; find the net cost.

Discount = $340 × .15 = **$51**

Net cost = $340 − **$51** = $289

7.1 Complements with respect to 1 (100%)

The complement is the number that must be added to a given discount to get 1 or 100%.

Find the complement of each with respect to 1.

Discount	Decimal Equivalent	Finding the Complement		Complement
20%	.2	1 − .2	=	.8
32%	.32	1 − .32	=	.68

7.1 Complements and series discounts

The complement of a discount is the percent paid. Multiply the complements of the series discounts to get the *net cost equivalent*.

Series discount, 10/20/10; find the net cost equivalent.

10/ 20/ 10
↓   ↓   ↓
.9 × .8 × .9 = **.648**

7.1 Net cost equivalent (percent paid) and net cost

Multiply the net cost equivalent (percent paid) by the list price to get the net cost.

List price, $280; series discount, 10/30/20; find the net cost.

10/ 30/ 20
↓   ↓   ↓
.9 × .7 × .8 = **.504** percent paid

Net cost = **.504** × $280 = $141.12

7.2 Single discount equivalent to a series discount

The single discount equivalent is found by multiplying the complements of the individual discounts to get the net cost equivalent, then subtracting from 1.

$$1 - \frac{\text{Net cost}}{\text{equivalent}} = \frac{\text{Single discount}}{\text{equivalent}}$$

Find the single discount equivalent to a 10/20/20 series discount.

10/ 20/ 20
↓   ↓   ↓
.9 × .8 × .8 = **.576**

1 − **.576** = .424 = 42.4%

7.2 Find list price, given the series discount and the net cost

First, find the net cost equivalent, percent paid. Then use the standard percent formula to find the list price (base).

$$B = \frac{P}{R}$$

New cost, $224; series discount, 20/20; find list price.

20/ 20
↓   ↓
.8 × .8 = .64

$$B = \frac{P}{R} = \frac{224}{.64} = \$350 \text{ list price}$$

CHAPTER 7 Quick Review

CONCEPTS	EXAMPLES

7.3 Number of days and dates
 	
 30-Day Months: April, June, September, November
 	
 31-Day Months: All the rest except February with 28 days (29 days in leap year)
	Date, July 24; find 10 days from date.
	 July 31 − 24 = 7 days remaining in July
	
	     10 total number of days
	   − 7 days remaining in July
	 August 3 (future date)

7.3 Ordinary dating and cash discounts
 	
 With ordinary dating, count days from the date of the invoice. Remember:
 	
   2/   10,   n/   30
   ↓   ↓   ↓   ↓
   %   days, net   days
	Invoice amount $182; terms 2/10, n/30; find the cash discount and amount due.
	
	 Cash discount: $182 × .02 = $3.64
	 Amount due: $182 − $3.64 = $178.36

7.4 Cash discounts with end-of-month dating (EOM or proximo dating) The final discount date and the net date are counted from the end of the month. If the invoice is dated the 26th or after, add the entire following month when determining the dates. If not stated, the net date is 20 days beyond the discount date.	Terms, 2/10 EOM; invoice date, Oct. 18; find the final discount date and the net payment date. Final discount date: **November 10, which is 10 days from the end of October** Net payment date: **November 30, which is 20 days beyond the discount date**
7.4 Receipt-of-goods (ROG) dating and cash discounts Time is counted from the date goods are received to determine the final cash discount date and the net payment date. If not stated, the net date is 20 days beyond the discount date.	Terms, 3/10 ROG; invoice date, March 8; goods received, May 10; find the final discount date and the net payment date. Final discount date: **May 20 (May 10 + 10 days)** Net payment date: **June 9 (May 20 + 20 days)**
7.4 Extra dating and cash discounts Extra dating adds extra days to the usual cash discount period, so 3/10–20 x is the equivalent to 3/30. If not stated, the net date is 20 days beyond the discount date.	Terms, 3/10–20 x, invoice date, January 8; find the final discount date and the net payment date. Final discount date: **February 7 (23 days in January + 7 days in February = 30)** Net payment date: **February 27 (February 7 + 20 days)**
7.4 Partial payment credit When only a portion of an invoice amount is paid within the cash discount period, credit will be given for the partial payment. Use the standard percent formula $$B = \frac{P}{R}$$ where the credit given is the base, the partial payment is the part, and (100% − cash discount) is the rate.	Invoice, $4000; terms, 2/10, n/30; invoice date, Oct. 10; partial payment of $2000 on Oct. 15. Find credit given for partial payment and the balance due on the invoice. $$B = \frac{P}{R} = \frac{\$2000}{100\% - 2\%} = \frac{\$2000}{.98} = \$2040.82$$ Credit = $2040.82 Balance due = $4000 − $2040.82 = $1959.18

Case Study

GEORGE FOREMAN

George E. Foreman grew up on the tough streets of Houston's Fifth Ward area, and as a young kid he was always getting into trouble. He joined the Job Corps and his life changed. One of the counselors noticed that he was always into fights, so he decided that George should put all of his energy into something positive. Thus began a successful boxing career, after which George started a business selling grills. If there is one thing George Foreman knows, it is good cooking!

1. Kitchen Crafters purchased George Foreman grills listed at $82.50 with a trade discount of 20/10. Find the cost of one dozen grills.

2. George Foreman's best grill lists for $119.99. The manufacturer gives a 25/10 trade discount and offers a cash discount of 3/15, net/30. Find the cost to Kitchen Crafters if both discounts are earned and taken.

3. The list price of a *George Foreman Grill Cookbook* is $16.95. If the cookbook is offered at a reduced price of $13.95 when purchased in bulk, what is the percent of markdown? Round to the nearest tenth of a percent.

INVESTIGATE

Talk with the manager of any retail business that sells products and ask: Are trade discounts common when buying from suppliers? Do you take advantage of all discounts? Do you use the Web when buying from suppliers?

Case in Point Summary Exercise

DISCOUNTS AT BED BATH & BEYOND

www.bedbathandbeyond.com

Facts:

- 1971: First formed
- 1985: Opened first superstore
- 1992: Listed on public stock exchange
- 2018: More than 1500 stores and sales of over $12 billion

Jack Williams works in a very active department called purchasing. He works with store managers, marketing, and supply companies. The people in his department try to have the right product, in the right amount, at the right place, and at the lowest cost. It is a tough job!

In early September, Williams placed an order with one vendor for products having a combined list price of $27,393. The vendor offered trade discounts of 20/10/10. The invoice arrived through the computer system the next day. It was dated September 4, had terms of 3/15 EOM, and showed a shipping charge of $748.38. Find the following.

1. The total amount of the invoice excluding shipping

2. The final discount date

3. The net payment date

4. The amount necessary to pay the invoice in full on October 11, including the shipping

5. Suppose that on October 11 the invoice is not paid in full, but a partial payment of $10,000 is made instead. Find the credit given for the partial payment and the balance due on the invoice including shipping.

1. _____

2. _____

3. _____

4. _____

5. _____
   _____

Discussion Question: Have you thought about a career in purchasing? Every company that handles products has people working in this field. See what you can find out about this career on the Web. You may wish to search for the American Purchasing Society.

Source: DISCOUNTS AT BED BATH & BEYOND, www.bedbathandbeyond.com.

Chapter 7 Test

To help you review, the numbers in brackets show the section in which the topic was discussed.

Find the net cost (invoice amount) for the following. Round to the nearest cent. [7.1]

1. List price: $348.22 less 10/20/10

2. List price: $1308 less 20/25

Find (a) the net cost equivalent and (b) the single discount equivalent for the following series discounts. [7.2]

3. 30/10

4. 20/10/20

Find the final discount date for the following. [7.4]

Invoice Date	Terms	Date Goods Received	Final Discount Date
5. Feb. 10	4/15 EOM	Feb. 16	
6. May 8	2/10 ROG	May 20	
7. Dec. 8	4/15 prox.	Jan. 5	
8. Oct. 20	2/20–40 extra	Oct. 31	

9. The following invoice was paid on November 15. Find (a) the invoice total, (b) the amount that should be paid after the cash discount, and (c) the total amount due, including shipping and insurance. [7.1–7.4]

GOURMET KITCHEN WHOLESALER

Terms: 2/10, 1/15, n/60   November 6

Quantity	Description	Unit Price	Extension Total
16	tablecloths, linen	@ 35.00 ea.	
8	rings, napkin	@ 6.50 ea.	
4 cases	cups, ceramic	@ 25.30 ea.	
12	bowls, 1 qt. stainless	@ 6.30 ea.	
		(a) Invoice Total	
		Cash Discount	
		(b) Due after Cash Discount	
		Shipping and Insurance	$38.75
		(c) Total Amount Due	

Solve the following application problems involving cash and trade discounts. Round to the nearest cent.

10. Dick's Sporting Goods made purchases at a net cost of $46,746 after a series discount of 20/20/20. Find the list price. **[7.2]**

    10. _____

11. An invoice of $3168 from Scottish Importers has cash terms of 4/20 EOM and is dated June 5. Find **(a)** the final date on which the cash discount may be taken and **(b)** the amount necessary to pay the invoice in full if the cash discount is earned. **[7.4]**

    (a) _____
    (b) _____

12. Mel's Diner purchased paper products list priced at $696 less series discounts of 10/20/10, with terms of 3/10–50 extra. If the retailer paid the invoice within 60 days, find the amount paid. **[7.4]**

    12. _____

13. The Fireside Shop offers chimney caps for $120 less 25/10. The same chimney cap is offered by Builders Supply for $111 less 25/5. Find **(a)** the firm that offers the lower price and **(b)** the difference in price. **[7.1]**

    (a) _____
    (b) _____

14. The amount of an invoice from Cloverdale Creamery is $1780 with terms of 2/10, 1/15, net 30 with shipping charges of $120.39. The invoice is dated March 8. **(a)** What amount should be paid on March 20? **(b)** What amount should be paid on April 3? **[7.3]**

    (a) _____
    (b) _____

15. Diamond Consulting receives an invoice dated November 23 for $2514 with terms of 3/15 EOM with shipping and insurance charges of $88.50. If the invoice is paid on December 14, find the amount necessary to pay the invoice in full. **[7.4]**

    15. _____

16. Bicycle Sports Shop receives an invoice amounting to $2916 with cash terms of 3/10 prox. and dated June 7. If a partial payment of $1666 is made on July 8, find **(a)** the credit given for the partial payment and **(b)** the balance due on the invoice. **[7.4]**

    (a) _____
    (b) _____

Mathematics of Selling

8

CHAPTER CONTENTS

8.1 Markup on Cost
8.2 Markup on Selling Price
8.3 Markdown
8.4 Turnover and Valuation of Inventory

CASE IN POINT

JAMES SMELTER works in inventory management at REI, which sells products for outdoor activities including cycling, camping, hiking, mountain climbing, skiing, snowboarding, and kayaking. It was founded by 23 mountain climbers, and employees still design many of the more than 40,000 products carried in inventory and listed for sale on the website. With roughly $500 million in inventory and sales of over $2 billion a year, the managing of inventory is a full-time and very important job for many REI employees.

Most companies seek to make substantial profits. REI is different! It is a cooperative owned by its 6+ million customers. At the end of every year, every REI customer gets a refund of about 10% on all purchases he or she made during the year. REI also donates money to nonprofits working to restore or protect the outdoor places that its customers love. The company is consistently listed as one of the "100 Best Companies to Work For" by *Fortune* magazine. It is one you may want to work for someday.

What price should we charge for this toaster?

The success of a retail business depends on many things. Managers must understand their customers, competitors, and trends. They must manage a supply chain from suppliers all the way to retail stores, or to a customer's home for Internet sales. They also must control costs and know how to price products. This section is about pricing products so that firms can generate enough revenue to pay for the items purchased and to pay overhead expenses such as wages and rent. They also want to have a little left over for profit.

8.1 Markup on Cost

OBJECTIVES

1. Recognize the terms used in selling.
2. Use the basic formula for markup.
3. Calculate markup based on cost.
4. Apply percent to markup problems.

CASE IN POINT James Smelter determines the selling price for ski and snowboard equipment. A price that is too high results in products that cannot be sold since customers buy elsewhere. A price that is too low results in the company not having enough revenue to pay bills.

OBJECTIVE 1

Recognize the terms used in selling

The terms used in markup are summarized here.

Cost is the amount paid to the manufacturer or supplier after trade and cash discounts have been taken. Shipping and insurance charges are included in cost.

Selling price is the price at which merchandise is sold to the public.

Markup, **margin**, or **gross profit** is selling price minus cost.

Operating expenses, or **overhead**, include the expenses of operating the business, such as wages, rent for buildings and equipment, utilities, insurance, and advertising.

Net profit (**net earnings**) is gross profit minus operating expenses.

OBJECTIVE 2 Use the basic formula for markup. Managers mark up the cost of an item before selling it. For example, REI pays $549.20 to buy a set of skis and marks the price up by $126.70. The basic **markup formula** that follows shows that the selling price is the sum of the cost and the markup.

$$\text{Selling price} = \text{Cost} + \text{Markup}$$
$$S = C + M$$
$$= \$549.20 + \$126.70$$
$$= \$675.90$$

The markup should be large enough to pay all associated operating expenses such as wages, utilities, marketing, management, and rent. It should also be large enough to have some left over for profit. However, smaller markups are sometimes used to help move inventory and to encourage customers to buy.

The information on the skis is shown here in a convenient table:

Cost	$549.20		C $549.20
+ Markup	$126.70	or	+ M $126.70
Selling price	$675.90		S $675.90

Cost is the amount paid to a supplier and includes shipping costs. Any markup remaining after paying operating expenses is profit.

Using the Basic Markup Formula

EXAMPLE 1 REI received the following items used by snowboarders. Use the basic markup formula to find the unknowns. C represents cost, M markup, and S selling price.

(a) C $34.48
    +M $13.40
    ─────────
    S $

(b) C $ 83.82
    +M $
    ─────────
    S $124.99

(c) C $
    +M $ 68.17
    ─────────
    S $227.24

SOLUTION

CASE IN POINT

(a) C $34.48
    +M $13.40
    ─────────
    S   **$47.88**

(b) C $ 83.82
    +M **$ 41.17**
    ─────────
    S $124.99

(c) C   **$159.07**
    +M $ 68.17
    ─────────
    S $227.24

QUICK CHECK 1

Find the selling price in part (a), the markup in (b), and the cost in (c).

(a) C $25
    +M $ 8
    ─────
    S $

(b) C $72
    +M $
    ─────
    S $98

(c) C $
    +M $ 35
    ─────
    S $118

OBJECTIVE 3 Calculate markup based on cost. Manufacturers often discuss inventories in terms of cost and use **markup on cost**, where markup is a stated as a percent of cost. Cost is the base, or 100% since the markup is ON COST. This is an application of the basic percent equation shown at left.

Finding Markup on Cost

$$\text{Markup on cost} = \frac{\text{Amount of markup}}{\text{Cost}} \quad \text{State as a percent.}$$

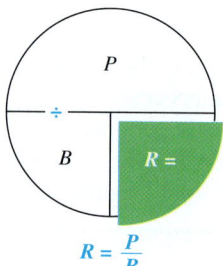

A retailer pays $90 for an MP3 player and then sells it for $126. Markup is on cost, which is the base, or 100%. The numbers 1, 2, and 3 in the following table reference the numbered calculations below.

```
           100%   C   $ 90
2 ────▶      %   M   $       ◀──── 1
3 ────▶      %   S   $126
```

Find the missing values as follows using cost as the base for all percents.

1. Selling price = Cost + Markup   or

   Markup = Selling price − Cost
          = $126 − $90 = $36

2. Markup as a percent of Cost = Markup ÷ Cost
                               = $36 ÷ $90 = 40%

3. Selling price as a percent of Cost = Selling price ÷ Cost
                                      = $126 ÷ $90 = 140%

Here is the completed table:

```
           100%   C   $ 90
            40%   M   $ 36
           140%   S   $126
```

The percents in the left column add up to 140%, and the amounts in the right column add up to the selling price of $126. All percents are in terms of cost. In other words,

Markup = 40% **of cost** and
Selling price = 140% **of cost**

OBJECTIVE 4 Apply percent to markup problems.

Solving for Percent of Markup on Cost — **EXAMPLE 2** A discount store bought hiking boots manufactured in Mexico for $60 and sells them for $81 a pair. Find the percent of markup based on cost.

SOLUTION

Set up the problem using the table form given earlier. All percents *must be in terms of cost*.

$$
\begin{array}{rll}
100\% & C & \$60 \text{ base} \\
?\% & M & \$\ ? \\
\hline
?\% & S & \$81
\end{array}
$$

Find the unknown values as follows.

$$\text{Markup} = \text{Selling price} - \text{Cost}$$
$$= \$81 - \$60 = \$21$$

$$\text{Markup percent} = \text{Markup} \div \text{Cost}$$
$$= \$21 \div \$60 = 35\%$$

$$\text{Selling price percent} = 100\% + \text{Markup percent}$$
$$= 100\% + 35\% = 135\%$$

The completed table below shows that markup is 35% of cost and selling price is 135% of cost.

$$
\begin{array}{rll}
100\% & C & \$60 \text{ base} \\
35\% & M & \$21 \\
\hline
135\% & S & \$81
\end{array}
$$

🖩 The calculator solution to the first part of this example is as follows.

(81 − 60) ÷ 60 = .35

Note: Refer to Appendix B for calculator basics.

QUICK CHECK 2

A retail buyer purchased some pedometers at a cost of $12 and will sell them for $16. What is the percent of markup based on cost? Round to the nearest tenth of a percent.

In Example 2 the cost was given; in the next example it is unknown.

Finding Cost When Cost Is Base — **EXAMPLE 3** Dick's Sporting Goods puts a markup on a dumbbell set of $16, which is 50% of the firm's cost. Complete the table.

SOLUTION

Cost is the base. Set the table up knowing that the $16 markup is 50% of the cost, which is unknown.

$$
\begin{array}{rll}
100\% & C & \$?\ \text{ base} \\
50\% & M & \$16 \\
\hline
?\% & S & \$\ ?
\end{array}
$$

Use the fact that markup of $16 is 50% of cost to find cost.

$$\text{Markup} = 50\% \times \text{Cost}$$
$$\$16 = .5 \times C$$

Divide both sides of the equation by .5 to find the cost.

$$C = \$16 \div .5 = \$32$$

> **Quick TIP**
> It is important to first find the base (the cost in this problem) and then find the other values in the table.

Complete the table by adding the percent and dollar columns to find the totals.

$$
\begin{array}{rll}
100\% & C & \$32 \text{ base} \\
+\ 50\% & M & +\$16 \\
\hline
150\% & S & \$48
\end{array}
$$

The weight set costs the retailer $32 and is marked up by $16. The selling price of $48 is 150% of the cost of $32.

QUICK CHECK 3

A 6-foot billiard table has a markup of $84, which is 35% based on cost. Find the cost and the selling price.

Finding the Markup and the Selling Price

EXAMPLE 4 Find the markup and the selling price for a belt if the cost is $23.60 and the markup is 45% of cost.

SOLUTION

Use the information given to set up the problem.

$$\begin{array}{rl} 100\% \ C & \$23.60 \ \text{base} \\ 45\% \ M & \$ \ ? \\ \hline ?\% \ S & \$ \ ? \end{array}$$

The percent column totals to 145%. Use the basic percent equation to find the following.

$$M = 45\% \text{ of Cost} = .45 \times \$23.60 = \$10.62$$

Find the selling price either by adding the cost of $23.60 to the markup of $10.62, or as follows:

$$S = 145\% \text{ of Cost} = 1.45 \times \$23.60 = \$34.22$$

The table shows that the selling price of the belt is $34.22.

$$\begin{array}{rl} 100\% \ C & \$23.60 \ \text{base} \\ 45\% \ M & \$10.62 \\ \hline 145\% \ S & \$34.22 \end{array}$$

This calculator solution uses the percent add-on feature found on many calculators.

23.6 [+] 45 [%] [=] 34.22

QUICK CHECK 4

The cost of some women's slippers is $45. If the markup is 22% of cost, find the markup and the selling price.

Finding Cost When Cost Is Base

EXAMPLE 5 REI sells a down sleeping bag rated for very cold weather for $308. If the markup on cost is 40%, find the amount that REI pays for one sleeping bag. Then find the markup.

SOLUTION

CASE IN POINT The cost is 100% and the markup is 40%, so the selling price is 140% of cost.

$$\begin{array}{rl} 100\% \ C & \$ \ ? \ \text{base} \\ 40\% \ M & \$ \ ? \\ \hline 140\% \ S & \$308 \end{array}$$

Since the base (cost) is not known, use the basic percent equation to find it.

$$\textbf{140\% of Cost} = \textbf{Selling price}$$
$$1.40 \times C = S$$
$$1.40 \times C = \$308 \quad \text{Substitute \$308 for } S.$$
$$C = \frac{\$308}{1.40} \quad \text{Divide both sides by 1.40.}$$
$$C = \$220$$

So, REI must pay $220 for the sleeping bag.

$$\begin{aligned} \text{Markup} &= \text{Selling price} - \text{Cost} \\ &= \$308 - \$220 \\ &= \$88 \end{aligned}$$

All values are given in the table.

$$
\begin{array}{rl}
100\% & C \quad \$220 \quad \text{base} \\
40\% & M \quad \$\ 88 \\
\hline
140\% & S \quad \$308
\end{array}
$$

Check the answer by making sure that both column totals are correct.

> **QUICK CHECK 5**
>
> Target sells a 2-pack of cotton socks for $4.99. If the markup is 30% on cost, how much can they afford to pay for the socks (cost)?

Finding the Cost and the Markup **EXAMPLE 6** The retail price of a 54-inch portable basketball system is $549.99. The retailer has operating expenses of 29.5% and wants a 5.5% profit, both based on cost, on this item. First find the total percent of markup on cost, then find the other values in the table.

SOLUTION

Find the percent markup required by the retailer by adding the operating expense and profit percents.

$$
\begin{aligned}
\text{Markup on cost} &= \text{operating expense} + \text{profit} \\
&= \quad\quad 29.5\% \quad\quad + 5.5\% \\
&= 35\%
\end{aligned}
$$

Now set the problem up in table form.

$$
\begin{array}{rl}
100\% & C \quad \$\ ? \quad\quad \text{base} \\
35\% & M \quad \$\ ? \\
\hline
?\ \% & S \quad \$549.99
\end{array}
$$

Quick TIP
When calculating markup on cost, *cost is always the base* and 100% always goes next to cost.

The percent column total is 135%. Find the base (cost) as follows.

$$
\text{Cost} = \frac{\text{Selling Price}}{\text{Rate}} = \frac{\$549.99}{1.35} = \$407.40
$$

$$
\begin{aligned}
\text{Markup} &= \text{Selling price} - \text{Cost} \\
&= \$549.99 - \$407.40 \\
&= \$142.59
\end{aligned}
$$

The final table is shown.

$$
\begin{array}{rl}
100\% & C \quad \$407.40 \quad \text{base} \\
35\% & M \quad \$142.59 \\
\hline
135\% & S \quad \$549.99
\end{array}
$$

The cost is $407.40 and the markup is $142.59.

> **QUICK CHECK 6**
>
> The selling price of a telescope is $64.99. If the markup is 40% of cost, find the cost and the markup to the nearest cent.

8.1 Exercises // MyLab Math

The shaded sections below contain solutions to help you get a QUICK START on the various types of exercises.

Solve for the missing numbers. Markup is based on cost. Round dollar amounts to the nearest cent. (See Examples 1–6.)

1. 100% C $12.40
    40% M $ 4.96
    140% S $ 17.36

2. 100% C $5.40
    25% M $1.35
    125% S $6.75

3.   % C $
     % M $
    120% S $32.60

286 CHAPTER 8 Mathematics of Selling

4.
%	C	$
50%	M	$ 50.00
%	S	$

5.
%	C	$
30%	M	$ 50.40
%	S	$

6.
100%	C	$78.00
%	M	$17.94
%	S	$

Find the missing numbers. Round rates to the nearest tenth of a percent and dollar amounts to the nearest cent. (See Examples 1–6.)

	Cost Price	Markup	% Markup on Cost	Selling Price
7.	$9.00	$2.70	30%	$11.70
8.	$36.00	$7.20	20%	$43.20
9.	$12.00	$7.20		
10.			100%	$68.98
11.	$153.60			$215.04
12.		$54.38	50%	
13.		$8.45		$42.25
14.	$640.70		22.3%	

15. Explain the factors that go into determining the markup. Do you think the salary of the chief operating officer (COO) of a company affects the markup on an inexpensive item? (See Objective 2 and Example 6.)

16. Write the markup formula in vertical form. Define each term. (See Objective 2.)

Solve the following application problems, using cost as a base. Round rates to the nearest tenth of a percent and dollar amounts to the nearest cent.

C// 17. EXERCYCLE REI pays $330.30 for a 6-person tent and the markup is 45% of cost. Find the markup.

100%	C	$330.30	base
45%	M	$?	
145%	S	$	

P = B × R = $330.30 × .45 = 148.635 = $148.64

17. $148.64

18. SKI JACKETS REI offers ski jackets, sizes S, M, and L, for $138. If the markup is 35% of cost, find the cost.

18. _____

19. RESPONDER BACKPACK The cost to a retailer of a responder backpack used in emergency situations is $84.96. The retailer uses a markup of 40% based on cost. Find the selling price.

19. _____

C// indicates an exercise that is related to the Case in Point feature.

20. **NEW TOYOTA** A dealer marks up the price of a new Toyota Prius to a selling price of $26,850. Find the cost of the car to the dealer if the markup is 8.5% of cost.

20. _____

21. **TABLE-TENNIS TABLES** Target purchases TIGA table-tennis tables at a cost of $180 each. If the company's operating expenses are 16% of cost, and a net profit of 7% of cost is desired, find the selling price of one table-tennis table.

21. _____

22. **TEXTBOOKS** A college bookstore pays $128.50 for the business math textbook you are currently reading. The bookstore is small and has a fairly high operating expense of 25.4%. The manager of the bookstore also wants a profit of 8.1%. Find the selling price per book and total selling price for 50 books.

22. _____

23. **OUTDOOR LIGHTING** Patios Plus sold an outdoor lighting set for $119.95. The markup on the set was $23.99. Find **(a)** the cost, **(b)** the markup percent on cost, and **(c)** the selling price as a percent of cost.

(a) _____
(b) _____
(c) _____

24. **GOLF CLUBS** Dick's Sporting Goods has a markup of $46.64 on golf clubs sold for $222.64. Find **(a)** the cost, **(b)** the markup percent on cost, and **(c)** the selling price as a percent of cost.

(a) _____
(b) _____
(c) _____

25. **TRACTOR PARTS** Bismark Tractor put a markup of 26% on cost on a part for which it paid $450. Find **(a)** the selling price as a percent of cost, **(b)** the selling price, and **(c)** the markup.

(a) _____
(b) _____
(c) _____

26. **CUSTOM-MADE JEWELRY** A jewelry dealer sold custom-made necklaces at a selling price that was 250% of his cost. If the markup is $135, find **(a)** the markup percent on cost, **(b)** the cost, and **(c)** the selling price.

(a) _____
(b) _____
(c) _____

QUICK CHECK ANSWERS

1. (a) $33 (b) $26 (c) $83
2. 33.3%
3. $240 cost; $324 selling price
4. $9.90 markup; $54.90 selling price
5. $3.84
6. $46.42 cost; $18.57 markup

8.2 Markup on Selling Price

OBJECTIVES

1. Understand the phrase *markup based on selling price*.
2. Solve markup problems when selling price is the base.
3. Use the markup formula to solve variations of markup problems.
4. Determine the percent markup on cost and the equivalent percent markup on selling price.
5. Convert markup percent on cost to markup percent on selling price.
6. Convert markup percent on selling price to markup percent on cost.
7. Find the selling price for perishables.

CASE IN POINT REI faces stiff competition from retailers such as Dick's Sporting Goods, as well as from smaller chains that specialize in road bicycles or skis, for example. Competition is often very serious for many firms.

Tour du France

OBJECTIVE 1 Understand the phrase *markup based on selling price*. Manufacturers often use markup on cost. However, retailers often prefer to use **markup on selling price**, meaning that selling price is the base, or 100%.

> **Markup on Cost:** Cost is the base, or 100%.
> **Markup on Selling Price:** Selling price is the base, or 100%.

> **Finding Markup on Selling Price**
>
> $$\text{Markup on selling price} = \frac{\text{Amount of markup}}{\text{Selling price}}$$

The same basic markup formula is used when using markup on selling price: $C + M = S$.

OBJECTIVE 2 Solve markup problems when selling price is the base. The table form used for these problems is identical to that of the previous section. However, *selling price is now the base* rather than cost. Since selling price is the base, put 100% adjacent to selling price, as shown.

? %	C $	?
%	M $	
100%	S $	base

Solving for Markup on Selling Price

EXAMPLE 1 During a sale, REI sells one model of kids sunglasses for $39.99. They pay $35 for each pair and calculate markup on selling price. Find the amount of markup, the percent of markup on selling price, and the percent of cost on selling price.

SOLUTION

Set up the problem.

$$
\begin{array}{rll}
?\% & C & \$35.00 \\
?\% & M & \$? \\
\hline
100\% & S & \$39.99 \quad \text{base}
\end{array}
$$

Markup = Selling price − Cost
       = $39.99   − $35 = $4.99

Solve for either of the rates, noting that the selling price is the base, or 100%. We will solve for the markup rate first.

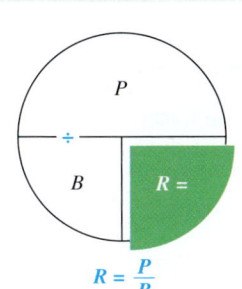

$$
\text{Rate} = \frac{\text{Markup}}{\text{Selling price}}
$$

$$
= \frac{\$4.99}{\$39.99}
$$

$$
= 12.5\% \quad (\text{rounded})
$$

Cost as a percent of selling price can be found either by subtracting 100% − 12.5%, or by dividing the cost of $35 by the selling price of $39.99.

$$
\begin{array}{rll}
87.5\% & C & \$35.00 \\
12.5\% & M & \$\ 4.99 \\
\hline
100.0\% & S & \$39.99 \quad \text{base}
\end{array}
$$

Here, selling price is the base and is associated with 100%. The markup in this example is very low—REI will probably take a loss on these sunglasses, but managers hope the low price will bring customers into the store.

QUICK CHECK 1

Walmart's cost of plain cotton gym shorts made in China is $13.50 and the selling price is $19.99. Find **(a)** the amount of markup, **(b)** the percent of markup on selling price, and **(c)** the percent of cost on selling price.

Markups vary widely due to the different costs of merchandise, operating costs, local competition, and the rate at which inventory is sold or turned over. The following table gives some average markups.

Average Markups for Retail Stores (Markup on Selling Price)

Type of Store	Markup	Type of Store	Markup
General merchandise stores	29.97%	Furniture and home furnishings	35.75%
Grocery stores	22.05%	Bars	52.49%
Motor vehicle dealers (new)	12.83%	Restaurants	56.35%
Gasoline service stations	14.47%	Drug and proprietary stores	30.81%
Other automotive dealers	29.57%	Liquor stores	20.19%
Apparel and accessories	37.64%	Sporting goods and bicycle shops	29.72%

(*Source:* Sole-proprietorship income tax returns, U.S. Treasury Dept., Internal Revenue Service, Statistics Division.)

OBJECTIVE 3 Use the markup formula to solve variations of markup problems. The same formula is used for markup on selling price as for markup on cost problems: Selling price = Cost + Markup. However, be careful to first identify the base; then put the 100% adjacent to it in the table.

Finding Cost When Selling Price Is Base

EXAMPLE 2 A Walmart employee needs a 35% markup on selling price in order to have a markup of $5.16 on a bottle of aspirin. How much can Walmart pay per bottle?

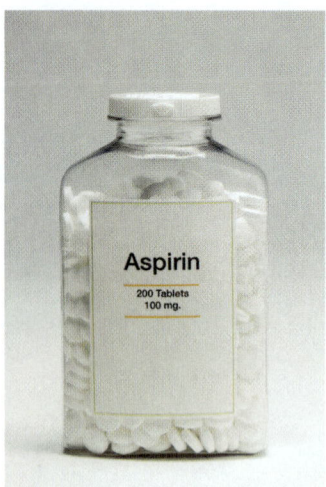

SOLUTION

Cost as a percent of selling price is found by subtracting 35% from 100% to find 65%.

$$
\begin{array}{lll}
65\% & C & \$\ ? \\
35\% & M & \$5.16 \\
\hline
100\% & S & \$\ ? \quad \text{base}
\end{array}
$$

Find the selling price as follows.

$$\text{Markup} = 35\% \text{ of selling price}$$
$$\$5.16 = .35 \times S$$
$$\frac{\$5.16}{.35} = \frac{.35 \times S}{.35} \quad \text{Divide both sides by .35.}$$
$$S = \$14.74 \quad \text{Selling price is the base.}$$

Finally, find cost by subtracting.

$$C = S - M = \$14.74 - \$5.16 = \$9.58$$

The final table is shown here.

$$
\begin{array}{lll}
65\% & C & \$\ 9.58 \\
35\% & M & \$\ 5.16 \\
\hline
100\% & S & \$14.74
\end{array}
$$

QUICK CHECK 2

A manager needs a 15% markup on selling price to obtain a markup of $13.50 on a pair of running shoes. How much can she afford to pay for each pair?

Finding Markup When Selling Price Is Given

EXAMPLE 3 Marilyn Westby must calculate the markup on a Wilson youth tennis racket. The selling price of the tennis racket is $15.99, and the markup is 20% of selling price. Find the markup.

SOLUTION

Cost as a percent of selling price is found by subtracting 20% from 100% to find 80%.

$$
\begin{array}{lll}
80\% & C & \$\ ? \\
20\% & M & \$\ \underline{} \\
\hline
100\% & S & \$15.99 \quad \text{base}
\end{array}
$$

Solve for the missing values using $P = R \times B$:

$$\text{Markup} = \$15.99 \times .2 = \$3.20$$

The markup is $3.20. Cost can be found by finding 80% of the selling price.

QUICK CHECK 3

The selling price of a three-pack of tennis balls is $9.99, and the markup is 25% of the selling price. Find the markup.

Finding Markup When Cost Is Given

EXAMPLE 4 Find the markup on a dartboard made in England if the cost is $27.45 and the markup is 25% of selling price.

SOLUTION

Subtract 25% from 100% to find that cost is 75% of selling price.

$$
\begin{array}{lll}
75\% & C & \$27.45? \\
25\% & M & \$\ ? \\
\hline
100\% & S & \$\ ? \quad \text{base}
\end{array}
$$

First, solve for selling price.

$$\text{Cost} = 75\% \text{ of Selling price}$$
$$\$27.45 = .75 \times S$$
$$S = \$27.45 \div .75 \quad \textcolor{blue}{\text{Divide both sides by .75.}}$$
$$S = \$36.60$$

Finally:

$$\text{Markup} = \text{Selling price} - \text{Cost} = \$36.60 - \$27.45$$
$$= \$9.15$$

Here is the completed table.

75%	C	$27.45
25%	M	$ 9.15
100%	S	**$36.60** base

Quick TIP
When calculating markup on selling price, *selling price* is *always the base* and 100% always goes next to selling price.

QUICK CHECK 4
Find the markup on a sleeping bag if the cost is $40.80 and the markup is 32% of selling price.

OBJECTIVE 4 Determine the percent markup on cost and the equivalent percent markup on selling price. Sometimes a markup based on cost must be compared with a markup based on selling price. For example, a salesperson who sells to both manufacturers which use markup on cost and to retailers which use markup on selling price might have to make conversions from one markup method to the other.

Determining Equivalent Markups

EXAMPLE 5 A manufacturer makes and sells fishing lures. One lure has a cost of $2.10 and is sold to distributors for $3.20. Find the percent markup on cost and also the percent markup on selling price.

SOLUTION

First set up the problem using *cost as the base*, or 100%.

100%	C	$2.10 base
? %	M	$?
? %	S	$3.20

Markup = S − C = $3.20 − $2.10 = $1.10
Markup as a percent of cost = $1.10 ÷ $2.10 = 52.4%   (rounded)
Selling price as a percent of cost = 100% + 52.4% = 152.4%

Quick TIP
Costs are marked up at each link in a supply chain. Manufacturers mark up the price when selling to distributors, who then mark up the price when selling to retailers.

Next, set up the table with *selling price as the base*, or 100%. The markup of $1.10 remains the same and can be entered in the table.

? %	C	$2.10
? %	M	$1.10
100%	S	$3.20 base

Cost as a percent of selling price = $2.10 ÷ $3.20 = 65.6%   (rounded)
Markup as a percent of selling price = 100% − 65.6% = 34.4%

Both completed tables are shown below.

Markup on Cost

100%	C	$2.10 base
52.4%	M	$1.10
152.4%	S	$3.20

Markup on Selling Price

65.6%	C	$2.10
34.4%	M	$1.10
100%	S	$3.20 base

A 52.4% markup on cost is the same as a 34.4% markup on selling price.

QUICK CHECK 5
The cost of a kick scooter is $45 and it sells for $59.99. Find **(a)** the percent of markup on cost and **(b)** the percent of markup on selling price.

Some managers talk in terms of markup on cost, others in terms of markup on selling price. Neither approach is any better than the other – managers simply use both methods for the same concept.

292 CHAPTER 8 Mathematics of Selling

OBJECTIVE 5 Convert markup percent on cost to markup percent on selling price.
Another way to convert between markup on cost and markup on selling price is to use **conversion formulas**.

> **Converting Markup Percent on Cost to Markup Percent on Selling Price**
>
> $$\frac{\%\text{ markup on cost}}{100\% + \%\text{ markup on cost}} = \%\text{ markup on selling price}$$

Converting Markup on Cost to Markup on Selling Price

EXAMPLE 6 Convert a markup of 25% on cost to its equivalent markup on selling price.

SOLUTION
Use the formula for converting markup percent on cost to markup percent on selling price.

$$\frac{\%\text{ markup on cost}}{100\% + \%\text{ markup on cost}} = \%\text{ markup on selling price}$$

$$\frac{25\%}{100\% + 25\%} = \frac{25\%}{125\%} = \frac{.25}{1.25} = .20 = 20\%$$

As shown, a markup of 25% on cost is equivalent to a markup of 20% on selling price.

🖩 The markup on cost (25%) is divided by 100% plus the markup on cost.

25 % ÷ (100 % + 25 %) = 0.2

QUICK CHECK 6
Convert a markup of 100% on cost to a markup on selling price.

OBJECTIVE 6 Convert markup percent on selling price to markup percent on cost.

> **Converting Markup Percent on Selling Price to Markup Percent on Cost**
>
> $$\frac{\%\text{ markup on selling price}}{100\% - \%\text{ markup on selling price}} = \%\text{ markup on cost}$$

Converting Markup on Selling Price to Markup on Cost

EXAMPLE 7 Convert a markup of 20% on selling price to its equivalent markup on cost.

SOLUTION
Use the formula for converting markup percent on selling price to markup percent on cost.

$$\frac{\%\text{ markup on selling price}}{100\% - \%\text{ markup on selling price}} = \%\text{ markup on cost}$$

$$\frac{20\%}{100\% - 20\%} = \frac{20\%}{80\%} = \frac{.2}{.8} = .25 = 25\%$$

A markup of 20% on selling price is equivalent to a markup of 25% on cost.

QUICK CHECK 7
Convert a markup of 40% on selling price to a markup on cost.

The table shows common markups and their equivalents.

Quick TIP
Retail managers use both methods and expect employees to understand both.

Markup Equivalents

Markup on Cost	Markup on Selling Price
20%	$16\frac{2}{3}\%$
25%	20%
50%	$33\frac{1}{3}\%$
75%	$42\frac{6}{7}\%$
100%	50%

8.2 Markup on Selling Price

OBJECTIVE 7 Find the selling price for perishables. Perishable items such as donuts or meat spoil and are then thrown out. This must be taken into account when finding the selling price of perishable items.

Several of the top ten retailers in the chart sell perishables, and so take spoilage into consideration when determining selling price.

Numbers in the News

Annual Sales of Top Ten Retail Giants
(in billions of dollars)

1. Walmart Stores — $353
2. The Kroger Co. — $104
3. Costco — $84
4. The Home Depot — $79
5. Walgreens — $77
6. Target — $73
7. CVS Health — $72
8. Amazon.com — $62
9. Albertsons — $58
10. Lowe's Companies — $58

Source: **Financial statements.**

Finding Selling Price for Perishables

EXAMPLE 8 New York Bagels bakes 60 dozen bagels at a cost of $6.48 per dozen. Generally an average of 5% of the bagels remain unsold at the end of the day and are donated to a homeless shelter. If a markup of 50% on selling price is needed, find the selling price per dozen.

SOLUTION

Step 1 Find the total cost to make all 60 dozen.

$$\text{Cost} = 60 \text{ dozen} \times \$6.48 = \$388.80$$

Step 2 Use a markup of 50% on selling price to find the selling price. Since selling price is the base, put 100% beside the S in the table.

50%	C	$388.80
50%	M	$
100%| S	$?	base

$$\text{Base} = \frac{\text{Part}}{\text{Rate}} = \frac{\$388.80}{.5} = \$777.60 \quad \text{revenue from all sales}$$

Step 3 Find the number of dozens that will be sold. Since 5% will not be sold, 100% − 5% or 95% will be sold.

$$95\% \times 60 \text{ dozen} = \mathbf{57} \text{ dozen bagels will be sold}$$

A total revenue from sales of $777.60 must be received from the 57 dozen sold.

Step 4 Find the selling price per dozen by dividing the total revenue by the number of dozens sold.

$$\frac{\$777.60}{\mathbf{57 \text{ dozen sold}}} = \mathbf{\$13.64} \text{ selling price per dozen}$$

A selling price of $13.64 per dozen results in the desired markup of 50% on selling price, assuming 5% of the bagels are donated to the homeless shelter. Managers often modify the price to a round number such as $13.75 per dozen.

294 CHAPTER 8 Mathematics of Selling

> **QUICK CHECK 8**
>
> The Cookie Jar bakes 80 dozen cookies at a cost of $1.08 per dozen. If a markup of 60% on selling price is needed and 10% of the cookies remain unsold and will be donated, find the selling price per dozen cookies.

8.2 Exercises // MyLab Math

The shaded sections below contain solutions to help you get a **QUICK START** *on the various types of exercises.*

Solve for the missing numbers. Markup is based on selling price. Round dollar amounts to the nearest cent. (See Examples 1–4.)

1.	75%	C	**$21.00**	**2.**	60%	C	$18.60	**3.**	C $145.00
	25%	M	**$ 7.00**		40%	M	$12.40		M
	100%	S	**$28.00**		100%	S	$31.00		100% S $250.00
4.	$66\frac{2}{3}\%$	C		**5.**	50%	C	$2025	**6.**	65% C
		M	$ 89.00			M			M $ 527.80
		S				S			S

Find the missing quantities by first computing the markup on one base and then computing the markup on the other. Round rates to the nearest tenth of a percent and dollar amounts to the nearest cent. (See Example 5.)

	Cost	Markup	Selling Price	% Markup on Cost	% Markup on Selling Price
7.	**$1920**	$480.00	**$2400.00**	25%	20%
8.	**$357.52**	$78.48	$436.00	**22%**	**18%**
9.	$13.80				38%
10.	$33.75		$67.50		
11.		$300.00		40%	
12.	$5.15		$15.45		

Find the equivalent markups on either cost or selling price using the appropriate formula. Round to the nearest tenth of a percent. (See Examples 6 and 7.)

	Markup on Cost	Markup on Selling Price		Markup on Cost	Markup on Selling Price
13.	100%	**50%**	14.		20%
	$\frac{100\%}{100\% + 100\%} = \frac{1}{2} = .5 = 50\%$				
15.	18%		16.	50%	

17. Use the table on page 289 to find the three types of retail stores with the lowest markups. Why do markups differ so much from one type of retail store to another?

18. Explain the difference between markup on cost and markup on selling price. Why are both used by managers? (See Sections 8.1 and 8.2.)

C// indicates an exercise that is related to the Case in Point feature.

Solve the following application problems. Round rates to the nearest tenth of a percent and dollar amounts to the nearest cent.

19. **HOME-WORKOUT EQUIPMENT** Dick's Sporting Goods has a markup of $437.50 on a stair-stepper. If this is a 35% markup on selling price, find **(a)** the selling price, **(b)** the cost, and **(c)** the cost as a percent of selling price.

    (a) _____
    (b) _____
    (c) _____

20. **SWIMMING POOL PUMP** Leslie's Pool Supply pays $187.19 for a new pool pump. If the markup is 28% on the selling price, find **(a)** the cost as a percent of selling price, **(b)** the selling price, and **(c)** the markup.

    (a) _____
    (b) _____
    (c) _____

21. **BLOUSES** Jana's Boutique purchased 20 blouses for $930. It sold 12 of the blouses for $85 each, 5 blouses for $68 each, and the remaining blouses for $49 each. Find **(a)** the total amount received for the blouses, **(b)** the total markup, **(c)** the markup percent on selling price, and **(d)** the equivalent markup percent on cost.

    (a) _____
    (b) _____
    (c) _____
    (d) _____

22. **RIVER-RAFT SALES** Olympic Sports purchased 380 river rafts for $7600. They sold 158 of the rafts at $45 each, 74 at $35 each, 56 at $30 each, and the remainder at $25 each. Find **(a)** the total amount received for the rafts, **(b)** the total markup, **(c)** the markup percent on selling price, and **(d)** the equivalent markup percent on cost.

    (a) _____
    (b) _____
    (c) _____
    (d) _____

23. **SELLING BANANAS** The produce manager at Tom Thumb Market anticipates that 10% of the bananas purchased will spoil and have to be thrown out. She buys 500 pounds of bananas at $.56 per pound. Find the selling price per pound if she wants a markup of 45%.

    23. _____

24. **PIZZA** The Pizza Shoppe makes and sells a lot of pizzas, but finds that 8% of them typically do not sell. If they produce 100 pizzas at a cost of $4.50 each and desire a markup of 70% on selling price, find the selling price per pizza.

    24. _____

QUICK CHECK ANSWERS

1. (a) $6.49 (b) 32.5%
   (c) 67.5%
2. $76.50
3. $2.50
4. $19.20
5. (a) 33.3%
   (b) 25%
6. 50%
7. 66.7%
8. $3 per dozen

Supplementary Application Exercises on Markup

Solve each of the following application problems. Round rates to the nearest tenth of a percent and dollar amounts to the nearest cent.

1. **DRUMS** Baytown Music pays $1040 for a Yamaha extreme electronic drum set. Find the markup if the markup percent on cost is 53.8%.

   1. _____

2. **BED IN A BAG** A complete comforter and sheet set has a cost of $132 to the retailer and is marked up 45% on cost. Find the markup.

   2. _____

3. **SNOWBOARD** What percent of markup on cost must be used if an Atlantis Snowboard costing $335 is sold for $399?

   3. _____

4. **WINE** Benjamin's Discount Wines sells a particular bottle of wine for $15.95. If it pays an average of $9.97 per bottle, find the percent markup on cost.

   4. _____

5. **AUTOMOTIVE SUPPLIES** An auto-parts dealer pays $41.88 per dozen cans of Chevron Techron fuel-injector cleaner, and the markup is 50% on selling price. Find the selling price per can.

   5. _____

6. **DIGITAL THERMOMETERS** A local pharmacy pays $102.50 for twenty-five digital thermometers. Find the selling price to the pharmacy per thermometer if the markup is 50% of selling price.

   6. _____

7. **CEILING FANS** The Fan Gallery's cost for a ceiling fan is $92.82, and its markup is 22% on the selling price. Find the selling price.

   7. _____

C// 8. **FLY-FISHING** REI pays $62.40 for a fly rod and marks it up 35% on selling price. Find the selling price.

   8. _____

9. **GOLF CLUBS** A golf shop pays $178.60 for a set of golf clubs. Using a markup of 40% on selling price, find **(a)** the cost as a percent of selling price, **(b)** the selling price, and **(c)** the markup.

   (a) _____
   (b) _____
   (c) _____

C// indicates an exercise that is related to the Case in Point feature.

Supplementary Application Exercises on Markup

10. DOUBLE-PANE WINDOWS Building Supply pays $3808 for all the double-pane windows needed for a small home. If the markup on the windows is 15% on selling price, what is **(a)** the cost as a percent of selling price, **(b)** the selling price, and **(c)** the markup?

(a) _____
(b) _____
(c) _____

11. COMMUNICATION EQUIPMENT A store purchased Blu-ray players at a cost of $288 per dozen. If the store needs 20% of cost to cover operating expenses and 15% of cost for the net profit, what are **(a)** the selling price of a DVD player and **(b)** the percent of markup on selling price?

(a) _____
(b) _____

12. BOWLING EQUIPMENT The Bowlers Pro-Shop determines that operating expenses are 23% of selling price and desires a net profit of 12% of selling price. If the cost of a team shirt is $29.25, what are **(a)** the selling price and **(b)** the percent of markup on cost?

(a) _____
(b) _____

13. MOUNTAIN BIKE REI advertises a fat-tire mountain bike for $199.90. If the store's cost is $2100 per dozen, what are **(a)** the markup per bicycle, **(b)** the percent of markup on selling price, and **(c)** the percent of markup on cost?

(a) _____
(b) _____
(c) _____

14. FLASH DRIVE Office Depot advertises 256 GB USB 3.0 flash drives for $49.99. Their cost is $449.91 per dozen. Find **(a)** the markup per flash drive, **(b)** the percent of markup on selling price, and **(c)** the percent of markup on cost.

(a) _____
(b) _____
(c) _____

15. LONG-STEMMED ROSES An Albertsons buys 144 dozen long-stemmed roses for $1890. If 25% of the roses cannot be sold and a markup of 100% on cost is needed, find the regular selling price per dozen roses.

15. _____

16. SPORTSWEAR Costco buys 2000 baseball caps at $5.00 per hat. If a markup of 50% on selling price is needed and 5% of the caps are damaged and cannot be sold, what is the selling price of each cap?

16. _____

8.3 Markdown

OBJECTIVES

1. Define the term *markdown* when applied to selling.
2. Calculate markdown, reduced price, and percent of markdown.
3. Define the terms associated with loss.
4. Determine the break-even point and operating loss.
5. Determine the amount of a gross or absolute loss.

CASE IN POINT James Smelter keeps a close eye on inventory at REI. One model of ski boot is not selling, so he marks the price down to move this model of boot out of the store. It is important to stock only items that are selling.

OBJECTIVE 1 Define the term *markdown* when applied to selling. The price of merchandise that is not selling well or is no longer in fashion is often marked down to move it out of the store. The difference between the original selling price and the reduced selling price is called the **markdown**. The selling price after the markdown is called the **reduced price**, **sale price**, or **actual selling price**. The basic **formula for markdown** is as follows.

Finding Reduced Price

$$\text{Reduced price} = \text{Original price} - \text{Markdown}$$

Finding the Reduced Price

EXAMPLE 1 Dick's Sporting Goods has reduced, or marked down, the price of a home gym. Find the reduced price if the original price was $2879 and the markdown is 30%.

SOLUTION

The markdown is 30% of $2879, or .3 × $2879 = $863.70. Find the reduced price as follows.

OBJECTIVE 2 Calculate markdown, reduced price, and percent of markdown.

$$\begin{array}{rl}
\$2879.00 & \text{original price} \\
- 863.70 & \text{markdown } (.30 \times \$2879) \\
\hline
\$2015.30 & \text{reduced price (70\% of original price)}
\end{array}$$

Quick TIP

The original selling price is the base or 100%, and the percent of markdown is calculated on the original selling price.

The calculator solution to this example uses the complement, with respect to 1, of the discount.

2879 × (1 − .3) = 2015.3

QUICK CHECK 1

Ski boots originally priced at $289.99 were marked down 40%. Find the reduced price.

The next example shows how to find **percent of markdown**.

Calculating the Percent of Markdown

EXAMPLE 2 The total inventory of coffee mugs at a gift shop has a retail value of $785. If the mugs were sold at reduced prices that totaled $530, what is the percent of markdown on the original price?

SOLUTION

First find the amount of the markdown.

$$\begin{array}{rl}
\$785 & \text{original price} \\
-530 & \text{reduced price} \\
\hline
\$255 & \text{markdown}
\end{array}$$

Find the percent markdown based on the original price, which is the base and goes in the denominator.

$$\text{Rate} = \frac{\text{Part}}{\text{Base}} = \frac{255}{785} = .3248 = 32.5\% \text{ markdown (rounded)}$$

The mugs were sold at a markdown of 32.5%.

QUICK CHECK 2
The entire inventory of decorations remaining after Christmas has a retail value of $1836. If the decorations are sold at reduced prices totaling $459, what is the percent of markdown on the original price?

Finding the Original Price

EXAMPLE 3 Target offers a child's car seat at a reduced price of $63 after a 25% markdown from the original price. Find the original price.

SOLUTION

After the 25% markdown, the reduced price of $63 represents 75% of the original price. The original price, or base, must be found.

$$\text{Base} = \frac{\text{Part}}{\text{Rate}} = \frac{\$63}{.75} = \$84 \text{ original price}$$

The original price of the car seat was $84.
  Check the answer by subtracting 25% of $84 from $84: $84 - (.25 \times \$84) = \$63$.

QUICK CHECK 3
Some winter gloves are sold at a reduced price of $17.99 after a markdown of 40% from the original price. Find the original price.

OBJECTIVE 3 Define the terms associated with loss. Markdowns are used to stimulate sales. They may be used when an item is not selling well, or perhaps when a firm has financial problems and needs to convert inventory to cash. However, markdowns are also used as part of a marketing strategy based on the fact that most people like the idea of getting a good deal by buying something "on sale."

The **break-even point** is the selling price that just covers the cost of the item plus overhead, which includes rent, utilities, marketing, accounting, etc. A company does not make or lose money on items sold at the break-even point. A **reduced net profit** occurs when an item is marked down, but still sold above the break-even point. An **operating loss** occurs when the selling price is below the break-even point. Finally, an **absolute loss** (**gross loss**) occurs if the selling price is less than the amount paid for the item.

Quick TIP
Think about a sweater for sale at a retailer. Management expects:

Cost of sweater to retailer	$65
Operating expenses	$25
Desired profit	$10
Original selling price	$100

If the sweater does not sell, the price is marked down. The retailer's break-even point is the cost of the sweater plus operating expenses: $65 + $25 = $90.

Price Marked Down To

$96	Reduced profit of $6
$90	Break-even (no profit)
$78	Operating loss of $12
$45	Absolute loss of $20

These formulas may help:

> Break-even point = Cost + Operating expenses
> Operating loss = Break-even point − Reduced selling price
> Absolute loss = Cost − Reduced selling price

Determining a Profit or a Loss

EXAMPLE 4 Appliance Giant paid $1600 for a 75-inch LCD flat-panel HDTV. If operating expenses are 30% of cost and the television is sold for $2000, find the amount of profit or loss.

SOLUTION

Operating expenses are 30% of cost.

$$\text{Operating expenses} = .30 \times \$1600 = \$480$$

The break-even point for the LCD HDTV is

$$\begin{aligned}\text{Break-even point} &= \text{Cost} + \text{Operating expenses} \\ &= \$1600 + (.3 \times \$1600) \\ &= \$1600 + \$480 \\ &= \$2080\end{aligned}$$

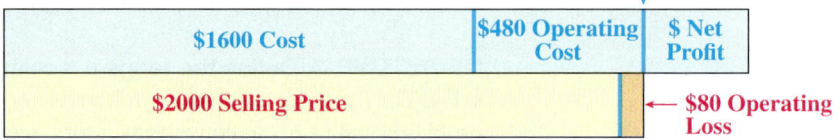

A profit is made if the television sells for more than the break-even point of $2080. A loss occurs if it sells for less than $2080. It sells for $2000, so

$$\text{Loss} = \$2080 - \$2000 = \$80$$

Since the selling price is less than the break-even point but greater than the cost to the retailer of $1600, the $80 is an operating loss—which is a real loss for the company.

 The calculator solution to this example follows.

1600 [+] [(] .3 [×] 1600 [)] [−] 2000 [=] 80

QUICK CHECK 4

Big Chime Electronics paid $480 for a flat-screen television set. If operating expenses are 35% of cost and the television is sold for $600, find the amount of the operating loss.

Determining the Operating Loss and the Absolute Loss

EXAMPLE 5 A ping pong table that normally sells for $360 at Dick's Sporting Goods is marked down 30%. If the cost of the table is $260 and the operating expenses are 20% of cost, find (a) the operating loss and (b) the absolute loss.

SOLUTION

(a) $\begin{aligned}\text{Break-even point} &= \text{Cost} + \text{Operating expenses} \\ &= \$260 + 20\% \text{ of } \$260 \\ &= \$260 + \$52 \\ &= \$312\end{aligned}$

$\text{Reduced price} = \$360 - (.3 \times \$360) = \$360 - \$108 = \$252$
$\text{Operating loss} = \$312 \text{ break-even point} - \$252 \text{ reduced price} = \$60$

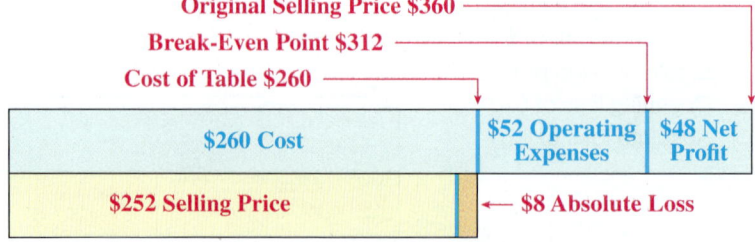

(b) The absolute loss is the cost of the table minus the selling price.

$$\text{Absolute Loss} = \$260 \text{ Cost} - \$252 \text{ Reduced Price} = \$8$$

> **QUICK CHECK 5**
>
> A propane heater normally selling for $290 is marked down 25%. If the cost of the heater is $220 and the operating expenses are 15% of cost, find **(a)** the operating loss and **(b)** the absolute loss.

The following bar graph shows the percent of adults who get an emotional high from making certain purchases. Customers love buying things—especially when they are on sale. This is valuable information to manufacturers, retailers, and merchandisers. In fact, companies aggressively track your preferences and target you with advertising focused on your interests.

Quick TIP

Try an experiment on your computer. Search for something you want to buy such as a pair of shoes. Then go on other websites and watch ads pop up for shoes.

Companies write information on your computer called *cookies* that allow them to hit you with focused advertising. Companies such as Google (Alphabet) make a lot of money doing this.

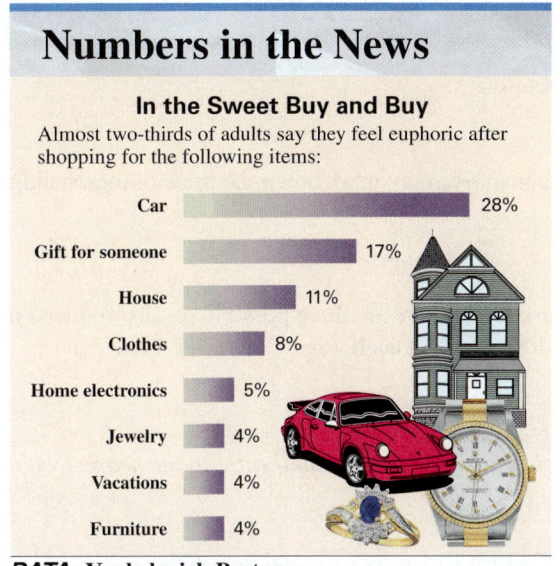

Numbers in the News

In the Sweet Buy and Buy

Almost two-thirds of adults say they feel euphoric after shopping for the following items:

- Car — 28%
- Gift for someone — 17%
- House — 11%
- Clothes — 8%
- Home electronics — 5%
- Jewelry — 4%
- Vacations — 4%
- Furniture — 4%

DATA: Yankelovich Partners.

8.3 Exercises // MyLab Math

The shaded sections below contain solutions to help you get a **QUICK START** *on the various types of exercises.*

Find the missing quantities. Round rates to the nearest whole percent and dollar amounts to the nearest cent. (See Examples 1–3.)

	Original Price	% Markdown	$ Markdown	Reduced Price
1.	$860	25%	$215	$645
	$R = \frac{P}{B} = \frac{\$215}{\$860} = .25 = 25\%; \$860 - \$215 = \$645$			
2.	$240	40%	$96	$144
	$B = \frac{P}{R} = \frac{\$144}{.6} = \$240; \$240 - \$144 = \$96$			
3.	$61.60	_____	_____	$43.12
4.	_____	$66\frac{2}{3}\%$	_____	$3.10
5.	$6.50	_____	$1.30	_____
6.	_____	50%	$65.25	_____

302 CHAPTER 8 Mathematics of Selling

Complete the following. If there is no operating loss or absolute loss, write "none." (See Examples 4 and 5.)

	Cost	Operating Expense	Break-even Point	Reduced Price	Operating Loss	Absolute Loss
7.	$96	$24	$120	$100	$20	none
	$96 + $24 = $120; $120 − $100 = $20					
8.	$25	$8	$33	$22	$11	$3
	$25 + $8 = $33; $33 − $22 = $11; $25 − $22 = $3					
9.	$50	_____	$66	$44	_____	_____
10.	$12.50	_____	$16.50	$11	_____	_____
11.	$310	$75	_____	_____	$135	_____
12.	$156	$44	_____	_____	$60	_____

13. Give five reasons a manager may mark down the price of merchandise. (See Objective 3.)

14. As a result of a markdown, there are three possible results: reduced net profit, operating loss, and absolute loss. Explain each. (See Objectives 4 and 5.)

Solve the following application problems. Round rates to the nearest whole percent and dollar amounts to the nearest cent.

15. **GPS SYSTEMS** If a Best Buy prices its total inventory of TomTom portable Global Positioning Systems at $133,509. If the original price of the inventory was $226,284, find the percent of markdown on the original price.
$226,284 − $133,509 = $92,775;
$R = \frac{P}{B} = \frac{\$92,775}{\$226,284} = .409 = 41\%$

15. 41% _____

16. **OAK DESK** An oak desk originally priced at $837.50 is reduced to $686.75. Find the percent of markdown on the original price.

16. _____

17. **ELLIPTICAL TRAINER** Dick's Sporting Goods paid $360 for a ProForm elliptical trainer. The operating expenses are $33\frac{1}{3}\%$ of cost. If they sell the elliptical trainer at a clearance price of $449.99, find the amount of profit or loss.

17. _____

18. **KAYAK** REI has an end-of-season sale, during which it sells an ocean kayak for two people for $865. If the cost was $641 and operating expenses are 28% of cost, find the amount of profit or loss.
;

18. _____

C// indicates an exercise that is related to the Case in Point feature.

19. TRUCK ACCESSORIES Pep Boys Automotive paid $208.50 for a pickup truck bedliner. The original selling price was $291.90, but this was marked down 35%. If operating expenses are 28% of the cost, find **(a)** the operating loss and **(b)** the absolute loss.

(a) _____
(b) _____

20. ANTIQUES American Antiques paid $153.49 for a fern stand. The original selling price was $208.78, but this was marked down 46% in order to make room for incoming merchandise. If operating expenses are 14.9% of cost, find **(a)** the operating loss and **(b)** the absolute loss.

(a) _____
(b) _____

21. CAR SEATS A firm pays $68 for a car seat and wants a 24% markup for operating expense as well as a 6% markup for profit. However, customers think the car seats are ugly, so they are not selling. Management reduces the price of each car seat by $30 to get rid of them. Find **(a)** the selling price, **(b)** the reduced price, **(c)** the percent of markdown to the nearest tenth of a percent, and **(d)** any operating or absolute loss.

(a) _____
(b) _____
(c) _____
(d) _____

22. MOTORCYCLES A Yamaha dealer pays $7250 for a new motorcycle model and marks up the price using 18% overhead and 9% profit. However a recession hits the area after a large employer goes bankrupt and lays off a lot of employees, so the motorcycle is not selling. The dealer reduces the price by $800 to get it out of the showroom. Find **(a)** the selling price, **(b)** the reduced price, **(c)** the percent of markdown to the nearest tenth of a percent, and **(d)** any operating or absolute loss.

(a) _____
(b) _____
(c) _____
(d) _____

QUICK CHECK ANSWERS

1. $173.99
2. 75%
3. $29.98
4. $48 operating loss
5. (a) $35.50 operating loss
   (b) $2.50 absolute loss

8.4 Turnover and Valuation of Inventory

OBJECTIVES
1. Determine average inventory.
2. Calculate inventory turnover.
3. Methods for tracking inventory.
4. Use the specific identification method to value inventory.
5. Determine inventory value using the weighted-average method.
6. Use the FIFO method to value inventory.
7. Use the LIFO method to value inventory.
8. Estimate inventory value using the retail method.

CASE IN POINT The marketing and inventory managers at REI work together to make sure they order only products that will sell. The last thing either wants is for inventory to build up needlessly in the company stores and warehouses.

Managers must constantly track the amount of inventory on hand and in transit. This allows them to determine profitability, understand what is selling and what is not, and complete tax returns and financial statements.

OBJECTIVE 1 Determine average inventory. **Average inventory** is found by finding the average of all inventories taken during a specific interval of time. For example, the average inventory for a quarter (3 months) is found by averaging the inventory at the beginning and at the end of the quarter, or by averaging inventory at the end of each of the three months of the quarter.

Determining Average Inventory

EXAMPLE 1 Inventory at a Footlocker store was $285,672 on April 1 and $198,560 on April 30. Find the average inventory for the month.

SOLUTION

$$\text{Average inventory} = \frac{\$285{,}672 + \$198{,}560}{2}$$
$$= \$242{,}116$$

QUICK CHECK 1

Inventory on September 1 was $176,840 and on September 30 it was $153,210. Find the average inventory for the month.

GET HUGE STOREWIDE SAVINGS!

YEAR-END SALE!

No Money Down & No Payment 'til June

Inventory must be carefully monitored, as it can quickly grow too large or fall too low. Companies often try to reduce inventories at the end of their accounting year using a year-end sale as indicated by the ad.

OBJECTIVE 2 Calculate inventory turnover. One measure of how well a company is doing financially is how fast inventory is being sold or turned over. **Inventory turnover** (also called **inventory turns** and **stock turnover**) is the number of times average inventory is sold during a year. Inventory is valued either at retail or at cost, so inventory turnover can be found using either; both are used. Inventory turnover differs greatly from one industry to another and from one firm to another. For example, firms with perishable items such as flowers must turn over inventory rapidly to avoid spoilage, whereas jewelry stores usually have a lower turnover.

Measuring Inventory Turnover

$$\text{Turnover at retail} = \frac{\text{Retail sales}}{\text{Average inventory at retail}}$$

$$\text{Turnover at cost} = \frac{\text{Cost of goods sold}}{\text{Average inventory at cost}}$$

The turnover ratio may be identical using both methods. However, **turnover at retail** is usually slightly lower than **turnover at cost**. This is due to **inventory shrinkage,** which is caused by theft, spoilage, breakage, clerical errors, or products that can no longer be sold.

Finding Stock Turnover at Retail

EXAMPLE 2 Last year, Red Clown Children's Apparel had retail sales of $559,320 and an average retail inventory of $49,601. Find the stock turnover at retail.

SOLUTION

$$\text{Turnover at retail} = \frac{\text{Retail sales}}{\text{Average inventory at retail}} = \frac{\$559,320}{\$49,601} = 11.28 \text{ at retail} \quad (\text{rounded})$$

The store turned over its average inventory 11.3 times during the year.

QUICK CHECK 2

Painters Supply had an average retail inventory of $1,120,400 and retail sales of $1,833,080. Find the stock turnover at retail.

Finding Stock Turnover at Cost

EXAMPLE 3 If the average inventory value at cost for Red Clown Children's Apparel in Example 2 was $27,260 and the cost of goods sold was $341,340, find the stock turnover at cost.

SOLUTION

$$\text{Turnover at cost} = \frac{\text{Cost of goods sold}}{\text{Average inventory at cost}} = \frac{\$341,340}{\$27,260} = 12.52 \text{ at cost} \quad (\text{rounded})$$

The turnover ratios found in Examples 2 and 3 are similar, although the turnover at retail is slightly lower, as expected.

QUICK CHECK 3

If the cost of goods sold by Painters Supply in Quick Check 2 was $1,204,300 and the average inventory at cost was $761,872, find the stock turnover at cost.

Inventory turnover ratios can be used to compare one department with another, one store with another, one quarter with another, or even one product line with another.

OBJECTIVE 3 Methods for tracking inventory. Managers constantly track inventory moving along the supply chain, sitting in warehouses, and on shelves in stores. Only by tracking inventory can they cut down on theft, figure out what is not selling, prevent the waste of a buildup in unneeded inventory, and move inventory quickly to locations where it is needed. In addition, the valuation of inventory helps determine profits, taxes, and financial statements, so managers are careful when choosing a method to evaluate inventory and when tracking inventory.

In the past, **periodic inventory** systems that required frequent **actual physical counts** of all inventory were used. Today, most firms use a **perpetual** or **continuous inventory** system, which uses reading devices, computers, and networks to track inventory minute by minute. These systems have several advantages but still require an occasional physical count.

Many technologies exist to track inventory. The **universal product code (UPC)** requires that a tag with black stripes called a **barcode** be placed on each item. The tag is scanned at registers or using other readers. Light is used to scan the barcodes, so the UPC must be visible and directed toward the machine reading the code. You have probably been stuck at a cash register as the attendant repeatedly tries to get the machine to read the UPC code of an item you are buying.

UPC Code

Many firms use **radio frequency identification (RFID)** chips placed on items that are read by **electronic product code (EPC)** devices. Walmart was the first retailer to require all suppliers to use RFID chips. One advantage of this method is that the chips do not have to be pointed at the machine reading the information; another is the low cost. Someday, this technology may allow customers to simply move an entire cart of purchases out the door without stopping at a cashier. The idea is that electronic systems read the RFID chips on all products in the cart very quickly, find a total amount due, add the sales tax, and then charge that amount to a known customer's charge or debit card. If this becomes commonplace, fewer cashiers will be needed.

RFID Chip

306 CHAPTER 8 Mathematics of Selling

All five of the following methods are used to value inventory: the specific identification method; the weighted-average method; the first-in, first-out method; the last-in, first-out method; and the retail method.

OBJECTIVE 4 Use the specific identification method to value inventory. The **specific identification method** is useful when items are easily identified and costs do not fluctuate. Each item is coded with cost information from which ending inventory is calculated. Since several items in stock may have been purchased at different times and have different costs, many managers prefer to use inventory at retail. In this case, the retail value of all identical items is the same, making it easier to value inventory.

OBJECTIVE 5 Determine inventory value using the weighted-average method. The **weighted average (average cost)** of inventory involves finding the average cost of an item and then multiplying the number of items remaining in inventory by the average cost per item.

Using Weighted Average (Average Cost) Inventory Valuation

EXAMPLE 4

CASE IN POINT

REI made the following purchases of one type of backpack during the year.

Beginning inventory	20 backpacks at $70
January	50 backpacks at $80
March	100 backpacks at $90
July	60 backpacks at $85
October	40 backpacks at $75

At the end of the year, there are 75 backpacks in inventory. Use the weighted-average method to find the inventory value.

SOLUTION

Find the total cost of all the backpacks.

Beginning inventory	20 × $70 = $1400
January	50 × $80 = $4000
March	100 × $90 = $9000
July	60 × $85 = $5100
October	40 × $75 = $3000
Total	270 $22,500

Find the average cost per backpack by dividing this total cost by the number purchased.

$$\frac{\$22{,}500}{270} = \$83.33$$

Since the average cost is $83.33 and 75 backpacks remain in inventory, the weighted-average method gives the inventory value of the remaining backpacks as $83.33 × 75 = $6249.75.

The calculator solution to this example has several steps. First, find the total number of backpacks purchased and place the total in memory.

20 [+] 50 [+] 100 [+] 60 [+] 40 [=] 270 [STO]

Next, find the total cost of all the backpacks purchased and divide by the number stored in memory. This gives the average cost per backpack.

20 [×] 70 [+] 50 [×] 80 [+] 100 [×] 90 [+] 60 [×] 85 [+] 40 [×] 75 [=] [÷] [RCL] [=] 83.3333

Finally, round the average cost to the nearest cent and multiply by the number of backpacks in inventory to get the weighted average inventory value.

83.33 [×] 75 [=] 6249.75

QUICK CHECK 4

Dick's Sporting Goods made the following purchases of the Iron Horse BMX bicycle during the year.

Beginning inventory	20 bicycles at $115
February	30 bicycles at $95
April	50 bicycles at $100
June	40 bicycles at $110
August	80 bicycles at $105
November	60 bicycles at $130

At the end of the year, there are 85 bicycles in inventory. Use the weighted-average method to find the inventory value.

The selling price of an item is determined by the cost of the item, the quantity purchased, overhead, product quality, and the location of the customer. The following graph shows that the average price of a tennis racket varies greatly by country.

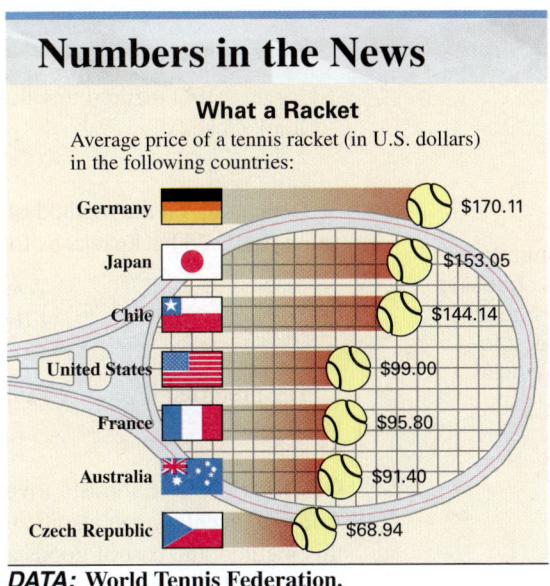

DATA: World Tennis Federation.

OBJECTIVE 6 Use the FIFO method to value inventory. The **first-in, first-out (FIFO) method** of inventory valuation assumes that the first goods to arrive are the first to be sold, and the last items purchased are assumed to be the last items remaining in inventory.

Using FIFO to Determine Inventory Valuation

EXAMPLE 5 Use the FIFO method to find the inventory value of the 75 backpacks from REI in Example 4.

SOLUTION

CASE IN POINT With the FIFO method, the 75 remaining backpacks are assumed to consist of the 40 backpacks bought in October and 35 (75 − 40 = 35) backpacks from the previous purchase in July. The value of the inventory is:

October	40 backpacks at $75 = $3000	value of last 40
July	35 backpacks at $85 = $2975	value of previous 35
	75 valued at $5975	

The value of the backpack inventory is $5975 using the FIFO method.

QUICK CHECK 5

Use the FIFO method to find the inventory value of the 85 bicycles from Dick's Sporting Goods in Quick Check 4.

OBJECTIVE 7 Use the LIFO method to value inventory. The **last-in, first-out (LIFO) method** of inventory valuation assumes that the last goods received are the first ones sold. Those remaining in inventory are those that were purchased first.

Using LIFO to Determine Inventory Valuation

EXAMPLE 6 Use the LIFO method to value the 75 backpacks in inventory at REI in Example 4.

SOLUTION

CASE IN POINT The calculation starts with the beginning inventory and moves through the year's purchases, resulting in 75 backpacks still in stock. The beginning inventory and January purchases come to 70 backpacks, so the cost of 5 more (75 − 70 = 5) backpacks from the March purchase is needed.

Beginning inventory	20 backpacks at $70 = $1400	value of first 20
January	50 backpacks at $80 = $4000	value of next 50
March	5 backpacks at $90 = $ 450	value of last 5
Total	75 valued at $5850	

The value of the backpack inventory is $5850 using the LIFO method.

QUICK CHECK 6

Use the LIFO method to value the 85 bicycles in inventory at Dick's Sporting Goods in Quick Check 4.

Quick TIP
Managers and accountants study carefully before choosing an inventory method, because inventory valuations affects income, taxes, and financial statements.

Depending on the method of valuing inventories that is used, REI may show the inventory value of the 75 backpacks as follows.

Average cost method	$6249.75
FIFO	$5975
LIFO	$5850

The preferred inventory valuation method is determined by management on the advice of an accountant.

OBJECTIVE 8 Estimate inventory value using the retail method. An estimate of the value of inventory can also be found using the **retail method of estimating inventory**. With this method, the cost of goods available for sale is found as a percent of the retail value of the goods available for sale during the same period. This percent is then multiplied by the retail value of inventory at the end of the period. The result is an estimate of the inventory at cost.

Estimating Inventory Value Using the Retail Method

EXAMPLE 7 The inventory on December 31 at a booth that sells phones in the mall was $129,200 at cost and $171,000 at retail. Purchases during the next three months were $165,400 at cost, $221,800 at retail, and net sales were $168,800. Use the retail method to estimate the value of inventory at cost on March 31.

SOLUTION

Step 1   Find the value of goods available for sale (inventory) at cost and at retail.

	At cost	At retail	
	$129,200	$171,000	beginning inventory
	+ 165,400	+ 221,800	purchases
	$294,600	$392,800	goods available for sale
		− 168,800	net sales
		$224,000	March 31 inventory at retail

Step 2   Find the retail value of current inventory.

Step 3   Now find the percent of the value of goods available for sale at cost to goods available for sale at retail (cost ratio).

$$\frac{\$294{,}600 \text{ goods available for sale at cost}}{\$392{,}800 \text{ goods available for sale at retail}} = .75 = 75\% \text{ (cost ratio)}$$

Step 4   Finally, the estimated inventory value at cost on March 31 is found by multiplying inventory at retail on March 31 by 75% (cost ratio).

Ending inventory at retail × % (cost ratio) = Inventory at cost

$224,000 × .75 = $168,000 March 31 inventory at cost

> **QUICK CHECK 7**
>
> At the end of June, Solar Solutions had an inventory of $87,500 at cost and $125,000 at retail. During the next three months, there were purchases of $103,200 at cost, $147,600 at retail, and net sales were $185,000. Use the retail method to estimate the value of inventory at cost at the end of September.

8.4 Exercises // MyLab Math

The shaded sections below contain solutions to help you get a **QUICK START** *on the various types of exercises.*

Find the average inventory in each of the following. (See Example 1.)

Date	Inventory Amount at Retail	Average Inventory	Date	Inventory Amount at Retail	Average Inventory
1. July 1	$18,300		2. January 1	$42,312	
October 1	$26,580		July 1	$38,514	
December 31	$23,139	$22,673	December 31	$30,219	$37,015
$68,019 total of inv. ÷ 3 = $22,673			$111,045 total of inv. ÷ 3 = $37,015		
3. January 1	$65,430		4. January 31	$69,480	
April 1	$58,710		April 30	$55,860	
July 1	$53,410		July 31	$80,715	
October 1	$78,950		October 31	$88,050	
December 31	$46,340	_____	January 31	$63,975	_____

Find the stock turnover at cost and at retail in each of the following. Round to the nearest hundredth. (See Examples 2 and 3.)

	Average Inventory at Cost	Average Inventory at Retail	Cost of Goods	Retail Sales	Turnover at Cost	Turnover at Retail
5.	$84,320	$127,690	$1,104,600	$1,782,300	13.10	13.96
	$1,104,600 ÷ $84,320 = 13.10; $1,782,300 ÷ $127,690 = 13.96					
6.	$284,300	$342,950	$5,040,500	$6,810,970	17.73	19.86
	$5,040,500 ÷ $284,300 = 17.73; $6,810,970 ÷ $342,950 = 19.86					
7.	$68,080	$125,240	$573,330	$951,900	_____	_____
8.	$38,074	$48,550	$260,420	$330,060	_____	_____
9.	$180,600	$256,700	$846,336	$1,196,222	_____	_____
10.	$411,580	$780,600	$1,905,668	$3,559,536	_____	_____

C// indicates an exercise that is related to the Case in Point feature.

CHAPTER 8 Mathematics of Selling

Find the inventory values using (a) the weighted-average method, (b) the FIFO method, and (c) the LIFO method for each of the following. Round to the nearest cent if necessary. (See Examples 4–6.)

Purchases	Now in Inventory	Weighted-Average Method	FIFO Method	LIFO Method
11. Beginning inventory: 10 units at $8 June: 25 units at $9 August: 15 units at $10	20 units	$182	$195	$170

($455 purchases ÷ 50) × 20 = $9.10 × 20 = $182 average cost method

```
   15 × $10 = $150              10 × $8 = $ 80
 + 5 × $ 9 = $ 45             +10 × $9 = $ 90
   20        $195 FIFO          20       $170 LIFO
```

12. Beginning inventory: 80 units at $14.50
July: 50 units at $15.80
October: 70 units at $13.90   90 units   _____ _____ _____

13. Beginning inventory: 50 units at $30.50
March: 70 units at $31.50
June: 30 units at $33.25
August: 40 units at $30.75   75 units   _____ _____ _____

14. Beginning inventory: 700 units at $1.25
May: 400 units at $1.75
August: 500 units at $2.25
October: 600 units at $3.00   720 units   _____ _____ _____

15. List three businesses you think would have a high inventory turnover rate, and three others that you think would have a low rate of inventory turnover. (See Objective 2.)

16. Which departments in a grocery store do you think have the highest turnover? Which ones have the lowest turnover? Why do you think this is true?

Solve the following application problems. Round stock turnover to the nearest hundredth.

17. STOCK TURNOVER AT COST The Glass Works has an average inventory at cost of $62,920, and cost of goods sold for the same period is $341,648. Find the stock turnover at cost.

$\frac{\$341,648}{\$62,920} = $ **5.43 turnover at cost**

17. 5.43 turnover at cost

18. STOCK TURNOVER AT RETAIL One Albertson's store has an average canned-fruit inventory of $4484 at retail. Retail sales of canned fruit for the year were $98,669. Find the stock turnover at retail.

18. _____

19. SPRAY-PAINT INVENTORY Hobby House made purchases of assorted colors of spray paint during the year as follows.

Beginning inventory 200 cans at $2.20
March 400 cans at $2.40
May 700 cans at $2.00
August 500 cans at $2.30
November 300 cans at $2.60

At the end of the year, they had 450 cans of spray paint in stock.
(a) Find the inventory value using the weighted-average method.
(b) Find the inventory value using the FIFO method.
(c) Find the inventory value using the LIFO method.

(a) _____
(b) _____
(c) _____

20. T-SHIRTS Adams Import made the following purchases of T-shirts made in Taiwan: beginning inventory was 650 shirts at $3.80 each; June, 500 shirts at $4.20 each; September, 450 shirts at $3.95 each; and December, 600 shirts at $4.05 each. An inventory at the end of the year shows that 775 T-shirts remain. (a) Find the inventory value using the weighted-average method. (b) Find the inventory value using the FIFO method. (c) Find the inventory value using the LIFO method.

(a) _____
(b) _____
(c) _____

21. WOOL SOCKS REI made the following purchases of socks.

Beginning inventory 200 packages at $3.10
May 250 packages at $3.50
August 300 packages at $4.25
October 280 packages at $4.50

An inventory at the end of October shows that 320 packages remain. (a) Find the inventory value using the weighted-average method. (b) Find the inventory value using the FIFO method. (c) Find the inventory value using the LIFO method.

(a) _____
(b) _____
(c) _____

22. ATHLETIC SHOES The beginning inventory and purchases of shoes made by The Sports Authority this year are as shown.

Beginning inventory 300 pairs at $21.60
March 400 pairs at $24.00
August 450 pairs at $24.30
November 350 pairs at $22.50

An inventory at the end of December shows that 530 pairs of shoes remain. (a) Find the inventory value using the weighted-average method. (b) Find the inventory value using the FIFO method. (c) Find the inventory value using the LIFO method.

(a) _____
(b) _____
(c) _____

312 CHAPTER 8 Mathematics of Selling

23. **PIANO REPAIR** The September 30 inventory at Liverpool Piano Repair was $43,750 at cost and $62,500 at retail. Purchases during the next three months were $51,600 at cost, $73,800 at retail, and net sales were $92,500. Use the retail method to estimate the value of the inventory at cost on December 31.

23. _____

24. **EVALUATING INVENTORY** Cell Phones Plus had an inventory of $27,000 at cost and $45,000 at retail on March 31. During the next three months, they made purchases of $108,000 at cost and $180,000 at retail and had net sales of $162,000. Use the retail method to estimate the value of inventory at cost on June 30.

24. _____

25. What are the benefits of using RFID chips to electronically track inventory? (See Objective 3.)

26. Give several reasons managers are careful when choosing a method to track inventory and also to value inventory. (See Objectives 3 and 7.)

QUICK CHECK ANSWERS

1. $165,025
2. 1.64 at retail
3. 1.58 at cost
4. $9334.70 weighted average
5. $10,425 FIFO
6. $8650 LIFO
7. $61,320 inventory at cost

Chapter 8 Quick Review

Chapter Terms Review the following terms to test your understanding of the chapter. For each term you do not know, refer to the page number found next to that term.

absolute loss [p. 299]
actual physical counts [p. 305]
actual selling price [p. 298]
average inventory [p. 304]
barcode [p. 305]
break-even point [p. 299]
continuous inventory [p. 305]
conversion formulas [p. 292]
cost [p. 281]
electronic product code (EPC) [p. 305]
first-in, first-out (FIFO) method [p. 307]

formula for markdown [p. 298]
gross loss [p. 299]
gross profit [p. 281]
inventory shrinkage [p. 304]
inventory turnover [p. 304]
inventory turns [p. 304]
last-in, first-out (LIFO) method [p. 308]
margin [p. 281]
markdown [p. 298]
markup [p. 281]
markup formula [p. 281]
markup on cost [p. 282]

markup on selling price [p. 288]
net earnings [p. 281]
net profit [p. 281]
operating expenses [p. 281]
operating loss [p. 299]
overhead [p. 281]
percent of markdown [p. 298]
periodic inventory [p. 305]
perpetual inventory [p. 305]
radio frequency identification (RFID) [p. 305]
reduced net profit [p. 299]
reduced price [p. 298]

retail method of estimating inventory [p. 308]
sale price [p. 298]
selling price [p. 281]
specific identification method [p. 306]
stock turnover [p. 304]
turnover at cost [p. 304]
turnover at retail [p. 304]
universal product code (UPC) [p. 305]
weighted average (average cost) method [p. 306]

CONCEPTS

8.1 Finding the markup on cost

	100%	Cost	base
	% +	Markup	?
	%	Selling price	

Cost is base. Use the basic percent formula.

$P = B \times R$

8.1 Calculating the percent of markup

	100%	C	$	base
rate	?%	M	$	
	%	S	$	

Solve for rate.

8.1 Finding the cost and the selling price

	100%	C	$?	base
	%	M	$	
	%	S	$?	

Solve for base.

8.2 Finding the markup on selling price

	%	C	$	
	%	M	$?	part
	100%	S	$	base

Solve for part.

EXAMPLES

	100%	C	$160	base
rate	25%	M	$?	
	%	S	$	

$P = B \times R$
$P = \$160 \times .25$
$P = \$40$ markup

	100%	C	$420	base
rate	?%	M	$	
	%	S	$546	

$546 - \$420 = \$126$ markup

$R = \dfrac{P}{B} = \dfrac{126}{420}$

$R = 30\%$

	100%	C	$?	base
rate	50%	M	$56	
	%	S	$?	

$B = \dfrac{P}{R} = \dfrac{56}{.5}$

$B = \$112$ cost

$\$112 + \$56 = \$168$ selling price

	%	C	$	
rate	25%	M	$?	part
	100%	S	$6.00	base

$P = B \times R$
$P = \$6.00 \times .25$
$P = \$1.50$

CHAPTER 8 Mathematics of Selling

CONCEPTS	EXAMPLES

8.2 Finding the cost

	%		C	$?	part
			M	$	
	100%		S	$	base

Example:

	%		C	$?	part
rate	35%		M	$87.50	
	100%		S	$	base

$$B = \frac{P}{R} = \frac{87.5}{.35} = \$250 \text{ selling price}$$

$250 − $87.50 = $162.50 cost

8.2 Calculating the selling price and the markup

	%		C	$	
			M	$?	
	100%		S	$?	base

Example:

	%		C	$150	
rate	25%		M	$?	
	100%		S	$?	base

100% − 25% = 75% cost

$$B = \frac{P}{R} = \frac{150}{.75} = \$200 \text{ selling price}$$

$200 − $150 = $50 markup

8.2 Converting markup on cost to markup on selling price

Use the formula

$$\% \text{ markup on selling price} = \frac{\% \text{ markup on cost}}{100\% + \% \text{ markup on cost}}$$

Convert 25% markup on cost to markup on selling price.

$$\% \text{ markup on selling price} = \frac{25\%}{100\% + 25\%}$$

$$= \frac{.25}{1.25} = .2 = 20\%$$

8.2 Converting markup on selling price to markup on cost

Use the formula

$$\% \text{ markup on cost} = \frac{\% \text{ markup on selling price}}{100\% - \% \text{ markup on selling price}}$$

Convert 20% markup on selling price to markup on cost.

$$\% \text{ markup on cost} = \frac{20\%}{100\% - 20\%}$$

$$= \frac{.2}{.8} = .25 = 25\%$$

8.2 Finding the selling price for perishables

1. Find total cost and selling price.
2. Subtract the quantity not sold to find the number of items that are sold.
3. Divide the total sales by the number of sellable units to get selling price per unit.

60 mini donuts cost 25¢ each; 10 are not sold; 50% markup on selling price; find selling price per mini donut.

Cost = 60 × $.25 = $15

	%		C	$15	part
rate	50%		M	$	
	50%				
	100%		S	$?	base

$$B = \frac{P}{R} = \frac{15}{.5} = \$30$$

60 − 10 = 50 mini donuts sold
$30 ÷ 50 = $.60 per mini donut

8.3 Finding the percent of markdown

Markdown is always a percent of the original price.
Use the formula

$$R = \frac{P}{B}$$

$$\text{Markdown percent} = \frac{\text{Markdown amount}}{\text{Original price}}$$

Original price, $76; markdown, $19; find the percent of markdown.

$$R = \frac{P}{B} = \frac{19}{76} = .25$$

R = 25% markdown

8.3 Calculating the break-even point

The cost plus operating expenses equals the break-even point.

Cost, $54; operating expenses, $16; find the break-even point.

$54 cost + **$16 operating expenses** = $70 break-even point

CONCEPTS	EXAMPLES
8.3 Finding operating loss The difference between the break-even point and the reduced price (when below the break-even point) is the operating loss.	Break-even point, $70; reduced price, $58; find the operating loss. $\quad\$70\quad$ break-even point $\underline{-\$58}\quad$ reduced price $\quad\$12\quad$ operating loss
8.3 Finding absolute loss (gross loss) When the reduced price is below cost, the difference between the cost and reduced price is the absolute loss.	Cost, $54; reduced price, $48; find the absolute loss. $54 cost − $48 reduced price = $6 absolute loss
8.4 Determining average inventory Inventory is taken two or more times. Totals are added together, then divided by the number of inventories taken to get the average.	Inventories, $22,635, $24,692, and $18,796; find the average inventory. $$\frac{\$22{,}635 + \$24{,}692 + \$18{,}796}{3}$$ $$= \frac{\$66{,}123}{3} = \$22{,}041 \text{ average inventory}$$
8.4 Finding turnover at retail Use the formula $$\text{Turnover} = \frac{\text{Retail sales}}{\text{Average inventory at retail}}$$	Quarterly retail sales, $893,724; average inventory at retail, $87,620; find turnover at retail. $$\frac{\$893{,}724}{\$87{,}620} = 10.2 \text{ at retail}$$
8.4 Finding turnover at cost Use the formula $$\text{Turnover} = \frac{\text{Cost of goods sold}}{\text{Average inventory at cost}}$$	Quarterly cost of goods sold, $257,747; average inventory at cost, $32,960; find turnover at cost. $$\frac{\$257{,}747}{\$32{,}960} = 7.82 \text{ at cost} \quad \text{(rounded)}$$
8.4 Using specific identification to value inventory Each item is cost coded, and the cost of each of the items is added to find total inventory.	Individual cost of each item in inventory is: item 1, $593; item 2, $614; item 3, $498; find total value of inventory. $593 + $614 + $498 = $1705 total value of inventory
8.4 Using weighted-average (average cost) method of inventory valuation This method values items in an inventory at the average cost of buying them.	Beginning inventory of 20 at $75; purchases of 15 at $80; 25 at $65; 18 at $70; 22 remain in inventory. Find the inventory value. $\quad$ 20 × $75 = $1500 $\quad$ 15 × $80 = $1200 $\quad$ 25 × $65 = $1625 $\quad$ 18 × $70 = $1260 Total 78 $\quad\quad\quad$ $5585 $$\frac{\$5585}{78} = \$71.60 \text{ average cost} \quad \text{(rounded)}$$ $71.60 × 22 = $1575.20 weighted-average method inventory value
8.4 Using first-in, first-out (FIFO) method of inventory valuation The first items in are the first sold. Inventory is based on cost of last items purchased.	Beginning inventory of 25 items at $40; purchased on August 7, 30 items at $35; 35 remain in inventory. Find the inventory. $\quad$ 30 × $35 = $1050 $\quad$ value of last 30 $\quad\underline{\;5 \times \$40 = \$\;200}\quad$ value of previous 5 $\quad$ 35 $\quad\quad\quad\;\;$ $1250 $\quad$ value of inventory using FIFO

CONCEPTS	EXAMPLES
8.4 Using last-in, first-out (LIFO) method of inventory valuation The items remaining in inventory are those items that were first purchased.	Beginning inventory of 48 items at $20 each; purchase on May 9, 40 items at $25 each; 55 remain in inventory. Find the inventory value. $48 \times \$20 = \$\ 960$ value of first 48 $\underline{7 \times \$25 = \$\ 175}$ value of last 7 $55 = \$1135$ value of inventory using LIFO
8.4 Estimating inventory value using the retail method $\dfrac{\text{Goods available for sale at cost}}{\text{Goods available for sale at retail}} = \%\,(\text{cost ratio})$ $\dfrac{\text{Ending inventory}}{\text{at retail}} \times \%\,(\text{cost ratio}) = \dfrac{\text{Inventory}}{\text{at cost}}$	Use the retail method to estimate the inventory value at cost. **Cost** **Retail** beginning inventory $9,000 $15,000 purchases +36,000 +60,000 goods available for sale **$45,000** **$75,000** net sales −54,000 ending inventory $21,000 $\dfrac{\text{goods available for sale at cost}}{\text{goods available for sale at retail}} \quad \dfrac{\$45,000}{\$75,000} = .6 = 60\%$ $\$21,000 \times .6 = \$12,600$ inventory value at cost

Case Study

MARKDOWN: REDUCING PRICES TO MOVE MERCHANDISE

Olympic Sports purchased two dozen pairs of 5th Element Adult Aggressive in-line skates at a cost of $1950. Operating expenses for the store are 25% of cost, while total markup on this type of product is 35% of selling price. Only 6 pairs of the skates sell at the original price, and the manager decides to mark down the remaining skates. The price is reduced 25% and 6 more pairs sell. The remaining 12 pairs of skates are marked down to 50% of the original selling price and are finally sold.

1. Find the original selling price of each pair of skates.   1. _____

2. Find the total of the selling prices of all the skates.   2. _____

3. Find the operating loss.   3. _____

4. Find the absolute loss.   4. _____

INVESTIGATE

Talk with the manager of a retail store. Is it important for the store to monitor inventory? Do they use markup based on cost or on retail? Do they ever use markdowns, and if so why? Ask the manager if she has ever marked a price down so much that an absolute loss resulted.

Case in Point Summary Exercise

RECREATIONAL EQUIPMENT INC. (REI)

www.REI.com

Facts:

- 1938: Established by 23 mountaineers
- 2001: Exceeded 2 million members
- 2005: Annual sales passed $1 billion
- 2018: About $200 million returned to members
  Three-quarters of profit shared with outdoor community

REI is a cooperative that focuses on the great outdoors. Although anyone can shop at REI, the firm is owned by the members who pay a one-time fee of $20 and receive annual dividends based on company profit and the amount of each member's annual purchases.

A marketing manager was interested in a new line of off-piste skis for skiing over rugged terrain and jumping off cliffs. The manager works with James Smelter in inventory to place the following orders for two different skis.

	Model Name	Number	Total Cost
1st order	Steep Alpine	15 pairs	$4395
2nd order	Cliff Hoppers	22 pairs	$7194

1. Find the cost per pair of skis for both models.

2. If the markup on cost for each pair of skis is 38%, find the list price for each.

3. The skis were received on January 1, and a physical inventory at the end of January showed that 7 pairs of Steep Alpine and 14 pairs of Cliff Hoppers remained. Find the average inventory at cost.

4. Use the data for these skis to find the turnover at cost for the month of January.

5. The managers are not happy with the quality of the skis and sell the remaining skis at a markdown of 40%. Find the price per pair of skis after being marked down.

6. Assume that operating expenses at REI are 22% of cost and determine the amount of any profit on these skis. If there was no profit, find the amount of the operating loss and/or the absolute loss.

1. _____

2. _____

3. _____

4. _____

5. _____

6. _____

Discussion Question: *Do you think firms often lose money on new products? What do you think would happen to a company that often loses money on new products? Explain.*

Source: RECREATIONAL EQUIPMENT INC. (REI), www.REI.com.

Chapter 8   Test

To help you review, the numbers in brackets show the section in which the topic was discussed.

Solve for (a), (b), and (c).   **[8.1 and 8.2]**

1. 100%   C $64.00
    (a)%   M $12.80
    ─────────────────
    (b)%   S $(c)

2. 100%   C $(b)
    38%   M $(c)
    ─────────────────
    (a)%   S $504.39

3. (a)%   C $134.40
    (b)%   M $(c)
    ─────────────────
    100%   S $168.00

4. (a)%   C $(c)
    (b)%   M $ 6.15
    ─────────────────
    100%   S $24.60

Find the equivalent markup on either cost or selling price, using the appropriate formula. Round to the nearest tenth of a percent.   **[8.2]**

	Markup on Cost	Markup on Selling Price		Markup on Cost	Markup on Selling Price
5.	25%	_____	6.	100%	_____

Complete the following. If there is no operating loss or absolute loss, write "none."   **[8.3]**

	Cost	Operating Expense	Break-even Point	Reduced Price	Operating Loss	Absolute Loss
7.	$160	$40	_____	$186	_____	_____
8.	$225	_____	$297	$198	_____	_____

Find the stock turnover at retail and at cost for the following. Round to the nearest hundredth.   **[8.4]**

	Average Inventory at Cost	Average Inventory at Retail	Cost of Goods Sold	Retail Sales	Turnover at Cost	Turnover at Retail
9.	$14,120	$25,572	$81,312	$146,528	_____	_____

Solve the following application problems.

10. Smart Phones, Inc. buys one model of smartphone, paying $2370 per case of 20 phones. Find the asking price for each assuming a 40% markup on cost.

    10. _____

11. It cost a singer-songwriter $67,300 for all of the expenses to engineer, record, mix, and promote a new music disc. The cost also included the manufacturing of 5000 discs. Assume each disc sells for $20. Find **(a)** the markup in dollars, **(b)** the markup as a percent of cost to the nearest tenth of a percent, and **(c)** the markup as a percent of selling price to the nearest tenth of a percent.

    (a) _____
    (b) _____
    (c) _____

12. REI buys jogging shorts manufactured in Indonesia for $195.00 per dozen pair. Find the selling price per pair if the retailer maintains a markup of 35% on selling price.   **[8.2]**

    12. _____

13. Restaurant Supply sells a walk-in refrigerator for $5250 while using a markup of 25% on cost. Find the cost.   **[8.1]**

    13. _____

C// indicates an exercise that is related to the Case in Point feature.

CHAPTER 8 Mathematics of Selling

14. The Computer Center sells a DeskJet print cartridge for $37.50. If the print cartridge costs the store $22.50, find the markup as a percent of selling price. **[8.2]**

14. _____

15. REI offers a men's soft-shell waterproof jacket for $199.95. If the jackets cost $1943.52 per dozen, find **(a)** the markup, **(b)** the percent of markup on selling price, and **(c)** the percent of markup on cost. Round to the nearest tenth of a percent. **[8.1 and 8.2]**

(a) _____

(b) _____

(c) _____

16. A motorcycle originally priced at $13,875 is marked down to $9990. Find the percent of markdown on the original price. **[8.3]**

16. _____

17. Leslie's Pool Supply, a retailer, pays $285 for a diving board. The original selling price was $399, but it was marked down 40%. If operating expenses are 30% of cost, find **(a)** the operating loss and **(b)** the absolute loss. **[8.3]**

(a) _____

(b) _____

18. Carpets Plus had an inventory of $117,328 on January 1, $147,630 on July 1, and $125,876 on December 31. Find the average inventory. **[8.4]**

18. _____

Round to the nearest dollar.

19. Ferrell Gas made the following purchases of fuel tanks during the year: 25 at $270 each, 40 at $330 each, 15 at $217 each, and 30 at $284 each. An inventory shows that 45 fuel tanks remain. Find the inventory value using the weighted-average method. **[8.4]**

19. _____

20. Find the value of the inventory listed in Exercise 19 using **(a)** the FIFO method and **(b)** the LIFO method. **[8.4]**

(a) _____

(b) _____

Chapters 5–8   Cumulative Review

CHAPTERS 5–8

The following credit-card transactions were made at the Patio Store. Answer Exercises 1–5 using this information. Round to the nearest cent. **[5.2]**

Sales			Credits
$428.80	$733.18	$22.51	$76.15
$316.25	$38.00	$162.15	$118.44
$68.95	$188.36		$13.86

1. Find the total amount of the sales slips.   1. _____

2. What is the total amount of the credit slips?   2. _____

3. Find the total amount of the deposit.   3. _____

4. Assuming that the bank charges the retailer a $1\frac{1}{4}\%$ discount charge, find the amount of the discount charge at the statement date.   4. _____

5. Find the amount of the credit given to the retailer after the fee is subtracted.   5. _____

Solve the following application problems.

6. Shaundra Brown worked 7 hours on Monday, 10 hours on Tuesday, 8 hours on Wednesday, 9 hours on Thursday, and 10 hours on Friday. Her regular hourly pay is $12.80. Find her gross earnings for the week if Brown is paid time and a half for all hours over 8 worked in a day.   **[6.1]**   6. _____

7. Last month, the employees of Jena's Bakery paid a total of $968.50 in Social Security tax, $223.50 in Medicare tax, and $1975.38 in federal withholding tax. Find the total amount the employer must send to the Internal Revenue Service.   **[6.4]**   7. _____

Find the net cost (invoice amount) for each of the following. Round to the nearest cent. **[7.1]**

8. List price $475.50, less 20/20 _____   9. List price $375, less 25/10/5 _____

Find the single discount equivalent for each of the following series discounts. **[7.2]**

10. 10/20 _____   11. 30/40/10 _____

Find the discount date and the net payment date for each of the following. The net payment date is 20 days after the final discount date. **[7.4]**

	Invoice Date	Terms	Date Goods Received	Final Discount Date	Net Payment Date
12.	May 27	2/10 ROG	June 5	_____	_____
13.	Oct. 9	3/15 EOM		_____	_____
14.	June 24	4/10–30 ex.		_____	_____

CHAPTER 8 Mathematics of Selling

Complete the following. If there is no operating loss or absolute loss, write "none." [8.3]

	Cost	Operating Expense	Break-even Point	Reduced Price	Operating Loss	Absolute Loss
15.	$312	$88	____	____	$120	____
16.	____	____	____	$220	$112	$32

Solve the following application problems.

17. The list price of a bike at Olympic Sports is $149.99. Find the dealer's cost if given a 20/20 trade discount and a 3/20, n/30 cash discount. Assume that the dealer earns the maximum cash discount. **[7.1 and 7.3]**

17. _____

18. Target purchases a top-selling video game for Xbox 360 paying $245 for 10 copies of the game. If the store wants a markup of 63.2% on the selling price, find the selling price per copy. **[8.2]**

18. _____

19. Digital Products has an average inventory of $37,568 at cost. If the cost of goods sold for the year was $483,876, find the stock turnover at cost. Round to the nearest hundredth. **[8.4]**

19. _____

20. Inventory at a local store was taken at retail value four times and was found to be $53,820; $49,510; $60,820; and $56,380. Sales during the same period were $252,077. Find the stock turnover at retail. Round to hundredths. **[8.4]**

20. _____

21. Thunder Manufacturing made the following purchases of rivet drums during the year: 25 at $135 each, 40 at $165 each, 15 at $108.50 each, and 30 at $142 each. An inventory shows that 45 rivet drums remain. Find the inventory value, using the weighted-average method. **[8.4]**

21. _____

22. Refer to Exercise 21. Find the inventory value using **(a)** the FIFO method and **(b)** the LIFO method. **[8.4]**

(a) _____
(b) _____

Simple Interest

CHAPTER CONTENTS

9.1 Basics of Simple Interest
9.2 Finding Principal, Rate, and Time
9.3 Simple Discount Notes
9.4 Discounting a Note Before Maturity

CASE IN POINT

APPLE INC. was founded by Steven Wozniak and Steven Jobs in 1976. It has revolutionized the computer, music, phone, movie, book and watch world with these devices: Macintosh, iPod, iPhone, iPad, and Apple watch. It sells more than 50 million devices per quarter and there are currently over 1 billion active Apple devices in the world.

While completing her degree in Information Systems at a local community college, Jessica Hernandez worked part-time in one of Apple's retail stores. She continued to work at the store after graduating and receiving a promotion to assistant manager. However, she has always wanted to own her own business and finally started a business to develop web pages for businesses.

CHAPTER 9 Simple Interest

Lending occurs everywhere. Individuals borrow to buy a car, pay for college, or buy a house. Businesses borrow to make payroll, expand, build a new building, develop a new product, or pay taxes. The interest on loans can be expensive. In fact, a common cause of **bankruptcy** is too much debt and payments that are too high. The interest rate at which the financially strongest firms borrow is called the **prime rate**. Most other firms pay a higher interest rate. Interestingly, the Koran forbids Muslims from charging interest. Instead, Islamic banks take an ownership interest in cars, houses, or firms until they are repaid.

Simple interest applies only to the **principal**, or the original amount borrowed, and it is usually used for loans lasting less than 1 year. Simple interest is discussed in this chapter. **Compound interest** requires interest to be paid on the principal and *also* on previously earned interest. It is covered in Chapter 10.

> **Quick TIP**
> The cost of borrowing money is called *interest*.

9.1 Basics of Simple Interest

OBJECTIVES
1. Solve for simple interest.
2. Calculate maturity value.
3. Use a table to find the number of days from one date to another.
4. Use the actual number of days in a month to find the number of days from one date to another.
5. Find exact and ordinary interest.
6. Define the basic terms used with notes.
7. Find the due date of a note.

CASE IN POINT Jessica Hernandez knew that expenses on a new business would begin immediately, yet revenues would be slow coming in at first. So she realized that she would need to borrow money to start her company.

Interest rates have a huge effect on costs. The graph shows the average interest rate on **consumer installment loans**, which are loans paid over time with a fixed number of payments. You may have purchased a car or a refrigerator on an installment loan. High interest rates make the monthly payments higher, potentially much higher. Low interest rates result in lower monthly payments.

Interest rates have varied greatly in the past. As you can see from the graph, interest rates were very high in 1980 but have generally fallen until recently. As interest rates increase, monthly payments increase and consumers have more difficulty financing and paying for things such as a new car or a house.

Interest Rate on Consumer Installment Loans

A major recession began in 2009 and it may still be affecting our economy since it affected so many people. The government responded by keeping interest rates very low for many years. Low interest rates reduce costs to borrowers, so keeping them low was an effort to encourage both individuals and firms to buy more. As consumers buy more, businesses become more profitable and hire more employees. Effectively, the government used low interest rates to try to increase the number of jobs and level of economic activity. However, economists now expect interest rates to increase as the effects of the 2009 recession finally fade away.

> **Quick TIP**
> Banks make a lot of simple interest loans but these types of loans are also used by others.

OBJECTIVE 1 Solve for simple interest. Simple interest is interest charged on the entire principal for the entire length of the loan. It is found using the formula that follows. **Principal** is the loan amount, **rate** is the annual interest rate, and **time** is the length of the loan *in years*.

9.1 Basics of Simple Interest

Finding Simple Interest

$$\text{Simple interest} = \text{Principal} \times \text{Rate} \times \text{Time}$$
$$I = P \times R \times T$$

1. Rate (R) must first be changed to a decimal or fraction.
2. Time (T) must first be converted to years.

Quick TIP
You can think of interest as rent. A person must pay rent to live in an apartment or house. Interest is "the rent," or the cost, of borrowing money.

So, for example, a rate of 7.5% should be changed to .075 and a time of 6 months should be changed to $\frac{6}{12} = .5$ year before using $I = PRT$.

Finding Simple Interest — EXAMPLE 1

Jessica Hernandez needs to borrow $85,000 for 9 months. Her bank would not lend her the money since she has no experience or assets. She found an individual who would lend her the money at 18.5%. However, her uncle agreed to go to the bank and **cosign** on a loan to her, which means he will have to repay the loan if Jessica fails to do so. On this basis, the bank agreed to lend her the money at 10% simple interest. Find the interest at **(a)** 18.5% and **(b)** 10%. **(c)** Then find the amount saved using the lower interest rate.

CASE IN POINT

SOLUTION

(a) First, convert 18.5% to .185 and 9 months to $\frac{9}{12}$ year. Then substitute values into $I = PRT$ to find the interest. The principal (P) is the amount of the loan.

$$I = PRT$$
$$I = \$85{,}000 \times .185 \times \tfrac{9}{12} \quad \text{Convert 18.5\% to .185}$$
$$I = \$11{,}793.75 \quad \text{simple interest}$$

(b)
$$I = PRT$$
$$I = \$85{,}000 \times .10 \times \tfrac{9}{12} \quad \text{Convert 10\% to .10 (or .1).}$$
$$I = \$6375 \quad \text{simple interest}$$

(c) Difference = $11,793.75 − $6375 = $5418.75

Hernandez quickly learned an important lesson: Interest costs can be very high. She was delighted that her uncle had agreed to cosign for her. It saved her nearly $5500 in interest in only 9 months.

The calculator solution for part **(a)** follows:

85000 × 18.5 % × 9 ÷ 12 = 11793.75

Note: Refer to Appendix B for calculator basics.

QUICK CHECK 1

Find the interest on a loan of $14,680 for 6 months at 9%.

OBJECTIVE 2 Calculate maturity value. The amount that must be repaid when the loan is due is the **maturity value** of the loan. Find this value by adding principal and interest.

Finding Maturity Value

$$\text{Maturity value} = \text{Principal} + \text{Interest}$$
$$M = P + I$$

326 CHAPTER 9 Simple Interest

Finding Maturity Value **EXAMPLE 2** Tom Swift needs to borrow $28,300 to remodel his bookstore so that he can serve coffee to customers as they browse or sit at their computers. He borrows the funds for 10 months at an interest rate of 9.25%. Find the interest due on the loan and the maturity value at the end of 10 months.

SOLUTION

Interest due is found using $I = PRT$, where T must be in years (10 months = $\frac{10}{12}$ year).

$$Interest = PRT$$

$$I = \$28{,}300 \times .0925 \times \frac{10}{12} = \$2181.46$$

$$Maturity\ value = P + I$$

$$M = \$28{,}300 + \$2181.46 = \$30{,}481.46$$

> **QUICK CHECK 2**
>
> Find the maturity value of a loan of $48,600 at 9% for 8 months.

OBJECTIVE 3 Use a table to find the number of days from one date to another. The **term** or **length of a loan** is often given in months, but it can be given in days. Or a loan may be due on a certain date such as July 24. To find how many days there are from one date to another, you might use the large table that appears on the next page.

For example, suppose it is June 11 and you want to know how many days until Christmas (December 25). In the table, June 11 is day 162 of the year and December 25 is day 359 of the year. Subtract to find the number of days until Christmas.

```
December 25 is day      359
June 11 is day         -162
                        197 days from June 11 to December 25
```

There are 197 days from June 11 to December 25.

Finding the Number of Days from One Date to Another, Using a Table **EXAMPLE 3** Use the table to find the number of days from **(a)** March 24 to July 22, **(b)** April 4 to October 10, **(c)** November 8 to February 17 of the following year, and **(d)** December 2 to January 17 of the following year. Assume that it is not a leap year.

SOLUTION

(a) July 22 is day   203
    March 24 is day   - 83
                          120 days from March 24 to July 22

(b) October 10 is day   283
    April 4 is day   - 94
                          189 days from April 4 to October 10

(c) November 8 is day 312, so there are $365 - 312 = 53$ days from November 8 to the end of the year. Add days until the end of the year plus days into the next year to find the total.

    November 8 to end of year   53
    February 17 is day   + 48
                                 101 days from November 8 to February 17 of next year

(d) December 2 is day 336, so there are $365 - 336 = 29$ days to the end of the year. Add days until the end of the year plus days into the next year to find the total.

    December 2 to end of year   29
    January 17 is day   + 17
                                 46 days from December 2 to January 17 of the next year

> **QUICK CHECK 3**
>
> Find the number of days from **(a)** July 7 to November 7 and **(b)** August 25 to January 20 of the following year.

NUMBER OF EACH DAY OF THE YEAR*

Day of Month	Jan.	Feb.	Mar.	Apr.	May	June	July	Aug.	Sept.	Oct.	Nov.	Dec.	Day of Month
1	1	32	60	91	121	152	182	213	244	274	305	335	1
2	2	33	61	92	122	153	183	214	245	275	306	336	2
3	3	34	62	93	123	154	184	215	246	276	307	337	3
4	4	35	63	94	124	155	185	216	247	277	308	338	4
5	5	36	64	95	125	156	186	217	248	278	309	339	5
6	6	37	65	96	126	157	187	218	249	279	310	340	6
7	7	38	66	97	127	158	188	219	250	280	311	341	7
8	8	39	67	98	128	159	189	220	251	281	312	342	8
9	9	40	68	99	129	160	190	221	252	282	313	343	9
10	10	41	69	100	130	161	191	222	253	283	314	344	10
11	11	42	70	101	131	162	192	223	254	284	315	345	11
12	12	43	71	102	132	163	193	224	255	285	316	346	12
13	13	44	72	103	133	164	194	225	256	286	317	347	13
14	14	45	73	104	134	165	195	226	257	287	318	348	14
15	15	46	74	105	135	166	196	227	258	288	319	349	15
16	16	47	75	106	136	167	197	228	259	289	320	350	16
17	17	48	76	107	137	168	198	229	260	290	321	351	17
18	18	49	77	108	138	169	199	230	261	291	322	352	18
19	19	50	78	109	139	170	200	231	262	292	323	353	19
20	20	51	79	110	140	171	201	232	263	293	324	354	20
21	21	52	80	111	141	172	202	233	264	294	325	355	21
22	22	53	81	112	142	173	203	234	265	295	326	356	22
23	23	54	82	113	143	174	204	235	266	296	327	357	23
24	24	55	83	114	144	175	205	236	267	297	328	358	24
25	25	56	84	115	145	176	206	237	268	298	329	359	25
26	26	57	85	116	146	177	207	238	269	299	330	360	26
27	27	58	86	117	147	178	208	239	270	300	331	361	27
28	28	59	87	118	148	179	209	240	271	301	332	362	28
29	29		88	119	149	180	210	241	272	302	333	363	29
30	30		89	120	150	181	211	242	273	303	334	364	30
31	31		90		151		212	243		304		365	31

*Add 1 to each date after February 29 for a leap year.

OBJECTIVE 4 Use the actual number of days in a month to find the number of days from one date to another. Another method for finding the number of days between specific dates is to use the number of days in each month as shown in the next table.

Number of Days in Each Month

31 Days	30 Days	28 Days
January	August	April	February
March	October	June	(29 days in leap year)
May	December	September	
July		November	

Two more ways of remembering the number of days in each month are the rhyme method and the knuckle method, as seen below.

Rhyme Method:

30 days hath September,
April, June, and November.
All the rest have 31, except
February, which has 28 and
in a leap year 29.

Knuckle Method:

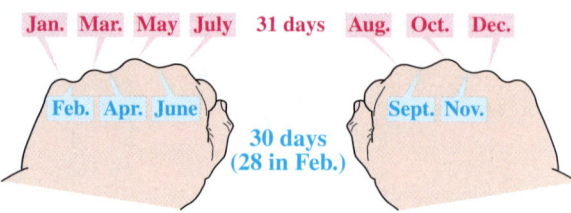

Finding the Number of Days from One Date to Another, Using Actual Days

EXAMPLE 4 Find the number of days from (a) June 3 to August 14 and (b) November 4 to February 21.

SOLUTION

(a) June has 30 days, so there are $30 - 3 = 27$ days from June 3 to the end of June.

June 3 to the end of June	27
31 days in July	31
14 days in August	+ 14
	72 days from June 3 to August 14

(b) November has 30 days, so there are $30 - 4 = 26$ days from November 4 to the end of November.

November 4 to end of November	26
31 days in December	31
31 days in January	31
21 days in February	+ 21
	109 days from November 4 to February 21

Quick TIP

To find the number of days from one date to another, do not count the day the loan was made, but do count the day the loan is paid.

QUICK CHECK 4

Find the number of days from March 14 to September 9.

OBJECTIVE 5 Find exact and ordinary interest. A simple interest rate is given as an annual rate, such as 7% per year. Since the rate is per year, time must also be given in years or fraction of a year when using $I = PRT$. If time is given in number of days, first change it to a fraction of a year.

Finding Time in Fraction of a Year

$$T = \frac{\text{Number of days in the loan period}}{\text{Number of days in a year}}$$

Exact interest calculations require the use of the exact number of days in the year, 365 or 366 if a leap year. **Ordinary interest**, or **banker's interest**, calculations require the use of 360 days. Many institutions, the government, and the Federal Reserve Bank use the exact number of days in a year in interest calculations. However, some banks and financial institutions still use 360 days.

Finding Exact and Ordinary Interest

For **exact** interest: Use 365 days (or 366 days if a leap year).

$$T = \frac{\text{Number of days in a loan period}}{365}$$

For **ordinary**, or **banker's**, interest: Use 360 days for the number of days.

$$T = \frac{\text{Number of days in a loan period}}{360}$$

Example 5 shows that **ordinary interest produces more interest** for the lending institution than does exact interest.

Finding Exact and Ordinary Interest

EXAMPLE 5 Radio station KOMA borrowed $148,500 on May 12 with interest due on August 27. If the interest rate is 10%, find the interest on the loan using (a) exact interest and (b) ordinary interest.

SOLUTION

Either the table method or the method of the number of days in a month can be used to find that there are 107 days from May 12 to August 27.

(a) The exact interest is found from $I = PRT$ with $P = \$148{,}500$, $R = .10$, and $T = \frac{107}{365}$.

$$I = PRT$$
$$I = \$148{,}500 \times .1 \times \frac{107}{365} \quad \text{Use 365 days.}$$
$$I = \$4353.29$$

(b) Find ordinary interest with the same formula and values, except $T = \frac{107}{360}$.

$$I = PRT$$
$$I = \$148{,}500 \times .1 \times \frac{107}{360} \quad \text{Use 360 days.}$$
$$I = \$4413.75$$

In this example, the ordinary interest is $\$4413.75 - \$4353.29 = \$60.46$ more than the exact interest.

Quick TIP
Ordinary interest results in slightly more interest than exact interest.

QUICK CHECK 5

Find the exact and ordinary interest for a 200-day loan of $19,500 at 9% to the nearest cent. Then find the difference between the two interest amounts.

> Use ordinary or banker's interest throughout the remainder of the book unless stated otherwise.

OBJECTIVE 6 Define the basic terms used with notes. Simple interest loans are commonly written using a **promissory note**, which is a legal document by which one person or firm agrees to pay a certain amount to another person or firm on a certain date.

This type of promissory note is called a **simple interest note**, since simple interest calculations involving $I = PRT$ are used. Here is the language of a simple interest promissory note.

> **Maker** or **payer:** The person borrowing the money. (Madeline Sullivan in the sample note)
> **Payee:** The person who loaned the money and who will receive the payment (Charles D. Miller in the sample note)
> **Term:** The length of time until the note is due (90 days in the sample note)
> **Face value** or **principal:** The amount being borrowed ($27,500 in the sample note)
> **Maturity value:** The face value plus interest, also the amount due at maturity ($27,500 in the sample note)
> **Maturity date** or **due date:** The date the loan must be paid off with interest (June 4 in the sample note)

Find the interest and the maturity value on the loan in the sample note above.

$$\text{Interest} = \text{Face value} \times \text{Rate} \times \text{Time}$$

$$\text{Interest} = \$27{,}500 \times .09 \times \frac{90}{360} = \$618.75$$

$$\text{Maturity value} = \text{Face value} + \text{Interest}$$

$$\text{Maturity value} = \$27{,}500 + \$618.75 = \$28{,}118.75$$

Madeline Sullivan must pay $28,118.75 to Charles D. Miller on June 4, the maturity date of the note.

Banks and financial institutions lend money only to individuals and firms they believe will repay the loan with interest. Even then, banks often require **collateral** or assets such as an automobile or stock in order to make a loan. If the loan is not repaid, the bank **forecloses** on the collateral, takes ownership, and sells or liquidates it. Funds from the sale of the collateral are first used to pay off the note and the expenses of the foreclosure. Any excess is returned to the maker of the note (person who borrowed the money).

Quick TIP
Interest rates are often higher when borrowing from individuals than from large banks. However, some people or firms do not have the credit history or collateral to borrow from a bank.

OBJECTIVE 7 Find the due date of a note. Some promissory notes give the term in months. In that case, the loan is due after the given number of months, but on the same day of the month as the original loan was made, as is shown next.

Date Made	Length of Loan	Date Due
March 12	5 months	August 12
April 24	7 months	November 24
October 7	9 months	July 7
January 31	3 months	April 30

When a calculated due date does not actually exist—"February 30" or "November 31," for example—the loan is considered to be due on the last day of the calculated month. For example, a loan made on January 31 for 3 months would normally be due on April 31. However, there are only 30 days in April, so the loan is due on the last day of April, or April 30.

Finding Due Date, Interest, and Maturity Value

EXAMPLE 6 Find the due date, interest, and maturity value for a $600,000 loan made to Benson Automotive on July 31 for 7 months at 7.5% interest.

SOLUTION

Interest and principal are due 7 months from July 31, or on February 31, which *does not* exist. Since February has only 28 days (unless it is a leap year), interest and principal are due on the last day of February, or February 28. If it were a leap year, the maturity value would be due on February 29.

$$I = PRT = \$600{,}000 \times .075 \times \frac{7}{12} = \$26{,}250$$

$$M = P + I = \$600{,}000 + \$26{,}250 = \$626{,}250$$

A total of $626,250 must be repaid on February 28.

Quick TIP
Do not convert the period of a loan from months to days to find the due date.

QUICK CHECK 6

Find the due date for a 6-month loan made on March 31.

9.1 Exercises // MyLab Math

The shaded sections below contain solutions to help you get a QUICK START on the various types of exercises.

Find simple interest and maturity value to the nearest cent. (See Examples 1 and 2.)

	Interest	Maturity Value
1. $3800 at 11% for 6 months $I = \$3800 \times .11 \times \frac{6}{12} = \$209$ $M = \$3800 + \$209 = \$4009$	$209	$4009
2. $10,200 at 9.5% for 10 months	_____	_____
3. $5500 at 8% for 1 year	_____	_____
4. $10,800 at 9% for 8 months	_____	_____

Find the exact number of days from the first date to the second. None of the years are leap years. (See Examples 3 and 4.)

5. February 15 to April 24
   From the table, February 15 is day 46; April 24 is day 114
   Number of days = 114 − 46 = 68 days

   5. __68__

6. May 22 to August 30

   6. _____

7. December 1 to March 10 of the following year

   7. _____

8. October 12 to February 22 of the following year

   8. _____

Find (a) the exact interest and (b) the ordinary interest for each of the following to the nearest cent. Then find (c) the amount by which the ordinary interest is larger. (See Example 5.)

9. $52,000 at $8\frac{3}{4}$% for 200 days
   (a) Exact interest = $\$52,000 \times .0875 \times \frac{200}{365} = \$2493.15$
   (b) Ordinary interest = $\$52,000 \times .0875 \times \frac{200}{360} = \$2527.78$
   (c) Ordinary is larger by $2527.78 − $2493.15 = $34.63

   (a) __$2493.15__
   (b) __$2527.78__
   (c) __$34.63__

C// indicates an exercise that is related to the Case in Point feature.

332 CHAPTER 9 Simple Interest

10. $185,000 at 7.5% for 180 days
    (a) _____
    (b) _____
    (c) _____

11. $38,750 at 10% for 240 days
    (a) _____
    (b) _____
    (c) _____

12. $52,610 at $8\frac{1}{2}$% for 82 days
    (a) _____
    (b) _____
    (c) _____

Identify each of the following from the promissory note shown. (See Objective 6.)

PROMISSORY NOTE

Jackson, Mississippi _____

_____ after date, _____ promise to pay to the order of
_____
_____ Dollars with interest at _____
_____ , payable at _____
Due _____

13. Maker _____
14. Payer _____
15. Payee _____
16. Face value _____
17. Term of loan _____
18. Date loan was made _____
19. Date loan is due _____
20. Maturity value _____

Find the date due, the amount of interest, and the maturity value. Use banker's interest. (See Example 6.)

Date Loan Was Made	Face Value	Term of Loan	Rate	Date Loan Is Due	Maturity Value
21. Mar. 12	$4800	220 days	9%	Oct. 18	$5064
I = $4800 × .09 × 220/360 = $264; M = $4800 + $264 = $5064					
22. Jan. 3	$12,000	100 days	9.8%	_____	_____
23. Nov. 10	$6300	180 days	$9\frac{1}{4}$%	_____	_____
24. July 14	$20,400	90 days	8%	_____	_____

Solve the following application problems. Round dollar amounts to the nearest cent.

25. **INVENTORY** Benson Automotive borrows $2,000,000 at $9\frac{1}{4}$% from a bank to buy land to build a building for a new dealership. Given that the loan is for 9 months, find (a) the interest and (b) the maturity value.
    (a) I = $2,000,000 × .0925 × 9/12 = $138,750
    (b) M = $2,000,000 + $138,750 = $2,138,750

    (a) $138,750
    (b) $2,138,750

26. **LOANS BETWEEN BANKS** Wells Fargo Bank borrows $25,000,000 at 4% for 90 days from a bank in Chicago. Find **(a)** the interest and **(b)** the maturity value.

(a) _____
(b) _____

27. **ROAD PAVING** Montoya Construction needs to borrow $375,000 to build a road and install utilities in a small subdivision. It borrows the funds at 8% for 90 days. When interest rates were high in 1980, the interest rate would have been 22%. Find the difference in interest charges between the two rates.

27. _____

28. **INTERNATIONAL BUSINESS** Lesly Pacas borrows 300,000 pesos for 90 days at 12% per year to remodel her hair salon. She lives in Guadalajara, Mexico, where the rate would have been 35% a decade ago. Find the difference in the interest charges based on the different rates.

28. _____

29. **CAPITAL IMPROVEMENT** Elizabeth Barton borrowed $13,700 to install a small rock fountain and fish pond in front of her flower shop. She signed a 90-day note on July 5 at $9\frac{1}{4}$% interest. Find **(a)** the due date and **(b)** the maturity value of the note.

(a) _____
(b) _____

30. **WEB DESIGN** On September 10, Jessica Hernandez signed a promissory note with a face value of $32,500 to help her pay the salaries of employees during a period of slow sales. The 90-day note is at 7.5% interest. Find **(a)** the due date and **(b)** maturity value of the note.

(a) _____
(b) _____

31. **HEALTH FOOD** On March 10, the owner of The Granary borrowed $80,000 on a 180-day promissory note at 10.5% interest. Find **(a)** the due date and **(b)** the maturity value of the note.

(a) _____
(b) _____

32. **CORPORATE FINANCE** On October 15, IBM borrows $45,000,000 at 6% from a bank in San Francisco and agrees to repay the loan in 120 days using ordinary interest. Find **(a)** the due date and **(b)** the maturity value.

(a) _____
(b) _____

33. **PENALTY ON UNPAID PROPERTY TAX** Joe Simpson's property tax is $3416.05 and is due on April 15. He does not pay until July 23. The county adds a penalty of 9.3% simple interest on his unpaid tax. Find the penalty using exact interest.

33. _____

334 CHAPTER 9 Simple Interest

34. PENALTY ON UNPAID INCOME TAX On January 5, Helen Terry made a property tax payment that was due on September 15. The penalty was at an annual rate of 7.5% simple interest per year on the unpaid tax of $4100. Find the penalty using exact interest.

34. _____

35. PAINT STORE On January 31, Jackson Paints & Supplies borrowed $128,000 to build a metal building for use as a warehouse. The 8-month loan has a rate of 9.5%. Find **(a)** the due date and **(b)** the maturity value of the loan.

(a) _____
(b) _____

36. CRIMINAL JUSTICE Harris County administrators borrow $1,850,000 to set up a branch forensic lab. The 9 month, 4.5% simple interest loan is made on the last day of May. Find **(a)** the due date and **(b)** the maturity value.

(a) _____
(b) _____

37. Explain the difference between exact interest and ordinary, or banker's, interest. (See Objective 5.)

38. List three companies that you have purchased products or services from in the past. List two reasons each of the companies may have needed to borrow money in the past.

QUICK CHECK ANSWERS

1. $660.60
2. $51,516
3. **(a)** 123 days **(b)** 148 days
4. 179 days
5. $961.64 exact interest; $975 ordinary interest; difference of $13.36
6. September 30

9.2 Finding Principal, Rate, and Time

OBJECTIVES
1. Find the principal.
2. Find the rate.
3. Find the time.

Principal (P), rate (R), and time (T) were given for all problems in **Section 9.1**, and we calculated interest. In this section, interest is given, and we solve for principal, rate, or time.

OBJECTIVE 1 Find the principal. The principal (P) is found by dividing both sides of the simple interest equation $I = PRT$ by RT. See Chapter 4 for a review of algebra if needed.

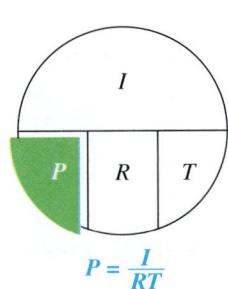

$$I = PRT$$
$$\frac{I}{RT} = \frac{PRT}{RT} \quad \text{Divide both sides by } RT.$$
$$\frac{I}{RT} = P \quad \text{or} \quad P = \frac{I}{RT}$$

$$P = \frac{I}{RT}$$

The various forms of the simple interest equation can be remembered using the circle sketch shown above. In the sketch, I (interest) is in the top half of the circle, with P (principal), R (rate), and T (time) in the bottom half of the circle. Find the formula for any one variable by covering the letter in the circle and then reading the remaining letters, noticing their position. For example, cover P and you are left with $\frac{I}{RT}$.

$$\text{Principal} = \frac{\text{Interest}}{\text{Rate} \times \text{Time (in years)}} \quad \text{or} \quad P = \frac{I}{RT}$$

Note: Use banker's interest with 360 days for all problems in this section.

Finding Principal Given Interest in Days **EXAMPLE 1** Gilbert Construction Company borrows funds at 10% for 54 days to finish building a home. Find the principal that results in interest of $1560.

SOLUTION

Write the rate as .10, the time as $\frac{54}{360}$, and then use the formula for principal.

$$P = \frac{I}{RT}$$

$$P = \frac{\$1560}{.10 \times \frac{54}{360}}$$

$$.10 \times \frac{54}{360} = .015 \qquad \text{Simplify the denominator.}$$

$$P = \frac{\$1560}{.015} = \$104{,}000 \qquad \text{Divide.}$$

Quick TIP
Remember that time must be in years or fraction of a year.

The principal is $104,000.
  Check the answer using $I = PRT$. The principal is $104,000, the rate is 10%, and the time is $\frac{54}{360}$ year. The interest should be, and is, $780.

$$I = \$104{,}000 \times .10 \times \frac{54}{360} = \$1560$$

The calculator approach to finding the principal uses parentheses so that the numerator is divided by the entire denominator.

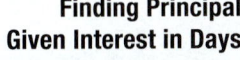

1560 ÷ (.10 × 54 ÷ 360) = 104,000

336 CHAPTER 9 Simple Interest

> **QUICK CHECK 1**
>
> A 90-day loan with a rate of 12% results in interest of $285. Find the principal.

Finding Principal Given Length of Loan

EXAMPLE 2 On February 2, Ebony Johnson took out a short-term loan to pay her college tuition. The loan is due to be repaid on April 15, when Johnson expects to receive an income tax refund. The interest on the loan is $93.60 at a rate of 6.5%. Find the principal, i.e. the amount borrowed.

SOLUTION

First find the number of days.

$$\begin{array}{rl} 26 & \text{days remaining in February} \\ 31 & \text{March} \\ +15 & \text{April} \\ \hline 72 & \text{days from February 2 to April 15} \end{array}$$

So $T = \frac{72}{360}$. Next find the principal.

$$P = \frac{I}{RT}$$

$$P = \frac{\$93.60}{.065 \times \frac{72}{360}} \qquad \text{Substitute values.}$$

$$= \frac{\$93.60}{.013} \qquad \text{Calculate denominator.}$$

$$= \$7200$$

Thomas borrowed $7200 for tuition. You can use the formula for simple interest to make sure the answer is correct.

$$I = \$7200 \times .065 \times \frac{72}{360} = \mathbf{\$93.60}$$

> **QUICK CHECK 2**
>
> A loan made on May 12 must be repaid on December 18. Find the principal given that the rate is 9% and the interest at maturity is $1551.

OBJECTIVE 2 Find the rate. Solve the formula $I = PRT$ for rate (R) by dividing both sides of the equation by PT. The rate found in this manner will be the annual interest rate.

$$I = PRT$$

$$\frac{I}{PT} = \frac{PRT}{PT} \qquad \text{Divide both sides by } PT.$$

$$\frac{I}{PT} = R \quad \text{or} \quad R = \frac{I}{PT}$$

$$R = \frac{I}{PT}$$

$$\boxed{\text{Rate} = \frac{\text{Interest}}{\text{Principal} \times \text{Time (in years)}} \quad \text{or} \quad R = \frac{I}{PT}}$$

9.2 Finding Principal, Rate, and Time

Finding Rate Given Length of Loan

EXAMPLE 3 An exchange student from the United States living in Paris deposits $10,000 in U.S. currency in a French bank for 45 days, when she has to pay her tuition and fees for a year. Find the rate if the interest is $18.13 in U.S. currency.

SOLUTION

$$\text{Rate} = \frac{I}{PT}$$

$$R = \frac{\$18.13}{\$10,000 \times \frac{45}{360}} \qquad \text{Substitute values.}$$

$$= \frac{\$18.13}{\$1250} \qquad \text{Simplify the denominator.}$$

$$= .014504 \text{ or } 1.45\%$$

Check the answer using the simple interest formula.

Solve using a calculator as follows. Notice that parentheses set off the denominator.

$\$18.13 \div (\$10,000 \times 45 \div 360) = .014504$

QUICK CHECK 3
A 120-day loan for $15,000 has interest of $412.50. Find the rate.

Finding Rate Given Length of Loan

EXAMPLE 4 Blaine Plumbing kept extra cash of $86,500 in an account from June 1 to August 16. Find the rate if the company earned $365.22 in interest during this period of time. Round to the nearest tenth of a percent.

SOLUTION
Find the number of days using the table on page 327.

August 16 is day   228
June 1 is day   − 152
              76 days

There are 76 days from June 1 to August 16.

$$T = \frac{76}{360}$$

$$\text{Rate} = \frac{I}{PT}$$

$$R = \frac{\$365.22}{\$86,500 \times \frac{76}{360}} = .0200$$

The rate of interest is 2.0%.

QUICK CHECK 4
A loan of $37,000 made on February 4 results in interest of $770.83. If the loan is due on May 15, find the rate to the nearest tenth of a percent.

OBJECTIVE 3 Find the time. The time (T) is found by dividing both sides of the simple interest equation $I = PRT$ by PR. Note that time will be in years.

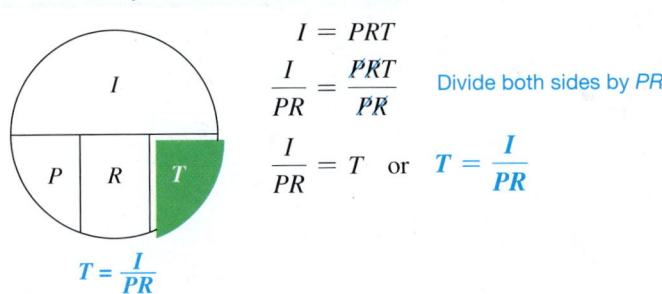

$$I = PRT$$
$$\frac{I}{PR} = \frac{PRT}{PR} \qquad \text{Divide both sides by } PR.$$
$$\frac{I}{PR} = T \text{ or } T = \frac{I}{PR}$$

$T = \dfrac{I}{PR}$

Finding Time

$$\text{Time (in years)} = \frac{\text{Interest}}{\text{Principal} \times \text{Rate}} \quad \text{or} \quad T = \frac{I}{PR}$$

The preceding formula gives time in years, but we often need time in days or months. Find these as follows.

Finding Time

$$\text{Time in days} = \frac{I}{PR} \times 360$$

$$\text{Time in months} = \frac{I}{PR} \times 12$$

Finding Time in Days Given Principal and Rate

EXAMPLE 5 Roberta Sanchez deposited $18,600 in an account paying 3% and earned $217 in interest. Find the number of days that the deposit earned interest.

SOLUTION

$$T \text{ in days} = \frac{I}{PR} \times 360$$

$$T = \frac{\$217}{\$18{,}600 \times .03} \times 360 = 140 \text{ days} \quad \text{(rounded)}$$

The money was on deposit for 140 days.

▦ Use parentheses around the denominator of the fraction to make sure that the calculations are done in the correct order.

217 ÷ (18600 × .03) × 360 = 140

QUICK CHECK 5

A loan for $22,000 results in interest of $1283.33 at 10.5%. Find the time to the nearest day.

Simple Interest Formulas

Interest	$I = PRT$
Principal	$P = \dfrac{I}{RT}$
Rate	$R = \dfrac{I}{PT}$
Time	$T \text{ (in years)} = \dfrac{I}{PR}$
	$T \text{ (in months)} = \dfrac{I}{PR} \times 12$
	$T \text{ (in days)} = \dfrac{I}{PR} \times 360$

All of these are modifications of the formula $I = PRT$.

9.2 Exercises // MyLab Math

The shaded sections below contain solutions to help you get a QUICK START on the various types of exercises.

Find the principal in each of the following. (See Example 1.)

	Rate	Time (in days)	Interest	Principal
1.	$7\frac{3}{4}\%$	90	$271.25	$14,000

$$P = \frac{\$271.25}{.0775 \times \frac{90}{360}} = \$14,000$$

	Rate	Time (in days)	Interest	Principal
2.	9.5%	120	$186	$5873.68

$$P = \frac{\$186}{.095 \times \frac{120}{360}} = \$5873.68$$

	Rate	Time (in days)	Interest	Principal
3.	10%	80	$112	_____
4.	5%	300	$1250	_____
5.	$8\frac{1}{2}\%$	120	$4420	_____
6.	10.5%	140	$87.20	_____

Find the rate in each of the following. Round to the nearest tenth of a percent. (See Example 3.)

	Principal	Time	Interest	Rate
7.	$7600	200 days	$498.22	11.8%

$$R = \frac{\$498.22}{\$7600 \times \frac{200}{360}} = 11.8\%$$

	Principal	Time	Interest	Rate
8.	$15,600	90 days	$312	_____
9.	$42,800	60 days	$677.67	_____
10.	$20,000	180 days	$850	_____
11.	$8000	4 months	$200	_____
12.	$4800	5 months	$197.60	_____

C// indicates an exercise that is related to the Case in Point feature.

340 CHAPTER 9 Simple Interest

Find the time in each of the following. In Exercises 13–16, round to the nearest day; in Exercises 17 and 18, round to the nearest month. (See Example 5.)

Principal	Rate	Interest	Time
13. $74,000	9.5%	$2343.33	120 days

$$T = \frac{\$2343.33}{\$74,000 \times .095} \times 360 = 119.9998 \text{ or } 120 \text{ days}$$

14. $36,000	9%	$585	_____
15. $24,000	8.75%	$700	_____
16. $20,000	8%	$1200	_____
17. $3500	$10\frac{1}{4}$%	$143.50	_____
18. $8400	$7\frac{1}{4}$%	$357	_____

In each of the following application problems, find principal to the nearest cent, rate to the nearest tenth of a percent, or time to the nearest day.

19. SHORT-TERM SAVINGS Hoyt Axton earned $244.80 interest in 9 months from an investment that paid 3.2% per year. Use the simple interest formula to estimate the amount initially invested.

$$P = \frac{\$244.80}{.032 \times \frac{9}{12}} = \$10,200$$

19. __$10,200__

20. BANK LOAN Citizens Bank earned $12,250 interest in 45 days from a bank loan with 5.6% interest per year. Find the amount initially invested.

20. _____

21. INVESTING IN BONDS Joan Gretz invested $7200 in a mutual fund containing bonds. Find the rate if she earned $237.50 in interest in 250 days.

21. _____

22. LAW ENFORCEMENT The Smith County Police Department borrowed $120,000 for 200 days to purchase new radar-detection equipment to detect speeders. Find the rate if the interest was $4067.

22. _____

23. PAYROLL Jessica Hernandez borrowed $45,000 so that she could meet payroll for the quarter. She agreed to pay interest of $1881.25 in 140 days. Find the rate.

23. _____

24. SAVING FOR RETIREMENT Mike Jordan deposited $2000 into a Roth Individual Retirement Account (IRA) investing in a mutual fund containing corporate bonds. Find the rate if he has $2192.50 in the account 15 months later.

24. _____

25. RETIREMENT ACCOUNT Over a period of 300 days, Shawna Johnson earned $450 interest in a retirement account paying interest at a rate of 5%. Find **(a)** the principal at the beginning of the 300 days and **(b)** the amount in the account at the end of 300 days.

(a) _____
(b) _____

26. INTEREST EARNINGS Patterson Plumbing had an account that earned $214.67 interest in 280 days. If the interest rate was 4%, find **(a)** the principal at the beginning of the 280 days and **(b)** the amount in the account at the end of the 280 days.

(a) _____
(b) _____

27. TIME OF DEPOSIT Benson Automotive earned $69.46 interest on a $9400 deposit in an account paying 3.5%. Find the number of days that the funds were on deposit.

27. _____

28. TIME OF DEPOSIT Find how long Quinlan Enterprises must deposit $7500 at 6% in order to earn $243.75 interest.

28. _____

29. INTEREST RATE Hot Air Balloon Tours, Inc. must pay the bank $23,515.27 in interest 300 days after making a loan of $328,120 to purchase hot air balloons.

29. _____

30. PENALTY ON LATE PAYMENT Juanita Comey pays her $1800 credit-card bill 70 days after it is due. She is charged a penalty of $59.50. Find the rate of interest that was charged as a penalty. (*Note:* Penalty rates are frequently quite high.)

30. _____

31. LAPTOPS Shanghai Computers purchased 20 laptops from a Chinese computer manufacturer. Shanghai paid the bill after 45 days, paying a finance charge of $150. If the Chinese company charges 10% interest, find **(a)** the cost of the 20 laptops excluding the interest and **(b)** the cost per laptop.

(a) _____
(b) _____

32. LAWN-MOWER Yard Mowers, Inc., bought 15 self-propelled, riding lawn mowers from Green Lawns, Ltd. The company paid after 90 days and was charged an annual finance charge of 12%, or $126. Find **(a)** the cost of the 15 lawn mowers excluding the interest and **(b)** the cost per mower.

(a) _____
(b) _____

33. PROMISSORY NOTE Jan Reus signed a promissory note for $15,000 at 7.5% interest that resulted in interest charges of $650. Find the term of the note to the nearest day.

33. _____

342 CHAPTER 9 Simple Interest

34. TIME OF DEPOSIT The Frampton Chamber of Commerce earns $682.71 interest on a $16,385 investment at 5.5%. Find the length of time of the investment to the nearest day.

34. _____

35. HOME CONSTRUCTION Gilbert Construction Company needs to borrow $350,000 for 1 year to build three homes. They can borrow from either of two banks. Interest charges from Bank One would amount to $23,650, whereas interest charges from First National Bank would amount to $25,000. Find the interest rates associated with a loan from (a) Bank One and (b) First National Bank.

(a) _____

(b) _____

36. INVENTORY PURCHASE Forest Nursery has had some problems with credit history in the past and expects high interest rates. But it needs to borrow $9500 on February 1 to buy additional inventory and will repay the loan on July 15. Interest charges for State Bank and First National Bank are $480 and $443.60, respectively. Find the rate for (a) the State Bank loan and (b) the First National Bank loan.

(a) _____

(b) _____

37. A retired couple receives $14,000 per year from Social Security and an additional $18,000 in interest from retirement plans and lifetime savings. They need all of their income to pay expenses, including medical bills. What will happen if the interest rate on their retirement plans and lifetime savings decreases significantly?

38. How would the formula for calculating time in days (given principal, interest, and rate) change if exact interest were used rather than ordinary interest?

QUICK CHECK ANSWERS

1. $9500
2. $28,200
3. 8.25%
4. 7.5%
5. 200 days

9.3 Simple Discount Notes

OBJECTIVES

1. Define the basic terms used with simple discount notes.
2. Find the bank discount and proceeds.
3. Find the face value.
4. Find the effective interest rate.
5. Understand U.S. Treasury bills.

CASE IN POINT Jessica Hernandez does not like to borrow money, but she had to when a large firm delayed payment on a bill. She has bills that must be paid now.

The clipping shows that the federal government has the power to change interest rates. Higher interest rates make it more expensive to borrow and slow economic growth. Lower interest rates make it cheaper to borrow and increase economic growth. The federal government uses interest rates as a tool *to control the growth rate of the economy*. The goal is to keep the economy growing fast enough to generate jobs but not so fast as to cause inflation, which is discussed in the next chapter.

Interest Rates Set to Go Up

All eyes are on the Fed chairwoman as the Federal Reserve Board meets. Economists expect rates to increase slightly. They have been at record lows since the recession began, but labor and material costs are rising.

In this section, we discuss **simple discount notes**, which are simply a different way to set up a promissory note. Any note that uses simple interest calculations with a lump-sum payment can be set up *either* as a simple interest note or as a simple discount note. One type of note is *not* better than the other type of note. They merely represent two different ways to discuss the same thing. We study both because some banks use simple interest notes while others use simple discount notes.

OBJECTIVE 1 Define the basic terms used with simple discount notes. Simple interest notes involve principal (face value or loan amount), interest rate, time, interest, and maturity value.

Simple discount notes involve **proceeds (loan amount)**, **discount rate**, time, **bank discount (interest)**, and **face value (or maturity value)**. Face value in a simple interest note is the amount loaned to the borrower, but it is the maturity value in a simple discount note. Simple discount notes are also called **interest-in-advance notes**, since interest is subtracted before funds are given to the borrower. A basic difference between the two types of notes is that simple interest is calculated based on principal, whereas simple discount is calculated based on maturity value. College students sometimes borrow money from the government using **Stafford loans**, which are simple discount notes.

Simple Interest versus Simple Discount Notes

Type of Note	Loan Amount		Interest		Repayment Amount
Simple interest	Face value (Principal)	+	Interest	=	Maturity value
Simple discount	Proceeds	+	Discount (Interest)	=	Face value (Maturity value)

Note: Simple interest is calculated on the *principal*, while simple discount is calculated on the *maturity value*.

Quick TIP Remember that these are just two different ways to set up a loan. One is no better than the other and both are common.

OBJECTIVE 2 Find the bank discount and proceeds. The formula for finding the bank discount is a form of the basic percent equation. The formula is similar to the one used to calculate simple interest, but different letters are used since the ideas differ slightly.

Calculating Bank Discount

Bank discount = Face value × Discount rate × Time

$$B = M \times D \times T$$

Then, if P is the proceeds to the borrower,

Proceeds (loan amount) = Face value − Bank discount   or $P = M - B$

Stated in another way,

Face value = Proceeds (loan amount) + Bank discount   or $M = P + B$

Note: Time must be given in years.

Finding Discount and Proceeds

EXAMPLE 1 Jim Peterson signs a simple discount note with a face or maturity value of $35,000 so that he can purchase a truck with plow for his snow removal business. The banker discounts the 10-month note at 9%. Find the amount of the discount and the proceeds.

SOLUTION

Peterson *does not* receive $35,000 from the bank—that is the amount he must repay when the loan matures. Use $M = \$35,000$, $D = 9\%$, and $T = \frac{10}{12}$ in the formula $B = MDT$ to find the discount, which is the interest that must be paid at maturity.

$$\text{Bank discount} = M \times D \times T$$
$$B = \$35,000 \times .09 \times \frac{10}{12} = \mathbf{\$2625}$$

The discount of $2625 is the interest charge on the loan. The proceeds that Peterson actually receives when making the loan is found using $P = M - B$.

$$P = M - B$$
$$P = \$35,000 - \$2625 = \$32,375$$

Peterson receives $32,375 but must repay $35,000 when the loan matures in 10 months. The difference is the interest.

QUICK CHECK 1

A simple discount loan has a maturity value of $15,800, discount rate of 9%, and time of 180 days. Find the bank discount and proceeds.

Finding the Proceeds **EXAMPLE 2** To finance a new electronic sign to put in front of a retail store, Mustang Auto signs a 6-month, simple discount note with a face value of $4500. Find the proceeds if the discount rate is 9.5%.

SOLUTION

The bank discount (B) is not known, but we do know that $B = MDT$. Therefore, we can substitute MDT in place of B.

$$P = M - B$$
$$P = M - MDT \qquad \text{Substitute } MDT \text{ in place of } B.$$
$$P = \$4500 - \left(\$4500 \times .095 \times \frac{6}{12}\right) \qquad \text{Substitute values.}$$
$$= \$4286.25$$

Mustang Auto receives $4286.25 but must pay back $4500 in 6 months.

QUICK CHECK 2

A 220-day loan to a firm with poor credit has a face value of $40,000 with a discount rate of 12%. Find the proceeds.

OBJECTIVE 3 Find the face value. If the loan amount (proceeds) of a simple discount note is known, use the following formula to find the corresponding face value.

Calculating Face Value to Achieve Desired Proceeds

$$M = \frac{P}{1 - DT}$$

where

M = Face value of the simple discount note
P = Proceeds received by the borrower
D = Discount rate used by the bank
T = Time of the loan (in years)

Note: The symbol D is the discount *rate*, not the bank discount.

Finding the Face Value **EXAMPLE 3** // Tina Watson purchased a classic 1961 Corvette and plans to rebuild it. She needs to borrow $18,000 for 180 days. Find the face value of the 10% simple discount note that would result in proceeds of $18,000 to Watson.

SOLUTION

Use the formula.

$$M = \frac{P}{1 - DT}$$

Replace P with $18,000, D with .10, and T with $\frac{180}{360}$.

$$M = \frac{\$18,000}{1 - \left(.10 \times \frac{180}{360}\right)} = \$18,947.37$$

The face value of the note is $18,947.37. However, Watson receives only $18,000 from the bank when the note is signed. She must repay $18,947.37 to the bank in 180 days.

🖩 The problem

$$\frac{\$18,000}{1 - \left(.10 \times \frac{180}{360}\right)}$$

can be solved using a calculator by first finding the denominator.

1 [−] [(] .1 [×] 180 [÷] 360 [)] [=] .95 [STO]
18000 [÷] [RCL] [=] 18947.37

QUICK CHECK 3

A 300-day note has proceeds of $48,000 and a discount rate of 8.8%. Find the maturity value.

Comparing Discount Notes and Simple Interest Notes **EXAMPLE 4** // The owner of a medical supplies store has been offered loans from two different banks. Each note has a face value of $75,000 and a time of 90 days. One note has a simple interest rate of 10%, and the other a simple discount rate of 10%. Identify the better deal.

SOLUTION

Find the interest owed on each.

Simple Interest Note	Simple Discount Note
$I = PRT$	$B = MDT$
$I = \$75,000 \times .10 \times \frac{90}{360}$	$B = \$75,000 \times .10 \times \frac{90}{360}$
$I = \$1875$	$B = \$1875$

The amount of interest is the same in both notes. Now find the amount the borrower would receive.

Simple Interest Note	Simple Discount Note
Face value = $75,000	Proceeds = $M - B$
	= $75,000 − $1875
	= $73,125

Quick TIP
A simple interest note is NOT always better than a simple discount note, or vice versa. You must compare specifics to discover which is better.

The borrower has the use of $75,000 with the simple interest note, but only $73,125 with the simple discount note. Yet the amount of interest is identical. So the simple interest note is the better deal in this situation. Find the maturity value of each note as follows.

Simple Interest Note	Simple Discount Note
$M = P + I$	Maturity = Face value
$= \$75,000 + \$1875$	$= \$75,000$
$= \$76,875$	

The differences between these two notes can be summarized as follows.

	Simple Interest Note	Simple Discount Note
Face value	$75,000	$75,000
Interest	$1875	$1875
Amount available to borrower	$75,000	$73,125
Maturity value	$76,875	$75,000

QUICK CHECK 4

Two notes both have face values of $24,000 and a time of 180 days. The first note has a simple interest rate of 9%, and the second has a simple discount rate of 9%. Find the maturity value of each.

OBJECTIVE 4 Find the effective interest rate. The federal **Truth in Lending Act** requires that interest rates be given in a form that can easily be compared.

The **effective rate of interest** is the interest rate that is calculated based on the actual amount of money received by the borrower. It is also called the **annual percentage rate**, the **APR**, and the **true rate**.

The discount rate of 10% in Example 4 is the **stated rate** or **nominal rate**, since it is the rate written on the note. It is not the effective rate, since the 10% applies to the maturity value of $75,000 and *not* to the proceeds of $73,125 actually received by the borrower. The next example shows how to find the effective rate for Example 4.

Finding the Effective Interest Rate

EXAMPLE 5 Find the effective rate of interest (APR) for the simple discount note of Example 4.

SOLUTION

Quick TIP
The discount rate is not an interest rate to be applied to proceeds (loan amount)—rather it is applied to face value.

Find the effective rate (APR) by using the formula for simple interest: $I = PRT$. In this case, $I = \$1875$ (the discount), $P = \$73,125$ (the proceeds), and $T = \frac{90}{360}$.

$$R = \frac{I}{PT}$$

$$R = \frac{\$1875}{\$73,125 \times \frac{90}{360}} = .1026 = 10.26\% \quad \text{(rounded)}$$

Thus, the 10.26% effective rate of the simple discount note is higher than the 10% effective rate of the simple interest note showing that the simple interest note is better for the borrower in this situation.

QUICK CHECK 5

Find the effective rate (APR) for a loan with a loan amount of $31,000, a time of 90 days, and interest of $891.25.

The interest rate 10.26% in Example 5 is the **effective rate of interest**. Federal regulations require that rates be rounded to the nearest quarter of a percent when communicated to a borrower. Since 10.26% is closer to 10.25% than to 10.50%, an APR of 10.25% must be reported to the borrower in Example 5.

The table below shows two identical loans, one a simple interest loan and the other a simple discount loan. Proceeds, interest charges, terms of the loans, and maturity values are all identical.

	Simple Interest Loan	Simple Discount Loan
Proceeds	$11,100	$11,100
Interest	$900	$900
Maturity Value	$12,000	$12,000
Face Value	$11,100	$12,000
Time	10 months	10 months
Interest Rate	9.73%	—
Discount Rate	—	9%

Quick TIP
The two loans in the table are identical in terms of proceeds, interest, and term. But they are set up differently.

Since the loans are identical, a 9.73% simple interest loan for 10 months is equivalent to a 9% simple discount note for 10 months. In both situations, the borrower receives $11,100 and must repay $12,000 ten months later. Therefore, the two loans have the *same effective rate*.

OBJECTIVE 5 Understand U.S. Treasury bills. The United States government borrows very large amounts of money from banks, various financial institutions, pension funds, wealthy individuals, and even foreign governments. In fact, the U.S. national debt is projected to pass $18 trillion by 2018, much of it owed to China and Japan.

$$\$18 \text{ trillion} = \$18,\underset{\uparrow\text{billions}}{000},\overset{\downarrow\text{trillions}}{000},000,000$$

If 18 trillion $1 bills could be stacked on top of one another, the stack would rise from the surface of the earth far past the moon. However, debt is not really in $1 bills. Think of debt as numbers written on contracts of one form or another and stored in computers. Our monetary system is not really based on paper currency. Instead, it is based on the government's backing of the dollar and the many contractual promises to repay loans that are everywhere in our society.

The U.S. government uses **U.S. Treasury bills**, or **T-bills**, to borrow for less than one year. An individual can buy a T-bill directly from the government or indirectly through a broker such as Merrill Lynch. T-bills use discount interest.

$18 Trillion One-Dollar Bills Stacked

Finding Facts About T-Bills

EXAMPLE 6 The president of a small bank in Mexico is worried that the peso is going to fall in value. So he purchases $1,000,000 in U.S. T-bills for safety. The T-bills are at a 4% discount rate for 26 weeks. Find (a) the total purchase price, (b) the total maturity value, (c) the interest earned, and (d) the effective rate of interest.

SOLUTION
$M = \$1,000,000; D = .04; T = \frac{26}{52}$

(a) Bank discount = Face value × Discount rate × Time
$= \$1,000,000 \times .04 \times \frac{26}{52} = \$20,000$

Purchase price = Face value − Bank discount
$= \$1,000,000 - \$20,000 = \$980,000$

(b) Maturity value = Face value
$= \$1,000,000$

(c) Interest = Bank discount
$= \$20,000$

(d) Effective rate $= \dfrac{\text{Interest earned}}{\text{Purchase price (proceeds)} \times \text{Time}}$

$= \dfrac{\$20,000}{\$980,000 \times \frac{26}{52}} = .04081 = 4.08\%$

348 CHAPTER 9 Simple Interest

> **QUICK CHECK 6**
>
> Find **(a)** the purchase price, **(b)** the maturity value, **(c)** the interest, and **(d)** the effective rate of interest (to the nearest hundredth) for a $1,000,000, 13-week T-bill with a discount rate of 5.2%. Round to the nearest hundredth of a percent.

9.3 Exercises // MyLab Math

The shaded sections below contain solutions to help you get a **QUICK START** *on the various types of exercises.*

Find the discount to the nearest cent, then find the proceeds. (See Example 1.)

	Face Value	Discount Rate	Time (Days)	Discount	Proceeds or Loan Amount
1.	$7800	9%	120	$234	$7566
	$B = \$7800 \times .09 \times \frac{120}{360} = \$234$; $P = \$7800 - \$234 = \$7566$				
2.	$15,000	10.25%	90	_____	_____
3.	$19,000	10%	180	_____	_____
4.	$12,500	11%	150	_____	_____
5.	$22,400	$8\frac{3}{4}\%$	75	_____	_____
6.	$18,050	8%	80	_____	_____

Find the maturity date and the proceeds for the following. (See Examples 1 and 2.)

	Face Value	Discount Rate	Date Made	Time (Days)	Maturity Date	Proceeds or Loan Amount
7.	$64,000	9.5%	Mar. 22	90	Jun. 20	$62,480
	$P = \$64,000 - (\$64,000 \times .095 \times \frac{90}{360}) = \$62,480$					
8.	$50,000	7.5%	Oct. 12	100	_____	_____
9.	$10,000	$10\frac{1}{4}\%$	July 12	150	_____	_____
10.	$18,500	$9\frac{1}{4}\%$	May 1	220	_____	_____
11.	$24,000	10%	Dec. 10	60	_____	_____
12.	$8000	10.5%	Nov. 4	165	_____	_____

C// indicates an exercise that is related to the Case in Point feature.

Solve each of the following application problems. Round rate to the nearest tenth of a percent, time to the nearest day, and money to the nearest cent.

13. **BOAT PURCHASE** An ExxonMobil employee borrowed $6000 from First National Bank to purchase a boat. He plans to repay the loan with a Christmas bonus he will receive in 120 days. If he borrowed the money at a discount rate of 8%, find (a) the discount and (b) the proceeds.

    (a) $B = \$6000 \times .08 \times \frac{120}{360} = \$160$
    (b) $P = \$6000 - \$160 = \$5840$

    (a) $160
    (b) $5840

14. **INCOME TAX PAYMENT** A manager signs a $48,000 simple discount note for six months for funds to pay corporate income taxes. If the discount rate is 8.5%, find (a) the discount and (b) the proceeds.

    (a) _____
    (b) _____

15. **CHRISTMAS TREE FARM** Danny Esposito signed a note with a face value of $40,000 at an 8.5% discount rate for funds to plant Christmas trees. Find the length of the loan in days if the discount is $1888.89.

    15. _____

16. **WEB-PAGE DESIGN** Jessica Hernandez was unable to collect funds owed her from a customer that declared bankruptcy. The shortage of cash forced Hernandez to sign a $12,200 note at a discount rate of 11% to pay her bills. She was told the interest would be $931.94. Find the length of the loan in days.

    16. _____

17. **CASINO** Wyatt Construction borrowed $157.25 million during the construction phase of adding a wing to a casino in Las Vegas. Management signed a 270-day note with a face value of $170 million. Find the discount rate.

    17. _____

18. **HARDWARE STORE** Kim Lee wanted to open a hardware store in a small town, so his uncle agreed to put up collateral for the loan. Lee signed a 200-day simple discount note with a face value of $125,000 and proceeds of $118,195. Find the discount rate.

    18. _____

19. **LAWSUIT** Benjamin Plumbing was sued for poor workmanship of a subcontractor, and the court ruled that the plumbing company had to pay $62,700. The bank lent them the funds at a discount rate of 8% for 240 days. Find the face value of the simple discount note.

    19. _____

20. **TRUCKS & SPRAYERS** A regional manager at Trugreen, Inc. authorizes the borrowing of $98,300 for trucks and sprayers needed to spray yards with fertilizers and pesticides. The simple discount note has a 9.25% rate and matures in 150 days. Find the face value of the loan needed.

    20. _____

21. **POOR CREDIT** Cathy Cox has poor credit but she found a bank that will lend her $4200. Still, the bank charges a 12% discount rate. Find (a) the proceeds if the note is for 10 months and (b) the effective interest rate charged by the bank.

    (a) _____
    (b) _____

350 CHAPTER 9 Simple Interest

22. BAD CREDIT HISTORY Tim Garcia has a poor credit history partly due to a divorce. A finance company finally agrees to lend him funds based on a note with a face value of $9400, but it requires him to use his truck as collateral. Even then, the bank charges him a high 16% discount rate. Find **(a)** the proceeds and **(b)** the effective rate if the note is for 7 months.

(a) _____
(b) _____

23. INTERNATIONAL FINANCE A business owner in England signs a 10% discount note for 40,000 English pounds (£40,000) with a bank in London. If the proceeds are £38,833.33, find the time of the note in days.

23. _____

24. GLOBAL TRADE A shipping company in Hong Kong needs to borrow proceeds of $65 million from a U.S. bank to build a cargo ship. It signs a note with a face value of $68 million at a discount rate of 7.3%. Find the time of the note in days.

24. _____

25. EARTHQUAKE DAMAGE A bridge in Japan was damaged by an earthquake. The firm repairing the bridge needs proceeds of 165 million Japanese yen (¥165,000,000) for 30 days to pay wages and buy supplies. A bank lends the funds at an 8% discount rate. Find **(a)** the face value and **(b)** the effective rate.

(a) _____
(b) _____

26. INTERNATIONAL FINANCE A Malaysian electric company requires proceeds of $720,000 (local currency) and borrows from a bank in Thailand at a 12% discount rate for 45 days. Find **(a)** the face value of the note and **(b)** the effective interest rate.

(a) _____
(b) _____

27. RACE HORSES Robert Johnson owns a farm and breeds race horses. To purchase three thoroughbreds, he signs a 180-day note with a maturity value of $265,000 and proceeds of $253,737.50. Find **(a)** the discount and **(b)** the true (or effective) rate.

(a) _____
(b) _____

28. PIZZA To remodel a restaurant, Two Brothers Pizza signs a 300-day note with proceeds of $63,159.72 and a maturity value of $68,000. Find **(a)** the discount and **(b)** the APR.

(a) _____
(b) _____

The following exercises apply to U.S. Treasury bills, discussed at the end of this section. (Assume 52 weeks per year for each exercise, and round to the nearest hundredth of a percent.) (See Example 6.)

29. **PURCHASE OF T-BILLS** A British investment firm purchases $25,000,000 in U.S. T-bills at a 6% discount rate for 13 weeks. Find **(a)** the purchase price of the T-bills, **(b)** the maturity value of the T-bills, **(c)** the interest earned, and **(d)** the effective rate.

(a) _____
(b) _____
(c) _____
(d) _____

30. **T-BILLS** Nina Horn buys a $50,000 T-bill at a 5.8% discount rate for 26 weeks. Find **(a)** the purchase price of the T-bill, **(b)** the maturity value, **(c)** the interest earned, and **(d)** the effective rate of interest.

(a) _____
(b) _____
(c) _____
(d) _____

31. Explain the main differences between simple interest notes and simple discount notes. (See Objective 1.)

32. As a borrower, would you prefer a simple interest note with a rate of 11% or a simple discount note at a rate of 11%? Explain using an example. (See Example 4.)

QUICK CHECK ANSWERS

1. $711; $15,089
2. $37,066.67
3. $51,798.56
4. $25,080; $24,000
5. 11.5%
6. (a) $987,000
   (b) $1,000,000
   (c) $13,000
   (d) 5.27%

9.4 Discounting a Note Before Maturity

OBJECTIVES

1. Understand the concept of discounting a note.
2. Find the proceeds when discounting simple interest notes.
3. Find the proceeds when discounting simple discount notes.

A note is a *legal responsibility* for one individual or firm to pay a specific amount on a specific date to another individual or firm. Notes can be bought and sold just as an automobile can be bought and sold. The clipping taken from a newspaper shows firms that buy notes. This section shows how to find the value of a note that is sold before its maturity date.

OBJECTIVE 1 Understand the concept of discounting a note. Businesses sometimes help their customers purchase products or services by accepting a promissory note rather than requiring an immediate cash payment. For example, a company that manufactures boats, a retailer that sells the boats, and a bank may do business as follows:

1. Boat manufacturer sells boats to a retailer and accepts a promissory note instead of cash.
2. Boat manufacturer needs cash and sells the note to a bank before it matures.
3. Retailer pays the maturity value of the note to the bank when due.

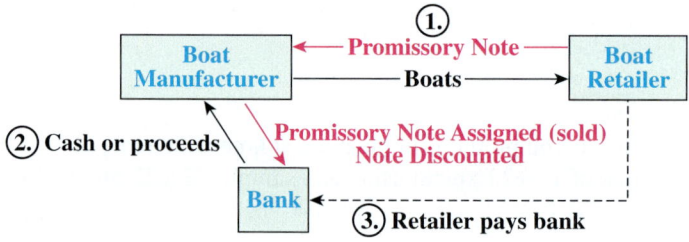

A bank makes money by charging interest. So it will not pay the full maturity value for a note that it buys. It will deduct a fee from the maturity value when figuring out what to pay for it, a process called "discounting a note." The fee is effectively interest for the number of days that the bank will hold the note until it is due. Both simple interest and simple discount notes can be sold at a discount before maturity. Here are some important definitions:

> **discounting the note**—finding the value of the note before it matures
> **discount period**—the number of days until the note matures
> **bank discount** (or **discount**)—the reduction from the maturity price paid by the bank
> **discount rate**—the percentage rate used by the bank to find the amount of the discount
> **proceeds**—the amount received by the seller of the note

The discount rate used by a bank depends on several things, including the **risk** that the borrower will NOT repay the bank. The discount rate will be low for a company with abundant assets and a great credit history, such as Apple. The discount rate will be significantly higher for a company with few assets, a shortage of cash coming in, and/or a lot of bills that have to be paid. Banks rarely buy notes made by a borrower with a questionable ability to repay. Managers work hard to minimize the risk of lending to someone who ends up not paying.

OBJECTIVE 2 Find the proceeds when discounting simple interest notes. The boat manufacturer from above who sells the note receives an amount called the **proceeds**. The bank then collects the maturity value from the maker of the note (the retailer).

These types of notes are often sold with **recourse**. This means that the bank receives reimbursement from the manufacturer if the retailer does not pay the bank when the note matures. Thus the bank is protected against loss.

9.4 Discounting a Note Before Maturity

Calculate the Proceeds When Discounting a Simple Interest Note

1. First, understand the simple interest note by finding:
   (a) the **due date** of the original note and
   (b) the **maturity value** of the original note ($M = P + I$, where $I = PRT$).
2. Then discount the simple interest note.
   (a) Find the **discount period**, which is the time (e.g., number of days) from the sale of the note to the maturity date of the note.
   (b) Find the **discount** using the formula
   $$B = M \times D \times T$$
   $$= \text{Maturity value} \times \text{Discount rate} \times \text{Discount period}$$
   (c) Find the **proceeds** after discounting the original note using $P = M - B$.

Finding Proceeds **EXAMPLE 1** Jameson Plumbing takes a simple interest, 180-day note from a contractor with a face value of $64,750 and a rate of 10.5%. The company sells the note to a bank 50 days later at a discount rate of 12%. Find the proceeds to the plumbing company.

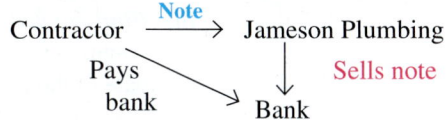

SOLUTION

Step 1 *Find the maturity value.* The face value equals the proceeds because this is a simple interest note.

Maturity value = Principal + Interest on the simple interest note
Maturity value = $64,750 + PRT   Since $I = PRT$
Maturity value = $64,750 + $64,750 \times .105 \times \dfrac{180}{360} = \$68,149.38$

Step 2 The note is discounted after 50 days, so the discount period is $180 - 50 = 130$ days. So the buyer of the note will own it for 130 days before the note is paid off.

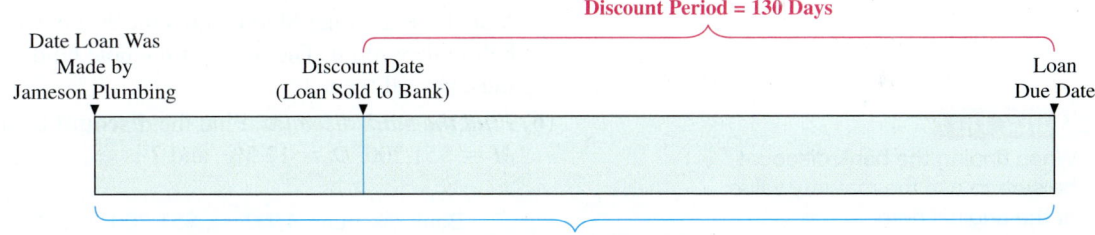

Use the formula $B = MDT$, with $M = \$68,149.38$, $D = .12$, and $T = \dfrac{130}{360}$ to find the discount.

$$\text{Bank discount} = MDT = \$68,149.38 \times .12 \times \dfrac{130}{360} = \$2953.14$$

Proceeds = Maturity value of simple interest note − Bank discount
Proceeds = $68,149.38 − $2953.14 = $65,196.24

To summarize:
1. A contractor signs a 180-day simple interest note with a face value of $64,750 to Jameson Plumbing.
2. After 50 days, Jameson Plumbing sells the note to a bank and receives $65,196.24.
3. The bank receives $68,149.38 on the maturity date of the loan.

QUICK CHECK 1

A simple interest note has a face value of $28,000, a rate of 9%, and a time to maturity of 240 days. It is discounted after 80 days at a rate of 11%. Find the maturity value of the simple interest note and the proceeds at the time of the discount.

Finding Proceeds EXAMPLE 2 // Matador Recording Studio holds a 200-day simple interest note from a rock group that agreed to pay them to record an album. The 12% simple interest note is dated March 24 and has a face value of $48,000. The recording studio wishes to convert the note to cash, so they sell it to a bank on August 15. Although the rock group is well known and has had some success with prior albums, the bank still considers the loan risky and requires a high discount rate of 12.5%. Find the proceeds to the recording studio.

SOLUTION //

Go through the two steps of discounting a note.

Step 1   *Find the maturity value.* The note is dated March 24 and is due in 200 days.

Due date: day 83 (March 24) + 200 days = day 283 (October 10)

Since this is a simple interest note, the proceeds are given but the maturity value must be found. First find the **interest** on the note if held until maturity.

$$I = PRT = \$48{,}000 \times .12 \times \frac{200}{360} = \$3200$$

Maturity value: $48,000 + $3200 = $51,200.

Step 2   Now discount this simple interest note.

(a) *Find the discount period.* The **discount period** is the number of days from August 15, which is the date the note is discounted (sold) to the bank, to the due date of the note (October 10).

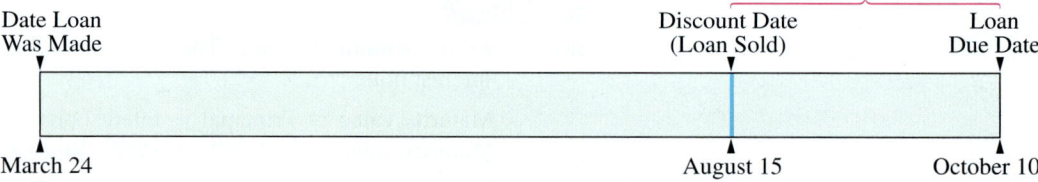

October 10 is day	283
August 15 is day	− 227
Discount period	**56 days**

Matador Recording Studio holds the 200-day note for 200 − 56 = 144 days before they sell it. The buyer of the note holds it for 56 days before the rock group must pay off the note.

Quick TIP //
When finding the bank discount, be sure to use the maturity value of the original note.

(b) *Find the bank discount.* Find the **discount** by using the formula $B = MDT$, where $M = \$51{,}200$, $D = 12.5\%$, and T is $\frac{56}{360}$.

$$\text{Bank discount: } MDT = \$51{,}200 \times .125 \times \frac{56}{360} = \$995.56$$

(c) *Find the proceeds.* **Proceeds** are found by subtracting the bank discount from the maturity value.

$$P = M - B$$
$$P = \$51{,}200 - \$995.56 = \$50{,}204.44$$

Date	Transaction
March 24	Rock group signs 200-day simple interest note for $48,000.
August 15	Matador Recording Studio sells note to bank for $50,204.44.
October 10	Bank receives $51,200 from payer (rock group).

QUICK CHECK 2 //

On March 27, Dayton Finance loans Jorge Rivera $9200 for 150 days at 11% simple interest. The finance company sells the note on April 24. Find the maturity value of the simple interest note and the proceeds to Dayton Finance if the note is sold at a discount rate of 12%.

It is also common for a business needing cash to sell part of its accounts receivable (money owed to the company) before it is due. The process is called **factoring**, and those who buy the accounts receivable are called **factors**. The calculations involved in factoring are the same as those for finding the discount discussed in this section.

OBJECTIVE 3 Find the proceeds when discounting simple discount notes. Two different discounts occur in problems of this type. The first occurs when the original simple discount note is signed. The second occurs later when the note is sold, which occurs before it matures.

> **Calculate the Proceeds When Discounting a Simple Discount Note**
> 1. First, understand the simple discount note by finding:
>   (a) the **due date** of the original note,
>   (b) the **discount** of the original note using $B = MDT$, and
>   (c) the **proceeds** from the original note using $P = M - B$.
>
>   The **maturity value (face value)** of the note is written on the note itself and is the value needed in step 2(b) below.
>
> 2. Then discount the simple discount note.
>   (a) Find the **discount period**, which is the time (e.g., number of days) from the sale of the note to the maturity date of the note.
>   (b) Find the **discount** using the formula $B = MDT$.
>   (c) Find the **proceeds** after discounting the original note using $P = M - B$.

Look carefully at the next example.

Finding the Proceeds

EXAMPLE 3 Garcia BMW uses excess cash to purchase a $100,000 Treasury bill with a term of 26 weeks at a 3.5% simple discount rate. However, the firm needs cash exactly 8 weeks later and sells the T-bill. During the 8 weeks, market interest rates changed slightly so that the bill was sold at a 3% discount rate. Find **(a)** the initial purchase price of the T-bill, **(b)** the proceeds received by the firm at the subsequent sale of the T-bill, and **(c)** the effective interest rate.

SOLUTION

(a) *Find the discount and proceeds.* The discount that Garcia BMW receives when buying the T-bill is found as follows.

$$B = MDT = \$100{,}000 \times .035 \times \frac{26}{52} = \$1750$$

The cost to the company is the maturity value minus the discount.

$$P = M - B = \$100{,}000 - \$1750 = \$98{,}250$$

Therefore, the U.S. government receives $98,250 from the sale of the T-bill.

(b) *Find the discount period, discount, and proceeds.* Now follow the steps in the table to find the proceeds Garcia BMW receives for selling the T-bill. The discount period is 18 weeks, since the T-bill is sold $26 - 8 = 18$ weeks before its due date.

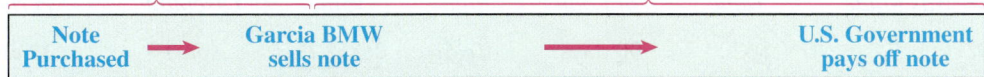

The discount at the time of the sale is as follows.

$$B = MDT = \$100{,}000 \times .03 \times \frac{18}{52} = \$1038.46$$

Finally, the proceeds equal the maturity value of the T-bill ($100,000) less the discount at the time of the sale.

$$P = M - B = \$100{,}000 - \$1038.46 = \$98{,}961.54$$

(c) Garcia BMW paid $98,250 to buy the T-bill and received $98,961.54 for it 8 weeks later.

$$\text{Interest received} = \$98{,}961.54 - \$98{,}250 = \$711.54$$

$$R = \frac{\$711.54}{\$98{,}250 \times \frac{8}{52}} = 4.71\% \text{ (rounded)}$$

356 CHAPTER 9 Simple Interest

Garcia BMW would have earned 3.5% on the T-bill had it kept the Treasury bill to maturity. Instead, the company sold it after market interest rates rose, but before the T-bill matured. So it ended up with an effective interest rate of 4.71% on the 8-week investment. This is higher than the 3.5% rate associated with the original purchase, but it resulted in less interest because the investment was for a shorter period than originally planned.

> **QUICK CHECK 3**
>
> A 240-day discount note has a maturity value of $24,000 and a discount rate of 8%. It is sold after 100 days at a discount rate of 10.5%. Find the maturity value of the original discount note and the proceeds at the time of the sale.

9.4 Exercises // MyLab Math

The shaded sections below contain solutions to help you get a **QUICK START** *on the various types of exercises.*

Find the discount period for each of the following. (See Examples 2 and 3, Step 2.)

Date Loan Was Made	Length of Loan	Date of Discount	Discount Period
1. Apr. 29	200 days	July 31	107 days
2. July 28	120 days	Sept. 20	
3. May 28	74 days	June 18	
4. Sept. 17	130 days	Jan. 13	

Find the proceeds to the nearest cent when each of the following is discounted. (See Examples 1 and 2.)

Maturity Value	Discount Rate	Discount Period	Proceeds
5. $10,400	8.5%	90 days	$10,179

$B = \$10,400 \times .085 \times \frac{90}{360} = \$221; P = \$10,400 - \$221 = \$10,179$

6. $4800	10.3%	200 days	
7. $25,000	9%	30 days	
8. $52,000	7.5%	250 days	

Find the maturity value of each of the following simple interest notes. Each note is then discounted at 12%. Find the discount period, the discount, and the proceeds after discounting. (See Examples 1 and 2.)

Date Loan Was Made	Face Value	Length of Loan	Rate	Maturity Value	Date of Discount	Discount Period	Discount	Proceeds
9. Feb. 7	$6200	90 days	10½%	$6362.75	Apr. 1	37 days	$78.47	$6284.28

$I = \$6200 \times .105 \times \frac{90}{360} = \$162.75; M = \$6200 + \$162.75 = \$6362.75$
Feb. 7 is day 38; Apr. 1 is day 91; 91 − 38 = 53
Discount period = 90 − 53 = 37 days
$B = \$6362.75 \times .12 \times \frac{37}{360} = \$78.47$
Proceeds = $6362.75 − $78.47 = $6284.28

10. June 15	$9200	140 days	12%		Oct. 22			

C// indicates an exercise that is related to the Case in Point feature.

	Date Loan Was Made	Face Value	Length of Loan	Rate	Maturity Value	Date of Discount	Discount Period	Discount	Proceeds
11.	July 10	$2000	72 days	11%	_____	Aug. 2	_____	_____	_____
12.	May 29	$5500	80 days	10%	_____	July 8	_____	_____	_____

First, find the initial proceeds of each of the following simple discount notes. Each note is then discounted at 11%. Find the discount period, the discount, and the proceeds after discounting. (See Example 3.)

	Date Loan Was Made	Maturity Value	Length of Loan	Rate	Initial Proceeds	Date of Discount	Discount Period	Discount	Proceeds at Time of Sale
13.	Jan. 12	$17,800	90 days	10%	$17,355	Mar. 1	42 days	$228.43	$17,571.57

$B = MDT = \$17,800 \times .10 \times \frac{90}{360} = \$445; P = M - B = \$17,800 - \$445 = \$17,355$
Jan. 12 is day 12; Due date is 12 + 90 = 102 or Apr. 12
Mar. 1 is day 60; Discount period is 102 − 60 = 42 days
$B = MDT = \$17,800 \times .11 \times \frac{42}{360} = \$228.43$
$P = M - B = \$17,800 - \$228.43 = \$17,571.57$

	Date Loan Was Made	Maturity Value	Length of Loan	Rate	Initial Proceeds	Date of Discount	Discount Period	Discount	Proceeds at Time of Sale
14.	Aug. 4	$24,000	120 days	10.5%	_____	Oct. 8	_____	_____	_____
15.	May 4	$32,100	150 days	9.5%	_____	July 10	_____	_____	_____
16.	Apr. 30	$22,000	200 days	9%	_____	July 12	_____	_____	_____

Solve the following application problems. Round interest and discount to the nearest cent.

17. **ROCK CRUSHER** First Bank loaned $360,000 for 180 days to a company purchasing a rock-crushing machine. The bank sold the 7% simple interest note 120 days later at an 8% discount rate. Find **(a)** the bank discount and **(b)** the proceeds.

    (a) $M = \$360,000 + (\$360,000 \times .07 \times \frac{180}{360}) = \$372,600$ (b) $P = \$372,600 - \$4968 = \$367,632$
    Discount period = 180 days − 120 days = 60 days
    $B = \$372,600 \times .08 \times \frac{60}{360} = \$4968$

    (a) $4968
    (b) $367,632

358 CHAPTER 9 Simple Interest

18. TRACTOR PURCHASE Cook and Daughters Farm Equipment accepts a $5800 simple interest note at 12% for 100 days, for a small used tractor. The note is dated May 12. On June 17, the firm discounts the note to a loan broker, at a 13% discount rate. Find **(a)** the bank discount and **(b)** the proceeds.

(a) _____
(b) _____

19. JEWELRY Hanson's Jewelry signed a 180-day simple discount note with a face value of $250,000 and a rate of 9% on March 19. The lender sells the note at an 8% discount rate on June 14. Find **(a)** the proceeds of the original note, **(b)** the discount period, **(c)** the discount, and **(d)** the proceeds at the sale of the note on June 14.

(a) _____
(b) _____
(c) _____
(d) _____

20. TANNING SALON Kathy Bates needs to buy new tanning beds, and agrees to a 10% simple discount note with a maturity value of $18,500 on July 30. The 120-day note is sold by the lender at a 12% discount rate on September 2. Find **(a)** the proceeds of the original note to Bates, **(b)** the discount period, **(c)** the discount, and **(d)** the proceeds at the sale of the note on September 2.

(a) _____
(b) _____
(c) _____
(d) _____

21. FINANCING CONSTRUCTION To build a new warehouse, Alco Fence Co. signed a $300,000 simple interest note at 9% for 150 days with National Bank on November 20. On February 6, National Bank sold all of its notes to Bank One. Find **(a)** the maturity value of the note and **(b)** the proceeds to National Bank given a discount rate of 10.5%.

(a) _____
(b) _____

22. BATTERY STORE A National Tire and Battery outlet borrowed $48,500 on a 200-day simple interest note to expand the battery store. The note was signed on December 28 and carried an interest rate of 9.8%. The note was then sold on March 17 at a discount rate of 10%. Find **(a)** the maturity value of the note and **(b)** the proceeds to the seller of the note on May 17.

(a) _____
(b) _____

23. PURCHASE OF A T-BILL Elizabeth Barton bought a $25,000, 26-week T-bill at a discount rate of 6.8% on August 7. She sold it 10 weeks later at a discount rate of 7%. Find **(a)** Barton's purchase price, **(b)** the discount 10 weeks later when she sold it, **(c)** the proceeds to Barton, and **(d)** the effective interest rate rounded to the nearest hundredth of a percent for the time Barton held the note.

(a) _____
(b) _____
(c) _____
(d) _____

24. PURCHASE OF A T-BILL Tina Klein bought a $10,000, 1.5%, 52-week T-bill on June 29 and sold it 26 weeks later at a discount rate of 1%. Find Klein's **(a)** purchase price for the T-bill, **(b)** the discount at time of sale, **(c)** the proceeds to Klein, and **(d)** the effective interest rate rounded to the nearest hundredth of a percent.

(a) _____
(b) _____
(c) _____
(d) _____

25. Give several reasons why a firm may need funds and decide to sell a note. Explain the steps to discount a simple interest note. (See Objective 2.)

26. Explain the effect of a rise in general market interest rates on an investor who is holding notes. Give an example. (See Example 3.)

QUICK CHECK ANSWERS

1. $29,680; $28,228.98
2. $9621.67; $9230.39
3. $24,000; $23,020

Supplementary Application Exercises on Simple Interest and Simple Discount

The shaded sections below contain solutions to help you get a **QUICK START** *on the various types of exercises.*

There are similarities and differences between simple interest and simple discount calculations. This exercise set compares these two important concepts. First, the key similarities between the two are as follows.

1. Both types of notes involve lump sums repaid with a single payment at the end of a stated period of time.
2. The length of time is generally 1 year or less.

The following table compares simple interest and simple discount notes.

	Simple Interest Note	Simple Discount Note
Variables	I = Interest P = Principal (face value) R = Rate of interest T = Time, in years or fraction of a year M = Maturity value	B = Discount P = Proceeds D = Discount rate T = Time, in years or fraction of a year M = Maturity value (face value)
Face value	Stated on note	Same as maturity value
Interest charge	$I = PRT$	$B = MDT$
Maturity value	$M = P + I$	Same as face value
Amount received by borrower	Face value or principal	Proceeds: $P = M - B$
Identifying phrases	Interest at a certain rate Maturity value greater than face value	Discount at a certain rate Proceeds Maturity value equal to face value
Effective interest rate	Same as stated rate, R	Greater than stated rate, D

Quick TIP

The variable P is used for *principal or face value* in simple interest notes, but P is used for *proceeds* in simple discount notes. P represents the amount received by the borrower.

Solve the following application problems. Round rates to the nearest tenth of a percent, time to the nearest day, and money to the nearest cent.

1. The owner of Redwood Furniture, Inc., signed a 120-day note for $18,000 at 11% simple interest. Find **(a)** the interest and **(b)** the maturity value.
   (a) I = PRT = $18,000 × .11 × $\frac{120}{360}$ = $660
   (b) M = $18,000 + $660 = $18,660

   (a) $660
   (b) $18,660

2. Bill Travis signed a note with a finance company for $18,500 so that he can begin repairing cars in his garage. The note is due in 300 days and has a discount rate of 14%. Travis hopes that a bank will refinance the note for him at a lower rate after he has been in business for 300 days. Find the proceeds.
   B = $18,500 × .14 × $\frac{300}{360}$ = $2158.33;
   P = $18,500 − $2158.33 = $16,341.67

   2. $16,341.67

c/ indicates an exercise that is related to the Case in Point feature.

3. Jessica Hernandez borrowed money to remodel a retail space she had leased for her computer training and web-design business. She signed a note with a 10% simple interest rate, interest of $4800, and time of 180 days. Find the principal.

3. _____

4. Bill Abel signed a simple interest note at a rate of only 6% because he had excellent collateral: a $100,000 certificate of deposit at the same bank. If the loan matures in 300 days and the interest is $9000, find the principal.

4. _____

5. Crazy-Good Ice Cream signed a $150,000 note at a simple discount rate of 10.5% and a discount of $8750. Find the length of the loan in days.

5. _____

6. A loan to a solar energy company was for $1,290,000 with a maturity value of $1,327,410 and a rate of 6%. Find the time.

6. _____

7. Jane Barber loaned her nephew $20,000 for 150 days at 9% simple interest. Find (a) the interest and (b) the maturity value.

(a) _____
(b) _____

8. John O'Neill borrowed $24,000 for 250 days at 7% simple interest. Find (a) the interest and (b) the maturity value.

(a) _____
(b) _____

9. BlueWater Pools signed a 5-month, $145,000 note at an 11.5% discount rate. Find the effective rate of interest.

9. _____

10. James Taylor signed an 80-day, $82,000 note at a 12% discount rate. Find the effective rate of interest.

10. _____

11. On October 14, Citibank loaned $10,000,000 to Fleet Mortgage Company for 180 days at a 10.5% discount rate. Find (a) the due date and (b) the proceeds.

(a) _____
(b) _____

12. On December 24, Junella Martin signed a 100-day note for $80,000 for a Jaguar. Given a discount rate of 11%, find (a) the due date and (b) the proceeds.

(a) _____
(b) _____

362   CHAPTER 9   Simple Interest

13. Lupe Galvez has a serious problem: two of her more energetic preschoolers keep getting out of the yard of her childcare center. She signs a note with interest charges of $670.83 to reinforce the fence around the entire yard. The simple interest note is for 140 days at 11.5%. Find the principal to the nearest dollar.

13. _____

14. On May 25, Drones-with-Cameras accepted a 250-day note for $82,000 from a firm that does city planning. The interest rate on the note is 9.5% simple interest. The note was discounted at 10.75% when sold to a bank on August 7. Find the proceeds.

14. _____

15. On November 19, a firm accepts an $18,000, 150-day note with a simple interest rate of 9%. The firm discounts the note at 12% on February 2. Find the proceeds.

15. _____

16. Barton's Flowers accepted a $16,000, 150-day note from Wedded Bliss Catering. The note had a simple interest rate of 11% and was accepted on May 12. The note was then discounted at 13% on July 20. Find the proceeds to Barton's Flowers.

16. _____

17. Herbert Leon signed a 220-day, 10% simple interest note with a face value of $28,000. In turn, the bank he borrowed the money from sold the note 90 days later at an 11% discount rate. Find (a) the interest, (b) the maturity value, (c) the discount period, (d) the discount, and (e) the proceeds to the bank.

(a) _____
(b) _____
(c) _____
(d) _____
(e) _____

18. Janice Dart signed a 140-day simple discount note at a rate of 9.9% with a maturity value of $82,000. The bank she borrowed the funds from sold the note 40 days later at a 10% discount rate. Find (a) the discount on the original note, (b) the proceeds of the original note, (c) the discount period, (d) the discount at the time of sale, and (e) the proceeds to the bank at the time of sale.

(a) _____
(b) _____
(c) _____
(d) _____
(e) _____

Supplementary Application Exercises on Simple Interest and Simple Discount   363

19. James and Tiffany Paterson need a 220-day loan for $68,000 to expand their auto repair shop. Bank One agrees to a simple interest note with a loan amount of $68,000 at $9\frac{1}{4}\%$ interest. Union Bank agrees to a simple discount note with proceeds of $68,000 and a 9.5% simple discount rate. Find **(a)** the interest for the simple interest note, **(b)** the maturity value of the discount note, **(c)** the interest for the discount note, and **(d)** the savings in interest charges of the simple interest note over the discount note.

(a) _____
(b) _____
(c) _____
(d) _____

20. Gilbert Construction Company needs to borrow $380,000 for $1\frac{1}{2}$ years to purchase some land to subdivide. One bank offers the firm a simple interest note with a principal of $380,000 and a rate of 12%. A second bank offers the company a discount note with proceeds of $380,000 and an 11% discount rate. **(a)** Which note produces the lower interest charges? **(b)** What is the difference in interest?

(a) _____
(b) _____

21. Show with an example that the effective interest rate is higher than the discount rate stated on a note.

22. Explain the difference in the meaning of the variable P (principal) in a simple interest note and the variable P (proceeds) in a simple discount note.

23. What is interest? Why is interest used?

24. Why might a bank use ordinary interest rather than exact interest?

Chapter 9 Quick Review

Chapter Terms Review the following terms to test your understanding of the chapter. For each term you do not know, refer to the page number found next to that term.

annual percentage rate (APR) [p. 346]
bank discount [p. 352]
banker's interest [p. 328]
bankruptcy [p. 324]
collateral [p. 330]
compound interest [p. 324]
consumer installment loan [p. 324]
cosign [p. 325]
discount [p. 353]
discounting the note [p. 352]
discount period [p. 352]

discount rate [p. 352]
due date [p. 353]
effective rate of interest [p. 346]
exact interest [p. 328]
face value [p. 329]
factoring [p. 355]
factors [p. 355]
foreclose [p. 330]
interest [p. 354]
interest-in-advance notes [p. 343]
length of a loan [p. 326]
loan amount [p. 343]

maker [p. 329]
maturity date [p. 329]
maturity value [p. 353]
nominal rate [p. 346]
ordinary interest [p. 328]
payee [p. 329]
payer [p. 329]
prime rate [p. 324]
principal [p. 329]
proceeds [p. 352]
promissory note [p. 329]
rate [p. 324]
recourse [p. 352]

risk [p. 352]
simple discount note [p. 343]
simple interest [p. 324]
simple interest note [p. 329]
Stafford loan [p. 343]
stated rate [p. 346]
T-bills [p. 347]
term [p. 326]
time [p. 324]
true rate [p. 346]
Truth in Lending Act [p. 346]
U.S. Treasury bills [p. 347]

CONCEPTS

9.1 Finding the simple interest when time is expressed in years

1. Use the formula $I = PRT$.
2. Express R in decimal form.
3. Express time in years.
4. Substitute values for P, R, and T and multiply.

EXAMPLES

A loan of $28,000 is made for $1\frac{1}{4}$ years at 10% per year. Find the simple interest.

$$I = PRT$$
$$I = \$28{,}000 \times .10 \times 1.25 = \$3500$$

The simple interest is $3500.

9.1 Finding the simple interest when time is expressed in months

1. Use the formula $I = PRT$.
2. Express R in decimal form.
3. Express time in years by dividing the number of months by 12.
4. Substitute values for P, R, and T and multiply.

Find the simple interest on $24,000 for 8 months at 10%.

$$I = PRT$$
$$I = \$24{,}000 \times .10 \times \frac{8}{12} = \$1600$$

The simple interest is $1600.

9.1 Finding the maturity value of a loan

1. Find I using the formula $I = PRT$.
2. Find the maturity value using the formula $M = P + I$.

A loan of $8500 is made for 1 year at 9%. Find the maturity value of the loan.

$$I = PRT$$
$$I = \$8500 \times .09 \times 1 = \$765$$
$$M = P + I$$
$$M = \$8500 + \$765 = \$9265$$

The maturity value is $9265.

9.1 Finding the number of days from one date to another using a table

1. Find the day corresponding to the final date using the table.
2. Find the day corresponding to the initial date.
3. Subtract the smaller number from the larger number.

Find the number of days from February 15 to July 28.
1. July 28 is day 209.
2. Feb. 15 is day 46.
3. Number of days is

$$\begin{array}{r} 209 \\ -\ 46 \\ \hline 163 \end{array}$$

There are 163 days from February 15 to July 28.

CHAPTER 9 Quick Review

CONCEPTS	EXAMPLES
9.1 Finding the number of days from one date to another using actual number of days in a month Add the actual number of days in each month or partial month from initial date to final date.	Find the number of days from April 20 to June 27. April 20 to April 30 10 days May 31 days June 27 days 68 days
9.1 Finding the exact interest Use the formula $$I = PRT$$ with $T = \dfrac{\text{Number of days of loan}}{365}$	Find the exact interest on a \$9000 loan at 8% for 140 days. $$I = PRT$$ $$I = \$9000 \times .08 \times \frac{140}{365} = \$276.16$$ The exact interest is \$276.16.
9.1 Finding the ordinary, or banker's, interest Use the formula $$I = PRT$$ with $T = \dfrac{\text{Number of days of loan}}{360}$	Find the ordinary interest on a loan of \$14,000 at 7% for 120 days. $$I = PRT$$ $$I = \$14,000 \times .07 \times \frac{120}{360} = \$326.67$$ The ordinary, or banker's, interest is \$326.67.
9.1 Finding the due date, interest, and maturity value of a simple interest promissory note when the term of the loan is in months 1. Add the number of months in the term of the note to the initial date of the note. 2. Use the formula $I = PRT$ to find interest. 3. Find the maturity value as follows. **Maturity value = Principal + Interest**	Find the due date, the interest, and the maturity value of a loan made on February 15 for 7 months at 8% with a face value of \$9400. September 15 is 7 months from February 15, so the note is due on September 15. $$I = PRT$$ $$I = \$9400 \times .08 \times \frac{7}{12} = \$438.67$$ $$M = \text{Principal} + \text{Interest}$$ $$M = \$9400 + \$438.67 = \$9838.67$$
9.1 Finding the due date of a promissory note when the term of the loan is expressed in days Use either a table or the actual number of days in each month.	A loan is made on August 14 and is due in 80 days. Find the due date. August 14 to August 31 17 days September 30 days October 31 days 78 days The loan is for 80 days, which is 2 days more than 78. Therefore, the loan is due on November 2.
9.2 Finding the principal given the interest, interest rate, and time Use the formula $$P = \frac{I}{RT}$$	Find the principal that produces interest of \$240 at 9% for 60 days. $$P = \frac{I}{RT}$$ $$P = \frac{\$240}{.09 \times \frac{60}{360}} = \$16,000$$ The principal is \$16,000.

CONCEPTS

9.2 Finding the rate of interest given the principal, interest, and time

Use the formula

$$R = \frac{I}{PT}$$

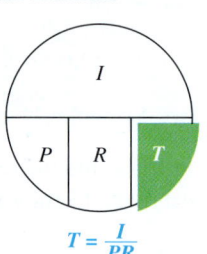

$$R = \frac{I}{PT}$$

9.2 Finding the time given the principal, rate of interest, and interest

To find the time in days, use the formula

$$T \text{ (in days)} = \frac{I}{PR} \times 360$$

To find the time in months, use the formula

$$T \text{ (in months)} = \frac{I}{PR} \times 12$$

$$T = \frac{I}{PR}$$

9.3 Finding the proceeds of a simple discount note

Calculate the bank discount using the formula $B = MDT$. Then calculate the proceeds or loan amount using the formula $P = M - B$.

9.3 Finding the face value of a simple discount note

Use the formula

$$M = \frac{P}{1 - DT}$$

9.3 Finding the effective interest rate

Find the interest (B) from the formula

$$B = MDT$$

Find proceeds from the formula

$$P = M - B$$

Then use the formula

$$R = \frac{I}{PT}$$

EXAMPLES

A principal of $8000 deposited for 250 days earns interest of $416.67. Find the rate of interest.

$$R = \frac{I}{PT}$$

$$R = \frac{\$416.67}{\$8000 \times \frac{250}{360}} = .075$$

Rate of interest = 7.5%.

Tom Jones invested $8000 at 8% and earned interest of $320. Find the time of the loan in days.

$$T = \frac{I}{PR} \times 360$$

$$T = \frac{\$320}{\$8000 \times .08} \times 360 = 180 \text{ days}$$

The loan was for 180 days.

Karen Pattern borrows $12,000 for 120 days at a discount rate of 9%. Find the proceeds.

$$B = MDT$$

$$B = \$12,000 \times .09 \times \frac{120}{360} = \$360$$

$$P = M - B$$

$$P = \$12,000 - \$360 = \$11,640$$

Sam Spade needs $15,000 for new equipment for his restaurant. Find the face value of a note that will provide the $15,000 in proceeds if he plans to repay the note in 180 days and the bank charges an 11% discount rate.

$$M = \frac{P}{1 - DT}$$

$$M = \frac{\$15,000}{1 - \left(.11 \times \frac{180}{360}\right)} = \$15,873.02$$

A 150-day, 11% simple discount note has a face value of $12,400. Find the effective rate to the nearest tenth of a percent.

$$B = \$12,400 \times .11 \times \frac{150}{360} = \$568.33$$

$$P = \$12,400 - \$568.33 = \$11,831.67$$

$$R = \frac{\$568.33}{\$11,831.67 \times \frac{150}{360}} = 11.5\%$$

CONCEPTS	EXAMPLES

9.4 Finding the proceeds to an individual or firm that discounts a simple interest note
 	
1. If necessary, find
    (a) the **due date** of the original note and
    (b) the **maturity value** of the original note ($M = P + I$, where $I = PRT$).
 	
2. (a) Find the **discount period**, which is the time (e.g., number of days) from the sale of the note to the maturity date of the note.
    (b) Find the **discount** using the formula $B = MDT$.
    (c) Find the **proceeds** using $P = M - B$.
	Moe's Ice Cream holds a 150-day note dated March 1 with a face value of $15,000 and a simple interest rate of 9%. Moe sells the note at a discount on June 1. Assume a discount rate of 11%. Find the proceeds.
	
	1. Due date = day 60 (March 1) + 150 days = day 210 or July 29
	$$I = PRT = \$15{,}000 \times .09 \times \frac{150}{360} = \$562.50$$
	$$M = P + I = \$15{,}000 + \$562.50 = \$15{,}562.50$$
	
	2. (a) The discount period is 58 days.
	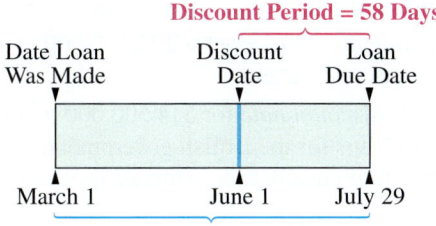
	Length of Loan: 150 Days
	
	(b) Bank discount = MDT
	$$B = \$15{,}562.50 \times .11 \times \frac{58}{360} = \$275.80$$
	(c) Proceeds = $M - B$
	$$D = \$15{,}562.50 - \$275.80 = \$15{,}286.70$$

9.4 Finding the proceeds to an individual or firm that discounts a simple discount note
 	
1. If necessary, find
    (a) the **due date** of the original note,
    (b) the **discount** of the original note using $B = MDT$, and
    (c) the **proceeds** from the original note using $P = M - B$.
   The **maturity value (face value)** of the note is written on the note itself.
 	
2. (a) Find the **discount period**, which is the time (e.g., number of days) from the sale of the note to the maturity date of the note.
    (b) Find the **discount** using the formula $B = MDT$.
    (c) Find the **proceeds** using $P = M - B$.
	On May 10, Applecrest Orchards signed a 120-day note for $22,000 at a simple discount rate of 10%. The note was sold on June 30 at a discount rate of 10.5%. Find **(a)** the proceeds from the original note and **(b)** the proceeds at the time of sale.
	
	(a) Due date is day 130 (May 10) + 120 days = day 250 or Sept. 7
	$$B = MDT = \$22{,}000 \times .10 \times \frac{120}{360} = \$733.33$$
	$$P = M - B = \$22{,}000 - \$733.33 = \$21{,}266.67$$
	(b) June 30 is day 181
	Discount period is $250 - 181 = 69$ days
	$$B = MDT = \$22{,}000 \times .105 \times \frac{69}{360} = \$442.75$$
	(c) $P = M - B = \$22{,}000 - \$442.75 = \$21{,}557.25$

Case Study

BANKING IN A GLOBAL WORLD: HOW DO LARGE BANKS MAKE MONEY?

Bank of America borrowed $80,000,000 at 5% interest for 180 days from a Japanese investment house. At the same time, the bank made the following three loans, each for the exact same 180-day period:

1. A 7% *simple interest note* for $38,000,000 to a Canadian firm that extracts oil from Canadian tar sands;
2. An 8.2% *simple discount note* for $27,500,000 to a European contractor building a factory in South Africa; and
3. An 8% *simple discount note* for $14,500,000 to a Louisiana company building minesweepers in New Orleans for the British government.
    (a) Find the difference between interest received and interest paid by the bank on these funds.

    (a) _____

    (b) The bank did not loan out all $80,000,000. Find the amount it actually loaned out.

    (b) _____

    (c) Find the effective rate of interest to the nearest hundredth of a percent.

    (c) _____

This seems like a low rate. However, Bank of America did this using funds borrowed from the Japanese investment house. So it did not even use its own money to make a little over one million dollars.

The very idea of interest is not acceptable in Muslim countries where a bank takes partial ownership of a company when it lends to a company, at least until funds are repaid. Even in countries that do allow interest, interest rates vary considerably. Use financial newspapers, magazines, or the World Wide Web to find interest rates in three different countries and compare them to those in the United States.

Case in Point Summary Exercise

APPLE, INC.

www.apple.com

Facts:

- **1976:** Founded
- **1984:** Introduced Macintosh
- **2001:** Introduced iPod
- **2003:** Opened iTunes Music Store
- **2007:** Introduced iPhone
- **2010:** Introduced iPad
- **2015:** Introduced Apple Watch

Jessica Hernandez wanted to open a retail outlet in a mall where she could sell iPhones. But Apple would not allow it, so she focused on building web pages for businesses. As her business started to grow, she realized she would soon need to hire one or two people and that she needed to buy some computers. In other words she could not grow her business without borrowing.

1. Hernandez decided to borrow $85,000 for 10 months. She found that banks would lend to her only if she had a cosigner on the note—fortunately her uncle was a successful business owner and he agreed to cosign. Bank One offered the funds at a 10% simple discount. Find the maturity value of the loan and the discount.

    1. _____

2. Once Hernandez's uncle agreed to cosign on a loan, Union Bank offered to lend Hernandez $85,000 at 10.5% simple interest. Find the interest and maturity value.

    2. _____

3. Find the loan with the lower interest and find the difference in interest.

    3. _____

4. Find the effective interest rate for both loans to the nearest hundredth of a percent.

    4. _____

Bank	Interest	Loan Amount	Effective Rate
Bank One	$7727.27	$85,000	
Union Bank	$7437.50	$85,000	

Discussion Question: Assume Hernandez has successfully managed her business for several years. List five reasons she may still need to borrow from time to time.

Source: Apple, Inc., www.apple.com.

Chapter 9 Test

To help you review, the numbers in brackets show the section in which the topic was discussed.

Find the simple interest for each of the following. Round to the nearest cent. [9.1]

1. $12,500 at $10\frac{1}{2}$% simple interest for 280 days

   1. _____

2. $8250 at $9\frac{1}{4}$% simple interest for 8 months

   2. _____

3. A loan of $47,000 at 8.5% simple interest made on June 8 and due on August 22

   3. _____

4. A promissory note for $4500 at 10.3% simple interest made on November 13 and due March 8

   4. _____

Solve the following application problems.

5. Joan Davies signed a 140-day simple interest note for $12,500 with a bank that uses *exact* interest. If the rate is 10.7%, find the maturity value. [9.1]

   5. _____

6. A French bakery borrowed $98,300 for new commercial equipment. The 8.25% simple interest loan was repaid in 6 months. Find the maturity value. [9.1]

   6. _____

7. Glenda Pierce plans to borrow $14,000 for a new hot tub and deck for her home. She has decided on a term of 200 days at 10.5% simple interest. However, she has a choice of two lenders. One calculates interest using a 360-day year and the other uses a 365-day year. Find the amount of interest Pierce will save by borrowing from the lender using the 365-day year. [9.1]

   7. _____

8. Lupe Gonzalez plans to lend $6500 to her friend so she can expand her business. Find the simple interest rate required for the fund to grow to $7247.50 in 15 months. [9.2]

   8. _____

9. Hilda Heinz lends $1200 to her sister Olga at a rate of 9%. Find how long it will take for her investment to earn $100 in interest. (Round to the nearest day.) [9.2]

   9. _____

10. A woman invested money received from an insurance settlement for 7 months at 5% simple interest. If she received $1254.17 interest on her investment during this time, find the amount that she invested. (Round to the nearest dollar.) [9.2]

    10. _____

11. Mike Fagan needs $25,000 to expand his flower shop. Find the face value of a simple discount note that will provide the $25,000 in proceeds if he plans to repay the note in 240 days and the bank charges a 9% discount rate. [9.3]

    11. _____

Find the discount and the proceeds for the following simple discount notes. **[9.3]**

	Face Value	Discount Rate	Time (Days)	Discount	Proceeds
12.	$9800	11%	120	_____	_____
13.	$10,250	9.5%	60	_____	_____

14. Barbara Waters signed a simple discount note for $15,000 for 120 days at a rate of 9%. Find **(a)** the proceeds and **(b)** the effective interest rate based on the proceeds received by Waters. **[9.3]**

(a) _____
(b) _____

15. Lizabeth Neault needed funds to open a law office. She borrowed $28,400 at 8.5% simple interest for 150 days on July 7. The bank she borrowed from sold the note at a 9% discount on August 20. Find the proceeds to the bank. **[9.4]**

15. _____

16. A 90-day simple discount promissory note for $9200 with a simple discount rate of 11% was signed on January 25. It was discounted on March 2 at 12%. Find the proceeds at the time of the sale. **[9.4]**

16. _____

17. A $20,000 T-bill is purchased at a 3.75% discount rate for 13 weeks. Find **(a)** the purchase price of the T-bill, **(b)** the maturity value, **(c)** the interest earned, and **(d)** the effective rate of interest to the nearest hundredth of a percent. **[9.3]**

(a) _____
(b) _____
(c) _____
(d) _____

The following note was discounted by a finance company at $12\frac{1}{2}\%$. Find the discount period, the discount, and the proceeds. **[9.4]**

	Date Loan Was Made	Face Value	Length of Loan	Rate	Date of Discount	Discount Period	Discount	Proceeds
18.	Jan. 25	$9200	90 days	10%	Mar. 12	_____	_____	_____

19. Jan Guerra lends $9000 to her second cousin using a 180-day 10% simple interest note that was signed on October 30. Guerra subsequently has a car accident and desperately needs money, so she sells the note at a discount of 15% on January 3 to an investor. Find **(a)** the discount, **(b)** the proceeds, and **(c)** the amount of money Guerra gains or loses. **[9.4]**

(a) _____
(b) _____
(c) _____

Compound Interest and Inflation

CHAPTER CONTENTS

- **10.1** Compound Interest
- **10.2** Interest-Bearing Bank Accounts and Inflation
- **10.3** Present Value and Future Value

Great Depression (1929–39)

CASE IN POINT

The Great Depression of 1929–39 was a very difficult time. The stock market dropped and people became afraid, so they bought less and stood for hours at banks to withdraw their funds. The banks had lent out most of the depositors' money and did not have enough cash for those wanting it. Fear increased, so people spent even less and kept any extra cash they had "under the mattress" rather than in a bank. As demand for nearly everything fell, companies laid off employees, and unemployment grew to around 25% of the workforce. Approximately 9000 banks went bankrupt and families suffered greatly! Many went hungry.

Many economists believe that the financial crisis of 2007–09 might have been as bad as the Great Depression of the 1930s were it not for economists' having studied that event in great detail and implementing ideas to avoid its worst effects. For example, to prevent banks from going bankrupt, the federal government loaned large amounts to the big banks and protected many smaller ones; stimulated demand by providing funds for jobs programs and keeping interest rates very low to encourage firms and people to buy; and supported the takeover of troubled financial firms by larger and more stable firms. For example, the government supported the merger of troubled home lender Countrywide and investment firm Merrill Lynch into Bank of America.

During the 2007–09 crisis, average unemployment soared around the world, rising to over 10% in 2009 in the United States, and even higher among minorities and younger workers. However, it did not reach the extreme levels of the Great Depression, and far fewer firms went bankrupt than in the 1930s. People were able to get cash out of the banks when they asked, which greatly boosted confidence in the banking system.

Despite all this, the effects of the crisis of 2007–09 lasted until 2017. Thankfully, major financial crises are rare.

Loans are at the core of world financial systems. People and firms with excess money lend to other people and firms who need money. *Interest* is the price of borrowing money, and so it is a fundamental aspect of life today. The **simple interest** loans discussed in Chapter 9 require interest to be calculated and credited only once during the life of the loan. On the other hand, the **compound interest** loans discussed in this chapter require interest to be calculated and credited more than once during the life of a loan or investment.

10.1 Compound Interest

OBJECTIVES

1. Use the simple interest formula $I = PRT$ to calculate compound interest.
2. Identify interest rate per compounding period and number of compounding periods.
3. Use the formula $M = P(1 + i)^n$ to find compound amount.
4. Use the table to find compound amount.

Present value is the value of an investment today, right now. Money left in an investment usually grows over time. The amount in an investment at a specific future date is called the **future amount**, **compound amount**, or **future value**. The future value depends not only on the amount initially invested, it also depends on the following:

1. **Compound interest—Compound interest results in a greater future value than simple interest.**
2. **Interest rate—A higher rate results in a greater future value.**
3. **Length of investment—An investment held longer usually results in a greater future value.**

To see the effects of these, compare the future values of a $10,000 investment using the following table:

Compare Investments:	To See That:
A and B	Compound interest results in more interest than simple interest.
B and C	Higher interest rates result in more interest.
C and D	Longer-term investments result in more interest.

	Investment	Term	Annual Rate	Interest	Future Value
A.	Simple interest	6 years	5%	$3,000	$13,000
B.	Compound interest	6 years	5%	$3,401	$13,401
C.	Compound interest	6 years	8%	$5,869	$15,869
D.	Compound interest	10 years	8%	$11,589	$21,589

Most wealthy people are over 50 because it takes years for compound interest to really work. The graph here shows the power of compound interest over time starting with an investment of $10,000 earning 5% and 8% per year. The $10,000 investment grows to over $100,000 in 30 years at 8%. This illustrates why it is so important to begin saving early. People save for an automobile, for college, and for retirement, and also so that they will not have to depend on a job for income someday.

"The most powerful force in the universe is compound interest."— Albert Einstein

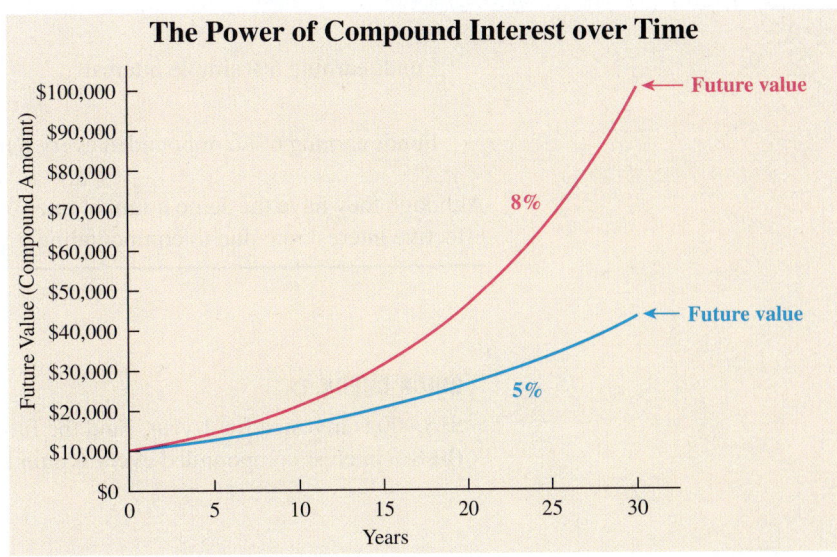

OBJECTIVE 1 Use the simple interest formula $I = PRT$ to calculate compound interest. Compound interest is interest calculated on previously credited interest in addition to the original principal. Compound interest calculations require that interest be calculated and credited to an account more than once each year. Examples 1 and 2 show how the simple interest formula can be used to find compound interest.

> **Finding Future Value (Compound Amount)**
> 1. Use $I = PRT$ to find simple interest for the period.
> 2. Add principal at the end of the previous period to the interest for the current period to find the principal at the end of the current period.

Comparing Simple to Compound Interest

EXAMPLE 1 Regina Foster wants to compare simple interest to compound interest on a $2000 investment.
(a) Find the interest if funds earn 6% simple interest for 1 year.
(b) Find the interest if funds earn 6% interest compounded every 6 months for 1 year.
(c) Find the difference between the two.
(d) Find the effective rate for both.

SOLUTION

(a) Simple interest on $2000 at 6% *for 1 year* is found as follows.
$$I = PRT = \$2000 \times .06 \times 1 = \$120$$

(b) Interest compounded every 6 months means that interest must be calculated at the end of each 6-months using $I = PRT$. Add interest to principal before proceeding.

$$\text{Interest for first 6 months} = PRT = \$2000 \times .06 \times \tfrac{1}{2} = \$60$$
$$\text{Principal at end of first 6 months} = \text{Original principal} + \text{Interest}$$
$$= \$2000 + \$60 = \$2060$$

The new principal of $2060 earns interest for the second 6 months.

$$\text{Interest for second 6 months} = PRT = \$2060 \times .06 \times \tfrac{1}{2} = \$61.80$$
$$\text{Principal at end of 1 year} = \$2060 + \$61.80 = \$2121.80$$

The interest earned in the second 6 months ($61.80) is larger than that earned in the first 6 months ($60), since the first interest amount of $60 is also earning interest during the second 6 months.

$$\text{Total compound interest} = \$60 + \$61.80 = \$121.80$$

(c) Difference in interest = $121.80 − $120 = $1.80.

Compound interest results in more interest. The difference here of $1.80 seems trivial, but compound interest results in *huge differences* over time as you will see shortly!

> **Quick TIP**
> Compound interest results in huge differences over time.

(d) The effective interest rate is the interest for the year divided by the original investment.

	Effective Interest Rate
Funds earning 6% simple interest	$\dfrac{\$120}{\$2000} = 6\%$
Funds earning 6% compounded every 6 months	$\dfrac{\$121.80}{\$2000} = 6.1\%$ (rounded)

Although they have the same nominal rate (6%), the compound interest investment has a larger effective interest rate due to compounding.

> **QUICK CHECK 1**
> $15,000 is invested for 1 year. Find the future value based on (a) simple interest of 8% and (b) 8% interest compounded every 6 months. (c) Then find the difference between the two.

Finding Compound Interest

EXAMPLE 2 The Simpsons want to pay $5000 down in 4 years on a new car. Today they invest $3800 in an investment that pays 6% interest compounded annually. (a) Find the excess of compound interest over simple interest at the end of 4 years. (b) Will they have enough money to meet their goal?

SOLUTION

First calculate interest using $I = PRT$ and round to the nearest cent. Then find the new principal by adding the interest earned to the preceding principal.

(a)
Year	P × R × T	Interest	P + I	Compound Amount
1	$3800.00 × .06 × 1	= $228.00	$3800.00 + $228.00	= $4028.00
2	$4028.00 × .06 × 1	= $241.68	$4028.00 + $241.68	= $4269.68
3	$4269.68 × .06 × 1	= $256.18	$4269.68 + $256.18	= $4525.86
4	$4525.86 × .06 × 1	= $271.55	$4525.86 + $271.55	= $4797.41

Compound interest = $4797.41 − $3800 = **$997.41**
Simple interest = PRT = $3800 × .06 × 4 = $912
Difference = **$997.41** − $912 = $85.41

(b) No, but almost! They will be short of their goal by $5000 − $4797.41 = **$202.59**.

Quick TIP
Round interest amounts to the nearest cent each time interest is calculated.

QUICK CHECK 2

Find the future amount at the end of 2 years for an $80,000 investment that earns 7% per year.

OBJECTIVE 2 Identify interest rate per compounding period and number of compounding periods. The **compounding period** is the time over which interest is calculated and added to principal. The **number of compounding periods** is the number of compounding periods, either in each year or in the life of a loan or investment.

Interest Compounded	Compound at the End of Every	Number of Compounding Periods in 1 Year
Annually	year	1
Semiannually	6 months	2
Quarterly	3 months	4
Monthly	1 month	12
Daily	1 day	365*

*Leap year has 366 compounding periods.

The **interest rate per compounding period** is the interest rate applied to each compounding period. It is found by dividing the annual interest rate by the number of compounding periods in a year. The total number of compounding periods is the number per year times the number of years. So, a loan compounded semiannually for 4 years will be compounded every 6 months for 4 years, or 8 times.

Rate	Compounded	Number of Compounding Periods per Year	Term	Rate per Compounding Period	Total Number of Compounding Periods
8%	semiannually	2	4 years	$\frac{8\%}{2} = 4\%$	4 years × 2 per year = 8
6%	monthly	12	$2\frac{1}{2}$ years	$\frac{6\%}{12} = .5\%$	$2\frac{1}{2}$ years × 12 per year = 30
4%	quarterly	4	5 years	$\frac{4\%}{4} = 1\%$	5 years × 4 per year = 20

376 CHAPTER 10 Compound Interest and Inflation

Finding the Interest Rate per Compounding Period and the Number of Compounding Periods

EXAMPLE 3 Find the interest rate per compounding period and the number of compounding periods for each.

(a) 5% compounded semiannually, 3 years
(b) 6% per year, compounded monthly, $2\frac{1}{2}$ years
(c) 2% per year, compounded quarterly, 5 years

SOLUTION

(a) 5% compounded semiannually is $\frac{5\%}{2}$ = 2.5% credited at the end of each 6 months. There are 3 years × 2 periods per year = 6 compounding periods in 3 years.
(b) 6% per year, compounded monthly, results in $\frac{6\%}{12}$ = .5% credited at the end of each month. There are 2.5 years × 12 periods per year = 30 compounding periods in 2.5 years.
(c) 2% per year, compounded quarterly, results in $\frac{2\%}{4}$ = .5% credited at the end of each quarter. There are 5 years × 4 periods per year = 20 compounding periods in 5 years.

> **QUICK CHECK 3**
>
> A loan requires that the 8% interest be compounded quarterly for 6 years. Find the interest rate per compounding period and the number of compounding periods.

OBJECTIVE 3 Use the formula $M = P(1 + i)^n$ to find compound amount. The **formula for compound interest** uses **exponents**, which is a short way of writing repeated products. For example,

$$2 \times 2 \times 2 = 2^3$$

Exponent: Multiply the base by itself 3 times.
base

Also, $4^2 = 4 \times 4 = 16$, and $5^4 = 5 \times 5 \times 5 \times 5 = 625$.

Assume that P dollars are deposited at a rate of interest i per compounding period for n periods. Then the compound amount and the interest are found as follows.

Formulas for Compounding Interest

Maturity value = $M = P(1 + i)^n$ where P = initial investment
Interest = $I = M - P$ n = total number of compounding periods
                                                      i = interest rate per compounding period

---

It is important to realize that in the formulas above

$$i = \frac{\text{Interest rate per year}}{\text{Number of compounding periods in a year}} \quad \text{and}$$

$$n = \text{Number of compounding periods per year} \times \text{Number of years}$$

Finding Compound Interest

EXAMPLE 4 An investment managed by Bank of America pays 7% interest per year compounded semiannually. Given an initial deposit of $4500, (a) use the formula to find the compound amount after 5 years, and (b) find the compound interest.

CASE IN POINT

SOLUTION

(a) Interest is compounded at $\frac{7\%}{2}$ = 3.5% every 6 months for 5 years × 2 periods per year = 10 periods. Therefore, 3.5% is the interest rate per compounding period (i) and 10 is the number of compounding periods (n).

$$M = P(1 + i)^n$$
$$= \$4500 \times (1 + .035)^{10}$$
$$= \$4500 \times (1.035)^{10}$$
$$= \$6347.69 \quad \text{(rounded)}$$

> **Quick TIP**
> Interest is such an important concept that many parents believe their kids should learn about it in high school.

The compound amount is $6347.69.

(b)
$$I = M - P$$
$$= \$6347.69 - \$4500 = \$1847.69$$

The interest is $1847.69.

🖩 The calculator solution for part **(a)** is as follows.

$$4500 \times (1 + .035) y^x 10 = \$6347.69 \quad \text{(rounded)}$$

Note: Refer to Appendix B for calculator basics.

> **QUICK CHECK 4**
>
> Use the formula for maturity value to find the compound amount and interest on a $9000 investment at 2% compounded semiannually for 5 years.

OBJECTIVE 4 Use the table to find compound amount. The value of $(1 + i)^n$ in the formula $M = P(1 + i)^n$ can be calculated using a calculator, or it can be found in the compound interest table below. The interest rate i at the top of the table is the interest rate *per compounding period*. The value of n down the far left (or far right) column of the table is *the total number of compounding periods*. The value in the body of the table is the compound amount, or maturity value, for each $1 in principal.

> **Finding Compound Amount (Future Value)**
>
> Compound amount = Principal × Number from compound interest table

Compound Interest Table

Interest Rate per Compounding Period

Period	½%	1%	1½%	2%	2½%	3%	4%	5%	6%	8%	Period
1	1.00500	1.01000	1.01500	1.02000	1.02500	1.03000	1.04000	1.05000	1.06000	1.08000	1
2	1.01003	1.02010	1.03023	1.04040	1.05063	1.06090	1.08160	1.10250	1.12360	1.16640	2
3	1.01508	1.03030	1.04568	1.06121	1.07689	1.09273	1.12486	1.15763	1.19102	1.25971	3
4	1.02015	1.04060	1.06136	1.08243	1.10381	1.12551	1.16986	1.21551	1.26248	1.36049	4
5	1.02525	1.05101	1.07728	1.10408	1.13141	1.15927	1.21665	1.27628	1.33823	1.46933	5
6	1.03038	1.06152	1.09344	1.12616	1.15969	1.19405	1.26532	1.34010	1.41852	1.58687	6
7	1.03553	1.07214	1.10984	1.14869	1.18869	1.22987	1.31593	1.40710	1.50363	1.71382	7
8	1.04071	1.08286	1.12649	1.17166	1.21840	1.26677	1.36857	1.47746	1.59385	1.85093	8
9	1.04591	1.09369	1.14339	1.19509	1.24886	1.30477	1.42331	1.55133	1.68948	1.99900	9
10	1.05114	1.10462	1.16054	1.21899	1.28008	1.34392	1.48024	1.62889	1.79085	2.15892	10
11	1.05640	1.11567	1.17795	1.24337	1.31209	1.38423	1.53945	1.71034	1.89830	2.33164	11
12	1.06168	1.12683	1.19562	1.26824	1.34489	1.42576	1.60103	1.79586	2.01220	2.51817	12
13	1.06699	1.13809	1.21355	1.29361	1.37851	1.46853	1.66507	1.88565	2.13293	2.71962	13
14	1.07232	1.14947	1.23176	1.31948	1.41297	1.51259	1.73168	1.97993	2.26090	2.93719	14
15	1.07768	1.16097	1.25023	1.34587	1.44830	1.55797	1.80094	2.07893	2.39656	3.17217	15
16	1.08307	1.17258	1.26899	1.37279	1.48451	1.60471	1.87298	2.18287	2.54035	3.42594	16
17	1.08849	1.18430	1.28802	1.40024	1.52162	1.65285	1.94790	2.29202	2.69277	3.70002	17
18	1.09393	1.19615	1.30734	1.42825	1.55966	1.70243	2.02582	2.40662	2.85434	3.99602	18
19	1.09940	1.20811	1.32695	1.45681	1.59865	1.75351	2.10685	2.52695	3.02560	4.31570	19
20	1.10490	1.22019	1.34686	1.48595	1.63862	1.80611	2.19112	2.65330	3.20714	4.66096	20
21	1.11042	1.23239	1.36706	1.51567	1.67958	1.86029	2.27877	2.78596	3.39956	5.03383	21
22	1.11597	1.24472	1.38756	1.54598	1.72157	1.91610	2.36992	2.92526	3.60354	5.43654	22
23	1.12155	1.25716	1.40838	1.57690	1.76461	1.97359	2.46472	3.07152	3.81975	5.87146	23
24	1.12716	1.26973	1.42950	1.60844	1.80873	2.03279	2.56330	3.22510	4.04893	6.34118	24
25	1.13280	1.28243	1.45095	1.64061	1.85394	2.09378	2.66584	3.38635	4.29187	6.84848	25
26	1.13846	1.29526	1.47271	1.67342	1.90029	2.15659	2.77247	3.55567	4.54938	7.39635	26
27	1.14415	1.30821	1.49480	1.70689	1.94780	2.22129	2.88337	3.73346	4.82235	7.98806	27
28	1.14987	1.32129	1.51722	1.74102	1.99650	2.28793	2.99870	3.92013	5.11169	8.62711	28
29	1.15562	1.33450	1.53998	1.77584	2.04641	2.35657	3.11865	4.11614	5.41839	9.31727	29
30	1.16140	1.34785	1.56308	1.81136	2.09757	2.42726	3.24340	4.32194	5.74349	10.06266	30

378 CHAPTER 10 Compound Interest and Inflation

Finding Compound Interest

EXAMPLE 5 In each case, find the interest earned on a $2000 deposit.
(a) For 3 years, compounded annually at 4%
(b) For 5 years, compounded semiannually at 6%
(c) For 6 years, compounded quarterly at 8%
(d) For 2 years, compounded monthly at 6%

SOLUTION

(a) In 3 years, there are $3 \times 1 = 3$ compounding periods. The interest rate per compounding period is $4\% \div 1 = 4\%$. Look across the top of the compound interest table above for 4% and down the side for 3 periods to find **1.12486**.

$$\text{Compound amount} = M = \$2000 \times 1.12486 = \$2249.72$$
$$\text{Interest earned} = I = \$2249.72 - \$2000 = \$249.72$$

(b) In 5 years, there are $5 \times 2 = 10$ semiannual compounding periods. The interest rate per compounding period is $6\% \div 2 = 3\%$. In the compound interest table, look at 3% at the top and 10 periods down the side to find **1.34392**.

$$\text{Compound amount} = M = \$2000 \times 1.34392 = \$2687.84$$
$$\text{Interest earned} = I = \$2687.84 - \$2000 = \$687.84$$

(c) Interest compounded quarterly is compounded 4 times a year. In 6 years, there are $6 \times 4 = 24$ quarters, or 24 periods. Interest of 8% per year is $\frac{8\%}{4} = 2\%$ per quarter. In the compound interest table, locate 2% across the top and 24 periods at the left, finding the number **1.60844**.

$$\text{Compound amount} = M = \$2000 \times 1.60844 = \$3216.88$$
$$\text{Interest earned} = I = \$3216.88 - \$2000 = \$1216.88$$

(d) In 2 years, there are $2 \times 12 = 24$ monthly compounding periods. Interest of 6% per year is $\frac{6\%}{12} = .5\%$ per month. Look in the compound interest table for .5% and 24 periods to find **1.12716**.

$$\text{Compound amount} = M = \$2000 \times 1.12716 = \$2254.32$$
$$\text{Interest earned} = I = \$2254.32 - \$2000 = \$254.32$$

QUICK CHECK 5

Find the interest earned on a $5000 investment for 4 years at 6% compounded semiannually.

USING A FINANCIAL CALCULATOR

We now show how to work Example 5(d) using a financial calculator. See Appendix C for a more detailed discussion of financial calculators and the notation used with them. But here are a few important things to keep in mind:

1. Present value (PV) is the amount of money today.
2. Future value (FV) is the amount of money at a specific date in the future.
3. The interest rate i refers to the interest rate per compounding period.
4. The symbol n refers to the number of compounding periods in the term of the investment.

Most financial calculators use the following convention: A negative number is used for an outflow of cash from an investor, and a positive number is used for an inflow of cash to an investor. Use a financial calculator by entering the three known values; then press the key for the unknown to find its value. Since financial calculators differ greatly, you should look at the instruction booklet that comes with your calculator.

Financial Calculator Solution Example 5(d) involves a $2000 deposit, which is the present value (PV). The unknown is the future value (FV). Since the original $2000 investment is an outflow to the investor, enter -2000 for PV. Then enter 24 (compounding periods) for n and .5% (per month) for the interest rate per compounding period. Finally, press FV to find the unknown future value.

$$-2000 \boxed{PV}\ 24 \boxed{n}\ .5 \boxed{i}\ \boxed{FV}\ \$2254.32$$

Answers may differ by very slight amounts due to the rounding of the table values.

The more often interest is compounded, the greater the amount of interest earned. You can use a financial calculator or the compound interest formula to confirm the numbers in the following table.

Compound interest makes a BIG difference!

Interest on $1000 at 8% per Year for 10 Years

Compounded	Interest
Not at all (simple interest)	$ 800.00
Annually	$1158.92
Semiannually	$1191.12
Quarterly	$1208.04
Monthly	$1219.64
Daily	$1225.35

Notice that daily compounding results in about 50% more interest over 10 years compared to simple interest.

Finding Compound Interest

EXAMPLE 6 John Smith sold his truck for $15,000, which he deposited in a retirement account that pays interest compounded semiannually. How much will he have after 15 years if the funds grow at
(a) 6%?   (b) 8%?   (c) 10%?

SOLUTION

In 15 years, there are $15 \times 2 = 30$ semiannual periods. The semiannual interest rates are
(a) $\frac{6\%}{2} = 3\%$ (b) $\frac{8\%}{2} = 4\%$ (c) $\frac{10\%}{2} = 5\%$

The compound amount is found using both the table and the compound amount formula since either can be used.

My truck goes down in value every year, but this investment will go up every year!

Using factors from the table:
(a) $15,000 × **2.42726** = $36,408.90
(b) $15,000 × **3.24340** = $48,651
(c) $15,000 × **4.32194** = $64,829.10

Using the formula for compound amount:
$M = \$15{,}000 \times (1 + .03)^{30} = \$36{,}408.94$
$M = \$15{,}000 \times (1 + .04)^{30} = \$48{,}650.96$
$M = \$15{,}000 \times (1 + .05)^{30} = \$64{,}829.14$

The small difference in compound amount using the two approaches is due to rounding in the table values. The formula for compound amount results in the exact answer, but we round the exact answer to the nearest penny.

The graph shows the growth in compound amount at these different interest rates. Importantly, annual interest earned becomes larger as the amount invested grows.

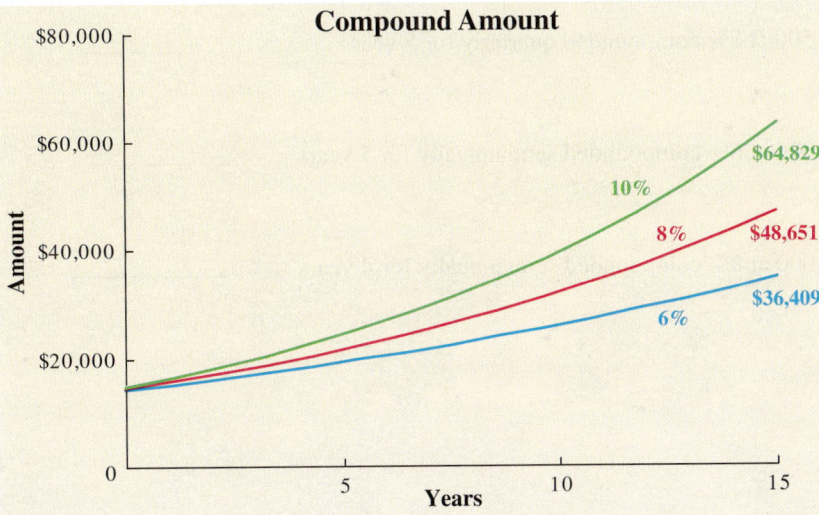

Quick TIP

Many people save during their working years so that they will have income from investments over and above Social Security payments when they are older.

QUICK CHECK 6

A bank offers a certificate of deposit that earns 2% compounded semiannually for 3 years. Find the compound amount for an investment of $3500.

10.1 Exercises // MyLab Math

The shaded sections below contain solutions to help you get a QUICK START on the various types of exercises.

Use the formula for compound amount, not the table, to find the compound amount and interest. (See Examples 3 and 4.)

	Compound Amount	Interest
1. $12,000 at 8% compounded annually for 4 years Compound interest is 8% per year for 4 years. $M = \$12{,}000 \times (1 + .08)^4 = \$12{,}000 \times 1.08 \times 1.08 \times 1.08 \times 1.08 = \$16{,}325.87$ $I = \$16{,}325.87 - \$12{,}000 = \$4325.87$	$16,325.87	$4325.87
2. $14,800 at 6% compounded semiannually for 4 years	_____	_____
3. $28,000 at 10% compounded quarterly for 1 year	_____	_____
4. $20,000 at 5% compounded quarterly for $\frac{3}{4}$ year	_____	_____

Use values from the compound interest table on page 377 to find both the compound amount and the compound interest. (See Examples 3–6.)

	Compound Amount	Interest
5. $32,350 at 6% compounded annually for 4 years Compound interest is 6% per year for 4 years. $M = \$32{,}350 \times 1.26248 = \$40{,}841.23$; $I = \$40{,}841.23 - \$32{,}350 = \$8491.23$	$40,841.23	$8491.23
6. $24,000 at 4% compounded annually for 10 years	_____	_____
7. $12,300 at 3% compounded semiannually for 4 years	_____	_____
8. $12,500 at 8% compounded quarterly for 5 years	_____	_____
9. $45,000 at 6% compounded semiannually for 5 years	_____	_____
10. $82,000 at 8% compounded semiannually for 4 years	_____	_____

C// indicates an exercise that is related to the Case in Point feature.

First find simple interest, then find compound interest assuming interest is compounded annually. Finally, find the excess of compound interest over simple interest.

	Principal	Rate	Number of Years	Simple Interest	Compound Interest	Difference
11.	$5400	6%	4	$1296	$1417.39	$121.39
12.	$9200	5%	6			
13.	$1200	8%	15			
14.	$4625	4%	10			

Use the table to solve the following application problems.

15. **CREDIT UNION** Bill Jensen deposits $8500 with Bank of America in an investment paying 5% compounded semiannually. Find **(a)** the compound amount and **(b)** the interest in 6 years.
Compound interest is $\frac{5\%}{2}$ = 2.5% and there are 6 × 2 = 12 compounding periods.
(a) M = $8500 × 1.34489 = $11,431.57; (b) I = $11,431.57 − $8500 = $2931.57

(a) $11,431.57
(b) $2931.57

16. **SAVINGS** Vickie Ewing deposits her savings of $2800 in an investment paying 6% compounded quarterly and she leaves it there for 5 years. Find **(a)** the compound amount and **(b)** the interest.

(a) _____
(b) _____

17. **SAVINGS** Tom Blasting invested $4500 in an investment paying 8% compounded quarterly for 3 years. Find **(a)** the compound amount and **(b)** the interest.

(a) _____
(b) _____

18. **INVESTMENT** Josh Crandall deposited $6000 in an annuity expected to pay 5% compounded semiannually for 4 years. Find **(a)** the compound amount and **(b)** the interest.

(a) _____
(b) _____

19. **INTERNATIONAL FINANCE** Chi Tang, a businessperson from Taiwan, deposits 25,000 yuan in a mutual fund with bonds managed by Bank of America that is expected to pay 6% compounded semiannually. Find **(a)** the balance in the account after 4 years and **(b)** the interest.

(a) _____
(b) _____

20. **UNITED KINGDOM** A firm in the UK places £42,000 (forty-two thousand pounds) in a bond paying 6% compounded quarterly and leaves it there as collateral for a loan. Find **(a)** the balance in the account after 1 year and **(b)** the interest.

(a) _____
(b) _____

21. **INVESTMENT DECISION** Bill Baxter has $25,000 to invest for a year. He can lend it to his sister, who has agreed to pay 10% simple interest for the year. Or, he can invest it in a retirement plan expected to pay 6% compounded quarterly for a year. How much additional interest would the simple interest loan to his sister generate? Find the effective interest rate for both investments to the nearest hundredth of a percent.

21. _____

CHAPTER 10 Compound Interest and Inflation

22. MAXIMIZING PROFIT Managers at Bank of America are debating about whether to lend $900,000 to a contractor at 7% simple interest or to a factory at 6% compounded monthly. Either loan would be for 9 months. Identify the loan that would pay more interest and how much additional interest it would pay. Then find the effective interest rate for the simple interest loan and the compound interest loan to the nearest tenth of a percent.

22. _____

23. INTERNATIONAL FINANCE JP Morgan lends $3,800,000 for 2 years at 6% compounded monthly to an Indonesian company that manufactures ships. Find **(a)** the future value and **(b)** interest.

(a) _____
(b) _____

24. CORPORATE FINANCE Key Bank lends $4,500,000 for $1\frac{1}{2}$ years at 8% compounded quarterly to Rengen Biomedical to fund a clinical trial on a new cancer drug. Find **(a)** the future value and **(b)** the interest.

(a) _____
(b) _____

25. INVESTING Sharon Nguyen has $25,000 to invest and believes that she can earn 6% compounded semiannually. Find the compound amount if she invests **(a)** for 2 years and **(b)** for 10 years. **(c)** Then find the additional amount earned due to the longer period.

(a) _____
(b) _____
(c) _____

26. WHICH INVESTMENT? Jan Reus sold her home and has $18,000 to invest. She believes she can earn 8% compounded quarterly. Find the compound amount if she invests for **(a)** 3 years and **(b)** 6 years. **(c)** Then find the additional amount earned due to the longer time period.

(a) _____
(b) _____
(c) _____

27. TIME OR RATE? Becky Hilton has a choice for her investment of $7500. She can invest it in funds she believes will earn either **(a)** 8% compounded quarterly for 2 years or **(b)** 6% compounded semiannually for 5 years. Find the future value of both. **(c)** Which is larger?

(a) _____
(b) _____
(c) _____

28. TIME OR RATE? Manager Isat Riyadh has a choice to make regarding short-term, excess corporate cash of $330,000. He can invest it **(a)** at 6% compounded quarterly for 6 months or **(b)** at 5% compounded semiannually for 1 year. Find the future amount of both. **(c)** Which is larger?

(a) _____
(b) _____
(c) _____

29. Explain the difference between simple interest and compound interest. (See Objectives 1 and 3.)

30. Explain the difference between 8% compounded monthly for 1 year and 8% simple interest for 1 year. Which is better for the lender? For the borrower? (See Objectives 2 and 3.)

31. Use an original investment of $10,000 to show the effect of interest rate and the term on future value. Do this by using two different interest rates and two different terms. (See Objective 3.)

32. List five companies that borrow money and state why they might need funds.

33. Explain why most wealthy people tend to be older. (See Objectives 3 and 4.)

QUICK CHECK ANSWERS

1. (a) $16,200 (b) $16,224 (c) $24
2. $91,592
3. 2%; 24 periods
4. $9941.60; $941.60
5. $1333.85
6. $3715.32

10.2 Interest-Bearing Bank Accounts and Inflation

OBJECTIVES

1. Define savings and other interest-bearing accounts.
2. Find interest compounded daily.
3. Define time deposit accounts.
4. Define inflation and the consumer price index.
5. Examine the effect of inflation on spendable income.
6. Understand the role of the government related to inflation.

CASE IN POINT Individuals, businesses, states, and even countries have money on deposit at Bank of America. These deposits can be in many forms, including checking accounts, savings accounts, money market accounts, and time deposits, such as certificates of deposits.

People sometimes think that banks have huge vaults of stored cash, but that is rarely the case. They usually have only enough cash in the vaults to meet customers' needs for the next few days. By far, most bank assets are in the loans to their many customers, rather than in cash.

OBJECTIVE 1 Define savings and other interest-bearing accounts. **Savings accounts, money market accounts,** and other interest-bearing accounts are offered by banks and credit unions and are usually a safe place to deposit money. These accounts are commonly insured by the Federal Deposit Insurance Corporation (FDIC) on deposits up to $250,000. Many of these accounts require a minimum balance. Interest rates paid on these accounts have varied from less than 1% per year to more than 5% per year and are often compounded daily. The Truth in Savings Act of 1991 resulted in Regulation DD, which requires that interest on savings accounts be paid based on the *exact* number of days.

Interest-bearing checking accounts are also offered by many banks and credit unions. These accounts often have a minimum balance, such as $1500, that must be maintained, although checks can be written on the account. They can be a good way to earn a little interest on your money as long as your balance does not fall too low, in which case the bank commonly charges a fee.

OBJECTIVE 2 Find interest compounded daily. Interest on savings accounts and other interest-bearing checking accounts is found using compound interest. It is common for banks to pay interest **compounded daily** so that interest is credited for every day that the money is on deposit.

The formula to find future value when interest is compounded daily is the same as given in Section 10.1:

$$M = P(1 + i)^n$$

where

$$i = \frac{\text{Annual interest rate}}{365 \text{ days in a year}} \quad \text{and}$$

$$n = \text{the number of days}$$

Quick TIP
It pays to earn interest whenever you can. Even small amounts add up.

You can use either the formula, the table in the margin at the left, or the table at the top of the next page. Since interest rates often vary, we will use $3\frac{1}{2}\%$ in this section, which is close to the historical average for accounts of this type. Even with daily compounding, interest is usually credited to accounts only at the end of each quarter.

The four quarters in a year begin on January 1, April 1, July 1, and October 1. Although some quarters have 91 or 92 days in them, we assume 90-day quarters for convenience in calculation. Assuming daily compounding and a compounding period expressed in days or quarters, compound amount and interest are found as follows.

Interest by Quarter for $3\frac{1}{2}\%$ Compounded Daily Assuming 90-Day Quarters

Number of Quarters	Value of $(1 + i)^n$
1	1.008667067
2	1.017409251
3	1.026227205
4	1.035121585

Finding Compound Amount and Interest

Compound amount = Principal × Number from table

Interest = Compound amount − Principal

Find the value of $(1 + i)^n$ using the table in the margin at the left of the prior page if time is given in number of quarters. Use the table below if time is 90 days or less.

Values of $(1 + i)^n$ for $3\frac{1}{2}$% Compounded Daily

Number of Days n	Value of $(1 + i)^n$	n	Value of $(1 + i)^n$	n	Value of $(1 + i)^n$	n	Value of $(1 + i)^n$	n	Value of $(1 + i)^n$
1	1.000095890	19	1.001823491	37	1.003554076	55	1.005287650	73	1.007024219
2	1.000191790	20	1.001919556	38	1.003650307	56	1.005384048	74	1.007120783
3	1.000287699	21	1.002015631	39	1.003746548	57	1.005480454	75	1.007217357
4	1.000383617	22	1.002111714	40	1.003842797	58	1.005576870	76	1.007313939
5	1.000479544	23	1.002207807	41	1.003939056	59	1.005673296	77	1.007410531
6	1.000575480	24	1.002303909	42	1.004035324	60	1.005769730	78	1.007507132
7	1.000671426	25	1.002400021	43	1.004131602	61	1.005866174	79	1.007603742
8	1.000767381	26	1.002496141	44	1.004227888	62	1.005962627	80	1.007700362
9	1.000863345	27	1.002592271	45	1.004324184	63	1.006059089	81	1.007796990
10	1.000959318	28	1.002688410	46	1.004420489	64	1.006155560	82	1.007893628
11	1.001055300	29	1.002784558	47	1.004516803	65	1.006252041	83	1.007990276
12	1.001151292	30	1.002880716	48	1.004613127	66	1.006348531	84	1.008086932
13	1.001247293	31	1.002976882	49	1.004709460	67	1.006445030	85	1.008183598
14	1.001343303	32	1.003073058	50	1.004805802	68	1.006541538	86	1.008280273
15	1.001439322	33	1.003169243	51	1.004902153	69	1.006638056	87	1.008376958
16	1.001535350	34	1.003265438	52	1.004998513	70	1.006734583	88	1.008473651
17	1.001631388	35	1.003361641	53	1.005094883	71	1.006831119	89	1.008570354
18	1.001727435	36	1.003457854	54	1.005191262	72	1.006927665	90	1.008667067

Note: The value of $(1 + i)^n$ for $3\frac{1}{2}$% compounded daily for a quarter with 91 days is 1.008763788 and for a quarter with 92 days is 1.008860519.

Finding Daily Interest   **EXAMPLE 1** Becky Gonzales received $12,500 from a divorce settlement. She plans to use the money for a down payment on a new Toyota Camry but decides to wait 60 days until the new models are out. She puts her money in a savings account earning $3\frac{1}{2}$% interest compounded daily for the 60 days. Find the amount of interest she will earn.

SOLUTION

The table value for 60 days is **1.005769730**.

$$\text{Compound amount} = \$12{,}500 \times \mathbf{1.005769730} = \mathbf{\$12{,}572.12}$$
$$\text{Interest} = \mathbf{\$12{,}572.12} - \$12{,}500 = \$72.12$$

The additional $72.12 isn't much money to Gonzales, but she is happy to earn some interest.

QUICK CHECK 1

Find the interest if $1200 is invested in a money market account earning 3.5% compounded daily for 90 days.

The next two examples show how interest is calculated when there are several deposits and/or withdrawals within a short period of time.

386 CHAPTER 10 Compound Interest and Inflation

Finding Interest on Multiple Deposits

EXAMPLE 2 Tom Blackmore is a private investigator who keeps his extra cash in a savings account to earn interest. On January 10, he deposited $2463 in a savings account paying $3\frac{1}{2}\%$ compounded daily. He deposited an additional $1320 on February 18 and $840 on March 3. Find the interest earned through April 10.

SOLUTION

Treat each deposit separately. The $2463 was in the account for 90 days (21 days in January, 28 days in February, 31 days in March, and 10 days in April). The value for 90 days from the table is **1.008667067**.

Compound amount = $2463 × **1.008667067** = **$2484.35**   first deposit plus interest

The $1320 deposited on February 18 was in the account for 51 days (10 days in February, 31 days in March, and 10 days in April).

Compound amount = $1320 × **1.004902153** = **$1326.47**   second deposit plus interest

The $840 was in the account for 38 days (28 days in March and 10 days in April).

Compound amount = $840 × **1.003650307** = **$843.07**   final deposit plus interest

The total amount in the account on April 10 is found by adding the three compound amounts.

Total in account = **$2484.35** + **$1326.47** + **$843.07** = **$4653.89**

To summarize:

Deposit	Number of Days Left in Account	Compound Amount
$2463	90	$2484.35
$1320	51	$1326.47
$ 840	38	$ 843.07
		$4653.89 **Total in account at end of quarter.**

The interest earned is the total amount in the account less the deposits.

Interest earned = **$4653.89** − ($2463 + $1320 + $840) = $30.89

QUICK CHECK 2

A money market account is opened with an $8500 deposit on April 10, and another $1500 is deposited on May 5. Find the total in the account on June 30 if funds earn $3\frac{1}{2}\%$ compounded daily. Also find the interest earned.

Finding Interest for the Quarter

EXAMPLE 3 Beth Gardner owns Blacktop Paving, Inc. She needs a place to keep extra cash, a place that will earn interest but that will allow her to get funds when needed. She opened a money market account on July 20 with a $24,800 deposit. She then withdrew $3800 on August 29 for an unexpected truck repair, and she made another withdrawal of $8200 on September 29 for payroll. Find the interest earned through October 1, given interest at $3\frac{1}{2}\%$ compounded daily.

SOLUTION

Of the original $24,800, a total of $24,800 − $3800 − $8200 = $12,800 earned interest from July 20 to October 1 or for 274 − 201 = 73 days. Find the factor **1.007024219** from the table.

Compound amount = $12,800 × **1.007024219** = $12,889.91
Interest = $12,889.91 − $12,800 = **$89.91**

The withdrawn $3800 earned interest from July 20 to August 29 or for 241 − 201 = 40 days.

Compound amount = $3800 × **1.003842797** = $3814.60
Interest = $3814.60 − $3800 = **$14.60**

Finally, the withdrawn $8200 earned interest from July 20 to September 29 or for 272 − 201 = 71 days.

Compound amount = $8200 × **1.006831119** = $8256.02
Interest = $8256.02 − $8200 = **$56.02**

Quick TIP
This example can also be solved using the formula for compound amount, $M = P(1 + i)^n$, or a financial calculator.

The total interest earned is ($89.91 + $14.60 + $56.02) = **$160.53**. The total in the account on October 1 is found as follows.

Deposits + Interest − Withdrawals = Balance on October 1
$24,800 + **$160.53** − ($3800 + $8200) = $12,960.53

10.2 Interest-Bearing Bank Accounts and Inflation

In summary:

Date	Deposit	Withdrawal	Balance
July 20	$24,800	—	$24,800
Aug. 29	—	$3800	$21,000
Sept. 29	—	$8200	$12,800
Oct. 1	$160.53 (Int.)	—	$12,960.53

QUICK CHECK 3

An account is opened with a deposit of $4000, but $3000 is withdrawn 40 days later. Find the amount in the account and the interest earned at the end of 90 days from the original deposit if interest is $3\frac{1}{2}\%$ compounded daily.

OBJECTIVE 3 Define time deposit accounts. Banks usually pay higher interest rates on funds left on deposit for *longer time periods* in **time deposits**. A **certificate of deposit (CD)** requires a minimum amount of money, such as $1000, to be on deposit for a minimum period of time, such as 1 year. Find the compound amount of a time deposit as follows.

Finding Compound Amount and Interest

$$\text{Compound amount} = \text{Principal} \times \text{Number from the table}$$
$$\text{Interest} = \text{Compound amount} - \text{Principal}$$

Compound Interest for Time Deposit Accounts Compounded Daily

Number of Years	1%	2%	3%	4%	5%	Number of Years
1	1.01005003	1.02020078	1.03045326	1.04080849	1.05126750	1
2	1.02020106	1.04080963	1.06183393	1.08328232	1.10516335	2
3	1.03045411	1.06183480	1.09417024	1.12748944	1.16182231	3
4	1.04081020	1.08328469	1.12749129	1.17350058	1.22138603	4
5	1.05127038	1.10516789	1.16182708	1.22138937	1.28400343	5
10	1.10516940	1.22139607	1.34984217	1.49179200	1.64866481	10

Quick TIP
A wide range of interest rates is given, since rates were very low but they are increasing at the time this edition was written.

Note: This compound interest table assumes daily compounding. The compound interest table on page 377 of Section 10.1 *does not*.

Finding Interest and Compound Amount for Time Deposits

EXAMPLE 4 Tony Sanchez plans to purchase tools for his auto-repair shop. Bank of America requires $20,000 in collateral before making the loan. Therefore, Tony deposits $20,000 with the bank in a 2-year certificate of deposit yielding 2% compounded daily. Find the compound amount and interest.

CASE IN POINT

SOLUTION
Look at the table for 2% and 2 years to find **1.04080963**.

$$\text{Compound amount} = \$20{,}000 \times 1.04080963 = \mathbf{\$20{,}816.19}$$
$$\text{Interest} = \mathbf{\$20{,}816.19} - \$20{,}000 = \$816.19$$

The compound amount formula $M = P(1 + i)^n$ can also be used to find the compound amount, but be careful as compounding is daily here:

Interest rate = $\frac{2\%}{365}$ per day and

Number of compounding periods is 2 years × 365 days per year = 730 days

$$M = \$20{,}000 \times \left(1 + \frac{.02}{365}\right)^{730} = \$20{,}816.19$$

The same future value is found using both methods.

> **QUICK CHECK 4**
>
> Find the compound amount and interest on $10,000 invested in a 4-year CD earning 2% compounded daily.

OBJECTIVE 4 Define inflation and the consumer price index. **Inflation** results in a continual rise in the price of goods and services. Another way to think about inflation is that a dollar is worth less each year. Either way, the end result is that inflation means you have to earn more each year to buy the same goods and services.

Inflation is important! For example, inflation of 2.5% per year will drive the cost of an automobile from $25,000 today to about $41,000 in 20 years. You need to increase your income regularly to keep up.

The following graph shows the increase in the average annual cost of tuition, fees, and room and board at public, two-year colleges. Costs have clearly increased a lot since 2000. Is it just that costs are going up, or is the dollar going down in value due to inflation? The answer is both in this case. Annual costs are increasing AND the dollar has slowly gone down in value.

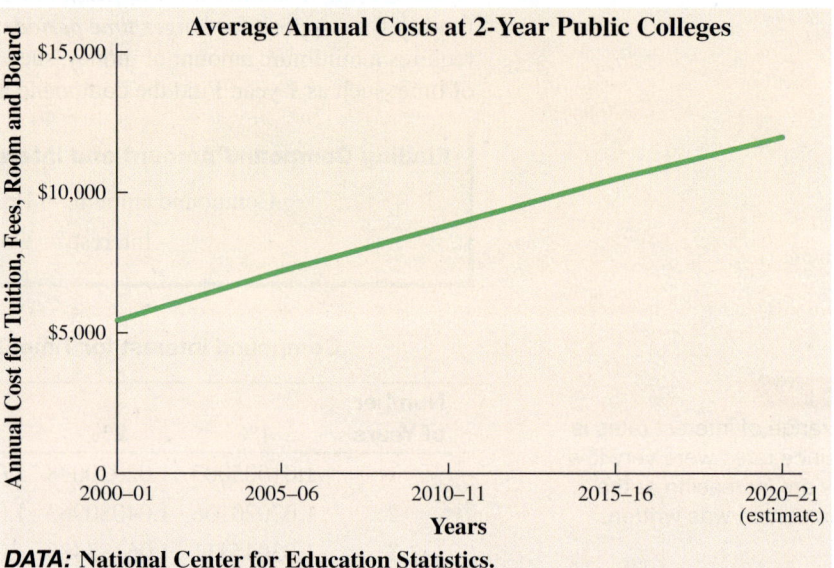

DATA: National Center for Education Statistics.

The **consumer price index (CPI)** or **cost of living index** calculated by the government is often used to track inflation. It measures the average change in prices from one year to the next for a common select group of goods and services, including food, housing, fuel, utilities, apparel, transportation, insurance, health care, and even pet care. The Bureau of Labor Statistics has historical data on inflation if you have further interest.

OBJECTIVE 5 Examine the effect of inflation on spendable income. Inflation can reduce the ability of a family to buy products and services. The next example shows what happens when a pay raise is less than the amount of inflation. It isn't pretty!

Estimating the Effects of Inflation

EXAMPLE 5 Inflation from one year to the next was 3% as measured by the CPI.

(a) Find the effect of the increase on a family with an after-tax annual income and budget of $39,600 (after taxes).

(b) What is the overall effect if the family members receive only a 1% (after tax) increase in pay for the year?

SOLUTION

(a) This is a percent problem. The cost of the goods and services that this family buys, if they buy the common bundle of goods and services, went up by 3% as measured by the CPI.

$$.03 \times \$39{,}600 = \mathbf{\$1188}$$

Therefore, next year these same goods and services will cost the family about

$$\$39{,}600 + \$1188 = \mathbf{\$40{,}788}$$

(b) The family's income went up 1% after taxes, or by

$$.01 \times \$39{,}600 = \mathbf{\$396}$$

Thus,

$$\text{New income} = \$39{,}600 + \$396 = \$39{,}996$$
$$\text{Family's loss in purchasing power} = \$40{,}788 - \$39{,}996 = \$792$$

QUICK CHECK 5

A family with an income of $32,000 receives a raise of 1.5% in a year when inflation is 3.5%. Find the decrease in purchasing power.

Example 5 can also be solved as follows.

Inflation rate	3.0%
− Raise	1.0%
Loss	2.0%

Loss in purchasing power = 2% × $39,600 = **$792**.

Example 5 shows that inflation can slowly erode purchasing power. Imagine the effect of losing purchasing power every year for 10 years. Inflation can erode purchasing power even though pay raises are received every year. Retired people are particularly concerned with inflation, since they live off of Social Security payments and their estate. Some retired people have had to go back to work because they did not plan for inflation.

The prior example shows how inflation can affect income. The next example shows how inflation can affect savings.

Estimating the Effects of Inflation

EXAMPLE 6 Joan Stringer has $48,500 in a retirement account. It is invested in a CD at a bank that pays 2% per year compounded daily during a year when the CPI increases by 2.7%. Find the gain or loss in purchasing power.

SOLUTION

Use the Compound Interest for Time Deposit Accounts Compounded Daily table on page 387 to find the factor for 2% and one year: **1.02020078**.

$$\text{Compound Amount} = \$48{,}500 \times \mathbf{1.02020078} = \mathbf{\$49{,}479.74}$$

Stringer must earn 2.7% for the year to keep up with inflation, so she is going to lose purchasing power for the year.

$$\text{Needed to keep up with inflation} = \$48{,}500 \times 1.027 = \$49{,}809.50$$
$$\text{Loss in purchasing power} = \$49{,}809.50 - \$49{,}479.74 = \mathbf{\$329.76}.$$

The purchasing power of Stringer's retirement account actually went down by $329.76 even though her funds earned interest and the amount in her account went up. The total funds in her retirement account at the end of the year will buy less than they would have at the beginning of the year.

QUICK CHECK 6

Tador Roofing deposits $50,000 in an account earning $3\frac{1}{2}\%$ compounded daily for 1 year. Find the loss in purchasing power in a year with an increase in the CPI of 4%.

Quick TIP

Spain, Greece, and Venezuela have all suffered from deflation recently, which has driven unemployment levels up in all three countries.

The opposite of inflation is **deflation**, which is a decrease in the general prices of goods and services. Deflation is relatively rare but potentially very serious. During the Great Depression of the 1930s, the prices of goods and services fell sharply in a deflationary spiral. People became worried about their savings and jobs. So, people postponed purchases, demand fell, and sales at firms plummeted. Managers responded by laying off even more workers, causing higher unemployment and even more fear. It was a vicious cycle.

OBJECTIVE 6 Understand the role of the government related to inflation. Both inflation that is too high (prices increasing rapidly) and deflation (prices falling) can be very harmful to families and businesses. As a result, the federal government works to control inflation using tools such as interest rates, over which they have some control.

When inflation is too high, the government increases interest rates, making it more expensive to borrow and thereby slowing lending and growth. When economic growth is too slow or there is a threat of deflation, the government decreases interest rates to encourage borrowing and, therefore, encourage business activity. Controlling inflation and deflation is very difficult to do, but it is a very important issue.

10.2 Exercises // MyLab Math

The shaded sections below contain solutions to help you get a QUICK START on the various types of exercises.

Find the interest earned by the following. Assume $3\frac{1}{2}$% interest compounded daily. (See Examples 1–3.)

	Amount	Date Deposited	Date Withdrawn	Interest Earned
1.	$4800	July 6	September 30	$39.75

There are $(31 - 6) + 31 + 30 = 86$ days.
Interest is $\$4800 \times 1.008280273 - \$4800 = \$39.75$.

	Amount	Date Deposited	Date Withdrawn	Interest Earned
2.	$3850	January 5	February 9	$12.94

There are $(31 - 5) + 9 = 35$ days.
Interest is $\$3850 \times 1.003361641 - \$3850 = \$12.94$.

	Amount	Date Deposited	Date Withdrawn	Interest Earned
3.	$8200	October 4	December 7	_____
4.	$2830	May 4	June 23	_____
5.	$17,958	September 9	November 7	_____
6.	$12,000	December 3	February 20	_____

Find the compound amount for each of the following certificates of deposit. Assume daily compounding. (See Example 4.)

	Amount Deposited	Interest Rate	Time in Years	Compound Amount
7.	$3900	2%	1	$3978.78

$\$3900 \times 1.02020078 = \$3978.78$

	Amount Deposited	Interest Rate	Time in Years	Compound Amount
8.	$8000	4%	1	_____
9.	$12,900	3%	10	_____
10.	$3600.40	1%	3	_____

11. Give three reasons to save and invest. Discuss why it is difficult to save.

12. List five ways in which inflation affects your family. (See Objective 4.)

C// indicates an exercise that is related to the Case in Point feature.

Solve the following application problems. If no interest rate is given, assume $3\frac{1}{2}\%$ interest compounded daily.

13. **SAVINGS ACCOUNT** Hilda Worth opened a savings account at Bank of America on April 1 with a $2530 deposit. She then deposited $150 on May 8 and $580 on May 24. Find **(a)** the balance on June 30 and **(b)** the interest earned through that date.

    (a) The $2530 (for 90 days) becomes $2530 × 1.008667067 = $2551.93
    The $150 (for 53 days) becomes $150 × 1.005094883 = $ 150.76
    The $580 (for 37 days) becomes $580 × 1.003554076 = $ 582.06
                                                         Total is $3284.75
    (b) Interest earned is $3284.75 − ($2530 + $150 + $580) = $24.75

    (a) **$3284.75**
    (b) **$24.75**

14. **PRINT SHOP** The manager of Quick Printing, Inc., is trying to get the most out of his assets, including cash that has been sitting in a checking account that does not pay interest. He opened a savings account with a deposit of $8765 on January 4. On February 11, he deposited $936. Then, on March 21, he deposited a tax refund check for $650. Find **(a)** the balance on March 31 and **(b)** the interest earned through that date.

    (a) _____
    (b) _____

15. **SAVINGS ACCOUNT** On April 1, MVP Sports opened a savings account at Bank of America with a deposit of $17,500. A withdrawal of $5000 was made 21 days later, another withdrawal of $980 was made 12 days before July 1. Find **(a)** the balance on July 1 and **(b)** the interest earned through that date. (*Hint:* See the footnote to the table on page 385.)

    (a) _____
    (b) _____

16. **SAVINGS ACCOUNT** The owner of International Magic opened a savings account for the extra cash in the firm. The initial deposit of $7800 was made on July 7. A withdrawal of $1500 was made 46 days later, and an additional withdrawal of $1000 was made 30 days before October 1. Find **(a)** the balance on October 1 and **(b)** the interest earned through that date.

    (a) _____
    (b) _____

17. **TIME DEPOSIT** Wes Cockrell has $4000 to deposit in a certificate of deposit, but he is debating whether to leave it there for 2 years or for 3 years. Assume 2% compounded daily in both situations, and find the compound amount in each case.

    17. _____

18. **COMPOUND INTEREST** Discount Auto Parts has $15,000 to deposit in a certificate of deposit for either 3 or 5 years. Assume 2% compounded daily and find the compound amount for both terms.

    18. _____

19. **LOAN COLLATERAL** An Italian firm deposited $800,000 in a 2-year time deposit earning 3% compounded daily with a New York bank as partial collateral for a loan. Find **(a)** the compound amount and **(b)** the interest earned.

    (a) _____
    (b) _____

CHAPTER 10 Compound Interest and Inflation

20. PUTTING UP COLLATERAL Joni Perez needs to borrow $20,000 to open a welding shop, but the bank will not lend her the money. Joni's uncle agrees to put up collateral for the loan with a $20,000, 4-year certificate of deposit paying 4% compounded daily. This means that the bank will take all or part of his deposit if Perez should fail to repay the loan. Find **(a)** the compound amount earned by her uncle and **(b)** the interest earned by her uncle.

(a) _____
(b) _____

21. RETIREMENT INCOME The Walters accumulated $235,000 during 40 years of work. They originally deposited this money in a 5-year time deposit earning 3% and used the income for living expenses. On renewing the time deposit, they found that interest rates on a 5-year time deposit had fallen and that they were going to receive only 2%. Find the difference in their *annual income* due to the decline in interest rates. (*Hint:* Don't use the compound interest table.)

21. _____

22. INHERITANCE Jessica Thompson inherited $80,000 and decided to put the money in one of two 4-year time deposits. The first time deposit yielded 3%, but the second yielded only 1%. Find the difference in the annual income. (*Hint:* Don't use the compound interest table.)

22. _____

23. PURCHASING POWER A family with an income and spending budget of $32,400 receives an increase in income of 1% after taxes in a year when inflation is 2.8%. Find the net gain or loss in purchasing power.

23. _____

24. INFLATION AND RETIREMENT Ben and Martha Wheeler are retired and they have $184,500 in a savings account at Bank of America paying $3\frac{1}{2}$% compounded daily. What is their gain or loss in purchasing power from interest in a year in which inflation is 2.5%?

24. _____

25. CORPORATE SAVINGS Dayton Tires has $180,000 to invest for 1 year. Find the future value if it earns **(a)** 2% per year compounded daily and **(b)** 3% per year compounded daily. **(c)** Then find the difference between the two.

(a) _____
(b) _____
(c) _____

26. EMERGENCY CASH After working long days in her small business for years, Kaitlyn Plank was finally making money. Since she was worried about emergencies, she decided to put $85,000 in an investment for 4 years. Find the future value if it earns **(a)** 2% per year compounded daily and **(b)** 4% per year compounded daily. **(c)** Then find the difference between the two.

(a) _____
(b) _____
(c) _____

27. Define inflation and discuss the government's role in controlling inflation. (See Objectives 5 and 6.)

28. Point out several differences between inflation and deflation. (See Objective 5.)

29. Define deflation and discuss how it can lead to severe problems. (See Objective 6.)

30. Describe the advantages to a family of having four months' income in a savings account. (See Objectives 1–6.)

QUICK CHECK ANSWERS

1. $10.40
2. $10,074.35; $74.35
3. $1020.20; $20.20
4. $10,832.85; $832.85
5. $640 loss
6. $243.92

10.3 Present Value and Future Value

OBJECTIVES
1. Define the terms *future value* and *present value*.
2. Calculate present value.
3. Use future value and present value to estimate the value of a business.

OBJECTIVE 1 Define the terms *future value* and *present value*. **Future value** is the amount available at a specific date in the future. It is the amount available after an investment has earned interest. All of the values found in Sections 10.1 and 10.2 were future values.

In contrast, **present value** is the amount needed today so that the desired future value will be available when needed. For example, an individual may need to know the present value that must be invested today in order to have a down payment for a new car in 3 years. Or a firm may need to know the present value that must be invested today in order to have enough money to purchase a new computer system in 20 months.

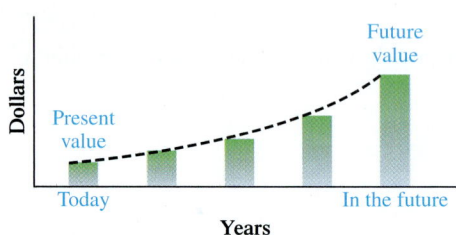

In this section the future value, interest rate and term are given. The present value that must be invested today to reach the future value is unknown and must be found.

Quick TIP
People save so they can eventually stop working, or can make a large purchase such as a car or go to college.

OBJECTIVE 2 Calculate present value. The present value needed at a given interest rate for a specific term can be found using a different form of the compound amount formula:

$$M = P(1 + i)^n$$
$$\frac{M}{(1 + i)^n} = P \quad \text{Divide both sides by } (1 + i)^n.$$
$$P = \frac{M}{(1 + i)^n} \quad \text{Exchange both sides of the equation.}$$

So, Present value = Compound amount (future value) divided by $(1 + i)^n$.

The present value can also be found by finding the appropriate value in the present value of a dollar table on the next page and substituting it into the following formula.

Finding Present Value

Present value (PV) = Future value $\times$ Table value

Finding Present Value EXAMPLE 1

Betty Clark needs to replace two pumps at her gas station in 3 years at an estimated cost of $12,000. What lump sum deposited today at 5% compounded annually must she invest to have the needed funds? How much interest will she earn?

SOLUTION

Step 1 The interest rate is 5% per compounding period for 3 compounding periods (years in this case). Look across the top of the table for 5% and down the left column for 3 to find **.86384**.

Present value = $12,000 × .86384 = **$10,366.08***

Step 2 Interest earned = $12,000 − **$10,366.08** = $1633.92.

*The present value can also be found using the equation Present value = future value divided by $(1 + i)^n$

$$P = \frac{M}{(1 + i)^n}$$
$$= \frac{\$12,000}{(1 + .05)^3} = \$10,366.05$$

The small difference is due to rounding.

Step 3 Check the answer by finding the future value of an investment of $10,366.08 in an account earning 5% compounded annually for 3 years. Use the table on page 377 to find **1.15763**.

$$\text{Future value} = \$10{,}366.08 \times 1.15763 = \$12{,}000.09$$

The reason it is not exactly $12,000 is rounding in the table value.

QUICK CHECK 1

Find the lump sum that must be deposited today to have a future value of $25,000 in 5 years if funds earn 6% compounded annually.

Present Value of a Dollar Table

Period	1%	1½%	2%	2½%	3%	4%	5%	6%	8%	10%	Period
1	.99010	.98522	.98039	.97561	.97087	.96154	.95238	.94340	.92593	.90909	1
2	.98030	.97066	.96117	.95181	.94260	.92456	.90703	.89000	.85734	.82645	2
3	.97059	.95632	.94232	.92860	.91514	.88900	.86384	.83962	.79383	.75131	3
4	.96098	.94218	.92385	.90595	.88849	.85480	.82270	.79209	.73503	.68301	4
5	.95147	.92826	.90573	.88385	.86261	.82193	.78353	.74726	.68058	.62092	5
6	.94205	.91454	.88797	.86230	.83748	.79031	.74622	.70496	.63017	.56447	6
7	.93272	.90103	.87056	.84127	.81309	.75992	.71068	.66506	.58349	.51316	7
8	.92348	.88771	.85349	.82075	.78941	.73069	.67684	.62741	.54027	.46651	8
9	.91434	.87459	.83676	.80073	.76642	.70259	.64461	.59190	.50025	.42410	9
10	.90529	.86167	.82035	.78120	.74409	.67556	.61391	.55839	.46319	.38554	10
11	.89632	.84893	.80426	.76214	.72242	.64958	.58468	.52679	.42888	.35049	11
12	.88745	.83639	.78849	.74356	.70138	.62460	.55684	.49697	.39711	.31863	12
13	.87866	.82403	.77303	.72542	.68095	.60057	.52032	.46884	.36770	.28966	13
14	.86996	.81185	.75788	.70773	.66112	.57748	.50507	.44230	.34036	.26333	14
15	.86135	.79985	.74301	.69047	.64186	.55526	.48102	.41727	.31524	.23939	15
16	.85282	.78803	.72845	.67362	.62317	.53391	.45811	.39365	.29189	.21763	16
17	.84438	.77639	.71416	.65720	.60502	.51337	.43630	.37136	.27027	.19784	17
18	.83602	.76491	.70016	.64117	.58739	.49363	.41552	.35034	.25025	.17986	18
19	.82774	.75361	.68643	.62553	.57029	.47464	.39573	.33051	.23171	.16351	19
20	.81954	.74247	.67297	.61027	.55368	.45639	.37689	.31180	.21455	.14864	20
21	.81143	.73150	.65978	.59539	.53755	.43883	.35894	.29416	.19866	.13513	21
22	.80340	.72069	.64684	.58086	.52189	.42196	.34185	.27751	.18394	.12285	22
23	.79544	.71004	.63416	.56670	.50669	.40573	.32557	.26180	.17032	.11168	23
24	.78757	.69954	.62172	.55288	.49193	.39012	.31007	.24698	.15770	.10153	24
25	.77977	.68921	.60953	.53939	.47761	.37512	.29530	.23300	.14602	.09230	25
26	.77205	.67902	.59758	.52623	.46369	.36069	.28124	.21981	.13520	.08391	26
27	.76440	.66899	.58586	.51340	.45019	.34682	.26785	.20737	.12519	.07628	27
28	.75684	.65910	.57437	.50088	.43708	.33348	.25509	.19563	.11591	.06934	28
29	.74934	.64936	.56311	.48866	.42435	.32065	.24295	.18456	.10733	.06304	29
30	.74192	.63976	.55207	.47674	.41199	.30832	.23138	.17411	.09938	.05731	30

Finding Present Value **EXAMPLE 2** The local Harley-Davidson shop has seen business grow rapidly. The owners plan to remodel the shop in one year at a cost of $280,000. How much should be invested in an investment earning 6% compounded semiannually to have the funds needed?

SOLUTION

The interest rate per compounding period is $\frac{6\%}{2} = 3\%$, and the number of compounding periods is 1 year × 2 periods per year = 2. Use the table to find **.94260**.

$$\text{Present value} = \$280{,}000 \times .94260 = \$263{,}928$$

The difference between the $280,001.22 and the desired $280,000 is due to rounding.

QUICK CHECK 2

In 5 years, Great Lakes Dairy estimates it will need $350,000 for a down payment to purchase a nearby farm. Find the amount that should be invested today to meet the down payment if funds earn 8% compounded quarterly.

Applying Present Value  **EXAMPLE 3** Radiux Inc. wishes to partner with a Korean company to purchase a satellite in 3 years. Radiux plans to make a cash down payment of 40% of its anticipated $8,000,000 cost and borrow the remaining funds from a bank. Find the amount Radiux should invest today in an investment earning 6% compounded annually to have the down payment needed in 3 years.

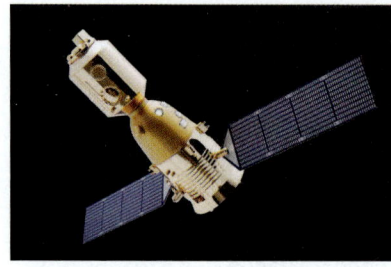

SOLUTION

First find the down payment to be paid in 3 years.

$$\text{Down payment} = .40 \times \$8{,}000{,}000 = \$3{,}200{,}000$$

This is the future value needed exactly 3 years from now. Using the present value of a dollar table on page 395 with 3 periods and 6% per period gives

$$\$3{,}200{,}000 \times .83962 = \$2{,}686{,}784$$

Radiux must invest $2,686,784 today at 6% interest compounded annually to have the required down payment of $3,200,000 in 3 years.

Financial Calculator Solution Using data from Example 3, the future value is $3,200,000. The number of periods is 3 and the interest rate per period is 6%. Enter these values into the financial calculator, with the future value being a positive number. Then press PV for the present value. It will be a negative number, indicating the outflows from Radiux.

  3200000 [FV] 3 [n] 6 [i] [PV] − 2686782 (rounded to the nearest dollar)

The difference from the value found in the example is due to rounding.

QUICK CHECK 3

Mom and Pop Jenkins plan to buy a new car in 2 years and want to make a down payment of 25% of the estimated purchase price of $32,000. Find the amount they need to invest to make the down payment if funds earn 6% compounded quarterly.

OBJECTIVE 3 Use future value and present value to estimate the value of a business. Sometimes a business has such a strong growth opportunity that it is valued at more than it would be if it had normal growth. In this case, the strong growth opportunity increases the market price of the business. To estimate the value of a business with strong growth, first estimate the future value of the business 2 to 5 years in the future. Then find the present value of this amount using the appropriate discount rate.

Evaluating a Business  **EXAMPLE 4** Brianna McGruder and Tanya Zoban own Extreme Sports, Inc., whose value is $120,000 today assuming normal growth. However, the partners believe the value will grow at 15% per year for the next four years. They want to take this rapid growth into consideration when valuing the business for a potential sale.

(a) Find the future value of the business in 4 years.

(b) Estimate the value of the retail store by finding the present value of the amount found in part (a) at 6% compounded quarterly.

SOLUTION

(a) The partners expect the business to grow at 15% per year for the next 4 years. There is no 15% column in the compound interest table of Section 10.1, so we use the formula $(1 + i)^n$, where $i = .15$ and $n = 4$.

$$\begin{aligned}\text{Future value} &= \$120{,}000 \times (1 + i)^n \\ &= \$120{,}000 \times (1 + .15)^4 = \mathbf{\$209{,}881 \text{ (rounded)}}\end{aligned}$$

This is an estimate of the value of the store in 4 years.

(b) Now find the present value of $209,881 assuming 6% compounded quarterly for 4 years or at $\frac{6\%}{4} = 1.5\%$ per quarter for $4 \times 4 = 16$ compounding periods. The value from the present value of a dollar table on page 395 is **.78803**.

Present value = $209,881 × **.78803** = $165,393 (**rounded**)

Thus, the partners should ask $165,400 for their business. The rapid growth rate adds about $165,400 − $120,000 = $45,400 to the value of the business.

QUICK CHECK 4

Assuming normal growth, a clothing store is worth $100,000. But the owners believe it will grow at 9% per year for the next three years. Estimate a reasonable selling price for the business by finding the present value at 5% per year compounded semiannually.

10.3 Exercises // MyLab Math

The shaded sections below contain solutions to help you get a QUICK START on the various types of exercises.

Find the present value and interest earned for each. (See Examples 1 and 2.)

	Amount Needed	Time (Years)	Interest	Compounded	Present Value	Interest Earned
1.	$12,300	3	6%	annually	$10,327.33	$1972.67
	P = $12,300 × .83962 = $10,327.33; I = $12,300 − $10,327.33 = $1972.67					
2.	$14,500	$2\frac{1}{2}$	8%	quarterly	$11,895.08	$2604.92
	P = $14,500 × .82035 = $11,895.08; I = $14,500 − $11,895.08 = $2604.92					
3.	$9350	4	5%	semiannually		
4.	$269,000	5	8%	semiannually		
5.	$18,853	11	6%	quarterly		
6.	$20,984	9	4%	quarterly		

Solve the following application problems.

7. **DIVORCE SETTLEMENT** Part of Abernathy's divorce settlement involves setting aside money today for college tuition for their daughter who enters college in 7 years. They estimate that the cost of four years' tuition and fees at the state university their daughter will attend will be $40,000. Find **(a)** the lump sum that must be invested at 4% compounded semiannually and **(b)** the amount of interest earned.
   (a) Lump sum = P = $40,000 × .75788 = $30,315.20
   (b) I = $40,000 − $30,315.20 = $9684.80

   (a) **$30,315.20**
   (b) **$9684.80**

8. **SELF-EMPLOYMENT** Samantha Garcia needs $25,000 in 4 years to start her own daycare business. **(a)** What lump sum should be invested today at 5%, compounded semiannually, to produce the needed amount? **(b)** How much interest will be earned?

   (a) _____
   (b) _____

9. **FINANCING COLLEGE EXPENSES** Mrs. Lorez wants all of her grandchildren to go to college and decides to help financially. How much must she give to each child at birth if they are to have $10,000 on entering college 18 years later, assuming 6% interest compounded annually?

   9. _____

10. BAKERY Carlos Mora recently immigrated to the United States from Central America. His family has agreed to help him set aside the cash needed to open a small bakery in 2 years once he completes a program at a culinary institute. Find the amount they must deposit today in an investment account expected to yield 4% compounded quarterly if he needs $95,000 to open the shop in 2 years.

10. _____

11. EXPANDING MANUFACTURING OPERATIONS Quantum Logic recently expanded its operations at a cost of $450,000. Management expects that the value of the investment will grow at a rate of 12% per year compounded annually for the next 5 years. **(a)** Find the future value of the investment. **(b)** Find the present value of the amount found in part **(a)** at a rate of 6% compounded annually. Round to the nearest dollar at each step.

(a) _____
(b) _____

12. BUSINESS EXPANSION Village Hardware expands its business at a cost of $20,000. They expect that the investment will grow at a rate of 10% per year compounded annually for the next 4 years. **(a)** Find the future value of the investment. **(b)** Find the present value of the amount found in part **(a)** at a rate of 6% compounded annually. Round to the nearest dollar at each step.

(a) _____
(b) _____

13. VALUE OF A BUSINESS Jessie Marquette believes her hair salon is worth $20,000 and estimates that its value will grow at 10% per year compounded annually for the next 3 years. If she sells the business, the funds will be invested at 8% compounded quarterly. **(a)** Find the future value if she holds onto the business. **(b)** What price should she insist on now if she sells the business? Round to the nearest dollar.

(a) _____
(b) _____

14. VALUE OF A BUSINESS John Fernandez figures his bike shop is worth $88,000 if sold today and that it will grow in value at 8% per year compounded annually for the next 6 years. If he sells the business, the funds will be invested at 5% compounded semiannually. **(a)** Find the future value of the shop. **(b)** What price should he insist on at this time if he sells the business?

(a) _____
(b) _____

15. Explain the difference between future value and present value. (See Objective 1.)

16. Explain how to value a rapidly growing business. (See Objective 3.)

QUICK CHECK ANSWERS

1. $18,681.50
2. $235,539.50
3. $7101.68
4. $111,670.35

Chapter 10 Quick Review

Chapter Terms *Review the following terms to test your understanding of the chapter. For each term you do not know, refer to the page number found next to that term.*

CD [p. 387]
CPI [p. 388]
certificate of deposit [p. 387]
compound amount [p. 373]
compound interest [p. 373]
compounded daily [p. 384]
compounding period [p. 375]

consumer price index [p. 388]
cost of living index [p. 388]
deflation [p. 389]
exponents [p. 376]
formula for compound interest [p. 376]
future amount [p. 373]

future value [p. 373]
inflation [p. 388]
interest-bearing checking accounts [p. 384]
interest rate per compounding period [p. 375]
money market accounts [p. 384]

number of compounding periods [p. 375]
present value [p. 373]
savings accounts [p. 384]
simple interest [p. 373]
time deposit [p. 387]

CONCEPTS

10.1 Finding compound amount and compound interest

Find the number of compounding periods (n) and the interest rate per period (i).
Use the compound interest table to find the interest on $1.
Multiply the table value by the principal to obtain the compound amount.
Subtract principal from compound amount to obtain the interest.

10.2 Finding the interest earned when the interest is compounded daily

Find the number of days that the deposit earns interest.
Use the 90-day or 1-quarter table to calculate interest on $1.
Find compound amount using the formula

   Compound amount = Principal × Table value

Find interest earned using the formula

   Interest = Compound amount − Principal

10.2 Finding the interest on time deposits

Use the compound interest for time deposit accounts table to find the interest on $1 compounded daily.
Find the compound amount using the formula

   Compound amount = Principal × Table value

Find interest using the formula

   Interest = Compound amount − Principal

10.2 Finding the effect of inflation on a pay raise

Find the new salary by multiplying the old salary by $(1 + \text{percent increase})$.
Find the salary needed to offset inflation by multiplying the old salary by $(1 + \text{inflation rate})$.
Find the gain or loss by subtracting.

EXAMPLES

Tom Jones invested $3000 at 6% compounded quarterly for 7 years.
There are $7 \times 4 = 28$ quarters or compounding periods in 7 years.
Interest of 6% per year = $\frac{6\%}{4} = 1\frac{1}{2}\%$ per period.
Find $1\frac{1}{2}\%$ across the top of the compound interest table and 28 down the left side to find **1.51722**.

   Compound amount = $3000 × **1.51722** = $4551.66
   Interest = $4551.66 − $3000 = $1551.66

Mary Carver deposits $1000 at $3\frac{1}{2}\%$ compounded daily on May 15. She withdraws the money on July 17. Find the compounded amount and interest earned.

May 15–May 31	16 days
June	30 days
July 1–July 17	17 days
	63 days

   Table value = **1.006059089**
   Compound amount = $1000 × **1.006059089** = **$1006.06**
   Interest = **$1006.06** − $1000 = $6.06

Susan Barbee invests $50,000 in a certificate of deposit paying 3% compounded daily. Find the amount after 4 years.

   Table value for 4 years at 3% = **1.12749129**
   Compound amount = $50,000 × **1.12749129** = $56,374.56
   Interest = $56,374.56 − $50,000 = $6374.56

Leticia Jaramillo earns $45,000 per year and gets a raise of 1% in a year in which inflation is 3%. Ignoring taxes, find the effect on her purchasing power.

   New salary = $45,000 × 1.01 = **$45,450**

Salary needed to offset
   inflation = $45,000 × 1.03 = **$46,350**

   Loss in purchasing power = $46,350 − $45,450 = $900

CONCEPTS	EXAMPLES
10.3 Finding the present value of a future amount Determine the number of compounding periods (n). Determine the interest per compounding period (i). Use the values of n and i to determine the table value from the present value table. Find present value from the following formula. $\text{Present value} = \text{Future value} \times \text{Table value}$	Sue York must pay a lump sum of \$4500 in 6 years. What lump sum deposited today at 6% compounded quarterly will amount to \$4500 in 6 years? Number of compounding periods $= 6 \times 4 = 24$ Interest per compounding period $= \dfrac{6\%}{4} = 1\dfrac{1}{2}\%$ **per period** Table value $= .69954$ Present value $= \$4500 \times .69954 = \$3147.93$
10.3 Finding the value of a business First estimate the future value a few years in the future. Then find the present value of that future value at the given discount rate.	A business worth \$180,000 is projected to grow at 12% per year for 4 years. Use the basic compound amount formula to find the future value. Then use a different form of the same equation to find the present value, assuming funds are invested at 5% compounded quarterly. $M = P(1 + i)^n = \$180{,}000(1 + .12)^4$ $= \$283{,}233.48$ **Future value** 5% compounded quarterly for 4 years is $\dfrac{5\%}{4} = 1.25\%$ per quarter for 20 quarters $P = \dfrac{M}{(1 + i)^n} = \dfrac{\$283{,}233.48}{(1 + .0125)^{20}} = \$220{,}924.54$ **Present value of firm**

Case Study //

VALUING A CHAIN OF MCDONALD'S RESTAURANTS

James and Mary Watson own a small chain of McDonald's restaurants that is valued at $2,300,000. They believe that the chain will grow in value at 12% per year compounded annually for the next 5 years. If they sell the chain, the funds will be invested at a rate of 6% compounded semiannually. They expect inflation to be 3% per year for the next 5 years. Ignore taxes, and answer the following, rounding answers to the nearest dollar at each step.

1. Find the future value of the chain after 5 years. Then find the price they should sell the chain for if they wish to have the same future value at the end of 5 years.

   1. _____

2. Find the future value of the chain if it grows at only 2% per year for 5 years. Then find the price they should ask for the chain given a 2% growth rate per year.

   2. _____

3. Use the compound amount formula to answer the following: What would the future value of the chain be if it grew at the expected rate of inflation? Find the price they should ask for the chain if it grows at that rate.

   3. _____

4. Complete the following table.

Growth Rate	Future Value	Market Value Today
2%	_____	_____
3% (inflation)	_____	_____
12%	_____	_____

   The value of the chain varies by more than one million dollars, depending on the rate of growth assumed for the business for the next 5 years.

INVESTIGATE

The interest rates that a bank pays depend on whether the money is in a checking account, money market account, savings account, or time deposit. Visit a local bank, and find the different interest rates that the bank will pay. Identify the conditions such as the minimum amount required, the minimum deposit, and the length of time the money must be on deposit to earn each interest rate.

Case in Point Summary Exercise

BANK OF AMERICA

www.bankofamerica.com

Facts:

- 1904: Founded by son of Italian immigrants
- 1930s: Survived the Great Depression
- 2008: Acquired Countrywide Financial (home loans)
- 2009: Acquired Merrill Lynch (investments)
- 2018: More than 5000 branches and operations in over 35 countries

The financial crisis that began in 2008 was caused by a speculative bubble in home prices amidst far too much debt. As home prices fell, the excessive debt of families, banks, and other financial institutions became obvious. People spent less, and corporations cut costs by laying off workers and rapidly slowing the inventory flowing in global supply chains. Home builders built far fewer homes and tried to quickly reduce debt. Banks made fewer loans, and credit became very difficult to obtain. As the losses mounted, stock markets fell sharply. By the end of 2008, many governments around the world reacted to prevent another Great Depression similar to the one that occurred in the 1930s.

To keep the economic systems working, the U.S. government spent several trillion dollars to support failing financial institutions and banks, extend unemployment benefits, increase employment, and support the sales of cars and homes. When it appeared that Countrywide Financial (home loans) and Merrill Lynch (investments) were in serious trouble, the U.S. government helped broker deals in which Bank of America took over both firms with the help of government loans and guarantees.

To demonstrate the effect of the recent financial crisis, we will use a small custom home builder called Horizon Homes, which used Bank of America.

1. Find the gross profit for each year by multiplying gross sales by the profit margin.

Year	Gross Sales	Profit Margin	Profit or Loss	
2007	$92,080,000	4.6%	_____	
2008	$64,160,000	−12.3%	_____	Financial crisis begins
2009	$28,034,000	−30.5%	_____	Firm struggles to survive
2010	$24,000,000	2.0%	_____	
2011	$32,060,000	1.6%	_____	

2. In 1996, managers at Horizon Homes had anticipated eventual financial problems and they invested $2,500,000 earning 6% per year (ignore taxes). Find the future value of this investment in 2008 when the firm had its first big loss. Was the value of the investment enough to offset the 2008 loss?

2. _____

3. (a) Find the combined loss by adding the losses in 2008 and 2009. (b) Find the present value that needed to be set aside in 1996 assuming 6% compounded annually to offset the total losses for 2008 and 2009. *Hint:* For simplicity, assume that the total of the two years losses occurred in 2009.

(a) _____
(b) _____

4. Businesses incur risk, the possibility that bad things happen without warning. For example, a company may be adversely affected by a competitor that brings a better product to the market. Discuss what managers can do to better prepare for risk.

Discussion Question: Do you and/or your family face financial risk? Discuss what you can do to better prepare for risk in your life.

Chapter 10 Test

To help you review, the numbers in brackets show the section in which the topic was discussed.

Find the compound amount and the interest earned for the following. [10.1]

	Amount	Rate	Compounded	Time (Years)	Compound Amount	Interest Earned
1.	$8700	4%	annually	8	_____	_____
2.	$12,000	6%	semiannually	5	_____	_____
3.	$9800	6%	semiannually	5	_____	_____
4.	$22,500	8%	quarterly	3	_____	_____

Find the interest earned by the following. Assume $3\frac{1}{2}\%$ interest compounded daily. [10.2]

5. $6400 deposited September 24 and withdrawn December 15   5. _____

6. $63,340 deposited December 5 and withdrawn March 2   6. _____

7. $37,650 deposited December 12 and withdrawn on February 29 (leap year)   7. _____

Find the present value of the following. [10.3]

	Amount Needed	Time (Years)	Rate	Compounded	Present Value
8.	$35,000	20	8%	annually	_____
9.	$15,750	7	6%	quarterly	_____
10.	$56,900	10	4%	semiannually	_____

Solve the following application problems.

11. A local branch of Gamestop, Inc., deposited $12,500 in a savings account on July 3 and then deposited an additional $3450 in the account on August 5. Find the balance on October 1 assuming an interest rate of $3\frac{1}{2}\%$ compounded daily. [10.2]   11. _____

12. Discount Auto Insurance deposited $1800 in a savings account paying $3\frac{1}{2}\%$ compounded daily on January 1 and deposited an additional $2300 in the account on March 12. Find the balance on April 1. [10.2]   12. _____

13. Mike George deposits $4000 in a certificate of deposit for 5 years. Find the compound amount if the interest rate is 3% compounded daily. [10.2]   13. _____

14. Benton Signs places $35,000 in a 2-year certificate of deposit yielding 2% compounded daily and uses it for collateral for a loan. Find the compound amount. [10.2]   14. _____

15. Liz Mulig earns $52,000 per year as a philosophy professor. She receives a raise of 2.5% in a year in which the CPI increases by 3.8%. Ignoring taxes, find the effect of the two increases on her purchasing power. [10.2]   15. _____

16. Peter Wong makes $65,000 per year as an editor for a publisher. He was notified of a 1.5% raise in a year in which the CPI increased by 3%. Find the gain or loss in his purchasing power. [10.2]   16. _____

CHAPTER 10 Compound Interest and Inflation

17. A note for $3500 was made at 8% per year compounded annually for 3 years. Find **(a)** the maturity value and **(b)** the present value of the note assuming 5% per year compounded semiannually. [10.3]

    (a) _____
    (b) _____

18. Computers, Inc., accepted a 2-year note for $12,540 in lieu of immediate payment for computer equipment sold to a local firm. Find **(a)** the maturity value given a 10% rate compounded annually and **(b)** the present value of the note at 6% per year compounded semiannually. [10.3]

    (a) _____
    (b) _____

19. A business worth $180,000 is expected to grow at 12% per year compounded annually for the next 4 years. **(a)** Find the expected future value. **(b)** If funds from the sale of the business today would be placed in an account yielding 8% compounded semiannually, what would be the minimum acceptable price for the business at this time? [10.3]

    (a) _____
    (b) _____

20. A corporation worth $40 million is expected to grow at 8% per year compounded annually for 5 years. **(a)** Find the future value to the nearest million. **(b)** The owners then propose to sell the firm and invest the proceeds in a new venture that should grow at 12% compounded annually for 4 years. Beginning with the future value from part **(a)** rounded to the nearest million, find the expected future value to the nearest million at the end of 4 additional years. [10.1]

    (a) _____
    (b) _____

Chapters 9–10 Cumulative Review

CHAPTERS 9 AND 10

Round money amounts to the nearest cent, time to the nearest day, and rates to the nearest tenth of a percent.

Find the value of the unknown quantity using simple interest. Use banker's interest.   [9.1–9.2]

	Interest	Principal	Rate	Time
1.	_____	$6800	8%	6 months
2.	_____	$6200	9.7%	250 days
3.	$165.28	_____	7%	100 days
4.	$4750	_____	6.5%	180 days
5.	$249.38	$10,500	_____	90 days
6.	$733.33	$12,000	_____	200 days
7.	$202.22	$9100	10%	_____
8.	$915	$18,300	6%	_____

1. _____
2. _____
3. _____
4. _____
5. _____
6. _____
7. _____
8. _____

Find the discount and the proceeds.   [9.3]

	Face Value	Discount Rate	Time (Days)	Discount	Proceeds
9.	$9000	12%	90	_____	_____
10.	$8750	$6\frac{1}{2}$%	210	_____	_____

Find the net proceeds when each of the following is discounted.   [9.4]

	Maturity Value	Discount Rate	Discount Period	Net Proceeds
11.	$5000	10%	90 days	_____
12.	$12,000	9%	150 days	_____

Find the compound amounts for the following.   [10.1]

13. $1000 at 4% compounded annually for 5 years

13. _____

14. $3520 at 8% compounded annually for 10 years

14. _____

Find the interest earned and compound amounts for each of the following. Assume $3\frac{1}{2}\%$ interest compounded daily. [10.2]

	Amount	Date Deposited	Date Withdrawn	Interest Earned	Compound Amount
15.	$12,600	March 24	June 3	_____	_____
16.	$7500	Nov. 20	Feb. 14	_____	_____

Find the present value and the amount of interest earned for the following. Round to the nearest cent. [10.3]

	Amount Needed	Time (Years)	Interest	Compounded	Present Value	Interest
17.	$1000	7	8%	annually	_____	_____
18.	$19,000	9	5%	semiannually	_____	_____

Solve the following application problems. Use a 360-day year where applicable.

19. Cathy Cockrell signed a 180-day simple discount note with a rate of 10% and a face value of $25,000. Find (a) the interest and (b) the proceeds. [9.3]

(a) _____
(b) _____

20. As a project manager, Regina Foster received a bonus of $18,000 for completing a difficult project on time. She invests it at 6% compounded quarterly for 5 years. Find the future value. [10.1]

20. _____

21. A divorce settlement requires Samantha James to pay her ex-spouse $12,000 in 2 years. What lump sum can be invested today at 5% compounded semiannually so that enough will be available for the payment? [10.3]

21. _____

22. Tom Davis owes $7850 to a relative. He has agreed to repay the money in 5 months, at an interest rate of 6%. One month before the loan is due, the relative discounts the loan at the bank. The bank charges a 7.92% discount rate. How much money does the relative receive? [9.4]

22. _____

23. The owner of Jessica's Cookies has an extra $3200 that she puts into a savings account paying $3\frac{1}{2}\%$ per year compounded daily. Find the interest if the funds are left there for 65 days. [10.2]

23. _____

Annuities, Stocks, and Bonds 11

CHAPTER CONTENTS

11.1 Annuities and Retirement Accounts

11.2 Present Value of an Ordinary Annuity

11.3 Sinking Funds (Finding Annuity Payments)

11.4 Stocks and Mutual Funds

11.5 Bonds

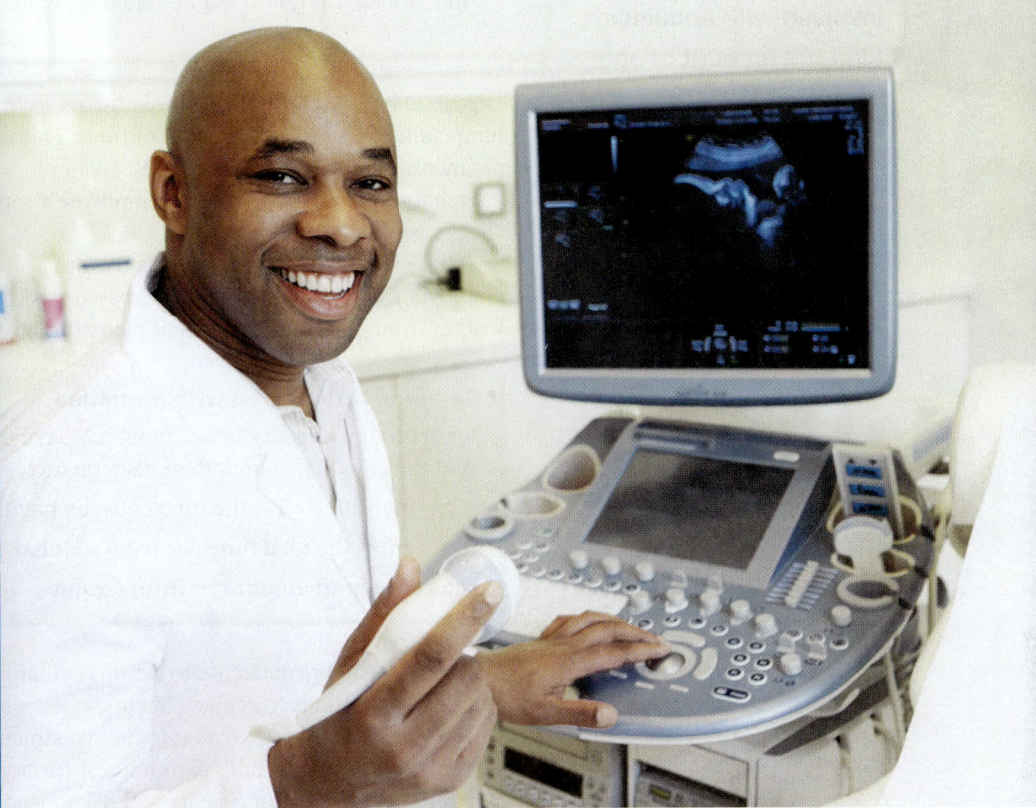

CASE IN POINT

DARNELL JOHNSON earned his two-year degree at a community college and then went to work as a technician at a hospital: Mayo Clinic. On his first day of work, he meet with Sella in Human Resources, who told him about the health insurance, life insurance, disability insurance, and retirement plan benefits he would have as a full-time employee.

He was surprised at all the benefits, including the fact that the hospital would contribute 4% of his salary into his retirement plan each month. In addition, the hospital would match any contribution he made into his retirement plan each month up to 3% of his salary. Contributions made into his retirement plan would be tax-deductible, meaning it would reduce the amount of income tax he would have to pay.

Samuel jotted down his thoughts, along with his first reactions, as he struggled to decide how much to contribute:

1. Hospital puts 4% of my salary into a retirement plan—great, no cost to me!
2. I can put in an additional 3% of my salary and the hospital will match it—cool!
3. So, up to 10% of my salary can be put into my retirement plan—seems like a LOT!
4. Any contributions I make are tax-deductible which reduces my income taxes—all right!
5. I can't get money out of my plan for many years—ugh, I don't know about this . . .

After thinking about it for a few days, Samuel decided to contribute 1% of his salary into the plan every month, so that he would at least get a matching 1% from the hospital in addition to the 4% the hospital automatically put in. As a result, 6% of his salary would be deposited into his plan every month.

11.1 Annuities and Retirement Accounts

OBJECTIVES

1. Define the basic terms involved with annuities.
2. Find the amount of an annuity.
3. Find the amount of an annuity due.
4. Understand different retirement accounts.

CASE IN POINT The benefits coordinator at the hospital asked Darnell Johnson if he preferred an annuity paying a guaranteed interest or one invested in a mutual fund containing stocks. Darnell knew he had a lot to learn—they were talking about his family's future.

OBJECTIVE 1 Define the basic terms involved with annuities. Chapter 10 was about lump sums invested for a period of time. This chapter discusses an **annuity**, which is a series of payments made at regular intervals. Examples of an annuity are monthly payments on a home, payments by a company into an employee's retirement account, and monthly Social Security checks paid to a retired person.

Annuities can involve payments used to accumulate funds, such as regular payments by an employer into a retirement plan, or payments from an accumulated sum of money to a retired worker from that same retirement plan after he or she has retired. Here are some important terms.

> **Common terms used with annuities**
>
> An **ordinary annuity** is one in which payments are made at the end of each period such as at the end of each month or each quarter.
>
> A **payment period** is the time between payments.
>
> The **term** is the total time needed for all payments to be made.
>
> The **compound amount** or **future value** is the value of the annuity at some future date.

Managers at a firm decide to set up regular payments into an account (an annuity) so that they will have funds to replace a work truck when needed. They deposit $3000 AT THE END OF EACH YEAR for 6 years into an investment in bonds that earns 8% compounded annually. The first payment will only earn interest for 5 years (see figure). Use the compound interest table in Section 10.1 for 5 years and 8% to find the future value of the first payment:*

$$\$3000 \times 1.46933 = \$4407.99$$

The future value of the annuity is *the sum* of the compound amounts of all six payments. The annuity ends on the day of the last payment. Therefore, the last payment, which is made at the end of year 6, earns no interest.

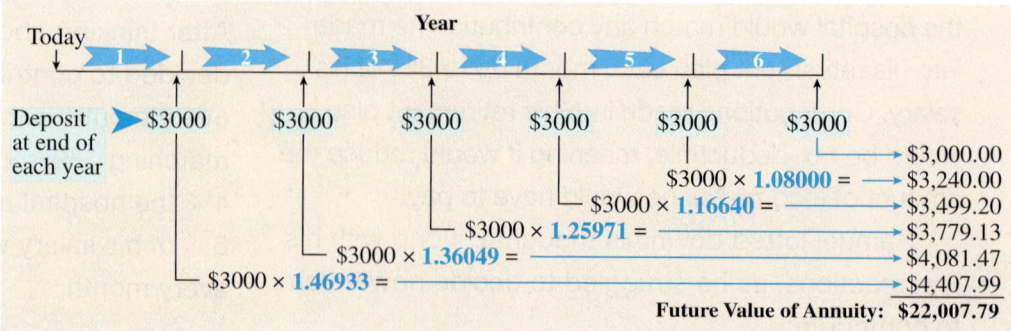

*Rather than using the tables in Section 10.1 to find the future value, you can use the compound amount formula: $M = P(1 + i)^n$. The first $3000 payment is compounded annually at 8% for 5 years, so $M = \$3000(1 + .08)^5 = \$4407.98$, which differs by one cent from the value found using the table due to rounding.

The future value of the annuity is $22,007.79, which will almost be enough to buy a new work pickup when needed. Find the total amount deposited in the annuity and interest earned as follows:

$$\text{Total deposits} = 6 \text{ years} \times \$3000 \text{ per year} = \mathbf{\$18,000}$$
$$\begin{aligned}\text{Interest earned} &= \text{Future value of annuity} - \textbf{Total deposits}\\&= \$22,007.79 \quad\quad - \quad \mathbf{\$18,000}\\&= \$4007.79\end{aligned}$$

The managers plan ahead so that interest earnings help pay a future cost.

OBJECTIVE 2 Find the amount of an annuity. The amount or future value of an annuity can also be found using the Amount of an Annuity Table given on page 410. The numbers in the table are the future value of an annuity with a regular payment of $1. Find the amount of an annuity with regular payments as follows.

> **Finding Amount of an Annuity**
>
> Amount = Payment × Number from amount of an annuity table

As a check, reconsider the annuity of $3000 at the end of each year for 6 years at 8% compounded annually. Locate 8% at the top of the Amount of an Annuity Table and 6 periods in the far left column to find **7.33593.**

$$\text{Amount} = \$3000 \times \mathbf{7.33593} = \$22{,}007.79$$

This amount is identical to the amount calculated earlier, but sometimes the values found using the table differ slightly from those found using a calculator.

Finding the Value of an Annuity and Interest Earned

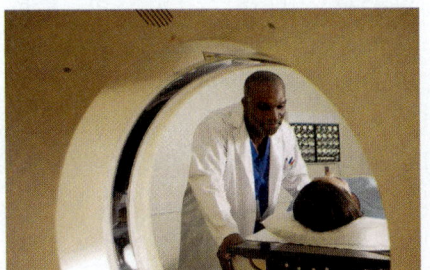

EXAMPLE 1 Darnell Johnson's employer contributes 4% of Darnell's $38,400 salary into his retirement plan. Additionally, the hospital matches the 1% of his salary that Darnell himself puts into his own plan. Find the future value in 10 years assuming (a) funds earn 5% compounded semiannually and (b) funds earn 8% compounded semiannually. (c) Then find the difference between the two. (*Hint:* Assume all contributions into the plan occur at the end of semiannual periods.)

CASE IN POINT

SOLUTION
First, find the amount put into the retirement plan at the end of every semiannual period.
Salary per semiannual period = $38,400 ÷ 2 = $19,200
Total semiannual contribution = Darnell's contribution + **Employer's contribution**

$$= .01 \times \$19{,}200 + \mathbf{.05 \times \$19{,}200}$$
$$= \$1152 \quad \textbf{invested at end of every six-months}$$

(a) Interest of $\frac{5\%}{2} = 2\frac{1}{2}\%$ per semiannual period for $10 \times 2 = 20$ six-month periods. Look across the top of the table for 2.5% and down the side for 20 periods to find **25.54466**.

$$\text{Amount} = \$1152 \times \mathbf{25.54466} = \mathbf{\$29{,}427.45}$$

(b) Interest of $\frac{8\%}{2} = 4\%$ per semiannual period for $10 \times 2 = 20$ six-month periods. Look across the top of the table for 4% and down the side for 20 periods to find **29.77808**.

$$\text{Amount} = \$1152 \times 29.77808 = \mathbf{\$34{,}304.35}$$

(c) Difference = $34,304.35 − $29,427.45 = **$4876.90**

> **QUICK CHECK 1**
>
> At the end of every quarter, $2000 is put into a retirement plan that earns 6% compounded quarterly. Find the future value in 5 years.

Amount of an Annuity Table

Period n	1%	1½%	2%	2½%	3%	4%	5%	6%	8%	10%	12%	Period n
1	1.00000	1.00000	1.00000	1.00000	1.00000	1.00000	1.00000	1.00000	1.00000	1.00000	1.00000	1
2	2.01000	2.01500	2.02000	2.02500	2.03000	2.04000	2.05000	2.06000	2.08000	2.10000	2.12000	2
3	3.03010	3.04522	3.06040	3.07562	3.09090	3.12160	3.15250	3.18360	3.24640	3.31000	3.37440	3
4	4.06040	4.09090	4.12161	4.15252	4.18363	4.24646	4.31013	4.37462	4.50611	4.64100	4.77933	4
5	5.10101	5.15227	5.20404	5.25633	5.30914	5.41632	5.52563	5.63709	5.86660	6.10510	6.35285	5
6	6.15202	6.22955	6.30812	6.38774	6.46841	6.63298	6.80191	6.97532	7.33593	7.71561	8.11519	6
7	7.21354	7.32299	7.43428	7.54743	7.66246	7.89829	8.14201	8.39384	8.92280	9.48717	10.08901	7
8	8.28567	8.43284	8.58297	8.73612	8.89234	9.21423	9.54911	9.89747	10.63663	11.43589	12.29969	8
9	9.36853	9.55933	9.75463	9.95452	10.15911	10.58280	11.02656	11.49132	12.48756	13.57948	14.77566	9
10	10.46221	10.70272	10.94972	11.20338	11.46388	12.00611	12.57789	13.18079	14.48656	15.93742	17.54874	10
11	11.56683	11.86326	12.16872	12.48347	12.80780	13.48635	14.20679	14.97164	16.64549	18.53117	20.65458	11
12	12.68250	13.04121	13.41209	13.79555	14.19203	15.02581	15.91713	16.86994	18.97713	21.38428	24.13313	12
13	13.80933	14.23683	14.68033	15.14044	15.61779	16.62684	17.71298	18.88214	21.49530	24.52271	28.02911	13
14	14.94742	15.45038	15.97394	16.51895	17.08632	18.29191	19.59863	21.01507	24.21492	27.97498	32.39260	14
15	16.09690	16.68214	17.29342	17.93193	18.59891	20.02359	21.57856	23.27597	27.15211	31.77248	37.27971	15
16	17.25786	17.93237	18.63929	19.38022	20.15688	21.82453	23.65749	25.67253	30.32428	35.94973	42.75328	16
17	18.43044	19.20136	20.01207	20.86473	21.76159	23.69751	25.84037	28.21288	33.75023	40.54470	48.88367	17
18	19.61475	20.48938	21.41231	22.38635	23.41444	25.64541	28.13238	30.90565	37.45024	45.59917	55.74971	18
19	20.81090	21.79672	22.84056	23.94601	25.11687	27.67123	30.53900	33.75999	41.44626	51.15909	63.43968	19
20	22.01900	23.12367	24.29737	25.54466	26.87037	29.77808	33.06595	36.78559	45.76196	57.27500	72.05244	20
21	23.23919	24.47052	25.78332	27.18327	28.67649	31.96920	35.71925	39.99273	50.42292	64.00250	81.69874	21
22	24.47159	25.83758	27.29898	28.86286	30.53678	34.24797	38.50521	43.39229	55.45676	71.40275	92.50258	22
23	25.71630	27.22514	28.84496	30.58443	32.45288	36.61789	41.43048	46.99583	60.89330	79.54302	104.60289	23
24	26.97346	28.63352	30.42186	32.34904	34.42647	39.08260	44.50200	50.81558	66.76476	88.49733	118.15524	24
25	28.24320	30.06302	32.03030	34.15776	36.45926	41.64591	47.72710	54.86451	73.10594	98.34706	133.33387	25
26	29.52563	31.51397	33.67091	36.01171	38.55304	44.31174	51.11345	59.15638	79.95442	109.18177	150.33393	26
27	30.82089	32.98668	35.34432	37.91200	40.70963	47.08421	54.66913	63.70577	87.35077	121.09994	169.37401	27
28	32.12910	34.48148	37.05121	39.85980	42.93092	49.96758	58.40258	68.52811	95.33883	134.20994	190.69889	28
29	33.45039	35.99870	38.79223	41.85630	45.21885	52.96629	62.32271	73.63980	103.96594	148.63093	214.58275	29
30	34.78489	37.53868	40.56808	43.90270	47.57542	56.08494	66.43885	79.05819	113.28321	164.49402	241.33268	30
31	36.13274	39.10176	42.37944	46.00027	50.00268	59.32834	70.76079	84.80168	123.34587	181.94342	271.29261	31
32	37.86901	40.68829	44.22703	48.15028	52.50276	62.70147	75.29883	90.88978	134.21354	201.13777	304.84772	32
33	38.86901	42.29861	46.11157	50.35403	55.07784	66.20953	80.06377	97.34316	145.95062	222.25154	342.42945	33
34	40.25770	43.93309	48.03380	52.61289	57.73018	69.85791	85.06696	104.18375	158.62667	245.47670	384.52098	34
35	41.66028	45.59209	49.99448	54.92821	60.46208	73.65222	90.32031	111.43478	172.31680	271.02437	431.66350	35

Finding the Amount of an Annuity and Interest Earned

EXAMPLE 2 At the birth of her grandson, Samantha Jones commits to help pay for his college education. She decides to make deposits of $600 at the end of each 6 months into an account for 17 years. Find the **amount of the annuity** and the interest earned, assuming 6% compounded semiannually.

SOLUTION

Interest of $\frac{6\%}{2} = 3\%$ is earned each semiannual period. There are $17 \times 2 = 34$ semiannual periods in 17 years. Find 3% across the top and 34 periods down the side of the table for **57.73018**.

$$\text{Amount} = \$600 \times \mathbf{57.73018} = \$34{,}638.11$$
$$\text{Interest} = \$34{,}638.11 - (\mathbf{34 \times \$600}) = \mathbf{\$14{,}238.11}$$

Jones knows that a college education will cost a lot more in 17 years than it does now, but she also knows that $34,638.11 will be of great help to her grandson.

Financial Calculator Solution

In this example, payment ($600), interest rate per compounding period (3%), and number of compounding periods (34) are known. Future value is the unknown. As described in the appendix on financial calculators, enter the payment as a negative number since it is an outflow of cash that Samantha Jones pays each month. Finally, press the $\boxed{FV}$ key to find the future value, which is a positive value since it will be an inflow of cash to her grandson.

$-600\ \boxed{PMT}\ 3\ \boxed{i}\ 34\ \boxed{n}\ \boxed{FV}\ 34638.11$

QUICK CHECK 2

Bob Nelson deposits $250 into a retirement account at the end of every month for 30 months. The fund holds international stocks and Nelson optimistically thinks it may yield 12% compounded monthly. Find the future amount.

Quick TIP
Payments are made at the BEGINNING OF EACH PERIOD in an ANNUITY DUE.

OBJECTIVE 3 Find the amount of an annuity due. Payments were made at the *end of each period* in the ordinary annuities discussed previously. In contrast, an annuity in which payments are made at the *beginning of each time period* is called an **annuity due**.

Finding the Amount of an Annuity Due

Step 1 Add 1 to the number of periods.
Step 2 Find: Amount = Payment × Number from amount of an annuity table.
Step 3 Subtract 1 payment.

Finding the Amount of an Annuity Due

EXAMPLE 3 Mr. and Mrs. Thompson set up an investment program using an *annuity due* with payments of $500 at the *beginning of each quarter*. Find **(a)** the amount of the annuity and **(b)** the interest if they make payments for 7 years into an investment expected to pay 8% compounded quarterly.

SOLUTION

Quick TIP
For an annuity due, be sure to add 1 period to the number of compounding periods and subtract 1 payment from the amount calculated.

(a) Step 1 There are 7 × 4 = 28 periods in 7 years. Since it is an annuity due, first add 1 to the 28 periods, making 29 periods.

Step 2 Interest of $\frac{8\%}{4}$ = 2% is earned each quarter. Look across the top of the table for 2% and down the side for 29 periods to find **38.79223**.

$$\$500 \times 38.79223 = \$19{,}396.12$$

Step 3 Now subtract one payment to find the amount of the annuity due.

Amount of annuity due = **$19,396.12** − $500 = $18,896.12

(b) Subtract the 28 payments (7 years × 4 payments per year) of $500 each to find the interest.

Interest = $18,896.12 − (28 × $500) = $4896.12

The calculator solution to finding the interest in part **(b)** follows.

18896.12 $\boxed{-}$ 28 $\boxed{\times}$ 500 $\boxed{=}$ 4896.12

Note: Refer to Appendix B for calculator basics.

QUICK CHECK 3

If $1000 is deposited at the beginning of every six months into an account that earns 5% compounded semiannually, find the amount after 8 years.

OBJECTIVE 4 Understand different retirement accounts. Many students see retirement as something very far away and mistakenly think they can wait until they are much older (say 50—ouch, that's old!) to worry about it. However, it takes a long time to save enough to retire and Social Security payments are not enough for most retirees. The clipping shows what can happen to someone who has not saved enough.

Retired Without Enough Assets . . .

After paying off the mortgage on his small home and saving some money, Tom Wheat retired at 65. He is 74 now and receives $1028 each month from Social Security after deductions are taken out for Medicare.

His health issues related to diabetes have gotten much worse, and he can no longer work. He struggles financially every month to get by and is worried. He may have to sell his home—but where will he live? He wishes he had saved more and even worked several more years before retiring.

Given that it takes years to accumulate a large amount of money, it is a good idea to start saving as soon as possible. Here are a few great features about retirement plans:

1. many firms make contributions into your plan at no cost to you;
2. your contributions are tax deductible;
3. funds in a retirement plan grow income-tax free; and
4. funds in a retirement plan can be withdrawn early in the event of emergencies (but there may be a penalty).

> Retirement plans are for everyone, *especially* recent college graduates.

The federal government sponsors a retirement and disability program for workers called Social Security. You may have noticed payroll deductions for FICA, which fund the program that will eventually pay your Social Security benefits.

Many employers have either **401(k)** or **403(b)** retirement plans into which they make regular contributions on behalf of each employee. The 401(k) plans are for those working in the private sector and the 403(b) are for various government and tax-exempt organizations. Employee contributions to either of these plans may be deductible and all funds grow income-tax free until funds are withdrawn.

Individuals who are not covered by an employer-sponsored plan may be eligible to contribute to an **individual retirement account (IRA)**. There are two types of IRAs:

> **Regular IRA** – Deposits are usually tax deductible and grow income-tax free. Funds are subject to income taxes when withdrawn.
>
> **Roth IRA** – Deposits are not tax deductible, but grow income-tax free. Funds are NOT subject to income taxes when withdrawn.

Quick TIP
Generally you must wait until you are $59\frac{1}{2}$ to withdraw funds from either type of IRA with no penalty. However, there are exceptions, resulting in penalty-free withdrawals for first-time home buyers, college costs, medical costs, etc.

Many individuals set up IRAs at firms such as Schwab, Fidelity, or Vanguard and manage the funds themselves. The maximum allowable contribution into either type of IRA is $5500 if you are under 50, or $6500 if you are 50 or older.

WHICH IRA IS BEST?	REGULAR IRA	ROTH IRA
Tax deductible?	If you qualify	No
Taxable at withdrawal?	Yes	No
Penalty for early withdrawal?	Yes, prior to age 59½	Yes
Mandatory withdrawal age?	70½	None

There are numerous rules related to each type of IRA. Research carefully before deciding which is best for you.

EXAMPLE 4 Finding the Value of an IRA

At 27, Joann Gretz sets up an IRA with online broker Charles Schwab, where she plans to deposit $2000 at the end of each year until age 60. Find the amount of the annuity if she invests in (a) a bond fund that has historically yielded 6% compounded annually versus (b) a stock fund that has historically yielded 10% compounded annually. Assume that future yields equal historical yields.

SOLUTION

Age 60 is 60 − 27 = **33 years away**, so she will make deposits at the end of each year for 33 years.

(a) Bond fund: Look down the left column of the amount of an annuity table on page 410 for 33 years and across the top for 6% to find **97.34316**.

$$\text{Amount} = \$2000 \times 97.34316 = \$194{,}686.32$$

(b) Stock fund: Look down the left column of the table for 33 years and across the top for 10% to find **222.25154**.

$$\text{Amount} = \$2000 \times 222.25154 = \$444{,}503.08$$

The differences in the two investments are shown in the figure. Gretz wants the larger amount, but she is worried she might lose money in the stock fund. See *Exercise 20* at the end of this section to find her investment choice.

Quick TIP
There are risks when investing. For example, stocks usually increase in value over the long term, but they can fall sharply in a recession.

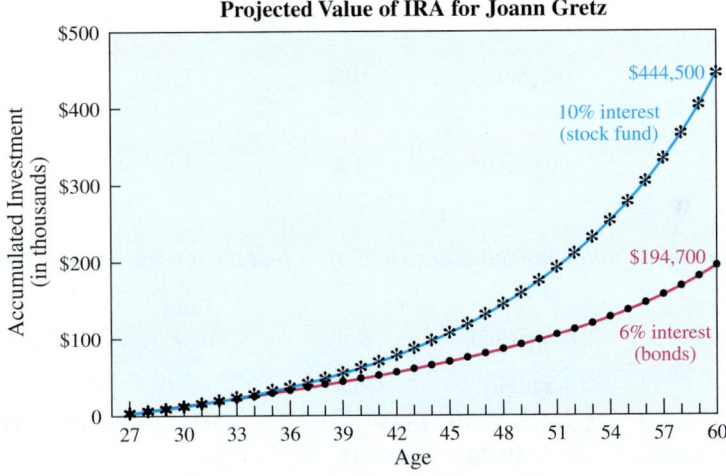
Projected Value of IRA for Joann Gretz

Quick TIP
Roughly 70% of families nearing retirement have saved less than $50,000. Most of these people will struggle during their retirement years, since Social Security will pay only part of what they need.

QUICK CHECK 4

Bill James plans to deposit $2500 in a regular IRA at the end of each six-month period for the next 17 years until he retires. Find the future value if funds earn (a) 5% compounded semiannually and (b) 8% compounded semiannually.

About 60 million people benefit every month from an annuity of payments from Social Security. Workers can retire as early as age 62 and get a smaller monthly payment. However, it is almost always better to work until age 68 or 70 to retire, when your Social Security checks are much larger. The bar graph shows the importance of Social Security.

Quick TIP
Work until you are 70 to maximize Social Security payments. Retiring before 70 may result in too little income in a few years.

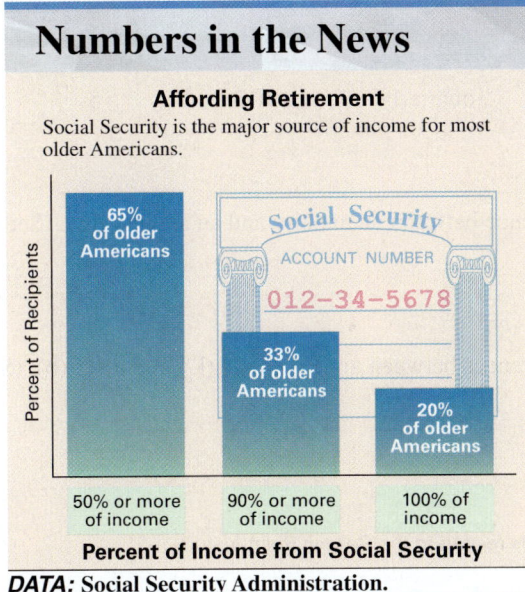
Numbers in the News
Affording Retirement
Social Security is the major source of income for most older Americans.
DATA: Social Security Administration.

11.1 Exercises // MyLab Math

The shaded sections below contain solutions to help you get a QUICK START on the various types of exercises.

Find the amount of the following ordinary annuities, then find the interest earned. (See Examples 1 and 2.)

	Amount of Each Deposit	Deposited	Rate	Time (Years)	Amount of Annuity	Interest Earned
1.	$900	annually	5%	18	$25,319.14	$9119.14

$900 × 28.13238 = $25,319.14; *I* = $25,319.14 − (18 × $900) = $9119.14

2.	$2900	annually	8%	5	$17,013.14	$2513.14

$2900 × 5.86660 = $17,013.14; *I* = $17,013.14 − (5 × $2900) = $2513.14

3.	$7500	semiannually	6%	10	_____	_____
4.	$9200	semiannually	8%	5	_____	_____
5.	$3500	quarterly	10%	7	_____	_____
6.	$6900	quarterly	4%	4	_____	_____

Find the amount of the following annuities DUE, then find the interest earned. (See Example 3.)

	Amount of Each Deposit	Deposited	Rate	Time (Years)	Amount of Annuity	Interest Earned
7.	$1200	annually	8%	5	$7603.12	$1603.12

Look up 8%, 5 + 1 = 6 periods, finding 7.33593. $1200 × 7.33593 − $1200 = $7603.12
Interest = $7603.12 − 5 × $1200 = $1603.12

8.	$400	annually	6%	6	$2957.54	$557.54

Look up 6%, 6 + 1 = 7 periods, finding 8.39384. $400 × 8.39384 − $400 = $2957.54
Interest = $2957.54 − 6 × $400 = $557.54

9.	$9500	semiannually	4%	9	_____	_____
10.	$1800	semiannually	5%	6	_____	_____
11.	$3800	quarterly	8%	3	_____	_____
12.	$10,200	quarterly	10%	5	_____	_____

13. Explain the difference between an annuity and an annuity due. (See Objectives 2 and 3.)

14. Describe the differences between an IRA, a 401(k), and a 403(b). (See Objective 4.)

C// indicates an exercise that is related to the Case in Point feature.

Solve the following application problems.

15. RETIREMENT PLANNING Darnell Johnson wants to know if he can retire in 35 years at age 60, when he plans to do a lot of fishing. Assume the deposit into his retirement account averages $3800 at the end of each year and that the funds earn 6% per year. Find **(a)** the amount of the annuity and **(b)** the interest earned.
(a) Amount = $3800 × 111.43478 = $423,452.16
(b) Interest = $423,452.16 − 35 × $3800 = $290,452.16

(a) $423,452.16
(b) $290,452.16

16. SAVING FOR A HOME Jim and Betty Collins need an additional $6500 for a down payment on a home they hope to buy in 2 years. They invest $800 at the end of each quarter in an account earning 6% compounded quarterly. Find **(a)** the amount of the annuity and **(b)** the interest earned.

(a) _____
(b) _____

17. CHILDCARE PAYMENTS Monique Chaney places $250 of her quarterly child support check into an annuity for the education of her child. She does this at the beginning of each quarter for 8 years into an account paying 8% per year, compounded quarterly. Find **(a)** the amount of the annuity and **(b)** the interest earned.

(a) _____
(b) _____

18. RETIREMENT Jason Horton works for Chevron as a welder on offshore drilling rigs. His retirement plan contributions are $3800 at the beginning of each 6-month period. Assume that the account grows at 6% compounded semiannually for 15 years. Find the **(a)** future value of the annuity and **(b)** the interest earned.

(a) _____
(b) _____

19. MUTUAL FUND INVESTING Sandra Gonzales deposits $1000 into a mutual fund containing international stocks at the end of each semiannual period for 12 years. Assume the fund earns 10% interest compounded semiannually and find the future value.

19. _____

20. T-BILL AND STOCK INVESTING Joann Gretz (see Example 4, page 413) decides to place half of her $2000 deposit at the end of each year into the bond fund and half into the stock fund. Assume the bond fund earns 6% compounded annually and the stock fund earns 10% compounded annually. Find the amount available in 33 years.

20. _____

QUICK CHECK ANSWERS

1. $46,247.34
2. $8696.22
3. $19,864.73
4. (a) $131,532.23 (b) $174,644.78

11.2 Present Value of an Ordinary Annuity

OBJECTIVES

1. Define the present value of an ordinary annuity.
2. Use the formula to find the present value of an ordinary annuity.
3. Find the equivalent cash price of an ordinary annuity.

OBJECTIVE 1 Define the present value of an ordinary annuity. This chapter is about annuities that involve regular payments. In Section 11.1, you learned how to find the future value of an annuity when payments were accumulating. In this section, you will learn how to find the present value of an annuity. There are two ways to think about the present value of an annuity: one replaces a stream of payments going into a fund that is growing in value, and the other replaces a stream of payments to pay a debt.

> **Two ways to think about the present value of an annuity**
>
> 1. *A single lump-sum deposit that results in the same future value as making regular payments for a specific amount of time.* For example, say a firm needs $100,000 in 5 years. It can achieve that goal either by making a single, lump-sum deposit (the present value) or by making regular payments (see Example 1).
> 2. *The lump sum needed today to fund all regular payments that need to be made.* For example, a judge requires a divorced man to make a $1500 child support payment at the end of each quarter until his son turns 18. This can be achieved by depositing a lump sum today (the present value) that will generate the necessary payments. (See Example 2.)

OBJECTIVE 2 Use the formula to find the present value of an ordinary annuity. The present value of an annuity with periodic payments at the end of each period is found using values from the table.

> **Finding Present Value of an Annuity**
>
> $$\text{Present value of annuity} = \text{Payment} \times \text{Number from the present value of an annuity table}$$

Finding the Present Value of an Annuity

EXAMPLE 1 The Daily News plans to accumulate funds ahead of time to purchase a computer network. It deposits $4325 into an account at the end of each quarter for 5 years. The account pays 6% compounded quarterly. **(a)** Find the lump sum that must be deposited today to accumulate the funds needed (the present value). **(b)** Use the concepts of Section 11.1 to find the future value of the annuity.

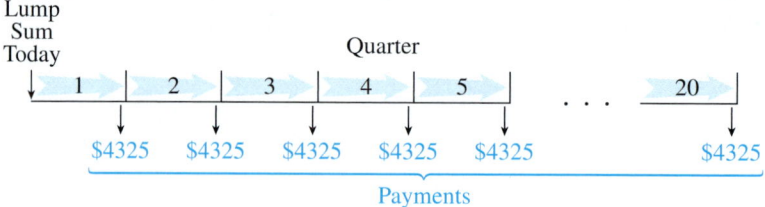

SOLUTION

$\frac{6\%}{4} = 1.5\%$ per quarter and 5 years $\times$ 4 quarters per year $= 20$ quarters

(a) To find the lump sum (present value) of the payments, use the present value of an annuity table with 1.5% per period and 20 periods to find **17.16864**.

$$\text{Present value} = \$4325 \times 17.16864 = \$74{,}254.37$$

(b) To find the future value, use the amount of an annuity table with a payment of $4325 for 20 periods in Section 11.1 to find **23.12367**.

$$\text{Future value} = \$4325 \times 23.12367 = \$100{,}009.87$$

The future value of $100,009.87 can be achieved with a single lump-sum deposit of $74,254.37 today, OR by making 20 end-of-quarter payments of $4325 into a fund. In both cases, we assume funds grow at 6% compounded quarterly. These two possibilities are shown by the two bar graphs on the next page.

11.2 Present Value of an Ordinary Annuity

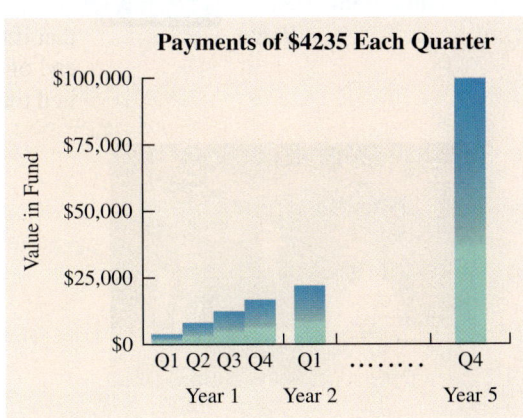

Actually, all three of the following have the exact same value assuming 6% compounded quarterly and ignoring income taxes.

1. 20 end-of-quarter deposits of $4325
2. A future value at the end of 5 years of $100,009.87
3. A present value on hand today of $74,254.37

QUICK CHECK 1

Walter and Beth Bates save $1200 at the end of each quarter for 7 years. Assume 8% compounded quarterly and find both **(a)** the future value and **(b)** the present value.

The three items listed at the end of the solution to Example 1 reinforce a very important idea in finance. The value of money depends on when it is available. Generally, money available today is worth more than money available in (say) 10 years. Lenders and investors use this idea every day to create profit. Suppose a bank lends you money today. It expects the interest on the debt to be enough to pay expenses, make up for bad loans, and also make a profit. So, you have to pay the bank back more than you borrowed.

Finding the Present Value **EXAMPLE 2** Tom and Brandy Barrett recently divorced. The judge gave custody of their 8 year-old son to Brandy and ruled that Tom must pay $1500 in child support to Brandy at the end of each quarter until the son turns 16. Find the lump sum that Tom must put into an account earning 6% compounded quarterly to cover the periodic payments. Find the interest earned.

SOLUTION

Payments must be made for $16 - 8 = 8$ years, or for $8 \times 4 = 32$ quarters. The interest rate per quarter is $\frac{6\%}{4} = 1.5\%$ per quarter. Look across the top of the Present Value of an Annuity Table for 1.5% and down the side for 32 payments to find 25.267145.

Quick TIP
Although the $1500 withdrawals to Brandy are at the end of each quarter, the original lump sum must be deposited at the beginning of the first year.

Present value of annuity = $1500 × 25.26714 = **$37,900.71**

Assuming 6% compounded quarterly, a deposit of $37,900.71 today is enough money to make 32 end-of-quarter payments of $1500 each. Interest earned is the sum of all payments less the original lump sum.

Interest = (32 × $1500) − $37,900.71 = **$10,099.29**

QUICK CHECK 2

Find the lump sum that must be set aside today to make end-of-month payments of $1000 for 2 years, assuming 12% compounded monthly. Then find the interest earned.

Finding the Present Value

EXAMPLE 3 An American company hires a project manager to work in Saudi Arabia. The contract states that if the manager works there for 5 years, he will receive an extra benefit of $15,000 at the end of each semiannual period for the 8 years that follow. Find the lump sum that can be deposited today to satisfy the contract, assuming 6% compounded semiannually.

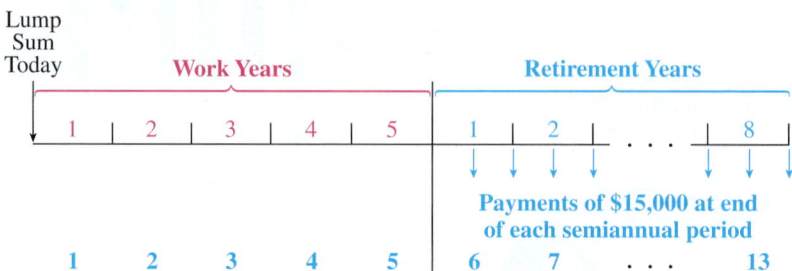

SOLUTION

The project manager works from years 1 to 5. He then receives two $15,000 annuity payments each year during years 6 through 13. Solve this problem by first finding the present value of the $15,000 payments at the moment the project manager leaves. Then find the present value of the amount needed today.

1. **Find the present value of the $15,000 payments at the moment manager leaves.**
   Use $\frac{6\%}{2} = 3\%$ per compounding period and $8 \times 2 = 16$ compounding periods to find **12.56110** in the present value of an annuity table.

   Present value of annuity = $15,000 × **12.56110** = $188,416.50

   This is the *present value* of the annuity needed at the beginning of year 6 to fund payments in years 6 through 13. But it is also the *future value* needed for the investment made today that will fund the eventual payments.

2. **Find the future value needed today to accumulate the $188,416.50 by the end of year 5.**
   Use the table showing present value of a dollar in Section 10.3 (page 395) with $\frac{6\%}{2} = 3\%$ per compounding period and $5 \times 2 = 10$ compounding periods to find **.74409**.

   Present value needed today = $188,416.50 × **.74409** = $140,198.83

A lump sum of $140,198.83 today will grow to $188,416.50 in 5 years. The $188,416.50 at the end of year 5 is enough to make 16 semiannual payments of $15,000 each during years 6 through 13.

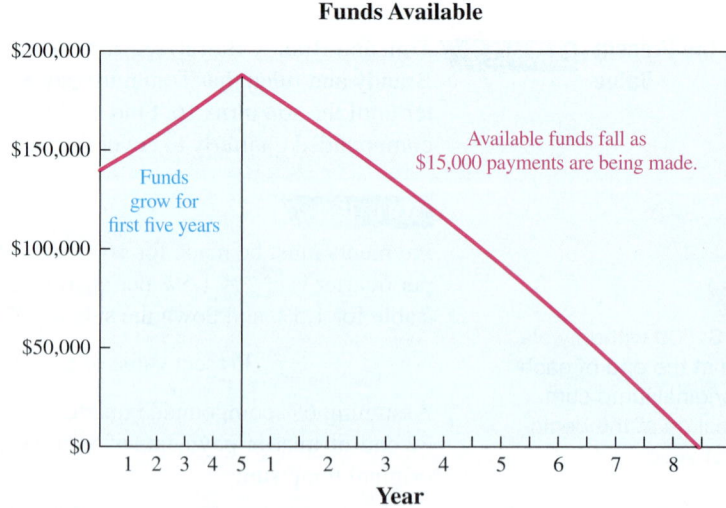

11.2 Present Value of an Ordinary Annuity

> **QUICK CHECK 3**
> A project manager signs a contract that will pay him a bonus of $20,000 at the end of each year for 5 years, beginning in 4 years. Assuming 4% per year, find the amount that must be set aside today to fund this benefit.

Present Value of an Annuity Table

Interest Rate per Period

Period	1%	1½%	2%	2½%	3%	4%	5%	6%	8%	10%	Period
1	.99010	.98522	.98039	.97561	.97087	.96154	.95238	.94340	.92593	.90909	1
2	1.97040	1.95588	1.94156	1.92742	1.91347	1.88609	1.85941	1.83339	1.78326	1.73554	2
3	2.94099	2.91220	2.88388	2.85602	2.82861	2.77509	2.72325	2.67301	2.57710	2.48685	3
4	3.90197	3.85438	3.80773	3.76197	3.71710	3.62990	3.54595	3.46511	3.31213	3.16987	4
5	4.85343	4.78264	4.71346	4.64583	4.57971	4.45182	4.32948	4.21236	3.99271	3.79079	5
6	5.79548	5.69719	5.60143	5.50813	5.41719	5.24214	5.07569	4.91732	4.62288	4.35526	6
7	6.72819	6.59821	6.47199	6.34939	6.23028	6.00205	5.78637	5.58238	5.20637	4.86842	7
8	7.65168	7.48593	7.32548	7.17014	7.01969	6.73274	6.46321	6.20979	5.74664	5.33493	8
9	8.56602	8.36052	8.16224	7.97087	7.78611	7.43533	7.10782	6.80169	6.24689	5.75902	9
10	9.47130	9.22218	8.98259	8.75206	8.53020	8.11090	7.72173	7.36009	6.71008	6.14457	10
11	10.36763	10.07112	9.78685	9.51421	9.25262	8.76048	8.30641	7.88687	7.13896	6.49506	11
12	11.25508	10.90751	10.57534	10.25776	9.95400	9.38507	8.86325	8.38384	7.53608	6.81369	12
13	12.13374	11.73153	11.34837	10.98318	10.63496	9.98565	9.39357	8.85268	7.90378	7.10336	13
14	13.00370	12.54338	12.10625	11.69091	11.29607	10.56312	9.89864	9.29498	8.24424	7.36669	14
15	13.86505	13.34323	12.84926	12.38138	11.93794	11.11839	10.37966	9.71225	8.55948	7.60608	15
16	14.71787	14.13126	13.57771	13.05500	12.56110	11.65230	10.83777	10.10590	8.85137	7.82371	16
17	15.56225	14.90765	14.29187	13.71220	13.16612	12.16567	11.27407	10.47726	9.12164	8.02155	17
18	16.39827	15.67256	14.99203	14.35336	13.75351	12.65930	11.68959	10.82760	9.37189	8.20141	18
19	17.22601	16.42617	15.67846	14.97889	14.32380	13.13394	12.08532	11.15812	9.60360	8.36492	19
20	18.04555	17.16864	16.35143	15.58916	14.87747	13.59033	12.46221	11.46992	9.81815	8.51356	20
21	18.85698	17.90014	17.01121	16.18455	15.41502	14.02916	12.82115	11.76408	10.01680	8.64869	21
22	19.66038	18.62082	17.65805	16.76541	15.93692	14.45112	13.16300	12.04158	10.20074	8.77154	22
23	20.45582	19.33086	18.29220	17.33211	16.44361	14.85684	13.48857	12.30338	10.37106	8.88322	23
24	21.24339	20.03041	18.91393	17.88499	16.93554	15.24696	13.79864	12.55036	10.52876	8.98474	24
25	22.02316	20.71961	19.52346	18.42438	17.41315	15.62208	14.09394	12.78336	10.67478	9.07704	25
26	22.79520	21.39863	20.12104	18.95061	17.87684	15.98277	14.37519	13.00317	10.80998	9.16095	26
27	23.55961	22.06762	20.70690	19.46401	18.32703	16.32959	14.64303	13.21053	10.93516	9.23722	27
28	24.31644	22.72672	21.28127	19.96489	18.76411	16.66306	14.89813	13.40616	11.05108	9.30657	28
29	25.06579	23.37608	21.84438	20.45355	19.18845	16.98371	15.14107	13.59072	11.15841	9.36961	29
30	25.80771	24.01584	22.39646	20.93029	19.60044	17.29203	15.37245	13.76483	11.25778	9.42691	30
31	26.54229	24.64615	22.93770	21.39541	20.00043	17.58849	15.59281	13.92909	11.34980	9.47901	31
32	27.26959	25.26714	23.46833	21.84918	20.38877	17.87355	15.80268	14.08404	11.43500	9.52638	32
33	27.98969	25.87895	23.98856	22.29188	20.76579	18.14765	16.00255	14.23023	11.51389	9.56943	33
34	28.70267	26.48173	24.49859	22.72379	21.13184	18.41120	16.19290	14.36814	11.58693	9.60857	34
35	29.40858	27.07559	24.99862	23.14516	21.48722	18.66461	16.37419	14.49825	11.65457	9.64416	35

Planning for Retirement

EXAMPLE 4 Tish Baker plans to retire from nursing at age 65 and hopes to withdraw $25,000 per year from her retirement plan until she is 90. **(a)** If money earns 8% per year compounded annually, how much will she need at age 65? **(b)** If she deposits $2000 per year into her retirement plan beginning at age 32, and if the retirement plan earns 8% per year compounded annually, will her retirement account have enough for her to meet her goals?

SOLUTION

(a) The amount needed at age 65 is the present value of an annuity of $25,000 per year for $90 - 65 = 25$ years with interest of 8% compounded annually. The present value of an annuity table is used to find the following.

$$\text{Present value} = \$25{,}000 \times 10.67478 = \$266{,}869.50$$

Baker will need $266,869.50 at age 65. This sum, at 8% compounded annually, will permit withdrawals of $25,000 per year until age 90.

(b) Baker makes payments of $2000 at the end of each year for $65 - 32 = 33$ years, at 8% compounded annually. These payments form a regular annuity. The amount of an annuity table in Section 11.1 is used to find the following.

$$\text{Future value} = \$2000 \times 145.95062 = \$291{,}901.24$$

The $291,901.24 in the retirement account at age 65 is slightly more than enough to fund 25 yearly withdrawals of $25,000 each. To have a chance of earning 8% per year, Tish would have to invest in places other than bank accounts, such as stocks and bonds where there are no guaranteed returns.

In addition, $25,000 per year will not be very much money when Tish turns 65, due to the probable 33 years' worth of inflation. So she will need quite a bit more than $266,869.50 to retire at 65. On the other hand, during those 33 years, her salary will have increased a lot due to pay raises, enabling her to save more.

> Still, saving early in life is VERY IMPORTANT since it allows compound interest to WORK FOR YOU over many years!

Financial Calculator Solution This problem can readily be solved using a financial calculator, although the numbers differ very slightly from those above due to rounding errors.

(a) The present value (PV) is unknown. The interest rate (*i*) is 8%, the payment (PMT) is $25,000, and the number of years (*n*) is 25. Enter the payment as a positive number since it is an inflow of cash to Tish Baker.

25000 [PMT] 8 [*i*] 25 [*n*] [PV] −266,869.40   (rounded)

(b) The unknown is the future value (FV). The interest rate (*i*) is 8%, the payment (PMT) is $2000, and the number of years (*n*) is 33. Enter the payment as a negative number since it is a cash outflow to be paid by Tish Baker.

−2000 [PMT] 8 [*i*] 33 [*n*] [FV] $291,901.24

Since the projected future value of $291,901.24 is more than the needed amount of $266,869.40, Baker should have enough.

QUICK CHECK 4

The Smiths want to plan for retirement using an annual income of $35,000 at the end of each year for 25 years based on a rate of 5% per year. **(a)** Find the present value needed when they retire to generate this income. **(b)** If they save $20,000 at the end of each year for 15 years, will they have enough?

OBJECTIVE 3 Find the equivalent cash price of an ordinary annuity. Payments made at different times cannot be compared directly to one another to see which is better. Rather, it is necessary to first find the present value of each payment (or series of payments) and compare these. The one that results in the larger present value is *the better of the two*. The present value that is equivalent to a payment (or series of payments) is called the **equivalent cash price**. Thus, the way to find the better deal is to compare the equivalent cash prices, which are the present values.

11.2 Present Value of an Ordinary Annuity

Comparing Methods of Investment

EXAMPLE 5 Jean Braddock offers some land to two different real estate developers. Kapton Homes offers $200,000 in cash today for the land. RealProperty offers $80,000 now as a down payment and payments of $10,000 at the end of each quarter for 4 years. Assume that money can be invested at 8% per year compounded quarterly. Which offer is better?

Which is the better offer?

SOLUTION

Since payments from the two real estate developers occur over different time periods, the present value of each offer must first be found to determine which is better.

The present value of Kapton Homes' offer is **$200,000**, since that payment is made now. The present value of RealProperty's offer is the sum of the down payment plus the present value of the series of payments. Use $\frac{8\%}{4} = 2\%$ per compounding period for 4 years $\times$ 4 quarters per year = 16 compounding periods. Use the present value of an annuity table to find **13.57771**.

Present value of payments = $10,000 \times$ **13.57771** = $135,777.10
Down payment + $80,000.00
Present value of RealProperty's offer **$215,777.10**

Therefore, $215,777.10 is the equivalent cash price to $80,000 down plus 16 quarterly payments of $10,000 each.

RealProperty's offer is the better of the two by the following amount.

$$\$215{,}777.10 - \$200{,}000 = \$15{,}777.10$$

QUICK CHECK 5

Ben James has two different offers for a lot: (1) $48,000 cash and (2) $12,000 down and $2659 per quarter for 12 quarters. Assuming 6% compounded quarterly, find the present value of both and determine the better of the two.

As shown in the next example, we can use the concepts in this section to estimate the present value of Social Security payments at the time of retirement. To do so, first simplify the problem by assuming that payments are made at the end of each year and that the payments remain constant. Note that increases in Social Security payments are due to inflation, so Social Security payments effectively remain constant in terms of buying power.

Finding the Present Value of Social Security Payments

EXAMPLE 6 On retiring at age 68, Shaunika Brown expects to receive $20,940 per year in Social Security payments. Assume that she receives these payments for 28 years and use a rate of 6% per year. Find the present value of her retirement payments.

SOLUTION

Use the present value of an annuity table with 28 years and 6% per year to find **13.40616**.

Present value = $20,940 $\times$ **13.40616** = $280,724.99

Her Social Security retirement payments have a present value of $280,724.99. In other words, the Social Security payments equate to having cash of $280,724.99 at the time of retirement.

Quick TIP

Some consider FICA withholding a cost, but it is really a way the government forces workers to save for their retirement (or disability).

QUICK CHECK 6

Since it will result in larger monthly checks, Ben Thomson works until age 70 before claiming Social Security benefits. Find the present value of his estimated $26,000 per year in payments assuming 5% per year and payments until his 85th birthday.

11.2 Exercises // MyLab Math

The shaded sections below contain solutions to help you get a QUICK START on the various types of exercises.

Find the present value of the following annuities. Round to the nearest cent. (See Examples 1–3.)

	Amount per Payment	Payment at End of Each	Time (Years)	Rate of Investment	Compounded	Present Value
1.	$1800	year	18	10%	annually	$14,762.54
	Present value = $1800 × 8.20141 = $14,762.54					
2.	$4100	year	7	6%	annually	$22,887.76
	Present value = $4100 × 5.58238 = $22,887.76					
3.	$2000	6 months	12	8%	semiannually	_____
4.	$1700	6 months	14	5%	semiannually	_____
5.	$894	quarter	6	4%	quarterly	_____
6.	$7500	quarter	5	10%	quarterly	_____

7. Explain the two different ways to think of the present value of an annuity. (See Objective 1.)

8. Explain the meaning of equivalent cash price. (See Objective 3.)

Solve the following application problems.

9. **INJURY LAWSUIT** The court ruled that Locan Corporation was liable in the death of an employee. The settlement called for the company to pay the employee's widow $65,000 at the end of each year for 20 years. Find the amount the company must set aside today, assuming 5% compounded annually.
   Present value of annuity = $65,000 × 12.46221 = $810,043.65

   9. $810,043.65

10. **COMPUTER REPLACEMENT** The hospital unit on which Darnell Johnson works sets aside an annual payment of $35,000 per year for 5 years so it will have funds to replace the personal computers, servers, and printers when needed. Assuming 5% compounded annually, what lump sum deposited today would result in the same future value?

    10. _____

11. **COLLEGE EXPENSES** In addition to his scholarship, Benjamin Wink needs $8000 every 6 months for living expenses and tuition at his university. It will take 5 years to complete his engineering degree. Assume funds earn 5% per year and find (a) the lump sum that must be deposited to meet this need and (b) the interest earned.

    (a) _____
    (b) _____

12. **DISASTER RELIEF** After a terrible cyclone in Bangladesh, an international disaster relief organization agreed to help support several families who lost everything with a payment of $25,000 every quarter for 5 years. Find (a) the lump sum that must be deposited to meet this need and (b) the interest earned assuming 6% per year, compounded quarterly.

    (a) _____
    (b) _____

C// indicates an exercise that is related to the Case in Point feature.

13. **PAYING FOR COLLEGE** Gabriel Martinez estimates that his daughter's college needs, beginning in 8 years, will be $3600 at the end of each quarter for 4 years. **(a)** Find the total amount needed in 8 years assuming 8% compounded quarterly. **(b)** Will he have enough money available in 8 years if he invests $700 at the end of each quarter for the next 8 years at 8% compounded quarterly?

(a) _____
(b) _____

14. **VAN PURCHASE** In 4 years, Jennifer Videtto will need to purchase a delivery van for her plumbing company. She estimates it will require a down payment of $10,000 with payments of $950 per month for 36 months. **(a)** Find the total amount needed in 4 years assuming 12% compounded monthly. **(b)** Will she have enough if she invests $2200 at the end of every quarter for 4 years and earns 6% compounded quarterly?

(a) _____
(b) _____

15. **SELLING A RESTAURANT** Anna Stanley has two offers for her pizza business. The first offer is a cash payment of $85,000, and the second is a down payment of $25,000 with payments of $3500 at the end of each quarter for 5 years. **(a)** Identify the better offer assuming 8% compounded quarterly. **(b)** Find the difference in the present values.

(a) _____
(b) _____

16. **GROCERY STORE** Adolf Hegman has two offers for his Canadian grocery company. The first offer is a cash payment of $540,000, and the second is a down payment of $240,000 with payments of $65,000 at the end of each semiannual period for 4 years. **(a)** Identify the better offer assuming 10% compounded semiannually. **(b)** Find the difference in the present values.

(a) _____
(b) _____

17. **SOCIAL SECURITY** Jessica Thames expects to receive $18,400 per year based on her deceased husband's contributions to Social Security. Assume that she receives payments for 14 years and a rate of 4% per year, and find the present value of this annuity.

17. _____

18. **SOCIAL SECURITY** Warren and Bernice White's combined Social Security payments add up to $35,400 per year. Assume payments for 20 years and a rate of 6% per year, and find the present value.

18. _____

QUICK CHECK ANSWERS

1. (a) $44,461.45 (b) $25,537.52
2. $21,243.39; $2756.61
3. $76,108.31
4. (a) $493,287.90
   (b) no, short by $61,716.70
5. $48,000; $41,003.07; $48,000 cash is the better offer.
6. $269,871.16

11.3 Sinking Funds (Finding Annuity Payments)

OBJECTIVES

1. Understand the basics of a sinking fund.
2. Set up a sinking fund table.

CASE IN POINT Darnell Johnson is excited! The hospital where he works is setting up a sinking fund to accumulate the money needed for a new building that will include a gymnasium and indoor swimming pool that he will be able to use.

OBJECTIVE 1 Understand the basics of a sinking fund. Individuals and businesses often need to raise a certain amount of money for use *at some fixed time in the future*. For example, a hospital administrator may need $1.8 million in 3 years to buy a new MRI machine that can be used to scan patients for cancer. This section shows how to find the periodic payment needed to achieve a specific future value on a particular date.

A fund set up to receive periodic payments is called a **sinking fund**. Sinking funds are used to provide money *to pay off a loan* in one lump sum *or to accumulate money* to build new factories, buy equipment, and so on. The payment needed at the end of each period to accumulate a fixed amount in a sinking fund at a specific future date is found as follows.

> **Finding Payment Needed to Accumulate a Specific Amount**
> Payment = Future value × Number from sinking fund table

Finding Periodic Payments

EXAMPLE 1 In 5 years, hospital administrators plan to erect a building that includes a gym and a swimming pool that can be used for therapy, at a cost of $16,500,000. To prepare for the expense, they decide to make end-of-quarter deposits into a sinking fund expected to earn 6% compounded quarterly. Find (a) the amount of each quarterly payment and (b) the interest earned.

CASE IN POINT

SOLUTION

(a) Use $\frac{6\%}{4}$ = 1.5% per compounding period for 5 × 4 years = 20 compounding periods in the sinking fund table on page 426 to find **.04325**.

Quarterly payment = $16,500,000 × **.04325** = **$713,625**

Twenty end-of-quarter payments of $713,625 at 6% compounded quarterly will grow to $16,501,629 using the table in Section 11.1.

(b) Interest is the future value minus the payments.

Interest = $16,501,629 − (20 × **$713,625**) = $2,229,129

> **QUICK CHECK 1**
> A utility company needs $300 million in 6 years to build a natural gas power plant. What amount does it need to deposit at the end of each quarter into a sinking fund if funds earn 8% compounded quarterly?

Finding the Periodic Payments

EXAMPLE 2 First Christian Church sold $100,000 worth of bonds that must be paid off in 8 years. It sets up a sinking fund to accumulate the necessary $100,000 to pay off the debt. Find the amount of each payment into a sinking fund if the payments are made at the end of each year and the fund earns 10% compounded annually. Find the amount of interest earned.

SOLUTION

Look along the top of the sinking fund table for 10% and down the side for 8 periods to find **.08744**.

Payment = $100,000 × **.08744** = $8744

The church must deposit $8744 at the end of each year for 8 years into an account paying 10% compounded annually to accumulate $100,000. The interest earned is the future value less all payments.

Interest = $100,000 − (8 × $8744) = $30,048

11.3 Sinking Funds (Finding Annuity Payments)

> **QUICK CHECK 2**
>
> A charter airline sold $1,000,000 in bonds to buy a new aircraft. It chose to make deposits at the end of each semiannual period to accumulate the funds needed to pay off the bonds in 4 years. Use 5% compounded semiannually and find (a) the payment and (b) the interest.

OBJECTIVE 2 Set up a sinking fund table. A **sinking fund table** is used to show the interest earned and the accumulated amount of a sinking fund at the end of each period.

Setting up a Sinking Fund Table  First Christian Church in Example 2 deposited $8744 at the end of each year for 8 years into a sinking fund that earned 10% compounded annually. Set up a sinking fund table for these deposits. After each calculation, round each answer to the nearest cent before proceeding.

SOLUTION

The sinking fund account contains no money until the end of the first year, when a single deposit of $8744 is made. Since the deposit is made at the end of the year, no interest is earned in the first year.

At the end of the second year, the account contains the original $8744 plus the interest earned on the $8744. This interest is found by the formula for simple interest.

$$I = \$8744 \times .10 \times 1 = \$874.40$$

An additional deposit is also made at the end of the second year, so that the sinking fund then contains the following total.

$$\$8744 + \$874.40 + \$8744 = \$18{,}362.40$$

Continue to get the following sinking fund table.

	Beginning of Period		End of Period	
Period	Accumulated Amount	Periodic Deposit	Interest Earned	Accumulated Amount
1	$0	$8744.00	$0	$8,744.00
2	$8,744.00	$8744.00	$874.40	$18,362.40
3	$18,362.40	$8744.00	$1836.24	$28,942.64
4	$28,942.64	$8744.00	$2894.26	$40,580.90
5	$40,580.90	$8744.00	$4058.09	$53,382.99
6	$53,382.99	$8744.00	$5338.30	$67,465.29
7	$67,465.29	$8744.00	$6746.53	$82,955.82
8	$82,955.82	$8748.60	$8295.58	$100,000.00

> **Quick TIP**
>
> Adjust the last payment as needed so that the future value exactly equals the desired amount. This explains why the last payment often differs slightly from the others.

The last payment is found as follows:

Desired Accumulated Amount − Interest Earned during Last Period − Accumulated Amount at end of Period 7
$100,000 − $8295.85 − $82,955.82

> **QUICK CHECK 3**
>
> The payments into a sinking fund are $2400 at the end of each quarter for 1 year. Assume interest of 6% compounded quarterly and construct a sinking fund table.

Frequently, an item costs more if its purchase is delayed a few years. The next example shows how to estimate the cost of a large purchase at a future date. It then shows how to find the payment needed to accumulate the necessary funds.

CHAPTER 11 Annuities, Stocks, and Bonds

Sinking Fund Table

Interest Rate per Compounding Period

Period	1%	1½%	2%	2½%	3%	4%	5%	6%	8%	10%	Period
1	1.00000	1.00000	1.00000	1.00000	1.00000	1.00000	1.00000	1.00000	1.00000	1.00000	1
2	.49751	.49628	.49505	.49383	.49261	.49020	.48780	.48544	.48077	.47619	2
3	.33002	.32838	.32675	.32514	.32353	.32035	.31721	.31411	.30803	.30211	3
4	.24628	.24444	.24262	.24082	.23903	.23549	.23201	.22859	.22192	.21547	4
5	.19604	.19409	.19216	.19025	.18835	.18463	.18097	.17740	.17046	.16380	5
6	.16255	.16053	.15853	.15655	.15460	.15076	.14702	.14336	.13632	.12961	6
7	.13863	.13656	.13451	.13250	.13051	.12661	.12282	.11914	.11207	.10541	7
8	.12069	.11858	.11651	.11447	.11246	.10853	.10472	.10104	.09401	.08744	8
9	.10674	.10461	.10252	.10046	.09843	.09449	.09069	.08702	.08008	.07364	9
10	.09558	.09343	.09133	.08926	.08723	.08329	.07950	.07587	.06903	.06275	10
11	.08645	.08429	.08218	.08011	.07808	.07415	.07039	.06679	.06608	.05396	11
12	.07885	.07668	.07456	.07249	.07046	.06655	.06283	.05928	.05270	.04676	12
13	.07241	.07024	.06812	.06605	.06403	.06014	.05646	.05296	.04652	.04078	13
14	.06690	.06472	.06260	.06054	.05853	.05467	.05102	.04758	.04130	.03575	14
15	.06212	.05994	.05783	.05577	.05377	.04994	.04634	.04296	.03683	.03147	15
16	.05794	.05577	.05365	.05160	.04961	.04582	.04227	.03895	.03298	.02782	16
17	.05426	.05208	.04997	.04793	.04595	.04220	.03870	.03544	.02963	.02466	17
18	.05098	.04881	.04670	.04467	.04271	.03899	.03555	.03236	.02670	.02193	18
19	.04805	.04588	.04378	.04176	.03981	.03614	.03275	.02962	.02413	.01955	19
20	.04542	.04325	.04116	.03915	.03722	.03358	.03024	.02718	.02185	.01746	20
21	.04303	.04087	.03878	.03679	.03487	.03128	.02800	.02500	.01983	.01562	21
22	.04086	.03870	.03663	.03465	.03275	.02920	.02597	.02305	.01803	.01401	22
23	.03889	.03673	.03467	.03270	.03081	.02731	.02414	.02128	.01642	.01257	23
24	.03707	.03492	.03287	.03091	.02905	.02559	.02247	.01968	.01498	.01130	24
25	.03541	.03326	.03122	.02928	.02743	.02401	.02095	.01823	.01368	.01017	25
26	.03387	.03173	.02970	.02777	.02594	.02257	.01956	.01690	.01251	.00916	26
27	.03245	.03032	.02829	.02638	.02456	.02124	.01829	.01570	.01145	.00826	27
28	.03112	.02900	.02699	.02509	.02329	.02001	.01712	.01459	.01049	.00745	28
29	.02990	.02778	.02578	.02389	.02211	.01888	.01605	.01358	.00962	.00673	29
30	.02875	.02664	.02465	.02278	.02102	.01783	.01505	.01265	.00883	.00608	30
31	.02768	.02557	.02360	.02174	.02000	.01686	.01413	.01179	.00811	.00550	31
32	.02667	.02458	.02261	.02077	.01905	.01595	.01328	.01100	.00745	.00497	32
33	.02573	.02364	.02169	.01986	.01816	.01510	.01249	.01027	.00685	.00450	33
34	.02484	.02276	.02082	.01901	.01732	.01431	.01176	.00960	.00630	.00407	34
35	.02400	.02193	.02000	.01821	.01654	.01358	.01107	.00897	.00580	.00369	35

Finding Periodic Payments and Interest Earned

EXAMPLE 4 Managers at Baton Chemicals frequently travel between the corporate offices and manufacturing plants and mines in South America. They plan to purchase a private jet in 4 years. The jet costs $14,800,000 today, and its cost is expected to grow at 5% per year. Find the quarterly payments needed to accumulate the necessary funds in a sinking fund in 4 years if the firm earns 6% compounded quarterly.

SOLUTION

First, find the cost (future value) of the jet in 4 years:
Use a rate of 5% per year and four compounding periods (years) in the compound interest table on page 377 to find **1.21551**.

Future cost of jet = $14,800,000 × **1.21551** = **$17,989,548**

Next, find the quarterly payment needed to accumulate $17,989,548 in 4 years:

Use a rate of $\frac{6\%}{4} = 1.5\%$ per quarter and $4 \times 4 = 16$ compounding periods in the sinking fund table to find **.05577**.

$$\text{Quarterly payment} = \$17,989,548 \times .05577 = \$1,003,277.09$$

Payments of $1,003,277.09 at the end of each quarter for 4 years into a sinking fund earning 6% compounded quarterly will result in the funds needed to purchase the jet.

QUICK CHECK 4

The $19,000 cost of a compressor is increasing by 5% per year. First find the amount needed to buy the compressor in 3 years. Then find the necessary semiannual payments into a sinking fund for the purchase if funds earn 10% compounded semiannually.

Quick TIP
Banks make money by charging more interest on money loaned out than they pay on funds deposited at the bank.

Two different interest rates are involved in Example 4. The price is increasing at 5% per year compounded annually, but deposits in the sinking fund earn 6% compounded quarterly. **Interest rate spreads** such as these are common. For example, banks use the difference between interest rates on deposits and interest rates charged on loans to make money.

11.3 Exercises // MyLab Math

The shaded sections below contain solutions to help you get a **QUICK START** *on the various types of exercises.*

Find the amount of each payment needed to accumulate the indicated amount in a sinking fund. (See Examples 1–3.)

1. $12,000, money earns 5% compounded annually, 4 years
   Payment = $12,000 × .23201 = $2784.12

   1. $2784.12

2. $125,000, money earns 6% compounded annually, 25 years
   Payment = $125,000 × .01823 = $2278.75

   2. $2278.75

3. $8200, money earns 6% compounded semiannually, 5 years

   3. _____

4. $12,000, money earns 10% compounded semiannually, 3 years

   4. _____

5. $50,000, money earns 4% compounded quarterly, 5 years

   5. _____

6. $32,000, money earns 6% compounded quarterly, 3 years

   6. _____

7. $78,500, money earns 6% compounded semiannually, 5 years

   7. _____

8. $29,804, money earns 12% compounded monthly, 2 years

   8. _____

9. Explain the difference between a sinking fund (see Objective 1) and the present value of an annuity discussed in Section 11.2.

C// indicates an exercise that is related to the Case in Point feature.

10. What is a sinking fund table? Who would use one? (See Objective 2.)

Solve each application problem.

11. STUDENT UNION A college needs $920,000 in 3 years to remodel the student union. It decides to make payments into a sinking fund at the end of each semiannual period.
(a) Find the amount of each payment, assuming 5% per year compounded semiannually.
(b) Find the total interest earned.
(a) Payment = $920,000 × .15655 = $144,026
(b) Interest = $920,000 − (6 × $144,026) = $55,844

(a) $144,026
(b) $55,844

12. SCUBA DIVING The owner of Emerald Diving plans to buy all new scuba diving equipment to rent to divers in 5 years at a cost of $122,300. He believes that he can earn 4% compounded quarterly. Find (a) the amount of each of the quarterly payments needed and (b) the total interest earned.

(a) _____
(b) _____

13. ACCUMULATING $1,000,000 Kyle Anderson is 25 years old and wants to accumulate $1,000,000 by the time he is 60. He believes he can earn 8% per year by investing in stocks and bonds. Find (a) the amount of the annual payment needed and (b) the total interest earned.

(a) _____
(b) _____

14. ALLIGATOR HUNTING Cajun Jack needs $45,000 in 4 years for boats used to hunt alligators. (a) Find the amount of each payment if payments are made at the end of each quarter with interest at 6% compounded quarterly. (b) Find the total amount of interest earned.

(a) _____
(b) _____

15. X-RAY EQUIPMENT Emergency Vet Clinic will need a new x-ray machine in 3 years. The owners decide to set up a sinking fund to accumulate the $82,400 needed at that time. Find the payment into the fund needed at the end of each year if money earns 2½% compounded annually. Round to the nearest dollar.

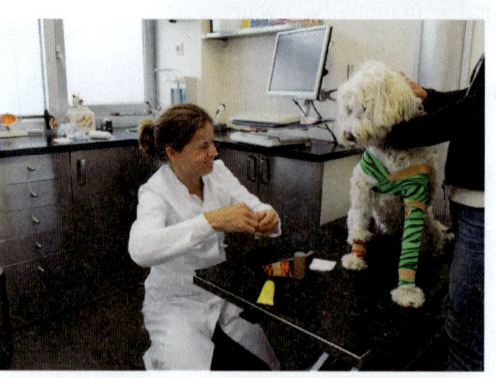

15. _____

16. NEW AUDITORIUM The membership of the Greenacres Baptist Church is large and growing rapidly. The leaders of the church are planning to remodel the auditorium with special features for their televised broadcasts at a cost of $2,800,000 in 5 years. The membership has set up a sinking fund with the idea of making a payment at the end of each quarter. Find the payment needed if money earns 8% compounded quarterly.

16. _____

17. A NEW SHOWROOM A Ford dealership wants to build a new showroom costing $2,300,000. It sets up a sinking fund with end-of-the-month payments in an account earning 5% compounded semiannually. Find the amount that should be deposited in this fund each semiannual period if the dealership wishes to build the showroom in **(a)** 3 years and **(b)** 4 years.

(a) _____
(b) _____

18. AIRPORT IMPROVEMENTS A city near Chicago sold $9,000,000 in bonds to pay for improvements to an airport. It sets up a sinking fund with end-of-the-quarter payments in an account earning 8% compounded quarterly. Find the amount that should be deposited in this fund each quarter if the city wishes to pay off the bonds in **(a)** 5 years and **(b)** 8 years.

(a) _____
(b) _____

19. LAND SALE Helen Spence sells a lot in Nevada. She will be paid a lump sum of $60,000 in 4 years. Until then, the buyer pays 8% simple interest every quarter. **(a)** Find the amount of each quarterly interest payment. **(b)** The buyer sets up a sinking fund so that enough money will be present to pay off the $60,000. The buyer wants to make semiannual payments into the sinking fund. The account pays 8% compounded semiannually. Find the amount of each payment into the fund. **(c)** Prepare a table showing the amount in the sinking fund after each deposit.

(a) _____
(b) _____

Payment Number	Amount of Deposit	Interest Earned	Total in Account

20. RARE STAMPS Jeff Reschke bought a rare stamp for his collection. He agreed to pay a lump sum of $4000 after 5 years. Until then, he pays 6% simple interest every 6 months. **(a)** Find the amount of each semiannual interest payment. **(b)** Reschke sets up a sinking fund so that money will be available to pay off the $4000. He wants to make annual payments into the fund. The account pays 8% compounded annually. Find the amount of each payment into the fund. **(c)** Prepare a table showing the amount in the sinking fund after each deposit.

(a) _____
(b) _____

Payment Number	Amount of Deposit	Interest Earned	Total in Account

21. SPORTS COMPLEX Prepare a sinking fund table for the first four payments for the hospital building with gym and pool described in Example 1.

Payment Number	Amount of Deposit	Interest Earned	Total in Account

CHAPTER 11 Annuities, Stocks, and Bonds

22. COMMERCIAL BUILDING Jeremy Painter plans to make a down payment of $70,000 on a commercial building for his plumbing company in 5 years. Construct a sinking fund table given semiannual payments of $6106.10 at the end of each period and an interest rate of 6% compounded semiannually.

Payment Number	Amount of Deposit	Interest Earned	Total in Account

23. SAVING FOR COLLEGE Barbara Funicello hopes to go to a private college where tuition is $22,500 per year. She believes that tuition will increase at 8% for the 4 years until she plans to enter college. Find the end-of-quarter payments needed to accumulate funds to pay the first year's tuition if funds earn 6% compounded quarterly. Round to the nearest dollar at each step.

23. _____

24. FIRE TRUCK A volunteer fire department anticipates purchasing a fire engine in 3 years. Today it would cost $625,000, but the cost is increasing at 4% per year. Find the semiannual payments needed to accumulate funds to purchase the truck if funds earn 6% compounded semiannually. Round to the nearest dollar at each step.

24. _____

25. NEW ROOF The manager of an apartment complex estimates she will need a new roof on one building in 5 years. Today it would cost $52,000 to reroof the building, but the cost is increasing at 8% per year. Find the semiannual payments needed to accumulate the necessary funds at 5% compounded semiannually. Round to the nearest dollar at each step.

25. _____

26. RESTAURANT The manager of Boston Sea Foods must remodel two bathrooms in 2 years. The estimated cost today is $90,000, but the costs are going up at 4% per year. Find the semiannual payments needed to a sinking fund if funds earn 5% compounded semiannually. Round to the nearest dollar.

26. _____

QUICK CHECK ANSWERS

1. $9,861,000
2. (a) $114,470 (b) $84,240
3.

| Period | Beginning of Period | | End of Period | |
	Accumulated Amount	Periodic Deposit	Interest Earned	Accumulated Amount
1	$0.00	$2400.00	$0.00	$2400.00
2	$2400.00	$2400.00	$36.00	$4836.00
3	$4836.00	$2400.00	$72.54	$7308.54
4	$7308.54	$2400.00	$109.63	$9818.17

4. $21,994.97; $3233.70

Supplementary Application Exercises on Annuities and Sinking Funds

Solve the following application problems. Round to the nearest cent.

1. Bill Carter deposits $500 at the end of each quarter for 6 years into a mutual fund that he believes will grow at 8% compounded quarterly. Find **(a)** the future value and **(b)** the interest.
   (a) Future value = $500 × 30.42186 = $15,210.93
   (b) Interest = $15,210.93 − (24 × $500) = $3210.93

   (a) $15,210.93
   (b) $3210.93

2. For 6 years, Jessica Savage deposits $1000 at the end of each quarter into a mutual fund earning 8% per year compounded quarterly. Find **(a)** the future value and **(b)** the interest.

   (a) _____
   (b) _____

3. Mr. and Mrs. Thompson deposit $2000 at the beginning of each year for 20 years into a retirement account earning 6% compounded annually. Find **(a)** the future value and **(b)** the interest.

   (a) _____
   (b) _____

4. Jaime Navarro deposits $1000 at the end of every 6 months into a Roth IRA for 8 years, where he optimistically hopes to earn 10% compounded semiannually. Find **(a)** the future value and **(b)** the interest.

   (a) _____
   (b) _____

5. Solectron needs to purchase new equipment for its production line in 3 years. The company has been advised to deposit $135,000 at the end of each quarter into an account that managers believe will yield 6% per year compounded quarterly. Find the lump sum that could be deposited today that will grow to the same future value.

   5. _____

6. Abel Plumbing saves $12,000 at the end of every semiannual period in an account earning 6% compounded semiannually to replace several of its trucks in 5 years. Find the lump sum that could be deposited today that will grow to the same future value.

   6. _____

7. Katherine Wysong was injured when she fell on ice at work. Her employer's workers compensation insurance paid for her medical bills and must also pay her $5000 per quarter for the next 6 years. Find the lump sum that must be deposited into an investment earning 4% per year compounded quarterly needed to make the payments.

   7. _____

8. Carl and Amy Glaser recently divorced. As part of the divorce settlement, Carl must pay Amy $1000 at the end of every quarter for 8 years. Find the lump sum he must deposit into an account earning 8% per year compounded quarterly to make the payments.

   8. _____

9. Houston Petrochemicals needs to purchase a new distillation tower in 3 years at a cost of $870,000. Find the annual payment the firm must make if funds are deposited into an account earning 8% compounded annually. Then set up a sinking fund table.

   9. _____

Payment Number	Amount of Deposit	Interest Earned	Total in Account

10. Ajax Coal sets up a sinking fund to purchase a new strip mining tractor costing $3,200,000 in 2 years. Find the semiannual payment the company must make into a sinking fund account earning 6% compounded semiannually. Then set up a sinking fund table.

Payment Number	Amount of Deposit	Interest Earned	Total in Account

11. Ben Jamison is 45. He wants to retire at age 65 and draw $25,000 per year from his retirement plan until he turns 85. He assumes that he can earn 8% per year *before* retiring but that he would invest more conservatively and earn 6% per year *after retiring*. **(a)** Find the amount needed at 65 to fund his retirement. **(b)** Then find the end-of-year payment into a sinking fund needed to accumulate this amount.

(a) _____
(b) _____

12. An accountant is only 28, but he hates his job and is dreaming of retiring early. He figures he will live to be 87 and wants $100,000 per year during retirement, which gives him a little extra for inflation. Assume funds earn 6% compounded annually. **(a)** Find the amount needed at 52 to fund his retirement. **(b)** Then find the end-of-year payment into a retirement fund needed to accumulate this amount. **(c)** Do you think he will be able to retire as planned? Why or why not?

(a) _____
(b) _____
(c) _____

13. Work a problem similar to Exercise 12, but use numbers that interest YOU! Decide at what age you would like to quit working and retire, as well as how much you think you would need and the number of years you will live after retiring. How much do you need to put into a retirement plan every year to achieve your goals?

11.4 Stocks and Mutual Funds

OBJECTIVES

1. Learn the basics about stocks.
2. Read stock tables.
3. Find the current yield on a stock.
4. Find the stock's PE ratio.
5. Define three common stock indexes.
6. Define mutual funds and exchange traded funds.

CASE IN POINT On his first day on his new job, Darnell Johnson was given a choice of investing his retirement account funds either in a fund paying a fixed interest or in a fund that contains stocks and bonds. Which would you choose? Why?

Many businesses are set up as **corporations**. For example, Microsoft, Apple, Nike, McDonald's, and Toyota are actually corporations. **Publicly held corporations** are those owned by the public; their stocks are traded daily in markets called stock markets. **Privately held corporations** are owned by one or a few individuals, and their stock is not traded on a market. For example, your medical doctor or plumber may have organized her small business as a privately held corporation.

A corporation is a form of business that gives the owners (the stockholders) **limited liability**. You do not have to worry that lawsuits will be filed against you or your family just because you own stock in General Motors, Inc. The owners of corporations will never lose more than they have invested in the corporation.

A corporation is set up with money, or **capital**, raised through the sale of shares of **stock**. A share of stock represents partial ownership of a corporation. If one million shares of stock are sold to establish a new firm, the owner of one share will own one-millionth of the corporation. In the past, stock ownership was indicated by **stock certificates** such as the one shown here. Today, very few companies issue stock certificates on paper. Proof of ownership is based on records maintained by brokerage firms and corporations.

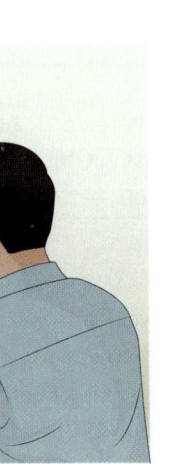

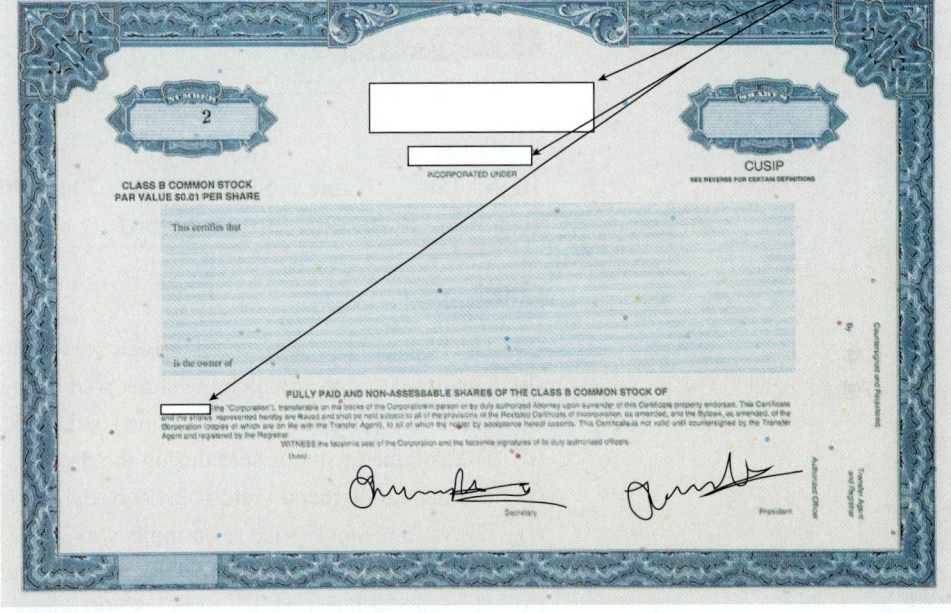

Corporations hold an **annual meeting** that is open to all **stockholders** (owners of stock) where management discusses financial results, problems, and future plans. The stockholders elect a **board of directors**, which is a group of people who represent the stockholders and oversee top management. The board of directors effectively hires the **executive officers** such as the president and possibly others. The board is usually involved in long-term major decisions, such as whether to enter a new market or offshore factory production to (say) China.

OBJECTIVE 1 Learn the basics about stocks. The two types of stock normally issued are **preferred stock** and **common stock**. As the name suggests, preferred stockholders *have certain rights* over common stockholders. For example, owners of preferred stock must be paid dividends *before* any dividends can be paid to owners of common stock. Also, corporate debt and preferred shareholders must be paid *before* common shareholders receive anything in the event that a corporation declares bankruptcy.

The shares of **publicly held corporations** are typically owned by many different individuals and institutions. Share prices of these firms are determined by supply and demand in public markets called **stock exchanges**. The New York Stock Exchange (NYSE) is the largest of the several exchanges in the United States. This exchange is located on Wall Street in New York City. Many foreign countries, including Japan, Taiwan, Germany, England, Canada, and Mexico, have their own stock exchanges.

434 CHAPTER 11 Annuities, Stocks, and Bonds

Businesses and individuals buy stock to make money. Stock prices usually go up over time, but not always since recessions drive stock prices lower and companies can go **bankrupt**. Many, but not all, publicly held corporations pay **dividends**—regular payments to stockholders usually made every quarter.

The long-term average return on funds invested in the Standard & Poor's index of 500 large companies is about 10% per year, including price appreciation and dividends. You can think of this as 10% compounded annually. However, this average return varies greatly from year to year, and the return on funds invested is sometimes negative, i.e., it goes down in value. The stock prices of individual companies may differ substantially from one year to the next.

Although some individuals believe they can do better than the overall market by buying stocks in companies they choose themselves, there is a lot of research showing this is very rarely true. Since you can lose a lot of money owning the stock of one particular company and no one can predict the future accurately, many professionals suggest that you own small amounts of stock in many different firms. Spreading out your investment dollars reduces risk. Stocks are best used for long-term investments. You would not want to put money you need next year into stocks, only to see the value fall sharply during the year. However, stocks can be a great place for investments of more than five years, such as retirement accounts or sinking funds.

OBJECTIVE 2 Read stock tables. Daily stock prices are readily found in many places on the Internet, such as finance.yahoo.com or wsj.com. Information on the stock of many widely held companies such as Apple or Nike is also found in the weekly financial magazine *Barron's*.

Reading the Stock Table

EXAMPLE 1 After receiving his first paycheck, Darnell Johnson went to Best Buy to look at iPhones and MP3 players. He liked the company so well that he decided to do some research on it. Analyze the information about Best Buy stock given in the table.

CASE IN POINT

52-Wk High	52-Wk Low	Name	Tick Sym	Vol. 100s	Yld.	P/E	Week's Last	Week's Chg.	Earnings Latest Year	Earnings This Year	Earnings Next Year	Div Amt.
49.40	25.31	BestBuy	BBY	242901	2.4	14	47.00	−2.00	2.56	3.26	3.47	.28

SOLUTION

(a) The highest price at which the stock sold during the year was $49.40.
(b) The lowest price at which the stock sold during the year was $25.31.
(c) The ticker symbol under which the stock trades on the market is BBY.
(d) The volume of shares sold during the day was 100 × 242901 = 24,290,100 shares.
(e) The annual dividend yield is 2.4% of the current stock price.
(f) The ratio of stock price to earnings was 14.
(g) The stock closed at $47.00 per share at the end of the week.
(h) The price of the stock was down $2.00 compared to the close at the end of the prior week.
(i) Best Buy earned $2.56 per share last year. It is expected to earn $3.26 per share this year and $3.47 per share next year.
(j) The most recent dividend was $.28 per share. We know it is a quarterly dividend, since total annual dividend (.28 × 4 = $1.12) equates to the 2.4% annual dividend yield noted in (e) above.

QUICK CHECK 1

Find the information in the stock table about Ball.

Individual stocks can be purchased using **stockbrokers** or individuals who make their living by advising investors. Stocks can also be purchased directly by individuals through any of the numerous **discount brokers** such as Schwab or Fidelity in a process similar to buying something on Amazon.

Finding the Cost of Stocks

EXAMPLE 2 Ignoring commissions, find the cost for each of the following purchases.

(a) 500 shares of Bank of America (BAC) at the close for the week
(b) 800 shares of Bladex (BLX) at the low for the year
(c) Then find the combined *annual dividend* from these two investments.

SOLUTION

(a) 500 shares × $22.66 per share = $11,330
(b) 800 shares × $19.63 per share = $15,704
(c) Quarterly dividend from Bank of America: $.075 × 500 = $37.50
   Quarterly dividend from Bladex: $.385 × 800 = $308.00
                      Total quarterly dividend = $345.50
   Total annual dividend = $345.50 × 4 quarters in a year = $1382.

QUICK CHECK 2

Ignore commissions and find the cost of 300 shares of BakerHughes (BHI) at the low for the year. Then find the *annual* dividend these shares will pay.

52-WK High	52-WK Low	Name	Tick Sym	Vol. 100s	Yld	P/E	Week's Last	Week's Chg.	Earnings Latest Year	Earnings This Year	Earnings Next Year	Div Amt.
37.28	27.01	BP	BP	321748	6.6	dd	36.59	+0.58	−2.12	1.08	2.40	.60
33.33	11.29	BP Prudhoe	BPT	10945	11.0	12	24.60	+0.25	5.86	…	…	.6793
18.12	11.05	BRF	BRFS	72494	…	30	14.01	−0.75	1.11	.38	.85	.2023
8.25	5.41	BRT Realty	BRT	z42456	…	7	8.16	+0.27	2.23	…	…	…
36.04	21.61	BT Group	BT	35632	2.5	11	23.27	+0.69	2.25	1.88	2.00	.2953
40.66	26.89	BWX Tech	BWXT	45847	.9	27	38.19	−0.83	1.22	1.65	1.82	.09
23.99	12.90	Babcock&Wilcox	BW	26213	…	dd	16.79	−0.46	.36	.70	1.39	…
39.36	26.40	BadgerMeter	BMI	8141	1.2	35	37.65	−0.85	1.80	1.14	1.28	.115
68.59	37.58	BakerHughes	BHI	219430	1.0	dd	66.39	+0.07	−4.49	−1.99	.47	.17
82.24	62.30	Ball	BLL	89713	.7	42	76.66	−0.18	1.99	3.44	4.28	.13
23.24	10.93	BancofCalifornia	BANC	82691	3.1	9	16.65	+0.25	1.34	1.85	1.92	.13
7.82	5.14	BancoBilbaoViz	BBVA	162001	5.1	11	6.80	+0.01	.43	.60	.67	.0869
10.15	3.77	BancoBradesco	BBDO	z14720	.7	…	8.33	+0.40	…	…	…	.0759
72.79	54.98	BancodeChile	BCH	2478	4.2	14	69.41	−1.12	5.08	5.36	5.59	2.9008
30.51	19.63	Bladex	BLX	9378	5.2	12	29.82	−0.26	2.67	2.66	3.06	.385
83.18	52.58	BancoMacro	BMA	6986	1.2	7	63.85	−2.47	…	7.61	8.38	.7488
8.77	3.02	BancSanBrasil	BSBR	74452	…	…	7.88	−0.06	…	.51	.54	.0411
23.50	15.69	BcoSantChile	BSAC	14416	4.9	16	22.12	−1.36	1.45	1.58	1.74	1.077
5.25	3.60	BancoSantander	SAN	362829	…	12	5.15	+0.03	.44	.45	.48	.0498
42.58	24.69	BanColombia	CIB	27625	3.2	11	36.97	+1.23	3.98	3.31	3.76	.2963
30.45	18.69	BancorpSouth	BXS	40197	1.7	24	29.75	−0.50	1.33	1.49	1.65	.125
23.39	10.99	BankofAmerica	BAC	f78177	1.3	19	22.66	−0.43	1.31	1.45	1.64	.075

(continued)

OBJECTIVE 3 Find the current yield on a stock. There is no certain way of choosing stocks that will increase in price. However, two **stock ratios** that people commonly look at before buying shares of a company are the **current yield** and the **price–earnings ratio**. It is used to compare the dividends paid by stocks selling at different prices. The result is commonly rounded to the nearest tenth of a percent.

Finding Current Yield

$$\text{Current yield} = \frac{\text{Annual dividend per share}}{\text{Closing price per share}}$$

436 CHAPTER 11 Annuities, Stocks, and Bonds

52-WK							Week's		Earnings			
High	Low	Name	Tick Sym	Vol. 100s	Yld	P/E	Last	Chg.	Latest Year	This Year	Next Year	Div Amt.
11.15	3.01	BankofAmWtA	BAC/WS/A	46742	...	...	10.65	−0.29	...	...	...	...
1.29	0.06	BankofAmWtB	BAC/WS/B	54227	...	...	1.11	−0.11	...	...	...	...
31.92	23.75	BankofButterfield	NTB	22930	...	24	31.77	+1.96	1.30	2.17	2.66	.10
89.42	54.55	BankofHawaii	BOH	11445	2.2	21	87.49	−1.61	3.70	4.16	4.26	.48
73.88	47.54	BankofMontreal g	BMO	28676	3.6	11	72.62	−0.37	6.94	7.75	8.24	.6549
49.54	32.20	BankNY Mellon	BK	316358	1.6	16	47.56	−1.06	2.71	3.15	3.48	.19
58.97	35.01	BkNovaScotia g	BNS	33290	3.9	10	56.94	−1.38	5.80	6.40	6.88	.5485
13.91	6.59	Bankrate	RATE	22355	...	dd	10.35	...	−.13	.63	.69	...
38.47	27.85	BankUnited	BKU	46916	2.3	18	37.28	−0.88	2.35	2.03	2.34	.21
13.42	6.76	Barclays	BCS	259529	1.9	dd	11.29	−0.45	−.12	.68	1.02	.0528
239.43	172.21	Bard CR	BCR	29506	.5	33	221.03	+4.39	1.77	10.26	11.36	.26
12.84	8.50	Barnes&NobleEduc	BNED	19595	...	dd	12.83	+0.84	...	.27	.45	...
13.63	7.25	Barnes&Noble	BKS	34902	4.8	97	12.60	−0.50	−.49	.57	.82	.15
49.36	30.07	BarnesGroup	B	11637	1.1	22	48.80	+0.74	2.19	2.52	2.69	.13
26.69	9.44	BarracudaNtwks	CUDA	28345	...	cc	22.42	−0.47	−.08	.69	.74	...
23.47	7.02	BarrickGold	ABX	f13097	.6	dd	14.28	−1.17	−2.44	.67	.96	.02

DATA: *Barron's*, December 19, 2016.

Finding the Current Yield **EXAMPLE 3** Use data in the stock table to estimate the current yield for Bankrate (RATE) and Barclays (BCS) rounded to the nearest tenth of a percent.

SOLUTION

$$\text{Current yield for Bankrate} = \frac{\text{Annual dividend}}{\text{Closing price}} = \frac{\$0}{\$10.35} = 0\%$$

$$\text{Current yield for Barclays} = \frac{\text{Annual dividend}}{\text{Closing price}} = \frac{\$.0528 \text{ per quarter} \times 4}{\$11.29} = 1.9\%$$

As you can see, the current yield is actually shown in the stock table under Yld.

QUICK CHECK 3

Find the current yield for ExxonMobil to the nearest tenth of a percent if the closing price is $90.87 and the annual dividend is $3.

Note that a company such as Bankrate in Example 3 may not pay a dividend since it may be having financial problems and needs to keep its cash, or because it may be growing rapidly and needs the cash to finance its future growth.

OBJECTIVE 4 Find the stock's PE ratio. One number that some people use to help decide which stock to buy is the **price–earnings ratio**, also called the **PE ratio**. It is often rounded to the nearest whole number.

Finding PE ratio

$$\text{PE ratio} = \frac{\text{Closing price per share}}{\text{Annual net income per share}}$$

Finding the PE Ratio **EXAMPLE 4** Find the PE ratio for each of the following corporations and round to the nearest whole number.

(a) Apple, Inc. (AAPL) with a closing price of $116.52 and annual earnings of $8.31
(b) Nike, Inc. (NKE) with a closing price of $51.91 and annual earnings of $2.22

SOLUTION

(a) PE ratio for Apple, Inc. $= \dfrac{\text{Closing price per share}}{\text{Annual net income per share}} = \dfrac{\$116.52}{\$8.31} = 14$

(b) PE ratio for Nike, Inc. $= \dfrac{\text{Closing price per share}}{\text{Annual net income per share}} = \dfrac{\$51.91}{\$2.22} = 23$

QUICK CHECK 4

Pfizer Inc (PFE) has a closing price of $32.48 and earnings of $1.00. Find the PE ratio to the nearest whole number.

Investors are often willing to pay more for rapidly growing companies because these companies may generate even higher profits in the near future. As a result, the PE ratios of rapidly growing companies are often higher than those of slow-growing companies.

A low PE ratio may indicate that a company is growing slowly or that it is having financial problems. It is best to compare the PE ratios of similar companies, such as oil giants ExxonMobil and Chevron. It usually is not worthwhile to compare the PE ratios of companies in very different businesses such as Walmart (retail) and ExxonMobil (energy).

OBJECTIVE 5 Define three common stock indexes. Three very common stock indexes are reported daily: the **Dow Jones Industrial Average**, **Standard & Poor's 500**, and the **NASDAQ Composite Index**. They are used to show the movements of the prices of select groups of stocks:

> **Dow Jones Industrial Average**—based on the prices of the stocks of 30 very large firms that trade on the New York Stock Exchange, including ExxonMobil, General Electric, and Boeing
>
> **Standard & Poor's 500**—based on the prices of stocks of 500 large firms that trade on either the New York or the NASDAQ stock exchange, including Apple, Amazon, and Chevron
>
> **NASDAQ Composite Index**—based on the prices of many firms that trade on the NASDAQ stock exchange, including mostly technology companies such as Apple, Facebook, and Twitter

The indexes show what specific groups of stocks are doing collectively. They do not indicate what is happening to the price of any one stock. The graph on the next page of the Dow Jones Industrial Average over time, labeled with major events such as the birth of the Internet.

In spite of the Great Depression and wars, you can clearly see that the trend of stock prices has been up over time. Although there have been periods when stocks have done very poorly for a period (e.g., the 2007 Great Recession), funds invested in stocks have usually done better than funds invested in savings accounts, certificates of deposit, or bonds. Of course, the timing of an investment is important. Investing in stocks at the beginning of the Great Recession in 2007 would have resulted in sharp losses by 2009 when the market bottomed. But investors who held onto stocks saw the market recover a few years later. Financial planners agree that stocks should be part of a long-range investment plan because they tend to do well over longer periods.

Many professionals consider Standard & Poor's 500 Index the best of the three indexes discussed earlier. It is not limited to 30 companies like the Dow Jones Industrial Average, and it is not focused only on technology companies like the NASDAQ Composite. You might think of the Standard & Poor's 500 as representing the most successful 500 publicly held companies in the United States. These companies all do business internationally, so together they are effectively a barometer of world economic growth and financial activity.

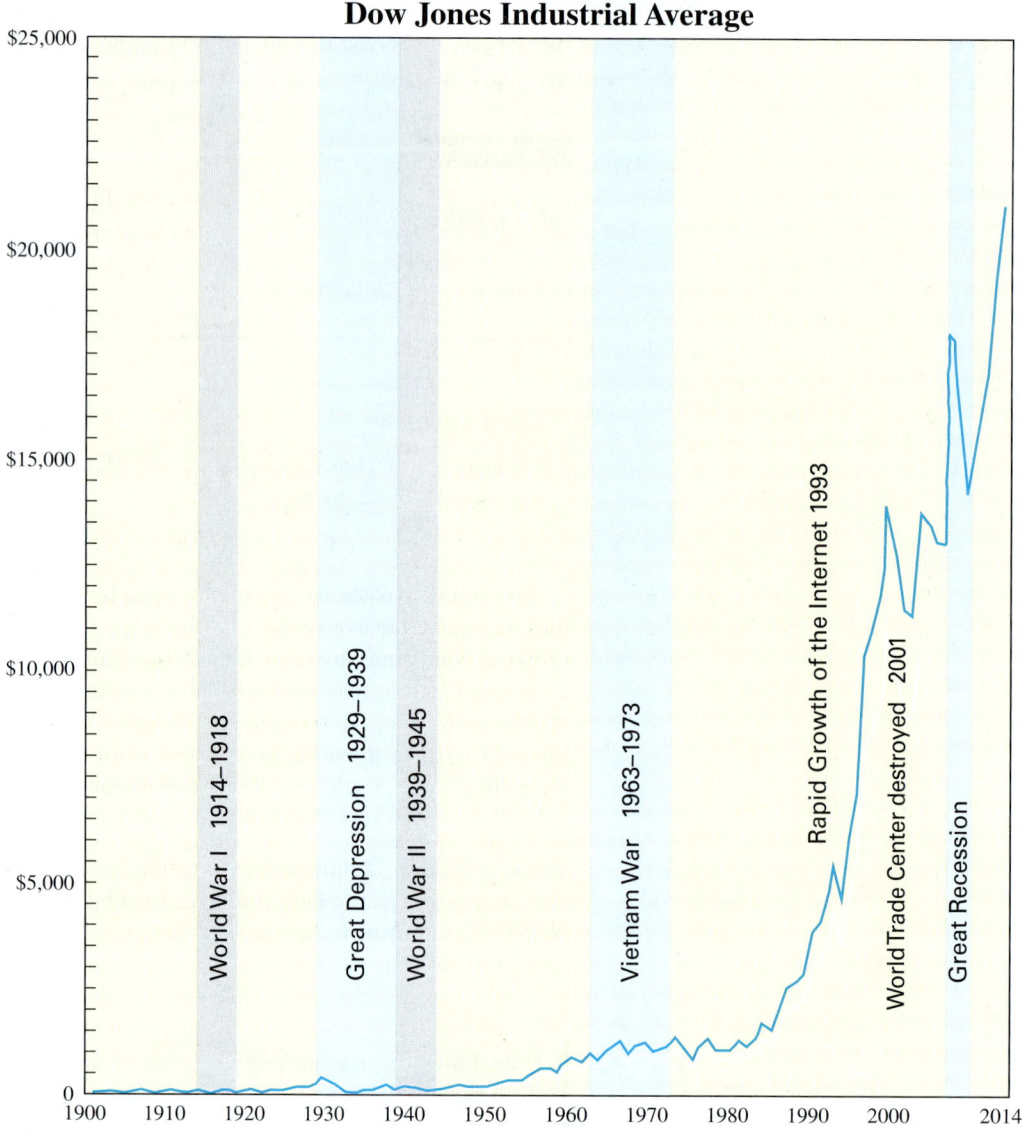

OBJECTIVE 6 Define mutual funds and exchange-traded funds. Owning the shares of just one company involves **risk**. Something unexpected may happen and the company's stock price may plunge, resulting in a big loss. Experts recommend you **diversify**, i.e. own stocks in different industries. Research shows that diversifying reduces the risk of a big loss in any year and usually yields better results.

One way to diversify is to put your retirement funds into one or more mutual funds. A **mutual fund** receives funds from many investors and uses them to buy and hold stock in many different companies. Having $1000 in a mutual fund means that you own a tiny portion of many different companies. In fact, one of the most successful investors in the world (Warren Buffet) recommends putting most of your money into a mutual fund that tracks the Standard & Poor's 500. By doing so, you are investing in 500 of the best firms in the world and which are managed by some of the best managers in the world. How can one person do any better by choosing individual stocks? The answer is that they almost never do better.

Mutual funds are actively managed, meaning that managers buy and sell the stocks of certain companies as they think best. **Exchange-traded funds (ETFs)** are similar in that they own a basket of stocks; however, they are usually not actively managed, reducing the costs of ownership for investors. ETFs buy a group of stocks that match certain indexes, such as Standard & Poor's 500 or some market sector such as energy companies. Essentially, ETFs can be a low-cost way to diversify your investments.

The table on the next page shows that some funds specialize by focusing investments in one area. For example, they may specialize in large-cap (large companies), small-cap (small companies), foreign countries (e.g., Japan), or other areas (real estate, energy, banking, etc.). Funds that specialize like this tend to be riskier than those that track the Standard & Poor's 500.

Numbers in the News

Exchange-traded fund (Ticker)	Invests in
Vanguard Total Stock Market VIPERs (VTI)	U.S. stocks, all sizes
iShares S&P 500 (IVV)	Large-cap U.S. stocks
iShares S&P Midcap 400 (IJH)	Mid-cap U.S. stocks
Midcap SPDRs (MDY)	Mid-cap U.S. stocks
iShares Russell 2000 (IWM)	Small-cap U.S. stocks
iShares MSCI Japan (EWJ)	Japanese stocks
iShares MSCI Pacific Ex-Japan (EPP)	Pacific Rim, but not Japan
iShares MSCI EAFE (EFA)	Europe and Pacific Rim
iShares MSCI Emerging Markets (EEM)	Emerging markets

Comparing Investment Alternatives

EXAMPLE 5 Cynthia Peck wants to know whether she should invest her retirement monies in certificates of deposit or in a mutual fund containing stock. Assume payments of $2000 per year for 30 years and **(a)** a certificate of deposit paying 4% compounded annually or **(b)** a mutual fund that tracks the Standard and Poor's 500 and has returned 8% per year. Find the future value for both and **(c)** compare the two investments.

SOLUTION

(a) Use 4% per year and 30 years in the table in Section 11.1 to find **56.08494**.

$$\text{Future value} = \$2000 \times 56.08494 = \$112{,}169.88$$

(b) Use 8% per year and 30 years in the table in Section 11.1 to find **113.28321**.

$$\text{Future value} = \$2000 \times 113.28321 = \$226{,}566.42$$

(c) Difference = $226,566.42 − $112,169.88 = $114,396.54

Stocks yield more but have higher risk. Cynthia Peck will need to decide how much risk she will accept before making a decision.

Quick TIP
Positive investment returns on stocks are not guaranteed. However, stocks usually return more than bank accounts over long periods of time.

QUICK CHECK 5
The owner of Termites Inc. plans to deposit $15,000 at the end of each year for 10 years into an investment account. One investment would pay 5% per year, and he assumes a stock fund would continue to yield 10% per year. Find the future value of both.

11.4 Exercises // MyLab Math

The shaded sections below contain solutions to help you get a **QUICK START** *on the various types of exercises.*

Find the following from the stock table on pages 435–436. (See Example 1.)

1. Low for the year for Bankrate (RATE) — **1. $6.59**
2. High for the year for Bard CR (BCR) — **2. $239.43**
3. Most recent dividend for Barnes&Noble (BKS) — 3. _____
4. Most recent dividend for Bank of Hawaii (BOH) — 4. _____
5. Volume for Bladex (BLX) — 5. _____
6. Volume for BarracudaNtwks (CUDA) — 6. _____
7. PE ratio for BT Group (BT) — 7. _____
8. PE ratio for BankofButterfield (NTB) — 8. _____
9. Estimated earnings per share for this year for Barrick Gold (ABX) — 9. _____
10. Estimated earnings per share for this year for Bank of Hawaii (BOH) — 10. _____

11. Dividend yield for BankNY Mellon (BK) 11. _____
12. Dividend yield for Baker Hughes (BHI) 12. _____
13. Estimated earnings next year for BP (BP) 13. _____
14. Estimated earnings next year for BadgerMeter (BMI) 14. _____
15. Change from previous week for BanColumbia (CIB) 15. _____
16. Change from previous week for Babcock&Wilcox (BW) 16. _____
17. Closing price for the week for BRF (BRFS) 17. _____
18. Closing price for the week for BancoSantander (SAN) 18. _____

Ignore commissions and find the cost for the following stock purchases at the closing price for the week. Then find the annual dividend that will be paid on those shares.

Stock	Number of Shares	Cost	Dividend
19. Baker Hughes (BHI) 800 × $66.39 = $53,112; 800 × $.17 × 4 = $544	800	$53,112	$544
20. Barnes & Noble (BKS)	200	_____	_____
21. Barclays (BCS)	500	_____	_____
22. Barrick Gold (ABX)	100	_____	_____
23. Bank of Nova Scotia (BNS)	700	_____	_____
24. British Petroleum (BP)	300	_____	_____

25. Explain corporations, shares, and shareholders. (See Objective 1.)

26. Use the chart of the Dow Jones Industrial Average and estimate the years in which stocks fell by more than 10%. (See Objective 5.)

Find the current yield for each of the following stocks. Round to the nearest tenth of a percent. (See Example 3.)

Stock	Current Price per Share	Annual Dividend	Current Yield
27. Coca-Cola (KO) $1.40 ÷ $41.60 = 3.4%	$41.60	$1.40	3.4%
28. Facebook (FB)	$117.27	$0	_____
29. McDonald's (MCD)	$123.14	$3.76	_____
30. Apple Inc. (AAPL)	$116.52	$2.28	_____
31. Nike (NKE)	$51.91	$.72	_____
32. Walmart (WMT)	$69.54	$2.00	_____

11.4 Exercises

Find the PE ratio for each of the following. Round all answers to the nearest whole number. (See Example 4.)

	Stock	Current Price per Share	Annual Net Earnings per Share	PE Ratio
33.	Coca-Cola (KO)	$41.60	$1.65	25
	$41.60 ÷ $1.65 = 25			
34.	Facebook (FB)	$117.27	$2.09	_____
35.	McDonald's (MCD)	$123.14	$5.32	_____
36.	Apple Inc. (AAPL)	$116.52	$8.31	_____
37.	Nike (NKE)	$51.91	$2.22	_____
38.	Walmart (WMT)	$69.54	$4.61	_____

Stock prices on consecutive days for a stock are shown next. Find the increase (decrease) in the price of each stock as a number and the percent increase (decrease) rounded to the nearest tenth of a percent.

39. $34.35, 35.20

40. $46.50, 45.90

Solve the following application problems.

41. **STOCK PURCHASE** Patsy Bonner buys 200 shares of Target at $56.30 and 100 shares of Pepsico at $68.73. Find the total cost ignoring commissions.
200 × $56.30 + 100 × $68.73 = $18,133

41. $18,133

42. **WRITING A WILL** In her will, Barbara Bains stated that the trustee should purchase 300 shares of McDonald's and 200 shares of Walmart and give the stock to her grandson on his 25th birthday. If the stocks are selling for $87.91 per share and $72.02 per share, respectively, find the total amount paid, ignoring broker's commissions.

42. _____

43. **CDS OR GLOBAL STOCKS** Stan Walker is comparing CDs currently yielding 2% compounded semiannually to a mutual fund with stocks that he believes will yield 8% compounded semiannually. Find the future value of an annuity with deposits of $600 every 6 months for 10 years for **(a)** the CDs and **(b)** the mutual fund. **(c)** Find the difference.

(a) _____
(b) _____
(c) _____

CHAPTER 11 Annuities, Stocks, and Bonds

44. FIXED RATE OR STOCKS Jesica Tate plans to contribute $2500 per year to a retirement plan and is debating the use of a certificate of deposit that pays 3% per year versus an Asian stock fund that she believes will yield 10% per year. Find the future value after 12 years for **(a)** the certificate of deposit and **(b)** the stock fund. **(c)** Find the difference.

(a) _____

(b) _____

(c) _____

QUICK CHECK ANSWERS

1. (a) The highest price at which the stock sold during the year was $82.24.
   (b) The lowest price the stock sold for during the year was $62.30.
   (c) The ticker symbol under which the stock trades on the market is BLL.
   (d) The volume of shares sold during the day was $100 \times 89713 = 8,971,300$ shares.
   (e) The annual dividend yield is .7% of the current price.
   (f) The ratio of stock price to annual earnings is 42.
   (g) The stock closed at $76.66 at the end of last week.
   (h) The price of the stock was down $.18 this week compared to last week.
   (i) Ball earned $1.99 per share last year. It is expected to earn $3.44 per share this year and $4.28 per share next year.
   (j) The most recent quarterly dividend was $.13 per share.
2. $11,274; $204
3. 3.3%
4. 32
5. $188,668.35; $239,061.30

11.5 Bonds

OBJECTIVES

1. Define the basics of bonds.
2. Read bond tables.
3. Find the commission charge on bonds and the cost of bonds.
4. Understand how mutual funds containing bonds are used for monthly income.

Quick TIP
The U.S. government borrows using bonds. China and Japan each own more than one trillion in U.S. bonds, meaning that the U.S. owes each country over $1,000,000,000,000.

CASE IN POINT Darnell Johnson has decided to include mutual funds holding stocks in his retirement plan, but he doesn't know about bonds. What are bonds? Should he invest in them?

OBJECTIVE 1 Define the basics of bonds. Corporations need money for many purposes, including to make payroll or pay for a new building or equipment. The figure below shows different methods firms use to raise funds. **Bonds** are loans to firms, consisting of legally binding contracts between the lenders and the borrowers. They often require interest payments every year, and repayment at a specific date in the future. Unlike shareholders, bondholders do not own part of the corporation. Countries, cities, and even churches also borrow money using bonds.

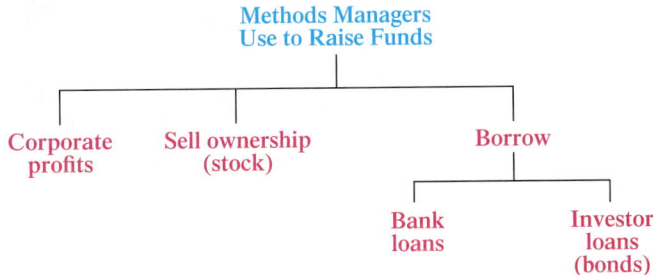

Common terms used with bonds

coupon rate – The interest rate stated on the bond.

face value – The amount to be repaid when the loan matures. Corporate bonds usually have a face value of $1000. Face value is sometimes referred to as **maturity value** or **par value**.

interest – Interest is found using the face value of the bond (usually $1000), the coupon rate, and the compounding period. Most bonds require interest to be paid every six months.

issuer – The issuer is the corporation or organization that is making the loan (selling the bond).

maturity date – Date at which the bond matures and the issuer repays the face value to the current owner of the bond.

term – The length of time until the bond matures. Terms on individual bonds are determined before the bond is issued, but the terms are often in the range of 1 to 30 years.

The prices of bonds vary according to the prevailing interest rate.

A company is said to go **bankrupt** if it can no longer meet its financial obligations to suppliers, banks, bondholders, and others. Bankruptcy is a complex process involving management, creditors, lawyers, and courts. Generally bankruptcy lawyers are paid first. Remaining assets are then used to pay off debt, including bonds. Shareholders *do not receive anything* unless there are assets remaining after all debts have been paid. Shareholders often receive very little or nothing from a bankruptcy, and bondholders often receive only 10 to 50 cents on every dollar originally loaned.

Corporations frequently use substantial amounts of debt to build factories, expand operations, or buy other companies. The interest that must be paid on that debt is the *cost of having debt*. The larger the debt, the greater the amount of revenue the company must set aside to pay interest. As interest rates increase, companies must set aside additional money to pay interest, leaving less for other purposes including profits.

OBJECTIVE 2 Read bond tables. The **face value**, or **par value**, of a bond is *the original amount of money* borrowed by a company. Most public corporations issue bonds with a par value of $1000. Principal and any interest due must be paid when a bond **matures**. Suppose that the owner of a bond needs money before the **maturity date** of the bond. In that event, the bond can be sold through a bond dealer, such as Merrill Lynch. The value of a bond is strongly influenced by the *credit history of the firm* and *market conditions at the time of the sale*. Suppose a company develops financial problems and investors start to worry that it might not be able to pay interest or debt when due. The price investors are willing to pay for each bond will immediately fall sharply. If the financial problems of the firm are severe, no one may want the bonds, in which case the price of the bond will fall drastically.

Market *interest rates fluctuate* from year to year, yet any one bond pays exactly the same dollar amount of interest each year. If interest rates rise, investors will pay less for a bond because they want the new, higher interest yield. If interest rates fall, investors will pay more for a bond because they are satisfied with the lower yield. As a result, the price of a bond fluctuates in the opposite direction of interest rates. A bond may have a face value of $1000, but it often trades at a value different than $1000, as shown in Example 1.

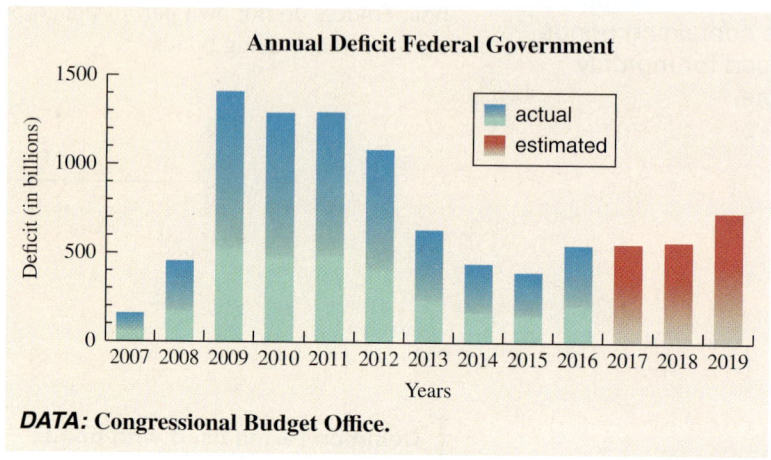

DATA: Congressional Budget Office.

It is difficult to forecast federal deficits, particularly just after a new president has taken office, as occurred in early 2017. So the estimated deficits for 2017, 2018, and 2019 in the bar graph are rough estimates.

The graph shows that U.S. government spending has exceeded revenues since 2007. The **budget deficits** soared in 2009–12 as the government spent significant funds during the Great Recession. The government finances deficits by borrowing from countries, pension and investment funds, and wealthy investors. Although some expected that the large deficits of 2009–12 would result in high inflation, it did not happen, partly because unemployment was so high that wages did not increase. The growth in wages is a major component of inflation.

Working with the Bond Table

EXAMPLE 1 Brandy Barrett was in an automobile accident that put her in the hospital for 3 weeks and required months of rehabilitation. The other driver was at fault, and his insurance company paid Barrett the liability limits on his policy of $50,000. Barrett needs monthly income and is thinking about investing the funds in Wells Fargo (WFC) bonds that mature in 2046. Analyze the data in the table.

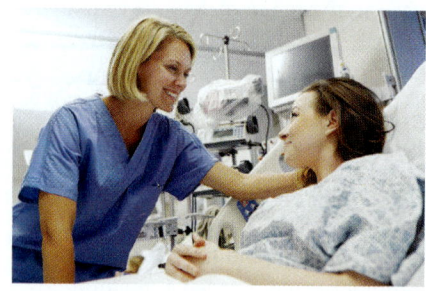

Company (Ticker)	Coupon	Maturity	Last Price	Last Yield	Est $ Vol (000's)
Wells Fargo (WFC)	4.750	December 7, 2046	99.004	4.813	171,626

SOLUTION

(a) The ticker symbol for Wells Fargo is WFC.
(b) Annual interest paid per bond = 4.75% of $1000 face value = **$47.50.**
(c) The bond matures on December 7, 2046, when Wells Fargo must repay the $1000 face value to the owner of the bond.
(d) The last price at which the bond traded for the day was 99.004% of the face value of $1000.

$$\text{Last price} = .99004 \times \$1000 = \textbf{\$990.04}$$

(e) An investor holding the bond until maturity in 2046 would earn 4.813% per year on the investment. This is often called the **yield to maturity**.
(f) The estimated face value of all the Wells Fargo bonds that sold during the week was

$$\text{Volume} = 171{,}626 \times \$1000 = \textbf{\$171{,}626{,}000}$$

QUICK CHECK 1

Analyze the data in the bond table for Bank of America (BAC) that matures in 2027.

Corporate Bonds

Company (Ticker)	Coupon	Maturity	Last Price	Last Yield	Est $ Vol (000's)
Noble Holding International Ltd (NE)	7.750	Jan 15, 2024	97.313	8.259	367,495
Viacom (VIA)	3.450	Oct 04, 2026	90.983	4.604	317,916
Viacom (VIA)	4.375	Mar 15, 2043	78.817	5.985	310,905
Citigroup (C)	3.200	Oct 21, 2026	94.411	3.889	309,316
Anheuser-Busch Inbev Finance (ABIBB)	4.900	Feb 01, 2046	105.409	4.559	301,132
Anheuser-Busch Inbev Finance (ABIBB)	3.650	Feb 01, 2026	99.567	3.706	278,699
Teva Pharm. Finance Netherlands Iii (TEVA)	3.150	Oct 01, 2026	90.551	4.345	261,095
Wells Fargo (WFC)	3.000	Oct 23, 2026	93.962	3.739	224,381
Ford Motor Co (F)	5.291	Dec 08, 2046	98.996	5.359	219,115
Deutsche Bank Ag (DB)	4.250	Oct 14, 2021	98.544	4.589	215,891
Metropolitan Life Global Funding I (MET)	3.450	Dec 18, 2026	99.648	3.492	215,256
Bank of America (BAC)	4.183	Nov 25, 2027	98.523	4.354	204,137
Actavis Funding Scs (AGN)	4.750	Mar 15, 2045	96.791	4.962	203,131
Citigroup (C)	2.900	Dec 08, 2021	99.095	3.098	196,479
JPMorgan Chase (JPM)	2.950	Oct 01, 2026	94.323	3.645	185,342
Wells Fargo (WFC)	4.750	Dec 07, 2046	99.004	4.813	171,626
Ensco International Incorporated (ESV)	4.700	Mar 15, 2021	98.125	5.197	168,876
Hsbc Holdings (HSBC)	4.375	Nov 23, 2026	99.147	4.482	159,976
Kraft Heinz Foods Co (KHC)	3.000	Jun 01, 2026	92.488	3.961	158,690
Apple (AAPL)	3.850	Aug 04, 2046	93.292	4.250	158,021
Anheuser-Busch Inbev Finance (ABIBB)	2.650	Feb 01, 2021	99.886	2.679	157,667
Oracle (ORCL)	1.900	Sep 15, 2021	96.953	2.587	157,638
Conocophillips Co (COP)	4.950	Mar 15, 2026	109.052	3.754	155,779
Unitedhealth Group (UNH)	3.450	Jan 15, 2027	100.050	3.444	155,153
Roper Technologies (ROP)	3.800	Dec 15, 2026	99.539	3.856	151,581

DATA: *Barron's*, December 16, 2016.

Using the Bond Table

EXAMPLE 2 Find the estimated volume sold during the prior week and the last sale price for the following bonds.

(a) Oracle (ORCL) maturing in 2021
(b) Apple (AAPL) maturing in 2046
(c) Ford Motor Company (F) maturing in 2046

SOLUTION

Company	Volume Sold	Last Sale Price per Bond
(a) Oracle	$157,638,000	$969.53
(b) Apple	$158,021,000	$932.92
(c) Ford Motor Co.	$219,115,000	$989.96

Notice the large volume of weekly sales. The global bond market is very active.

QUICK CHECK 2

Find the volume sold and the last sale price for Metropolitan Life Global Funding I (MET) bonds maturing in 2026.

446 CHAPTER 11 Annuities, Stocks, and Bonds

OBJECTIVE 3 Find the commission charge on bonds and the cost of bonds. The cost of buying and selling bonds has fallen sharply due to competition among brokers, and it varies widely depending on the volume of bonds being sold or bought. A common charge to buy or sell a few bonds is $1 per bond, but commissions are much lower for large transactions.

Commission costs must be taken into account when calculating the effective interest rate of a bond. It is important to remember that the effective interest rate is not the same as the last yield. Effective interest rate is calculated by dividing the total interest the investor earns by the total cost of the bond including commissions. Last yield takes into consideration the maturity value of the bond, the time to maturity, and all interest payments, but not commissions.

Finding the Cost to Buy Bonds

EXAMPLE 3 Assume that the sales charge is $1 per bond, and find the following for Kraft Heinz Foods Co (KHC) maturing in 2026.

(a) The total cost of purchasing 20 bonds
(b) The total annual interest paid on these bonds
(c) The effective interest rate to the buyer, including the cost of buying the bonds

SOLUTION

(a) Total cost = (Price per bond + Sales charge per bond) × Number of bonds
   = ($924.88 + $1) × 20 = **$18,517.60**

(b) Annual interest = Coupon rate × Par value of bond × Number of bonds
   = (0.03 × $1000) × 20 bonds = $600

(c) Effective rate = $\dfrac{\text{Total interest}}{\text{Total cost of bonds}} = \dfrac{\$600}{\$18,517.60} = 3.2\%$

QUICK CHECK 3

Find the total cost, annual interest, and effective rate for 10 Citigroup (C) bonds maturing in 2026 if the sales charge is $1 per bond.

Finding the Net Amount from the Sale of Bonds

EXAMPLE 4 Find the amount received from the sale of 50 Viacom (VIA) bonds maturing in 2043. Assume a commission of $1 per bond.

SOLUTION

Amount received = (Sales price of a bond − Sales charge per bond) × Number of bonds
   = ($788.17 − $1) × 50 = $39,358.50

QUICK CHECK 4

Find the amount received from the sale of 30 Conoco Phillips Co (COP) bonds maturing in 2026. Assume a commission of $1 per bond.

OBJECTIVE 4 Understand how mutual funds containing bonds are used for monthly income. Stock prices vary greatly but have higher average returns than bonds over time. As a result, financial planners recommend more stock investments for younger people and fewer for people near retirement. Since bonds generate a regular income and the risk of losing principal is usually less, bonds may be a better investment for those close to retiring.

Stocks that pay dividends are also used for income, but dividends are not guaranteed and those stocks usually pay less than bonds.

Using a Bond Fund for Income

EXAMPLE 5 Physicians have told Brandy Barrett in Example 1 that she may have long-lasting effects from the car accident. She puts the $50,000 received from the insurance company into a mutual fund containing bonds since there is less risk than investing in the bond of one company. **(a)** Find her annual income if the fund yields 6.5% per year. **(b)** How much would Barrett need to invest in the fund to earn $10,000 per year?

> **SOLUTION**
>
> (a) Use the formula for simple interest: $I = PRT$.
>
> $$\text{Interest} = \$50,000 \times .065 = \$3250$$
>
> (b) Again use the formula for simple interest, but now the principal (P) is unknown. Divide both sides of $I = PRT$ by RT to find the following form of the equation.
>
> $$\text{Principal} = P = \frac{I}{RT} = \frac{\$10,000}{.065 \times 1} = \$153,846.15$$

> **QUICK CHECK 5**
>
> James Corporation wants $80,000 per year in interest. Find the amount the firm must invest in a bond fund yielding 6.125% to attain this annual income.

11.5 Exercises // MyLab Math

The shaded sections below contain solutions to help you get a **QUICK START** *on the various types of exercises.*

Use the bond table on page 445 to find the following for Anheuser-Busch Inbev Finance (ABIBB) maturing in 2046. *(See Examples 1 and 2.)*

1. Price per bond   1. **$1054.09**

2. Volume of bonds sold during the week   2. _____

3. Date when bonds must be paid off by Anheuser-Busch   3. _____

4. Annual interest paid   4. _____

5. Last yield or yield to maturity   5. _____

6. Price to buy 50 of these bonds, including sales charge of $1 per bond   6. _____

Find the cost, including sales charges of $1 per bond, for each of the following purchases. *(See Example 3.)*

Bond	Maturity	Number Purchased	Cost
7. Noble Holding Int. Ltd (NE) ($973.13 + $1) × 20 bonds = $19,482.60	Jan. 15, 2024	20	**$19,482.60**
8. Deutsche Bank Ag (DB)	Oct. 4, 2021	50	_____
9. Teva Pharmaceutical (TEVA)	Oct. 1, 2026	100	_____
10. Unitedhealth Group (UNH)	Jan. 15, 2027	80	_____
11. Ensco Int. (ESV)	Mar. 15, 2021	30	_____
12. JPMorgan Chase (JPM)	Oct. 1, 2026	200	_____

13. What are bonds and why are they used? *(See Objectives 1 and 2.)*

C// indicates an exercise that is related to the Case in Point feature.

CHAPTER 11 Annuities, Stocks, and Bonds

14. Explain how a bondholder can estimate the effective interest rate on the total cost of the investment, including commissions. (See Example 3.)

Solve each application problem. Assume a sales commission of $1 per bond, unless indicated otherwise, and use the table on page 445. Round the rate to the nearest tenth of a percent.

15. BOND FUND Pete Chong manages bonds for a fund company. He purchases 4000 Actavis Funding Scs (AGN) bonds that mature in 2045. Since this is a large purchase, the commission is only $0.20 per bond. Find **(a)** the total cost of the purchase including commissions, **(b)** the annual interest, and **(c)** the effective interest rate based on total cost including commissions.

(a) Total cost = ($967.91 + $.20) × 4000 = $3,872,440
(b) Annual interest = (.0475 × $1000) × 4000 = $190,000
(c) Effective interest rate = $190,000 ÷ $3,872,440 = 4.9%

(a) _____
(b) _____
(c) _____

16. BOND PURCHASE New York purchased 10,000 bonds issued by Citigroup (C) bonds maturing in 2021. The commission is $.20 per bond due to the large volume. Find **(a)** the total cost including commissions, **(b)** the annual interest and **(c)** the effective interest rate.

(a) _____
(b) _____
(c) _____

17. BOND PURCHASE Charles Smitten needed more income than offered by money market funds, so he purchased 25 Oracle (ORCL) bonds maturing in 2021. Find **(a)** the total cost including commissions, **(b)** the annual interest and **(c)** the effective interest rate.

(a) _____
(b) _____
(c) _____

18. RETIREMENT FUNDS United Worker's pension administrator purchased 90 Apple (AAPL) bonds maturing in 2046. Find **(a)** the total cost including commissions, **(b)** the annual interest and **(c)** the effective interest rate.

(a) _____
(b) _____
(c) _____

19. BOND FUND Bernice Clarence places $45,000 in a bond fund that is currently yielding 8% compounded annually. **(a)** Find interest for the first year. **(b)** She decides to let all interest payments remain in the account. Find the amount in the account after 10 years if the fund continues to earn 8% compounded annually.

(a) _____
(b) _____

20. BOND FUND The hospital where Darnell Johnson works has an endowment funded by alumni and business owners in the community. The manager of the endowment invested $500,000 in a bond fund yielding 6% compounded semiannually. Find **(a)** the interest for the first year and **(b)** the future value of the account in 8 years.

(a) _____
(b) _____

QUICK CHECK ANSWERS

1. Ticker symbol: BAC; annual interest: $41.83; matures on November 25, 2027; last price = $985.23; yield to maturity = 4.354%; face value of bonds sold during the week = $204,137,000
2. $215,256,000; $996.48
3. $9451.10; $320; 3.4%
4. $32,685.60
5. $1,306,122.45

Chapter 11 Quick Review

Chapter Terms Review the following terms to test your understanding of the chapter. For each term you do not know, refer to the page number found next to that term.

401(k) [p. 412]
403(b) [p. 412]
amount of the annuity [p. 410]
annual meeting [p. 433]
annuity [p. 408]
annuity due [p. 411]
bankrupt [p. 434]
board of directors [p. 433]
bond [p. 443]
budget deficits [p. 444]
capital [p. 433]
common stock [p. 433]
compound amount of the annuity [p. 408]
corporation [p. 433]
coupon rate [p. 443]
current yield [p. 435]
discount brokers [p. 434]
diversify [p. 438]
dividends [p. 434]
Dow Jones Industrial Average [p. 437]
equivalent cash price [p. 420]
ETF [p. 438]
exchange-traded funds (ETFs) [p. 438]
executive officers [p. 433]
face value [p. 443]
future value of the annuity [p. 408]
individual retirement account (IRA) [p. 412]
interest [p. 443]
interest rate spreads [p. 427]
IRA [p. 412]
issuer [p. 443]
limited liability [p. 433]
mature [p. 443]
maturity date [p. 443]
mutual fund [p. 438]
NASDAQ Composite Index [p. 437]
ordinary annuity [p. 408]
par value [p. 443]
payment period [p. 408]
PE ratio [p. 436]
preferred stock [p. 433]
present value [p. 416]
price–earnings (PE) ratio [p. 435]
privately held corporation [p. 433]
publicly held corporation [p. 433]
regular IRA [p. 412]
risk [p. 438]
Roth IRA [p. 412]
sinking fund [p. 424]
sinking fund table [p. 425]
Standard and Poor's 500 [p. 437]
stock [p. 433]
stockbrokers [p. 434]
stock certificates [p. 433]
stock exchanges [p. 433]
stockholders [p. 433]
stock ratios [p. 435]
term [p. 408]
yield to maturity [p. 444]

CONCEPTS

11.1 Finding the amount of an ordinary annuity

Determine the number of periods in the annuity (n) and the interest rate per annuity period (i).
Use n and i in the annuity table to find the value of $1 at the term of annuity.
Find the value of an annuity using the formula.

$$\text{Amount} = \text{Payment} \times \text{Number from table}$$

EXAMPLES

Ed Navarro deposits $800 at the end of each quarter for 7 years into an IRA. Given interest of 6% compounded quarterly, find the future value.

$n = 7 \times 4 = 28$ periods; $i = \frac{6\%}{4} = 1.5\%$ per period

Number from table is **34.48148**.

Amount = $800 \times$ **34.48148** = **$27,585.18**

11.1 Finding the amount of an annuity due

Determine the number of periods in the annuity. Add 1 to the value and use this as the value of n.
Determine the interest rate per annuity period, and use the table to find the value of $1 at term of annuity.

The amount of the annuity is

$$\text{Payment} \times \text{Number from table} - 1 \text{ payment}$$

Find the amount of an annuity due if payments of $700 are made at the beginning of each quarter for 3 years in an account paying 8% compounded quarterly.

$n = 3 \times 4 + 1 = 13$; $i = \frac{8\%}{4} = 2\%$

Number from table is **14.68033**.

Amount = $700 \times$ **14.68033** − **$700** = $9576.23

11.2 Finding the present value of an annuity

Determine the payment per period.
Determine the number of periods in the annuity (n).
Determine the interest rate per period (i).
Use the values of n and i to find the number in the present value of an annuity table.

The present value of an annuity is

$$\text{Present value} = \text{Payment} \times \text{Number from table}$$

What lump sum deposited today at 8% compounded annually will yield the same total as payments of $600 at the end of each year for 10 years?

Payment = $600; $n = 10$

Interest = 8%

Number from table is **6.71008**.

Present value = $600 \times$ **6.71008** = **$4026.05**

CONCEPTS

11.2 Finding the equivalent cash price

Determine the amount of the annuity payment.
Determine the number of periods in the annuity (n).
Determine the interest rate per annuity period (i).
Use n and i in the present value of an annuity table.
Add the present value of the annuity to the down payment to obtain today's equivalent cash price.

11.3 Determining the payment into a sinking fund

Determine the number of payments (n). Determine the interest rate per period (i). Find the value of the payment needed to accumulate $1 from the sinking fund table.

Calculate the payment using

$$\text{Payment} = \text{Future value} \times \text{Number from table}$$

11.3 Setting up a sinking fund table

Determine the required payment into the sinking fund.
Calculate the interest at the end of each period.
Add the previous total, next payment, and interest to determine the total.
Repeat these steps for each period.

11.4 Reading the stock table

Use the stock table in Section 11.4 to find the following information for British Petroleum Prudhoe (BPT).

EXAMPLES

A buyer offers to purchase a business for $75,000 down and payments of $4000 at the end of each quarter for 5 years. Money is worth 8% compounded quarterly. Find the present value or equivalent cash price of the offer.

Payment = $4000; $n = 5 \times 4 = 20$

Interest = $\frac{8\%}{4} = 2\%$

Number from table is **16.35143**.

Present value = $4000 \times $ **16.35143** $ = $ **$65,405.72**

Equivalent cash value = $75,000 + **$65,405.72**
$= $140,405.72$

No-Leak Plumbing plans to accumulate $500,000 in 4 years in a sinking fund for a new building. Find the amount of each semiannual payment if the fund earns 6% compounded semiannually.

$n = 2 \times 4 \text{ years} = 8 \text{ periods}; i = \frac{6\%}{2} = 3\%$ per period

Number from the table is **.11246**.

Payment = $500,000 \times $ **.11246** $ = $56,230$

A company wants to set up a sinking fund to accumulate $10,000 in 4 years. It wishes to make semiannual payments into the account, which pays 8% compounded semiannually. Set up a sinking fund table.

$$n = 4 \times 2 = 8; \quad i = \frac{8\%}{2} = 4\%$$

Number from table is **.10853**.

Payment = $10,000 \times $ **.10853** $ = 1085.30

Payment	Amount of Deposit	Interest Earned	Total
1	$1085.30	$0	$1085.30
2	$1085.30	$43.41	$2214.01
3	$1085.30	$88.56	$3387.87
4	$1085.30	$135.51	$4608.68
5	$1085.30	$184.35	$5878.33
6	$1085.30	$235.13	$7198.76
7	$1085.30	$287.95	$8572.01
8	$1085.11	$342.88	$10,000.00

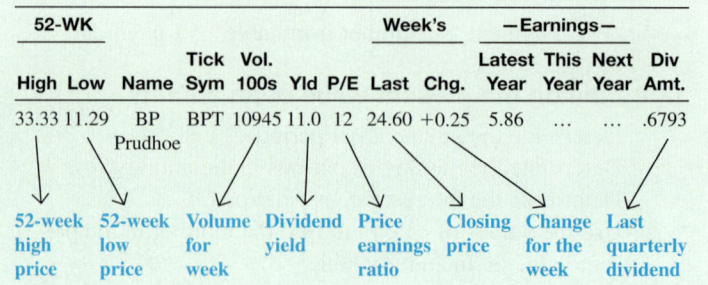

52-WK						Week's		—Earnings—		
High	Low	Name	Tick Sym	Vol. 100s	Yld	P/E	Last	Chg.	Latest Year	This Year	Next Year	Div Amt.
33.33	11.29	BP Prudhoe	BPT	10945	11.0	12	24.60	+0.25	5.86	...	...	.6793

52-week high price | 52-week low price | Volume for week | Dividend yield | Price earnings ratio | Closing price | Change for the week | Last quarterly dividend

CHAPTER 11 Quick Review

CONCEPTS

11.4 Finding the current yield on a stock

To determine the current yield, use the formula

$$\text{Current yield} = \frac{\text{Annual dividend}}{\text{Closing price}}$$

11.4 Finding the price–earnings (PE) ratio

To find the PE ratio, use the formula

$$\text{PE ratio} = \frac{\text{Price per share}}{\text{Annual net income per share}}$$

11.5 Reading the bond table

Use the bond table in Section 11.5 to find the following information for Roper Technologies (ROP) maturing in 2026.

11.5 Determining the cost of purchasing bonds

First locate the bond in the table. Find the price of the bond and multiply by $1000. Add $1 to this amount for commissions, then multiply by the number of bonds.

11.5 Determining the amount received from the sale of bonds

First locate the bond in the table. Find the price of the bond and multiply by $1000. Subtract $1 from this amount for commissions, then multiply by the number of bonds.

11.5 Finding the effective yield of a bond

Find the cost of the bond after commission. Find the interest paid on the bond using the coupon rate. Finally, divide the interest by the cost of the bond.

EXAMPLES

Find the current yield for a stock if the purchase price is $35 and the annual dividend is $.64.

$$\text{Current yield} = \frac{\$.64}{\$35} = 1.8\% \text{ (rounded)}$$

Find the PE ratio for a stock priced at $42.50 with earnings of $2.11.

$$\text{PE Ratio} = \frac{\$42.50}{\$2.11} = 20 \text{ (rounded)}$$

Company (Ticker)	Coupon	Maturity	Last Price	Last Yield	Est $ Vol (000's)
Roper Technologies (ROP)	3.800	Dec 15, 2026	99.539	3.856	151,581

- Pays 3.8% of $1000 per year
- Matures Dec. 15, 2026
- Closed at $995.39
- Yield to maturity of 3.856%
- Estimated volume for the week of $151,581,000

Including sales charges of $1 per bond, find the cost of 50 Apple (AAPL) bonds maturing in 2046.

$$(\$932.92 + \$1) \times 50 = \$46{,}696$$

Find the amount received after the sales charge from the sale of 20 Citigroup (C) bonds maturing in 2021.

$$(\$990.95 - \$1) \times 20 = \$19{,}799$$

Find the effective interest rate, to the nearest tenth, for a bond with a coupon rate of 3.8% and selling at 98.25.

Price of bond $= .9825 \times \$1000 + \$1 = \mathbf{\$983.50}$

Interest $= .038 \times \$1000 = \$38$

$$\text{Effective rate} = \frac{\$38}{\$983.50} = 3.9\% \text{ (rounded)}$$

Case Study

FINANCIAL PLANNING

At age 37, Paul Li decides to plan for his retirement at age 67. He currently has a net worth of about $45,000 including the equity in his home. He assumes that his employer will contribute $3500 to his retirement plan at the end of each year for the next 30 years. He plans to put one-half of his money in a mutual fund containing stocks and the other one-half in a mutual fund containing bonds.

1. Estimate Li's net worth at age 67 if his net worth grows at 5% per year, his stock fund grows at 10% per year, and his bond fund grows at 6% per year.

    1. _____

2. Li is amazed that he will be able to accumulate over $600,000. However, he knows that inflation will increase his cost of living significantly in 30 years. He assumes 3% inflation and wants to find the income he needs at age 67 to have the same purchasing power as $40,000 today. (*Hint:* Look at inflation in Section 10.2 and use the compound interest table in Section 10.1.)

    2. _____

3. Li has read newspaper articles stating that Social Security benefits will be reduced in the years ahead. After some thought, he decides to be conservative and assume that Social Security will pay only the first $30,000 of the annual income he needs at age 67. Find the remaining income he will need beginning at age 67.

    3. _____

4. Li decides to plan funding for his retirement for 20 years, from ages 67 to 87. If funds earn 8% compounded annually, find the present value of the annual income that he needs at 67 based on the income from part 3. above.

    4. _____

5. Will his expected savings fund his retirement?

    5. _____

6. What could go wrong with his plans?

    6. _____

INVESTIGATE

There is some discussion about the ability of Social Security to pay retirement benefits. Do you think Social Security will be around to help you during your retirement? What percent of your retirement needs do you think Social Security will pay? Try to support your views with recent articles from newspapers, magazines, or the World Wide Web.

Case in Point Summary Exercise

MAYO CLINIC

www.mayoclinic.com

- 1889: Founded in Rochester, MN
- 1987: Expanded to Jacksonville, FL
- 2016: Over 50,000 employees
- 2018: More than a million patients

The Mayo Clinic is the largest integrated, not-for-profit medical group practice in the world. With main facilities in Minnesota, Arizona, and Florida, it is ranked among the best hospitals in the world. Its mission is focused on people:

> To inspire hope and contribute to health and well-being by providing the best care to every patient through integrated clinical practice, education and research.

After getting his two-year degree in radiation therapy and working for three years at a small hospital, Darnell Johnson went to work for the Mayo Clinic at a salary of $63,500. The clinic puts 6% of his salary into his retirement plan and also allows Darnell to contribute up to an additional 5% of his salary.

1. Johnson decides to put 5% of his salary into his retirement plan in addition to the amount put into the plan by his employer. Find the total annual contribution into the retirement plan.

   1. _____

2. Johnson decides to invest his retirement funds in a mutual fund with global stocks that he hopes will yield 8% per year. Find the future value in his account after 12 years.

   2. _____

3. Assume the Mayo Clinic has exactly 55,000 employees and that the average salary per person including physicians is $87,300 per year, which includes the cost of health insurance paid by the employer. Estimate the annual payroll.

   3. _____

4. Use the annual payroll to estimate the annual contribution the Mayo Clinic must make into (a) FICA (Social Security) and Medicare (7.65% of all wages) and (b) the retirement plans of all employees.

   (a) _____
   (b) _____

5. (a) Find the total labor costs by adding the amounts in Exercises 3 and 4 above and then (b) write the number using words. It is clear that the Mayo Clinic needs a LOT of revenue to pay employees and this does not even include the costs of buildings, equipment, supplies, etc.

   (a) _____
   (b) _____

   _____

6. The Mayo Clinic saved the life of a wealthy entrepreneur who had cancer. He decided to donate $400,000 per year for the next 5 years to the hospital. Find the present value of this gift, assuming 6% per year.

   6. _____

7. Managers have decided to build a new building in 7 years and will need $47,400,000. The chief financial officer (CFO) decides to fund this cost by making contributions into a sinking fund at the end of each 6-month period. Find the payment needed every 6 months if funds earn 5% per year.

   7. _____

Chapter 11 Test

To help you review, the numbers in brackets show the section in which the topic was discussed.

Find the amounts of the following annuities. **[11.1]**

	Amount of Each Deposit	Deposited	Rate per Year	Number of Years	Type of Annuity	Amount of Annuity
1.	$1000	annually	6%	8	ordinary	_____
2.	$4500	semiannually	4%	9	ordinary	_____
3.	$30,000	quarterly	8%	6	due	_____
4.	$2600	semiannually	5%	12	due	_____

5. James Rivera earned his degree in drafting at a community college and recently began his new career. He was happy to learn that his new employer will deposit $2500 into his 401(k) retirement account at the end of each year. Find the amount he will have accumulated in 15 years if funds earn 8% per year. **[11.1]**

5. _____

6. James Rivera from Exercise 5 has also decided to invest $1000 at the end of each 6 months in an IRA that grows tax deferred. Find the amount he will have accumulated if he does this for 15 years and earns 6% compounded semiannually. **[11.1]**

6. _____

Find the present value of the following annuities. **[11.2]**

	Amount per Payment	Payment at End of Each	Number of Years	Interest Rate	Compounded	Present Value
7.	$1000	year	9	6%	annually	_____
8.	$4500	6 months	6	10%	semiannually	_____
9.	$1500	quarter	5	6%	quarterly	_____
10.	$14,000	quarter	6	8%	quarterly	_____

11. Betty Yowski borrows money for a new swimming pool and hot tub. She agrees to repay the note with a payment of $1200 per quarter for 6 years. Find the amount she must set aside today to satisfy this capital requirement in an investment earning 8% compounded quarterly. **[11.2]**

11. _____

12. Dan and Mary Foster just divorced. The divorce settlement included $650 a month payment to Dan for the 4 years until their son turns 18. Find the amount Mary must set aside today in an account earning 6% compounded semiannually to satisfy this financial obligation. **[11.2]**

12. _____

Find the amount of the payment needed into a sinking fund for each of the following. **[11.3]**

	Amount Needed	Years Until Needed	Interest Rate	Interest Compounded	Amount of Payment
13.	$100,000	9	6%	annually	_____
14.	$250,000	10	8%	semiannually	_____
15.	$360,000	11	6%	quarterly	_____
16.	$800,000	12	10%	semiannually	_____

Solve the following application problems.

17. The owner of King BBQ plans to open a new restaurant in 4 years at a cost of $200,000. Find the required semiannual payment into a sinking fund if funds are invested in an account earning 6% per year compounded semiannually. [11.3]

17. _____

18. Lupe Martinez will owe her retired mother $45,000 for a piece of land. Find the required quarterly payment into a sinking fund if Lupe pays it off in 4 years and the interest rate is 10% per year compounded quarterly. [11.3]

18. _____

19. George Jones purchases 200 shares of ExxonMobil stock at $85.82 per share. Find **(a)** the total cost and **(b)** the annual dividend if the dividend per share is $2.28. [11.4]

(a) _____
(b) _____

20. Belinda Deal purchases 25 IBM bonds at 95.1. They have a coupon rate of 4.2%. Find **(a)** the total cost if commissions are $1 per bond, **(b)** the annual interest, and **(c)** the effective interest rate rounded to the nearest tenth. [11.5]

(a) _____
(b) _____
(c) _____

21. Explain the following. [11.4]

52-Wk		Name	Tick Sym.	Vol. 100s	Yld.	P/E	Week's		Earnings			Div Amt.
High	Low						Last	Chg.	Latest Year	This Year	Next Year	
65.90	20.10	BlackDeck	BDK	74515	.8	27	63.44	+1.23	5.47	2.53	2.89	.12

22. Explain the following. [11.5]

Company (Ticker)	Coupon	Maturity	Last Price	Last Yield	EST $ Vol (000s)
International Paper Co (IP)	7.950	Jun 15, 2032	117.389	5.369	209,183

Business and Consumer Loans

12

CHAPTER CONTENTS

- 12.1 Open-End Credit and Charge Cards
- 12.2 Installment Loans
- 12.3 Early Payoffs of Loans
- 12.4 Personal Property Loans
- 12.5 Real Estate Loans

CASE IN POINT

AFTER JACKIE WATERTON received her degree in business, she went to work for Citigroup, Inc. which is one of the largest banks in the world. She worked in electronic banking for two years and then moved to the department that manages consumer credit. Recently she was promoted and now heads a group that works with people who have major credit problems. She is surprised at how much debt some customers accumulate.

12.1 Open-End Credit and Charge Cards

Debt is extremely important to every modern economy and is used daily by individuals, firms, school districts, cities, governments, and others. Individuals borrow for small and large purchases. Firms borrow to make payroll, build buildings, and pay taxes. School districts borrow to build a new school. Cities borrow to build a sports complex, and state and federal governments borrow heavily and often. When used carefully, debt is healthy and helps everyone involved. However, too much debt is a problem for any person, firm, or governmental entity.

Student Debt Problems

According to the Federal Reserve Bank of New York, the average debt of students graduating from college in the northeast exceeds $37,000. Total student debt including debt owed by students who have already graduated is $1.4 trillion dollars. Students are borrowing more to pay higher tuitions at a time when there is more competition than ever for scholarships. Worse, students in some majors struggle to find good jobs, making it very difficult for them to repay loans.

OBJECTIVES

1. Define open-end credit.
2. Define revolving charge accounts.
3. Use the unpaid balance method.
4. Use the average daily balance method.
5. Define loan consolidation.

CASE IN POINT // Jackie Waterton has worked with many families who have serious debt problems. She often helps these families set up budgets and then tries to reduce their monthly debt payments by refinancing loans at lower interest rates when possible.

This chapter covers several different kinds of loans including credit and charge card loans, short-term installment loans, and loans on personal property and real estate.

OBJECTIVE 1 Define open-end credit. A common way of buying on credit, called **open-end credit**, has no fixed payments. The customer continues making payments until no outstanding balance is owed. With open-end credit, additional credit is often extended before the initial amount is paid off. Examples of open-end credit include most department-store charge accounts and charge cards. Individuals are given a **credit limit**, or a maximum amount that may be charged on these accounts. The lender determines the credit limit for each person based on income, assets, other debts, and credit history.

A sale paid for with a **debit card** authorizes the retailer's bank to debit the purchaser's checking account immediately upon receipt, so debit cards do not involve loans or credit. The funds either are, or are not, in your bank account when making a debit. A Visa or MasterCard can be either a credit or a debit card depending on the bank and the cardholder's preference.

OBJECTIVE 2 Define revolving charge accounts. Individuals often make several charges a month to a bank card or department-store account. The charges are accumulated throughout the month and billed at the end of the month. Sometimes people pay off the full amount owed on a card at the end of a month; but often they do not which results in an interest charge. This type of account is called a **revolving charge account**, because the full amount owed is often not paid. There are more than 3 billion credit cards in use worldwide. This results in a very large amount of credit extended to card holders. Banks earn a lot of interest on credit-card loans.

According to the Federal Trade Commission, credit and charge card fraud costs cardholders and issuers hundreds of millions of dollars each year. Individuals making charges while on vacation sometimes return home to find that someone has stolen their credit-card number and is using it to make purchases. It is important to immediately report any fraudulent use of your credit card to the bank issuing the card. By law, your maximum liability is limited to $50 once you have reported the theft.

Quick TIP // Interest charges are often very high on revolving charge accounts, especially on cash advances.

Sample Credit-Card Receipt

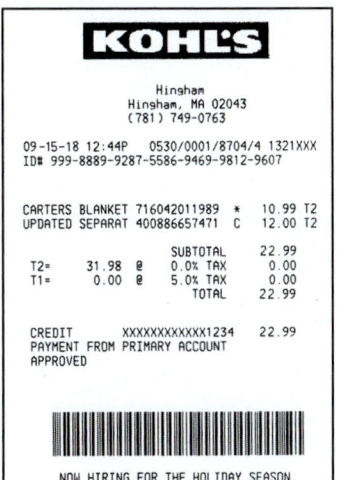

Prevent Credit-Card Losses

1. Keep a separate record of account number, phone number, and address of each bank that has issued you a credit card so that you can quickly notify them if a card is stolen.
2. Be sure to retrieve your card after making a purchase and before leaving the business.
3. Save your receipts and compare amounts to billing statements.
4. Open bills promptly and check the individual charges for any errors.
5. Check your account balance once or twice a month so you can note any fraudulent charges sooner.
6. Immediately report any questionable charges to the bank issuing the card.

458 CHAPTER 12 Business and Consumer Loans

There are billions of credit- and debit-card transactions every month, with transactions occurring all over the world. Merchants prefer debit-card sales due to the lower cost for them and the fact that funds are immediately transferred from the customer's bank account to the merchant's. On the other hand, banks generally prefer credit-card transactions due to the higher fees earned. Merchants and customers alike continue to move to the use of cards over cash for convenience. At the end of each billing cycle, each credit-card holder receives a statement showing details. For example, the statement below is an **itemized billing** for a Visa card that shows payments, credits, purchases, and interest charges made by the bank. These statements also show any fees related to purchases made in a foreign country. Notice that the due date of the payment is also shown on the statement—one has to make the payment by the due date or face penalties.

Finance charges may include interest, credit life insurance, a time-payment differential, and carrying charges. Interest charges can be avoided if the total balance is paid by the end of the **grace period**. Grace periods range between 15 and 30 days depending on the company. Often, there is no grace period on cash advances, so that finance charges are assessed beginning immediately. Lenders also charge **late fees** for payments that are received after the due date. **Over-the-limit fees** are charged by the lender when the borrower charges more than an approved maximum amount of debt.

OBJECTIVE 3 Use the unpaid balance method. Finance charges on open-end credit accounts may be calculated using the **unpaid balance method**. This method calculates finance charges based on the unpaid balance at *the end of the previous month*. Any purchases or returns during the current month are not used in calculating the finance charge, as you can see in the next example.

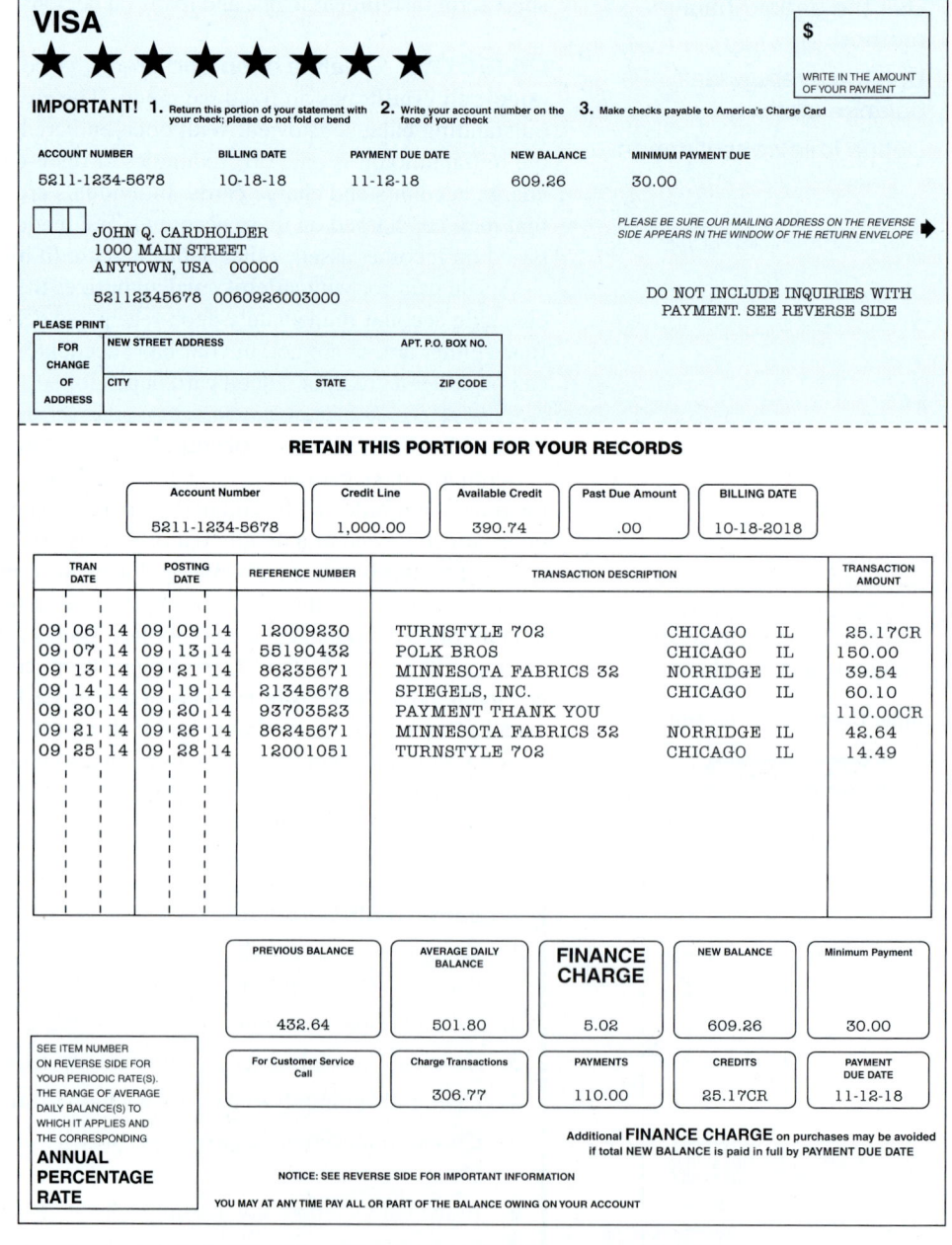

Finding Finance Charge Using the Unpaid Balance Method

EXAMPLE 1 (a) Peter Brinkman's MasterCard account had an unpaid balance of $870.40 on November 1. During November, he made a payment of $100 and used the card to purchase a puppy costing $150 for his son. Find the finance charge and the unpaid balance on December 1 if the bank charges 1.5% per month on the unpaid balance.

$$\text{Finance charge based on unpaid balance} = \$870.40 \times .015$$
$$= \$13.06$$

Find the unpaid balance on December 1 as follows.

$$\underbrace{\$870.40}_{\text{previous balance}} + \underbrace{\$13.06}_{\text{finance charge}} + \underbrace{\$150}_{\substack{\text{purchases during month}}} - \underbrace{\$100}_{\text{payment}} = \underbrace{\$933.46}_{\text{new balance}}$$

Quick TIP
The interest rate on charge card loans ranges from 12% to over 20% per year depending on credit history and other factors.

(b) During December, Brinkman made a payment of $50, charged $240.56 for Christmas presents, returned $35.45 worth of items, and took his family to dinner with charges of $92.45. Find his unpaid balance on January 1.

The finance charge calculated on the unpaid balance is $933.46 \times .015 = \$14.00$. The unpaid balance on January 1 follows.

$$\$933.46 + \$14.00 + \$240.56 + \$92.45 - \$35.45 - \$50 = \$1195.02$$

Month	Unpaid Balance at Beginning of Month	Finance Charge	Purchases During Month	Returns	Payment	Unpaid Balance at End of Month
November	$870.40	$13.06	$150.00	—	$100	$ 933.46
December	$933.46	$14.00	$333.01	$35.45	$ 50	$1195.02

The total finance charge during the 2-month period was $13.06 + \$14.00 = \$27.06$.

(c) Brinkman moves the balance to another charge card that charges only .8% interest per month. Find his savings in finance charges for January.

$$\text{Savings} = \underbrace{(\$1195.02 \times .015)}_{\text{old charge card}} - \underbrace{(\$1195.02 \times .008)}_{\text{new charge card}} = \$8.37$$

QUICK CHECK 1

The unpaid balance on a Visa card was $284.37. During the month, a payment of $200 was made and charges of $357.54 were added. If the finance charge is 1.2% per month on the unpaid balance, find (a) the finance charge for the month and (b) the new balance at the end of the month.

Suppose you decide to purchase a $1000 digital television set and charge it to a Visa card with finance charges of 1.5% per month on the unpaid balance. Further suppose that you make payments of $50 every month and don't charge anything else on the card. As shown in the table, it will take 24 months to pay off the television set. The $1000 television set will cost you an extra $197.83 in finance charges for a total cost of $1197.83, or nearly $1200.

Month	Unpaid Balance at Beginning of Month	Finance Charge	Payment	Unpaid Balance at End of Month
1	$1,000.00	$15.00	$50.00	$965.00
2	$965.00	$14.48	$50.00	$929.48
3	$929.48	$13.94	$50.00	$893.42
⋮	⋮	⋮	⋮	⋮
22	$143.53	$ 2.15	$50.00	$95.68
23	$ 95.68	$ 1.44	$50.00	$47.12
24	$ 47.12	$ 0.71	$47.83	$ 0.00
Totals		$197.83	$1197.83	

The nearly $200 in interest increases the cost of the television by nearly 20%. So the lender makes a lot of money at your expense! Here is how to buy the TV for less:

1. wait several months to buy it since the prices of technology items usually fall rapidly, and
2. save ahead of time for the TV in an account that pays you interest.

Doing this could result in you getting the $1000 TV for less than $850, rather than $1200 which includes interest.

OBJECTIVE 4 Use the average daily balance method. Most revolving charge plans calculate finance charges using the **average daily balance method**. First, the balance owed on the account is found at the end of each day during a month or billing period. All of these amounts are added, and the total is divided by the number of days in the month or billing period. The result is the average daily balance of the account, which is then used to calculate the finance charge.

Finding the Average Daily Balance

EXAMPLE 2 Beth Hogan's balance on a Visa card was $209.46 on March 3. Her activity for the next 30 days is shown in the table. **(a)** Find the average daily balance on April 3. Given finance charges based on $1\frac{1}{2}$% on the average daily balance, find **(b)** the finance charge for the month and **(c)** the balance owed on April 3.

Transaction Description		Transaction Amount
Previous balance $209.46		
March 3	Billing date	
March 12	Payment	$50.00 CR*
March 17	Walmart	$28.46
March 20	Mail order	$31.22
April 1	Auto parts	$59.10

*CR represents *credit*.

SOLUTION

(a)

Date	Unpaid Balance	Number of Days Until Balance Changes
March 3	$209.46	9
March 12	$159.46 = $209.46 − March 12 payment of $50	5
March 17	$187.92 = $159.46 + March 17 charge of $28.46	3
March 20	$219.14 = $187.92 + March 20 charge of $31.22	12
April 1	$278.24 = $219.14 + April 1 charge of $59.10	2
April 3	end of billing cycle …	31 total number of days in billing period

It is 9 days from March 3 to March 12, so the unpaid balance remains at $209.46 for 9 days.

Quick TIP

The billing period in Example 2 is 31 days, but some billing periods are for 30 or fewer days. Be sure to use the correct number of days.

There are 31 days in the billing period (March has 31 days). Find the average daily balance as follows:

Step 1 Multiply each unpaid balance by the number of days for that balance.
Step 2 Total these amounts.
Step 3 Divide by the number of days in that particular billing cycle (month).

Step 1

Unpaid Balance		Days		Total Balance
$209.46	×	9	=	$1885.14
$159.46	×	5	=	797.30
$187.92	×	3	=	563.76
$219.14	×	12	=	2629.68
$278.24	×	2	=	556.48
				$6432.36 ← Step 2

Step 3

$$\frac{\$6432.36}{31} = \$207.50 \text{ average daily balance}$$

Hogan will pay a finance charge based on the average daily balance of $207.50.

(b) The finance charge is $.015 \times \$207.50 = \$3.11$.

(c) The amount owed on April 3 is the beginning unpaid balance less any returns or payments, plus new charges and the finance charge.

$$\underset{\substack{\text{previous} \\ \text{balance}}}{\$209.46} - \underset{\text{payment}}{\$50} + \underset{\text{new charges}}{(\$28.46 + \$31.22 + \$59.10)} + \underset{\substack{\text{finance} \\ \text{charge}}}{\$3.11} = \$281.35$$

QUICK CHECK 2

The July 5 balance on a credit card was $494. A payment of $400 was made on July 22, and a charge of $258.67 was made on July 25. If the finance charge is based on $1\frac{1}{2}\%$, find (a) the average daily balance on August 5, (b) the finance charge for the month, and (c) the unpaid balance on August 5.

The finance charge in the previous example was small, only $3.11. Have you ever wondered how banks make money on charge cards? Let's explore this issue. Citigroup recently had a balance of $150 billion outstanding on credit cards. Many people pay off their loan balances every month, resulting in no interest charges. However, assume the bank earns an average of 1.2% on $50 billion of the outstanding debt. Then

$$\text{Interest} = \$50,000,000,000 \times .012 = \mathbf{\$600,000,000}$$

Quick TIP
Some banks make a lot of money from credit card debt!

The revenue to the bank for the month was $600,000,000 just from interest charges on outstanding credit-card debt. Of course, the bank has a lot of expenses related to generating the revenue, but that is still a lot of money.

If the finance charges are expressed on a per-month basis, find the **annual percentage rate** by multiplying the monthly rate by 12, the number of months in a year. For example, $1\frac{1}{2}\%$ per month is the same as:

$$1\tfrac{1}{2}\% \times 12 = 1.5\% \times 12 = 18\% \text{ per year}$$

OBJECTIVE 5 Define loan consolidation. Credit can be *very easy to get* for individuals who have a good credit history and a stable job. The clipping shows that too much spending and borrowing often creates problems.

Easy Credit–Big Worry

James Machler has a degree in nursing and a great job. In the past, he has been careful with his money, even paying off an extra $80 each month on his student loans and saving a little each month. But he finally splurged and bought his dream car, a BMW. Now his cash flow is tight and his employer has just announced there will be some layoffs. Suddenly, James is very worried.

Too many people get caught in a credit trap. They effectively spend more than they can afford by borrowing. They then end up with problems that can last for months or years. Below are some tips to gain control of your finances.

Quick TIP
Managers constantly apply the same basic ideas in businesses as they work to increase revenue and decrease costs.

Gaining Control of Your Finances

1. Increase your income by investing in yourself. Choose a career you enjoy, and get training and education.
2. Make a budget and stick to it.
3. Spend less. Here are some suggestions.
    (a) Make sure you can afford your rent or mortgage payment.
    (b) Eat out less often.
    (c) Drive that old automobile one or two more years.
    (d) Purchase a less expensive automobile or reduce the number of automobiles in your family by one.
    (e) Be careful with the amount you spend on entertainment, hobbies, and travel.
    (f) Don't buy on impulse. If you want something, write it down on a piece of paper and stick it on your refrigerator for 30 days. After 30 days, ask yourself if you actually need the item.

4. Try to pay cash for things the day you buy them rather than using credit.
5. Save more by paying yourself first. Do this by saving every month before you spend money on other things.
6. Unexpected costs always occur. Set aside money ahead of time so you are better prepared.
7. Contribute to a long-range retirement plan. Your employer may help by matching your contributions in addition to paying your salary.

Quick TIP
You may be able to consolidate your loans using an online company such as lendingclub.com.

Have you ever found yourself in a position where you cannot make all of your monthly payments? If so, you may be able to **consolidate your loans** into a single loan with one lower monthly payment. The new loan may have a lower interest rate and also a longer term, meaning that payments must be made for a longer period of time. This process can help you afford your monthly payments rather than defaulting on debt. **Defaulting on your debt**, or not making your payments, can mean repossession of your automobile or furniture, eviction from your apartment or house, and/or court appearances. Defaulting on your debt also ruins your credit history and can make it difficult to borrow money to buy a car or a home for years into the future.

Consolidating Loans **EXAMPLE 3** Bill and Jane Smith married two years ago. Both were happy when they had their first child, but they needed to buy several things on credit. They now have the monthly payments shown below. The Smiths are having difficulties making the payments and they sometimes argue over money. They bank online at Citibank and ask Jackie Waterton for help.

Revolving Accounts	Debt	Annual Percentage Rate	Minimum Monthly Payment	Other Payments	Monthly Payment
Target	$3880.54	18%	$150	Rent	$800
Home Depot	$1620.13	16%	$60	Jane's car payment	$315
MasterCard	$3140.65	14%	$100	Bill's truck payment	$268
Visa	$4920.98	18%	$200	Total	$1383
Total	$13,562.30		Total $510		

CASE IN POINT

SOLUTION

Jackie Waterton did the following:
1. Put the Smiths on a **strict monthly budget** so expenses are better balanced with income.
2. Consolidated their revolving account debts into one longer-term, low-interest loan (this required a loan guarantee from Bill's father).
3. Decreased one automobile payment by refinancing the loan over a longer term.

Here are their new monthly payments.

Account	Monthly Payment	New Status
Citigroup loan for $13,562.30	$337.50	Revolving loans were consolidated
Rent	$800.00	Unchanged
Jane's car payment	$247.50	Refinanced using a longer term
Bill's truck payment	$268.00	Unchanged
	Total $1653.00	

Reduction in payments = ($510 + $1383) − $1653 = **$240 per month**

Quick TIP
Individuals who consolidate their loans and then borrow even more can get into very serious financial difficulties.

The Smiths should be all right as long as they:
1. **Stay on their monthly budget.**
2. **Do not make additional credit purchases.**
3. **Continue to make all payments.**

The Smiths may end up with severe debt problems if they borrow more before the existing loan balances are significantly reduced. Borrowing more could force them to declare bankruptcy.

> **QUICK CHECK 3**
>
> A family refinanced the mortgage on their home, reducing the payment from $1269.45 to $1093.12. They also sold a car that had a monthly payment of $385.65 and negotiated a $50-per-month payment reduction on their second car. Find the decrease in monthly payments.

12.1 Exercises // MyLab Math

The shaded sections below contain solutions to help you get a QUICK START on the various types of exercises.

Find the finance charge on each of the following revolving charge accounts. Assume interest is calculated on the unpaid balance of the account. (See Example 1.)

	Unpaid Balance	Monthly Interest Rate	Finance Charge
1.	$6425.40	0.8%	**$51.40**
	$6425.40 × .008 = $51.40		
2.	$595.35	$1\frac{1}{2}\%$	_____
3.	$1201.43	$1\frac{1}{4}\%$	_____
4.	$2540.33	1.2%	_____

Complete the following tables, showing the unpaid balance at the end of each month. Assume an interest rate of 1.4% on the unpaid balance. (See Example 1.)

	Month	Unpaid Balance at Beginning of Month	Finance Charge	Purchases During Month	Returns	Payment	Unpaid Balance at End of Month
5.	October	$437.18	_____	$128.72	$27.85	$125	_____
	November	_____	_____	$291.64	—	$175	_____
	December	_____	_____	$147.11	$17.15	$150	_____
	January	_____	_____	$27.84	$127.76	$225	_____
6.	October	$2184.60	_____	$540.68	$50.87	$200	_____
	November	_____	_____	$890.27	—	$250	_____
	December	_____	_____	$1240.13	$89.69	$125	_____
	January	_____	_____	$384.20	—	$575	_____

7. Compare the unpaid balance method and the average daily balance method for calculating interest on open-end credit accounts. (See Objectives 3 and 4.)

8. Explain how consolidating loans may be of some advantage to the borrower. What are the disadvantages? (See Objective 5.)

C// indicates an exercise that is related to the Case in Point feature.

464 CHAPTER 12 Business and Consumer Loans

Find the finance charge for the following revolving charge accounts. Assume that interest is calculated on the average daily balance of the account. (See Example 2.)

	Average Daily Balance	Monthly Interest Rate	Finance Charge
9.	$1458.25 $1458.25 × .014 = $20.42	1.4%	$20.42
10.	$841.60 $841.60 × .015 = $12.62	$1\frac{1}{2}$%	$12.62
11.	$389.95	$1\frac{1}{4}$%	_____
12.	$6320.10	0.9%	_____
13.	$1235.68	1.4%	_____
14.	$4235.47	$1\frac{3}{4}$%	_____

Solve the following application problems.

15. HOT TUB PURCHASE Betty Thomas borrowed $6500 on her Visa card to install a hot tub with landscaping around it. The interest charges are 1.6% per month on the unpaid balance. **(a)** Find the monthly interest charges. **(b)** Find the interest charges if she moves the debt to a credit card charging 1% per month on the unpaid balance. **(c)** Find the monthly savings.

(a) _____
(b) _____
(c) _____

16. CREDIT-CARD BALANCE Alphy Jurarim used a credit card from Citibank to help pay for tuition expenses while in college and now owes $5232.25. The interest charges are 1.75% per month. **(a)** Find the interest charges. **(b)** Find the interest charges if he moves the debt to a credit card charging .8% per month on the unpaid balance. **(c)** Find the savings.

(a) _____
(b) _____
(c) _____

(a) Find the average daily balance for the following credit-card accounts. Assume one month between billing dates using the proper number of days in the month. (b) Then find the finance charge if interest is 1.5% per month on the average daily balance. (c) Finally, find the new balance. (See Example 2.)

17. Previous balance $2340.52

November 12	Billing date	
November 20	Payment	$1000
November 21	Road bicycle & equipment	$1440.30
November 30	Flowers	$65.40

Nov. 12 to Nov. 20 = 8 days at $2340.52 gives $18,724.16
Nov. 20 to Nov. 21 = 1 day at $1340.52 gives $1340.52
Nov. 21 to Nov. 30 = 9 days at $2780.82 gives $25,027.38
Nov. 30 to Dec. 12 = 12 days at $2846.22 gives $34,154.64
8 + 1 + 9 + 12 = 30 days
$18,724.16 + $1340.52 + $25,027.38 + $34,154.64 = $79,246.70

(a) Average daily balance = $\dfrac{\$79{,}246.70}{30}$ = $2641.56

(b) Finance charge = $2641.56 × .015 = $39.62
(c) New balance = $2340.52 + $39.62 − $1000 + $1440.30 + $65.40 = $2885.84

(a) $2641.56
(b) $39.62
(c) $2885.84

18. Previous balance $228.95

January 27	Billing date	
February 9	Walmart	$11.08
February 13	Returns	$26.54
February 20	Payment	$29
February 25	Restaurant	$71.19

(a) _____
(b) _____
(c) _____

19. Previous balance $312.78

June 11	Billing date	
June 15	Returns	$106.45
June 20	Coat	$115.73
June 24	Car rental	$74.19
July 3	Payment	$115

(a) _____
(b) _____
(c) _____

20. Previous balance $714.58

August 17	Billing date	
August 21	Mail order	$26.94
August 23	Returns	$25.41
August 27	Target	$31.82
August 31	Payment	$128.00
September 9	Returns	$71.14
September 11	Groceries	$110.00
September 14	Cash advance	$100.00

(a) _____
(b) _____
(c) _____

21. Previous balance $355.72

March 29	Billing date	
March 31	Returns	$209.53
April 2	Auto parts	$28.76
April 10	Pharmacy	$14.80
April 12	Returns	$63.54
April 13	Returns	$11.71
April 20	Payment	$72.00
April 21	Flowers	$29.72

(a) _____
(b) _____
(c) _____

12.2 Installment Loans

OBJECTIVES

1. Define installment loan.
2. Find the total installment cost and the finance charge.
3. Use the formula for approximate APR.
4. Use the table to find APR.

OBJECTIVE 1 Define installment loan. A loan is **amortized** if both principal and interest are paid off using a sequence of equal periodic payments. This type of loan is called an **installment loan**. People use installment loans for cars, boats, home improvements, and even for consolidating several loans into one affordable loan. Firms use installment loans to purchase equipment, computers, vehicles, mining equipment, etc.

The graphic shows the total interest that must be paid when financing a new Ford Escape over 3, 4, and 5 years. Financing it for 3 years results in $1568 in interest. Financing it for 5 years raises the interest costs to $2600. Financing it for 5 years increases the total cost to $25,000 + $2600 = $27,600, effectively increasing the cost by 10.4%.

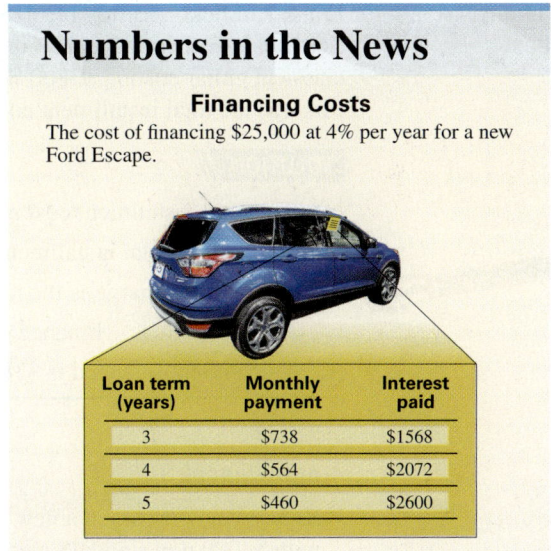

Numbers in the News

Financing Costs
The cost of financing $25,000 at 4% per year for a new Ford Escape.

Loan term (years)	Monthly payment	Interest paid
3	$738	$1568
4	$564	$2072
5	$460	$2600

Quick TIP

Shorter-term loans usually result in much less interest even though payments are sometimes higher!

The federal **Truth in Lending Act** (Regulation Z) of 1969 requires lenders to disclose their **finance charge** (the charge for credit) and **annual percentage rate (APR)** on installment loans. The federal government *does not* regulate rates. Each individual state sets the maximum allowable rates and charges.

The interest rate that is **stated** (in the newspaper, a marketing brochure, or a problem in a textbook) is also called the **nominal rate**. The nominal or stated rate can differ from the annual percentage rate or APR, which is based on the actual amount received by the borrower. The APR is the *true effective annual interest rate* for a loan. Information on two loans of $1000 each is shown below. The stated rates and interest on the loans are identical, but the terms differ, as do the APRs.

	Loan Amount	Stated Rate	Interest	Term	APR	
Loan 1	$1000	10%	$100	1 year	$R = \dfrac{I}{PT} = \dfrac{\$100}{\$1000 \times 1}$	$= 10\%$
Loan 2	$1000	10%	$100	9 months	$R = \dfrac{I}{PT} = \dfrac{\$100}{\$1000 \times \frac{9}{12}}$	$= 13.3\%$

The interest rates on the two loans are 10% and 13.3%, respectively, so they differ significantly. It pays to shop around for the lowest APR. Your credit history may affect your interest rate.

Credit history is a record of how responsible the borrower has been in terms of amounts borrowed and whether payments were made on time. Those with a poor credit history have more difficulty borrowing and are usually charged a higher interest rate, resulting in higher monthly payments. Being late on two or three payments of charge card debt can result in higher payments if you then purchase an automobile. However, if you make your payments regularly and do not borrow excessively, negative information will eventually age off your credit history report.

OBJECTIVE 2 Find the total installment cost and the finance charge. The total **installment cost** includes the down payment and all subsequent payments. The **finance charge** is the difference between the total installment cost and the price if paid in cash. So, the finance charge includes interest and any fees charged by the lender.

> **Finding the Total Installment Cost, Finance Charge, and Amount Financed**
> Total installment cost = Down payment + Payment amount × Number of payments
> Finance charge = Total installment cost − Cash price
> Amount financed = Cash price − Down payment

Finding the Total Installment Cost

EXAMPLE 1 Frank Kimlicko recently received his master's degree and began work at a community college as a music professor specializing in classical guitar. He purchased an exquisite-sounding classical guitar costing $3800 with $500 down and 36 monthly payments of $109.61 each. Find **(a)** the total installment cost, **(b)** the finance charge, and **(c)** the amount financed.

SOLUTION

(a) The total installment cost is the down payment plus the total of all monthly payments.

Total installment cost = $500 + ($109.61 × 36) = **$4445.96**

(b) The finance charge is the total installment cost less the cash price.

Finance charge = **$4445.96** − $3800 = $645.96

(c) The amount financed is $3800 − $500 = $3300.

QUICK CHECK 1
Robert Chu purchased a new Honda Civic costing $24,200, including taxes and licensing, with $4000 down and 48 payments of $488.25 each. Find **(a)** the total installment cost, **(b)** the finance charge, and **(c)** the amount financed.

Quick TIP
Try to borrow as little as possible. Remember you have to pay it all back along with interest!

Many students use an installment loan called a **Stafford loan** to help pay costs while in college. The government pays the interest on a *subsidized* Stafford loan while the student borrower is in school on at least a half-time basis. In contrast, the student is responsible for interest on *unsubsidized* Stafford loans. Repayment of a loan begins six months after the borrower ceases at least half-time enrollment. You can find information about Stafford loans at the financial aid office at your college or at a bank.

OBJECTIVE 3 Use the formula for approximate APR. The **approximate annual percentage rate (APR)** for a loan paid off in monthly payments can be found using the following formula.

$$\text{Approximate APR} = \frac{24 \times \text{Finance charge}}{\text{Amount financed} \times (1 + \text{Total number of payments})}$$

The formula is *only an estimate* of the APR. It is not accurate enough for the purposes of the federal Truth in Lending Act, which requires the use of tables.

Finding the Annual Percentage Rate

EXAMPLE 2 Florent Ze decides to buy a used car for $6400. She makes a down payment of $1200 and monthly payments of $169 for 36 months. Find the approximate annual percentage rate rounded to the nearest tenth of a percent.

SOLUTION
Use the steps outlined above.

Total installment cost = $1200 + ($169 × 36 months) = **$7284**
Finance charge = **$7284** − $6400 = $884
Amount financed = $6400 − $1200 = $5200

12.2 Installment Loans

Quick TIP
The precise APR can be found using a financial calculator as shown in examples in Appendix C.

Use the formula for approximate APR. Replace the finance charge with $884, the amount financed with $5200, and the number of payments with 36.

$$\text{Approximate APR} = \frac{24 \times \text{Finance charge}}{\text{Amount financed} \times (1 + \text{Total number of payments})}$$

$$= \frac{24 \times \$884}{\$5200 \times (1 + 36)}$$

$$= \frac{\$21{,}216}{\$192{,}400}$$

$$= .110 \text{ or } 11.0\% \text{ approximate APR}$$

The approximate annual percentage rate on this loan is 11%. Example 3 shows how to find the actual APR for this loan.

QUICK CHECK 2

Bob Drake purchases a Harley-Davidson motorcycle costing $26,500. He financed the purchase at his bank with a $5000 down payment and payments of $693.74 for 36 months. Estimate the annual percentage rate to the nearest tenth of a percent.

OBJECTIVE 4 Use the table to find APR. Special tables must be used to find annual percentage rates *accurate enough* to satisfy federal law which requires accuracy to the nearest quarter of a percent. These tables are available from a Federal Reserve Bank or the Board of Governors of the Federal Reserve System, Washington, DC 20551. The table on page 470 shows a small portion of these tables. The APR is found from the APR table as follows.

Finding the Annual Percentage Rate (APR)

Step 1 Multiply the finance charge by $100, and divide by the amount financed.

$$\frac{\text{Finance charge} \times \$100}{\text{Amount financed}}$$

The result is the finance charge per $100 of the amount financed.

Step 2 Read down the left column of the annual percentage rate table to the proper number of payments. Go across to the number closest to the number found in Step 1. Read the number at the top of that column to find the annual percentage rate.

Finding the Annual Percentage Rate — **EXAMPLE 3** In Example 2, a used car costing $6400 was financed at $169 per month for 36 months after a down payment of $1200. The total finance charge was $884, and the amount financed was $5200. Find the annual percentage rate.

SOLUTION

Quick TIP
When using the annual percentage rate table, select the column with the table number that is closest to the finance charge per $100 of amount financed.

Step 1 Multiply the finance charge by $100, and divide by the amount financed.

$$\frac{\$884 \times \$100}{\$5200} = \$17.00 \quad \text{Round to two decimal places for use in the table.}$$

This gives the finance charge per $100 financed.

Step 2 Read down the left column of the annual percentage rate table to the line for 36 months (the actual number of monthly payments). Follow across to the right to find the number closest to $17.00. Here, find **17.01**. Read the number at the top of this column of figures to find the annual percentage rate, 10.50%.

In this example, 10.50% is the annual percentage rate that must be disclosed to the buyer of the car. In Example 2, the formula for the approximate annual percentage rate gave an answer of 11%, which is not accurate enough to meet the requirements of the law.

QUICK CHECK 3

A refrigerator costing $1450 was financed with $100 down and 20 monthly payments of $74.95 each. Find **(a)** the finance charge, **(b)** amount financed, and **(c)** the annual percentage rate.

Annual Percentage Rate Table for Monthly Payment Plans

Annual Percentage Rate (Finance Charge Per $100 of Amount Financed)

Number of Payments	10.00%	10.25%	10.50%	10.75%	11.00%	11.25%	11.50%	11.75%	12.00%	12.25%	12.50%	12.75%	13.00%	13.25%	13.50%	13.75%	Number of Payments
12	5.50	5.64	5.78	5.92	6.06	6.20	6.34	6.48	6.62	6.76	6.90	7.04	7.18	7.32	7.46	7.60	12
14	6.36	6.52	6.69	6.85	7.01	7.17	7.34	7.50	7.66	7.82	7.99	8.15	8.31	8.48	8.64	8.81	14
16	7.23	7.41	7.60	7.78	7.97	8.15	8.34	8.53	8.71	8.90	9.08	9.27	9.46	9.64	9.83	10.02	16
18	8.10	8.31	8.52	8.73	8.93	9.14	9.35	9.56	9.77	9.98	10.19	10.40	10.61	10.82	11.03	11.24	18
20	8.98	9.21	9.44	9.67	9.90	10.13	10.37	10.60	10.83	11.06	11.30	11.53	11.76	12.00	12.23	12.46	20
22	9.86	10.12	10.37	10.62	10.88	11.13	11.39	11.64	11.90	12.16	12.41	12.67	12.93	13.19	13.44	13.70	22
24	10.75	11.02	11.30	11.58	11.86	12.14	12.42	12.70	12.98	13.26	13.54	13.82	14.10	14.38	14.66	14.95	24
26	11.64	11.94	12.24	12.54	12.85	13.15	13.45	13.75	14.06	14.36	14.67	14.97	15.28	15.59	15.89	16.20	26
28	12.53	12.86	13.18	13.51	13.84	14.16	14.49	14.82	15.15	15.48	15.81	16.14	16.47	16.80	17.13	17.46	28
30	13.43	13.78	14.13	14.48	14.83	15.19	15.54	15.89	16.24	16.60	16.95	17.31	17.66	18.02	18.38	18.74	30
32	14.34	14.71	15.09	15.46	15.84	16.21	16.59	16.97	17.35	17.73	18.11	18.49	18.87	19.25	19.63	20.02	32
34	15.25	15.65	16.05	16.44	16.85	17.25	17.65	18.05	18.46	18.86	19.27	19.67	20.08	20.49	20.90	21.31	34
36	16.16	16.58	17.01	17.43	17.86	18.29	18.71	19.14	19.57	20.00	20.43	20.87	21.30	21.73	22.17	22.60	36
38	17.08	17.53	17.98	18.43	18.88	19.33	19.78	20.24	20.69	21.15	21.61	22.07	22.52	22.99	23.45	23.91	38
40	18.00	18.48	18.95	19.43	19.90	20.38	20.86	21.34	21.82	22.30	22.79	23.27	23.76	24.25	24.73	25.22	40
42	18.93	19.43	19.93	20.43	20.93	21.44	21.94	22.45	22.96	23.47	23.98	24.49	25.00	25.51	26.03	26.55	42
44	19.86	20.39	20.91	21.44	21.97	22.50	23.03	23.57	24.10	24.64	25.17	25.71	26.25	26.79	27.33	27.88	44
46	20.80	21.35	21.90	22.46	23.01	23.57	24.13	24.69	25.25	25.81	26.37	26.94	27.51	28.08	28.65	29.22	46
48	21.74	22.32	22.90	23.48	24.06	24.64	25.23	25.81	26.40	26.99	27.58	28.18	28.77	29.37	29.97	30.57	48
50	22.69	23.29	23.89	24.50	25.11	25.72	26.33	26.95	27.56	28.18	28.80	29.42	30.04	30.67	31.29	31.92	50

Finding the Annual Percentage Rate

EXAMPLE 4 Two Brothers from Italy Pizza borrowed $48,000 to remodel their store. They agreed to a note with payments of $1565.78 per month for 36 months. Find the annual percentage rate.

SOLUTION

Total installment cost = $0 down payment + $1565.78 × 36 = **$56,368.08**
Finance charge = **$56,368.08** − $48,000 = $8368.08
Amount financed = $48,000 − $0 down payment = **$48,000**

Now use the formula for the APR.

$$\frac{\$8368.08 \times \$100}{\$48{,}000} = 17.4335$$

Find the row associated with 36 payments in the annual percentage rate table. Look to the right across that row to find the number closest to 17.4335, which is 17.43. Look to the top of that column to find **10.75%**. This is the APR to the nearest quarter of a percent.

QUICK CHECK 4

An insurance agent borrowed $22,500 for new hardware and software for her growing business. She agreed to a note with payments of $858.10 per month for 30 months and put a certificate of deposit up for collateral instead of making a down payment. Find the annual percentage rate.

12.2 Exercises // MyLab Math

The shaded sections below contain solutions to help you get a QUICK START on the various types of exercises.

Find the finance charge (FC) and the total installment cost (TIC) for the following. (See Example 1.)

	Amount Financed	Down Payment	Cash Price	Number of Payments	Amount of Payment	Total Installment Cost	Finance Charge
1.	$1400	$400	$1800	24	$68.75	$2050	$250
	TIC = $400 + (24 × $68.75) = $2050; FC = $2050 − $1800 = $250						
2.	$650	$125	$775	24	$32	$893	$118
	TIC = $125 + (24 × $32) = $893; FC = $893 − $775 = $118						
3.	$1500	none	$1500	12	$150	_____	_____
4.	$1200	none	$1200	20	$70	_____	_____
5.	$2525	$375	$2900	18	$176	_____	_____
6.	$6388	$380	$6768	60	$136	_____	_____

Find the approximate annual percentage rate using the approximate annual percentage rate formula. Round to the nearest tenth of a percent. (See Example 2.)

	Amount Financed	Finance Charge	No. of Monthly Payments	Approximate APR
7.	$11,500	$1200	30	8.1%
	Approx. APR = $\dfrac{24 \times \$1200}{\$11{,}500 \times (1 + 30)} = 8.1\%$			
8.	$2200	$346	36	_____
9.	$7542	$1780	48	_____
10.	$4500	$650	36	_____
11.	$1852	$111.98	18	_____
12.	$8046	$973	24	_____

C// indicates an exercise that is related to the Case in Point feature.

472 CHAPTER 12 Business and Consumer Loans

Find the annual percentage rate using the annual percentage rate table. (See Example 3.)

	Amount Financed	Finance Charge	No. of Monthly Payments	APR
13.	$1400	$185.68	24	12.25%
	$\frac{FC \times \$100}{AF} = \frac{\$185.68 \times \$100}{\$1400} = 13.26$; from 24-payment row, APR = 12.25%			
14.	$345	$24.62	12	_____
15.	$442	$28.68	14	_____
16.	$4690	$1019.61	48	_____
17.	$1450	$132.50	18	_____
18.	$650	$73.45	24	_____

19. Explain the difference between open-end credit and installment loans. (See Section 12.1 and Objective 1 of this section.)

20. Make a list of items that you have bought on an installment loan. Make another list of things you plan to buy in the next 2 years on an installment loan. (See Objective 1.)

Solve the following application problems. Use the formula on page 468 to estimate the APR, and round rates to the nearest tenth of a percent.

21. METAL LATHE Benson Fabrication purchased a new precision metal lathe for $74,800. The company made a down payment of 20% and financed the balance using 36 monthly payments of $1916.85. Find **(a)** the amount financed, **(b)** the total installment cost, and **(c)** the finance charge. **(d)** Then estimate the APR.
  (a) Down payment = .2 × $74,800 = $14,960
     Amount financed = $74,800 − $14,960 = $59,840
  (b) Total installment cost = $14,960 + 36 × $1916.85 = $83,966.60
  (c) Finance charge = $83,966.60 − $74,800 = $9166.60
  (d) Approximate APR = $\frac{24 \times \$9166.60}{\$59,840 \times (1 + 36)}$ = 9.9%

    (a) $59,840
    (b) $83,966.60
    (c) $9166.60
    (d) 9.9%

22. REMODEL Jiya Shin remodeled her den at a cost of $9400. She put 10% down and financed the balance with 24 monthly payments of $390.15. Find **(a)** the amount financed, **(b)** the total installment cost, and **(c)** the finance charge. **(d)** Then estimate the APR.

    (a) _____
    (b) _____
    (c) _____
    (d) _____

23. ELECTRIC GUITAR Yanni Benjamin purchased a good-quality electric guitar with amplifier and financed $3600 over 12 months. The finance charge was $260. **(a)** Estimate the APR, then **(b)** find the exact APR using the table.

    (a) _____
    (b) _____

24. CANNING MACHINE Aluminum Cans Inc. purchased a new machine to press aluminum into cans and financed $385,000 over 30 months. The finance charge was $54,411. (a) Estimate the APR, then (b) find the exact APR using the table.

(a) _____
(b) _____

Solve the following application problems and use the table to find the annual percentage rate.

25. CHIP FABRICATION A Chinese computer chip manufacturer borrowed $84 million worth of Chinese yuan to purchase some sophisticated equipment. The note required 24 monthly payments of $3.88 million each. Find the annual percentage rate.

$FC = 24 \times \$3.88 \text{ million} - \$84 \text{ million} = \$9.12 \text{ million}$

$\dfrac{FC \times \$100}{AF} = \dfrac{\$9.12 \times \$100}{\$84 \text{ million}} = 10.86;$ from 24-payment row, APR = 10.00%

25. 10.00%

26. REFRIGERATOR Sears sells a refrigerator for $1600 with no down payment, $214.88 in interest charges, and 30 equal payments. Find the annual percentage rate.

26. _____

27. TOYOTA Sarah Gonzales bought a Toyota Corolla for $20,800 including title and license. She made a down payment of $2000 and agreed to make 50 payments to Citibank of $463.57 each. Find (a) the amount financed, (b) the total installment cost, (c) the total interest paid, and (d) the APR.

(a) _____
(b) _____
(c) _____
(d) _____

28. SKI BOAT James Berry purchased a ski boat costing $12,800 with $500 down and loan payments to Citibank of $399 per month for 36 months. Find (a) the amount financed, (b) the total installment cost, (c) the total interest paid, and (d) the APR.

(a) _____
(b) _____
(c) _____
(d) _____

29. COMPUTER SYSTEM A contractor in Mexico City purchased a computer system for 650,000 pesos. After making a down payment of 100,000 pesos, he agreed to make payments of 26,342.18 pesos per month for 24 months. Find (a) the total installment cost and (b) the annual percentage rate.

(a) _____
(b) _____

30. TRACTOR PURCHASE An electrical contractor in Hiroshima, Japan, with poor credit, purchased a tractor costing 2,700,000 yen. He made a down payment of 1,000,000 yen and agreed to monthly payments of 54,855 yen for 36 months. Find (a) the total installment cost and (b) the annual percentage rate.

(a) _____
(b) _____

31. PECAN TREES Josefina Torres and her husband need $85,000 to plant pecan trees on their small farm. They pay $20,000 down and finance the balance with 40 payments of $1942.24. Find (a) the total installment cost and (b) the annual percentage rate.

(a) _____
(b) _____

474 CHAPTER 12 Business and Consumer Loans

32. GOAT CHEESE Toni Smith wants to produce goat cheese and needs $120,000 to purchase 400 goats. He pays $30,000 down and finances the balance with 24 monthly payments of $4253.44. Find **(a)** the total installment cost and **(b)** the annual percentage rate.

(a) _____

(b) _____

33. Should businesses be able to charge interest rates of over 20% to customers with very poor credit histories? Why or why not? If not, do you think any firm would lend to these customers?

34. Explain why it is important for the government to regulate the way in which interest rates are stated.

QUICK CHECK ANSWERS

1. (a) $27,436 (b) $3236 (c) $20,200
2. 10.5%
3. (a) $149 (b) $1350 (c) 12.25%
4. 10.75%

12.3 Early Payoffs of Loans

OBJECTIVES

1. Use the United States Rule for an early payment.
2. Find the amount due on the maturity date using the United States Rule.
3. Use the Rule of 78 when prepaying a loan.

OBJECTIVE 1 Use the United States Rule for an early payment. It is common for a payment to be made on a loan *before it is due*. This may occur when a person receives extra money or refinances a debt at a lower interest rate somewhere else. Prepayments of loans are discussed in this section.

The **United States Rule**, used by the U.S. government and most states and financial institutions, requires any loan payment to first be applied to any interest owed. The balance of the payment is then used to reduce the principal amount of the loan. We will continue to use 360-day years in the calculations of this section.

Using the United States Rule

- **Step 1** Find the simple interest due from the date the loan was made until the date the partial payment is made. Use the formula $I = PRT$.
- **Step 2** Subtract this interest from the amount of the payment.
- **Step 3** Any difference is used to reduce the principal.
- **Step 4** Treat additional partial payments in the same way, always finding interest on *only* the unpaid balance after the last partial payment.
- **Step 5** The remaining principal plus interest on this unpaid principal is then due on the due date of the loan.

OBJECTIVE 2 Find the amount due on the maturity date using the United States Rule. If the partial payment is not large enough to pay the interest due, the payment is held until enough money is available to pay the interest due. Thus, a partial payment for less than the interest due offers no advantage to the borrower since the money is held until enough is available to pay the interest owed.

Finding the Amount Due  **EXAMPLE 1** On August 14, Dr. Jane Ficker signed a 180-day note for $28,500 for a used x-ray machine for her dental office. The note has an interest rate of 8% compounded annually. On October 25, a payment of $8500 is made. **(a)** Find the balance owed on the principal after the payment. **(b)** If no additional payments are made, find the amount due at maturity of the loan.

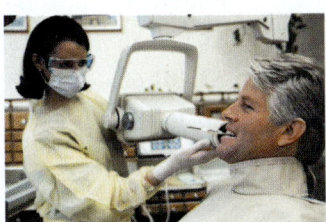

SOLUTION

(a) Step 1   Use $I = PRT$ to find interest from August 14 to October 25 (72 days).

$$\text{Interest} = \$28{,}500 \times .08 \times \frac{72}{360} = \$456$$

Step 2   Amount applied to principal = Payment − Interest
= $8500 − $456
= $8044

Step 3   New principal = Previous principal − Amount applied to principal
= $28,500 − $8044
= $20,456

(b) Step 4   Go on to Step 5 since there are no additional partial payments.

Step 5   The partial payment was made on day 72 of the 180-day note. Interest on the new principal of $20,456 will be for 180 − 72 = **108** days.

$$\text{Interest} = \$20{,}456 \times .08 \times \frac{108}{360} = \$490.94$$

Due at maturity = $20,456 − $490.94 = $19,965.06

QUICK CHECK 1

A 200-day note, signed on April 7 for $48,300, has a rate of 9% compounded annually. A payment of $12,000 was made on June 25. Find **(a)** the balance owed on the principal after the payment on June 25 and **(b)** the amount due at maturity.

Finding the Interest Paid and Amount Due

EXAMPLE 2 On March 1, Boston Dairy signs a promissory note for $38,500 to replace some milking equipment for their Holsteins. The note is for 180 days at a rate of 10%. The dairy makes the following partial payments: $6000 on June 9 and $3500 on July 11. Find the interest paid on the note and the amount due on the due date of the note.

SOLUTION

The first partial payment is on June 9 or, using the number of days in each month, after $(30 + 30 + 31 + 9) = 100$ days.

$$\text{Interest for 100 days} = \$38,500 \times .10 \times \frac{100}{360} = \$1069.44 \text{ (rounded)}$$

First partial payment $\quad\quad\quad\quad\quad$ $6000.00
Portion going to interest $\quad\quad\quad\quad$ − 1069.44
Portion going to reduce debt $\quad\quad$ **$4930.56**

Debt on June 9 *after 1st partial payment* = $38,500 − **$4930.56** = $33,569.44

The second partial payment occurs $21 + 11 = 32$ days later.

$$\text{Interest for 32 days} = \$33,569.44 \times .10 \times \frac{32}{360} = \$298.40 \text{ (rounded)}$$

Second partial payment $\quad\quad\quad\quad$ $3500.00
Portion going to interest $\quad\quad\quad\quad$ − 298.40
Portion going to reduce debt $\quad\quad$ **$3201.60**

Debt on July 11 *after 2nd partial payment* = $33,569.44 − **$3201.60** = $30,367.84

The first partial payment is made after 100 days, and the second partial payment is made after an additional 32 days. Thus, the due date of the note is $180 − 100 − 32 = 48$ days after the second partial payment.

$$\text{Interest for the last 48 days} = \$30,367.84 \times .10 \times \frac{48}{360} = \$404.90 \text{ (rounded)}$$

Amount due *at maturity* = $30,367.84 + **$404.90** = $30,772.74

Date Payment Made	Amount of Payement	Applied to Interest	Applied to Principal	Remaining Balance
June 9	$6,000.00	$1069.44	$4,930.56	$33,569.44
July 11	$3,500.00	$298.40	$3,201.60	$30,367.84
At maturity	$30,772.74	$404.90	$30,367.84	$0
	Total $40,272.74	$1772.74	$38,500.00	

Quick TIP
A spreadsheet such as Excel is an excellent tool for calculations of this type.

QUICK CHECK 2

A 120-day note has a face value of $7500 and a rate of 11.5%. A partial payment of $1700 is made after 40 days. Find **(a)** the balance owed on the principal after the partial payment and **(b)** the amount due at maturity.

OBJECTIVE 3 Use the Rule of 78 when prepaying a loan. A variation of the United States Rule, called the **Rule of 78**, is still used by many lenders for installment loans. This rule allows a lender *to earn more of the finance charge during the early months* of the loan compared with the United States Rule. Lenders typically use this rule *to protect against* early payoffs on small loans. Effectively, the lender will earn a higher rate of interest in the event of an early payoff under the Rule of 78 than under the United States Rule.

The Rule of 78 gets its name based on a loan of 12 months—the sum of the months $1 + 2 + 3 + \cdots + 12 = 78$. The finance charge for the first month is $\frac{12}{78}$ of the total charge, with $\frac{11}{78}$ in the second month, $\frac{10}{78}$ in the third month, and so on, with $\frac{1}{78}$ in the final month.

The Rule of 78 can be applied to loans *with terms other than 12 months*. For example, the sum of the months in a 6-month contract is $1 + 2 + 3 + 4 + 5 + 6 = 21$. The finance charge for the first month is $\frac{6}{21}$; $\frac{5}{21}$ for the second month, and so on. Similarly, the sum of the months in a 15-month contract is $1 + 2 + \cdots + 15 = 120$. The finance charge for the first month of a 15-month contract is $\frac{15}{120}$ or $\frac{1}{8}$, and so on.

Monthly Finance Charges

Term of Loan	Month 1	Month 2	Month 3	Month 4	Month 5	Month 6	Month 7	Month 8	Month 9	Month 10	Month 11	Month 12
12 months	$\frac{12}{78}$	$\frac{11}{78}$	$\frac{10}{78}$	$\frac{9}{78}$	$\frac{8}{78}$	$\frac{7}{78}$	$\frac{6}{78}$	$\frac{5}{78}$	$\frac{4}{78}$	$\frac{3}{78}$	$\frac{2}{78}$	$\frac{1}{78}$
6 months	$\frac{6}{21}$	$\frac{5}{21}$	$\frac{4}{21}$	$\frac{3}{21}$	$\frac{2}{21}$	$\frac{1}{21}$						

The total finance charge on an installment loan is calculated when a loan is first made. Early payoff of a loan results in a lower finance charge. The portion of the finance charge that *has not yet been earned* by the lender under the Rule of 78, called **unearned interest** or **refund**, is found as follows.

Finding unearned interest

$$U = F\left(\frac{N}{P}\right)\left(\frac{1+N}{1+P}\right)$$

where

U = unearned interest $\qquad F$ = finance charge
N = number of payments remaining $\qquad P$ = original number of payments

---

Finding Unearned Interest and Balance Due

EXAMPLE 3 Aaron Ortego borrowed $6000, which he is paying back in 24 monthly payments of $295 each. With 9 payments remaining, he decides to repay the loan in full. Find **(a)** the amount of unearned interest and **(b)** the amount necessary to repay the loan in full. Use the Rule of 78.

SOLUTION

(a) Total of all payments = 24 payments × $295 = **$7080**

Finance charge = **$7080** − $6000 = **$1080** (Amount borrowed)

Find the amount of unearned interest as follows. The finance charge is $1080, the scheduled number of payments is 24, and the loan is paid off with 9 payments left. Solve as follows.

$$\text{Unearned interest} = \$1080 \times \left(\frac{9}{24}\right) \times \left(\frac{1+9}{1+24}\right) = \$162$$

(b) When Ortego decides to pay off the loan, he has 9 payments of $295 left.

Sum of remaining payments = 9 payments × $295 = **$2655**

Ortego saves the unearned interest of $162 by paying off the loan early. Therefore, the amount needed to pay the loan in full is the sum of the remaining payments minus the unearned interest.

Amount needed to repay the loan in full = Remaining payments − Unearned interest
= **$2655** − $162 = **$2493**

QUICK CHECK 3

Simplot Maps has signed a note with a face value of $24,000 that requires 20 monthly payments of $1307.76. The manager pays the debt in full after 12 payments. Find **(a)** the finance charge, **(b)** the amount of unearned interest, and **(c)** the amount needed to repay the loan in full using the Rule of 78.

Finding Unearned Interest and Balance Due

EXAMPLE 4 The Smiths borrow $1200 for a new washer and dryer and agree to make 18 monthly payments of $74.30 each. After the 10th payment, they pay the loan in full. Find **(a)** the amount of unearned interest and **(b)** the amount needed to repay the loan in full using the Rule of 78.

SOLUTION

(a)
$$\text{Sum of payments} = 18 \text{ payments} \times \$74.30 = \$1337.40$$
$$\text{Finance charge} = \$1337.40 - \$1200 = \$137.40$$

Since there are $(18 - 10) = $ **8 payments remaining** when he pays the loan off, the unearned interest is found as follows.

$$\text{Unearned interest} = \$137.40 \times \frac{8}{18} \times \frac{(1+8)}{(1+18)} = \$28.93$$

(b) The amount needed to pay the loan in full is the sum of the remaining payments less the unearned interest.

$$\text{Sum of remaining 8 payments} = 8 \text{ payments} \times \$74.30 = \$594.40$$
$$\text{Amount needed to repay the loan in full} = \$594.40 - \$28.93 = \$565.47$$

QUICK CHECK 4

Tim O'Murphy borrows $2100 to buy a riding lawn mower and agrees to make 12 monthly payments of $186.58 each. He pays the debt in full after 6 months. Find **(a)** the finance charge, **(b)** the amount of unearned interest, and **(c)** the amount needed to repay the loan in full using the Rule of 78.

12.3 Exercises // MyLab Math

The shaded sections below contain solutions to help you get a **QUICK START** *on the various types of exercises.*

Find the balance due on the maturity date of the following notes. Find the total amount of interest paid on each note. Use the United States Rule. (See Examples 1 and 2.)

	Principal	Interest	Time (Days)	Partial Payments	Balance Due	Total Interest Paid
1.	$9800	$8\frac{1}{2}\%$	150	$1800 on day 50	$8307.31	$307.31

Interest for 50 days = $9800 \times .085 \times \frac{50}{360} = \$115.69$
Amount of 1st payment applied to reduce debt = $1800 - $115.69 = $1684.31
Debt after 1st payment = $9800 - $1684.31 = $8115.69
Interest due at end of 150-day note = $8115.69 \times .085 \times \frac{100}{360} = \$191.62$
Balance due on maturity date = $8115.69 + $191.62 = $8307.31
Total interest paid = $115.69 + $191.62 = $307.31

	Principal	Interest	Time (Days)	Partial Payments	Balance Due	Total Interest Paid
2.	$7200	7.5%	120	$3000 on day 45	_____	_____
3.	$15,000	10.5%	200	$6500 on day 100	_____	_____
4.	$76,900	11%	180	$31,250 on day 75	_____	_____
5.	$18,457	12%	120	$5978 on day 34 $3124 on day 55	_____	_____
6.	$39,864	9%	105	$8458 on day 43 $11,354 on day 88	_____	_____

C// indicates an exercise that is related to the Case in Point feature.

Each of the following loans is paid in full before the date of maturity. Find the amount of unearned interest. Use the Rule of 78. (See Example 3.)

Finance Charge	Total Number of Payments	Remaining Number of Payments When Paid in Full	Unearned Interest
7. $1050	24	11	$231
$1050 × $\frac{11}{24}$ × $\frac{(1+11)}{(1+24)}$ = $231			
8. $422	30	16	_____
9. $881	36	12	_____
10. $650	24	22	_____
11. $900	36	6	_____
12. $1250	60	12	_____

13. Explain why banks prefer the Rule of 78 to the United States Rule in the event of prepayment. (See Objective 3.)

14. Explain how an early payoff on a loan helps the borrower.

Solve the following application problems using the United States Rule.

15. LANDSCAPING Andrew Raring borrowed $8900 from Citibank to landscape his yard. The 240-day note had an interest rate of 8.5% compounded annually. He repaid the note in 140 days with his income-tax refund. Find **(a)** the interest due and **(b)** the total amount due.

(a) Loan is for 140 days
   Interest = $8900 × .085 × $\frac{140}{360}$ = $294.19
(b) Amount due = $8900 + $294.19 = $9194.19

(a) $294.19
(b) $9194.19

16. COMPUTER CONSULTANT The computer system at Genome Therapy crashed several times last year. On January 10, the company borrowed $125,000 at 9% compounded annually for 250 days to pay a consultant to work on the Novell network. However, they decide to pay the loan in full on July 1. Find **(a)** the interest due and **(b)** the total amount due.

(a) _____
(b) _____

17. PARTIAL PAYMENT Solar Technologies borrowed $92,000 on May 7, signing a note due in 90 days at 11.25% interest. On June 24, the company made a partial payment of $24,350. Find **(a)** the amount due on the maturity date of the note and **(b)** the interest paid on the note.

(a) _____
(b) _____

480 CHAPTER 12 Business and Consumer Loans

18. **REMODELING** The Second Avenue Butcher Shop financed a remodeling program by giving the builder a note for $32,500. The note was made on September 14 and is due in 120 days. Interest on the note is 9.75%. On December 9, the firm makes a partial payment of $9000. Find **(a)** the amount due on the maturity date of the note and **(b)** the interest paid on the note.

(a) _____
(b) _____

19. **INVENTORY** The Washington News signed a note on February 18, maturing on May 15. The face value of the note was $104,500, with interest of 11%. The firm made a partial payment of $38,000 on March 20 and a second partial payment of $27,200 on April 16. Find **(a)** the amount due on the maturity date of the note and **(b)** the amount of interest paid on the note.

(a) _____
(b) _____

20. **SURVEILLANCE CAMERAS** To help detect trespassers at night, a small security firm purchased some high-technology cameras using a note from Citibank for $32,000. The note was signed on July 26 and was due on November 20. The interest rate is 13%. The firm made a partial payment of $6000 on August 31 and a second partial payment of $11,700 on October 4. Find **(a)** the amount due on the maturity date of the note and **(b)** the interest paid on the note.

(a) _____
(b) _____

Solve the following application problems using the Rule of 78. (See Example 3.)

21. **ENGAGEMENT RING** Tom Stowe purchased a diamond engagement ring for $1150. He paid $100 down and agreed to 12 monthly payments of $95 each. After making 7 payments, he paid the loan in full. Find **(a)** the unearned interest and **(b)** the amount necessary to pay the loan in full.
   (a) Finance charge = $100 + (12 × $95) − $1150 = $90
   $U = \$90 \times \frac{5}{12} \times \frac{6}{13} = \$17.31$
   (b) (5 × $95) − $17.31 = $457.69

(a) $17.31
(b) $457.69

22. **GARBAGE TRUCK** Haul-it-Away, Inc., purchased a dump truck for $62,000. The owners made a down payment of $22,000 and financed the remainder with 36 payments of $1328.57 each. They paid off the note with 12 payments remaining. Find **(a)** the amount of unearned interest and **(b)** the amount necessary to pay the loan in full.

(a) _____
(b) _____

23. **PRINTING** BlackTop Printing made a $5000 down payment on a special copy machine costing $23,800. The loan agreement with Citibank called for 20 monthly payments of $1025 each. Find **(a)** the finance charge, **(b)** the unearned interest, and **(c)** the amount necessary to pay the loan in full after the 14th payment.

(a) _____
(b) _____
(c) _____

24. **MOVIE PROJECTORS** Movie 6, Inc., purchased two movie projectors at a total cost of $12,200 with a down payment of $1500. The company agreed to make 12 monthly payments of $945 each. Find **(a)** the finance charge, **(b)** the unearned interest, and **(c)** the amount necessary to pay the loan in full after the 8th payment.

(a) _____
(b) _____
(c) _____

25. **WEB DESIGN** Blackstone Web Design needed $76,800 to purchase computers, software, and network equipment. The owners paid $15,000 down and financed the balance with 30 monthly payments of $2423.89 each. Find **(a)** the total installment cost, **(b)** the finance charge, **(c)** the unearned interest, and **(d)** the amount needed to pay the loan in full after the 10th payment.

(a) _____
(b) _____
(c) _____
(d) _____

26. **ASPHALT CRUMB** Binston Asphalt borrowed $850,000 to purchase a machine that converts old tires into asphalt crumb for use as a base material under sports arenas. The company paid $200,000 of the cost up front and financed the balance with 36 monthly payments of $20,821.42 each. Find **(a)** the total installment cost, **(b)** the finance charge, **(c)** the unearned interest, and **(d)** the amount needed to pay the loan in full after the 25th payment.

(a) _____
(b) _____
(c) _____
(d) _____

QUICK CHECK ANSWERS

1. (a) $37,253.93 (b) $38,380.86
2. (a) $5895.83   (b) $6046.50
3. (a) $2155.20   (b) $369.46
   (c) $10,092.62
4. (a) $138.96   (b) $37.41
   (c) $1082.07

12.4 Personal Property Loans

OBJECTIVES

1. Define personal property and real estate.
2. Use the formula for amortization to find payment.
3. Set up an amortization schedule.
4. Find monthly payments.

CASE IN POINT Before making a personal property loan, Jackie Waterton carefully checks credit history and all sources of income as well as assets and debts. She does not want to approve a loan to someone who cannot repay with interest.

OBJECTIVE 1 Define personal property and real estate. Items that can be moved from one location to another, such as an automobile, a boat, or a stereo, are called **personal property**. In contrast, land and homes cannot be moved and are called **real estate** or **real property**. Personal property loans are discussed in this section, and real estate loans are discussed in the next section.

Banks, credit unions, finance companies, and many other types of companies make money through personal property loans. Typically, these loans are repaid, or **amortized**, using monthly payments. When a borrower does not make the payments as promised, the lender may **repossess** the personal property and sell it. Lenders do not want to go through this process, and only lend to individuals they believe will repay the debt with interest.

Interest rates for personal property loans vary significantly and are sometimes negotiable. Shop around for the best terms before borrowing.

OBJECTIVE 2 Use the formula for amortization to find payment. The periodic payment needed at the end of each period to amortize a loan with interest i per period, over n periods, is found by using the following formula.

> **Finding Periodic Payment**
>
> Payment = Loan amount × Number from amortization table

Amortizing a Loan **EXAMPLE 1** Sven Yarborough is a mechanic at a Ford dealership. He was so impressed with the quality of Fords that he purchased an SUV at a cost of $29,400, including tax, title, and license, after the rebate. He made a down payment of $3500 and financed the balance at a rate of 6% per year for 4 years. Find **(a)** the monthly payment, **(b)** the portion of the first payment that is interest, **(c)** the balance due after one payment, **(d)** the interest owed for the second month, and **(e)** the balance after the second payment.

Quick TIP
Be sure to round to the nearest cent after each calculation.

SOLUTION

(a) Amount financed = $29,400 − $3500 = $25,900.

Use $\frac{6\%}{12}$ = .5% per month and 4 years × 12 = 48 months in the table to find **.02349**.

Monthly payment = $25,900 × **.02349** = $608.39 (rounded)

(b) Interest for 1st month = $I = PRT$ = $25,900 × .06 × $\frac{1}{12}$ = **$129.50** (rounded)

Amount of 1st payment applied to principal = $608.39 − **$129.50** = **$478.89**

(c) Balance after 1st payment = $25,900 − $478.89 = $25,421.11.

(d) Interest for 2nd month = PRT = $25,421.11 × .06 × $\frac{1}{12}$ = **$127.11** (rounded)

Amount of 2nd payment applied to principal = $608.39 − **$127.11** = **$481.28**

(e) Balance after 2nd payment = $25,421.11 − **$481.28** = **$24,939.83**.

To summarize:

Payment Number	Payment	Interest	Applied to Principal	Loan Balance After Payment
1	$608.39	$129.50	$478.89	$25,421.11
2	$608.39	$127.11	$481.28	$24,939.83

Note that monthly payment and monthly interest are rounded to the nearest cent before going on to the other calculations.

QUICK CHECK 1

Marine Ltd. paid $450,000 down on a tugboat costing $1,800,000. The firm financed the balance at 6% per year for 4 years. Find **(a)** the monthly payment and **(b)** the balance after the first payment.

OBJECTIVE 3 Set up an amortization schedule.

Creating an Amortization Table

EXAMPLE 2 Ruth Jinseng needed $22,300 for equipment for her car wash. Due to issues with her credit history, the bank required her to put $5000 down and agreed to lend the balance at a relatively high 12% compounded quarterly with quarterly payments spread out over two years. **(a)** Find the quarterly payment, and **(b)** show the payments in a table called an **amortization schedule**.

SOLUTION

(a) The amount financed is $22,300 − $5000 = $17,300. Find the factor from the table using 2 × 4 = 8 quarters and $\frac{12\%}{4}$ = 3% per quarter. Then multiply the amount financed by the factor to find the payment.

$$\text{Payment} = \$17{,}300 \times .14246 = \$2464.56$$

(b) **Amortization Schedule**

Payment Number	Amount of Payment	Interest for Period	Portion to Principal	Principal at End of Period
0	—	—	—	$17,300.00
1	$2,464.56	$519.00	$1,945.56	$15,354.44
2	$2,464.56	$460.63	$2,003.93	$13,350.51
3	$2,464.56	$400.52	$2,064.04	$11,286.47
4	$2,464.56	$338.59	$2,125.97	$9,160.50
5	$2,464.56	$274.82	$2,189.74	$6,970.76
6	$2,464.56	$209.12	$2,255.44	$4,715.32
7	$2,464.56	$141.46	$2,323.10	$2,392.22
8	$2,463.99*	$71.77	$2,392.22	$0.00

*The last payment differs slightly so that the final debt ends up at exactly $0.

The table illustrates a very important point! Notice that interest is high at the time of the first payment but eventually goes to nearly zero. This is because the loan balance is high at first—the amount of interest depends on the amount owed.

QUICK CHECK 2

BlueLake Marina purchased a party barge costing $85,700 with a down payment of $20,000 and financed the balance at 8% for 4 quarters. Find the payment and construct an amortization schedule for the first 2 quarters.

Amortization Table

Interest Rate per Period

Period	½%	1%	1½%	2%	2½%	3%	4%	6%	8%	10%	Period
1	1.00500	1.01000	1.01500	1.02000	1.02500	1.03000	1.04000	1.06000	1.08000	1.10000	1
2	.50375	.50751	.51128	.51505	.51883	.52261	.53020	.54544	.56077	.57619	2
3	.33667	.34002	.34338	.34675	.35014	.35353	.36035	.37411	.38803	.40211	3
4	.25313	.25628	.25944	.26262	.26582	.26903	.27549	.28859	.30192	.31547	4
5	.20301	.20604	.20909	.21216	.21525	.21835	.22463	.23740	.25046	.26380	5
6	.16960	.17255	.17553	.17853	.18155	.18460	.19076	.20336	.21632	.22961	6
7	.14573	.14863	.15156	.15451	.15750	.16051	.16661	.17914	.19207	.20541	7
8	.12783	.13069	.13358	.13651	.13947	.14246	.14853	.16104	.17401	.18744	8
9	.11391	.11674	.11961	.12252	.12546	.12843	.13449	.14702	.16008	.17364	9
10	.10277	.10558	.10843	.11133	.11426	.11723	.12329	.13587	.14903	.16275	10
11	.09366	.09645	.09929	.10218	.10511	.10808	.11415	.12679	.14008	.15396	11
12	.08607	.08885	.09168	.09456	.09749	.10046	.10655	.11928	.13270	.14676	12
13	.07964	.08241	.08524	.08812	.09105	.09403	.10014	.11296	.12652	.14078	13
14	.07414	.07690	.07972	.08260	.08554	.08853	.09467	.10758	.12130	.13575	14
15	.06936	.07212	.07494	.07783	.08077	.08377	.08994	.10296	.11683	.13147	15
16	.06519	.06794	.07077	.07365	.07660	.07961	.08582	.09895	.11298	.12782	16
17	.06151	.06426	.06708	.06997	.07293	.07595	.08220	.09544	.10963	.12466	17
18	.05823	.06098	.06381	.06670	.06967	.07271	.07899	.09236	.10670	.12193	18
19	.05530	.05805	.06088	.06378	.06676	.06981	.07614	.08962	.10413	.11955	19
20	.05267	.05542	.05825	.06116	.06415	.06722	.07358	.08718	.10185	.11746	20
21	.05028	.05303	.05587	.05878	.06179	.06487	.07128	.08500	.09983	.11562	21
22	.04811	.05086	.05370	.05663	.05965	.06275	.06920	.08305	.09803	.11401	22
23	.04613	.04889	.05173	.05467	.05770	.06081	.06731	.08128	.09642	.11257	23
24	.04432	.04707	.04992	.05287	.05591	.05905	.06559	.07968	.09498	.11130	24
25	.04265	.04541	.04826	.05122	.05428	.05743	.06401	.07823	.09368	.11017	25
26	.04111	.04387	.04673	.04970	.05277	.05594	.06257	.07690	.09251	.10916	26
27	.03969	.04245	.04532	.04829	.05138	.05456	.06124	.07570	.09145	.10826	27
28	.03836	.04112	.04400	.04699	.05009	.05329	.06001	.07459	.09049	.10745	28
29	.03713	.03990	.04278	.04578	.04889	.05211	.05888	.07358	.08962	.10673	29
30	.03598	.03875	.04164	.04465	.04778	.05102	.05783	.07265	.08883	.10608	30
31	.03490	.03768	.04057	.04360	.04674	.05000	.05686	.07179	.08811	.10550	31
32	.03389	.03667	.03958	.04261	.04577	.04905	.05595	.07100	.08745	.10497	32
33	.03295	.03573	.03864	.04169	.04486	.04816	.05510	.07027	.08685	.10450	33
34	.03206	.03484	.03776	.04082	.04401	.04732	.05431	.06960	.08630	.10407	34
35	.03122	.03400	.03693	.04000	.04321	.04654	.05358	.06897	.08580	.10369	35
36	.03042	.03321	.03615	.03923	.04245	.04580	.05289	.06839	.08534	.10334	36
37	.02967	.03247	.03541	.03851	.04174	.04511	.05224	.06786	.08492	.10303	37
38	.02896	.03176	.03472	.03782	.04107	.04446	.05163	.06736	.08454	.10275	38
39	.02829	.03109	.03405	.03717	.04044	.04384	.05106	.06689	.08419	.10249	39
40	.02765	.03046	.03343	.03656	.03984	.04326	.05052	.06646	.08386	.10226	40
41	.02704	.02985	.03283	.03597	.03927	.04271	.05002	.06606	.08356	.10205	41
42	.02646	.02928	.03226	.03542	.03873	.04219	.04954	.06568	.08329	.10186	42
43	.02590	.02873	.03172	.03489	.03822	.04170	.04909	.06533	.08303	.10169	43
44	.02538	.02820	.03121	.03439	.03773	.04123	.04866	.06501	.08280	.10153	44
45	.02487	.02771	.03072	.03391	.03727	.04079	.04826	.06470	.08259	.10139	45
46	.02439	.02723	.03025	.03345	.03683	.04036	.04788	.06441	.08239	.10126	46
47	.02393	.02677	.02980	.03302	.03641	.03996	.04752	.06415	.08221	.10115	47
48	.02349	.02633	.02938	.03260	.03601	.03958	.04718	.06390	.08204	.10104	48
49	.02306	.02591	.02896	.03220	.03562	.03921	.04686	.06366	.08189	.10095	49
50	.02265	.02551	.02857	.03182	.03526	.03887	.04655	.06344	.08174	.10086	50

12.4 Personal Property Loans

OBJECTIVE 4 Find monthly payments. The loan payoff table that follows can be used as an alternative to the amortization table on the prior page. It shows higher interest rates and longer terms. Personal property loans often involve higher interest rates, since people are more likely to default on a loan for, say, a digital television set than on a loan for their home. Furthermore, individuals with poor or no credit history must often pay a higher interest rate.

This table has a different format from the table on the preceding page. The APR is down the left column, and the number of months is across the top of this table.

> **Finding Loan Payment**
>
> Payment = Loan amount × Number from loan payoff table

Loan Payoff Table

APR	18	24	30	36	42	48	54	60	APR
8%	.05914	.04523	.03688	.03134	.02738	.02441	.02211	.02028	8%
9%	.05960	.04568	.03735	.03180	.02785	.02489	.02259	.02076	9%
10%	.06006	.04615	.03781	.03227	.02832	.02536	.02307	.02125	10%
11%	.06052	.04661	.03828	.03274	.02879	.02585	.02356	.02174	11%
12%	.06098	.04707	.03875	.03321	.02928	.02633	.02406	.02225	12%
13%	.06145	.04754	.03922	.03369	.02976	.02683	.02456	.02275	13%
14%	.06192	.04801	.03970	.03418	.03025	.02733	.02507	.02327	14%
15%	.06238	.04849	.04018	.03467	.03075	.02783	.02558	.02379	15%
16%	.06286	.04896	.04066	.03516	.03125	.02834	.02610	.02432	16%
17%	.06333	.04944	.04115	.03565	.03176	.02885	.02662	.02485	17%
18%	.06381	.04993	.04164	.03615	.03226	.02937	.02715	.02539	18%
19%	.06428	.05041	.04213	.03666	.03278	.02990	.02769	.02594	19%
20%	.06476	.05090	.04263	.03716	.03330	.03043	.02823	.02649	20%

Finding Amortization Payments **EXAMPLE 3** After a trade-in, Vickie Ewing owes $17,400 on a new Harley-Davidson motorcycle and wishes to pay the loan off in 60 months. Her banker told her that the interest rate would be somewhere between 9% and 14% depending on her credit history.

(a) Find the monthly payment at both interest rates.

(b) Find the total finance charge at both interest rates.

(c) Find the extra cost of having poor credit.

SOLUTION

(a) Monthly payment at 9% = $17,400 × .02076 = $361.22.

Monthly payment at 14% = $17,400 × .02327 = $404.90.

(b) The finance charge is the sum of all of the payments minus the amount financed.

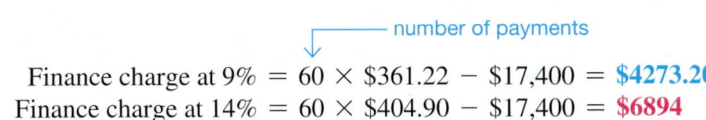

Finance charge at 9% = 60 × $361.22 − $17,400 = **$4273.20**
Finance charge at 14% = 60 × $404.90 − $17,400 = **$6894**

(c) Extra cost of poor credit = **$6894** − **$4273.20** = $2620.80.

Quick TIP
Clearly, a poor credit history can be very costly!

> **QUICK CHECK 3**
>
> After his down payment, Josh Crandall needs to borrow $12,800 for 48 months to buy a truck. He can finance it at 8% per year if he has excellent credit but, at 15% per year if he has poor credit. Find **(a)** the monthly payment at both interest rates, **(b)** the total finance charge for both rates, and **(c)** the extra cost of having poor credit.

12.4 Exercises — MyLab Math

The shaded sections below contain solutions to help you get a QUICK START on the various types of exercises.

Find the payment necessary to amortize the following loans using the amortization table. Round to the nearest cent if needed. (See Example 1.)

	Amount of Loan	Interest Rate	Payments Made	Number of Years	Payment
1.	$6800	8%	annually	5	$1703.13
	Payment = $6800 × .25046 = $1703.13				
2.	$7500	10%	annually	6	$1722.08
	Payment = $7500 × .22961 = $1722.08				
3.	$4500	8%	semiannually	$7\frac{1}{2}$	_____
4.	$12,000	6%	semiannually	8	_____
5.	$96,000	8%	quarterly	$7\frac{3}{4}$	_____
6.	$210,000	12%	quarterly	8	_____
7.	$4876	12%	monthly	3	_____
8.	$6800	6%	monthly	3	_____

Use the loan payoff table to find the monthly payment (MP) and finance charge (FC) for each of the following loans. (See Example 3.)

	Amount Financed	Number of Months	APR	Monthly Payment	Finance Charge
9.	$5300	42	9%	$147.61	$899.62
	MP = $5300 × .02785 = $147.61; FC = (42 × $147.61) − $5300 = $899.62				
10.	$4800	24	12%	$225.94	$622.56
	MP = $4800 × .04707 = $225.94; FC = (24 × $225.94) − $4800 = $622.56				
11.	$14,200	48	10%	_____	_____
12.	$8102	48	8%	_____	_____
13.	$11,750	60	11%	_____	_____
14.	$16,000	60	10%	_____	_____

C// indicates an exercise that is related to the Case in Point feature.

15. Explain the difference between personal property and real property. Why do you sometimes need to borrow to buy personal property? (See Objective 1.)

16. Explain why a loan officer always looks at a credit report before making a loan. What should you do if your credit report is not accurate?

Solve the following application problems using the amortization table.

17. **FORESTRY OPERATIONS** Blackstone Logging borrowed $62,400 to purchase a used truck to haul logs. The loan has a rate of 12% per year and requires 40 monthly payments. Find (a) the monthly payment and (b) the total interest paid.
    (a) Monthly payment = $62,400 × .03046 = $1900.70
    (b) Total interest = (40 × $1900.70) − $62,400 = $13,628

    (a) $1900.70
    (b) $13,628

18. **OPENING A RESTAURANT** Chuck and Judy Nielson opened a restaurant at a cost of $340,000. They paid $40,000 of their own money and agreed to pay the remainder in quarterly payments over 7 years at 12%. Find (a) the quarterly payment and (b) the total amount of interest paid over 7 years.

    (a) _____
    (b) _____

19. **PRINTER** An insurance firm pays $4000 for a new high-speed color printer. It amortizes the loan for the printer in 4 annual payments at 8%. Prepare an amortization schedule for this machine.

Payment Number	Amount of Payment	Interest for Period	Portion to Principal	Principal at End of Period

20. **TRACTOR PURCHASE** Long Haul Trucking purchases a tractor for pulling 18-wheel trailers on interstate highways at a cost of $72,000. It agrees to pay for it with a loan from Citibank that will be amortized over 9 annual payments at 8% interest. Prepare an amortization schedule for the truck.

Payment Number	Amount of Payment	Interest for Period	Portion to Principal	Principal at End of Period

488 CHAPTER 12 Business and Consumer Loans

Solve the following application problems. Use the loan payoff table. (See Objective 3.)

21. **ELECTRONIC EQUIPMENT** An engineering firm purchases specialized software for $24,500. The firm makes a down payment of $10,000 and amortizes the balance with monthly loan payments to Citibank of 11% for 4 years. Prepare an amortization schedule showing the first 5 payments.

Payment Number	Amount of Payment	Interest for Period	Portion to Principal	Principal at End of Period

22. **AMORTIZING A LOAN** Rebecca Reed just graduated from dental school and borrows $120,000 from Citibank to purchase equipment. She agreed to amortize the loan with monthly payments at 10% for 4 years. Prepare an amortization schedule for the first 5 payments.

Payment Number	Amount of Payment	Interest for Period	Portion to Principal	Principal at End of Period

23. **APPLIANCE REPAIR** Jessica Navarro needs $50,000 to expand her business. She has $15,000 and was forced to finance the balance at a high 14% interest rate for 36 months due to not having much credit history and also making a couple of late payments on a Mastercard. Prepare an amortization schedule showing the first 5 payments.

Payment Number	Amount of Payment	Interest for Period	Portion to Principal	Principal at End of Period

24. SCUBA EQUIPMENT Jessica Chien needed $280,000 for inventory for a scuba diving shop that she was opening. She had $40,000, and the bank loaned her the balance at 11% for 30 months. Prepare an amortization schedule showing the first 5 payments.

Payment Number	Amount of Payment	Interest for Period	Portion to Principal	Principal at End of Period

QUICK CHECK ANSWERS

1. (a) $31,711.50 (b) $1,325,038.50

2.
Payment Number	Amount of Payment	Interest for Period	Portion to Principal	Principal at End of Period
0	—	—	—	$65,700.00
1	$17,254.13	$1,314.00	$15,940.13	$49,759.87
2	$17,254.13	$995.20	$16,258.93	$33,500.94

3. (a) $312.45; $356.22 (b) $2197.60; $4298.56 (c) $2100.96

12.5 Real Estate Loans

OBJECTIVES

1. Determine monthly payments on a home.
2. Prepare a repayment schedule.
3. Define escrow accounts.
4. Define fixed and variable rate loans.
5. Understand your credit score.

CASE IN POINT Jackie Waterton's position at Citibank requires her to work with customers who have difficulty paying their bills. Since the house payment is often the largest expense for a family, she often looks to see if she can reduce the monthly payment on the mortgage.

OBJECTIVE 1 Determine monthly payments on a home. A home is *one of the most expensive purchases* that a person makes in his or her lifetime. The monthly payment for a home **mortgage** depends on the amount borrowed, the interest rate, and the term of the loan.

The bar graph shows that home **mortgage rates** have fallen from a high of 15% in 1982 to record low rates of below 4% recently. Interest rates are a very important factor in determining the monthly payment and therefore the affordability of a particular home. If interest rates fall, people **refinance** their home loans at lower rates to reduce their monthly payments.

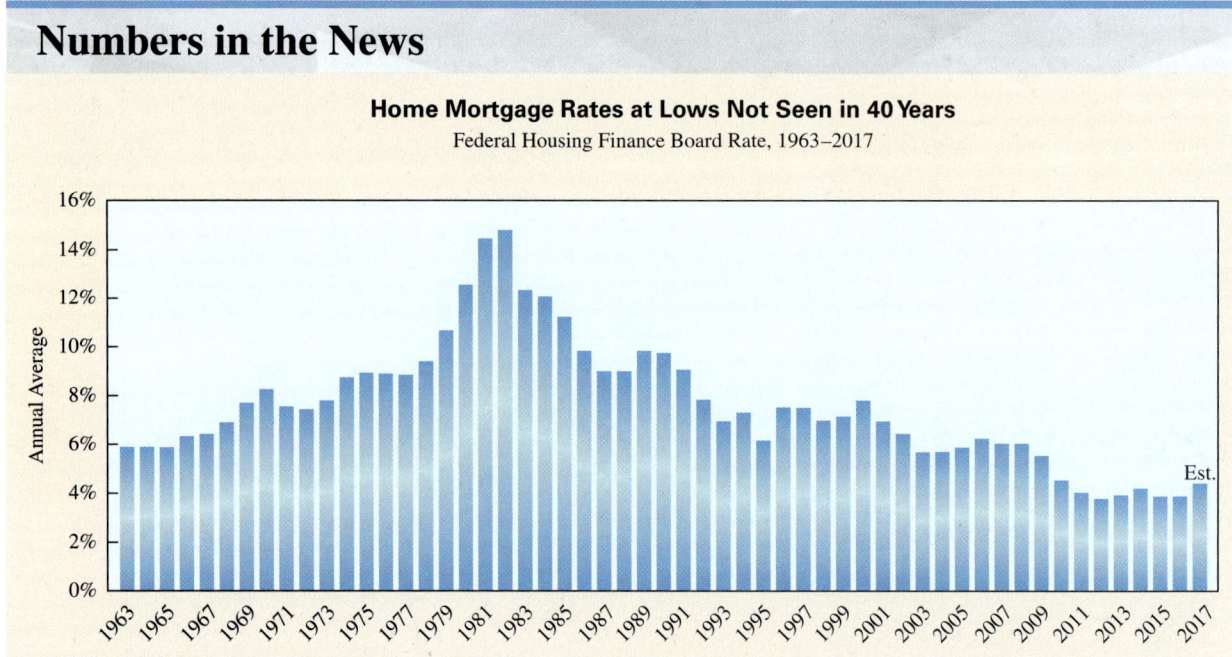

Numbers in the News

Home Mortgage Rates at Lows Not Seen in 40 Years
Federal Housing Finance Board Rate, 1963–2017

Monthly payments for real estate loans are found in the same way as those for the personal property loans found in the prior section. However, real estate loans have longer terms, so different tables are included in this section. The real estate amortization table below shows the monthly payment necessary to repay a $1000 loan for differing interest rates and lengths of repayment. To use the table, first find the amount to be financed in thousands by dividing the total amount to be borrowed by $1000. Then multiply this value by the appropriate number from the table.

Real Estate Amortization Table
(Principal and Interest per Thousand Dollars Borrowed)

Terms in Years	4%	4½%	5%	5½%	6%	6½%	7%	7½%	8%	Terms in Years
10	10.12	10.36	10.61	10.85	11.10	11.35	11.62	11.88	12.14	10
15	7.40	7.65	7.91	8.17	8.44	8.71	8.99	9.28	9.56	15
20	6.06	6.33	6.60	6.88	7.16	7.46	7.76	8.06	8.37	20
25	5.28	5.56	5.85	6.14	6.44	6.75	7.07	7.39	7.72	25
30	4.77	5.07	5.37	5.68	6.00	6.32	6.65	7.00	7.34	30

12.5 Real Estate Loans

Understanding the Effects of Rate and Term

EXAMPLE 1 After making a down payment, the Stringers need to borrow $140,000 to purchase a condominium. They want to know the effect of the interest rate and term of the loan on cost. **(a)** Find the monthly payment for both 15 and 30 years at 4% and 6%. Then find **(b)** the total cost of the home with each loan and **(c)** the finance charge for each loan.

SOLUTION

Quick TIP
Be sure to divide the loan amount by $1000 before calculating the monthly payment.

(a) The amount to be financed in thousands = $140,000 ÷ $1000 = 140. Multiply this value by the appropriate factor from the real estate amortization table.

	Monthly Payment
4% interest for 15 years = 140 × 7.40 =	**$1036.00**
6% interest for 15 years = 140 × 8.44 =	**$1181.60**
4% interest for 30 years = 140 × 4.77 =	**$667.80**
6% interest for 30 years = 140 × 6.00 =	**$840.00**

Monthly payments range from $667.80 to $1181.60, depending on the interest rate and length of the loan. The lower payments associated with 30-year terms above may look good at first, but look at **(b)** and **(c)** below.

(b) The total cost of the home is the sum of all payments over the respective months and years.

	Monthly Payment		Total Cost
4% interest for 15 years =	$1036.00 × 15 years × 12 months/yr	=	**$186,480**
6% interest for 15 years =	$1181.60 × 15 years × 12 months/yr	=	**$212,688**
4% interest for 30 years =	$667.80 × 30 years × 12 months/yr	=	**$240,408**
6% interest for 30 years =	$840.00 × 30 years × 12 months/yr	=	**$302,400**

(c) The finance charge, or interest cost, of each of the loans is the total cost found in part **(b)** minus the amount financed of $140,000.

Quick TIP
You can reduce your long-term cost of a mortgage by paying more than the required payment every month. Even an extra $60 per month can make a big difference over time.

Interest Cost of Purchasing the Condominium

Term of Loan	4%	6%
15 years	$46,480	$72,688
30 years	$100,408	$162,400

least expensive — *most expensive*

First, notice that interest adds quite a bit to the cost of the $140,000 loan. The longer term of 30 years results in a significantly lower house payment, but it also results in a much higher eventual cost. So, the longer term loan of 30 years is *not necessarily the best choice*. Clearly, higher interest rates and longer terms ADD HUGE amounts to the interest that must be paid to purchase a property.

QUICK CHECK 1

A couple plans to borrow $220,000 for 30 years to purchase a home. Find both the payment and the total 30-year cost assuming **(a)** a 4.5% rate and **(b)** a 7% rate.

Mortgages of 30 years are common. However, some people choose **accelerated mortgages** with terms of 15 or 20 years to reduce the total amount of interest paid to mortgage companies.

OBJECTIVE 2 Prepare a repayment schedule. Lenders use computers to calculate an **amortization schedule**, also called a **repayment schedule**. This schedule separates each payment into the portion going to interest and the portion reducing the debt (principal payment).

Preparing a Repayment Schedule

EXAMPLE 2 After making a large down payment, the Zinks purchase a new home by borrowing $195,000 at 7% for 30 years. Prepare a loan repayment schedule for this loan.

SOLUTION

First find the monthly payment, then use simple interest calculations for the first two months. Be sure to round to the nearest cent at each step.

$$\text{Monthly payment} = \$195 \times 6.65 = \$1296.75$$

First month:

$$\text{Interest} = PRT = \$195{,}000 \times .07 \times \tfrac{1}{12} = \$1137.50$$

Amount that payment reduces principal = $1296.75 − $1137.50 = **$159.25**
Remaining debt at end of 1st month = **$195,000** − **$159.25** = $194,840.75

Every time a payment is made, interest is first subtracted from the payment. As a result, only a small portion of the first payment is applied to reduce the principal.

Second month:

$$\text{Interest} = PRT = \$194{,}840.75 \times .07 \times \tfrac{1}{12} = \$1136.57$$

Amount that payment reduces principal = $1296.75 − $1136.57 = **$160.18**
Remaining debt at end of 2nd month = $194,840.75 − **$160.18** = $194,680.57

Data in the computer-generated table that follows show that interest is high at first when the debt is high. Every month, the principal goes down, reducing the interest slightly. It takes 262 months (nearly 22 years) to pay off one-half of the debt and only 98 months to pay off the rest.

Loan Repayment Schedule

Loan: $195,000; Term: 30 Years; Rate 7%; Payment $1296.75							
Payment Number	Interest Payment	Principal Payment	Remaining Balance	Payment Number	Interest Payment	Principal Payment	Remaining Balance
0	—	—	$195,000.00	258	$586.73	$710.02	$99,872.80
1	$1137.50	$159.25	$194,840.75	259	$582.59	$714.16	$99,158.64
2	$1136.57	$160.18	$194,680.57	260	$578.43	$718.32	$98,440.32
3	$1135.64	$161.11	$194,519.46	261	$574.24	$722.51	$97,717.81
4	$1134.70	$162.05	$194,357.41	262	$570.02	$726.73	$96,991.08
5	$1133.75	$163.00	$194,194.41	.	.	.	.
6	$1132.80	$163.95	$194,030.46	.	.	.	.
7	$1131.84	$164.91	$193,865.55	.	.	.	.
8	$1130.88	$165.87	$193,699.68	359	$19.14	$1277.61	$2,004.17
9	$1129.91	$166.84	$193,532.84	360	$11.69	$2004.17*	$0.00

*Due to rounding, the last payment needs to be a little larger to bring the debt to exactly $0.

QUICK CHECK 2

After a large down payment, a couple finances a home with a loan of $128,600 at $4\frac{1}{2}\%$ for 20 years. First find the monthly payment, then prepare a loan repayment schedule for the first three months.

OBJECTIVE 3 Define escrow accounts. Many lenders require **escrow accounts** (also called **impound accounts**) for people taking out a mortgage. With an escrow account, buyers pay $\frac{1}{12}$ of the total estimated property tax and insurance each month. The lender holds these funds until the taxes and insurance fall due and then *pays the bills for the borrower*. Many consumer groups oppose this practice, since the lender earns interest on the money while waiting for payments to come due. In fact, some states require that interest be paid on escrow accounts on any homes located in those states.

Finding the Total Monthly Payment **EXAMPLE 3** The Campbell family borrowed $322,450 for 25 years at 5% to purchase a home with 5 acres for the miniature horses they own. Annual insurance and taxes on the property are $940 and $7382, respectively. Find the total monthly payment.

SOLUTION

Use the real estate amortization table to find a factor of $5.85. Add monthly insurance and taxes to the monthly payment to find the total payment.

$$\text{Monthly payment} = (322.45 \times \$5.85) + \left(\frac{\$940 + \$7382}{12}\right)$$

$$= \$1886.33 + \$693.50 = \$2579.83 \text{ per month}$$

Hot Dog & Thor

QUICK CHECK 3

A 20-year mortgage is taken out with a debt of $147,000 and a rate of $6\frac{1}{2}\%$. Taxes are estimated to be $2400 per year, and a homeowner's insurance policy costs $647 per year. Find the total monthly payment.

OBJECTIVE 4 Define fixed and variable rate loans. Home loans with fixed, stated interest rates are called **fixed-rate loans**. Many borrowers prefer fixed-rate loans, since they know that the principal and interest portion of their monthly payments will be fixed until they either sell the house or pay off the debt. It is good for a borrower to lock in an interest rate on a home mortgage before interest rates go up, since it establishes a monthly payment. Many economists believe interest rates will increase in the future, so now might be a good time to buy a home using a fixed-rate loan.

Another type of home loan called an **adjustable rate mortgage, or ARM**, has a **variable interest rate**. This means that the interest rate is periodically reset by the lender to either a higher or a lower rate. Usually, there is a maximum limit to the increase in the rate during any one reset. Nonetheless, higher interest rates mean higher monthly payments for homeowners.

The Great Recession of 2008–2012 resulted in high unemployment and falling home prices. As a result, many people were unable to make their mortgage payments and lost their homes when lenders foreclosed. Some homeowners were **underwater**, meaning they owed more than the loan amount on their home. The graph here shows that the number of **foreclosures** has dropped significantly since 2010–11 as unemployment has fallen and house prices have gone back up. Can you use the graph to determine the year(s) when economic conditions improved substantially?

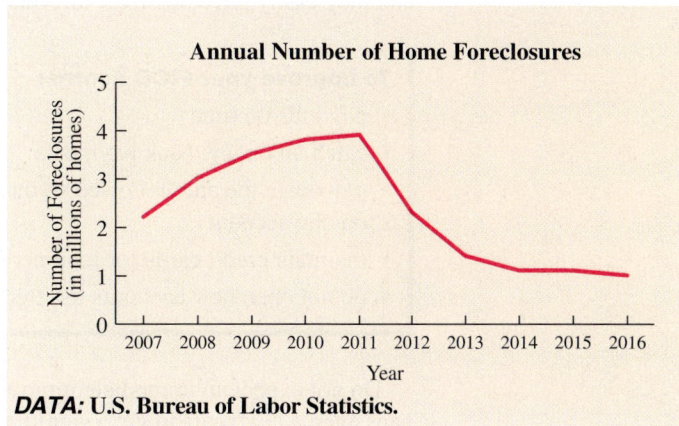

DATA: U.S. Bureau of Labor Statistics.

OBJECTIVE 5 Understand your credit score. Mortgage companies, banks, and other lending institutions determine your creditworthiness using a FICO® score. This **credit score** was established by **FICO** (Fair Isaac Corporation) in 1989, and it is based on numerical values ranging from 300 to 850.

Those with higher credit scores find it easier to borrow, and pay lower interest rates. Those with lower credit scores are charged higher interest rates, i.e., if they can get a loan. Three national credit-reporting agencies have databases that contain financial information about you: Equifax, TransUnion, and Experian. Each calculates a FICO score for you. The three scores may differ slightly. It is a good idea to contact at least one of the three credit reporting agencies annually to examine your credit history. You may be able to catch and correct mistakes on your credit history file.

Here is how one large lender uses FICO scores:

FICO Score	Creditworthiness
above 730	excellent
700–729	good
670–699	increased scrutiny
585–669	high credit risk
below 585	very high credit risk

FICO scores based on credit history have a large effect on borrowing costs. Here are some data based on a 30-year fixed-rate home loan of $250,000 taken out in 2018. Monthly payments are rounded to the nearest dollar.

FICO Score	Average Mortage Rate	Monthly Payment*	Total Interest over 30-Year Loan**
above 730	4.00%	$1194	$179,840
600–640	5.50%	$1419	$260,840

*The monthly payment does not include taxes and insurance.
**We assume the loan is never refinanced.

Quick TIP
It literally pays you to have a good FICO score.

Making a few late payments can reduce your credit score as much as 100 points. In turn, this could increase monthly payments on the home above by $225 and increase the total 30-year interest cost by $81,000. Plan ahead and try never to make a late payment.

Your FICO score is based on a weighted average of the following factors:

1. payment history,
2. amount owed,
3. length of credit history,
4. new credit, and
5. types of credit in use.

Quick TIP
Charging rates that depend on creditworthiness helps a lender ensure that its portfolio of loans remains profitable.

The largest weights in your FICO score are based on payment history and amount owed. You can control these factors, but you cannot control the third factor, length of credit history, since that depends on your age. New credit refers to the amount of new credit you have recently been given.

Higher FICO scores make it easier to borrow and result in a lower interest rate, saving you money. If you start making late payments or miss a payment or two, your score goes down, making it more difficult to borrow. To raise your FICO score, first make sure the credit agencies do not have any incorrect information. They handle billions of pieces of data every month and may easily have incorrect information about you. It also helps if you do the following:

To improve your FICO scores:
- pay bills on time
- catch up on any back payments
- pay down the amount owed so that you are not using more than 30% of the credit limit on any account
- maintain credit cards for long periods of time and
- do not open new accounts too frequently

Do not expect an immediate improvement in your credit score. It may take months or perhaps even a few years to get a significant improvement, depending on your credit history.

12.5 Exercises // MyLab Math

The shaded sections below contain solutions to help you get a QUICK START on the various types of exercises.

Use the real estate amortization table to find the monthly payment for the following loans. (See Example 1.)

	Amount of Loan	Interest Rate	Term of Loan	Monthly Payment
1.	$340,000	$4\frac{1}{2}$%	15 years	$2601.00
	Payment = 340 × $7.65 = $2601			
2.	$263,100	4%	30 years	$1254.99
	Payment = 263.1 × $4.77 = $1254.99			
3.	$168,500	4%	25 years	_____
4.	$132,000	$6\frac{1}{2}$%	25 years	_____
5.	$187,900	5%	20 years	_____
6.	$280,000	6%	15 years	_____

7. Explain how different interest rates can make a large difference in interest charges over a number of years. (See Example 1.)

8. Explain why long-term loans result in a lot of interest. Also explain the effects of shortening the term of a loan. (See Example 1.)

Find the total monthly payment, including taxes and insurance, for the following loans. (See Example 3.)

	Amount of Loan	Interest Rate	Term of Loan	Annual Taxes	Annual Insurance	Monthly Payment
9.	$198,000	6%	30 years	$2870	$825	$1495.92
	198 × $6.00 = $1188; $1188.00 + $\frac{\$2870 + \$825}{12}$ = $1495.92					
10.	$368,000	$4\frac{1}{2}$%	30 years	$4870	$940	$2349.93
	368 × $5.07 = $1865.76; $1865.76 + $\frac{\$4870 + \$940}{12}$ = $2349.93					
11.	$232,900	4%	15 years	$3250	$680	_____
12.	$195,000	5%	30 years	$3920	$850	_____

C// indicates an exercise that is related to the Case in Point feature.

CHAPTER 12 Business and Consumer Loans

Amount of Loan	Interest Rate	Term of Loan	Annual Taxes	Annual Insurance	Monthly Payment
13. $149,400	$5\frac{1}{2}$%	20 years	$2840	$665	_____
14. $283,000	5%	30 years	$3450	$850	_____

Solve the following application problems.

C// 15. HOME PURCHASE The Potters want to buy a small cottage costing $127,000 with annual insurance and taxes of $720 and $2300, respectively. They have saved $10,000 for a down payment, and they can get a 5%, 30-year mortgage from Citibank. They are qualified for a home loan as long as the total monthly payment does not exceed $1000. Are they qualified?

Loan amount = $127,000 − $10,000 = $117,000
Monthly payment = 117 × $5.37 = $628.29
Total payment = $628.29 + $\frac{\$720 + \$2300}{12}$ = $879.96; Yes, qualified

15. **Yes, qualified**

16. CONDOMINIUM PURCHASE The Polinki family wants to buy a condominium that costs $225,000 with annual insurance and taxes of $850 and $3200, respectively. They plan to pay $20,000 down and amortize the balance at $4\frac{1}{2}$% per year for 25 years. They are qualified for a loan as long as the payments do not exceed $1350. Are they qualified for the loan?

16. _____

17. HOME LOAN June and Bill Able borrow $122,500 on a duplex they own at $7\frac{1}{2}$% for 15 years. Prepare a repayment schedule for the first two payments. (See Example 2.)

Payment Number	Total Payment	Interest Payment	Principal Payment	Remaining Balance

18. ELDERLY HOUSING Gabriel Godwin purchases a tiny home for his elderly mother. After a large down payment, he finances $88,600 at 5% for 10 years. Prepare a repayment schedule for the first two payments. (See Example 2.)

Payment Number	Total Payment	Interest Payment	Principal Payment	Remaining Balance

---

QUICK CHECK ANSWERS

1. (a) $1115.40; $401,544 (b) $1463; $526,680
2. Monthly payment = $814.04

Payment Number	Interest Payment	Principal Payment	Remaining Balance
0	—	—	$128,600.00
1	$482.25	$331.79	$128,268.21
2	$481.01	$333.03	$127,935.18
3	$479.76	$334.28	$127,600.90

3. $1350.54

Chapter 12 Quick Review

Chapter Terms Review the following terms to test your understanding of the chapter. For each term you do not know, refer to the page number found next to that term.

accelerated mortgages [p. 491]
adjustable rate mortgages (ARM) [p. 493]
amortization schedule [p. 483]
amortize [p. 467]
annual percentage rate [p. 461]
approximate annual percentage rate (APR) [p. 468]
average daily balance method [p. 460]
consolidate loans [p. 462]
credit history [p. 467]
credit limit [p. 457]
credit score [p. 494]
debit cards [p. 457]
defaulting on debt [p. 462]
escrow accounts [p. 493]
FICO [p. 494]
finance charges [p. 467]
fixed-rate loans [p. 493]
foreclosure [p. 493]
grace period [p. 458]
impound accounts [p. 493]
installment cost [p. 468]
installment loan [p. 467]
itemized billing [p. 458]
late fees [p. 458]
mortgage [p. 490]
mortgage rates [p. 490]
nominal rate [p. 467]
open-end credit [p. 457]
over-the-limit fees [p. 458]
personal property [p. 482]
real estate [p. 482]
real property [p. 482]
refinance [p. 490]
refund of unearned interest [p. 477]
repayment schedule [p. 491]
repossess [p. 482]
revolving charge account [p. 457]
Rule of 78 [p. 476]
Stafford loan [p. 468]
stated rate [p. 467]
strict monthly budget [p. 462]
Truth in Lending Act [p. 467]
underwater [p. 493]
unearned interest [p. 477]
United States Rule [p. 475]
unpaid balance method [p. 458]
variable interest rate [p. 493]

CONCEPTS

12.1 Finding the finance charge on a revolving charge account, using the unpaid balance method

Start with the unpaid balance of the previous month. Then find the finance charge on the unpaid balance. Next, add the finance charge and any purchases. Finally, subtract any payments made.

12.1 Finding the finance charge on a revolving charge account, using the average daily balance method

First find the unpaid balance on each day of the month. Then add up the daily unpaid balances.

Next divide the total of the daily unpaid balances by the number of days in the billing period.

Finally, calculate the finance charge by multiplying the average daily balance by the finance charge.

EXAMPLES

Debbie Mahoney's MasterCard account had an unpaid balance of $385.65 on March 1. During March, she made a $100 payment and charged $68.92. Find the finance charge and the unpaid balance on April 1 if the bank charges 1.25% per month on the unpaid balance.

Finance charge = $385.65 × .0125 = $4.82

$\underset{\text{previous balance}}{\$385.65} + \underset{\text{finance charge}}{\$4.82} + \underset{\text{purchases}}{\$68.92} - \underset{\text{payment}}{\$100} = \underset{\text{new balance}}{\$359.39}$

The following is a summary of a credit-card account.

Previous balance		$115.45
November 1	Billing date	
November 15	Payment	$35.00
November 22	Charge	$45.00

Find the average daily balance and the finance charge if interest is 2% per month on the average daily balance.

Balance on Nov. 1 = $115.45

Nov. 1 − 15 = 14 days at $115.45
14 × $115.45 = **$1616.30**

Payment on Nov. 15 = $35.00
Balance on Nov. 15 = $115.45 − $35 = $80.45

Nov. 15 − Nov. 22 = 7 days at $80.45
7 × $80.45 = **$563.15**

Charge on Nov. 22 of $45.00
Balance on Nov. 22 = $80.45 + $45 = $125.45

Nov. 22 − Dec. 1 = 9 days at $125.45
9 × $125.45 = **$1129.05**

Daily balances

$1616.30 + $563.15 + $1129.05 = **$3308.50**

Average daily balance = $\dfrac{\$3308.50}{30}$ = $110.28

Finance charges = $110.28 × .02 = $2.21

CONCEPTS	EXAMPLES

12.2 Finding the total installment cost, finance charge, and amount financed
 	
Total installment cost
= Down payment
  + (Amount of each payment × Number of payments)
 	
Finance charge (interest)
  = Total installment cost − Cash price
 	
Amount financed (principal of loan)
  = Cash price − Down payment
	Joan Taylor bought a leather coat for $1580. She put $350 down and then made 12 payments of $115 each. Find the total installment cost, the finance charge, and the amount financed.
	
	Total installment cost = $350 + (12 × $115) = $1730
	Finance charge = $1730 − $1580 = $150
	Amount financed = $1580 − $350 = $1230

12.2 Determining the approximate APR using a formula First determine the finance charge. Then find the amount financed. Next approximate the APR using the formula. **Approximate APR** $$= \frac{24 \times \text{Finance charge}}{\text{Amt. fin.} \times (1 + \text{Total number of payments})}$$	Tom Jones buys a motorcycle for $8990. He makes a down payment of $1800 and then makes monthly payments of $230 for 36 months. Estimate the APR. **Total installment cost** = $1800 + ($230 × 36) = $10,080 **Finance charge** = $10,080 − $8990 = **$1090** **Amount financed** = $8990 − $1800 = $7190 **Approximate APR** = $\dfrac{24 \times \$1090}{\$7190 \times (1 + 36)}$ = 9.8% (rounded)
12.2 Finding the APR using a table First determine the finance charge per $100 of amount financed, using the formula $$\frac{\text{Finance charge} \times \$100}{\text{Amount financed}}$$ Then read down the left column of the annual percentage rate table to the proper number of payments. Go across to the number closest to the number found above. Look to the top of the column to find the annual percentage rate.	Lupe Torres has poor credit but is able to buy a used car for $6500. She makes a down payment of $1000 and agrees to make 24 monthly payments of $260.83. Use the table to find the APR. **Finance charge** = $1000 + ($260.83 × 24) − $6500 = $759.92 **Amount financed** = $5500 Finance charge per $100 $$\frac{\$759.92 \times \$100}{\$5500} = 13.82 \text{ (rounded)}$$ Use the 24-payment row in the table to find APR = 12.75%
12.3 Finding the amount due on the maturity date using the United States Rule First determine the simple interest due from the date the loan was made until the date of the partial payment. Then subtract this interest from the amount of the payment and reduce the principal by the difference. Next find the interest from the date of partial payment to the due date of the note, and add the unpaid balance and interest to find the amount due.	Sam Wiley signs a 90-day note on August 1 for $5000 at an interest rate of 12%. On September 15, he makes a payment of $1800. Find the balance owed on the principal. If no additional payments are made, find the amount due on the maturity date of the loan. From August 1 to September 15, there are 30 + 15 = 45 days. $I = \$5000 \times .12 \times \dfrac{45}{360}$ = **$75** interest due $1800 payment − 75 interest due $1725 applied to principal reduction $5000 amount owed − 1725 principal reduction **$3275 balance owed** Note is for 90 days; a partial payment was made after 45 days. Interest on $3275 will be charged for 90 − 45 = 45 days. $I = \$3275 \times .12 \times \dfrac{45}{360}$ = **$49.13** $3275.00 principal owed + 49.13 interest **$3324.13 amount due**

CHAPTER 12 Quick Review

CONCEPTS	EXAMPLES
12.3 Finding the unearned interest using the Rule of 78 First calculate the finance charge. Then find the unearned interest using the formula $$U = F\left(\frac{N}{P}\right)\left(\frac{1+N}{1+P}\right)$$ where U = unearned interest F = finance charge N = number of payments remaining P = total number of payments Next find the total of the remaining payments. Finally, subtract the unearned interest to find the balance remaining.	Jennifer Salas borrows $1500, which she pays back in 36 monthly installments of $52.75 each. With 10 payments remaining, she decides to pay the loan in full. Find **(a)** the amount of unearned interest and **(b)** the amount necessary to pay the loan in full. 36 payments of $52.75 each for a total repayment of 36 × $52.75 = $1899. **Finance charge** = $1899 − $1500 = **$399** **Unearned interest** = $399 × $\frac{10}{36}$ × $\frac{(1+10)}{(1+36)}$ = **$32.95** 10 payments of $52.75 are left. These payments total $52.75 × 10 = $527.50 $527.50 − $32.95 = $494.55 This is the amount needed to pay the loan in full.
12.4 Finding the periodic payment for amortizing a loan First determine the number of periods for the loan and the interest rate per period. The payment is found by multiplying the loan amount by the number from the amortization table.	Bob Smith agrees to pay $12,000 for a used truck. The amount will be repaid in monthly payments over 3 years at an interest rate of 12%. Find the amount of each payment. 3 × 12 = **36 periods (payments)** $\frac{12\%}{12}$ = **1% per period** Number from table is **.03321**. Payment = $12,000 × .03321 = **$398.52**
12.4 Setting up an amortization schedule First find the periodic payment. Then calculate the interest owed in the first period using the formula $I = PRT$. Next subtract the value of I from the periodic payment. This is the amount applied to the reduction of the principal. Then find the balance after the first periodic payment by subtracting the value of the debt reduction from the original amount. Now repeat the above steps until the original loan is amortized (paid off). You may have to adjust the last payment a small amount so that the final debt is exactly $0.	Terri Meyer borrows $1800. She will repay this amount in 2 years with semiannual payments at an interest rate of 8%. Set up an amortization schedule. 2 × 2 = **4 periods (payments); 4% per period** $\frac{8\%}{2}$ = **4% per year** Number from table is **.27549** Payment = $1800 × .27549 = **$495.88** $I = PRT$ Interest owed = $1800 × .08 × $\frac{1}{2}$ = **$72** **Debt reduction** = $495.88 − $72 = **$423.88** **Balance of loan** = $1800 − $423.88 = **$1376.12** | Payment Number	Amount of Payment	Interest for Period	Portion to Principal	Principal at End of Period
---	---	---	---	---
0	—	—	—	$1800.00
1	$495.88	$72.00	$423.88	$1376.12
2	$495.88	$55.04	$440.84	$ 935.28
3	$495.88	$37.41	$458.47	$ 476.81
4	$495.88	$19.07	$476.81	$ 0.00	

CONCEPTS

12.4 Finding monthly payments, total amount paid, and finance charge

First multiply the amount to be financed by the number from the amortization table or the loan payoff table to find the periodic payment. Then find the total amount repaid by multiplying the periodic payment by the number of payments. Finally, subtract the amount financed from the total amount repaid to obtain the finance charge.

12.5 Finding the amount of monthly home loan payments and total interest charges over the life of a home loan

Using the number of years and the interest rate, find the amortization value per thousand dollars from the real estate amortization table.

Next multiply the table value by the number of thousands in the principal to obtain the monthly payment.

Then find the total amount of the payments and subtract the original amount owed from the total payments to obtain interest paid.

EXAMPLES

Ben Apostolides purchased a new Toyota Camry and owes $16,400 after the trade-in. He decides on a term with 50 monthly payments. Assume an interest rate of 6% compounded monthly and find the amount of each payment, total amount paid, and the finance charge.

Monthly payment = $16,400 × .02265 = **$371.46**

Total amount paid = 50 × **$371.46** = $18,573

Finance charge = $18,573 − $16,400 = $2173

Tom Begay borrowed $195,850 to buy a house. It is a 15-year loan at an interest rate of $4\frac{1}{2}$% and taxes and insurance will be $3800 and $750, respectively. Find the total monthly payment and the interest charges over the life of the loan.

Payment = loan payment + payment into escrow for taxes and insurance

$$\text{Payment} = 195.85 \times \$7.65 + \frac{\$3800 + \$750}{12}$$

$$= \mathbf{\$1498.25} + \$379.17$$

$$= \$1877.42$$

Total payments = monthly loan payment × number of years × 12 months/year

$$= \$1498.25 \times 15 \text{ years} \times 12 \text{ months/year}$$

$$= \$269,685$$

Total interest = total payments − original loan amount

$$= \$269,685 \quad - \quad \$195,850$$

$$= \$73,835$$

Case Study

CONSOLIDATING LOANS

Roberto and Julie Hernandez are struggling to make their monthly payments. They have accumulated too much debt, which was easy to do with two young kids at home. Julie works 30 hours a week and takes care of the kids after day care. Their credit history was poor, resulting in high interest rates on loans. However, a few months ago Roberto added a second job, and they have been making payments regularly for 5 months.

1. Find the monthly payments on each of the following purchases and the total monthly payment.

Purchase	Original Loan Amount	Interest Rate	Term of Loan	Monthly Payment
Honda Accord	$18,800	12%	4 years	_____
Ford truck	$14,300	18%	4 years	_____
Home	$126,800	$6\frac{1}{2}$%	15 years	_____
2nd mortgage on home	$4,500	12%	3 years	_____
			Total	_____

2. These monthly expenses do not include car insurance ($215 per month), health insurance ($290 per month), or real estate taxes and insurance on their home ($3350 per year), among other expenses. Find their total monthly outlay for all of these expenses.

Expense	Monthly Outlay
Payments on debt from 1. above	_____
Car insurance	_____
Health insurance	_____
Real estate taxes and insurance on home	_____
Total	_____

3. A loan officer told them that they can (1) refinance the remaining $14,900 amount on the Honda Accord at 12% over 4 years, (2) refinance the remaining $8600 loan amount on the Ford truck at 12% over 3 years, (3) refinance the remaining $121,850 loan amount on their home at 5% over 30 years, and (4) reduce their car insurance payments by $28 per month. Complete the following table.

Item	Current Loan Amount	New Interest Rate	New Term of Loan	New Monthly Payment
Honda Accord				_____
Ford truck				_____
Home				_____
2nd mortgage on home				_____
Car insurance				_____
Health insurance				_____
Real estate taxes and insurance on home				_____
				_____

4. Find the reduction in their monthly payments. 4. _____

Part of the savings in the monthly payment came from reducing the interest rates. The remainder of the savings came from extending the loans further into the future, meaning that the Hernandez family will, in the long run, pay more interest. But at least their current bills are reduced by $715.11 per month. Their goal now is to avoid new debt and pay off some of the existing debt.

INVESTIGATE

The interest rate that you are charged for borrowing money differs depending on the bank you go to for a loan. Find current interest rates for financing a 2-year-old Toyota Camry from at least two banks in the area in which you live. Then go online and look for a lower interest rate.

Case in Point Summary Exercise

UNDERWATER ON A HOME

www.citigroup.com

Facts:

- **1812:** Founded in New York City
- **1914:** Opened first international branch, in Argentina
- **2008:** Lost $27.7 billion due to the Great Recession
- **2018:** Over 200 million customers in more than 160 countries

Citigroup Inc. (Citi) provides various banking, lending, insurance, and investment services to individual and corporate customers worldwide. It operates more than 7200 branches and 7000 ATMs. Citi has more than $150 billion in credit-card loans outstanding in addition to huge volumes of home loans and business loans.

An economic recession such as occurred in 2008–12 can create very serious problems for a large bank such as Citibank. To illustrate the nature of some of the problems, we use the example of one family: Tom and Marie Duston purchased their first home in early 2008 when home loans were still very easy to get. They bought a beautiful new 4-bedroom, $2\frac{1}{2}$-bath home with a down payment of only $8500.

1. They financed the loan balance of $306,500 using an adjustable rate mortgage (ARM). The monthly payment of $1100 did not include taxes and insurance. In fact, the monthly payment was all interest, meaning that nothing was applied against the debt each month. Find the monthly payment given taxes of $6400 per year and insurance of $980 per year.

    1. _____

2. The Dustons were told that the interest rate on their ARM loan would reset in 2011, so they knew the payments might increase. However, they were not worried since they assumed that their incomes and also the value of the house would be higher by then. But home prices fell across much of the country as did the value of their home. By 2011, an appraiser estimated that it was worth only 75% of the original loan balance of $306,500, which they still owed. They owed more on the home than it was worth, a situation often described as "being underwater." Find out what the house was worth in 2011 and the amount by which they were underwater.

    2. _____

3. The Dustons were shocked to find out that they would have to come up with $76,625 to pay off the bank loan to sell their home. They were further shocked to find out they would also need to come up with an additional $23,000 to pay various expenses, such as the real estate commission related to the sale of their home. Estimate the total amount they would have to pay to sell their home, rounded to the nearest thousand.

    3. _____

4. The Dustons did not have $100,000, so they asked about refinancing the loan balance of $306,500. At first, the bank wanted them to pay off the loan, but it finally agreed to refinance it. The Dustons felt trapped! It was difficult to understand that they were underwater by so much given that they had made every payment on time for 3 years. With the help of a government program designed to help underwater homeowners current on their mortgage payments, the bank agreed to refinance $285,000 on the home on a 30-year fixed mortgage at 5%. The difference between the debt of $306,500 and $285,000 was essentially forgiven due to the government program. Find the new home payment not including taxes and insurance.

4. _____

5. So the Dustons' monthly payment, not including taxes and insurance, increased from $1100 per month up to the figure found for #4 above. Find the increase in the monthly payment.

5. _____

From the Dustons' perspective: The Dustons were shocked about what had happened. They had financed their home using an ARM loan and made every payment on time for 3 years. However, the value of their home fell sharply, and they had to refinance their home in 2011. They did not have the money needed to sell the house and pay off the debt. After a stressful period for the family, the bank finally agreed to help by refinancing $285,000 on their home. The monthly payment increased significantly. But, importantly, the home was still worth quite a bit less than they owed on it! They were still underwater and might lose more on the house later. The good news is that they were able to stay in the house.

From the bank's perspective: The bank lent the Dustons $306,500 on an ultra-low interest ARM loan and received payments for 3 years. It was not the bank's fault that the value of the home fell by 25%. Managers essentially believed that the loss was the homeowner's responsibility. However, they found a government program that would allow the bank to forgive part of the Dustons' debt and refinance the remaining $285,000. This was still more than the house was worth. The managers were worried that the Dustons might still abandon the house. In that event, the bank would be forced to go through the lengthy and costly foreclosure process, which would probably result in further losses for the bank.

Discussion Question: *Are the Dustons right? Is the bank right? Make an argument for each side.*

Chapter 12 Test

To help you review, the numbers in brackets show the section in which the topic was discussed.

Solve for following problems.

1. A cruise line needs to update some sonar equipment on one of its luxury ships that sails the Caribbean. The cost of the equipment is $214,500. The company makes a down payment of $20,000 and agrees to 24 monthly payments of $8975 per month. Find the total finance charge. [12.1]

    1. _____

2. The balance on John Baker's MasterCard on November 1 is $680.45. In November, he charges an additional $337.32, has returns of $45.42, and makes a payment of $50. If the finance charges are calculated at 1.5% per month on the unpaid balance, find his balance on December 1. [12.1]

    2. _____

Use the annual percentage rate table to find the annual percentage rate. [12.2]

Amount Financed	Finance Charge	Number of Payments	APR
3. $5280	$1010.59	36	_____
4. $1130	$149.84	24	_____

Solve the following application problems.

5. Barton Springs Landscaping buys a used truck for $18,700 and agrees to make 36 payments of $612.25 each. Find the annual percentage rate on the loan. [12.2]

    5. _____

6. A note with a face value of $7000 is made on June 21. The note is for 90 days and carries interest of 10%. A partial payment of $2800 is made on July 17. Find the amount due on the maturity date of the note. [12.3]

    6. _____

7. Mock Construction bought a truck and financed $7400 with 48 monthly payments of $228.14 each. Suppose the firm pays the loan off with 12 payments left. Use the Rule of 78 to find (a) the amount of unearned interest and (b) the amount necessary to pay off the loan. [12.3]

    (a) _____
    (b) _____

Find the amount of each payment necessary to amortize the following loans. [12.4]

8. Jenson SawLogs borrows $34,500 to buy a new electric generator. The company agrees to make quarterly payments for 2 years at 10% per year. Find the amount of the quarterly payment.

    8. _____

9. Scented Candles remodeled its lobby at a cost of $36,000. It pays $6000 down and pays off the balance in payments made at the end of each quarter for 5 years. Interest is 10% compounded quarterly. Find the amount of each payment so that the loan is fully amortized.

    9. _____

CHAPTER 12 Business and Consumer Loans

Find the monthly payment necessary to amortize the following home mortgages. **[12.5]**

10. $236,000, $4\frac{1}{2}$%, 15 years

    10. _____

11. $134,560, 7%, 15 years

    11. _____

Work the following application problems. **[12.5]**

12. Mr. and Mrs. Zagorin plan to buy a small, unfinished one-room cabin for $90,000, paying 20% down and financing the balance at $5\frac{1}{2}$%, for 30 years. The taxes are $960 per year, with fire insurance costing $352 per year. Find the monthly payment (including taxes and insurance).

    12. _____

13. Billiards Galore purchases a commercial building for $680,000, pays 20% down, and finances the balance at 7% for 15 years. Taxes and insurance are $14,500 and $3200 per year, respectively. (a) Find the monthly payment. (b) Assume that insurance and taxes do not increase, and find the total cost of owning the building for 15 years, including the down payment.

    (a) _____
    (b) _____

14. Jerome Watson, owner of Watson Welding, has a metal building built for his business and makes a $25,000 down payment. He finances the balance of $122,500 for 20 years at 8%. (a) Find the total monthly payment given taxes of $3200 per year and insurance of $1275 per year. (b) Assume that insurance and taxes do not increase, and find the total cost of owning the building for 20 years (including the down payment).

    (a) _____
    (b) _____

Chapters 11–12 Cumulative Review

CHAPTERS 11 AND 12

Round money amounts to the nearest cent and rates to the nearest tenth of a percent.

Find the amount and interest earned for each of the following ordinary annuities. **[11.1]**

	Amount of Each Deposit	Deposited	Rate	Time (Years)	Amount of Annuity	Interest Earned
1.	$1000	annually	4%	8	_____	_____
2.	$2000	quarterly	6%	5	_____	_____

Find the amount of each annuity due and the interest earned. **[11.1]**

	Amount of Each Deposit	Deposited	Rate	Time (Years)	Amount of Annuity	Interest Earned
3.	$2500	annually	5%	6	_____	_____
4.	$1800	semiannually	8%	5	_____	_____

Find the present value of the following annuities. **[11.2]**

	Amount per Payment	Payment at End of Each	Time (Years)	Rate of Investment	Compounded	Present Value
5.	$925	6 months	11	8%	semiannually	_____
6.	$27,235	quarter	8	8%	quarterly	_____

Find the required payment into a sinking fund. **[11.3]**

	Future Value	Interest Rate	Compounded	Time (Years)	Payment
7.	$3600	8%	annually	7	_____
8.	$4500	10%	quarterly	7	_____

Solve the following application problems using 360-day years where applicable.

9. At 58, Thomas Jones knows that he needs to save more. He decides to invest $300 per quarter in a mutual fund he hopes will earn 10% compounded quarterly. Find the accumulated amount at age 65. **[11.1]**

9. _____

10. A public utility needs $60 million in 5 years for a major capital expansion. What annual payment must the firm place into a sinking fund earning 5% per year in order to accumulate the required funds? **[11.3]**

10. _____

CHAPTER 12 Business and Consumer Loans

11. Sarah Warren purchased 100 shares of stock at $23.45 per share. The company had earnings of $1.56 and a yearly dividend of $.35. Find **(a)** the cost of the purchase ignoring commissions, **(b)** the price–earnings ratio to the nearest whole number, and **(c)** the dividend yield. **[11.4]**

(a) _____
(b) _____
(c) _____

12. Martin Wicker buys 9000 GM bonds due in 2020 at 104.38 for the pension fund he manages. The coupon rate is 6.4%. Find **(a)** the cost to purchase the bonds if the commission is $.15 per bond, **(b)** the annual interest from all of the bonds, and **(c)** the effective interest rate. **[11.5]**

(a) _____
(b) _____
(c) _____

13. James Thompson purchased a large riding lawnmower costing $2800 with $500 down and payments of $108.27 per month for 24 months. Find **(a)** the total installment cost, **(b)** the finance charge, and **(c)** the amount financed. **(d)** Then use the table to find the annual percentage rate to the nearest quarter of a percent. **[12.2]**

(a) _____
(b) _____
(c) _____
(d) _____

14. Abbie Spring's unpaid balance on her Visa card on July 8 was $204.37. She made a payment of $100 on July 14 and had charges of $34.95 on July 16 and $95.12 on July 30. Assume an interest rate of 1.6% per month and find the balance on August 8 using **(a)** the unpaid balance method and **(b)** the average daily balance method. **[12.1]**

(a) _____
(b) _____

15. Mayberry Pets borrows to purchase a van to transport animals and supplies. They agree to make quarterly payments on the $22,400 debt for 3 years at a rate of 8% compounded quarterly. Find **(a)** the quarterly payment and **(b)** the total amount of interest paid. **[12.4]**

(a) _____
(b) _____

16. The Hodges purchase an older 3-bedroom, 1-bath home for $195,000 with 5% down. They finance the balance at 5% per year for 30 years. If insurance is $720 per year and taxes are $4140 per year, find the monthly payment. **[12.5]**

16. _____

17. On January 10, Bob Jones signed a 200-day note for $24,000 to finance some work on a roof. The note was at 9% per year simple interest. Due to an unexpected income tax refund, he was able to repay $10,000 on April 15. Use the United States Rule and **(a)** find the balance owed on the principal after the partial payment. **(b)** Then find the amount due at maturity of the loan. **[12.3]**

(a) _____
(b) _____

18. Karoline Jacobs borrowed $2200 for new kitchen appliances. She agreed to pay the loan back with 8 payments of $290.69 each. After 3 payments, she decides to go ahead and pay off the loan in full. Use the Rule of 78 to find (a) the amount of unearned interest and (b) the amount needed to repay the loan in full.   [12.3]

(a) _____

(b) _____

19. County Squire Electrical lost a lawsuit and must pay the injured party $35,000 at the end of each semiannual period for 2 years. If the funds earn 4% per year compounded semiannually, find the amount that needs to be set aside today to fulfill this obligation.   [11.2]

19. _____

20. James Booker signs an employment contract that should provide him $35,000 at the end of each year for 3 years when he retires in 4 years. If funds earn 8% per year, find the present value needed today to meet the eventual retirement income.   [10.3 and 11.2]

20. _____

21. Explain the terms *present value, future value,* and *annuity.*

22. Describe stocks and bonds, explaining similarities and differences.

Taxes and Insurance

13

CHAPTER CONTENTS

13.1 Property Tax
13.2 Personal Income Tax
13.3 Fire Insurance
13.4 Motor-Vehicle Insurance
13.5 Life Insurance

CASE IN POINT

MARTHA SPENCER OWNS The Doll House. She specializes in antique dolls from around the world and sells a lot of antique Barbie dolls. Her business was struggling, so she set up a website that increased sales. In particular, international sales really took off, as shown in the graph. Spencer had never realized Asians would buy so many antique dolls. She estimates that one-third of sales will be to overseas clients in 2018.

Spencer owns the building in which her business is located, and she has several employees. As a result, she must keep up with all the laws related to taxes and insurance. It is not fun to work through tax and insurance issues, but Spencer knows it must be done—and done well—in order for her small business to prosper.

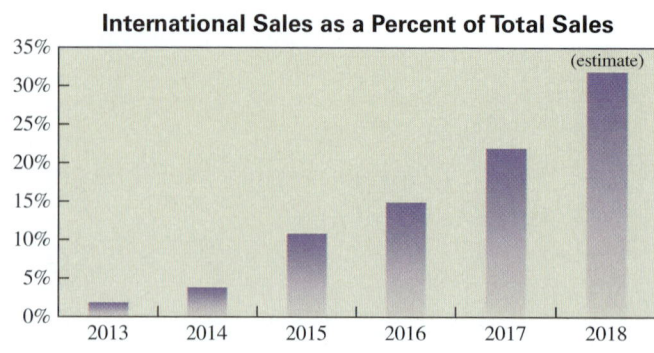

Both **taxes** and **insurance** are facts of life. In one form or another, we pay taxes to many different entities, including the federal government, states, counties, school districts, and cities. The pie chart shows the number of days per year the "average worker" works to pay various taxes. Property taxes and individual income taxes are discussed in this chapter.

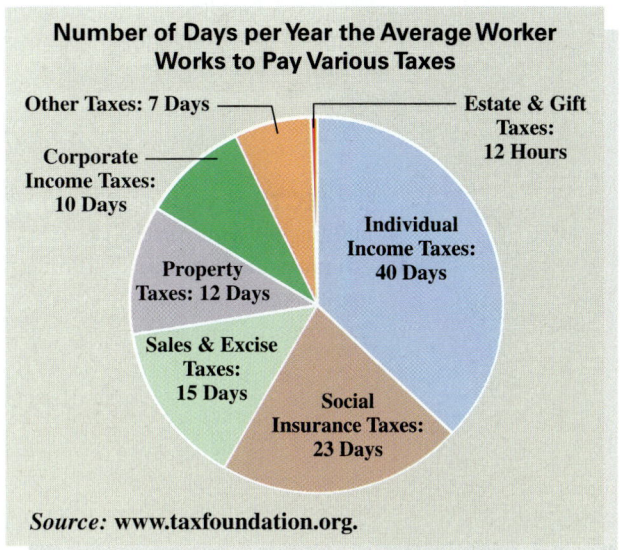

Oliver Wendell Holmes, Jr.
"Taxes are the price we pay to live in a civilized society."

Individuals buy insurance to protect from **risk**—damage to homes and automobiles, injury from car accidents, medical expenses, or expenses at the time of an unexpected death. Companies buy insurance to protect from potential losses to buildings and property, losses due to workers injured on the job, or even lawsuits from customers. Three common types of insurance are discussed in this chapter.

13.1 Property Tax

OBJECTIVES

1. Define *fair market value* and *assessed value*.
2. Find the tax rate.
3. Find the property tax.
4. Express tax rate in percent, in dollars per $100, in dollars per $1000, and in mills.
5. Find taxes given the assessed value and the tax rate.

CASE IN POINT Martha Spencer was surprised by the property tax bill on her commercial building this year. Her taxes went up significantly and she was determined to find out why. She also wanted to know where the money was going.

Property owners pay a property tax that is based on the value of their real estate. The money received from property taxes are used to pay for all of the following:

Taxes Pay for the Following:

- police protection and emergency response
- fire protection and emergency response
- 911 emergency service
- public schools
- street and sidewalk maintenance
- snow removal
- parks and recreational facilities
- activity programs for senior citizens
- libraries

Property taxes are very important sources of revenue for public schools and even many community colleges. The taxes paid by the property owners near where you live help fund public schools and many community colleges.

School District Battles Financial Problems

A local school district is struggling. The mission of the district is founded on a quality education for all, but the current financial picture is grim. Income from the state has gone down at the same time that enrollment is up slightly. The school board is asking for a small increase in the property tax, but board members are afraid they will not get it. Without an increase in funding, the district may have to lay off some teachers and administrators.

OBJECTIVE 1 Define *fair market value* and *assessed value*. The value of lands and buildings is determined by a **tax assessor**, who may either be appointed or elected. He/she estimates the **fair market value**, which is the price at which a property can be sold. The fair market value is multiplied by a percent called the **assessment rate** to find the **assessed value** of the property. The assessed value is used to find the property tax.

Fair Market Value ⇨ Assessed Value ⇨ Property Tax

Finding the Assessed Valuation of Property

EXAMPLE 1 Find the assessed value for the following pieces of property owned by Martha Spencer.
(a) Home: fair market value $298,000; assessment rate 40%
(b) Business property: fair market value $328,500; assessment rate 35%
(c) Commercial lot located in a different state: fair market value $123,800; assessment rate 60%

Quick TIP
The assessed value is used as a middle step to determine the property tax.

SOLUTION
Multiply the fair market value by the assessment rate.
(a) $298,000 × 40% = $119,200
(b) $328,500 × .35 = $114,975
(c) $123,800 × .60 = $74,280

QUICK CHECK 1
A commercial building has a fair market value of $1,480,000. If the assessment rate is 70%, find the assessed value.

Just because the assessment rate is higher in one area than another does not necessarily mean that the taxes are higher. The assessed value of the property must be multiplied by the tax rate to find the tax.

OBJECTIVE 2 Find the tax rate. A taxing authority such as a city or community college district first estimates the revenue needed for the year. It then finds the **property-tax rate** required to generate that amount of tax as follows.

Finding the Property-Tax Rate
Step 1   Estimate the amount of money needed by the taxing authority.
Step 2   Find the total fair market value of all real properties in the tax district.
Step 3   Find the total assessed value of all real properties in the tax district.
Step 4   Property tax rate = $\dfrac{\text{Total tax amount needed}}{\text{Total assessed value}}$

Finding the Tax Rate

EXAMPLE 2 Find the tax rate for the following school districts, rounded to the nearest hundredth of a percent.
(a) Amount needed = $14,253,000; total assessed value = $1,575,890,000
(b) Amount needed = $268,000,000; total assessed value = $36,190,000,000

SOLUTION

(a) Tax rate = $\dfrac{\$14{,}253{,}000}{\$1{,}575{,}890{,}000}$ = .90% (rounded)

(b) Tax rate = $\dfrac{\$268{,}000{,}000}{\$36{,}190{,}000{,}000}$ = .74% (rounded)

QUICK CHECK 2

A large community college district needs $22,350,000 this year. The taxing authority has determined that the total assessed value of all the real estate in the district is $3.48 billion. Find the property tax rate to the nearest hundredth of a percent.

OBJECTIVE 3 Find the property tax. The tax rate is applied to the assessed value to find the property tax due as follows.

Finding the Property Tax

Tax = Tax rate × Assessed value

OBJECTIVE 4 Express tax rate in percent, in dollars per $100, in dollars per $1000, and in mills. Tax rates are stated differently by different taxing entities. However, just because they are stated differently does not mean that the tax is either higher or lower in one area than in another. The table that follows describes methods that are in common use.

Methods to Calculate Property Tax

Percent. Some areas express tax rates as a percent of assessed value. The yearly tax on a piece of property with an assessed value of $174,000 at a tax rate of 2.17% follows.

Tax = .0217 × **$174,000** = **$3775.80**

Dollars per $100. In some areas, the tax rate is expressed as a number of dollars per $100 of assessed value. In this event, find the tax on a piece of land by first finding the number of hundreds in the assessed value and then multiplying the number of hundreds by the tax rate. For example, assume an assessed value of $56,300 and a tax rate of $11.42 per $100 of assessed value and find taxes for the year as follows.

$56,300 ÷ 100 = **563 hundreds**   Divide by 100 by moving the decimal point 2 places to the left.

Tax = $11.42 × 563 = **$6429.46**   Multiply by the tax rate to find tax.

Dollars per $1000. In other areas, the tax rate is expressed as a number of dollars per $1000 of assessed value. If the tax rate is $98.12 per $1000, a piece of property having an assessed value of $197,000 would be taxed as follows.

$197,000 = **197 thousands**   Move the decimal point 3 places to the left to divide by 1000.

Tax = $98.12 × 197 = **$19,329.64**

Mills. Other taxing authorities express tax rates in mills or one-thousandths of a dollar. For example, a tax rate might be expressed as 46 mills. Divide the 46 mills by 1000 to find .046 per dollar of assessed value. Assuming a tax rate of 46 mills, the tax on a house assessed at $81,000 is found as follows.

46 mills = .046

Tax = **.046** × $81,000 = **$3726**

The following chart shows the same tax rates written in the four different systems. Although expressed differently, the rates in each row of this chart are equivalent tax rates.

Percent	Per $100	Per $1000	In Mills
1.25%	$1.25	$12.50	12.5
3.2%	$3.20	$32	32
9.87%	$9.87	$98.70	98.7

OBJECTIVE 5 Find taxes given the assessed value and the tax rate. Property taxes are found by multiplying the tax rate by the assessed value.

Finding the Property Tax **EXAMPLE 3** Find the taxes on each of the following pieces of property. Assessed values and tax rates are given.

(a) $58,975; 8.4%
(b) $875,400; $7.82 per $100
(c) $129,600; $64.21 per $1000
(d) $221,750; 94 mills

SOLUTION

Multiply the tax rate by the assessed value.

(a) 8.4% = .084

$$\text{Tax} = \text{Tax rate} \times \text{Assessed value}$$
$$\text{Tax} = .084 \times \$58,975 = \$4953.90$$

(b) $875,400 = **8754 hundreds**

$$\text{Tax} = \$7.82 \times \mathbf{8754} = \$68,456.28$$

(c) $129,600 = **129.6 thousands**

$$\text{Tax} = \$64.21 \times \mathbf{129.6} = \$8321.62$$

(d) 94 mills = .094

$$\text{Tax} = .094 \times \$221,750 = \$20,844.50$$

> **Quick TIP**
>
> The annual property tax paid on a large commercial building such as a mall are huge.

> **QUICK CHECK 3**
>
> A home is assessed at $65,000. Find the property tax if the tax rate is (a) 3.4%, (b) $4.50 per $100, (c) $38.40 per $1000, and (d) 48.2 mills.

Not everyone pays at the same property-tax rates. For example, churches are usually exempt from property taxes. Many states give homeowners an exemption that reduces the property tax on their home. Still other states give property-tax exemptions to the elderly.

Comparing Tax Rates **EXAMPLE 4** Diego Garcia considers property tax as he decides where to build a 40-unit apartment house with an expected market value of $2,800,000. In one county, property is assessed at 50% of market value with a tax rate of 3.2%. In a second county, property is assessed at 80% with a tax rate of 35 mills. (a) Find the county with the lower tax. (b) Find the amount saved by building in the county with the lower tax.

SOLUTION

(a) Property tax 1st county = $2,800,000 × .5 × .032 = **$44,800**
Property tax 2nd county = $2,800,000 × .8 × .035 = **$78,400**
The first county has the lower tax.

(b) Amount saved = **$78,400** − **$44,800** = $33,600 per year

> **QUICK CHECK 4**
>
> One county has a tax rate of $2.40 per $100, and a second county has a tax rate of $26.60 per $1000. Once built, a home will have an assessed value of $290,000. (a) Find the county with the lower tax. (b) Find the amount saved by building in the county with the lower tax.

13.1 Exercises // MyLab Math

The shaded sections below contain solutions to help you get a QUICK START on the various types of exercises.

Find the assessed value for each of the following pieces of property. (See Example 1.)

	Fair Market Value	Rate of Assessment	Assessed Value		Fair Market Value	Rate of Assessment	Assessed Value
1.	$85,000	40%	$34,000	2.	$68,000	60%	$40,800
	$85,000 × .4 = $34,000				$68,000 × .6 = $40,800		
3.	$142,300	50%	_____	4.	$98,200	42%	_____
5.	$1,300,500	25%	_____	6.	$2,450,000	80%	_____

Find the tax rate for the following. Write the tax rate as a percent, rounded to the nearest hundredth of a percent. (See Example 2.)

	Total Tax Amount Needed	Total Assessed Value	Tax Rate		Total Tax Amount Needed	Total Assessed Value	Tax Rate
7.	$18,300,000	$6,850,000,000	.27%	8.	$7,600,000	$1,752,500,000	.43%
	$18,300,000 ÷ $6,850,000,000 = .0027 = .27% (rounded)				$7,600,000 ÷ $1,752,500,000 = .0043 = .43% (rounded)		
9.	$1,580,000	$197,500,000	_____	10.	$2,175,000	$480,300,000	_____
11.	$1,224,000	$816,000,000	_____	12.	$28,630,000	$12,350,000,000	_____

Write the given tax rate using the other three methods. (See Example 3.)

	Percent	Per $100	Per $1000	In Mills
13.	4.84%	(a) $4.84	(b) $48.40	(c) 48.4
14.	(a) _____	$6.75	(b) _____	(c) _____
15.	(a) _____	(b) _____	$70.80	(c) _____
16.	(a) _____	(b) _____	(c) _____	28

17. Explain the steps used to find property tax. Where are funds received from property tax spent? (See Objective 1.)

18. Explain the difference between fair market value, assessed value, and property tax. (See Objectives 1, 2, and 3.)

C// indicates an exercise that is related to the Case in Point feature.

516 CHAPTER 13 Taxes and Insurance

Find the property tax for the following. (See Example 3.)

	Assessed Value	Tax Rate	Tax		Assessed Value	Tax Rate	Tax
19.	$86,200	$6.80 per $100	$5861.60	**20.**	$41,300	$46.40 per $1000	
	862 × $6.80 = $5861.60						
21.	$128,200	42 mills		**22.**	$37,250	3.4%	

Solve the following application problems.

23. **REAL ESTATE TAXES** Martha Spencer owns the building used by The Doll House. The property has a fair market value of $328,500, the assessment rate is 35%, and the local tax rate is 5.2%. Find the property tax.
$328,500 × .35 = $114,975; $114,975 × .052 = $5978.70

23. $5978.70

24. **FOURPLEX** Chad LeCompte owns a four-unit apartment building with a fair market value of $248,000. Property in the area is assessed at 40% of market value, and the tax rate is 5.5%. Find the amount of the property tax.

24. _____

25. **RADIO STATION** A new radio station broadcasts from a building having a fair market value of $480,000. Property in the area is assessed at 25% of market value, and the tax rate is $65.30 per $1000 of assessed value. Find the property tax.

25. _____

26. **OFFICE COMPLEX** Huron Development just purchased a modern office complex for $12,380,000. The county assesses the property at 60% of market value and has a property tax of $18.40 per $1000 of assessed value. Find the property tax.

26. _____

27. **WALMART SUPERCENTER** A new Walmart Supercenter is expected to have a fair market value of $32,500,000. It will be assessed at 65% and the tax rate is $2.18 per $100 of assessed value. Estimate the annual property tax.

27. _____

28. **MOTORCYCLES** One Harley-Davidson store has property with a fair market value of $518,600. The property is located in an area that is assessed at 35% of market value. The tax rate is $7.35 per $100. Find the property tax.

28. _____

29. **PHARMACY** A building with a stand-alone pharmacy has a fair market value of $1,400,000 and a tax rate of 23 mills. It is in an area assessed at 30% of market value. Find the property tax.

29. _____

30. CONVENIENCE STORE A convenience store on 1.2 acres of land is located in a prime location. The owner has been offered $1,650,000 for the building and land. Property is assessed at 40%. Given a tax rate of 26.5 mills, find the property tax.

30. _____

31. COMPARING PROPERTY TAX RATES In one parish (county), property is assessed at 40% of market value, with a tax rate of 32.1 mills. In a second parish, property is assessed at 24% of market value, with a tax rate of 50.2 mills. A telephone company is trying to decide where to place a storage building with a fair market value of $95,000. **(a)** Which parish would charge the lower property tax? **(b)** Find the annual amount saved.

(a) _____
(b) _____

32. TAXES ON HOME Jacque Henri plans to build a home with a fair market value of $240,000 in one of two neighboring counties. In Smith County, property is assessed at 30% of market value, with a tax rate of 45.6 mills. In Justin County, property is assessed at 58% of market value with a tax rate of 38.5 mills. Find **(a)** which county has the lower property tax and **(b)** the annual amount saved if the home is built in that county.

(a) _____
(b) _____

QUICK CHECK ANSWERS

1. $1,036,000
2. .64%
3. (a) $2210 (b) $2925 (c) $2496 (d) $3133
4. (a) first county (b) $754 per year

13.2 Personal Income Tax

OBJECTIVES

1. List the four steps that determine income-tax liability.
2. Find the adjusted gross income.
3. Know the standard deduction amounts.
4. Find the taxable income and income tax.
5. List possible deductions.
6. Determine a balance due or a refund from the Internal Revenue Service.
7. Prepare a 1040A and a Schedule 1 federal tax form.

CASE IN POINT Martha Spencer is responsible for paying all appropriate taxes related to her business as well as her personal income taxes. Since she is busy and taxes are complex, she hires an accountant to do her taxes.

The diagram below is an estimate of the incoming revenue and outgoing expenses of the federal government for 2017. The largest source of revenue is individual **income taxes**. Individuals also pay about one-half of Social Security and other payroll taxes, which is the second-largest source of revenue (see Chapter 6). You can see that corporations pay much less than individuals in income taxes, although they do also pay about one-half of Social Security and other payroll taxes.

Defense spending was expected to be over $800 billion, but the larger category of spending is to individuals for Social Security, Medicare, Medicaid, and safety net programs. Effectively the government taxes individuals and corporations and spends much of that money on the elderly and the poor.

Notice that spending was expected to exceed revenue by about $500 billion. Consequently, the federal government was expected to borrow $500 billion to pay its bills. The total federal debt, which has accumulated over decades, was nearly $20 trillion ($20,000,000,000,000) in 2017, which is about $60,000 for each individual in the United States. The federal government was expected to pay $223 billion in interest on that debt in 2017. (The government has to pay interest on debt just as a student has to pay interest on student debt or a car loan.)

An ongoing debate is whether those with the highest incomes pay their fair share of tax. The top 50% of all earners pay more than 97% of all individual income taxes. Thus, the bottom 50% pay less than 3% of all income taxes, although they do pay Social Security, Medicare, property, gasoline, sales, and other taxes. Although the top 10% pay about $900 billion in income taxes, many wealthy individuals pay relatively little in income taxes due to various tax avoidance strategies allowed in the tax code, such as depreciation (which is covered in Chapter 14).

Income and wealth inequality has grown dramatically in the United States over the past 25 years. Earnings for families with lower incomes have simply not kept up with earnings of those with higher incomes. Most people believe this is a bad situation that will lead to more poverty and related social problems. Another factor affecting wealth inequality is that tax rates for higher earners have generally gone down, so that many of the wealthy pay a lower percentage of their incomes in taxes than in the past.

Discussion question: What ideas do you have to decrease income inequality?

Quick TIP
Payments to individuals (Social Security, Medicare, and Medicaid and other) were 59.1% of total money spent. Interest on the national debt was 6% of total money spent.

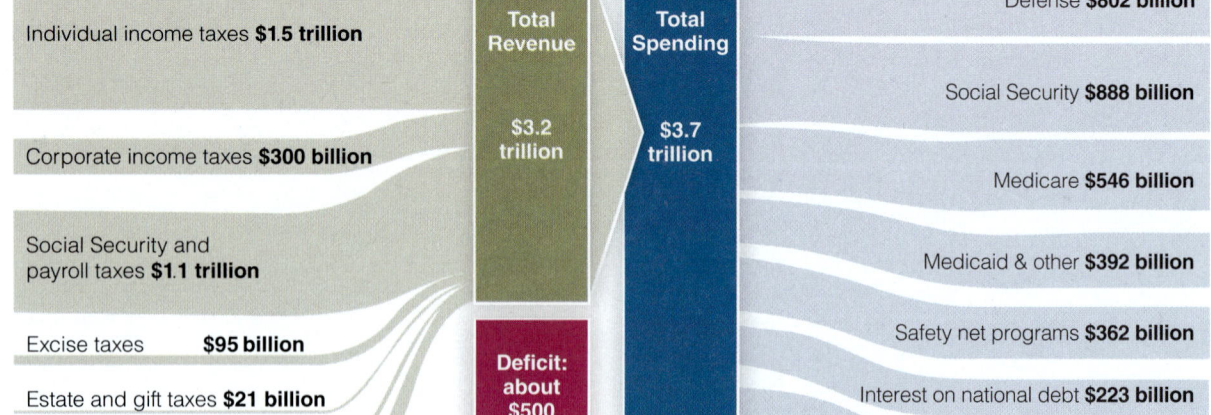

Source: Office of Management and Budget.

Most working adults in the United States are required to file an income tax return each year with the **Internal Revenue Service (IRS)**. The IRS is the branch of the U.S. government responsible for collecting income taxes. More than 150,000,000 tax returns are expected to be filed next year, at least 90% of them electronically.

13.2 Personal Income Tax

Quick TIP
Always go to the IRS for the most recent forms and laws.

There are thousands of rules about income tax preparation and many different forms, each with its own instructions. However, many people have relatively simple income tax returns that do not require professionals to complete. No matter whether you use a professional to file your income-tax return or do it yourself, you should carefully save all records related to income and expenses. These records should be kept until the statute of limitations runs out, which is 3 years from the date the tax return was filed or 2 years from the date the tax was paid, whichever is later. The IRS can audit you during this period and it requires proof of income and deductions. Many accountants recommend that you keep records for 7 years, but you should keep records on property such as your home as long as they are needed, often longer than 3 years.

OBJECTIVE 1 List the four steps that determine income-tax liability. Income tax calculations can be very complicated, but here are the four steps common to all personal tax returns. We will only look at a simple tax return in this section.

> **Finding Your Income Tax Liability**
> Step 1 Find the adjusted gross income (AGI) for the year.
> Step 2 Find the taxable income.
> Step 3 Find the tax.
> Step 4 Check to see if a refund is due or if money is owed to the government.

OBJECTIVE 2 Find the adjusted gross income. The first step in finding personal income tax is to find **adjusted gross income**. Adjusted gross income is the total of all income less certain adjustments. Employers are required to send out **W-2 forms** showing wages paid, federal income taxes withheld, Social Security tax withheld, and Medicare tax withheld. Other types of income such as interest, dividends, and self-employment income are shown on **1099 forms**, which are also mailed to each individual. Sample W-2 and 1099-INT forms are shown on the next page.

> **Finding Adjusted Gross Income**
> Step 1 Add amounts from all W-2 and 1099 forms along with dividends, capital gains, unemployment compensation, and tips or other employee compensation.
> Step 2 From this sum, subtract allowed adjustments such as contributions to a *regular* **individual retirement account (IRA)** or alimony payments.

Finding Adjusted Gross Income (AGI)

EXAMPLE 1 As an assistant manager at The Doll House, Jennifer Crum earned $24,738.41 during the 9 months she worked last year and $1624.01 in interest from her credit union (see her W-2 and 1099 forms). She had $1500 in regular IRA contributions. Find her adjusted gross income.

SOLUTION

$$\text{Adjusted gross income} = \text{Wages} + \text{Interest} - \textbf{IRA contribution}$$
$$= \$24{,}738.41 + \$1624.01 - \$1500$$
$$= \$24{,}862.42$$

QUICK CHECK 1
Last year, Marelli Food Distribution paid marketing manager Beth Hogan $95,000. Hogan also earned $6450 in interest and contributed $2500 to an IRA. Find her adjusted gross income.

When you file your income tax return, a copy of all W-2 forms must be sent to the Internal Revenue Service along with the completed tax forms. However, the IRS does not require copies of 1099 forms to be sent unless income tax was withheld. This is because the IRS receives copies of the 1099 forms from the employers.

Form W-2 Wage and Tax Statement

22222 Void ☐	**a** Employee's social security number 123–45–6789

For Official Use Only ▶
OMB No. 1545-0008

b Employer identification number (EIN) 94–1287319	**1** Wages, tips, other compensation $24,738.41	**2** Federal income tax withheld $3275.60
c Employer's name, address, and ZIP code *The Doll House* *1568 Liberty Heights Ave.* *Baltimore, MD 21230*	**3** Social security wages $24,738.41	**4** Social security tax withheld $1533.78
	5 Medicare wages and tips $24,738.41	**6** Medicare tax withheld $358.71
	7 Social security tips	**8** Allocated tips
d Control number	**9** Advance EIC payment	**10** Dependent care benefits
e Employee's first name and initial: *Jennifer* Last name: *Crum* Suff.	**11** Nonqualified plans	**12a** See instructions for box 12
2136 Old Road *Towson, MD 21285*	**13** Statutory employee ☐ Retirement plan ☐ Third-party sick pay ☐	**12b**
	14 Other	**12c**
		12d
f Employee's address and ZIP code		

15 State Employer's state ID number MD 600–5076	**16** State wages, tips, etc.	**17** State income tax	**18** Local wages, tips, etc.	**19** Local income tax	**20** Locality name

Form **W-2** Wage and Tax Statement **20__**

Department of the Treasury—Internal Revenue Service

Form 1099-INT Interest Income

9292 ☐ VOID ☐ CORRECTED

PAYER'S name, street address, city or town, state or province, country, ZIP or foreign postal code, and telephone no. *Employees Credit Union* *2572 Brookhaven Drive* *Dundalk, MD 21222*	Payer's RTN (optional)	OMB No. 1545-0112 **20__** Form **1099-INT**	**Interest Income**
	1 Interest income $		
	2 Early withdrawal penalty $		Copy 1
PAYER'S federal identification number 94–1287319	RECIPIENT'S identification number 123–45–6789	**3** Interest on U.S. Savings Bonds and Treas. obligations $1624.01	For State Tax Department
RECIPIENT'S name *Jennifer Crum*	**4** Federal income tax withheld $	**5** Investment expenses $	
	6 Foreign tax paid $	**7** Foreign country or U.S. possession	
Street address (including apt. no.) *2136 Old Road*	**8** Tax-exempt interest $	**9** Specified private activity bond interest $	
City or town, state or province, country, and ZIP or foreign postal code *Towson, MD 21285*	**10** Market discount $	**11** Bond premium $	
FATCA filing requirement ☐	**12** Bond premium on Treasury obligations $	**13** Bond premium on tax-exempt bond $	
Account number (see instructions)	**14** Tax-exempt and tax credit bond CUSIP no.	**15** State **16** State identification no.	**17** State tax withheld $ $

Form **1099-INT** www.irs.gov/form1099int Department of the Treasury - Internal Revenue Service

13.2 Personal Income Tax

OBJECTIVE 3 Know the standard deduction amounts. Most people are almost finished with their income tax return at this point. A taxpayer next subtracts the larger of either the itemized deductions or the standard deduction from adjusted gross income. Itemized deductions are described in Objective 5 in this section and are associated with several limitations based on adjusted gross income. The **standard deduction** amount is based on the taxpayer's filing status as follows.

Finding Standard Deduction

- $6,300 for single taxpayers
- $12,600 for married taxpayers filing jointly
- $6,300 for married taxpayers filing separately
- $9,300 for head of household

Married people can file either together on one tax return (jointly) or separately. A **head of household** is an unmarried person who provides a home for other people, such as a dependent child or even a dependent parent of the taxpayer.

Additional standard deductions are given for taxpayers and dependents who are blind or 65 years of age or older. The standard deduction amounts change from one year to the next. The most current amounts can be obtained from the IRS.

The next step is to find the number of **personal exemptions**. One exemption is allowed for yourself, another for your spouse if filing a joint return, and yet another exemption for each child or other dependent. A **dependent** is a child, spouse, parent, or certain other relative to whom the primary taxpayer contributes all or a major portion of necessary financial support. The number of exemptions does not depend on whether the filing status is single or married. The deduction for personal exemptions is $4050 times the number of exemptions. Here are some examples.

Filing Status and Exemptions	Total Deduction
Single individual (1 exemption)	1 × $4050 = $4050
Married with 3 children (5 exemptions)	5 × $4050 = $20,250
Single, head of household, 2 children (3 exemptions)	3 × $4050 = $12,150

Taxable income is found by subtracting the standard deduction amount and personal exemptions from adjusted gross income. Taxes are then calculated based on taxable income as shown next.

OBJECTIVE 4 Find the taxable income and income tax. Taxable income is used along with data from the tax rate schedule shown next to find the income tax. As you can see from the tax rate schedules, individual income tax rates range from a low of 10% to a high of 39.6% depending on income and filing status. Assume a single filer has a taxable income of $37,300. Use the table for single filer below to find:

$$\text{Federal income tax} = \$927.50 + \textbf{15\% of the amount over \$9275}$$
$$= \$927.50 + .15 \times (\$37,300 - \$9275)$$
$$= \$5131.25$$

Note that the 15% applies only to the excess over $9275 and not to the entire $37,300. Find taxes as a percent of taxable income as follows.

$$\text{Percent of taxable income paid in taxes} = \frac{\text{Amount paid in taxes}}{\text{Taxable income}}$$
$$= \frac{\$5131.25}{\$37,300} = 13.8\% \quad \text{(rounded)}$$

Therefore, this taxpayer must pay nearly 14% of taxable income in income taxes. This does not include Social Security or Medicare taxes, nor does it include any other taxes the individual has to pay such as sales taxes, property taxes, gasoline taxes, etc.

2016 Tax Rate Schedules

SINGLE

Taxable Income	Find the Tax
$0–$9,275	10%
$9,276–$37,650	$927.50 plus 15% of the amount over $9,275
$37,651–$91,150	$5,183.75 plus 25% of the amount over $37,650
$91,151–$190,150	$18,558.75 plus 28% of the amount over $91,150
$190,151–$413,350	$46,278.75 plus 33% of the amount over $190,150
$413,351–$415,050	$119,934.75 plus 35% of the amount over $413,350
$415,051 or more	$120,529.75 plus 39.6% of the amount over $415,050

MARRIED FILING JOINTLY

Taxable Income	Find the Tax
$0–$18,550	10%
$18,551–$75,300	$1,855 plus 15% of the amount over $18,550
$75,301–$151,900	$10,367.50 plus 25% of the amount over $75,300
$151,901–$231,450	$29,517.50 plus 28% of the amount over $151,900
$231,451–$413,350	$51,791.50 plus 33% of the amount over $231,450
$413,351–$466,950	$111,818.50 plus 35% of the amount over $413,350
$466,951 or more	$130,578.50 plus 39.6% of the amount over $466,950

Quick TIP
All of these figures change every year.

MARRIED FILING SEPARATELY

Taxable Income	Find the Tax
$0–$9,275	10%
$9,276–$37,650	$927.50 plus 15% of the amount over $9,275
$37,651–$75,950	$5,183.75 plus 25% of the amount over $37,650
$75,951–$115,725	$14,758.75 plus 28% of the amount over $75,950
$115,726–$206,675	$25,895.75 plus 33% of the amount over $115,725
$206,676–$233,475	$55,909.25 plus 35% of the amount over $206,675
$233,476 or more	$65,289.25 plus 39.6% of the amount over $233,475

HEAD OF HOUSEHOLD

Taxable Income	Find the Tax
$0–$13,250	10%
$13,251–$50,400	$1,325 plus 15% of the amount over $13,250
$50,401–$130,150	$6,897.50 plus 25% of the amount over $50,400
$130,151–$210,800	$26,835 plus 28% of the amount over $130,150
$210,801–$413,350	$49,417 plus 33% of the amount over $210,800
$413,351–$441,000	$116,258.50 plus 35% of the amount over $413,350
$441,001 or more	$125,936 plus 39.6% of the amount over $441,000

Find Taxable Income and the Income Tax Amount

EXAMPLE 2 Find the taxable income and income tax for each of the following.

(a) Herbert White, married filing jointly, 5 daughters, adjusted gross income $48,300
(b) Onita Fields, single, no dependents, adjusted gross income $28,400
(c) Imogene Griffin, single, head of household, 2 children, adjusted gross income $74,500
(d) Jeffy Norwood, married filing separately, 1 child, adjusted gross income $145,000

SOLUTION

(a) Herbert White + spouse + 5 daughters = 7 exemptions

Taxable income = $48,300 − $12,600 − (7 × $4050) = **$7350**
  (standard deduction) (deduction for exemptions) (taxable income)

The tax rate for married filing jointly with less than $18,550 in taxable income is 10%.

Income tax = .10 × **$7350** = $735

(b) Onita Fields = 1 exemption

Taxable income = $28,400 − $6300 − (1 × $4050) = **$18,050**
  (standard deduction) (deduction for exemptions) (taxable income)

Income tax = $927.50 + **15% of amount over $9275**
= $927.50 + .15 × ($18,050 − $9275)
= **$2243.75**

(c) Imogene Griffin + 2 children = 3 exemptions

$$\text{Taxable income} = \$74{,}500 - \underset{\substack{\text{standard}\\\text{deduction}}}{\$9300} - \underset{\substack{\text{deduction}\\\text{for exemptions}}}{(3 \times \$4050)} = \underset{\substack{\text{taxable}\\\text{income}}}{\$53{,}050}$$

$$\begin{aligned}\text{Income tax} &= \$6897.50 + \text{25\% of the amount over \$50,400}\\ &= \$6897.50 + .25 \times (\$53{,}050 - \$50{,}400)\\ &= \$7560\end{aligned}$$

Quick TIP
Notice that the tax rate and therefore the income taxes go up with increased taxable income.

(d) Jeffy Norwood + 1 child = 2 exemptions

$$\text{Taxable income} = \$145{,}000 - \underset{\substack{\text{standard}\\\text{deduction}}}{\$6300} - \underset{\substack{\text{deduction}\\\text{for exemptions}}}{(2 \times \$4050)} = \underset{\substack{\text{taxable}\\\text{income}}}{\$130{,}600}$$

$$\begin{aligned}\text{Income tax} &= \$25{,}895.75 + \text{33\% of the amount over \$115,725}\\ &= \$25{,}895.75 + .33 \times (\$130{,}600 - \$115{,}725)\\ &= \$30{,}804.50\end{aligned}$$

QUICK CHECK 2

John O'Neill files jointly with his wife. They have 1 child and an adjusted gross income of $62,300 and use the standard deduction. Find their taxable income and income tax.

OBJECTIVE 5 List possible deductions. Actually, taxpayers may deduct *the larger of* **itemized deductions** *or* the standard deduction from their adjusted gross income *before* finding taxable income. This is particularly applicable to individuals who are paying interest on a home loan, but sometimes others can use this to their advantage. The most common **tax deductions** are listed next.

Medical and dental expenses: Only medical and dental expenses exceeding 10% of your adjusted gross income, or 7.5% if you or your spouse are 65 or older, can be deducted. Expenses reimbursed by an insurance company are not deductible.

Taxes: State and local taxes, real estate taxes, and personal property taxes may be deducted (but not federal income or payroll taxes).

Interest: Interest paid on a debt such as a home mortgage or student loan is deductible. Other personal interest is not deductible.

Gifts to charity: Gifts to a qualified organization (such as a church) are deductible.

Casualty or theft losses: Losses due to a casualty (e.g., fire) or theft are deductible if not reimbursed by insurance.

Unreimbursed employee expenses, **tax preparation**, and some **miscellaneous deductions:** These expenses are deductible only if the total exceeds 2% of the taxpayer's adjusted gross income.

You can also deduct alimony paid to a former spouse, or a spouse from whom you are separated, if required under a divorce or separation decree, whether or not you itemize deductions.

EXAMPLE 3 Using Itemized Deductions to Find Taxable Income and Income Tax

Kristina Kelly is single, has 1 child, and had an adjusted gross income of $58,700 last year. She paid $3240 in real estate taxes, $7280 in home mortgage interest, and donated $1200 to her church. Find her taxable income and her income tax if she files as head of household.

SOLUTION

$$\text{Itemized deductions} = \$3240 + \$7280 + \$1200 = \$11{,}720$$

Her itemized deductions of $11,720 exceed the standard deduction of $6300 for a single person, so she uses the itemized deduction amount.

$$\text{Taxable income} = \$58{,}700 - \underset{\substack{\text{itemized}\\\text{deductions}}}{\$11{,}720} - \underset{\substack{\text{personal}\\\text{exemptions}}}{(2 \times \$4050)} = \$38{,}880$$

$$\begin{aligned}\text{Income tax} &= \$5183.75 + \text{25\% of the amount over \$37,650}\\ &= \$5183.75 + .25 \times (\$38{,}880 - \$37{,}650)\\ &= \$5491.25\end{aligned}$$

QUICK CHECK 3

Tom Garcia and his wife file jointly. They have no children, earned an adjusted gross income of $48,200, and paid the following: $2350 in other taxes, $6807.45 in home mortgage interest, and $500 as a charitable deduction to the Red Cross. Find their taxable income and income tax.

OBJECTIVE 6 Determine a balance due or a refund from the Internal Revenue Service. A taxpayer may have paid more to the IRS than is due. Add up the total amount of income tax paid using the W-2 forms. Usually, no taxes are withheld on 1099 forms. If the amount withheld is greater than the tax owed, the taxpayer is entitled to a refund. If the amount withheld is less than the tax owed, then the taxpayer must send the difference along with the tax return to the IRS.

Determining Tax Due or Refund

EXAMPLE 4 Jim Clark works as a petroleum engineer for Chevron, where he helps design deep, complex wells being drilled offshore Africa. His wife stays at home and takes care of their toddler. His salary last year was $123,500, and $1210 was withheld from each monthly paycheck. They file a joint return and use the standard deduction. Find the amount of income tax due to the IRS or the amount overpaid, as applicable.

SOLUTION

Adjusted gross income	$123,500
Standard deduction	−12,600 **married filing jointly**
Personal exemptions	12,150 **3 exemptions × $4050**
Taxable income	$98,750

Income tax = $10,367.50 + **25% of the amount over $75,300**
= $10,367.50 + .25 × ($98,750 − $75,300)
= $16,230

Amount withheld last year = $1210 × 12 = **$14,520**
Amount owed to IRS = $16,230 − **$14,520** = $1710

The Clarks must send in an additional $1710 to the IRS with their income tax return. Depending on the state in which they live, they may also owe some state income taxes, although a few states do not have income taxes.

QUICK CHECK 4

James Benson files as head of household and has 2 children. His adjusted gross income is $59,200, and he uses the standard deduction. Last year, his employer held $280 out of every bimonthly paycheck. Find **(a)** his taxable income, **(b)** tax due, **(c)** amount withheld by his employer last year, and **(d)** either the additional amount of income taxes he owes or the amount he overpaid.

Use the simplest IRS form possible when filing your income taxes. Here are basic guidelines starting with the simplest form. Be sure to check with the IRS before choosing the tax form—the rules change often!

> **1040EZ (general guidelines)**
> 1. Single or married filing jointly with no dependents
> 2. No adjustments to income
> 3. Cannot itemize
> 4. Limited sources of income and use of tax credits
> 5. Under age 65 with taxable income less than $100,000
> 6. Taxable interest not over $1500

> **1040A (general guidelines)**
> 1. Sources of income limited to wages, interest, capital gains, and a few other categories
> 2. Limited adjustments to income and use of tax credits
> 3. Cannot itemize
> 4. Taxable income less than $100,000

If neither the 1040EZ nor the 1040A form applies, then you must use the 1040 form. Partnerships and corporations require a completely different set of forms.

OBJECTIVE 7 Prepare a 1040A and a Schedule 1 federal tax form. The next example shows how to complete an income tax return using **Form 1040A** and **Schedule B (Form 1040A)**.

Preparing a 1040A and a Schedule 1

EXAMPLE 5 Jennifer Crum is single and claims one exemption. Her income appears on the W-2 and 1099 forms on page 520. She contributes $1500 to a regular IRA. Crum satisfies the requirements to use Form 1040A, but she must also fill out Schedule B (Form 1040A), since she has more than $1500 in interest. Round figures to the nearest dollar.

Form 1040A — U.S. Individual Income Tax Return (99) — 2016

Department of the Treasury—Internal Revenue Service

IRS Use Only—Do not write or staple in this space.

OMB No. 1545-0074

Your first name and initial: Jennifer
Last name: Crum
Your social security number: 1 2 3 4 5 6 7 8 9

If a joint return, spouse's first name and initial / Last name / Spouse's social security number

Home address (number and street). If you have a P.O. box, see instructions. 2136 Old Road

City, town or post office, state, and ZIP code. Towson, MD 21285

Make sure the SSN(s) above and on line 6c are correct.

Presidential Election Campaign — Check here if you, or your spouse if filing jointly, want $3 to go to this fund. Checking a box below will not change your tax or refund. ☑ You ☐ Spouse

Filing status
Check only one box.

1. ☑ Single
2. ☐ Married filing jointly (even if only one had income)
3. ☐ Married filing separately. Enter spouse's SSN above and full name here. ▶
4. ☐ Head of household (with qualifying person). (See instructions.) If the qualifying person is a child but not your dependent, enter this child's name here. ▶
5. ☐ Qualifying widow(er) with dependent child (see instructions)

Exemptions

6a ☑ **Yourself.** If someone can claim you as a dependent, **do not** check box 6a.
6b ☐ **Spouse**
6c **Dependents:**

(1) First name Last name	(2) Dependent's social security number	(3) Dependent's relationship to you	(4) ✓ if child under age 17 qualifying for child tax credit (see instructions)
			☐
			☐
			☐
			☐

If more than six dependents, see instructions.

Boxes checked on 6a and 6b: **1**
No. of children on 6c who:
• lived with you
• did not live with you due to divorce or separation (see instructions)
Dependents on 6c not entered above
Add numbers on lines above ▶ **1**

6d Total number of exemptions claimed.

Income

Attach Form(s) W-2 here. Also attach Form(s) 1099-R if tax was withheld.

If you did not get a W-2, see instructions.

Line	Description		Amount
7	Wages, salaries, tips, etc. Attach Form(s) W-2.	7	$24,738
8a	**Taxable** interest. Attach Schedule B if required.	8a	$1,624
8b	**Tax-exempt** interest. **Do not** include on line 8a.		
9a	Ordinary dividends. Attach Schedule B if required.	9a	
9b	Qualified dividends (see instructions).		
10	Capital gain distributions (see instructions).	10	
11a	IRA distributions.		
11b	Taxable amount (see instructions).	11b	
12a	Pensions and annuities.		
12b	Taxable amount (see instructions).	12b	
13	Unemployment compensation and Alaska Permanent Fund dividends.	13	
14a	Social security benefits.		
14b	Taxable amount (see instructions).	14b	
15	Add lines 7 through 14b (far right column). This is your **total income.** ▶	15	$26,362

Adjusted gross income

Line	Description		Amount
16	Educator expenses (see instructions).	16	
17	IRA deduction (see instructions).	17	$1,500
18	Student loan interest deduction (see instructions).	18	
19	Tuition and fees. Attach Form 8917.	19	
20	Add lines 16 through 19. These are your **total adjustments.**	20	$1,500
21	Subtract line 20 from line 15. This is your **adjusted gross income.** ▶	21	$24,862

For Disclosure, Privacy Act, and Paperwork Reduction Act Notice, see separate instructions. Cat. No. 11327A Form **1040A** (2016)

Form 1040A (2016) Page 2

Tax, credits, and payments

Standard Deduction for—
- People who check any box on line 23a or 23b **or** who can be claimed as a dependent, see instructions.
- All others:
Single or Married filing separately, $6,300
Married filing jointly or Qualifying widow(er), $12,600
Head of household, $9,300

Line	Description	Amount
22	Enter the amount from line 21 (adjusted gross income).	$24,862
23a	Check if: ☐ **You** were born before January 2, 1952, ☐ Blind / ☐ **Spouse** was born before January 2, 1952, ☐ Blind — Total boxes checked ▶ 23a	☐
b	If you are married filing separately and your spouse itemizes deductions, check here ▶ 23b	☐
24	Enter your **standard deduction**.	$6,300
25	Subtract line 24 from line 22. If line 24 is more than line 22, enter -0-.	$18,562
26	**Exemptions.** Multiply $4,050 by the number on line 6d.	$4,050
27	Subtract line 26 from line 25. If line 26 is more than line 25, enter -0-. This is your **taxable income.** ▶	$14,512
28	**Tax,** including any alternative minimum tax (see instructions). 28 $1,713	
29	Excess advance premium tax credit repayment. Attach Form 8962. 29	
30	Add lines 28 and 29.	$1,713
31	Credit for child and dependent care expenses. Attach Form 2441. 31	
32	Credit for the elderly or the disabled. Attach Schedule R. 32	
33	Education credits from Form 8863, line 19. 33	
34	Retirement savings contributions credit. Attach Form 8880. 34	
35	Child tax credit. Attach Schedule 8812, if required. 35	
36	Add lines 31 through 35. These are your **total credits.**	$0
37	Subtract line 36 from line 30. If line 36 is more than line 30, enter -0-.	$1,713
38	Health care: individual responsibility (see instructions). Full-year coverage ☐	
39	Add line 37 and line 38. This is your **total tax.**	$1,713
40	Federal income tax withheld from Forms W-2 and 1099. 40 $3,276	
41	2016 estimated tax payments and amount applied from 2015 return. 41	
42a	**Earned income credit (EIC).** 42a	
b	Nontaxable combat pay election. 42b	
43	Additional child tax credit. Attach Schedule 8812. 43	
44	American opportunity credit from Form 8863, line 8. 44	
45	Net premium tax credit. Attach Form 8962. 45	
46	Add lines 40, 41, 42a, 43, 44, and 45. These are your **total payments.** ▶	$3,276

If you have a qualifying child, attach Schedule EIC.

Refund

Direct deposit? See instructions and fill in 48b, 48c, and 48d or Form 8888.

47	If line 46 is more than line 39, subtract line 39 from line 46. This is the amount you **overpaid.**	$1,563
48a	Amount of line 47 you want **refunded to you.** If Form 8888 is attached, check here ▶ ☐ 48a	$1,563
b	Routing number ▶ c Type: ☐ Checking ☐ Savings	
d	Account number	
49	Amount of line 47 you want **applied to your 2017 estimated tax.** 49	$0

Amount you owe

50	**Amount you owe.** Subtract line 46 from line 39. For details on how to pay, see instructions. ▶	
51	Estimated tax penalty (see instructions). 51	

Third party designee

Do you want to allow another person to discuss this return with the IRS (see instructions)? ☐ **Yes.** Complete the following. ☐ **No**

Designee's name ▶   Phone no. ▶   Personal identification number (PIN) ▶

Sign here

Joint return? See instructions. Keep a copy for your records.

Under penalties of perjury, I declare that I have examined this return and accompanying schedules and statements, and to the best of my knowledge and belief, they are true, correct, and accurately list all amounts and sources of income I received during the tax year. Declaration of preparer (other than the taxpayer) is based on all information of which the preparer has any knowledge.

Your signature	Date	Your occupation	Daytime phone number
Jennifer Crum	4/14/2017	**Assistant Manager**	(410)-286-2594
Spouse's signature. If a joint return, **both** must sign.	Date	Spouse's occupation	If the IRS sent you an Identity Protection PIN, enter it here (see inst.)

Paid preparer use only

Print/type preparer's name	Preparer's signature	Date	Check ☐ if self-employed	PTIN
Firm's name ▶			Firm's EIN ▶	
Firm's address ▶			Phone no.	

Form **1040A** (2016)

SCHEDULE B
(Form 1040A or 1040)
(Rev. January 2017)
Department of the Treasury
Internal Revenue Service (99)

Interest and Ordinary Dividends

▶ Attach to Form 1040A or 1040.
▶ Information about Schedule B and its instructions is at *www.irs.gov/scheduleb*.

OMB No. 1545-0074

2016

Attachment Sequence No. **08**

Name(s) shown on return: **Jennifer Crum**

Your social security number: **123-45-6789**

Part I — Interest

(See instructions on back and the instructions for Form 1040A, or Form 1040, line 8a.)

Note: If you received a Form 1099-INT, Form 1099-OID, or substitute statement from a brokerage firm, list the firm's name as the payer and enter the total interest shown on that form.

Line	Description		Amount
1	List name of payer. If any interest is from a seller-financed mortgage and the buyer used the property as a personal residence, see instructions on back and list this interest first. Also, show that buyer's social security number and address ▶ **Employees Credit Union**	1	$1,624
2	Add the amounts on line 1	2	$1,624
3	Excludable interest on series EE and I U.S. savings bonds issued after 1989. Attach Form 8815	3	
4	Subtract line 3 from line 2. Enter the result here and on Form 1040A, or Form 1040, line 8a ▶	4	$1,624

Note: If line 4 is over $1,500, you must complete Part III.

Part II — Ordinary Dividends

(See instructions on back and the instructions for Form 1040A, or Form 1040, line 9a.)

Note: If you received a Form 1099-DIV or substitute statement from a brokerage firm, list the firm's name as the payer and enter the ordinary dividends shown on that form.

Line	Description		Amount
5	List name of payer ▶	5	
6	Add the amounts on line 5. Enter the total here and on Form 1040A, or Form 1040, line 9a ▶	6	

Note: If line 6 is over $1,500, you must complete Part III.

Part III — Foreign Accounts and Trusts

(See instructions on back.)

You must complete this part if you **(a)** had over $1,500 of taxable interest or ordinary dividends; **(b)** had a foreign account; or **(c)** received a distribution from, or were a grantor of, or a transferor to, a foreign trust.

		Yes	No
7a	At any time during 2016, did you have a financial interest in or signature authority over a financial account (such as a bank account, securities account, or brokerage account) located in a foreign country? See instructions		
	If "Yes," are you required to file FinCEN Form 114, Report of Foreign Bank and Financial Accounts (FBAR), to report that financial interest or signature authority? See FinCEN Form 114 and its instructions for filing requirements and exceptions to those requirements		
b	If you are required to file FinCEN Form 114, enter the name of the foreign country where the financial account is located ▶		
8	During 2016, did you receive a distribution from, or were you the grantor of, or transferor to, a foreign trust? If "Yes," you may have to file Form 3520. See instructions on back		

For Paperwork Reduction Act Notice, see your tax return instructions.   Cat. No. 17146N   Schedule B (Form 1040A or 1040) 2016

13.2 Exercises // MyLab Math

The shaded sections below contain solutions to help you get a QUICK START on the various types of exercises.

Find the adjusted gross income for each of the following people. (See Example 1.)

Name	Income from Jobs	Interest	Misc. Income	Dividend Income	Adjustments to Income	Adjusted Gross Income
1. R. Jacob	$22,840	$234	$1209	$48	$1200	$23,131
$22,840 + $234 + $1209 + $48 − $1200 = $23,131						
2. K. Chandler	$68,156	$285	$73	$542	$317	$68,739
$68,156 + $285 + $73 + $542 − $317 = $68,739						
3. P. Kalb	$98,600	$2300	$410	$4650	$2140	_____
4. The Jazwinskis	$33,650	$722	$375	$218	$473	_____
5. The Brashers	$38,643	$1020	$3820	$1050	$0	_____
6. The Claxtons	$93,680	$3247	$8115	$2469	$2800	_____
7. The Wheats	$21,370	$420	$0	$0	$0	_____
8. Tiny Peyton	$68,540	$1290	$480	$318	$1840	_____

Find the amount of taxable income and the tax owed for each of the following people. The letter following the names indicates the marital status, and all married people are filing jointly. (See Examples 2 and 3.)

Name	Number of Exemptions	Adjusted Gross Income	Total Deductions	Taxable Income	Tax Owed
9. R. Rodriguez, S	1	$36,840	$2460	$26,490	$3509.75
$36,840 − $6300 − $4050 = $26,490; $927.50 + .15 × ($26,490 − $9275) = $3509.75					
10. S. Simpson, S	1	$26,190	$1248	$15,840	$1912.25
$26,190 − $6300 − $4050 = $15,840; $927.50 + .15 × ($15,840 − $9275) = $1912.25					
11. The Pacas, M	2	$72,450	$6040	_____	_____
12. The Laytons, M	4	$65,290	$8040	_____	_____
13. The Jordans, M	3	$99,500	$14,320	_____	_____
14. P. Jong, Head of Household	4	$162,370	$15,800	_____	_____
15. K. Tang, Married filing separately	2	$85,332	$8170	_____	_____
16. G. Begay, Head of Household	4	$58,332	$9480	_____	_____

C// indicates an exercise that is related to the Case in Point feature.

CHAPTER 13 Taxes and Insurance

	Name	Number of Exemptions	Adjusted Gross Income	Total Deductions	Taxable Income	Tax Owed
17.	D. Kien, Head of Household	2	$82,650	$8100	_____	_____
18.	L. Dimon, M	2	$262,680	$18,350	_____	_____

Find the tax refund or tax due for the following people. The letter following the names indicates the marital status. Assume a 52-week year and that married people are filing jointly. (See Example 4.)

	Name	Taxable Income	Federal Income Tax Withheld from Checks	Tax Refund or Tax Due
19.	L. Karecki, S	$78,500	$1516 monthly	**$2795.75 tax refund**
	$18,192 − $15,396.25 = $2795.75 tax refund			
20.	K. Turner, S	$32,060	$347.80 monthly	_____
21.	M. Hunziker, S	$23,552	$72.18 weekly	_____
22.	The Fungs, M	$38,238	$119.27 weekly	_____
23.	The Todds, M	$202,100	$3200 monthly	_____
24.	S. Benson, HH	$160,300	$720 weekly	_____

25. Use the graphic at the front of this section and list where government money comes from and where it goes, including dollar amounts. Explain the deficit. (See Objective 1.)

26. List four possible tax deductions, and explain the effect that a tax deduction will have on taxable income and on income tax due. (See Objective 5.)

Find the tax in the following application problems.

27. **MARRIED—INCOME TAX** The Tobins had an adjusted gross income of $98,700 last year. They had deductions of $2820 for state income tax, $490 for city income tax, $4400 for property tax, $5800 in mortgage interest, and $1450 in contributions. They file a joint return and claim 5 exemptions.

27. _____

28. **SINGLE—INCOME TAX** Diane Bolton works at The Doll House and had an adjusted gross income of $34,975 last year. She had deductions of $971 for state income tax, $1864 for property tax, $3820 in mortgage interest, and $235 in contributions. Bolton claims one exemption and files as a single person.

28. _____

29. Carol Ridgeway has an adjusted gross income of $73,200 and files as head of household since she has a dependent, 9-year-old daughter. Her deductions are $7143.

29. _____

30. **MARRIED—INCOME TAX** The Hernandez family had an adjusted gross income of $48,260 last year. They had deductions of $1078 for state income tax, $253 for city income tax, $3240 for property tax, $5218 in mortgage interest, and $386 in contributions. They claim three exemptions and file a joint return.

30. _____

31. **HEAD OF HOUSEHOLD** Martha Spencer, owner of The Doll House, had wages of $73,800, dividends of $385, interest of $1672, and adjustments to income of $1058 last year. She had deductions of $877 for state income tax, $342 for city income tax, $4986 for property tax, $5173 in mortgage interest, and $1800 in contributions. She claims four exemptions and files as head of household.

31. _____

32. **HEAD OF HOUSEHOLD** John Walker had wages of $48,200, other income of $2892, dividends of $340, interest of $651, and a regular IRA contribution of $2000 last year. He had deductions of $1163 for taxes, $5350 in mortgage interest, and $540 in contributions. Walker claims two exemptions and files as head of household.

32. _____

33. **MARRIED** John and Vicki Karsten had combined wages and salaries of $64,280, other income of $5283, dividend income of $324, and interest income of $668. They have adjustments to income of $2484. Their itemized deductions are $7615 in mortgage interest, $2250 in state income tax, $3300 in real estate taxes, and $1219 in charitable contributions. The Karstens filed a joint return and claimed 3 exemptions.

33. _____

34. **HEAD OF HOUSEHOLD** Jayne Binyan is single with one dependent and files as head of household. She is the chief financial officer at the university where she works, and her salary is $173,400. She has other income of $2800, interest income of $8400, and an adjustment to income of $6000 for a 401K retirement plan at work. Her itemized deductions are $14,380 in mortgage interest, $4820 in state income tax, $7800 in real estate taxes, and $6800 in charitable contributions.

34. _____

QUICK CHECK ANSWERS

1. $98,950
2. $37,550; $4705
3. $27,500; $3197.50
4. (a) $37,750
   (b) $5000
   (c) $6720
   (d) overpaid by $1720

13.3 Fire Insurance

OBJECTIVES

1. Define the terms *policy, coverage, face value,* and *premium.*
2. Find the annual premium for fire insurance.
3. Use the coinsurance formula.
4. Understand multiple-carrier insurance.
5. List additional types of insurance coverage.

CASE IN POINT Martha Spencer owns the building in which her business is located. A fire in that building could ruin her financially. Although she hopes that there is never a fire, she carries fire insurance just in case.

Insurance protects against risk. For example, there is only a slight chance that a particular building or home will be damaged by fire in any year. However, the financial loss from a fire could be devastating to the owner. Therefore, people and companies pay a small fee each year to an insurance company to protect them against catastrophic losses. The insurance company collects money from many different people and companies that buy insurance and pays money to the few who suffer damages.

The Hernandezes are worried their house may burn so they buy insurance.

Quick TIP
Almost every adult has insurance. Even children are often on their parents' health insurance.

Quick TIP
There are 2.5 million employees working in the insurance industry in the United States. You may wish to consider working in this industry.

Insurance is important and common. For example, lenders require that fire insurance be purchased on a building or home before lending funds. Most states require automobile drivers to buy and carry a minimum amount of car insurance. Firms large and small buy liability insurance to protect against a lawsuit or a **catastrophic event** that would be financially damaging to the company. Parents buy life insurance to support their kids in the event of a parent's untimely death. People carry health insurance so they will not be overwhelmed by the very high costs associated with surgeries and hospital bills.

Individuals buy insurance to protect against losses due to fire, theft, illness or health problems, disability, car wrecks, lawsuits, and even death. Companies buy insurance to protect against losses due to fire, automobile accidents, employee illnesses, lawsuits, and worker accidents on the job. In this section, we talk about fire insurance.

OBJECTIVE 1 Define the terms *policy, coverage, face value,* and *premium.* The contract between the owner of a building and an insurance company is called a **policy** or an **insurance policy**. A basic fire policy provides **coverage** or protection for both the owner of the building and the company that holds the mortgage on the building. The owner of a building can also purchase coverage on the **contents** of the building and liability insurance in the event someone is injured while on the property. Homeowners often purchase a **homeowners' policy**, which includes all of these coverages. The graph on the next page shows that the cost of homeowners' insurance has increased partly due to the size of **claims** paid; however, inflation also contributes to price increases.

13.3 Fire Insurance

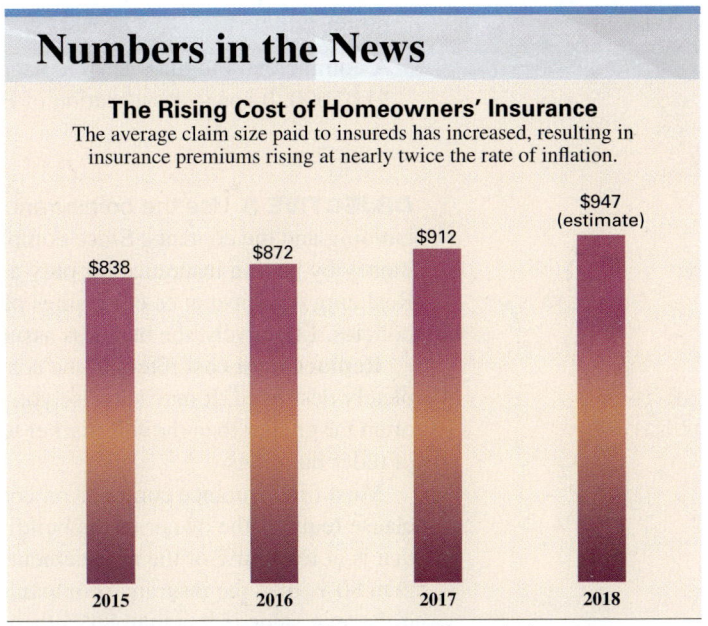

Numbers in the News

The Rising Cost of Homeowners' Insurance
The average claim size paid to insureds has increased, resulting in insurance premiums rising at nearly twice the rate of inflation.

- 2015: $838
- 2016: $872
- 2017: $912
- 2018: $947 (estimate)

DATA: National Association of Insurance Commissioners.

The dollar value of the insurance coverage provided on a building is called the **face value** of the policy. The annual cost of the policy is called the **premium**. The premium is calculated based on factors such as age of the building, materials used in the construction of the building, location, crime rate of the neighborhood, presence of any safety features such as a sprinkler system, security system, and history of previous insurance claims on the property.

OBJECTIVE 2 Find the annual premium for fire insurance. Although many factors are used to determine premiums, we will consider only the location (territory) and the age, construction material, and general condition of a building (building classification) in this chapter.

Building classifications are assigned to building types by insurance company employees called **underwriters**. Underwriters also assign ratings called **territorial ratings** to each area that describe the quality of fire protection in the area and the level of crime. Although fire insurance rates vary greatly, the rates in the following table are typical.

Quick TIP
Insurance company employees use computers to determine premiums.

Annual Rates for Each $100 of Insurance

Territorial Rating	Building Classification					
	A		B		C	
	Building	Contents	Building	Contents	Building	Contents
1	$.25	$.32	$.36	$.49	$.45	$.60
2	$.30	$.44	$.45	$.55	$.54	$.75
3	$.37	$.46	$.54	$.60	$.63	$.80
4	$.50	$.52	$.75	$.77	$.84	$.90
5	$.62	$.58	$.92	$.99	$1.05	$1.14

Finding the Annual Fire Insurance Premium

EXAMPLE 1 The Doll House is in a building rated class C. It is in territory 4. Find the annual premium if the replacement cost of the building is $640,000 and the contents are valued at $186,500.

SOLUTION

Building:

Replacement cost in hundreds = $640,000 ÷ 100 = **6400**
Insurance premium for the building = **6400** × $.84 = $5376

Contents:

Replacement cost in hundreds = $186,500 ÷ 100 = **1865**
Insurance premium for contents = **1865** × $.90 = $1678.50

Total premium: $5376 + $1678.50 = **$7054.50**

QUICK CHECK 1

A commercial building has a replacement cost of $1,480,000 and contents valued at $110,000. It has a classification of B and a territory rating of 5. Find the annual premium.

OBJECTIVE 3 Use the coinsurance formula. Most fires damage only a portion of a building and the contents. Since complete destruction of a building is rare, many owners save money by buying insurance for only a portion of the value of the building and contents. Realizing this, insurance companies place a **coinsurance clause** in almost all fire insurance policies. Effectively, the business assumes part of the risk of a loss under coinsurance.

Replacement cost refers to the cost to replace (rebuild) a building in the event it is completely destroyed. It may surprise you to learn that the replacement cost for an older building is often far greater than the fair market value, since new construction costs often exceed the value of older buildings.

Most fire insurance contracts on commercial buildings have an 80% coinsurance clause. This clause requires the owner of the building to have an insurance policy in effect with a face value that is at least 80% of the replacement cost of the building. If the policy has a face value greater than 80%, then the insurance companies pays for all losses caused by a fire. On the other hand, if the face value is less than 80%, then the insurance company will pay only a portion of any loss. The owner of the building must pay the balance. An owner in this situation is said to be **underinsured**.

Finding Amount Insurance Will Pay

$$\text{Amount insurance company will pay (assuming 80\% coinsurance)} = \text{Amount of loss} \times \frac{\text{Amount of policy}}{80\% \text{ of replacement cost}}$$

Using the Coinsurance Formula — EXAMPLE 2

Dayton Properties owns a small apartment building with a replacement cost of $760,000. The fire insurance policy has an 80% coinsurance clause and a face value of $570,000. A fire started in the kitchen of a tenant and swept through three apartments, resulting in $144,000 in losses. Find the amount of the loss that the insurance company will pay.

SOLUTION

The policy must have a face value of at least 80% of $760,000 or $608,000 in order to receive the payment for the entire loss. Since the face value of $570,000 is less than 80% of the replacement cost, the company will pay only the following portion of the loss.

$$\text{Amount insurance company pays} = \$144,000 \times \frac{\$570,000}{\$608,000} = \$135,000$$

$$\text{Amount not paid by insurance company} = \$144,000 - \$135,000 = \$9000$$

Dayton Properties is responsible for $9000.

The calculator solution to this example uses chain calculations and parentheses to set off the denominator. The result is then subtracted from the fire loss.

$$144{,}000 \times 570{,}000 \div (\ 80\ \% \times 760{,}000\ ) = 135{,}000$$
$$144{,}000 - 135{,}000 = 9000$$

Note: Refer to Appendix B for calculator basics.

Quick TIP
Work with an insurance agent to find a balance between premiums paid and amount of risk you take personally.

QUICK CHECK 2

A real estate investment trust owns an office building with a replacement cost of $8,400,000 that is insured for $5,600,000. Find the amount of the loss paid for by the insurance company if a fire causes $2,300,000 in losses and the policy has an 80% coinsurance feature.

13.3 Fire Insurance

Finding the Amount of Loss Paid by the Insurance Company

EXAMPLE 3 A Swedish investment group owns a warehouse with a replacement cost of $3,450,000. The company has a fire insurance policy with a face value of $3,400,000. The policy has an 80% coinsurance feature. If the firm has a fire loss of $233,500, find the portion of the loss paid by the insurance company.

SOLUTION

$$80\% \text{ of replacement cost} = .80 \times \$3,450,000 = \$2,760,000$$

The business has a fire insurance policy with a face value of more than 80% of the value of the store. Therefore, the insurance company pays the entire $233,500 loss.

QUICK CHECK 3

A plumbing company owns its own building with a replacement cost of $1,600,000. A fire results in damages of $445,000. Find the amount the insurance company will pay if the company has $1,400,000 in insurance coverage.

OBJECTIVE 4 Understand multiple-carrier insurance. A business may have fire insurance policies with several companies at the same time. Perhaps additional insurance coverage was purchased over a period of time, as new additions were made to a factory or building complex. Or perhaps the building is so large that one insurance company does not want to take the entire risk by itself, so several companies each agree to take a portion of the insurance coverage and thereby share the risk. In either event, the insurance coverage is divided among **multiple carriers**. When an insurance claim is made against multiple carriers, each insurance company pays its fractional portion of the total claim on the property.

Understanding Multiple-Carrier Insurance

EXAMPLE 4 Youngblood Apartments has an insured loss of $1,800,000 while having insurance coverage beyond its coinsurance requirement. The insurance is divided among Company A with $5,900,000 coverage, Company B with $4,425,000 coverage, and Company C with $1,475,000 coverage. Find the amount of the loss paid by each of the insurance companies assuming the apartments are insured for 100% of replacement cost.

SOLUTION

Start by finding the total face value of all three policies.

$$\$5,900,000 + \$4,425,000 + \$1,475,000 = \$11,800,000 \text{ total face value}$$

$$\text{Company A pays } \frac{\$5,900,000}{\$11,800,000} = \frac{1}{2} \text{ of the loss}$$

$$\text{Company B pays } \frac{\$4,425,000}{\$11,800,000} = \frac{3}{8} \text{ of the loss}$$

$$\text{Company C pays } \frac{\$1,475,000}{\$11,800,000} = \frac{1}{8} \text{ of the loss}$$

Since the insurance loss is $1,800,000, the amount paid by each of the multiple carriers is

$$\text{Company A: } \frac{1}{2} \times \$1,800,000 = \$900,000$$

$$\text{Company B: } \frac{3}{8} \times \$1,800,000 = \$675,000$$

$$\text{Company C: } \frac{1}{8} \times \$1,800,000 = \underline{\$225,000}$$

$$\text{Total loss} = \$1,800,000$$

QUICK CHECK 4

A fully insured bank has $780,000 in fire damage. The coverage is divided between Company A ($1,200,000) and Company B ($800,000). Find the amount of the loss paid by each company assuming the bank building is insured for 100% of replacement cost.

Understanding Partial Coverage and Multiple Carriers

EXAMPLE 5 The fire damage to a small shopping center with a replacement cost of $4,800,000 was limited to $420,000 thanks to an advanced sprinkler system. The insurance coverage is divided between Company A ($2,000,000) and Company B ($1,200,000). Find the amount of the loss paid by each, assuming management had full coverage.

SOLUTION

$$80\% \text{ of } \$4,800,000 = \$3,840,000$$

$$\text{Total insurance coverage} = \$2,000,000 + \$1,200,000 = \$3,200,000$$

Since the total insurance coverage is less than 80% of the replacement cost of the building, the insurance companies will pay only a portion of the total damages.

$$\text{Amount paid by insurance: } \$420,000 \times \frac{\$3,200,000}{\$3,840,000} = \$350,000$$

$$\text{Paid by Company A: } \frac{\$2,000,000}{\$3,200,000} \times \$350,000 = \$218,750$$

$$\text{Paid by Company B: } \frac{\$1,200,000}{\$3,200,000} \times \$350,000 = \$131,250$$

QUICK CHECK 5

An office building with a replacement cost of $800,000 has fire damage of $100,000. The insurance coverage is divided between Company 1 ($300,000) and Company 2 ($200,000). First **(a)** find the amount covered by insurance, then **(b)** find the amount paid by each company.

The amount of money paid for insurance premiums by businesses is often small compared with other business expenses. Likewise, the average household pays only a small portion of its budget for insurance premiums. The following chart shows the percent of total household spending going to pay for insurance coverage.

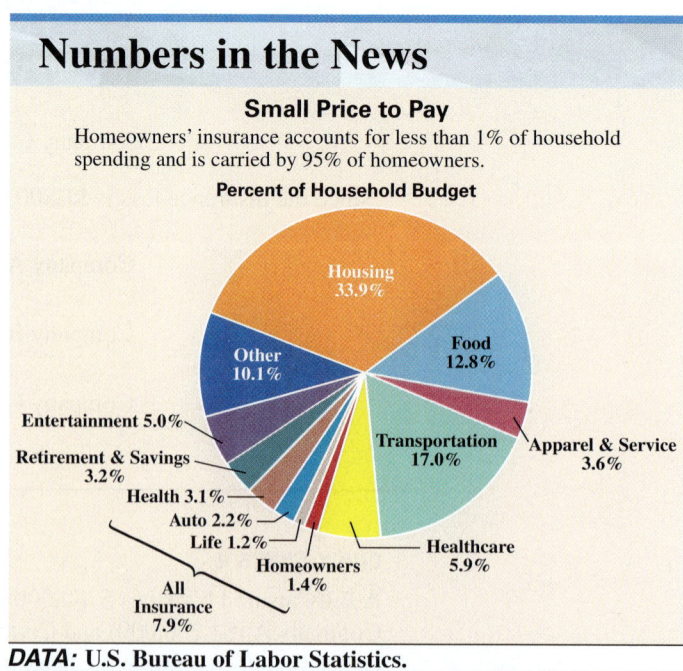

Numbers in the News

Small Price to Pay

Homeowners' insurance accounts for less than 1% of household spending and is carried by 95% of homeowners.

Percent of Household Budget

- Housing 33.9%
- Transportation 17.0%
- Food 12.8%
- Other 10.1%
- All Insurance 7.9%
- Healthcare 5.9%
- Entertainment 5.0%
- Apparel & Service 3.6%
- Retirement & Savings 3.2%
- Health 3.1%
- Auto 2.2%
- Homeowners 1.4%
- Life 1.2%

DATA: U.S. Bureau of Labor Statistics.

13.3 Exercises

OBJECTIVE 5 List additional types of insurance coverage.
Here are some other commonly purchased insurance coverages:

Disability insurance pays monthly payments in event of disability.

Homeowners' insurance pays a homeowner in the event of fire, theft, or accidents.

Liability insurance protects business owners against lawsuit.

Long-term care insurance pays for nursing home costs.

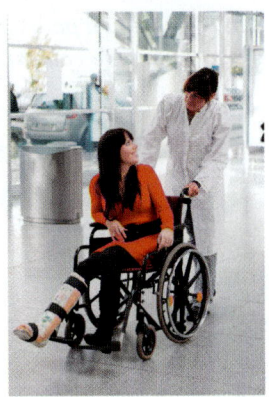

Quick TIP
Medicare and Medicaid are government-sponsored health insurance programs.

Health insurance pays for medical expenses.

Renter's insurance protects a renter against loss of his personal property.

Workers' compensation pays employees for injuries while on the job.

Financial planners recommend that you determine where you have risk and that you cover at least some of that risk using insurance. Do this by first figuring out where you could suffer a loss that would hurt you financially. Then, talk to an experienced insurance agent to understand how to manage that risk. It is also worthwhile to compare rates at different insurance companies.

13.3 Exercises // MyLab Math

The shaded sections below contain solutions to help you get a **QUICK START** *on the various types of exercises.*

Find the total annual premium for each of the following. Use the table on page 543. (See Example 1.)

	Territorial Rating	Building Classification	Building Value	Contents Value	Total Annual Premium
1.	2	B	$280,000	$80,000	$1700
2.	5	A	$220,500	$105,000	$1976.10
3.	1	C	$285,000	$152,000	_____
4.	4	C	$2,325,000	$111,500	_____
5.	5	B	$782,600	$212,000	_____
6.	3	A	$596,400	$206,700	_____

Find the amount to be paid by the insurance company in the following problems. Assume that each policy includes an 80% coinsurance clause. (See Examples 2 and 3.)

	Replacement Cost of Building	Face Value of Policy	Amount of Loss	Amount Paid
7.	$1,450,000	$850,000	$96,000	$70,344.83
	$1,450,000 × .8 = $1,160,000 (80%); $\frac{\$850,000}{\$1,160,000}$ × $96,000 = $70,344.83			
8.	$187,400	$140,000	$10,850	$10,132.07
	$187,400 × .8 = $149,920 (80%); $\frac{\$140,000}{\$149,920}$ × $10,850 = $10,132.07			
9.	$287,000	$232,500	$19,850	_____
10.	$780,000	$585,000	$10,400	_____
11.	$218,500	$195,000	$36,500	_____
12.	$750,000	$500,000	$56,000	_____

CHAPTER 13 Taxes and Insurance

Find the amount paid by each insurance company in the following problems involving multiple carriers. Assume that the coinsurance requirement is met. (See Example 4.)

	Insurance Loss	Companies	Coverage	Amount Paid
13.	$80,000	Company 1	$750,000	$60,000
		Company 2	$250,000	$20,000

$\frac{750,000}{1,000,000} \times \$80,000 = \$60,000;\ \frac{250,000}{1,000,000} \times \$80,000 = \$20,000$

	Insurance Loss	Companies	Coverage	Amount Paid
14.	$360,000	Company A	$1,200,000	_____
		Company B	$800,000	_____
15.	$650,000	Company 1	$1,350,000	_____
		Company 2	$1,200,000	_____
		Company 3	$450,000	_____
16.	$1,600,000	Company A	$4,800,000	_____
		Company B	$800,000	_____
		Company C	$2,400,000	_____

Find the annual fire insurance premium in each of the following application problems. Use the table on page 533.

17. **COMMERCIAL BUILDING** Home Depot owns a building with a replacement cost of $1,400,000 and with contents of $360,000. The building is class C with a territorial rating of 5.
    14,000 × 1.05 = $14,700; 3600 × $1.14 = $4104
    $14,700 + $4104 = $18,804

    17. **$18,804**

18. **FIRE INSURANCE PREMIUM** James Baker owns a class-B building with a replacement cost of $165,400. Contents are valued at $128,000. The territorial rating is 3.

    18. _____

19. **AIRPLANE HANGAR** Valley Crop Dusting owns an airplane hanger located in a class-B area. It has a replacement cost of $107,500. Contents are worth $39,800. The territorial rating is 2.

    19. _____

20. **INDUSTRIAL BUILDING INSURANCE** London's Dredging Equipment is in a class-C building with a territorial rating of 4. The building has a replacement cost of $305,000 and the contents are worth $682,000.

    20. _____

21. Determine where you have a risk of large financial loss—think in terms of after graduation and when you have a job and perhaps a family. What can you do to first minimize, and then manage the risk? (See Objective 1.)

22. Describe several factors that are used to determine the premium charged for fire insurance. (See Objective 2.)

C// indicates an exercise that is related to the Case in Point feature.

In the following application problems, find the amount of the loss paid by (a) the insurance company and (b) the insured. Assume an 80% coinsurance clause.

23. **FIRE LOSS** The Doll House stores supplies in a separate building with a replacement cost of $328,500, but Martha Spencer insured it for only $200,000 in order to save money on insurance premiums. An electrical short causes a fire that results in $180,000 in damage.
    (a) $328,500 × .8 = $262,800; $\frac{\$200,000}{\$262,800}$ × $180,000 = $136,986.30
    (b) $180,000 − $136,986.30 = $43,013.70

    (a) $136,986.30
    (b) $43,013.70

24. **FIRE** Bantam Art is in a building with a replacement cost of $395,000. The shop is insured for $280,000. Fire loss is $22,500.

    (a) _____
    (b) _____

25. **SALVATION ARMY LOSS** The main office of the Salvation Army suffers a loss from fire of $45,000. The building has a replacement cost of $550,000 and is insured for $300,000.

    (a) _____
    (b) _____

26. **TRIPLEX** Fang Li purchased a small triplex that has a replacement cost of $185,000. She insures it for $111,000 and has a fire loss of $28,000 when a smoker accidently starts a fire.

    (a) _____
    (b) _____

In the following application problems, find the amount paid by each of the multiple carriers. Assume that the coinsurance requirement has been met.

27. **COINSURED FIRE LOSS** C. Wood Plumbing had an insured fire loss of $548,000. It has insurance coverage as follows: Company A, $600,000; Company B, $400,000; and Company C, $200,000.
    A: $\frac{600,000}{1,200,000}$ × $548,000 = $274,000
    B: $\frac{400,000}{1,200,000}$ × $548,000 = $182,666.67
    C: $\frac{200,000}{1,200,000}$ × $548,000 = $91,333.33

    A: $274,000
    B: $182,666.67
    C: $91,333.33

28. **COINSURED FIRE LOSS** Beacon Cycles had an insured fire loss of $68,500. It has insurance as follows: Company 1, $60,000; Company 2, $40,000; and Company 3, $30,000.

    1: _____
    2: _____
    3: _____

29. **MAJOR FIRE LOSS** Gold's Gym had fire insurance coverage as follows: Company 1, $360,000; Company 2, $120,000; and Company 3, $240,000. The gym has an insured fire loss of $250,000.

    1: _____
    2: _____
    3: _____

30. **MULTIPLE INSURANCE CARRIERS** Tokyo International had an insured fire loss of $2,100,000. The company has insurance as follows: Company A, $2,000,000; Company B, $1,750,000; Company C, $1,250,000.

    A: _____
    B: _____
    C: _____

CHAPTER 13 Taxes and Insurance

Find the amount paid by each of the multiple carriers in the following two problems. Note that the coinsurance requirement has not been met.

31. UNDERINSURED A building with a replacement cost of $1,200,000 has fire damages of $420,000. The insurance coverage is split between Company A ($200,000) and Company B ($300,000). Find **(a)** the amount of the loss covered and **(b)** amount paid by each company.

(a) _____

(b) _____

32. UNDERINSURED Family Dollar owns a building with a replacement cost of $1,800,000 that has fire and smoke damages of $310,000. The insurance coverage is split between Company 1 ($1,000,000) and Company 2 ($400,000). Find **(a)** the amount of the loss covered and **(b)** the amount paid by each company.

(a) _____

(b) _____

33.

QUICK CHECK ANSWERS

1. $14,705 2. $1,916,666.67
3. $445,000
4. Company A—$468,000;
   Company B—$312,000

5. (a) $78,125
   (b) Company A—$46,875;
       Company B—$31,250

13.4 Motor-Vehicle Insurance

OBJECTIVES

1. Describe the factors that affect the cost of motor-vehicle insurance.
2. Define liability insurance.
3. Define property damage insurance.
4. Describe comprehensive and collision insurance.
5. Define no-fault and uninsured motorist insurance.
6. Apply youthful-operator factors.
7. Find the amounts paid by the insurance company and the insured.

CASE IN POINT Martha Spencer owns a van used by her employees. Knowing that she or an employee could have an accident resulting in serious costs, she insures the van. She carries higher liability limits than are required by the state, since she does not want one serious accident to result in the loss of her business.

OBJECTIVE 1 Describe the factors that affect the cost of motor-vehicle insurance. Almost everyone who owns a vehicle buys motor vehicle insurance. The insurance is required by firms making loans to buy vehicles and also by most states. Furthermore, vehicle insurance protects the vehicle owner from lawsuits and other expenses associated with vehicle accidents and can also help pay for injuries to a passenger during an accident.

Young drivers sometimes ask why their car insurance premiums are so high. The data in the graph show that young people are involved in far more accidents than drivers between the ages of 20 and 79. More accidents translates to higher costs. Young drivers with speeding tickets or who have been in accidents often have to pay very high premiums.

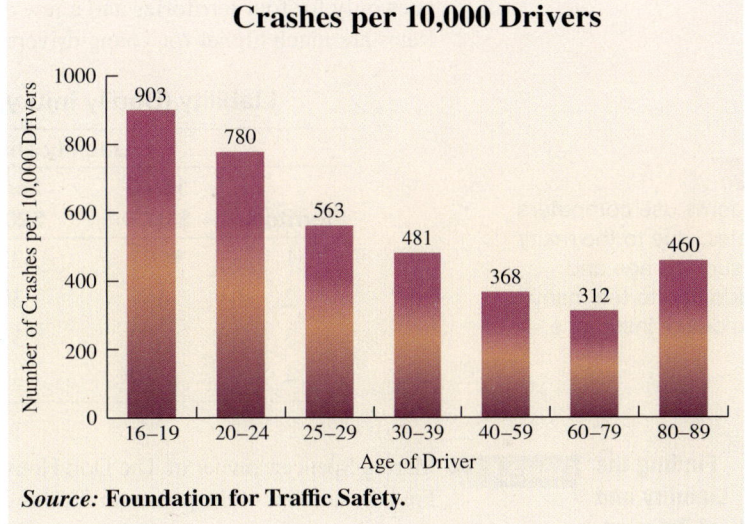

Source: **Foundation for Traffic Safety.**

The **premium**, or cost, of an insurance policy is determined by **actuaries** who work for the insurance company. Actuaries look at the frequency and severity of accidents based on several factors, including

- location of the insured vehicle,
- age and sex of the driver,
- miles driven, and
- driving history of the driver.

These factors help measure the risk of insuring a particular driver, which is used to determine the premium. For example, the insurance premium for a red sports car driven by a 17-year-old male will be very high, since the data show a high probability of such a driver being in an accident.

OBJECTIVE 2 Define liability insurance. **Liability** or **bodily injury coverage** protects the insured in case he or she injures someone with a car. Many states have minimum amounts of liability insurance coverage set by law. The amount of liability insurance is expressed as a fraction, such as 15/30. The fraction 15/30 means that the insurance company will pay up to $15,000 for injury to any one person, but the maximum payout for two or more people injured in the same accident is limited to $30,000.

542 CHAPTER 13 Taxes and Insurance

Experts recommend liability limits of at least 100/300, since a single accident can result in a lot of damages and costs. See the clipping reproduced below.

One Accident Affects Life

Thomas Garcia has driven for 10 years and has never had an accident or a speeding ticket. However, his life changed instantly when he accidently ran through a stop sign and hit a car driven by a nurse. She is now disabled and can no longer work. Garcia's policy limits of 25/50 are not enough to pay her bills, and her lawyer is taking Garcia to court. Garcia does not want to lose his savings and wishes he had carried higher limits.

The following table shows typical premium rates for various amounts of liability coverage. Included in the cost of the liability insurance is **medical insurance** for the driver and passengers in case of injury. For example, the table column 15/30 means that any passenger injured in the insured's vehicle can receive up to $1000 for medical expenses arising from an accident, no matter who was at fault. Insurance companies divide the nation into territories. We show rates only for four territories and a few different liability limits for 30- to 50-year-old drivers. Rates are much higher for young drivers, as shown later in this section.

> **Quick TIP**
> Insurance agents use computers to access rates, due to the many risk factors such as age and location in addition to the many possible choices of insurance coverage.

Liability (Bodily Injury) and Medical Insurance (per Year)

Territory	Liability and Medical Expense Limits				
	15/30 $1000	25/50 $2000	50/100 $3000	100/300 $5000	250/500 $10,000
1	$207	$222	$253	$282	$308
2	269	302	341	378	392
3	310	314	375	398	459
4	340	362	375	398	445

Finding the Liability and Medical Premium

EXAMPLE 1 Martha Spencer, owner of The Doll House, is in territory 2 and wants 100/300 liability coverage. Find the amount of the premium for this coverage and the amount of medical coverage included.

CASE IN POINT

SOLUTION

Look up territory 2 and 100/300 coverage in the liability and medical insurance table to find an annual premium of $378 just for liability and medical coverage. This cost includes $5000 medical coverage.

> **QUICK CHECK 1**
> Find the annual cost, for a person living in territory 4, for 250/500 liability coverage with $10,000 in medical insurance.

OBJECTIVE 3 Define property damage insurance. Liability coverage pays if you injure someone. **Property damage coverage** pays if you damage someone else's property such as an automobile or a building. The following table shows the annual cost for various **policy limits** on property damage. You are responsible for damages above the policy limit.

Property Damage Insurance (per Year)

Territory	Property Damage Limits			
	$10,000	$25,000	$50,000	$100,000
1	$88	$93	$97	$103
2	168	192	223	251
3	129	134	145	158
4	185	203	236	262

> **Quick TIP**
> Insurance rates vary greatly according to age, sex, location, type of vehicle, driving history, etc.

13.4 Motor-Vehicle Insurance

Finding the Premium for Property Damage

EXAMPLE 2 Find the annual premium if Martha Spencer, in territory 2, wants property damage coverage of $50,000.

CASE IN POINT

SOLUTION
Property damage coverage of $50,000 in territory 2 requires a premium of $223.

QUICK CHECK 2
Find the premium for $100,000 in property damage coverage for a person living in territory 4.

Would this be covered by comprehensive or collision?

OBJECTIVE 4 Describe comprehensive and collision insurance. **Comprehensive coverage** pays for damages to the insured's vehicle caused by a fire, by theft of the automobile, by vandalism (e.g., someone purposefully scratches off paint), by a tree falling onto the automobile, and by other similar events. In general, comprehensive insurance pays for damages that are not covered by collision insurance.

Collision coverage pays for repairs to the insured's vehicle when it is being driven and collides with an object.

You choose the **deductible** for both comprehensive and collision insurance when you purchase the insurance. Common deductibles for each coverage range anywhere from $200 to $2000. Higher deductibles result in a lower insurance premium. However, higher deductibles require you to pay more out of your pocket when the vehicle is damaged. So, be careful to choose a deductible that you can afford.

In the event of damages to your automobile, you pay the deductible and the insurance company pays the balance. For example, assume a tree falls on your car during a storm and that you have a $1000 deductible on your comprehension coverage. The amount that must be paid by the insured and the insurance company are shown.

Damages to your automobile	$3480
Deductible you must pay	−1000
Insurance company pays	**$2480**

Quick TIP
Lenders require owners to carry comprehensive and collision insurance.

The table that follows shows typical rates for comprehensive and collision insurance, each with a $2000 deductible. The rates are determined by actuaries based on the territory in which you live, the age group of your vehicle, and the symbol based on the cost to repair your specific vehicle. Age group 1 is a vehicle that is less than 2 years of age. Age group 2 is a vehicle that is at least 2 but less than 3 years of age, and so on. Age group 6 is a vehicle 6 years of age or older.

Quick TIP
Comprehensive and collision costs are much higher than these figures for those living in large cities with aggressive drivers and for individuals that choose lower deductibles.

Comprehensive and Collision Insurance (per Year)

		Comprehensive ($2000 Deductible)			Collision ($2000 Deductible)		
		Symbol			Symbol		
Territory	Age Group	6	7	8	6	7	8
1	1	$58	$64	$90	$153	$165	$184
	2, 3	50	56	82	135	147	171
	4, 5	44	52	76	116	128	147
	6	34	44	64	92	110	128
2	1	$26	$28	$40	$89	$95	$104
	2, 3	40	52	68	80	86	98
	4, 5	20	24	34	71	77	86
	6	30	36	45	95	103	115
3	1	$70	$78	$108	$145	$157	$174
	2, 3	60	66	90	128	139	162
	4, 5	52	64	92	111	122	139
	6	48	59	63	110	132	157
4	1	$42	$46	$66	$97	$104	$124
	2, 3	36	40	58	87	94	107
	4, 5	32	38	54	77	84	94
	6	60	68	84	140	158	166

Finding the Comprehensive and Collision Premiums

EXAMPLE 3 Martha Spencer, owner of The Doll House, is in territory 2 and has a 2-year-old minivan that has a symbol of 8. Use the comprehensive and collision insurance table to find the cost for (a) comprehensive coverage and (b) collision coverage.

CASE IN POINT

SOLUTION

(a) The cost of comprehensive coverage is $68.
(b) The cost of collision coverage is $98.

QUICK CHECK 3

Find the premium for (a) comprehensive and (b) collision coverage for a 3-year-old vehicle in territory 4 and symbol 8.

OBJECTIVE 5 Define no-fault and uninsured motorist insurance. Some states have **no-fault** laws. Under no-fault insurance, all medical expenses and costs associated with an accident are paid to each individual by *his or her own insurance company*, no matter who is at fault. Legislators and insurance companies argue that no-fault insurance removes lawyers, courts, and juries from the process and results in quicker, less costly settlements. Others (including trial lawyers) argue that no-fault insurance leaves accident victims unable to recover all of their damages.

Most states do not have no-fault laws, thereby requiring the insurance company of the person at fault to pay for damages. A potential problem in these states is that the motorist who caused an accident either has no insurance at all or has too little insurance for the damages that occurred. **Uninsured motorist coverage** protects a vehicle owner from financial liability when hit by a driver with no insurance. **Underinsured motorist coverage** provides protection to a vehicle owner when hit by a driver who has *too little* insurance. Typical costs for uninsured motorist insurance are shown in the table at the left. The cost of this coverage has increased a lot over the past few years, partly due to the fact that the number of uninsured and underinsured drivers has increased.

Uninsured Motorist Insurance (per Year)

Territory	Basic Limit
1	$242
2	$219
3	$317
4	$285

Determining the Premium for Uninsured Motorist Coverage

EXAMPLE 4 Martha Spencer, in territory 2, wants uninsured motorist coverage. Find the premium in the uninsured motorist insurance table.

CASE IN POINT

SOLUTION

The premium for uninsured motorist coverage in territory 2 is $219.

QUICK CHECK 4

Find the annual cost of uninsured motorist coverage in territory 1.

OBJECTIVE 6 Apply youthful-operator factors. Insurance companies base their rates on actual claim data for different age groups. Since younger drivers are far more likely to be involved in serious accidents, insurance companies charge higher rates for **youthful** than for **adult** drivers. Many excellent young drivers end up paying for the poor driving habits of a few others in their same age group. The age at which a youth becomes an adult varies from company to company. Generally, drivers under 25 are considered youthful drivers, and drivers 25 or older are considered adults.

The following table shows typical youthful-operator factors based on age and on whether or not the operator has taken driver's training. You can see that the factors are much higher for youthful operators who have not taken driver's training. The steps to apply these factors follow.

1. Determine the total premium for all coverages desired.
2. Multiply the total premium by the youthful-operator factor from the table.

13.4 Motor-Vehicle Insurance

Youthful-Operator Factor

Age	With Driver's Training	Without Driver's Training
20 or less	2.35	2.95
21–25	2.10	2.40

Using the Youthful-Operator Factor

EXAMPLE 5 Janet Ito lives in territory 4, is 22 years old, has had driver's training, and has no speeding tickets. She drives a 5-year-old car with a symbol of 7. She wants a 25/50 liability policy with these coverages: $2000 in medical payments, $10,000 property damage, comprehensive and collision, and uninsured motorist. Find her annual insurance premium using the tables in this section.

SOLUTION

1. Determine the total premium for all coverages desired.

25/50 liability insurance	$362
$10,000 property damage	185
Comprehensive insurance	38
Collision	84
Uninsured motorist	+$285
Subtotal	**$954**

2. Since she is younger than 25, multiply the premium by the youthful-operator factor from the table.

   $$\$954 \times 2.10 = \$2003.40$$

   The calculator solution to this example uses parentheses and chain calculations.

   (362 + 185 + 38 + 84 + 285) × 2.10 = 2003.4

QUICK CHECK 5

Scott Burch is 20, lives in territory 3, has not had driver's training, and drives a 3-year-old car with a symbol of 6. He wants the following coverages: 100/300 liability, $5000 medical, $50,000 property damage, uninsured motorist, and both comprehensive and collision insurance. Find the annual premium.

OBJECTIVE 7 Find the amounts paid by the insurance company and the insured. If you are at fault in an automobile accident and the damages *exceed the limits* on your insurance policy, you *may be personally liable* for the excess. You can help avoid this situation if you increase the liability and property damage limits on your policy. Sometimes, it does not cost much to increase your liability limits. For example, the additional cost of increasing liability coverage in territory 1 from 50/100 to 100/300 is only $29 per year ($282 − $253).

Finding the Amounts Paid by the Insurance Company and the Insured

EXAMPLE 6 James Benson has 25/50 liability limits, $25,000 property damage limits, and $500 deductible collision insurance. While on vacation, he was at fault in an accident that caused $5800 damage to his car, $3380 in damage to another car, and severe injuries to the other driver and his passenger. A subsequent lawsuit for injuries resulted in a judgment of $45,000 and $35,000, respectively, to the other parties. Find the amounts that the insurance company will pay for (a) repairing Benson's car, (b) repairing the other car, and (c) paying the court judgment resulting from the lawsuit. (d) How much will Benson have to pay the injured parties?

SOLUTION

(a) The insurance company will pay $4800 (**$5800 − $1000 deductible**) to repair Benson's car.
(b) Repairs on the other car will be paid to the property damage limits ($25,000). Here, the total repairs of $3380 are paid.
(c) Since more than one person was injured, the insurance company pays the limit of $50,000 ($25,000 to each of the two injured parties).
(d) Benson is liable for $30,000 (**$80,000 − $50,000**), the amount awarded over the insurance limits.

546 CHAPTER 13 Taxes and Insurance

> **QUICK CHECK 6**
> Janet Yahn has the following coverages: 50/100 liability, $3000 medical, $25,000 property damage, and $500 deductible collision. She runs through a stop sign and hits a van with a value of $12,200. The van is destroyed and the cost to repair Yahn's car is $8600. The driver of the van had medical costs of $68,000 and Yahn had medical costs of $2800. Find each cost Yahn must pay.

13.4 Exercises // MyLab Math

The shaded sections below contain solutions to help you get a QUICK START on the various types of exercises.

Find the annual premium for the following. (See Examples 1–5.)

	Name	Territory	Age	Driver Training?	Liability	Property Damage	Comprehensive Collision Age Group	Symbol	Uninsured Motorist?	Annual Premium
1.	Smyth	3	42	No	50/100	$25,000	2	7	Yes	$1031
	$375 + $134 + $66 + $139 + $317 = $1031									
2.	Thompson	4	20	Yes	25/50	$25,000	4	7	No	$1614.45
	$362 + $203 + $38 + $84 = $687; $687 × 2.35 = $1614.45									
3.	Shraim	3	52	No	250/500	$50,000	2	8	Yes	_____
4.	Waldron	2	67	No	50/100	$100,000	1	6	Yes	_____
5.	Applegate	4	22	No	100/300	$100,000	1	6	Yes	_____
6.	Rodriguez	1	24	Yes	50/100	$50,000	3	8	Yes	_____

7. Describe four factors that determine the premium on an automobile insurance policy. (See Objective 1.)

8. Explain in your own words the difference between liability (bodily injury) and property damage. (See Objectives 2 and 3.)

Solve the following application problems.

9. **ADULT AUTO INSURANCE** Bill Poole is 47 years old, lives in territory 4, and drives a 2-year-old car with a symbol of 7. He wants 250/500 liability limits, $100,000 property damage limits, comprehensive and collision insurance, and uninsured motorist coverage. Find his annual insurance premium.
$445 + $262 + $40 + $94 + $285 = $1126

9. $1126

C// indicates an exercise that is related to the Case in Point feature.

10. **ADULT AUTO INSURANCE** Martha Spencer, owner of The Doll House, is thinking about moving from a house in territory 2 to a house in territory 3 and wants to know the effect on the insurance costs for her new car. She currently lives in territory 2 and the car has symbol 7. Her coverages are 250/500 for liability, $100,000 for property damage, comprehensive, collision, and uninsured motorist. She is 53 years old. Find the change in annual cost.

10. _____

11. **YOUTHFUL-OPERATOR AUTO INSURANCE** Brandy Barrett is 23 years old, took a driver's education course, lives in territory 1, and drives a 4-year-old vehicle with a symbol of 6. She wants 50/100 liability limits, $25,000 property damage limits, comprehensive and collision insurance, and uninsured motorist coverage. Find her annual insurance premium.

11. _____

12. **YOUTHFUL OPERATOR—NO DRIVER'S TRAINING** Karen Roberts's father gave her a new Honda Accord to use at college. She is 17, has not had driver's training, lives in territory 1, and her vehicle has a symbol of 6. Her father asked her to carry 50/100 liability limits, $25,000 property damage limits, comprehensive and collision insurance, and uninsured motorist coverage. Find her annual insurance premium.

12. _____

13. **BODILY INJURY INSURANCE** Suppose your bodily injury policy has limits of 25/50 and you injure a person on a bicycle. The judge awards damages of $36,500 to the cyclist. **(a)** How much will the company pay? **(b)** How much will you pay?

(a) _____
(b) _____

14. **BODILY INJURY INSURANCE** Martha Spencer, owner of The Doll House, lost control of her car while trying to find her smartphone and forced another driver off the road. The court awarded $28,000 to the driver of the other car and $8000 to a passenger of the other car. Spencer had limits of 15/30. **(a)** Find the amount the insurance company paid. **(b)** Find the amount Spencer had to pay.

(a) _____
(b) _____

15. **MEDICAL EXPENSES AND PROPERTY DAMAGE** Wes Hanover accidentally backed into a parked car. He caused $4300 in damage to the car, and Hanover's passenger needed stitches in her forehead, which cost $850. Hanover had 15/30 liability limits, $1000 medical expense, and property damage of $10,000. Find the amount paid by the insurance company for **(a)** damages to the automobile and **(b)** medical expenses.

(a) _____
(b) _____

16. **MEDICAL EXPENSES AND PROPERTY DAMAGE** Jessica Wallace backed into a new Mercedes and caused $12,800 in damage to the car. She also injured the vertebrae in her neck, requiring surgery costing $48,200. She had 50/100 liability limits, $10,000 in property damage, and $3000 in medical expense coverage. Find the amount paid by the insurance company for **(a)** damages to the automobile and **(b)** medical expenses.

(a) _____
(b) _____

17. **INSURANCE COMPANY PAYMENT** A reckless driver caused Sandy Silva to collide with a car in another lane. Silva had 50/100 liability limits, $25,000 property damage limits, and collision coverage with a $100 deductible. Silva's car had damage of $1878, while the other car suffered $6936 in damages. The resulting lawsuit gave injury awards of $60,000 and $55,000, respectively, in damages for personal injury to the two people in the other car. Find the amount that the insurance company will pay for **(a)** repairing Silva's car, **(b)** repairing the other car, and **(c)** personal injury damages. **(d)** How much must Silva pay beyond her insurance coverage, including the collision deductible?

(a) _____
(b) _____
(c) _____
(d) _____

18. **INSURANCE PAYMENT** Bob Armstrong lost control of his car and crashed into another car. He had 15/30 liability limits, $10,000 property damage limits, and collision coverage with a $100 deductible. Damage to Armstrong's car was $2980; the other car, with a value of $22,800, was totaled. The results of a lawsuit awarded $75,000 and $45,000, respectively, in damages for personal injury to the two people in the other car.

(a) _____
(b) _____
(c) _____
(d) _____

Find the amount that the insurance company will pay for (a) repairing Armstrong's car, (b) repairing the other car, and (c) personal injury damages. (d) How much must Armstrong pay beyond his insurance coverage?

19. Explain why insurance companies charge a higher premium for auto insurance sold to a youthful operator. Do you think this is fair? (See Objective 6.)

20. Property damage pays for damage caused by you to the property of others. Since the average cost of a new car today is over $20,000, what amount of property damage coverage would you recommend to a friend who owns her own business?

QUICK CHECK ANSWERS

1. $445
2. $262
3. (a) $58 (b) $107
4. $242
5. $3091.60
6. $500 to repair her own car; $18,000 for medical expenses of other driver

13.5 Life Insurance

OBJECTIVES

1. Understand life insurance that does not accumulate cash value.
2. Understand life insurance that accumulates cash value.
3. Find the annual premium for life insurance.
4. Use premium factors with different modes of premium payment.

CASE IN POINT Martha Spencer owns the Doll House. Recently divorced, she worries about her two children should she become ill, disabled, or die. As a result, she carries medical insurance, disability insurance, and $250,000 in life insurance.

There is no doubt about it: Insurance is expensive! Yet most of us need insurance (car, home, medical, disability, and/or life insurance) because the risks of an accident, a fire, a health crisis, or an early death are so high and can be disastrous for the family. The graph on the left shows that about one out of every five adults believe they need more insurance. The figure on the right illustrates that many employers help insure their employees.

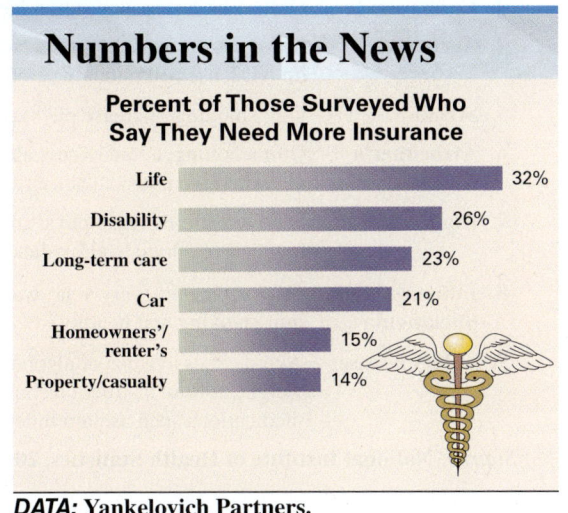

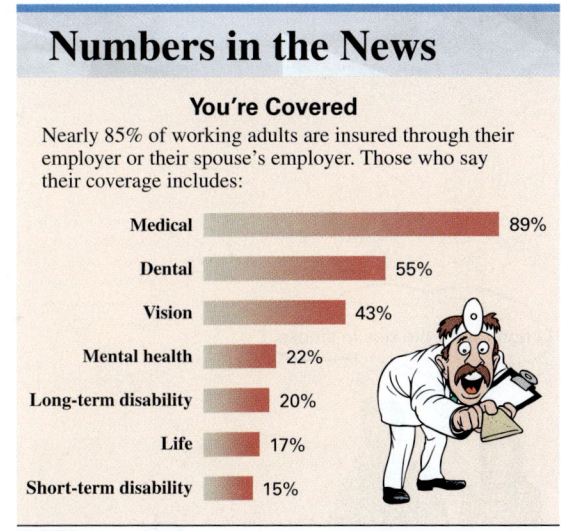

People buy **life insurance** to pay for their own burial expenses, to pay off a home mortgage or a car loan, to provide for a spouse and/or children, or to pay for their children's future college expenses. Some forms of life insurance build up a cash retirement value; other forms do not. Life insurance can also be important for the owner(s) of a business. Upon the death of the business owner, life insurance proceeds can provide a company with enough money to continue until it can be sold. Alternatively, life insurance proceeds can be used by one partner of a firm to buy out the ownership interest of a deceased partner.

Quick TIP

Proceeds from a life insurance policy are generally free of income taxes.

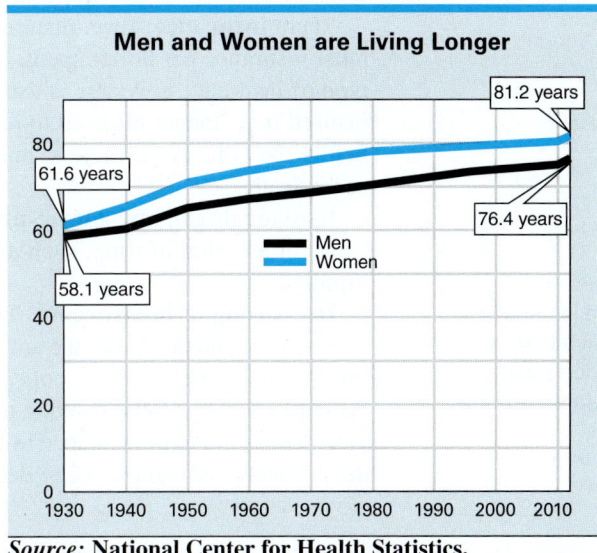

Life insurance rates have declined markedly over the past 30 or so years for two reasons: people are living longer and competition has reduced costs. The increase in competition is partly a result of sales of insurance over the Internet, which lowers costs and increases the availability of information about prices. Competition often, but not always, results in reduced prices to consumers.

Life insurance rates are determined on the basis of mortality data. The following table lists the most common causes of death in the United States and includes suggestions to reduce the risk of each.

Most Common Causes of Death

Cause	To Reduce Risk:
1. Heart disease	Quit smoking, lose weight, eat more fish and less red meat, increase physical activity
2. Cancer	Quit smoking, lose weight, eat healthy, increase physical activity
3. Lung disease	Quit smoking, reduce salt in diet
4. Accidents	Drive carefully (do not speed, stop at stop signs, do NOT use cell phone when driving, etc.)
5. Stroke	Quit smoking, increase physical activity, eat better
6. Alzheimer's	Quit smoking, avoid excess alcohol, eat a healthy diet, increase physical activity
7. Diabetes	Lose weight, increase physical activity, eat a low-carbohydrate healthy diet
8. Influenza and pneumonia	Get vaccinated every year, wash hands frequently, quit smoking, eat healthy
9. Kidney disease	Avoid excessive use of alcohol, maintain a healthy weight, follow instructions on over-the-counter medications such as pain relievers, quit smoking

Source: **National Institute of Health Statistics, 2016.**

Quick TIP //
Diabetes is a rapidly growing problem worldwide.

Quick TIP //
Mortgage insurance is usually much more expensive than a regular term insurance policy. Compare prices before you buy.

Look carefully at the table. Doing just a few things regularly will greatly reduce your chances of having any one of these very serious medical problems. Experts recommend that you quit smoking, eat healthy (generally more vegetables and less junk food and red meat), exercise more, avoid excess salt, drive carefully, maintain a healthy weight, avoid excessive use of alcohol, and follow instructions on over-the-counter medications such as pain relievers.

It is also very important that parents train their children to follow these guidelines. Patterns learned early in life tend to be followed later. So you help yourself and you help your family (even little sisters and brothers) if you quit smoking, eat healthy, etc.

OBJECTIVE 1 Understand life insurance that does not accumulate cash value.

Term insurance. Term insurance is the lowest-cost type of life insurance. It provides the most insurance per dollar spent, but it does not build up any cash values for retirement. This type of insurance coverage is usually renewable until some age, such as 70, when the insured is no longer allowed to renew it. As a result, most people discontinue term insurance before they die. Nevertheless, many prefer it because it is the cheapest protection against an early death.

Individuals can buy a **level-premium** term policy in which the premium remains constant for a period of time, such as 10 years or 20 years. Thereafter, premiums increase rapidly.

Decreasing term insurance. This is a type of term insurance with fixed premiums commonly to age 60 or 65, but the amount of life insurance decreases periodically. An example of this is a mortgage insurance policy on a home. The amount of life insurance on the owner decreases as the amount owed to the mortgage company decreases. Many large companies provide decreasing term insurance to employees as a benefit. The table provides an example of the death benefits of one particular decreasing term policy.

Example of Death Benefits for a Decreasing Term Policy with a Premium of $11 per Month

Age	Amount of Life Insurance
Under 29	$160,000
30–34	$140,000
35–39	$120,000
40–44	$100,000
45–49	$72,000
50–54	$44,000
55–59	$28,000
60–66	$16,000
67 and over	$0

OBJECTIVE 2 Understand life insurance that accumulates cash value.

Whole life (also called **straight life**, **ordinary life**, or **permanent**). This type of insurance provides a death benefit and a savings plan. The insured commonly pays a constant premium until death or retirement, whichever occurs first. If the policy is in force at the time of death, a death benefit is paid. Alternatively, the insured may choose to convert the accumulated **cash value** to a retirement benefit.

Universal life. This type of insurance provides the life insurance protection of term insurance plus a tax-deferred way to accumulate assets. It sometimes allows people to establish a permanent policy at a lower premium than they would have to pay under a whole life policy, and it gives the insured more flexibility. For example, universal life can help a family obtain more insurance when young children are at home and then help accumulate savings later after the children are grown.

Variable life. This type of insurance allows the policyholder to make choices among several investment options. It places the investment risk on the policyholder by allowing the insured to invest in any of the following: money market funds, bond funds, stock funds, or a combination of the three.

Limited-payment life insurance. Limited-payment life is similar to whole life insurance, except that premiums are paid for only a fixed number of years, such as 20. This type of insurance is thus often called 20-pay life, representing payments for 20 years. The premium for limited-payment life is higher than that for whole life policies. Limited-payment life is most appropriate for athletes, actors, and others whose income is likely to be high for several years and then decline.

Endowment policies are the most expensive type of policy since they accumulate cash more rapidly than the other types of life insurance. These policies guarantee payment of a fixed amount of money to a given individual, whether or not the insured lives. Endowment policies might be taken out by parents to guarantee a sum of money for their children's college education. Because of the high premiums, this is one of the least popular types of policies today.

> **Quick TIP**
> Limited-payment and endowment policies cost more, but they accumulate more money for use later in life.

OBJECTIVE 3 Find the annual premium for life insurance. Calculation of life insurance rates by actuaries is based on statistical data involving mortality tables, interest rates, and other factors. Women tend to live a few years longer than men, so a woman pays a lower life insurance premium than a man of the same age. Incidentally, women are more likely to be disabled than men and therefore have higher disability insurance rates than men. Use the actual age of a man to find the premium factor in the table below. However, subtract 5 from the age of a woman before finding the premium factor in the table.

Sample Annual Premium Rates* per $1000 of Life Insurance

Age	10-Year Level Premium Term	Whole Life	Universal Life	20-Pay Life
20	1.60	4.07	3.48	12.30
21	1.65	4.26	3.85	12.95
22	1.69	4.37	4.10	13.72
23	1.73	4.45	4.56	14.28
24	1.78	4.68	4.80	15.95
25	1.82	5.06	5.11	16.60
30	1.89	5.66	6.08	18.78
35	2.01	7.68	7.45	21.60
40	2.56	12.67	10.62	24.26
45	3.45	19.86	15.24	28.16
50	5.63	26.23	21.46	32.59
55	8.12	31.75	28.38	38.63
60	14.08	38.42	36.72	45.74

*For women, subtract 5 years from the actual age. For example, the rates for a 30-year-old woman are found on the age 25 row of the table.

Quick TIP
Life insurance premiums are higher for smokers, people with health problems, and even those with dangerous hobbies such as skydiving.

Finding Annual Premium

Annual premium = Number of thousands × Rate per $1000

Finding the Life Insurance Premium

EXAMPLE 1 Martha Spencer became the primary source of income for her family at age 35 after her divorce. At that time, she decided that she needed $250,000 in life insurance to pay off the mortgage on her home and to provide for her children. Find her annual premium for (a) a 10-year level premium term policy, (b) a whole life policy, (c) a universal life policy, and (d) a 20-pay life plan.

CASE IN POINT

SOLUTION
First, divide the desired amount of life insurance by $1000 to find the number of thousands.

$250,000 ÷ $1000 = 250 thousands

Since Spencer is a woman, subtract 5 from her actual age before using the table (35 − 5 = 30). Look in the table at age 30 for the rates for each type of insurance.

Quick TIP
Use the actual age of a man when using the table of premiums. However, subtract 5 from the age of a woman before using the table.

		Annual Premium
(a)	10-year level premium term	250 × **1.89** = $472.50
(b)	Whole life	250 × **5.66** = $1415
(c)	Universal life	250 × **6.08** = $1520
(d)	20-pay life	250 × **18.78** = $4695

Spencer wanted to buy universal life because of the savings feature, which would help her save for retirement. However, she purchased the level premium term instead, since her income was limited and the level term policy was much cheaper.

QUICK CHECK 1
James Liberty is 50 and wants to buy $300,000 in life insurance. Find the annual cost for (a) 10-year level premium term, (b) whole life, (c) universal life, and (d) 20-pay life.

OBJECTIVE 4 Use premium factors with different modes of premium payment. Many companies give the insured the option of paying the premium semiannually, quarterly, or monthly. For this convenience, the policyholder *pays an additional amount* that is determined by a **premium factor**. The following table shows typical premium factors.

Premium Factors	
Mode of Payment	**Premium Factor**
Semiannually	.51
Quarterly	.26
Monthly	.0908

Using a Premium Factor

EXAMPLE 2 The annual insurance premium on a $200,000 10-year level premium term life policy for Jane Rodriguez is $378. Use the premium factors table to find the amount of premium and the total annual cost if she pays (a) semiannually, (b) quarterly, or (c) monthly.

SOLUTION

	Premium	Annual Cost
(a) Semiannually:	$378 × .51 = $192.78	$192.78 × **2 payments/year** = $385.56
(b) Quarterly:	$378 × .26 = $98.28	$98.28 × **4 payments/year** = $393.12
(c) Monthly:	$378 × .0908 = $34.32	$34.32 × **12 payments/year** = $411.84

The annual cost increases when going from one annual premium, to semiannual, to quarterly, and finally to monthly.

QUICK CHECK 2

The annual premium on a life insurance policy is $470. Find the premium if payments are made (a) semiannually, (b) quarterly, and (c) monthly.

Pricing Life Insurance

EXAMPLE 3 Shauna Jones has decided to buy $100,000 in life insurance to make sure that her kids will have money if something happens to her. Jones is 28 and in good health. Find the monthly premium for (a) 10-year level premium term and (b) universal life.

SOLUTION

(a) 10-year level premium term:

                                          premium factor for monthly payment

Monthly premium = ($100 × 1.73) × .0908 = **$15.71**

(b) Universal life:

                                          premium factor for monthly payment

Monthly premium = ($100 × 4.56) × .0908 = **$41.40**

QUICK CHECK 3

Darryl Foster is 40 years old and wants to buy $400,000 of whole life insurance. Find the quarterly premium.

13.5 Exercises — MyLab Math

The shaded sections below contain solutions to help you get a QUICK START on the various types of exercises.

Find the annual premium, the semiannual premium, the quarterly premium, and the monthly premium for each of the following. (Note: Subtract 5 years for women.) Round to the nearest cent.

	Face Value of Policy	Age of Insured	Sex of Insured	Type of Policy	Annual Premium	Semi-annual Premium	Quarterly Premium	Monthly Premium
1.	$100,000	45	F	Term	$256	$130.56	$66.56	$23.24
	100 × $2.56 = $256; $256 × .51 = $130.56; $256 × .26 = $66.56; $256 × .0908 = $23.24							
2.	$60,000	30	M	Whole life	$339.60	$173.20	$88.30	$30.84
	60 × $5.66 = $339.60; $339.60 × .51 = $173.20; $339.60 × .26 = $88.30; $339.60 × .0908 = $30.84							
3.	$35,000	40	M	20-pay life				
4.	$60,000	50	F	20-pay life				
5.	$85,000	30	M	Universal life				
6.	$150,000	60	M	Term				
7.	$75,000	21	M	Whole life				
8.	$80,000	35	F	Term				
9.	$65,000	60	M	20-pay life				
10.	$50,000	45	F	Universal life				

11. What can the working parents of a young toddler do to decrease the risk of serious illness of either parent? Explain why the couple might choose to buy life insurance on the spouse with the higher income. (See the table above Objective 1.)

12. Explain the difference between term insurance and insurance that builds cash value. (See Objectives 1 and 2.)

C// indicates an exercise that is related to the Case in Point feature.

Solve the following application problems.

13. **LIFE INSURANCE** Bill Able married on his 25th birthday. He adopted the child his wife had from a previous marriage and wanted to care for the child as his own. To protect his new family against the loss of income in the event of his death, he purchased $200,000 in 10-year level premium term. Find the annual cost.

    $200,000 ÷ $1000 = 200
    200 × $1.82 = $364

    13. $364

14. **WHOLE LIFE INSURANCE** Jessica Smith buys a whole life policy with a face value of $100,000 at age 35. Find the annual premium.

    14. _____

15. **KEY EMPLOYEE INSURANCE** Martha Spencer owns the Doll House and has a 35-year-old key male employee whom she wants to insure for $50,000. Find the annual premium **(a)** for 10-year level term and **(b)** for whole life.

    (a) _____
    (b) _____

16. **20-PAY LIFE POLICY** Luan Lee buys a $100,000, 20-pay life policy at age 45. Her son Bryan is the beneficiary and will collect the face value of the policy. **(a)** Find the annual premium. **(b)** How much will Bryan get if his mother dies after making payments for 12 years?

    (a) _____
    (b) _____

17. **WHOLE LIFE** Rebecca Martinez is 35 and needs $100,000 in coverage. She opts for a whole life policy since it will also accumulate cash and force her to save. Find the annual premium and the total premiums paid if she keeps the policy to age 65.

    17. _____

18. **UNIVERSAL LIFE INSURANCE** Richard Gonsalves takes out a universal life policy with a face value of $50,000. He is 40 years old. Find the monthly premium.

    18. _____

19. **PREMIUM FACTORS** The annual premium for a whole life policy is $872. Using premium factors, find **(a)** the semiannual premium, **(b)** the quarterly premium, and **(c)** the monthly premium.

    (a) _____
    (b) _____
    (c) _____

20. **PREMIUM FACTORS** A universal life policy has an annual premium of $2012. Use premium factors to find **(a)** the semiannual premium, **(b)** the quarterly premium, and **(c)** the monthly premium.

    (a) _____
    (b) _____
    (c) _____

QUICK CHECK ANSWERS

1. (a) $1689 (b) $7869 (c) $6438
   (d) $9777
2. (a) $239.70 (b) $122.20 (c) $42.68
3. $1317.68

Chapter 13 Quick Review

Chapter Terms Review the following terms to test your understanding of the chapter. For each term you do not know, refer to the page number found next to that term.

1099 forms [p. 519]
actuaries [p. 541]
adjusted gross income [p. 519]
assessed value [p. 512]
assessment rate [p. 512]
bodily injury coverage [p. 541]
cash value [p. 551]
casualty or theft losses [p. 523]
catastrophic event [p. 532]
claims [p. 532]
coinsurance clause [p. 534]
collision coverage [p. 534]
comprehensive coverage [p. 534]
contents [p. 532]
coverage [p. 532]
decreasing term insurance [p. 550]
deductible [p. 543]
dependent [p. 521]
disability insurance [p. 537]
dollars per $100 [p. 513]
dollars per $1000 [p. 513]

endowment policies [p. 551]
face value [p. 533]
fair market value [p. 512]
Form 1040A [p. 525]
gifts to charity [p. 523]
health insurance [p. 537]
head of household [p. 521]
homeowners' policy [p. 532]
income tax [p. 518]
individual retirement account (IRA) [p. 519]
insurance [p. 511]
insurance policy [p. 532]
Internal Revenue Service [p. 518]
IRS [p. 518]
itemized deductions [p. 523]
level-premium term insurance [p. 550]
liability insurance [p. 537]
life insurance [p. 549]
limited-payment life [p. 551]
long-term care insurance [p. 537]

medical and dental expenses [p. 523]
medical insurance [p. 542]
mills [p. 513]
miscellaneous deductions [p. 523]
multiple carriers [p. 535]
no-fault [p. 544]
ordinary life [p. 551]
permanent life [p. 551]
personal exemptions [p. 521]
policy [p. 532]
policy limits [p. 542]
premium [p. 533]
premium factor [p. 552]
property damage coverage [p. 542]
property-tax rate [p. 512]
risk [p. 511]
renter's insurance [p. 537]
replacement cost [p. 534]
Schedule B (Form 1040A) [p. 525]

standard deduction [p. 521]
straight life [p. 551]
taxable income [p. 521]
tax assessor [p. 512]
tax deduction [p. 523]
taxes [p. 511]
tax preparation [p. 523]
term insurance [p. 550]
territorial ratings [p. 533]
underinsured [p. 534]
unreimbursed employee expenses [p. 523]
underwriters [p. 533]
uninsured motorist coverage [p. 544]
universal life [p. 551]
variable life [p. 551]
W-2 forms [p. 519]
whole life [p. 551]
workers' compensation [p. 537]
youthful [p. 544]

CONCEPTS

13.1 Fair market value and assessed value

The value of property is multiplied by a given percent to arrive at the assessed value.

Assessed value = **Assessment rate** × Market value

13.1 Tax rate

The tax rate formula is

$$\text{Tax rate} = \frac{\text{Total tax amount needed}}{\text{Total assessed value}}$$

13.1 Tax rates in different forms

1. **Percent**: multiply by assessed value.
2. **Dollars per $100**: move decimal point 2 places to the left in assessed value and multiply.
3. **Dollars per $1000**: move decimal point 3 places to the left in assessed value and multiply.
4. **Mills**: move decimal point 3 places to the left in rate and multiply by assessed value.

Use the formula

Property tax = Assessed value × **Tax rate**

EXAMPLES

The assessment rate is 30%; fair market value is $115,000; find the assessed value.

$$30\% \times \$115{,}000 = \$34{,}500$$

Tax amount needed: $3,864,400; total assessed value: $946,320,000; find the tax rate.

$$\frac{\$3{,}864{,}400}{\$946{,}320{,}000} = .00408 = .41\% \quad (\text{rounded})$$

Assessed value, $90,000; tax rate, 2.5%:

$$\$90{,}000 \times 2.5\% = \$2250$$

Tax rate, **$2.50** per $100:

$$900 \times \$2.50 = \$2250$$

Tax rate, **$25** per $1000:

$$90 \times \$25 = \$2250$$

Tax rate, **25 mills**:

$$\$90{,}000 \times .025 = \$2250$$

CONCEPTS	EXAMPLES
13.2 Adjusted gross income Adjusted gross income includes wages, salaries, tips, dividends, and interest. Subtract IRA contributions and alimony.	Salary, $32,540; interest income, $875; dividends, $315; find adjusted gross income. $$\$32{,}540 + \$875 + \$315 = \$33{,}730$$
13.2 Standard deduction amounts The majority of taxpayers use the standard deduction allowed by the IRS.	$6300 for single taxpayers $12,600 for married taxpayers filing jointly $6300 for married taxpayers filing separately $9300 for head of household
13.2 Taxable income The larger of either the total of itemized deductions or the standard deduction is subtracted from adjusted gross income along with $4050 for each personal exemption.	Single taxpayer, adjusted gross income, $38,500; itemized deductions, $3850; find taxable income. Note that itemized deductions are less than the standard deduction of $6300 and the personal exemption is $4050. Taxable income = $38,500 − $6300 − $4050 = $28,150
13.2 Tax rates There are seven tax rates: 10%, 15%, 25%, 28%, 33%, 35%, and 39.6%.	Single: 10%; 15% over $9275; 25% over $37,650; 28% over $91,150; 33% over $190,150; 35% over $413,350; 39.6% over $415,050 Married filing jointly: 10%; 15% over $18,550; 25% over $75,300; 28% over $151,900; 33% over $231,450; 35% over $413,350; 39.6% over $466,950 Married filing separately: 10%; 15% over $9275; 25% over $37,650; 28% over $75,950; 33% over $115,725; 35% over $206,675; 39.6% over $233,475 Head of household: 10%; 15% over $13,250; 25% over $50,400; 28% over $130,150; 33% over $210,800; 35% over $413,350; 39.6% over $441,000
13.2 Balance due or a refund from the IRS If the total amount withheld by employers is greater than the tax owed, a refund results. If the tax owed is the greater amount, a balance is due.	Tax owed, $1253; tax withheld, $113 per month for 12 months. Find balance due or refund. $113 withheld × 12 = $1356 withheld $1356 withheld − $1253 owed = $103 refund
13.3 Annual premium for fire insurance The building and territorial ratings are used to find the premiums per $100 for the building and contents. The two are added.	Building value, $180,000; contents, $35,000. Premiums are: building, $.75 per $100; contents, $.77 per $100. Find the annual premium. Building: 1800 (hundreds) × $.75 = $1350 Contents: 350 (hundreds) × $.77 = $269.50 Total premium: $1350 + $269.50 = $1619.50
13.3 Coinsurance formula Part of the risk of fire is taken by the insured. An 80% coinsurance clause is common. $$\text{Loss paid by insurance company} = \text{Amount of loss} \times \frac{\text{Policy amount}}{\text{80\% of replacement cost}}$$	Replacement cost, $250,000; policy amount, $150,000; fire loss, $40,000; 80% coinsurance clause; find the amount of loss paid by the insurance company. The face value ($150,000) is less than 80% of the replacement cost ($200,000). $$\$40{,}000 \times \frac{\$150{,}000}{\$200{,}000} = \$30{,}000 \text{ (amount insurance company pays)}$$

CONCEPTS	EXAMPLES
13.3 Multiple carriers Several companies insure the same property, which limits the risk of the insurance company, with each paying its fractional portion of any claim.	Insured loss, $500,000 Insurance is Company A with $1,000,000; Company B with $750,000; Company C with $250,000; find the amount of loss paid by each company. Total insurance: $$1{,}000{,}000 + 750{,}000 + 250{,}000 = 2{,}000{,}000$$ Company A: $$\frac{1{,}000{,}000}{2{,}000{,}000} \times \$500{,}000 = \$250{,}000$$ Company B: $$\frac{750{,}000}{2{,}000{,}000} \times \$500{,}000 = \$187{,}500$$ Company C: $$\frac{250{,}000}{2{,}000{,}000} \times \$500{,}000 = \$62{,}500$$
13.4 Annual auto insurance premium Most drivers are required to purchase automobile insurance. The premium is determined by the types of coverage selected, the type of car, geographic territory, past driving record, and other factors.	Determine the premium: territory, 2; liability, 50/100; property damage, $50,000; comprehensive and collision, 3-year-old car with a symbol of 8; uninsured motorist coverage; driver is age 23 with driver's training. $341 liability 223 property damage 68 comprehensive 98 collision 219 uninsured motorist $949 × 2.10 youthful-operator factor = **$1992.90**
13.5 Annual life insurance premium There are several types of life policies. Use the table and multiply by the number of $1000s of coverage. Subtract 5 years from the age of women. Premium = Number of thousands × **Rate per $1000**	Find the annual premiums on a $50,000 policy for a 30-year-old man. (a) 10-year level premium term: 50 × **$1.89** = $94.50 (b) Whole life: 50 × **$5.66** = $283 (c) Universal life: 50 × **$6.08** = $304 (d) 20-pay life: 50 × **$18.78** = $939
13.5 Premium factors Life insurance premiums may be paid semiannually, quarterly, or monthly. The annual premium is multiplied by the premium factor to determine the premium amount.	The annual life insurance premium is $740. Use the table to find the (a) semiannual, (b) quarterly, and (c) monthly premiums. (a) Semiannual: $740 × .51 = $377.40 (b) Quarterly: $740 × .26 = $192.40 (c) Monthly: $740 × .0908 = $67.19

Case Study

FINANCIAL PLANNING FOR PROPERTY TAXES AND INSURANCE

Baker's Pottery manufactures and sells ceramic pots of all types, shapes, and styles. Planning ahead, the company set aside $53,500 to pay property taxes, fire insurance premiums, and life insurance premiums on the company president. All of these premiums happen to be due in the same month. Find each of the following.

1. The company property has a fair market value of $1,990,000 and is assessed at 75% of this value. If the tax rate is $7.90 per $1000 of assessed value, find the annual property tax.

    1. _____

2. The building occupied by the company is a class-B building with a replacement cost of $1,730,000. The contents are worth $3,502,000 and the territorial rating is 4. Find the annual fire insurance premium.

    2. _____

3. The president of the company is a 50-year-old woman who lost the use of her legs in an automobile accident. She needs life insurance and the company buys a $250,000, 10-year level premium life insurance policy on her. Find the semiannual premium.

    3. _____

4. Find the total amount needed to pay property taxes, the fire insurance premium, and the semiannual life insurance premium.

    4. _____

5. How much more than the amount needed had the company set aside to pay these expenses?

    5. _____

INVESTIGATE

Estimate the amount of life insurance you need (or may need in 10 years if you prefer). Use the Web to find prices for term and whole life insurance for this face amount. Explain why whole life insurance has a much higher annual premium.

Case in Point Summary Exercise

MATTEL INC.—TAXES AND INSURANCE

www.mattel.com

Facts:

- 1945: Founded by Elliot and Ruth Handler
- 1959: Barbie introduced
- 2004: More than one billion Barbies sold
- 2018: Sales over $5 billion

Antique dolls representing adults from the 17th and 18th centuries have been found, but they are very rare. Individual craftsmen in England made most of these earliest dolls. The craftsmen carved the dolls of wood, painted their features, and also designed the costumes for the dolls. Some of these earliest dolls are valued at more than $40,000.

The Barbie doll is the most popular fashion doll ever created. If all the Barbie dolls that have been sold since 1959 were placed head-to-toe, the dolls would circle the earth more than seven times. The most popular Barbie ever sold was the Totally Hair Barbie, which was introduced in 1992. With hair from the top of her head to her toes, more than 10 million of these dolls were sold, resulting in revenue of $100 million. With annual retail sales at over $1 billion, Barbie is the #1 brand of doll for girls.

Martha Spencer sells dolls around the world through her website. Her business has grown over the years, but she still carefully watches costs. She knows that she must control costs to be successful. Help her figure out some of these costs.

1. Spencer is looking at a building that she hopes to buy for what she believes is the fair market value of $310,000. However, the replacement cost is estimated to be $420,000. The assessment rate is 30% and the tax rate is $64 per $1000 of assessed value. Find the property tax for the year.

    1. _____

2. Janet Chino, one of Spencer's employees, needs help estimating her income tax based on the following: wages—$28,410; interest—$212; ordinary dividends—$84; and IRA contribution—$500. Find the following: total income, adjusted gross income, taxable income, and income tax given that Chino is single and claims only 1 exemption. Use the tax rate schedules.

    2. _____

3. Find the cost of fire insurance if the building in Question 1 is in territory 5 and has a building classification of C. The contents are valued at $145,000.

    3. _____

4. Spencer purchases a new Toyota and buys 250/500 liability limits, $100,000 property damage, comprehensive, collision, and uninsured motorist. She is in territory 2 and the vehicle has symbol 8. Find the annual premium given that she is 35.

    4. _____

CHAPTER 13 Case in Point Summary Exercise   561

5. Spencer is a 35-year-old woman who is divorced and has two children. She decides to purchase $500,000 of life insurance to protect her children. Find the annual cost of 10-year level term.

5. _____

6. Find the total of all of the costs that Martha Spencer must pay. This total is only a small part of her total bills each year—clearly she has to sell a lot of dolls to pay her bills.

6. _____

Discussion Question: Is it important to pay property taxes and buy insurance? What should happen to those who do not pay property taxes? What, if anything, should happen to those not buying the required minimum amounts of insurance on their vehicles?

Chapter 13 Test

To help you review, the numbers in brackets show the section in which the topic was discussed.

Find the following property tax rates. **[13.1]**

	Percent	Per $100	Per $1000
1.	5.76%	_____	_____
2.	_____	_____	$93.50

Find the taxable income and the tax for each of the following people. The letter following the names indicates the marital status. Assume those married file jointly. **[13.2]**

	Name	Number of Exemptions	Adjusted Gross Income	Itemized Deductions	Taxable Income	Tax
3.	J. Spalding, S	2	$68,295	$5380	_____	_____
4.	The Bensons, M	4	$43,487	$8315	_____	_____

Find the tax owed in the following problems.

5. Bradkin's Toggery owns property with a fair market value of $209,200. Property in the area is assessed at 30% of fair market value with a tax rate of 3.65%. Find the annual tax. **[13.1]**

    5. _____

6. The Blakely family has an adjusted gross income of $98,316. They are married and file jointly with five exemptions and deductions of $8420. **[13.2]**

    6. _____

7. Kari Heen had an adjusted gross income of $44,600 last year. She had deductions of $1280 for state income tax, $3620 for property tax, $3540 in mortgage interest, and $1450 in contributions. Heen claims one exemption and files as a single person. **[13.2]**

    7. _____

Find the annual fire insurance premium for the following. Use the table on page 533. **[13.3]**

8. Southside Plating owns a class-B building with a replacement cost of $780,000. Contents are valued at $128,600. The territorial rating is 5.

    8. _____

9. A fourplex is valued at $220,000. The fire insurance policy (with an 80% coinsurance clause) has a face value of $150,000. If the building has a fire loss of $50,000, find the amount of the loss that the insurance company will pay.

    9. _____

10. Dave's Body and Paint has an insurable loss of $72,000, while having insurance coverage beyond coinsurance requirements. The insurance is divided among Company A with $250,000 coverage, Company B with $150,000 coverage, and Company C with $100,000 coverage. Find the amount of the loss paid by each of the insurance companies.

    A: _____
    B: _____
    C: _____

Find the annual motor-vehicle insurance premium for the following people. **[13.4]**

Name	Territory	Age	Driver Training?	Liability	Property Damage	Comprehensive Collision Age Group	Symbol	Uninsured Motorist?	Annual Premium
11. Ramos	3	18	Yes	15/30	$10,000	5	7	Yes	_____
12. Larik	1	42	No	50/100	$100,000	1	8	Yes	_____

Find the annual premium, the semiannual premium, the quarterly premium, and the monthly premium for each of the following life insurance policies. Use the tables in Section 13.5. **[13.5]**

	Annual	Semiannual	Quarterly	Monthly
13. Irene Chong, whole life, $28,000 face value, age 35, female	_____	_____	_____	_____
14. Gil Eckern, 20-pay life, $80,000 face value, age 40, male	_____	_____	_____	_____

Solve the following.

15. Betsy Monikens (age 28) and Jim Faber (age 30) recently married and need to buy both car and life insurance. They have one car, 4 years old, with symbol 7 and live in territory 2. Auto: 50/100 liability with $3000 in medical expense limits, $25,000 in property damage, comprehensive, $1000 deductible collision, and uninsured motorist. Life: $100,000 10-year level term on Jim and $50,000 universal life on Betsy. Find the annual cost of **(a)** the auto insurance and **(b)** the life insurance. **[13.4 and 13.5]**

(a) _____

(b) _____

16. Jessie Hernandez's truck spun out of control on an icy road, causing an accident with another driver. Damages to his truck were $6400, damages to the other driver's vehicle were $8200, and the other driver had medical expenses of $12,900. Hernandez was not hurt and his policy showed that he had liability limits of 15/30, medical expenses of $1000, property damage of $10,000, no comprehensive or collision, and no uninsured motorist. Identify each of the costs he must pay. **[13.4]**

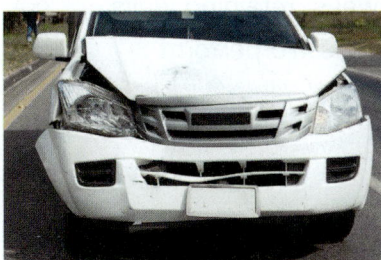

16. _____

Depreciation

14

CHAPTER CONTENTS

14.1 Straight-Line Method

14.2 Declining-Balance Method

14.3 Sum-of-the-Years'-Digits Method

14.4 Units-of-Production Method

14.5 Modified Accelerated Cost Recovery System

CASE IN POINT

CAPITAL CURB AND CONCRETE is owned and managed by John Goodby. The company does concrete work for building contractors and for the owners of new and existing residential and commercial properties. The company builds curbs and sidewalks, driveways and walkways, retaining walls, patios, and concrete mowing strips. To perform these services, Capital Curb and Concrete owns and maintains trucks, trailers, tractors, backhoes, trenchers, and concrete-pumping equipment. All of this equipment wears out with use and is depreciated when doing income taxes and preparing financial statements.

The federal government allows businesses to deduct expenses such as salaries, payroll taxes, rent, maintenance, and utilities from revenue when determining income used to calculate income tax. However, it does not allow firms to deduct the full cost of buildings, trucks, machines, improvements to buildings, and so on, in one year. The cost of items of this type that last for more than one year must be deducted over their **useful life**. The process of spreading the income tax deduction out over the useful life of an asset is called **depreciation**.

Over the years, several methods of computing depreciation have been used, including **straight-line**, **declining-balance**, **sum-of-the-years'-digits**, and **units-of-production**. These methods are used in keeping company accounting records and sometimes when preparing state income tax returns. Items purchased after 1981 are depreciated for federal income tax returns with the **accelerated cost recovery system** or the **modified accelerated cost recovery system**, discussed later.

A company might not use the same method of depreciation for everything. For example, some states do not allow the method of depreciation required on federal income taxes. Further, firms sometimes use one method of depreciation for income tax returns and another for financial statements.

Quick TIP
Depreciation can also be used by individuals owning rental property such as a rent house.

14.1 Straight-Line Method

OBJECTIVES

1. Understand the terms used in depreciation.
2. Use the straight-line method of depreciation to find the amount of depreciation each year.
3. Use the straight-line method to find the book value of an asset.
4. Use the straight-line method to prepare a depreciation schedule.

CASE IN POINT John Goodby and his accountant decide on which method of depreciation to use for the depreciable assets of Capital Curb and Concrete. A commonly used method is the straight-line method of depreciation.

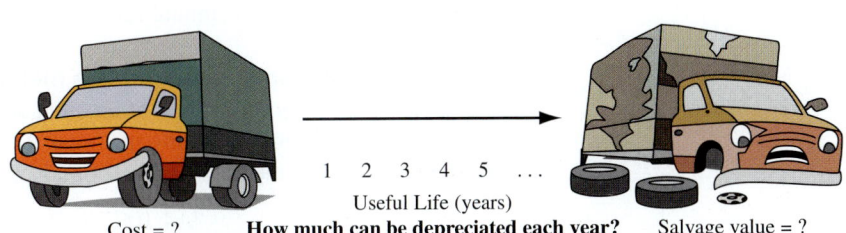

Cost = ?   How much can be depreciated each year?   Salvage value = ?
Useful Life (years): 1 2 3 4 5 ...

OBJECTIVE 1 Understand the terms used in depreciation. The physical assets of a company such as machinery, trucks, cars, and computers are **tangible assets**. Assets such as patents and copyrights, franchise fees, and customer lists are **intangible assets**. In general, either type of asset may be depreciated as long as its useful life can be determined.

> **Key terms related to depreciation**
>
> **Cost** is the basis for determining depreciation. It is the total amount paid for the asset.
>
> **Useful life** is the period of time during which the asset will be used. The **Internal Revenue Service** has guidelines for estimating the life of an asset used in a particular trade or business. However, useful life depends on the use of the asset, the repair policy, the replacement policy, obsolescence, and other factors.
>
> **Salvage value** or **scrap value** (sometimes called **residual value**) is the estimated value of an asset when it is retired from service, traded in, disposed of, or exhausted. An asset may have a salvage value of zero, or no salvage value.
>
> **Accumulated depreciation** is the amount of depreciation taken so far, a running balance of depreciation to date.
>
> **Book value** is the cost of an asset minus the total depreciation to date. The book value at the end of an asset's life is equal to the salvage value. The book value can never be less than the salvage value.

566 CHAPTER 14 Depreciation

OBJECTIVE 2 Use the straight-line method of depreciation to find the amount of depreciation each year. The simplest method of depreciation, **straight-line depreciation**, assumes that assets lose an equal amount of value during each year of life. For example, suppose a heavy equipment trailer is purchased by Capital Curb and Concrete at a cost of $14,100. The trailer has a useful life of 8 years and a salvage value of $2100, which is the estimated amount it can be sold for at the end of its useful life. Use the following formula to find the amount that can be depreciated:

> **Finding Amount to Be Depreciated**
> Amount to be depreciated = Cost − Salvage value

Here, the amount to be depreciated over the 8-year period is:

$$\begin{array}{rl} \$14,100 & \text{cost} \\ -\ 2,100 & \text{salvage value} \\ \hline \$12,000 & \text{amount to be depreciated} \end{array}$$

With the straight-line method, an equal amount of depreciation is taken each year over the 8-year life of the trailer.

> **Finding Annual Depreciation**
> $$\text{Depreciation} = \frac{\text{Depreciable amount}}{\text{Years of life}} = \frac{\$12,000}{8} = \$1500$$

Quick TIP
No money changes hands due to depreciation. Depreciation is used to calculate federal and state income taxes as well as financial statements.

Each year during the 8-year life of the trailer, the annual depreciation will be $1500, or $\frac{1}{8}$ of the depreciable amount. The annual rate of depreciation is $12\frac{1}{2}\%$ ($\frac{1}{8} = 12\frac{1}{2}\%$).

OBJECTIVE 3 Use the straight-line method to find the book value of an asset. The book value, or remaining value, of an asset at the end of a year is the original cost minus the depreciation up to and including that year (**accumulated depreciation**). The book value of the trailer at the end of the first year is found as follows:

$$\begin{array}{rl} \$14,100 & \text{cost} \\ -\ 1,500 & \text{first year's depreciation} \\ \hline \$12,600 & \text{book value at end of the first year} \end{array}$$

Book value is found with the following formula.

Quick TIP
Book value is the value of an asset that is carried on the financial statements (or books) of a firm.

> **Finding Book Value**
> Book value = Cost − Accumulated depreciation

Subtract the second year's depreciation from the book value at the end of year 1 to find the book value at the end of year 2, and so on.

Finding First-Year Depreciation and Book Value

EXAMPLE 1 A fast food restaurant purchased a new low-fat chicken cooker at a cost of $26,500. The estimated life of the fryer is 5 years, with a salvage value of $3500. Find **(a)** the annual rate of depreciation, **(b)** the annual amount of depreciation, and **(c)** the book value at the end of the first year.

SOLUTION

(a) The annual rate of depreciation is 20% (5-year life = $\frac{1}{5}$ **per year = 20%**).

(b) $$\begin{array}{rl} \$26,500 & \text{cost} \\ -\ 3,500 & \text{salvage value} \\ \hline \$23,000 & \text{depreciable amount} \end{array}$$

This $23,000 will be depreciated evenly over the 5-year life for an annual depreciation of $4600 ($23,000 × 20% = $4600).

(c) Since the annual depreciation is $4600, the book value at the end of the first year will be

$$\begin{array}{rl} \$26,500 & \text{cost} \\ -\ 4,600 & \text{depreciation in the first year} \\ \hline \$21,900 & \text{book value at the end of the first year} \end{array}$$

To solve Example 1 using a calculator, first use parentheses to find the depreciable amount. Next, divide to find depreciation. Finally, find the book value.

$$(\; 26{,}500 \; - \; 3500 \;) \; \div \; 5 \; = \; 4600$$
$$26{,}500 \; - \; 4600 \; = \; 21{,}900$$

Note: Refer to Appendix B for calculator basics.

QUICK CHECK 1

Class Printers purchased a new printing press at a cost of $18,400. The estimated life of the printing press is 8 years, with a salvage value of $3600. Find **(a)** the annual rate of depreciation, **(b)** the annual amount of depreciation, and **(c)** the book value at the end of the first year.

When using straight-line depreciation, the steps to find the book value at the end of any year are as follows:

Finding Book Value
Step 1 Find depreciation to date by multiplying the annual depreciation times the number of years.
Step 2 Subtract this value from cost to find the book value.

If an asset is expected to have **no salvage value** at the end of its expected life, the entire cost will be depreciated over its life. In Example 1, if the chicken cooker had been expected to have no salvage value at the end of 5 years, the annual amount of depreciation would have been $5300 ($26,500 × 20% = $5300).

Finding the Book Value at the End of Any Year

EXAMPLE 2 A lighted display case at Albertson's cost $3400, has an estimated life of 10 years, and has a salvage value of $800. Find the book value at the end of 6 years.

SOLUTION

The annual rate of depreciation is 10% (10-year life is $\frac{1}{10}$ or 10%).

$$\text{Depreciable amount} = \text{Cost} - \text{Salvage value}$$
$$= \$3400 - \$800 = \$2600$$
$$\text{Annual depreciation} = \$2600 \times 10\% = \$260$$
$$\text{Accumulated depreciation over 6 years} = \$260 \times 6 \text{ years}$$
$$= \$1560$$

Find the book value at the end of 6 years by subtracting the accumulated depreciation from the cost.

$3400 cost
− 1560 accumulated depreciation (6 years)
$1840 book value at the end of 6 years

After 6 years, this display case would be carried on the firm's books with a value of $1840.

Quick TIP
The book value helps the owner of a business estimate the value of the business, which is important when the owner is borrowing money or trying to sell the business.

QUICK CHECK 2

Two Brothers Alignment bought a new wheel alignment system for $15,900. If the salvage value of the equipment is $4500 and the estimated life is 12 years, find the book value at the end of 8 years.

OBJECTIVE 4 Use the straight-line method to prepare a depreciation schedule. A **depreciation schedule** is often used to show the annual depreciation, accumulated depreciation, and book value over the useful life of an asset.

Preparing a Depreciation Schedule

EXAMPLE 3 Capital Curb and Concrete bought a new pickup truck for $21,500. It is a small truck that will be used exclusively in the business. It has a useful life of 5 years, at which time it will have a salvage value (trade-in value) of $3500. Prepare a depreciation schedule using the straight-line method of depreciation.

SOLUTION

The annual rate of depreciation is 20% (5-year life = $\frac{1}{5}$ per year = 20%). Find the depreciable amount as follows.

$21,500   cost
− 3,500   salvage value
$18,000   depreciable amount

This $18,000 will be depreciated evenly over the 5-year life for an annual depreciation of $3600 ($18,000 × 20% = $3600). This depreciation schedule includes a year zero that represents the initial purchase of the truck.

Depreciation Schedule

Year	Computation	Amount of Depreciation	Accumulated Depreciation	Book Value
0	—	—	—	$21,500
1	(20% × $18,000)	$3600	$3,600	$17,900
2	(20% × $18,000)	$3600	$7,200	$14,300
3	(20% × $18,000)	$3600	$10,800	$10,700
4	(20% × $18,000)	$3600	$14,400	$7,100
5	(20% × $18,000)	$3600	$18,000	$3,500

The depreciation is $3600 each year. The accumulated depreciation at the end of 5 years is equal to the depreciable amount, and the book value at the end of 5 years is equal to the salvage value.

Quick TIP
The firm in this example cannot deduct the entire cost of the truck from income in the year purchased. It deducts $3600 every year for 5 years so that the cost of the truck is spread out over its useful life.

QUICK CHECK 3

A concrete extruding machine is purchased at a cost of $4300. The machine has a useful life of 4 years and a salvage value of $700. Prepare a depreciation schedule using the straight-line method of depreciation to find the book value at the end of each year of the asset's life.

To help you understand the differences between the depreciation methods presented in this chapter, we will depreciate the same pickup truck using straight-line depreciation in Example 3 and using double-declining-balance in Section 14.2 and sum-of-the-years'-digits methods in Section 14.3.

Depreciation is used with assets having a useful life of *more than one year*. The asset to be depreciated must have a predictable life. A truck can be depreciated because its useful life can be estimated, but land cannot be depreciated because its life is considered to be indefinite.

The graph below shows the remaining value of the pickup truck in Example 3 as it is depreciated over its useful life. The amount shown on the vertical axis in the graph represents the book value, or the value at which the truck is carried on the firm's financial statements.

Quick TIP
Sometimes the book value of an asset is close to the actual market value. At other times, the book value of an asset is either above or below the actual market value.

Book Value
Straight-Line Depreciation of Truck

Depreciable assets can be large and very expensive. For example, one utility company plans to depreciate the several-billion-dollar cost of a nuclear power plant shown in the photo at the left over its several-decade useful life. The income tax savings resulting from depreciation help the utility company afford the cost of building the power plant.

14.1 Exercises // MyLab Math

The shaded sections below contain solutions to help you get a QUICK START on the various types of exercises.

Find the annual straight-line rate of depreciation, given the following estimated lives. (See Example 1.)

	Life	Annual Rate		Life	Annual Rate
1.	5 years $\frac{1}{5} = 20\%$	20%	2.	4 years $\frac{1}{4} = 25\%$	25%
3.	8 years	_____	4.	10 years	_____
5.	20 years	_____	6.	25 years	_____
7.	15 years	_____	8.	30 years	_____
9.	80 years	_____	10.	40 years	_____
11.	50 years	_____	12.	100 years	_____

Find the annual amount of depreciation for the following, using the straight-line method. (See Examples 1 and 2.)

13. Cost: $9000
    Estimated life: 20 years
    Estimated scrap value: None
    Annual depreciation: **$450**
    $9000 × 5% = $450

14. Cost: $3400
    Estimated life: 4 years
    Estimated scrap value: $800
    Annual depreciation: **$650**
    $3400 − $800 = $2600
    $2600 × 25% = $650

15. Cost: $2700
    Estimated life: 3 years
    Estimated scrap value: $300
    Annual depreciation: _____

16. Cost: $8100
    Estimated life: 6 years
    Estimated scrap value: $750
    Annual depreciation: _____

17. Cost: $46,800
    Estimated life: 5 years
    Estimated scrap value: None
    Annual depreciation: _____

18. Cost: $2,730,000
    Estimated life: 10 years
    Estimated scrap value: $310,000
    Annual depreciation: _____

C// indicates an exercise that is related to the Case in Point feature.

Find the book value at the end of the first year for the following, using the straight-line method. (See Examples 1 and 2.)

19. Cost: $3200
Estimated life: 8 years
Estimated scrap value: $400
Book value: **$2850**

$3200 - $400 = $2800
$2800 \times 12.5\% = $350
$3200 - $350 = $2850

20. Cost: $35,000
Estimated life: 10 years
Estimated scrap value: $2500
Book value: _____

21. Cost: $5400
Estimated life: 12 years
Estimated scrap value: $600
Book value: _____

22. Cost: $4500
Estimated life: 5 years
Estimated scrap value: None
Book value: _____

Find the book value at the end of 5 years for the following, using the straight-line method. (See Examples 1 and 2.)

23. Cost: $4800
Estimated life: 10 years
Estimated scrap value: $750
Book value: **$2775**

$4800 - $750 = $4050
$4050 \times 10\% = $405
$405 \times 5 = $2025
$4800 - $2025 = $2775

24. Cost: $16,000
Estimated life: 20 years
Estimated scrap value: $2000
Book value: _____

25. Cost: $80,000
Estimated life: 50 years
Estimated scrap value: $10,000
Book value: _____

26. Cost: $660
Estimated life: 8 years
Estimated scrap value: $100
Book value: _____

Solve the following application problems.

27. MACHINERY DEPRECIATION Capital Curb and Concrete selects the straight-line method of depreciation for a Bobcat loader costing $12,000 with a 3-year life and an expected scrap value of $3000. Prepare a depreciation schedule.

Year	Computation	Amount of Depreciation	Accumulated Depreciation	Book Value
0	—	—	—	$12,000
1	$(33\tfrac{1}{3}\% \times \$9000)$	$3000	$3000	$9,000
2	$(33\tfrac{1}{3}\% \times \$9000)$	$3000	$6000	$6,000
3	$(33\tfrac{1}{3}\% \times \$9000)$	$3000	$9000	$3,000

28. LABORATORY EQUIPMENT Taconic Medical purchased new lab equipment costing $18,000, having an estimated life of 4 years and a salvage value of $1600. Prepare a depreciation schedule using the straight-line method of depreciation.

Year	Computation	Amount of Depreciation	Accumulated Depreciation	Book Value
0	—	—	—	$18,000
1				
2				
3				
4				

29. VEHICLE DEPRECIATION Capital Curb and Concrete paid $51,200 for a $1\frac{1}{2}$-ton, dual-axle flatbed truck with an estimated life of 6 years and a salvage value of $14,000. Prepare a depreciation schedule using the straight-line method of depreciation.

Year	Computation	Amount of Depreciation	Accumulated Depreciation	Book Value
0	—	—	—	$51,200
1				
2				
3				
4				
5				
6				

30. BUSINESS FIXTURES The Venture Center purchased a new commercial air conditioner at a cost of $7800 and estimates the useful life as 10 years, after which it will have no salvage value. Prepare a depreciation schedule, calculating depreciation by the straight-line method.

Year	Computation	Amount of Depreciation	Accumulated Depreciation	Book Value
0	—	—	—	$7800
1				
2				
3				
4				
5				
6				
7				
8				
9				
10				

31. Explain the purpose of depreciation. When is it used and who uses it?

32. Develop a single formula that will show how to find annual depreciation using the straight-line method of depreciation. (See Objective 1.)

572 CHAPTER 14 Depreciation

33. BARGE DEPRECIATION A Dutch petroleum company purchased a barge for $1,300,000. The estimated life is 20 years, at which time it will have a salvage value of $200,000. Find (a) the annual amount of depreciation using the straight-line method and (b) the book value at the end of 5 years.

(a) _____
(b) _____

34. DEPRECIATING EQUIPMENT The new digital automated loading system at Capital Curb and Concrete has a cost of $14,500, an estimated life of 8 years, and a scrap value of $2100. Find (a) the annual depreciation and (b) the book value at the end of 4 years using the straight-line method of depreciation.

(a) _____
(b) _____

35. DEPRECIATING MACHINERY A bottle-capping machine costs $88,000, has an estimated life of 8 years, and has a scrap value of $16,000. Use the straight-line method of depreciation to find (a) the annual rate of depreciation, (b) the annual amount of depreciation, and (c) the book value at the end of the first year.

(a) _____
(b) _____
(c) _____

36. DRILLING RIG Transocean Drilling spent $236,000,000 to purchase a new offshore drilling rig capable of drilling to 40,000 feet. The useful life is 20 years and the salvage value is estimated to be $12,000,000. Use the straight-line method of depreciation to find (a) the annual rate of depreciation, (b) the annual amount of depreciation, and (c) the book value at the end of 5 years.

(a) _____
(b) _____
(c) _____

QUICK CHECK ANSWERS

1. (a) 12.5% rate
   (b) $1850 depreciation
   (c) $16,550 book value
2. $8300
3. year 0 = $4300
   year 1 = $3400
   year 2 = $2500
   year 3 = $1600
   year 4 = $ 700

14.2 Declining-Balance Method

OBJECTIVES
1. Describe the declining-balance method of depreciation.
2. Find the double-declining-balance rate.
3. Use the double-declining-balance method to find the amount of depreciation and the book value for each year.
4. Use the double-declining-balance method to prepare a depreciation schedule.

CASE IN POINT Straight-line depreciation assumes that an asset loses an equal amount of value each year of its life. This is not realistic for most of the equipment owned by Capital Curb and Concrete. For example, a new tractor loses much more value during its first year of life than during its fifth year of life.

OBJECTIVE 1 Describe the declining-balance method of depreciation. Methods of **accelerated depreciation** more accurately reflect the rate at which assets actually lose value. One of the more common accelerated methods of depreciation is the **double-declining-balance method** or **200% method**. With this method, the **double-declining-balance rate** must first be found. This rate is multiplied by last year's book value to get this year's depreciation. Since the book value declines from year to year, the annual depreciation also declines, giving the origin of the name of this method.

OBJECTIVE 2 Find the double-declining-balance rate. Calculate depreciation using the double-declining-balance method by first finding the straight-line rate of depreciation. Then adjust the straight-line rate to the desired double-declining-balance rate by multiplying by 2.

Finding the 200% Declining-Balance Rate

EXAMPLE 1 Find the straight-line rate and the double-declining-balance (200%) rate for each of the following years of life.

SOLUTION

Years of Life	Straight-Line Rate	Double-Declining-Balance Rate
3	33.33% ($\frac{1}{3}$)	$\times 2 = 66.67\%$ ($\frac{2}{3}$)
4	25%	$\times 2 = 50\%$
5	20%	$\times 2 = 40\%$
8	12.5%	$\times 2 = 25\%$
10	10%	$\times 2 = 20\%$
20	5%	$\times 2 = 10\%$
25	4%	$\times 2 = 8\%$
50	2%	$\times 2 = 4\%$

QUICK CHECK 1

Find the straight-line rate and the double-declining-balance rate for (a) 40 years, (b) 6 years, (c) 15 years, and (d) 25 years.

OBJECTIVE 3 Use the double-declining-balance method to find the amount of depreciation and the book value for each year. Use the formula below to find the amount of depreciation using the double-declining-balance method. The declining balance is the total cost in the first year and the previous year's book value in the following years.

Finding Amount of Depreciation

Depreciation = Double-declining-balance rate × Declining balance

Quick TIP
Throughout the remainder of this section and chapter, money amounts will be rounded to the nearest dollar.

Do not subtract salvage value from cost when using this method of depreciation. Rather, simply multiply the appropriate percentage by the book value remaining at the end of the prior year. It is important to remember that *book value never goes below the salvage value*.

574 CHAPTER 14 Depreciation

Finding Depreciation and Book Value Using Double-Declining-Balance

EXAMPLE 2 Capital Curb and Concrete purchased a small portable storage building for $8100. It is expected to have a life of 10 years, at which time it will have no salvage value. Using the double-declining-balance method of depreciation, find the first and second years' depreciation and the book value at the end of the first and second years.

SOLUTION

The straight-line depreciation rate for property with a 10-year recovery is 10%. Therefore, the double-declining balance rate is twice that, or 10% × 2 = 20%.

First year:

$$\text{Depreciation} = 20\% \text{ of declining balance}$$
$$= 20\% \times 8100 \text{ cost}$$
$$= \$1620$$

Find the book value at the end of the first year as follows.

$8100   cost
−1620   depreciation to date
$6480   book value at the end of the first year

Second year:

$$\text{Depreciation} = 20\% \text{ of book value at end of last year}$$
$$= 20\% \times \$6480$$
$$= \$1296$$

Book value at end of second year is found as follows.

$6480   Book value at the end of year 1
−1296   Year 2 depreciation
$5184   Book value at the end of year 2

QUICK CHECK 2

A pet store purchased a saltwater fish aquarium for $6800. It is expected to have a life of 8 years, with no salvage value. Using the double-declining-balance method of depreciation, find **(a)** the first and second years' depreciation and **(b)** the book value at the end of the first and second years.

OBJECTIVE 4 Use the double-declining-balance method to prepare a depreciation schedule. The next example shows a depreciation schedule for the pickup truck discussed in Example 3 of Section 14.1. As this example shows, the same rate is used each year with the declining-balance method, and the rate is multiplied by the declining balance, which is last year's book value. Also, the amount of depreciation in a given year may have to be adjusted so that book value is never less than salvage value.

Preparing a Depreciation Schedule

EXAMPLE 3 Capital Curb and Concrete bought a new pickup truck at a cost of $21,500. It is estimated the truck will have a useful life of 5 years, at which time it will have a salvage value (trade-in value) of $3500. Prepare a depreciation schedule using the double-declining-balance method of depreciation.

Quick TIP

Never subtract the salvage value from the cost when calculating depreciation using the double-declining-balance method.

SOLUTION

The annual rate of depreciation is 40% **(20% straight-line × 2 = 40%)**. Do not subtract salvage value from cost before calculating depreciation. In year 1, the full cost is used to calculate depreciation.

Year	Computation	Amount of Depreciation	Accumulated Depreciation	Book Value
0	—	—	—	$21,500
1	(40% × $21,500)	$8600	$8,600	$12,900
2	(40% × $12,900)	$5160	$13,760	$7,740
3	(40% × $7,740)	$3096	$16,856	$4,644
4		$1144*	$18,000	$3,500
5		$0	$18,000	$3,500

*In year 4 of the table, 40% of $4644 is $1858. If this amount were subtracted from $4644, the book value would drop below the salvage value of $3500. Since book value never falls below salvage value, depreciation of $1144 ($4644 − $3500) is taken in year 4, so that book value equals salvage value. No further depreciation remains for year 5 or subsequent years.

> **QUICK CHECK 3**
>
> A concrete-extruding machine is purchased at a cost of $4300. The machine has a useful life of 4 years and a salvage value of $700. Prepare a depreciation schedule using the double-declining-balance method of depreciation to find the book value at the end of each year of the asset's life.

The following graph shows the remaining book value by year of the pickup truck in Example 3. It shows them using both the straight-line and double-declining-balance methods of depreciation. Notice that the book value at the end of the asset's useful life is the same (salvage value = $3500) using either method.

Quick TIP

Although total depreciation is the same under straight-line and double-declining-balance, depreciation occurs more rapidly under double-declining-balance depreciation. The result is that book value falls more rapidly under double-declining-balance.

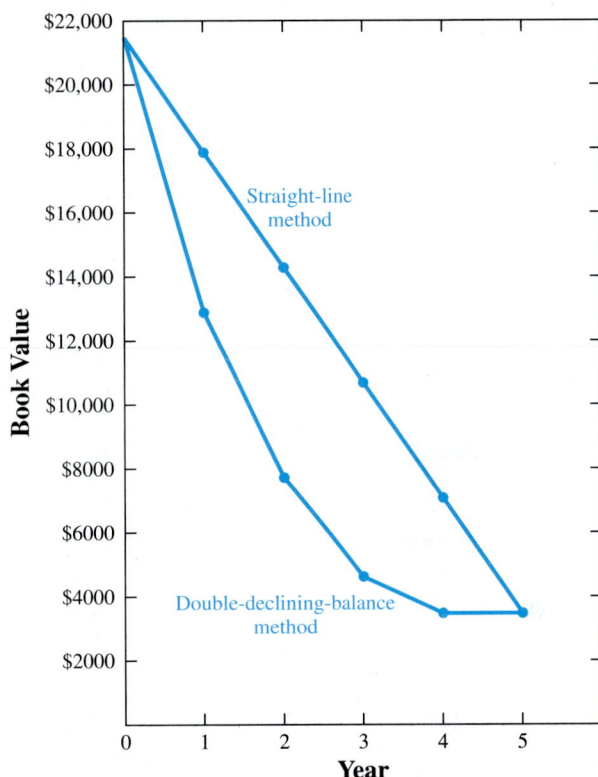

The advantage of using double-declining-balance over straight-line depreciation is that the deduction from income is greater in the early years of the useful life. In turn, this results in lower income tax in the early years. The disadvantage is that the deduction from income is much lower in the last years of the useful life. Managers are careful when choosing the depreciation methods, since they can have a large effect on income and, therefore, a significant effect on income taxes that must be paid.

14.2 Exercises / MyLab Math

The shaded sections below contain solutions to help you get a QUICK START on the various types of exercises.

Find the annual double-declining-balance (200% method) rate of depreciation, given the following estimated lives. (See Example 1.)

	Life	Annual Rate		Life	Annual Rate
1.	5 years	40%	2.	20 years	10%
	20% × 2 = 40%			5% × 2 = 10%	
3.	8 years	_____	4.	25 years	_____
5.	15 years	_____	6.	4 years	_____
7.	10 years	_____	8.	30 years	_____
9.	6 years	_____	10.	40 years	_____
11.	50 years	_____	12.	100 years	_____

Find the first year's depreciation for the following, using the double-declining-balance method of depreciation. (See Example 2.)

13. Cost: $15,000
    Estimated life: 10 years
    Estimated scrap value: $3000
    Depreciation (year 1): **$3000**
    $15,000 × 20% = $3000

14. Cost: $10,800
    Estimated life: 20 years
    Estimated scrap value: None
    Depreciation (year 1): **$1080**
    $10,800 × 10% = $1080

15. Cost: $22,500
    Estimated life: 5 years
    Estimated scrap value: $500
    Depreciation (year 1): _____

16. Cost: $243,000
    Estimated life: 40 years
    Estimated scrap value: $40,000
    Depreciation (year 1): _____

17. Cost: $3800
    Estimated life: 4 years
    Estimated scrap value: None
    Depreciation (year 1): _____

18. Cost: $11,400
    Estimated life: 6 years
    Estimated scrap value: $3500
    Depreciation (year 1): _____

Find the book value at the end of the first year for the following, using the double-declining-balance method of depreciation. Round to the nearest dollar. (See Examples 1 and 2.)

19. Cost: $4200
    Estimated life: 10 years
    Estimated scrap value: $1000
    Book value: **$3360**
    $4200 × 20% = $840
    $4200 − $840 = $3360 book value

20. Cost: $2500
    Estimated life: 6 years
    Estimated scrap value: $400
    Book value: _____

21. Cost: $1620
    Estimated life: 8 years
    Estimated scrap value: None
    Book value: _____

22. Cost: $11,280
    Estimated life: 5 years
    Estimated scrap value: $1600
    Book value: _____

C// indicates an exercise that is related to the Case in Point feature.

Find the book value at the end of 3 years for the following, using the double-declining-balance method of depreciation. Round to the nearest dollar. (See Examples 1 and 2.)

23. Cost: $16,200
Estimated life: 8 years
Estimated scrap value: $1500
Book value: **$6834**

$16,200 × 25% = $4050 dep. year 1
$16,200 − $4050 = $12,150
$12,150 × 25% = $3038 dep. year 2
$12,150 − $3038 = $9112
$9112 × 25% = $2278 dep. year 3
$9112 − $2278 = $6834 book value year 3

24. Cost: $8500
Estimated life: 10 years
Estimated scrap value: $1100
Book value: _____

25. Cost: $6000
Estimated life: 3 years
Estimated scrap value: $750
Book value: _____

26. Cost: $75,000
Estimated life: 50 years
Estimated scrap value: None
Book value: _____

Solve the following application problems.

27. WEIGHT-TRAINING EQUIPMENT Planet Fitness selects the double-declining-balance method of depreciation for some weight-training equipment costing $14,400. If the estimated life of the equipment is 4 years and the salvage value is zero, prepare a depreciation schedule.

Year	Computation	Amount of Depreciation	Accumulated Depreciation	Book Value
0	—	—	—	$14,400
1	(50% × $14,400)	$7200	$7,200	$7,200
2	(50% × $7,200)	$3600	$10,800	$3,600
3	(50% × $3,600)	$1800	$12,600	$1,800
4		$1800*	$14,400	$0

*To depreciate to $0 scrap value

28. SOUND SYSTEM A sound system for physical therapy offices cost $11,760, has a 3-year life, and a scrap value of $1400. Prepare a depreciation schedule using the double-declining-balance method of depreciation.

Year	Computation	Amount of Depreciation	Accumulated Depreciation	Book Value
0	—	—	—	$11,760
1				
2				
3				$1,400

578 CHAPTER 14 Depreciation

29. CONVEYOR SYSTEM Use the double-declining-balance method of depreciation to prepare a depreciation schedule for a new conveyor. Cost = $14,000; estimated life = 5 years; estimated scrap value = $2500. (Round to the nearest dollar.)

Year	Computation	Amount of Depreciation	Accumulated Depreciation	Book Value
0	—	—	—	$14,000
1				
2				
3				
4				
5				$2,500

30. ELECTRONIC ANALYZER Managers at a General Motors dealership decide to use the double-declining-balance method of depreciation on an electronic analyzer that was acquired at a cost of $25,500. If the estimated life of the analyzer is 8 years and the estimated scrap value is $3500, prepare a depreciation schedule. (Round to the nearest dollar.)

Year	Computation	Amount of Depreciation	Accumulated Depreciation	Book Value
0	—	—	—	$25,500
1				
2				
3				
4				
5				
6				
7				
8				$3,500

31. Another name for the double-declining-balance method of depreciation is the 200% method. Explain why the straight-line method of depreciation is often called the 100% method. (See Objective 2.)

32. Explain why the amount of depreciation taken in the last year of an asset's life may be zero when using the double-declining-balance method of depreciation. (See Objective 4.)

33. CARPET-CLEANING EQUIPMENT John Walker, owner of The Carpet Solution, purchased some truck-mounted carpet-cleaning equipment at a cost of $8200. The estimated life of the equipment is 8 years, and the expected salvage value is $1250. Use the double-declining-balance method of depreciation to find the depreciation in the third year.

33. _____

34. COMMERCIAL FISHING BOATS A commercial fishing boat costs $717,000 and has an estimated life of 10 years and a salvage value of $225,000. Find the depreciation in the second year using the double-declining-balance method of depreciation.

34. _____

35. FISHING BOAT Reef Tours purchased a fishing boat for $39,240. It has an estimated life of 5 years with no salvage value. Use the double-declining-balance method of depreciation to find the book value at the end of the third year.

35. _____

36. COMMUNICATION EQUIPMENT Laura Rogers purchased some communication equipment for her public relations firm at a cost of $19,700. She estimates the life of the equipment to be 8 years, at which time the salvage value will be $1000. Use the double-declining-balance method of depreciation to find the book value at the end of 5 years.

36. _____

37. CONSTRUCTION POWER TOOLS Capital Curb and Concrete purchased a portable cement mixer at a cost of $5800. The mixer has a life of 8 years and a scrap value of $1000. Use the double-declining-balance method of depreciation to find **(a)** the annual rate of depreciation, **(b)** the amount of depreciation in the first year, **(c)** the accumulated depreciation at the end of the fifth year, and **(d)** the book value at the end of the fifth year.

(a) _____
(b) _____
(c) _____
(d) _____

38. SCHOOL EQUIPMENT Gale Klein bought some white boards to write on for her reading clinic at a cost of $3620. The estimated life of the white boards is 5 years, with a salvage value of $400. Use the double-declining-balance method of depreciation to find **(a)** the annual rate of depreciation, **(b)** the amount of depreciation in the first year, **(c)** the accumulated depreciation at the end of the third year, and **(d)** the book value at the end of the third year.

(a) _____
(b) _____
(c) _____
(d) _____

QUICK CHECK ANSWERS

1. (a) $2\frac{1}{2}\%$, 5%
   (b) $16\frac{2}{3}\%$, $33\frac{1}{3}\%$
   (c) $6\frac{2}{3}\%$, $13\frac{1}{3}\%$
   (d) 4%, 8%
2. (a) $1700; $1275 (b) $5100; $3825
3. year 0 = $4300
   year 1 = $2150
   year 2 = $1075
   year 3 = $700
   year 4 = $700

14.3 Sum-of-the-Years'-Digits Method

OBJECTIVES

1. Understand the sum-of-the-years'-digits method of depreciation.
2. Find the depreciation fraction for the sum-of-the-years'-digits method.
3. Use the sum-of-the-years'-digits method to find the amount of depreciation for each year.
4. Prepare a depreciation schedule for the sum-of-the-years'-digits method.

OBJECTIVE 1 Understand the sum-of-the-years'-digits method of depreciation. The sum-of-the-years'-digits method of depreciation is another accelerated depreciation method. The double-declining-balance method of depreciation produces more depreciation than the straight-line method in the early years of an asset's life and less depreciation in the later years. The **sum-of-the-years'-digits method**, however, produces results in between the straight-line and the double-declining-balance methods—more than straight-line at the beginning and more than double-declining-balance at the end.

OBJECTIVE 2 Find the depreciation fraction for the sum-of-the-years'-digits method. The use of the sum-of-the-years'-digits method requires a **depreciation fraction** instead of the depreciation rate used earlier. The annual depreciation is this depreciation fraction multiplied by the depreciable amount (cost minus salvage value). The depreciation fraction decreases annually, as does the depreciation.

To find the depreciation fraction, first find the denominator, which remains constant for every year of the life of the asset. The denominator is the sum of all the years of the estimated useful life of the asset (sum of the years' digits). For example, if the life is 6 years, the denominator is 21, **since $1 + 2 + 3 + 4 + 5 + 6 = 21$**. The numerator of the fraction, which decreases each year, gives the number of years of life remaining at the beginning of that year.

Finding the Depreciation Fraction

EXAMPLE 1 Find the depreciation fraction for each year if the sum-of-the-years'-digits method of depreciation is used for an asset with a useful life of 6 years.

SOLUTION

First determine the denominator of the depreciation fraction. The denominator is 21 $(1 + 2 + 3 + 4 + 5 + 6 = 21)$. Next determine the numerator for each year. The number of years of life remaining at the beginning of any year is the numerator.

Quick TIP
It is common not to write these fractions in lowest terms, so that the year in question can be seen.

Year	Depreciation Fraction	
1	$\frac{6}{21}$	← 6 is the numerator in Year 1 since there are 6 years remaining in life of asset.
2	$\frac{5}{21}$	
3	$\frac{4}{21}$	
4	$\frac{3}{21}$	
5	$\frac{2}{21}$	
6	$\frac{1}{21}$	
21 ← sum of the years' digits	$\frac{21}{21}$	

Under the sum-of-the-years'-digits method, an asset having a life of 6 years is assumed to lose $\frac{6}{21}$ of its value the first year, $\frac{5}{21}$ the second year, and so on. The sum of the six fractions in the table is $\frac{21}{21}$, or 1, so that the entire depreciable amount is used over the 6-year life.

QUICK CHECK 1

Find the depreciation fraction for each year if the sum-of-the-years'-digits method of depreciation is to be used for an asset with a useful life of 10 years.

A quick method of finding the denominator of the sum of the years' digits is by the formula

$$\frac{n(n+1)}{2}$$

where n is the estimated life of the asset. Here are some examples of how to find the denominator.

Useful Life	$\frac{n(n+1)}{2}$	Denominator of Fraction
6	$\frac{6(6+1)}{2}$	$= 21$
8	$\frac{8(8+1)}{2}$	$= 36$
10	$\frac{10(10+1)}{2}$	$= 55$

OBJECTIVE 3 Use the sum-of-the-years'-digits method to find the amount of depreciation for each year. Use the following formula and multiply the depreciation fraction in any year by the depreciable amount to calculate the amount of depreciation in that year.

> **Finding Annual Depreciation**
>
> Depreciation = Depreciation fraction × Depreciable amount

Finding Depreciation Using the Sum-of-the-Years'-Digits Method

EXAMPLE 2 Capital Curb and Concrete purchases a DitchMaster trencher at a cost of $8940. The trencher has a useful life of 8 years and an estimated salvage value of $1200. Find the first and second years' depreciation using the sum-of-the-years'-digits method.

SOLUTION

The depreciation fraction has a denominator of 36 (or $1 + 2 + 3 + 4 + 5 + 6 + 7 + 8$). The numerator in the first year is 8. The first-year fraction, $\frac{8}{36}$, is multiplied by the amount to be depreciated, **$7740** ($8940 cost − $1200 salvage value).

$$\frac{8}{36} \times \$7740 = \$1720$$

The first year's depreciation is $1720.
 The depreciation fraction for the second year, $\frac{7}{36}$, is multiplied by the original depreciable amount of $7740.

$$\frac{7}{36} \times \$7740 = \$1505$$

The second year's depreciation is $1505.

> **QUICK CHECK 2**
>
> A dry cleaning plant purchases new cleaning equipment at a cost of $28,600. The equipment has a life of 6 years and a salvage value of $4600. Find the first and second years' depreciation using the sum-of-the-years'-digits method.

OBJECTIVE 4 Prepare a depreciation schedule for the sum-of-the-years'-digits method. For comparison, the next example uses the same truck as in Sections 14.1 and 14.2.

582 CHAPTER 14 Depreciation

Preparing a Depreciation Schedule

EXAMPLE 3 Capital Curb and Concrete bought a new pickup truck for $21,500. The truck is estimated to have a useful life of 5 years, at which time it will have a salvage value of $3500. Prepare a depreciation schedule using the sum-of-the-years'-digits method of depreciation.

SOLUTION

The depreciation fraction has a denominator of 15 (or $1 + 2 + 3 + 4 + 5$). The depreciable amount is $21,500 - $3500 = $18,000.

Year	Computation	Amount of Depreciation	Accumulated Depreciation	Book Value
0	—	—	—	$21,500
1	$\left(\frac{5}{15} \times \$18{,}000\right)$	$6000	$6,000	$15,500
2	$\left(\frac{4}{15} \times \$18{,}000\right)$	$4800	$10,800	$10,700
3	$\left(\frac{3}{15} \times \$18{,}000\right)$	$3600	$14,400	$7,100
4	$\left(\frac{2}{15} \times \$18{,}000\right)$	$2400	$16,800	$4,700
5	$\left(\frac{1}{15} \times \$18{,}000\right)$	$1200	$18,000	$3,500

Quick TIP

The straight-line and sum-of-the-years'-digits methods require that salvage value be subtracted from cost to find depreciable amount. The declining-balance method does not.

QUICK CHECK 3

A concrete-extruding machine is purchased at a cost of $4300. The machine has a useful life of 4 years and a salvage value of $700. Prepare a depreciation schedule using the sum-of-the-years'-digits method of depreciation to find the book value at the end of each year of the asset's life.

The following graph shows the book value at the end of each year for the truck in Example 3 using straight-line, double-declining-balance, and sum-of-the-years'-digits methods of depreciation. Notice that cost, useful life, salvage value, and book value at the end of useful life are exactly the same under all three methods.

In fact, book value at the end of the useful life is always the same as salvage value. Annual depreciation using the sum-of-the-years'-digits method falls between that for the straight-line and double-declining-balance methods.

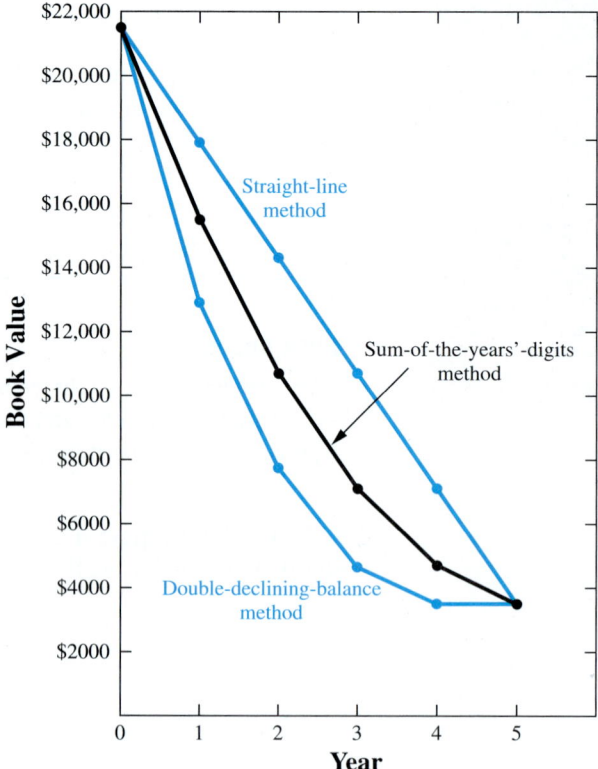

The amount of depreciation in any one year on one depreciable asset may not be large. However, depreciation on the many different depreciable assets owned by a company can amount to quite a bit. So the differences between depreciation methods can make a significant difference on income taxes.

14.3 Exercises // MyLab Math

The shaded sections below contain solutions to help you get a QUICK START on the various types of exercises.

Find the sum-of-the-years'-digits depreciation fraction for the first year given the following estimated lives. (See Example 1.)

	Life	First-Year Fraction		Life	First-Year Fraction
1.	4 years	$\frac{4}{10}$	**2.**	3 years	$\frac{3}{6}$
	$4 + 3 + 2 + 1 = 10; \frac{4}{10}$			$3 + 2 + 1 = 6; \frac{3}{6}$	
3.	6 years	_____	**4.**	5 years	_____
5.	7 years	_____	**6.**	8 years	_____
7.	10 years	_____	**8.**	20 years	_____

Find the first year's depreciation for the following, using the sum-of-the-years'-digits method of depreciation. Round to the nearest dollar. (See Example 2.)

9. Cost: $4800
Estimated life: 4 years
Estimated scrap value: $700
Depreciation (year 1): **$1640**
$4800 − $700 = $4100
$4100 × $\frac{4}{10}$ = $1640

10. Cost: $5600
Estimated life: 5 years
Estimated scrap value: $800
Depreciation (year 1): **$1600**
$5600 − $800 = $4800
$4800 × $\frac{5}{15}$ = $1600

11. Cost: $60,000
Estimated life: 10 years
Estimated scrap value: $5000
Depreciation (year 1): _____

12. Cost: $1440
Estimated life: 8 years
Estimated scrap value: None
Depreciation (year 1): _____

13. Cost: $18,500
Estimated life: 3 years
Estimated scrap value: $3500
Depreciation (year 1): _____

14. Cost: $97,400
Estimated life: 8 years
Estimated scrap value: $11,000
Depreciation (year 1): _____

Find the book value at the end of the first year for the following, using the sum-of-the-years'-digits method of depreciation. Round to the nearest dollar. (See Example 3.)

15. Cost: $9500
Estimated life: 8 years
Estimated scrap value: $1400
Book value: **$7700**
$8100 × $\frac{8}{36}$ = $1800
$9500 − $1800 = $7700

16. Cost: $14,800
Estimated life: 10 years
Estimated scrap value: None
Book value: **$12,109**
$14,800 × $\frac{10}{55}$ = $2691
$14,800 − $2691 = $12,109

C// indicates an exercise that is related to the Case in Point feature.

17. Cost: $3800
Estimated life: 5 years
Estimated scrap value: $500
Book value: _____

18. Cost: $15,650
Estimated life: 6 years
Estimated scrap value: $2000
Book value: _____

Find the book value at the end of 3 years for the following, using the sum-of-the-years'-digits method of depreciation. Round to the nearest dollar. (See Example 3.)

19. Cost: $2240
Estimated life: 6 years
Estimated scrap value: $350
Book value: **$890**

$\frac{6}{21} + \frac{5}{21} + \frac{4}{21} = \frac{15}{21}$ dep. in 3 years
$2240 - \$350 = \$1890$
$1890 \times \frac{15}{21} = \$1350$ dep.
$2240 - \$1350 = \$890$

20. Cost: $27,500
Estimated life: 10 years
Estimated scrap value: None
Book value: _____

21. Cost: $4500
Estimated life: 8 years
Estimated scrap value: $900
Book value: _____

22. Cost: $6600
Estimated life: 5 years
Estimated scrap value: $1500
Book value: _____

Solve the following application problems. Round to the nearest dollar.

23. DEPRECIATING OFFICE EQUIPMENT Capital Curb and Concrete uses the sum-of-the-years'-digits method of depreciation to prepare a depreciation schedule for an engineering workstation that costs $3900, has an expected life of 3 years, and has an estimated salvage value of $480.

Year	Computation	Amount of Depreciation	Accumulated Depreciation	Book Value
0	—	—	—	$3900
1	$\left(\frac{3}{6} \times \$3420\right)$	$1710	$1710	$2190
2	$\left(\frac{2}{6} \times \$3420\right)$	$1140	$2850	$1050
3	$\left(\frac{1}{6} \times \$3420\right)$	$570	$3420	$480

24. FACTORY Gliddens purchased a new tool for a paint mixer at a cost of $14,400. The expected life of the mixer is 4 years, at which time the salvage value is estimated to be $2400. Complete a depreciation schedule using the sum-of-the-years'-digits method of depreciation.

Year	Computation	Amount of Depreciation	Accumulated Depreciation	Book Value
0	—	—	—	$14,400
1				
2				
3				
4				

25. **COMMERCIAL FREEZER** Joe's Pizzeria purchased a new freezer at a cost of $10,800. The estimated life of the freezer is 6 years, at which time the salvage value is estimated to be $2400. Complete a depreciation schedule using the sum-of-the-years'-digits method of depreciation.

Year	Computation	Amount of Depreciation	Accumulated Depreciation	Book Value
0	—	—	—	$10,800
1				
2				
3				
4				
5				
6				

26. **FORKLIFT DEPRECIATION** Capital Curb and Concrete purchased a new forklift. Prepare a depreciation schedule using the sum-of-the-years'-digits method of depreciation. Cost = $15,000; estimated life = 10 years; estimated scrap value = $4000.

Year	Computation	Amount of Depreciation	Accumulated Depreciation	Book Value
0	—	—	—	$15,000
1				
2				
3				
4				
5				
6				
7				
8				
9				
10				

27. Write a description of how the depreciation fraction is determined in any year of an asset's life when using the sum-of-the-years'-digits method of depreciation. (See Objective 2.)

28. Discuss the differences between the three depreciation methods covered in this chapter. (See Sections 14.1, 14.2, and 14.3.)

29. **WIND TURBINE** Vulcan Utilities purchased several large wind turbines to generate electricity from the wind. Each turbine costs $3,200,000, has an estimated life of 20 years, and has an estimated scrap value of $300,000. Find the depreciation for the third year.

29. _____

CHAPTER 14 Depreciation

30. LANDSCAPE EQUIPMENT Capital Curb and Concrete purchased a new trailer at a cost of $32,000. The expected life of the unit is 8 years and the salvage value is expected to be $5000. Use the sum-of-the-years'-digits method of depreciation to determine the first year's depreciation.

30. _____

31. HOSPITAL EQUIPMENT Medical Equipment, Inc. uses the sum-of-the-years'-digits method of depreciation on all hospital rental equipment. If it purchases new hospital beds at a cost of $12,800 and estimates the life of the beds to be 10 years with no scrap value, find the book value at the end of the fourth year.

31. _____

32. COMMERCIAL TILE First Savings Bank has installed commercial tile at a cost of $25,200. It has a useful life of 6 years and no salvage value. Find **(a)** the first and **(b)** the second year's depreciation, using the sum-of-the-years'-digits method.

(a) _____
(b) _____

33. FAST FOOD RESTAURANTS In-N-Out Burgers purchased a new low-fat deep-fry unit at a cost of $12,420. The expected life of the unit is 8 years with a scrap value of $1800. Use the sum-of-the-years'-digits method of depreciation to find **(a)** the first year's depreciation fraction, **(b)** the amount of depreciation in the first year, **(c)** the accumulated depreciation at the end of the eighth year, and **(d)** the book value at the end of the fourth year.

(a) _____
(b) _____
(c) _____
(d) _____

34. SEWER DRAIN SERVICE Armour Drain bought a new sewer line root remover for $6725. The life of the machine is 10 years, and the scrap value is $1500. Use the sum-of-the-years'-digits method of depreciation to find **(a)** the first year's depreciation fraction, **(b)** the amount of depreciation in the first year, **(c)** the accumulated depreciation at the end of the tenth year, and **(d)** the book value at the end of the sixth year.

(a) _____
(b) _____
(c) _____
(d) _____

QUICK CHECK ANSWERS

1. year 1 = $\frac{10}{55}$; year 2 = $\frac{9}{55}$
   year 3 = $\frac{8}{55}$; year 4 = $\frac{7}{55}$
   year 5 = $\frac{6}{55}$; year 6 = $\frac{5}{55}$
   year 7 = $\frac{4}{55}$; year 8 = $\frac{3}{55}$
   year 9 = $\frac{2}{55}$; year 10 = $\frac{1}{55}$
2. year 1 = $6857 (rounded)
   year 2 = $5714 (rounded)
3. year 0 = $4300
   year 1 = $2860
   year 2 = $1780
   year 3 = $1060
   year 4 = $ 700

Supplementary Application Exercises on Depreciation

Round to the nearest dollar if necessary.

1. **DISTRIBUTION CENTER** Xienshang Electronics built a new distribution center in California for $2,600,000. It has an estimated life of 40 years and an estimated scrap value of $400,000. Use the straight-line method of depreciation to find the book value of the distribution center at the end of 10 years.

2. **COPY MACHINES** The UPS Store® purchased a new high-speed copy machine for $9480. Use the straight-line method of depreciation to find the amount of depreciation that should be charged off each year if the equipment has an estimated life of 4 years and a scrap value of $1500.

3. **WOODWORKING** Custom Cabinets bought a lathe for turning table and chair legs at a total cost of $18,500. The estimated life of the lathe is 5 years and there is no scrap value. Find the depreciation in the first year using the double-declining-balance method of depreciation.

4. **LANDSCAPING EQUIPMENT** Capital Curb and Concrete purchased a scraper at a cost of $22,000. If the estimated life of the scraper is 10 years, find the book value at the end of 3 years using the double-declining-balance method of depreciation.

5. **ENGRAVING** Joan Benson bought an engraving machine for her business for $2850. It has an estimated life of 6 years and salvage value of $600. Use the sum-of-the-years'-digits method of depreciation to find the book value of all the iPhones at the end of the third year.

588 CHAPTER 14 Depreciation

6. **BUSINESS SIGNAGE** The outdoor sign used by the Donut Palace cost $7375, has an estimated life of 10 years, and has a salvage value of $500. Use the sum-of-the-years'-digits method of depreciation to find the book value at the end of the third year.

6. _____

7. **JEWELRY CASES** The Diamond Center installed new shatterproof display cases at a cost of $45,600. Using the straight-line method of depreciation, find the amount of depreciation that should be charged off *each year* if the estimated life of the display cases is 10 years and the scrap value is $8000.

7. _____

8. **RECREATION EQUIPMENT** River Rentals purchased a sculling boat at a cost of $32,000. The estimated life of the boat is 15 years, and the scrap value is $6500. Find the book value at the end of 10 years using the straight-line method of depreciation.

8. _____

9. **BUSINESS SAFE** Alton Credit Union purchased a safe at a cost of $78,000. If the estimated life of the safe is 20 years, find the book value at the end of 2 years using the double-declining-balance method of depreciation.

9. _____

10. **THEATER SEATING** A movie theater just installed new seating at a total cost of $228,000. The estimated life of the seating is 10 years, there is no salvage value, and the double-declining-balance method of depreciation is used. Find the depreciation in the first year.

10. _____

11. **SHUTTLE VAN** Airport Parking, Inc. purchased a shuttle van for $38,600. It has an estimated life of 4 years and a scrap value of $4400. Use the sum-of-the-years'-digits method of depreciation to find the annual depreciation for each year.

11. _____
_____
_____
_____

12. **DIESEL TRACTOR** Using the sum-of-the-years'-digits method of depreciation, find the amount of depreciation to be charged off each year on a diesel tractor purchased by Capital Curb and Concrete. The tractor has a cost of $85,000, an estimated life of 5 years, and a scrap value of $13,000.

12. _____
    _____
    _____
    _____
    _____

13. **INDUSTRIAL FORKLIFT** Lumber Plus buys 5 industrial lifts at a cost of $14,825 each. The life of the lifts is estimated to be 10 years and the scrap value is $3000 each. Use the sum-of-the-years'-digits method of depreciation to find the total book value of all the lifts at the end of the fourth year.

13. _____

14. **CAR-WASH MACHINERY** Bubble Car Wash buys new pumps for an automatic car wash at a cost of $50,950. It has a scrap value of $10,000 and an estimated life of 6 years. Use the sum-of-the-years'-digits method of depreciation to find the book value at the end of the third year.

14. _____

15. **VIDEO EQUIPMENT** Video Productions purchased some studio equipment manufactured in Australia. The total cost of the equipment was $21,600. It has an estimated life of 5 years and a salvage value of $2400. Use the sum-of-the-years'-digits method of depreciation to find the book value of the equipment at the end of the second year.

15. _____

CHAPTER 14 Depreciation

16. **RESTAURANT TABLES** Little Italy bought new dining-room tables at a cost of $14,750. The estimated life of the tables is 8 years, at which time they will be worthless. Use the double-declining-balance method of depreciation to find the book value at the end of the third year.

16. _____

17. **REFRIGERATED DISPLAY CASE** Goldie's Delicatessen purchased a new refrigerated display case for $10,800. It has an estimated life of 10 years and a salvage value of $1500. Use the straight-line method of depreciation to find **(a)** the accumulated depreciation and **(b)** the book value at the end of the sixth year.

(a) _____
(b) _____

18. **SOFT-DRINK BOTTLING** A small soft-drink bottler purchased two automatic filling and capping machines at a cost of $185,000 each. The machines have a scrap value of $30,000 each and an estimated life of 8 years. Use the double-declining-balance method of depreciation to find **(a)** the accumulated depreciation and **(b)** the book value of the machines at the end of the third year.

(a) _____
(b) _____

14.4 Units-of-Production Method

OBJECTIVES

1. Describe the units-of-production method of depreciation.
2. Use the units-of-production method to find the depreciation per unit.
3. Calculate the annual depreciation using the units-of-production method.
4. Prepare a depreciation schedule using the units-of-production method.

CASE IN POINT Capital Curb and Concrete owns a stump chipper that is used to remove tree stumps when preparing for concrete work. The company's accountant suggests that the stump chipper be depreciated based on the number of hours used, rather than the number of years owned.

OBJECTIVE 1 Describe the units-of-production method of depreciation. An asset often has a useful life given in terms of **units of production**, such as hours or miles of service. For example, an airplane or truck may have a useful life given as hours of flying time or miles driven. A steel press or stamping machine may have a useful life given as the total number of units that it can produce. For these assets, the units-of-production method often is used.

The units-of-production depreciation method results in a constant amount of depreciation per unit. So, annual depreciation using units-of production depreciation will vary depending on number of units for the year.

OBJECTIVE 2 Use the units-of-production method to find the depreciation per unit. Find the depreciation per unit with the following formula.

Finding Depreciation per Unit

$$\text{Depreciation per unit} = \frac{\text{Depreciable amount}}{\text{Units of life}}$$

For example, suppose the stump chipper owned by Capital Curb and Concrete costs $7500, has a salvage value of $1500, and is expected to operate 700 hours. First, find the depreciable amount.

$$\begin{array}{rl} \$7500 & \text{cost} \\ -\ 1500 & \text{salvage value} \\ \hline \$6000 & \text{depreciable amount} \end{array}$$

OBJECTIVE 3 Calculate the annual depreciation using the units-of-production method. Then find the allowable depreciation per hour of use.

$$\frac{\$6000 \text{ depreciable amount}}{700 \text{ hours of life}} = \$8.57 \text{ (rounded) depreciation per hour}$$

Use the following formula and multiply the number of hours used during the year by the depreciation per unit to find the annual depreciation.

Finding Annual Depreciation

$$\text{Depreciation} = \text{Number of units (hours)} \times \text{Depreciation per unit (hour)}$$

Quick TIP
Depreciation is limited to the extent that it can never result in a book value that is less than the salvage value.

The annual depreciation for a year in which the stump chipper was used for 280 hours follows.

$$280 \times \$8.57 = \$2399.60 \text{ depreciation}$$

Heavy-Truck Makers Find the Good Times Are Rolling

The newspaper headline suggests that the big 18-wheelers that run day and night on the highways are busy. The rigs cost $85,000 or more. The useful life depends on the number of miles in the life of a truck, which can range up to 1,000,000 miles. Since many firms have a lot of money tied up in the rigs, managers want rapid depreciation to reduce income taxes. These trucks are sometimes operated by a team of two, who alternate driving and together drive over 100,000 miles in a year.

Using Units-of-Production Depreciation

EXAMPLE 1 American Trucking purchased a new Kenworth truck for $95,000. The truck has a salvage value of $15,000 and an estimated life of 500,000 miles. Find the depreciation for a year in which the truck is driven 128,000 miles.

SOLUTION

First find the depreciable amount.

$$\begin{array}{rl} \$95,000 & \text{cost} \\ -\ 15,000 & \text{scrap value} \\ \hline \$80,000 & \text{depreciable amount} \end{array}$$

Next find the depreciation per unit.

$$\frac{\$80,000 \text{ depreciable amount}}{500,000 \text{ miles of life}} = \$.16 \text{ depreciation per mile}$$

Multiply to find the depreciation for the year.

$$128,000 \text{ miles} \times \$.16 = \$20,480 \text{ depreciation for the year}$$

Quick TIP
The $.16 depreciation per mile is an estimate of the wear on the truck per mile driven. Traveling salespersons use this method to estimate wear on their work vehicles.

QUICK CHECK 1

Grayson Medical Clinic purchased an emergency generator at a cost of $19,600. The generator has a salvage value of $2500 and an estimated life of 7500 hours. Find the depreciation for a year in which the generator operated 840 hours.

Preparing a Depreciation Schedule

EXAMPLE 2 Bagel Boys purchased a packaging machine at a cost of $52,300. It has an estimated salvage value of $4000 and an expected life of 690,000 units. Prepare a depreciation schedule using the units-of-production method of depreciation. Use the following packaging schedule.

Year 1	240,000 units
Year 2	150,000 units
Year 3	90,000 units
Year 4	120,000 units
Year 5	90,000 units

SOLUTION

The depreciable amount is $48,300 ($52,300 − $4000). The depreciation per unit is

$$\frac{\$48,300}{690,000 \text{ units}} = \$.07 \text{ per unit}$$

The annual depreciation is found by multiplying the number of units packaged each year by the depreciation per unit.

		Annual Depreciation
Year 1	240,000 units × $.07 =	$16,800
Year 2	150,000 units × $.07 =	$10,500
Year 3	90,000 units × $.07 =	$6,300
Year 4	120,000 units × $.07 =	$8,400
Year 5	90,000 units × $.07 =	$6,300
Total	690,000 units	$48,300

depreciable amount

OBJECTIVE 4 Prepare a depreciation schedule using the units-of-production method. The results above are used in the preparation of the depreciation schedule.

Year	Computation	Depreciation	Accumulated Depreciation	Book Value
0	—	—	—	$52,300
1	(240,000 × $.07)	$16,800	$16,800	$35,500
2	(150,000 × $.07)	$10,500	$27,300	$25,000
3	(90,000 × $.07)	$6,300	$33,600	$18,700
4	(120,000 × $.07)	$8,400	$42,000	$10,300
5	(90,000 × $.07)	$6,300	$48,300	$4,000

The book value at the end of year 5 is the salvage value of $4000 since all the units of production (boxes packaged in this case) have been used.

QUICK CHECK 2

Assume that the cost of the machine in Example 2 was $58,200 and the salvage value was $3000. Use the units-of-production method of depreciation and the same production schedule to prepare a depreciation schedule for the machine. Show the book value for each year.

The newspaper clipping shows additional interest and funding being given to the study of entrepreneurship. Some community colleges offer courses in entrepreneurship that prepare a student to start a business. Most businesses use depreciation.

Entrepreneur Course Becoming Popular

More students are taking courses in entrepreneurship at local colleges. In the past, universities have prepared students to work for large corporations, but today many jobs are found in small companies and startups which value entrepreneurial skills such as preparing a business plan. The best courses in entrepreneurship also focus on innovation or new products and ways of doing things.

14.4 Exercises // MyLab Math

The shaded sections below contain solutions to help you get a QUICK START on the various types of exercises.

Find the depreciation per unit in the following. Round to the nearest thousandth of a dollar. (See Example 1.)

	Cost	Salvage Value	Estimated Life	Depreciation per Unit
1.	$16,800	$1800	20,000 units	$.75
	$16,800 − $1800 = $15,000; $15,000 ÷ 20,000 = $.75			
2.	$22,500	$1500	60,000 units	_____
3.	$3750	$250	120,000 units	_____
4.	$7500	$500	15,000 miles	_____
5.	$37,500	$7500	125,000 miles	_____

CHAPTER 14 Depreciation

6.	$300,000	$25,000	4000 hours	_____
7.	$175,000	$25,000	5000 hours	_____
8.	$125,000	$20,000	500,000 miles	_____

Find the amount of depreciation in each of the following. (See Example 1.)

	Depreciation per Unit	Units Produced	Amount of Depreciation
9.	$.46	55,000	$25,300
	55,000 × $.46 = $25,300		
10.	$.18	275,000	$49,500
	275,000 × $.18 = $49,500		
11.	$.54	32,000	_____
12.	$.73	16,500	_____
13.	$.185	15,000	_____
14.	$.032	73,000	_____
15.	$.14	22,200	_____
16.	$.075	110,000	_____

17. Describe the types of situations in which the units-of-production method of depreciation is most applicable.

18. Create your own example to demonstrate how the annual depreciation amount is found using the units-of-production method of depreciation. (See Objective 3.)

C// indicates an exercise that is related to the Case in Point feature.

Solve the following application problems. Round to the nearest dollar.

19. **DEEP FRYER** Michael's Diner purchased a new grilling station at a cost of $6800. The expected life is 5000 hours of production, at which time it will have a salvage value of $500. Using the units-of-production method, prepare a depreciation schedule given the following production: year 1: 1350 hours; year 2: 1820 hours; year 3: 730 hours; year 4: 1100 hours.

Year	Computation	Amount of Depreciation	Accumulated Depreciation	Book Value
0	—	—	—	$6800
1	(1350 × $1.26)	$1701	$1701	$5099
2	(1820 × $1.26)	$2293	$3994	$2806
3	(730 × $1.26)	$920	$4914	$1886
4	(1100 × $1.26)	$1386	$6300	$500

20. **HEAVY-DUTY TRUCK** Capital Curb and Concrete purchased a Kenworth truck at a cost of $87,000. It estimates that the truck will have a life of 300,000 miles and a salvage value of $15,000. Use the units-of-production method to prepare a depreciation schedule given the following production: year 1: 108,000 miles; year 2: 75,000 miles; year 3: 117,000 miles.

Year	Computation	Amount of Depreciation	Accumulated Depreciation	Book Value
0	—	—	—	$87,000

QUICK CHECK ANSWERS

1. $1915
2. year 1 = $39,000
   year 2 = $27,000
   year 3 = $19,800
   year 4 = $10,200
   year 5 = $ 3,000

14.5 Modified Accelerated Cost Recovery System

OBJECTIVES

1. Understand the modified accelerated cost recovery system (MACRS).
2. Determine the recovery period of different types of property.
3. Find the depreciation rate, given the recovery period and recovery year.
4. Use the MACRS to find the amount of depreciation.
5. Prepare a depreciation schedule using the MACRS.

CASE IN POINT Several different depreciation methods are used by Capital Curb and Concrete for financial statement and income tax purposes. However, the modified accelerated cost recovery system (MACRS) must be used for federal income tax purposes. This means that every depreciable asset owned by the company must have a MACRS depreciation schedule.

OBJECTIVE 1 Understand the modified accelerated cost recovery system (MACRS). A depreciation method known as the **accelerated cost recovery system (ACRS)** originated as part of the Economic Recovery Tax Act of 1981. It was modified by several tax acts and now applies to all property placed in service after 1986. The new method is known as the **modified accelerated cost recovery system (MACRS)**. The result is that there are now three systems for computing depreciation for *federal tax purposes*.

Federal Tax Depreciation Methods

1. The MACRS method of depreciation is used for all property placed in service after 1986.
2. The ACRS method of depreciation will continue to be used for all property placed in service from 1981 through 1986.
3. The straight-line, declining-balance, and sum-of-the-years'-digits methods continue to be used for property placed in service before 1981.

The MACRS system is designed for tax purposes (it is sometimes called the **income-tax method**), and businesses often use some alternative method of depreciation for financial accounting purposes. Some states do not allow the modified accelerated cost recovery system of depreciation for finding state income tax liability. This means that businesses must use the *MACRS* on the *federal tax return* and one of the other methods on those state tax returns.

OBJECTIVE 2 Determine the recovery period of different types of property. Under the modified accelerated cost recovery system, assets are placed in one of nine **recovery classes**, depending on whether the law assumes a 3-, 5-, 7-, 10-, 15-, 20-, 27.5-, 31.5-, or 39-year life for the asset. The following table gives brief descriptions of the types of items belonging to each recovery class. To help you interpret the table: a property with a recovery period of three years falls in the 3-year property recovery class.

Department of the Treasury
Internal Revenue Service

Publication 946
Cat. No. 13081F

How To Depreciate Property

MACRS Recovery Classes

3-year property	Tractor units for use over-the-road, any racehorse that is over 2 years old, any other horse that is over 12 years old, and qualified rent-to-own property
5-year property	Automobiles, taxis, trucks, buses, computers and peripheral equipment, office machinery, copiers, research equipment, breeding cattle, and dairy cattle
7-year property	Office furniture and fixtures (desks, files, safes), and any property not designated by law to be in any other class
10-year property	Vessels, barges, tugs, and similar water transportation equipment
15-year property	Improvements made directly to land, such as shrubbery, fences, roads, bridges, and any single-purpose agricultural or horticultural structure, and any tree or vine bearing fruits or nuts
20-year property	Certain farm buildings such as a storage shed
27.5-year property	Residential rental real estate such as rental houses, apartments, and mobile homes
31.5-year property	Nonresidential rental real estate such as office building, stores, and warehouses if placed in service before May 13, 1993
39-year property	Nonresidential property placed in service after May 12, 1993

14.5 Modified Accelerated Cost Recovery System

Finding the Recovery Period for Property

EXAMPLE 1 Capital Curb and Concrete owns the following assets. Determine the recovery period for each of them.

(a) computer equipment (b) an industrial warehouse (built after May 12, 1993)
(c) a pickup truck (d) office furniture (e) a farm building (storage shed)

CASE IN POINT

SOLUTION

Use the MACRS recovery class list above.

(a) 5 years (b) 39 years (c) 5 years (d) 7 years (e) 20 years

---

QUICK CHECK 1

Determine the recovery period for each of the following assets using the MACRS recovery classes.

(a) a drilling barge (b) a taxi cab (c) office landscaping (d) an apartment building

---

OBJECTIVE 3 Find the depreciation rate, given the recovery period and recovery year. Salvage value is ignored under MACRS, so depreciation is based on the entire cost of the asset. The depreciation rates used by MACRS are determined by applying various versions of double-declining-balance and straight-line depreciation methods to the different classes. The Internal Revenue Service provides these rates in a table similar to the "MACRS Depreciation Rates" table on the next page. To determine the rate of depreciation for any year of life, find the recovery year in the left-hand column and then read across to the allowable recovery period.

Notice in the table that the number of recovery years is one greater than the class life of the property. This is because only a half-year of depreciation is allowed for the first year the property is placed in service, regardless of when the property is placed in service during the year. This is known as the **half-year convention** and is used by most taxpayers.

Finding the Rate of Depreciation with MACRS

EXAMPLE 2 Find the rate of depreciation given the following recovery class and recovery year (year of depreciation within that recovery class).

	(a)	(b)	(c)	(d)
Recovery Class	3 years	10 years	5 years	27.5 years
Recovery Year	3	4	2	12

SOLUTION

(a) 14.81% (b) 11.52% (c) 32.00% (d) 3.637%

QUICK CHECK 2

Find the rate of depreciation given the following recovery class and recovery year.

	(a)	(b)	(c)	(d)
Recovery Class	7 years	20 years	31.5 years	15 years
Recovery Year	4	2	20	7

MACRS Depreciation Rates
Applicable Percent for the Class of Property

Recovery Year	3-Year	5-Year	7-Year	10-Year	15-Year	20-Year	27.5-Year	31.5-Year	39-Year
1	33.33	20.00	14.29	10.00	5.00	3.750	3.485	3.042	2.461
2	44.45	32.00	24.49	18.00	9.50	7.219	3.636	3.175	2.564
3	14.81	19.20	17.49	14.40	8.55	6.677	3.636	3.175	2.564
4	7.41	11.52	12.49	11.52	7.70	6.177	3.636	3.175	2.564
5		11.52	8.93	9.22	6.93	5.713	3.636	3.175	2.564
6		5.76	8.92	7.37	6.23	5.285	3.636	3.175	2.564
7			8.93	6.55	5.90	4.888	3.636	3.175	2.564
8			4.46	6.55	5.90	4.522	3.636	3.175	2.564
9				6.56	5.91	4.462	3.636	3.174	2.564
10				6.55	5.90	4.461	3.637	3.175	2.564
11				3.28	5.91	4.462	3.636	3.174	2.564
12					5.90	4.461	3.637	3.175	2.564
13					5.91	4.462	3.636	3.174	2.564
14					5.90	4.461	3.637	3.175	2.564
15					5.91	4.462	3.636	3.174	2.564
16					2.95	4.461	3.637	3.175	2.564
17						4.462	3.636	3.174	2.564
18						4.461	3.637	3.175	2.564
19						4.462	3.636	3.174	2.564
20						4.461	3.637	3.175	2.564
21						2.231	3.636	3.174	2.564
22							3.637	3.175	2.564
23							3.636	3.174	2.564
24							3.637	3.175	2.564
25							3.636	3.174	2.564
26							3.637	3.175	2.564
27							3.636	3.174	2.564
28							1.97	3.175	2.564
29								3.174	2.564
30								3.175	2.564
31								3.174	2.564
32								1.720	2.564
33–39									2.564
40									0.107

OBJECTIVE 4 Use the MACRS to find the amount of depreciation. MACRS is the method of depreciation that must be used on federal income taxes for property placed in service after 1986. There are three important points to remember.

1. Do not use salvage value.
2. Refer to the table to find the life of the asset for that class of property.
3. Find the depreciation amount by multiplying the appropriate percent from the table by the cost of the item.

14.5 Modified Accelerated Cost Recovery System

Finding the Amount of Depreciation with MACRS

EXAMPLE 3 Capital Curb and Concrete bought a new pickup truck for $21,500. Find the amount of depreciation for the pickup truck in the fourth year.

CASE IN POINT

SOLUTION

A pickup truck has a recovery period of 5 years. From the table on page 598, the depreciation rate in the fourth year of recovery of 5-year property is **11.52%**. Multiply this rate by the full cost of the property to determine the amount of depreciation.

$$11.52\% \times \$21{,}500 = \$2477 \text{ (rounded)}$$

The amount of depreciation is $2477.

QUICK CHECK 3

Find the depreciation in year 2 for the pickup truck in Example 3.

Preparing a Depreciation Schedule with MACRS

EXAMPLE 4 Omaha Insurance purchased new office furniture at a cost of $24,160. Prepare a depreciation schedule using the modified accelerated cost recovery system.

SOLUTION

No salvage value is used with MACRS. Office desks and chairs have a 7-year recovery period. The annual depreciation rates for 7-year properties are shown in the table at the side.

Recovery Year	Recovery Percent (Rate)
1	14.29%
2	24.49%
3	17.49%
4	12.49%
5	8.93%
6	8.92%
7	8.93%
8	4.46%

OBJECTIVE 5 Prepare a depreciation schedule using the MACRS. For the furniture in Example 4, multiply the appropriate percents by $24,160 to get the results shown in the following depreciation schedule.

Year	Computation	Amount of Depreciation	Accumulated Depreciation	Book Value
0	—	—	—	$24,160
1	(14.29% × $24,160)	$3452	$3,452	$20,708
2	(24.49% × $24,160)	$5917	$9,369	$14,791
3	(17.49% × $24,160)	$4226	$13,595	$10,565
4	(12.49% × $24,160)	$3018	$16,613	$7,547
5	(8.93% × $24,160)	$2157	$18,770	$5,390
6	(8.92% × $24,160)	$2155	$20,925	$3,235
7	(8.93% × $24,160)	$2157	$23,082	$1,078
8	(4.46% × $24,160)	$1078	$24,160	$0

Quick TIP

These tables can be quickly generated using spreadsheet software.

QUICK CHECK 4

The city of San Francisco purchased a trolley at a cost of $128,000. Use the MACRS method of depreciation to find the book value at the end of the fourth year.

The MACRS method of depreciation allows a rapid rate of investment recovery for items in classes with short lives, yet it is simple to use since there is no need to estimate the life of an asset or the salvage value.

14.5 Exercises // MyLab Math

The shaded sections below contain solutions to help you get a QUICK START on the various types of exercises.

Use the MACRS depreciation rates table to find the recovery percent (rate), given the following recovery years and recovery periods. (See Examples 1 and 2.)

	Recovery Year	Recovery Period	Percent (Rate)		Recovery Year	Recovery Period	Percent (Rate)
1.	3	5-year	**19.20%**	2.	5	7-year	**8.93%**
3.	9	10-year	_____	4.	1	3-year	_____
5.	1	5-year	_____	6.	5	20-year	_____
7.	15	27.5-year	_____	8.	11	31.5-year	_____
9.	6	5-year	_____	10.	4	27.5-year	_____
11.	14	39-year	_____	12.	4	31.5-year	_____

Find the first year's depreciation for each of the following using the MACRS method of depreciation and the MACRS depreciation rates table. Round to the nearest dollar. (See Example 3.)

13. Cost: $12,250
    Recovery period: 7 years
    Depreciation (year 1): **$1751**
    14.29% rate
    $12,250 × .1429 = $1751 depreciation

14. Cost: $8790
    Recovery period: 5 years
    Depreciation (year 1): **$1758**
    20% rate
    $8790 × .20 = $1758 depreciation

15. Cost: $430,500
    Recovery period: 10 years
    Depreciation (year 1): _____

16. Cost: $72,300
    Recovery period: 20 years
    Depreciation (year 1): _____

17. Cost: $48,000
    Recovery period: 10 years
    Depreciation (year 1): _____

18. Cost: $12,340
    Recovery period: 3 years
    Depreciation (year 1): _____

Find the book value at the end of the first year for each of the following using the MACRS method of depreciation and the MACRS depreciation rates table. Round to the nearest dollar. (See Example 4.)

19. Cost: $9380
    Recovery period: 3 years
    Book value: **$6254**
    $9380 × .3333 = $3126
    $9380 − $3126 = $6254 book value

20. Cost: $68,700
    Recovery period: 5 years
    Book value: _____

C// indicates an exercise that is related to the Case in Point feature.

14.5 Exercises

21. Cost: $18,800
Recovery period: 10 years
Book value: _____

22. Cost: $156,000
Recovery period: 27.5 years
Book value: _____

Find the book value at the end of 3 years for each of the following using the MACRS method of depreciation and the MACRS depreciation rates table. Round to the nearest dollar. (See Example 4.)

23. Cost: $9570
Recovery period: 5 years
Book value: $2756

20% + 32% + 19.2% = 71.2% rate 3 years
$9570 × .712 = $6813.84 = $6814 dep. 3 years
$9570 − $6814 = $2756 book value year 3

24. Cost: $18,800
Recovery period: 3 years
Book value: _____

25. Cost: $136,800
Recovery period: 27.5 years
Book value: _____

26. Cost: $390,400
Recovery period: 31.5 years
Book value: _____

Solve the following application problems. Use the MACRS depreciation rates table. Round to the nearest dollar. (See Example 4.)

27. STORAGE TANK New England Propane purchased a storage tank for $10,980. Prepare a depreciation schedule using the MACRS method of depreciation (3-year property).

Year	Computation	Amount of Depreciation	Accumulated Depreciation	Book Value
0	—	—	—	$10,980
1	(33.33% × $10,980)	$3660	$3,660	$7,320
2	(44.45% × $10,980)	$4881	$8,541	$2,439
3	(14.81% × $10,980)	$1626	$10,167	$813
4	(7.41% × $10,980)	$813*	$10,980	$0

*Due to rounding in prior years

28. COMPANY VEHICLES Capital Curb and Concrete purchased a pickup truck at a cost of $21,500. Prepare a depreciation schedule using the MACRS method of depreciation (5-year property).

Year	Computation	Amount of Depreciation	Accumulated Depreciation	Book Value
0	—	—	—	$21,500
1				
2				
3				
4				
5				
6				

CHAPTER 14 Depreciation

29. DRILLING EQUIPMENT Gulf Drilling purchased a pump to circulate drilling fluids for $122,700. Prepare a depreciation schedule using the MACRS method of depreciation (10-year property).

Year	Computation	Amount of Depreciation	Accumulated Depreciation	Book Value
0	—	—	—	$122,700
1				
2				
3				
4				
5				
6				
7				
8				
9				
10				
11				

30. RESIDENTIAL RENTAL PROPERTY Joseph Flores purchased some residential rental real estate for $990,000. Find the book value at the end of the tenth year using the MACRS method of depreciation (27.5-year property).

30. _____

Year	Computation	Amount of Depreciation	Book Value
0	—	—	$990,000
1			
2			
3			
4			
5			
6			
7			
8			
9			
10			

31. Explain why businesses sometimes use more than one method of depreciation. (See Objective 1.)

32. Why do you suppose the IRS publishes tables defining which items go into each recovery class? (See Objective 2.)

Use the MACRS depreciation rates table in the following application problems. Round to the nearest dollar.

33. **COMMERCIAL FISHING BOAT** Reef Fisheries purchased a new fishing boat for $74,125. Find the depreciation in year 8 using the MACRS method of depreciation.

33. _____

34. **SHOPPING CENTER** A new parking lot was added to the Oak Shopping Center at a cost of $118,000. Find the depreciation in year 12 using the MACRS method of depreciation.

34. _____

35. **LAPTOP COMPUTERS** Capital Curb Concrete purchased two laptop computers for the office for $1700. Find the book value at the end of the third year using the MACRS method of depreciation.

35. _____

36. **DENTAL OFFICE** Dr. Jill Owens Family Dentistry purchased new furniture for its patient reception area at a cost of $27,400. Find the book value at the end of the fifth year using the MACRS method of depreciation.

36. _____

37. **OFFICE BUILDING** James Bradley purchased an office building at a cost of $860,000. Find the amount of depreciation for each of the first five years using the MACRS method of depreciation (39-year property).

Year 1: _____
Years 2–5: _____

38. **COFFEE SHOP** Martha Rose bought a building for $620,000 to use for her business. The cost of the building does not include the cost of land, which cannot be depreciated. Find the amount of depreciation for each of the first five years using the MACRS method of depreciation (39-year property).

Year 1: _____
Years 2–5: _____

QUICK CHECK ANSWERS

1. (a) 10 years (b) 5 years
   (c) 15 years (d) 27.5 years
2. (a) 12.49% (b) 7.219%
   (c) 3.175% (d) 5.90%
3. $6880
4. $22,118

Chapter 14 Quick Review

Chapter Terms Review the following terms to test your understanding of the chapter. For each term you do not know, refer to the page number found next to that term.

200% method [p. 573]
accelerated cost recovery system (ACRS) [p. 596]
accelerated depreciation [p. 573]
accumulated depreciation [p. 565]
book value [p. 565]
cost [p. 565]
declining-balance [p. 565]

depreciable amount [p. 566]
depreciation [p. 565]
depreciation fraction [p. 580]
depreciation schedule [p. 567]
double-declining-balance rate [p. 573]
half-year convention [p. 597]
income tax method [p. 596]
intangible assets [p. 565]

Internal Revenue Service [p. 565]
modified accelerated cost recovery system (MACRS) [p. 596]
no salvage value [p. 567]
Publication 946 [p. 596]
recovery classes [p. 596]
recovery periods [p. 596]
residual value [p. 565]

salvage value [p. 565]
scrap value [p. 565]
straight-line depreciation [p. 566]
sum-of-the-years'-digits [p. 580]
tangible assets [p. 565]
units of production [p. 591]
useful life [p. 565]

CONCEPTS

14.1 Straight-line method of depreciation
The depreciation is the same each year.

$$\text{Depreciation} = \frac{\text{Depreciable amount}}{\text{Years of life}}$$

14.1 Book value
Book value is the remaining value at the end of the year.

Book value = Cost − Accumulated depreciation

14.2 Double-declining-balance rate
First find the straight-line rate, and then adjust it. For the 200% method, **multiply by 2**.

14.2 Double-declining-balance depreciation method
First find the double-declining-balance rate, and then multiply by the cost in year 1. The rate is then multiplied by the declining book value in the following years.

Depreciation =
Double-declining-balance rate × Declining balance

EXAMPLES

Cost, $4500; scrap value, $800; life, 8 years; find the annual amount of depreciation.

$4500   cost
− 800   scrap
$3700   depreciable amount

$$\frac{\$3700}{8} = \$462.50 \text{ depreciation each year}$$

Cost, $4000; scrap value, $1000; life, 3 years; find the book value at the end of the first year.

$4000   cost
− 1000   scrap value
$3000   depreciable amount

$$\frac{\$3000}{3} = \$1000 \text{ depreciation}$$

$4000   cost
− 1000   depreciation
$3000   book value year 1

The life of an asset is 10 years. Find the double-declining-balance (200%) rate.

$$10 \text{ years} = 10\% \left(\frac{1}{10}\right) \text{ straight-line}$$

$$2 \times 10\% = 20\% \text{ per year}$$

Cost, $1400; life, 5 years; find the depreciation in years 1 and 2.

$2 \times 20\%$ (straight-line rate) = 40%

Year 1: **40%** × $1400 = $560 depreciation year 1

$1400 − **$560** = $840 book value year 1

Year 2: **40%** × $840 = $336 depreciation year 2

CHAPTER 14 Quick Review

CONCEPTS	EXAMPLES
14.3 Sum-of-the-years'-digits depreciation fraction Add the years' digits to get the denominator. The numerator is the number of years of life remaining. The shortcut formula for finding the denominator is: $$\frac{n(n+1)}{2}$$	Useful life is 4 years. Find the depreciation fraction for each year. $$1 + 2 + 3 + 4 = 10 \quad \text{or} \quad \frac{4(4+1)}{2} = 10$$ | Year	Depreciation Fraction
1	$\frac{4}{10}$
2	$\frac{3}{10}$
3	$\frac{2}{10}$
4	$\frac{1}{10}$	
14.3 Sum-of-the-years'-digits depreciation method First find the depreciation fraction, and then multiply by the depreciable amount. **Depreciation = Depreciation fraction × Depreciable amount**	Cost, $2500; salvage value, $400; life, 6 years; find depreciation in year 1. Depreciation fraction = $\frac{6}{21}$ Depreciable amount = $2100 ($2500 − $400) Depreciation = $\frac{6}{21}$ × $2100 = $600
14.4 Units-of-production depreciation amount per unit Use the following formula. $$\text{Depreciation per unit} = \frac{\text{Depreciable amount}}{\text{Units of life}}$$	Cost, $10,000; salvage value, $2500; useful life, 15,000 units; find depreciation per unit. $10,000 − $2500 = $7500 depreciable amount Depreciation per unit = $\frac{\$7500 \text{ depreciable amount}}{15{,}000 \text{ units of life}}$ = $.50
14.4 Units-of-production depreciation method Multiply the number of units (hours) of production by the depreciation per unit (per hour). **Depreciation = Number of units (hours) × Depreciation per unit (hour)**	Cost, $25,000; salvage value, $2000; useful life, 100,000 units; production in year 1, 22,300 units; find the first year's depreciation. 1. $25,000 − $2000 = $23,000 depreciable amount 2. $\frac{\$23{,}000}{100{,}000}$ = $.23 depreciation per unit 3. 22,300 × $.23 = $5129 depreciation year 1
14.5 Modified accelerated cost recovery system (MACRS) Established in 1986 for federal income tax purposes. No salvage value. Recovery periods are: 3-year 5-year 7-year 10-year 15-year 20-year 27.5-year 31.5-year 39-year Find the proper rate from the table and then multiply by the cost to find depreciation.	Use the table, finding the recovery period column at the top of the table and the recovery year in the left-hand column. Cost, $4850; recovery class, 5 years; recovery year, 3; find the depreciation. Rate is **19.20%** from table. $4850 × .192 = $931 depreciation

Case Study

COMPARING DEPRECIATION METHODS

Trader Joe's purchased freezer cases at a cost of $285,000. The estimated life of the freezer cases is 5 years, at which time they will have no salvage value. The company would like to compare allowable depreciation methods and decides to prepare depreciation schedules for the fixtures using the straight-line, double-declining-balance, and sum-of-the-years'-digits methods of depreciation. Answer the following questions.

1. What is the book value at the end of 3 years using the straight-line depreciation method?

    1. _____

2. Using the double-declining-balance method of depreciation, what is the book value at the end of the third year?

    2. _____

3. With the sum-of-the-years'-digits method of depreciation, what is the accumulated depreciation at the end of 3 years?

    3. _____

4. What amount of depreciation will be taken in year 4 with each of the methods?

    4. _____
    _____
    _____

INVESTIGATE

Identify a store or business with which you are familiar and list six of its depreciable assets. Examples could be such items as buildings, computer equipment, vehicles, and fixtures. Using the information on the MACRS method of depreciation, give the recovery period and the first-year depreciation rate for each of the six depreciable assets you listed.

Case in Point Summary Exercise

FORD MOTOR COMPANY

www.ford.com

Facts:

- 1925: Price of first Ford truck was $281
- 1950s: Cars styled for more comfort
- 2003: Celebrates 100th anniversary
- 2018: More than 200,000 employees worldwide

Henry Ford started Ford Motor Company in 1903 with $28,000 in cash. His greatest contribution to automobile manufacturing was the moving assembly line that allowed individual workers to stay in one place and perform the same task on each vehicle as it passed by. The assembly line allowed Ford to cut costs and reduce the prices of new automobiles so that average working families could afford them. Today, the Ford Motor Company sells a variety of vehicles under the Ford and Lincoln brands.

Capital Curb and Concrete purchased a new Ford truck for $26,500 to be used by the sales manager, who often drives to construction sites to work with customers. The firm's accountant believes that the useful life of the truck is 5 years and that the estimated salvage value is $4500.

1. Using the information above, find the book value of the pickup truck after 3 years using the straight-line method of depreciation.

2. Find the book value of the truck at the end of 3 years using the double-declining-balance method of depreciation.

3. The company also spent $118,350 to purchase a truck that hauls concrete. Which method of depreciation must it use when calculating federal income taxes? What is the appropriate recovery class? Use this information for all of the following exercises.

4. Use MACRS to find annual depreciation for the first three years. Round to the nearest dollar.

5. Find the book value at the end of the third year.

6. Find the percent of the total depreciation that is taken in the first three years. Round to the nearest tenth of a percent.

7. What is the depreciation in the year after book value reaches salvage value?

Discussion Question: Explain depreciation in your own words. Why is depreciation important to business managers?

Chapter 14 Test

To help you review, the numbers in brackets show the section in which the topic was discussed.

Find the annual straight-line and double-declining-balance rates (percents) of depreciation and the sum-of-the-years'-digits fraction for the first year for each of the following estimated lives. [14.1–14.3]

Life	Straight-Line Rate	Double-Declining-Balance Rate	Sum-of-the-Years'-Digits Fraction
1. 4 years			
2. 5 years			
3. 8 years			
4. 20 years			

Solve the following application problems. Round to the nearest dollar.

5. Cloverdale Creamery purchased a soft-serve ice cream maker at a cost of $82,000. The machine has an estimated life of 10 years and a scrap value of $3000. Use the straight-line method of depreciation to find the annual depreciation. [14.1]

5. _____

6. Capital Curb and Concrete purchased a new dump truck for $38,000. If the estimated life of the dump truck is 8 years, find the book value at the end of 2 years using the double-declining-balance method of depreciation. [14.2]

6. _____

7. Feather River Youth Camp purchased a diesel generator for $8250. Use the sum-of-the-years'-digits method of depreciation to determine the amount of depreciation to be taken during *each of the 4 years* on the diesel generator that has a 4-year life and a scrap value of $1500. [14.3]

Year 1: _____
Year 2: _____
Year 3: _____
Year 4: _____

8. A private road costs $56,000 and has a 15-year recovery period. Find the depreciation in the third year using the MACRS method of depreciation. [14.5]

8. _____

9. One water filter at Micro Brew costs $74,000, has an estimated life of 20 years, and has an estimated scrap value of $12,000. Use the straight-line method of depreciation to find the book value of the machinery at the end of 10 years. [14.1]

9. _____

10. Gold's Gym has added paging and intercom features to the communication systems of four stores at a cost of $2800 per store. The estimated life of the systems is 10 years, with no expected salvage value. Using the sum-of-the-years'-digits method of depreciation, find the total book value of all the systems at the end of the third year. [14.3]

10. _____

C// indicates an exercise that is related to the Case in Point feature.

11. Table Fresh Foods purchased a machine to package its presliced garden salads. The machine costs $20,100 and has an estimated life of 30,000 hours and a salvage value of $1500. Use the units-of-production method of depreciation to find **(a)** the annual amount of depreciation and **(b)** the book value at the end of each year, given the following use information: year 1: 7800 hours; year 2: 4300 hours; year 3: 4850 hours; year 4: 7600 hours. [14.4]

(a) _____

(b) _____

12. The Rice Growers Cooperative paid $2,800,000 to build a new rice-drying plant. The recovery period is 39 years. Use the MACRS method of depreciation to find the book value of the rice-drying plant at the end of the fifth year. [14.5]

12. _____

Financial Statements and Ratios

15

CHAPTER CONTENTS

15.1 The Income Statement

15.2 Analyzing the Income Statement

15.3 The Balance Sheet

15.4 Analyzing the Balance Sheet

CASE IN POINT

APPLE, INC. gave the world the Mac computer, iPod, iPhone, iPad, iTunes, MacBook Pro, Apple Apps, and the Apple Watch. It was the most valuable company in the world by 2016, although its value sometimes falls below that of other giant firms such as Alphabet or ExxonMobil. Over 60% of its revenue comes from the sale of iPhones.

Apple's shares trade on the NASDAQ stock exchange under the symbol AAPL. You can become a part-owner of Apple by buying shares of Apple stock. Recent financial statements for this very successful and creative company are shown in this chapter.

4. Explain why lenders and investors look at the income statement before making a loan or an investment. (See Objective 1.)

5. Can a family prepare a personal income statement? Who would want to look at it? (See Objective 1.)

6. **DENTAL-SUPPLY COMPANY** New England Dental Supply is a regional wholesaler that had gross sales last year of $2,215,000. Returns totaled $26,000. Inventory on January 1 was $215,000. Goods purchased during the year totaled $1,123,000. Freight was $4000. Inventory on December 31 was $265,000. Wages and salaries were $326,000, rent was $59,000, advertising was $11,000, utilities were $12,000, taxes on inventory and payroll totaled $28,200, and miscellaneous expenses were $18,800. In addition, income taxes for the year amounted to $197,800. Complete the following income statement for this firm. (See Example 2.)

New England Dental Supply
Income Statement
Year Ending December 31

Gross Sales _____
Returns _____
Net Sales _____
   Inventory, January 1 _____
   Cost of Goods
     Purchased _____
     Freight _____

   Total Cost of Goods Purchased _____
   Total of Goods Available for Sale _____
   Inventory, December 31 _____
Cost of Goods Sold _____
Gross Profit _____
Expenses
   Salaries and Wages _____
   Rent _____
   Advertising _____
   Utilities _____
   Taxes on Inventory, Payroll _____
   Miscellaneous Expenses _____
   Total Expenses _____
Net Income before Taxes _____
   Income Taxes _____
Net Income _____

QUICK CHECK ANSWERS

1. $.6 million, or $600,000
2. $1.2 million

15.2 Analyzing the Income Statement

OBJECTIVES
1. Compare income statements using vertical analysis.
2. Compare income statements to published charts.
3. Compare income statements using horizontal analysis.

CASE IN POINT Investors look carefully at the income statement of Apple, or any other company, before investing in the stock of that company. They buy stock in the company only if they believe it is going to be profitable.

OBJECTIVE 1 Compare income statements using vertical analysis. A firm can find its net income for a given period of time by going through the steps presented in the preceding section. A question that might then be asked is "What happened to each part of the sales dollar?" The first step toward answering this question is to list each of the important items on the income statement as a percent of net sales. This process is called a **vertical analysis** of the income statement.

In a vertical analysis, each item on the income statement is found as a percent of the net sales.

$$R = \frac{P}{B} \quad \text{or} \quad R = \frac{\text{Particular item}}{\text{Net sales}}$$

For example, we can use data from the 2016 income statement of Apple shown on page 612. The percent cost of goods sold is found by dividing cost of goods sold divided by net sales.

$$\text{Percent cost of goods sold} = \frac{\$121{,}576}{\$215{,}091} = 56.5\% \quad (\text{rounded})$$

So the firm spent nearly 57% of total revenue for the cost of goods. A lot of this money went to Asia since many of the goods were produced at factories on that continent. Other companies may have either higher or lower percent cost of goods sold. For example, Microsoft Corporation writes computer software. The cost of labor to produce the software is very high (operating expense), but the cost of goods such as optical disks on which software is written is very low. So you would expect this same ratio to be much lower for Microsoft.

A **comparative income statement** is used to compare results from two or more years. It can be used to show how the company is doing over time.

Performing a Vertical Analysis  First, do a vertical analysis for the 2015 and 2016 income statements for Apple and then build a comparative income statement and show the results in a table.

Apple, Inc.
Consolidated Income Statement
(in millions of dollars)

Year Ending September 30	2015	2016
Gross Sales	$233,715	$215,091
Returns	− 0	− 0
Net Sales	$233,715	$215,091
Cost of Goods Sold	− 129,589	− 121,576
Gross Profit	$104,126	$93,515
Operating Expenses	− 31,611	− 32,143
Net Income before Taxes	$72,515	$61,372
Income Taxes	− 19,121	− 15,685
Net Income	$53,394	$45,687

SOLUTION

Calculate each value in the column labeled 2015 as a percent of gross sales for the year, rounded to the nearest tenth. Gross sales equals net sales for both 2015 and 2016 since there were no returns. Then do the same for 2016.

15.2 Analyzing the Income Statement

Quick TIP

Managers are so cost- and revenue-conscious that they make a budget of what they expect to spend ahead of time. They then compare actual expenditures to the budget, shaving expenses as needed and trying to boost revenue to meet the plan.

Comparative Income Statement for Apple, Inc.

	2015	2016
Percent Cost of Goods Sold	$\dfrac{\$129{,}589}{\$233{,}715} = 55.4\%$	$\dfrac{\$121{,}576}{\$215{,}091} = 56.5\%$
Percent Gross Profit	$\dfrac{\$104{,}126}{\$233{,}715} = 44.6\%$	$\dfrac{\$93{,}515}{\$215{,}091} = 43.5\%$
Percent Operating Expenses	$\dfrac{\$31{,}611}{\$233{,}715} = 13.5\%$	$\dfrac{\$32{,}143}{\$215{,}091} = 14.9\%$
Percent Net Income before Taxes	$\dfrac{\$72{,}515}{\$233{,}715} = 31.0\%$	$\dfrac{\$61{,}372}{\$215{,}091} = 28.5\%$
Percent Income Taxes	$\dfrac{\$19{,}121}{\$233{,}715} = 8.2\%$	$\dfrac{\$15{,}685}{\$215{,}091} = 7.3\%$
Percent Net Income after Taxes	$\dfrac{\$53{,}394}{\$233{,}715} = 22.8\%$	$\dfrac{\$45{,}687}{\$215{,}091} = 21.2\%$

The comparative income statement is important, as it allows managers to compare costs and other items from one quarter or year to the next. It shows that the percent cost of goods sold went up slightly from 55.4% of gross revenue in 2015 to 56.5% of gross revenue in 2016. Percent operating expenses also went up, from 13.5% to 14.9% of gross revenue. It is not good that both categories of cost went up, but this explains why the percent net income before taxes fell from 31% of gross revenue to 28.5% of gross revenue. Investors and bankers would likely watch these trends carefully to see if they continue. However, although Apple's percent net income after taxes went down from 22.8% to 21.2% of gross revenue, it is still a very healthy after-tax profit margin. The average profit margin of the 500 companies in Standard & Poor's 500 is less than 10%, so Apple's margins remain far above average.

Apple paid $19,121,000,000 in income taxes in 2015 and $15,685,000,000 in 2016. Some believe that corporations do not pay enough income tax, and they might have a point as the total income tax paid by publicly held corporations has fallen over the past 30 years. Of course, corporations also pay some payroll taxes on behalf of employees (see Chapter 6) and property taxes, and many pay health insurance for employees as well as contributing to employee retirement plans. But the amount of taxes that corporations pay remains in the public debate.

It is important to note that the percentages in the comparative income statement change from quarter to quarter and from year to year. Major investors watch these numbers carefully.

QUICK CHECK 1

An income statement shows the following figures in millions: Net Sales $23.5, Returns $0, Cost of Goods Sold $12.9, and Operating Expenses $3.8. Do a vertical analysis and round each percent to the nearest tenth.

OBJECTIVE 2 Compare income statements to published charts. If you own a business or are considering investing in one, you can use charts, published by the government, of industry averages for comparison purposes.

Average Ratios by Type of Business

Type of Business	Cost of Goods	Gross Profit	Total Expenses*	Net Income	Wages	Rent	Advertising
Supermarkets	82.7%	17.3%	13.9%	3.4%	6.5%	.8%	1.0%
Men's and women's apparel	67.0%	33.0%	21.2%	11.8%	8.0%	2.5%	1.9%
Women's apparel	64.8%	35.2%	23.4%	11.7%	7.9%	4.9%	1.8%
Shoes	60.3%	39.7%	24.5%	15.2%	10.3%	4.7%	1.6%
Furniture	68.9%	31.2%	21.7%	9.6%	9.5%	1.8%	2.5%
Appliances	66.9%	33.1%	26.0%	7.2%	11.9%	2.4%	2.5%
Drugs	67.9%	32.1%	23.5%	8.6%	12.3%	2.4%	1.4%
Restaurants	48.4%	51.6%	43.7%	7.9%	26.4%	2.8%	1.4%
Service stations	76.8%	23.2%	16.9%	6.3%	8.5%	2.3%	.5%

*Total Expenses represents the total of all expenses involved in running the firm. These expenses include, but are not limited to, wages, rent, and advertising.

Comparing Business Ratios

EXAMPLE 2 Gina Burton wishes to compare the business ratios of her shoe store to industry averages. Figures from her store and industry averages for shoe stores are shown in the table.

	Cost of Goods	Gross Profit	Total Expenses	Net Income	Wages	Rent	Advertising
Burton's Shoes	58.2%	41.8%	28.3%	13.5%	11.7%	5.6%	2.8%
Shoes (from previous chart)	60.3%	39.7%	24.5%	15.2%	10.3%	4.7%	1.6%

SOLUTION

Burton's expenses are higher than the average for other shoe stores and her net income is lower. Wages are higher—perhaps because her store is located in an area with high wages or perhaps the store is not large enough to efficiently utilize its employees. Burton also spends a higher percent than average for advertising. Thus, Burton may be able to find ways to enhance the performance of her business such as reducing advertising expenses without lowering sales.

QUICK CHECK 2

A furniture store has total expenses of 21.3% and cost of goods of 72%. Compare to averages.

OBJECTIVE 3 Compare income statements using horizontal analysis. Another way to analyze an income statement is to prepare a **horizontal analysis**. A horizontal analysis finds the percent of change (either increases or decreases) between the current time period and a previous time period. This comparison can expose unusual changes, such as a rapid increase in expenses or decline in net sales or profits.

Finding Percent Change

$$\% \text{ of change} = \frac{\text{Change}}{\text{Previous year's amount}}$$

Always use *last year* as the base.

Do a horizontal analysis by finding the amount of change from the previous year compared to the current year in dollars and as a percent. For example, the comparative income statement of Apple shows that net sales decreased from $233,715 (in millions) in 2015 to $215,091 (in millions) in 2016.

$$\text{Decrease in net sales} = \$233{,}715 - \$215{,}091 = \$18{,}624 \text{ millions}$$

$$\text{Percent increase in net sales} = \frac{\$18{,}624}{\$233{,}715} = 8.0\% \quad \text{(rounded)}$$

So Apple's sales actually fell by 8% from 2015 to 2016. The decrease in sales may have been a result of any of the following or even other reasons:

1. an economic recession,
2. competitors introduced great products and gained **market share,**
3. a well-publicized crisis in the firm such as a big lawsuit or defective products, and/or
4. no cool new products were introduced.

There was no recession in 2015–16, and there was no well-publicized crisis at Apple during those years. So perhaps competitors had strong products in the market and/or Apple had not introduced any cool new products recently. In fact, Apple only introduces a new iPhone every few years. It introduced the iPhone 7 in September 2016, too late to have much effect on revenue that year. Another reason may simply be that Apple's growth has slowed after a torrid growth rate for most of the previous 10 years. Let's look at Apple's stock price to see what the price did during 2016.

The following graph shows the price of Apple stock (AAPL) compared to the **NASDAQ** index, which is a composite of many technology companies. Apple stock generally fell or was flat during the 2016 fiscal year of September 2015–September 2016. This is generally what you would expect during a year in which gross sales and net profit fell. However, the price of Apple stock rose during 2017 probably due to sales of the new iPhone 7 introduced in September 2016. The iPhone 7 accounts for more than 60% of Apple's revenue.

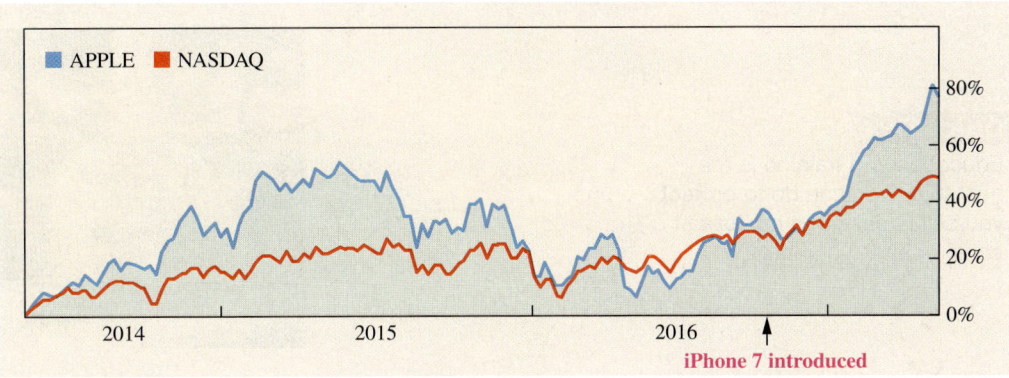

Performing a Horizontal Analysis

EXAMPLE 3 Do a horizontal analysis for the 2015 and 2016 income statements for Apple using the data given in Example 1.

SOLUTION

Find the increase by subtracting the 2015 figure from the 2016 figure. Then divide by the 2015 figure to find the percent increase and round to the nearest tenth of a percent.

Apple, Inc. Consolidated Statement Year Ending September 30 (in millions of dollars)				
	2015	2016	Increase (Decrease)	Percent
Net Sales	$233,715	$215,091	($18,624)	(8.0%)
Gross Profit	$104,126	$93,515	($10,611)	(10.2%)
Net Income before Taxes	$72,515	$61,372	($11,143)	(15.4%)
Income Taxes	$19,121	$15,685	($3,436)	(18.0%)
Net Income	$53,394	$45,687	($7,707)	(14.4%)

All of the numbers fell from 2015 to 2016 and the percents were all a percent decrease. Parentheses around a number or percent indicate a negative number. Apple clearly did not do as well in 2016 as it had in 2015. Although net sales fell by 8%, net income fell by 14.4%—something an investor would watch carefully. Stock prices do not go up much or for long when net income falls.

Similarly to most businesses, Apple operates in a competitive environment. Although it has competition in all of its markets, Apple has been well-managed and innovative enough in delivering quality new products that it has outgrown its competitors. This has not always been the case, as evidenced by its flat stock price from 2000 through 2006. In the future, management is likely to

1. further increase sales outside the United States,
2. develop additional products and services,
3. increase and improve marketing efforts, and
4. continue to work to decrease costs.

QUICK CHECK 3

Partial results for a company are: Gross Sales $58.6 million and Operating Expenses $12.1 million. The same results for 2013 were: Gross Sales $61.2 million and Operating Expenses $12.9 million. Do a *horizontal comparison*.

620 CHAPTER 15 Financial Statements and Ratios

During the past 50 years, many firms have moved manufacturing jobs to countries where wages are lower, such as Mexico, China, and India. Now there is evidence that service jobs (e.g., accountant, lawyer, and engineer) are being moved to lower-wage countries as well. Education is *the* fundamental tool you have to protect yourself and your family from a loss in job and income. It is important to continue to learn in life, no matter what your age, skills, level of education, or experience!

Quick TIP
Education and training is the best thing you can do to protect yourself in case of a job loss at some point in your life.

15.2 Exercises // MyLab Math

The shaded sections below contain solutions to help you get a **QUICK START** *on the various types of exercises.*

Prepare a vertical analysis for each of the following firms. Round percents to the nearest tenth of a percent. (See Example 1.)

1. **SCUBA SHOPPE** Reef Scuba, Inc., had net sales of $439,000, operating expenses of $143,180, and a cost of goods sold of $198,400.

   Percent cost of goods sold $= \frac{\$198,400}{\$439,000} = 45.2\%$

   Percent operating expenses $= \frac{\$143,180}{\$439,000} = 32.6\%$

   1. **45.2%; 32.6%**

2. **COFFEE SHOP** Tatum's Coffee and Books had operating expenses of $198,400, a cost of goods sold of $287,104, and net sales of $589,250.

   2. _____

3. **GUITAR SHOP** One quarter, the Guitar Shoppe had a cost of goods sold of $243,570, operating expenses of $140,450, and net sales of $480,300.

   3. _____

4. **FLOWER SHOP** Trisha's Flowers had net sales of $294,380, operating expenses of $68,650, and a cost of goods sold of $163,890.

   4. _____

The following charts show some figures from income statements. In each case, prepare a vertical analysis by expressing each item as a percent of net sales. Then write in the appropriate average percent from the table in the book. (See Objective 2.)

5.

Gooden Drugs

	Amount	Percent	Average Percent
Net Sales	$850,000	100%	100%
Cost of Goods Sold	$570,350		
Gross Profit	$279,650		
Wages	$106,250		
Rent	$21,250		
Advertising	$12,750		
Total Expenses	$209,100		
Net Income before Taxes	$70,550		

6.

Best Burgers

	Amount	Percent	Average Percent
Net Sales	$600,000	100%	100%
Cost of Goods Sold	$280,000		
Gross Profit	$320,000		
Wages	$160,600		
Rent	$15,000		
Advertising	$8,000		
Total Expenses	$255,000		
Net Income before Taxes	$65,000		

7. Explain the difference between a vertical and a horizontal analysis, and indicate uses for each. (See Objectives 1 and 3.)

8. Explain why a comparative income statement is a very important tool for managers. (See Objective 1.)

9. Complete the comparative income statement. Round to the nearest tenth of a percent.

Hernandez Nursery Comparative Income Statement

	This Year		Last Year	
	Amount	Percent	Amount	Percent
Gross Sales	$1,856,000		$1,692,000	
Returns	$6,000		$12,000	
Net Sales		100.0%		100.0%
Cost of Goods Sold	$1,102,000		$950,000	
Gross Profit	$748,000		$730,000	
Wages	$252,000		$248,000	
Rent	$82,000		$78,000	
Advertising	$111,000		$122,000	
Utilities	$32,000		$17,000	
Taxes on Inv., Payroll	$17,000		$18,000	
Miscellaneous Expenses	$62,000		$58,000	
Total Expenses	$556,000		$541,000	
Net Income before Taxes				

10. Complete the horizontal analysis for the Hernandez Nursery comparative income statement given in Exercise 9. Round to the nearest tenth of a percent.

Hernandez Nursery Horizontal Analysis

	This Year	Last Year	Increase or (Decrease) Amount	Percent
Gross Sales	$1,856,000	$1,692,000		
Returns	$6,000	$12,000		
Net Sales	$1,850,000	$1,680,000		
Cost of Goods Sold	$1,102,000	$950,000		
Gross Profit	$748,000	$730,000		
Wages	$252,000	$248,000		
Rent	$82,000	$78,000		
Advertising	$111,000	$122,000		
Utilities	$32,000	$17,000		
Taxes on Inv., Payroll	$17,000	$18,000		
Miscellaneous Expenses	$62,000	$58,000		
Net Income before Taxes	$192,000	$189,000		

The following table gives the percents for various items from the income statements of firms in two businesses. Complete these tables by including the appropriate percents from the table on page 617. Identify any areas that might require attention by management.

Type of Store	Cost of Goods	Gross Profit	Total Operating Expenses	Net Income	Wages	Rent	Advertising
11. Women's apparel	66.4%	33.6%	25.3%	8.3%	8.4%	6.5%	1.9%
	___	___	___	___	___	___	___
12. Drug store	71.2%	28.8%	26.5%	2.3%	12.9%	5.3%	2.0%
	___	___	___	___	___	___	___

13. Find a graph on the Web that shows the stock price of Apple, Inc. (symbol: AAPL) and print it out.

14. Use the Internet to find a graph of the stock-price history of a company with which you are familiar.

QUICK CHECK ANSWERS

1. Returns 2.1%; Cost of Goods sold 54.9%; Operating Expenses 16.2%
2. Total Expenses are below average of 21.7%; Cost of Goods is above average of 68.9%.
3. Gross Sales increased 4.4%; Operating Expenses increased 6.6%.

15.3 The Balance Sheet

OBJECTIVES

1. Understand the terms on a balance sheet.
2. Prepare a balance sheet.

CASE IN POINT It is also important to look at the balance sheet before deciding to buy stock in a company. The income statement shows information during an interval of time, such as a quarter or a year. In contrast, the balance sheet shows information at one point in time, such as the amount of cash Apple had on a specific day.

OBJECTIVE 1 Understand the terms on a balance sheet. An income statement shows the revenues and expenses of a firm for a period of time such as a month, quarter, or year. In contrast, a **balance sheet** describes the financial condition of a firm *at one point in time*, such as the last day of a year. A balance sheet shows the **assets** of a firm, which is the total value of everything owned by a business at a particular time. Assets include property, equipment, money owed to the company, cash, and securities owned. A balance sheet also shows the amounts owed by the business to others, called **liabilities**. Think of assets as property and investments owned by a company and liabilities as debts that must be paid at some point.

Both assets and liabilities are divided into two categories, **long-term** and **current (short-term)**. Long-term generally applies to assets or liabilities with a life of more than a year. Short-term applies when the time involved is less than a year.

Assets—property and investments owned by the firm
    Current assets—cash or items that can be converted into cash within a short period of time such as a year
        Cash—amount in checking and savings accounts and money market instruments
        Accounts receivable—funds owed by customers of the firm
        Notes receivable—value of all notes owed to the firm
        Inventory—cost of merchandise that the firm has for sale
    Plant and equipment—assets that are expected to be used for more than one year (also called **fixed assets** or **plant assets**)
        Land—book value of any land owned by the firm
        Buildings—book value of any buildings owned by the firm
        Equipment—book value of equipment, store fixtures, furniture, and similar items owned by the firm

Liabilities—debts that must be paid at some point
    Current liabilities—items that must be paid by the firm within a short period of time, usually one year
        Accounts payable—amounts that must be paid to other firms
        Notes payable—value of all notes owed by the firm
    Long-term liabilities—items that will be paid after one year
        Mortgages payable—total due on all mortgages
        Long-term notes payable—total of all other debts of the firm

The difference between the total of all assets and the total of all liabilities is called the **owners' equity**, which is also referred to as **net worth** or, for a corporation, **stockholders' equity**. The relationship among owners' equity, assets, and liabilities is shown in the fundamental formula below.

$$\text{Owners' equity} = \text{Assets} - \text{Liabilities}$$
$$\text{or}$$
$$\text{Assets} = \text{Liabilities} + \text{Owners' equity}$$

OBJECTIVE 2 Prepare a balance sheet.

Preparing a Balance Sheet

EXAMPLE 1 On September 30, 2016, Apple's balance sheet showed the assets and liabilities shown below. Complete the balance sheet and find the Stockholders' Equity.

SOLUTION

Apple, Inc. Consolidated Statement Year Ending September 30, 2016 (in millions of dollars)		
ASSETS:		
Current Assets:		
Cash and short-term investments	$67,155	
Accounts receivable and other	$29,299	
Inventory and other	$10,415	
Total Current Assets	**$106,869**	sum of all current assets
Other Assets:		
Net plant and property	$27,010	
Other assets	$187,807	
Total Assets	**$321,686**	total current and other assets
LIABILITIES:		
Current Liabilities:		
Loans and accounts payable	$48,899	
Other current liabilities	$30,107	
Total Current Liabilities	**$79,006**	sum of all current liabilities
Other Liabilities:		
Long-term debt	$75,427	
Other long-term liabilities	$39,004	
Total Liabilities	**$193,437**	total current and other liabilities
Stockholders' Equity:	**$128,249**	Total assets − Total liabilities
Total Liabilities and Equity	$321,686	

Quick TIP
Assets are cash, inventory, facilities, etc., whereas liabilities are debts to others. Banks ask people for a balance sheet when they seek to borrow money.

On September 30, 2016, Apple had about $67,155,000,000 in cash and short-term investments that could quickly be converted to cash. This balance sheet only provides information for September 30, 2016, and not any other day, so the amount of cash and short-term investments on a different day could be very different. It had $79,006,000,000 in current liabilities (short-term debt) and $75,427,000,000 in long-term debt.

Although Apple has historically kept inventories to very low levels, it had $10,415,000,000 in inventory on September 30, 2016. Perhaps products were selling more slowly than anticipated, so inventory may have built up more than management wanted. We noted in the prior section that sales fell during 2016. Management makes projections or forecasts of what they believe the company can sell and then contracts with suppliers in the supply chain to build that number of products months ahead of actual delivery. Thus, management may have been a little optimistic in their projections and contracted with sellers for those volumes. In turn, the sellers built and shipped the contracted number of products even though sales did not materialize at the expected level. So inventory levels may have increased a little, although they are certainly not excessive for a company of Apple's size.

Looking at inventory on shelves in stores, customers tend to think of inventory as always sitting around. However, managers view excess inventory as wasteful. They only want the minimum needed in stores at any one time. They tend to think of inventory as moving rapidly along global supply chains rather than sitting somewhere for any length of time.

Apple's Global Supply Chain Getting iPhones to Chicago

> **QUICK CHECK 1**
>
> A plumbing contractor's balance sheet shows the following assets: Cash, $398; Accounts Receivable and other, $789; Inventory, $481, Net Plant and Property, $1047; Other Assets; $212. The liabilities were: Loans and Accounts Payable, $659; Other Current Liabilities, $241; Long-Term Debt, $384, Other Long-Term Liabilities, $412. Find the Stockholders' Equity. All figures are in thousands.

Remember this important relationship:

> **Finding Stockholders' Equity**
>
> Stockholders' equity = Total assets − Total liabilities

15.3 Exercises // MyLab Math

Complete the balance sheets for the following business firms. (See Example 1.)

1. **GROCERY CHAIN** Brookshire's Grocery (all figures in millions): fixtures, $28; buildings, $290; land, $466; cash, $273; notes receivable, $312; accounts receivable, $264; inventory, $180; notes payable, $312; mortgages payable, $212; accounts payable, $63; long-term notes payable, $55.

Brookshire's Grocery Balance Sheet December 31 (in millions)

ASSETS	
Current Assets	
Cash	____
Notes Receivable	____
Accounts Receivable	____
Inventory	____
Total Current Assets	____
Plant Assets	
Land	____
Buildings	____
Fixtures	____
Total Plant Assets	____
Total Assets	____
LIABILITIES	
Current Liabilities	
Notes Payable	____
Accounts Payable	____
Total Current Liabilities	____
Long-Term Liabilities	
Mortgages Payable	____
Long-Term Notes Payable	____
Total Long-Term Liabilities	____
Total Liabilities	____
OWNERS' EQUITY	
Owners' Equity	____
Total Liabilities and Owners' Equity	____

626 CHAPTER 15 Financial Statements and Ratios

2. **JETSON AIRCRAFT SALES** Land is $8750; accounts payable total $49,230; notes receivable are $2600; accounts receivable are $37,820; cash is $14,800; buildings are $21,930; notes payable are $3780; owners' equity is $54,320; long-term notes payable are $18,740; mortgages payable total $26,330; inventory is $49,680; fixtures are $16,820. All figures are in thousands.

Jetson Aircraft Sales Balance Sheet—December 31 (in thousands)

ASSETS		
Current Assets		
Cash	_____	
Notes Receivable	_____	
Accounts Receivable	_____	
Inventory	_____	
Total Current Assets		_____
Plant Assets		
Land	_____	
Buildings	_____	
Fixtures	_____	
Total Plant Assets		_____
Total Assets		_____
LIABILITIES		
Current Liabilities		
Notes Payable	_____	
Accounts Payable	_____	
Total Current Liabilities		_____
Long-Term Liabilities		
Mortgages Payable	_____	
Long-Term Notes Payable	_____	
Total Long-Term Liabilities		_____
Total Liabilities		_____
OWNERS' EQUITY		
Owners' Equity		_____
Total Liabilities and Owners' Equity		_____

3. Compare a balance sheet to an income statement. (See Sections 15.1 and 15.3)

4. Use the Internet to find the amount of cash and equivalents at the end of the most recent fiscal year for GM and The Coca-Cola Company. Which company has more? Why do you suppose it has as much as it does?

QUICK CHECK ANSWER

1. $1,231,000

15.4 Analyzing the Balance Sheet

OBJECTIVES

1. Compare balance sheets using vertical analysis.
2. Compare balance sheets using horizontal analysis.
3. Find financial ratios.

OBJECTIVE 1 Compare balance sheets using vertical analysis. A balance sheet can be analyzed in much the same way as an income statement. In a **vertical analysis**, each item on the balance sheet is expressed as a percent of total assets. A **comparative balance sheet** shows the vertical analysis for two different years.

Comparing Balance Sheets  First, do a vertical analysis for the 2015 and 2016 balance sheets for Apple by calculating each value as a percent of the total assets for the respective year (round percents to the nearest tenth). Then compare the percents to identify the changes.

SOLUTION

Apple, Inc.
Consolidated Statement
Year Ending September 30, 2016
(in millions of dollars*)

	2015		2016	
Assets:	Amount	Percent	Amount	Percent
Current Assets:				
Cash and short-term investments	$41,601	14.3%	$67,155	20.9%
Accounts receivable and other	$30,343	10.4%	$29,299	9.1%
Inventory	$17,434	6.0%	$10,415	3.2%
Total Current Assets	$89,378	30.8%	$106,869	33.2%
Other Assets:				
Net plant and property	$22,471	7.7%	$27,010	8.4%
Other assets	$178,630	61.5%	$187,807	58.4%
Total Assets	$290,479	100.0%	$321,686	100.0%
Liabilities:	Amount	Percent	Amount	Percent
Current Liabilities:				
Loans and accounts payable	$46,489	16.0%	$48,899	15.2%
Other current liabilities	$34,121	11.7%	$30,107	9.4%
Total Current Liabilities	$80,610	27.8%	$79,006	24.6%
Other Liabilities:				
Long-term debt	$53,463	18.4%	$75,427	23.4%
Other long-term liabilities	$37,051	12.8%	$39,004	12.1%
Total Liabilities	$171,124	58.9%	$193,437	60.1%
Stockholders' Equity:	$119,355	41.1%	$128,249	39.9%
Total Liabilities and Equity	$290,479	100.0%	$321,686	100.0%

*Rounding may cause columns to not add to percent shown.

Notice that total current assets increased from $89,378,000,000 to $106,869,000,000 and they also increased to 33.2% of total assets. In particular, cash and short-term investments increased significantly. Current liabilities went down both in dollars and as a percent of total assets; however, total liabilities increased significantly. It would appear that management shifted some debt from short term to long term. Stockholder's equity increased nicely to $128,249,000,000.

Looking at both the balance sheets and the income statements from prior sections, a careful investor would want to know why total liabilities increased significantly even though the total income went down in 2016 from 2015. It is not a good trend for liabilities to be increasing as sales and profits fall. However, Apple has a very high profit margin, relatively little debt compared to assets, and hot new products coming out every few years.

> **QUICK CHECK 1**
>
> The manager at Timberon Boots was concerned about current liabilities. The 2016 balance sheet showed loans and accounts payable of $3,879,000 and total assets of $18,243,000. The 2017 balance sheet showed loans and accounts payable of $5,125,000 and assets of $18,572,000. Do a vertical analysis and compare.

OBJECTIVE 2 Compare balance sheets using horizontal analysis. Perform a **horizontal analysis** by finding the change, both in dollars and in percent, for each item on the balance sheet from one year to the next. As before, always use the previous year as a base when finding the percents.

Using Horizontal Analysis **EXAMPLE 2** Apple's cash and short-term investments increased from $41,601 million at the end of 2015 to $67,155 at the end of 2016. The percent increase is found as follows.

$$\text{Percent increase} = \frac{\$67,155 - \$41,601}{\$41,601} = 61.4\% \text{ (rounded)}$$

Apple's cash and short-term investments grew by an amazingly high 61.4% during the year. A movement that significant does not happen by accident, so managers must have focused on building cash and short-term investments during 2016. Perhaps they were saving for an investment to be made soon, or perhaps they wanted to increase liquidity to decrease risk. Complete a horizontal analysis of the current assets portion of Apple's balance sheet. Round percentages to the nearest tenth.

SOLUTION

	Comparative Analysis Apple, Inc. Year Ending September 30, 2016 (in millions of dollars)			
Current Assets:	2015	2016	Amount of Increase/Decrease	Percent
Cash and short-term investments	$41,601	$67,155	$25,554	61.4%
Accounts receivable and other	$30,343	$29,299	($1,044)*	(3.4%)
Inventory	$17,434	$10,415	($7,019)	(40.3%)
Total Current Assets	$89,378	$106,869	$17,491	19.6%

*Parentheses indicate a negative number.

Accounts receivable and other remained about the same, but inventory went down substantially even as cash and short investments increased substantially. So the decrease in inventory explains part of the increase in cash and short-term investments, but not all of it. However, current assets still grew by a 19.6%. This buildup in cash and short-term investments suggests that Apple is even better prepared to handle any crisis such as a rapid fall-off in sales, or an aggressive move by a competitor into one of its primary markets.

> **QUICK CHECK 2**
>
> Use the data from Quick Check 1 to do a partial horizontal analysis by finding the percent increase for 2017 over 2016.

OBJECTIVE 3 Find financial ratios. **Financial ratios** can be calculated from the balance sheet. These ratios can be compared against earlier years for the same company or against the same financial ratios of other firms in the same industry. A ratio that is far out of line may indicate either financial difficulties or perhaps that the firm is doing better than expected.

The first two ratios we discuss are designed to measure the **liquidity** of a firm. Liquidity refers to a firm's ability to raise cash quickly without being forced to sell assets at a big loss.

Find the **current ratio** by dividing current assets by current liabilities. This ratio is read as "current assets to current liabilities" and is sometimes called **banker's ratio**.

Finding the Current Ratio

$$\text{Current ratio} = \frac{\text{Current assets}}{\text{Current liabilities}}$$

15.4 Analyzing the Balance Sheet

Finding the Current Ratio — **EXAMPLE 3** — The September 30, 2016, balance sheet for Apple shows current assets of $106,869 and current liabilities of $79,006, both in millions. Find the current ratio to the nearest hundredth.

SOLUTION

$$\text{Current ratio} = \frac{\$106,869}{\$79,006} = 1.35 \quad \text{(rounded)}$$

The current ratio is often written as 1.35 to 1, or as 1.35:1.

QUICK CHECK 3

A firm has current liabilities of $2.9 billion, with current assets of $1.56 billion. Find the current ratio.

Lending institutions and bondholders look at current ratio before making a loan. A common rule of thumb (not necessarily applicable to all businesses) is that the current ratio should be at least 1:1. A current ratio lower than 1 indicates that the firm might have problems meeting its short-term liabilities; a current ratio much over 2 suggests that managers are having difficulty managing working capital.

Apple's current ratio falls in the desired range. But this alone is not enough to conclude that the company is doing well. It is important to look at many different issues, such as financial conditions, the company, the industry, social trends, management, and the economy, before investing or making loans.

One disadvantage of the current ratio is that inventory is included in current assets. In a period of financial difficulty, a firm might have trouble disposing of its inventory at a reasonable price. Some accountants think that the **acid test** of a firm's financial health is to consider only **liquid assets**. Liquid assets are those that can quickly be converted to cash and are listed as current assets on balance sheets. As a general rule, the acid-test ratio should be at least 1 to 1, so that liquid assets are at least enough to cover current liabilities.

The **acid-test ratio**, also called the **quick ratio**, is defined as follows.

Quick TIP
It is very difficult to choose a stock in which to invest. That is why most financial professionals recommend investing in mutual funds or in funds that track the Standard & Poor's 500 index.

Finding the Acid-test Ratio

$$\text{Acid-test ratio} = \frac{\text{Liquid assets}}{\text{Current liabilities}}$$

Finding the Acid-Test Ratio — **EXAMPLE 4** — Compare the acid-test ratios for Apple for 2015 and 2016. Round the ratio to the nearest hundredth.

SOLUTION

	Current Assets	−	Inventory		Current Liabilities	Acid-Test Ratio
2015	$89,378	−	$17,434	= $71,944	$80,610	.89
2016	$106,869	−	$10,415	= $96,454	$79,006	1.22

The acid-test ratio was below 1 in 2015 but was well over 1 in 2016, so it showed clear improvement.

QUICK CHECK 4

Find the acid-test ratio for a company with current assets of $745,600, inventory of $217,050, and current liabilities of $240,800.

A company with a large amount of capital invested should have a higher net income than a company with a small amount invested. To check on this, accountants often find the **ratio of net income after taxes to average owners' equity**. The **average owners' equity** is found by adding the owners' equity at the beginning and end of the year and dividing by 2.

Finding Average Owners' Equity

$$\text{Average owners' equity} = \frac{\text{Owners' equity at beginning} + \text{Owners' equity at end}}{2}$$

Then the ratio of net income after taxes to average owners' equity is found.

Finding Ratio of Net Income After Taxes to Average Owners' Equity

$$\text{Ratio of net income after taxes to average owners' equity} = \frac{\text{Net income after taxes}}{\text{Average owners' equity}}$$

Finding the Return on Average Equity

EXAMPLE 5 Find the 2016 ratio of net income after taxes to average owners' equity for Apple.

SOLUTION

Stockholders' equity increased from $119,355 in 2015 to $128,249 in 2016.

$$\text{Average owners' equity} = \frac{\$119{,}355 + \$128{,}249}{2} = \$123{,}802 \text{ (in millions)}$$

Quick TIP
The ratio of net income after taxes to average owners' equity requires data from both the income statement and the balance sheet.

Use the net income after taxes for 2016 from the income statement in Section 15.1 to complete the following calculation.

$$\text{Ratio of net income after taxes to average owners' equity} = \frac{\$45{,}687}{\$123{,}802} = 36.9\%$$

QUICK CHECK 5

Stockholders' equity for a firm increased from $3456 at the end of 2016 to $3720 at the end of 2017. Net income after taxes during the year was $402 (all figures in thousands). Find the ratio of net income after taxes to average owners' equity.

The ratio of net income after taxes to average owners' equity should be significantly higher than the interest rate paid on interest-bearing investments such as bonds. Otherwise, investors would sell the stock and buy bonds, which are usually less risky than stocks. Generally, investments with a higher risk (e.g., stocks) must offer the chance for higher returns than those with less risk (e.g., bonds) in order to attract investors.

The ratio of net income after taxes to average owners' equity far exceeds rates on government bonds for Apple. In fact, at nearly 37%, it is unusually high. But we again note that it is important to look at more than one ratio before investing. It takes a lot of detective work to determine whether a company will continue to grow profitably and to make an informed decision related to investment or lending. Investment advisors tend to have a lot of training and experience as they try to separate out the good investments from the mediocre or poor, and they are NOT always correct!

All the ratios discussed here are more meaningful when comparing a company with another within the same industry. For example, it does not make sense to compare ExxonMobil to Apple, since the two are in very different businesses with very different capital and inventory needs. Rather, it is best to compare the financial ratios for Apple to a **direct competitor**, or one that produces similar products and/or offers similar services. However, it is difficult to find a direct competitor to Apple (one selling products in most of the same markets), which shows how difficult financial analysis can be.

The financial ratios of this section can be summarized as follows.

Summary of Ratios

$$\text{Current ratio} = \frac{\text{Current assets}}{\text{Current liabilities}} \qquad \text{Acid-test ratio} = \frac{\text{Liquid assets}}{\text{Current liabilities}}$$

$$\text{Ratio of net income after taxes to average owners' equity} = \frac{\text{Net income after taxes}}{\text{Average owners' equity}}$$

15.4 Exercises — MyLab Math

1. **YACHT CONSTRUCTION** Complete this balance sheet using vertical analysis. Round to the nearest tenth of a percent. (See Example 1.)

Comparative Balance Sheet for Pleasure Yacht, Inc. (in thousands of dollars)

	Amount Last Year	Percent Last Year	Amount This Year	Percent This Year
Assets:				
Current Assets				
Cash	$42,000	_____	$52,000	_____
Notes Receivable	$6,000	_____	$8,000	_____
Accounts Receivable	$120,000	_____	$148,000	_____
Inventory	$120,000	_____	$153,000	_____
Total Current Assets	_____	_____	_____	_____
Plant Assets				
Land	$8,000	_____	$10,000	_____
Buildings	$11,000	_____	$14,000	_____
Fixtures	$13,000	_____	$15,000	_____
Total Plant Assets	_____	_____	_____	_____
Total Assets	_____	100%	_____	100%
Liabilities:				
Current Liabilities				
Accounts Payable	$4,000	_____	$3,000	_____
Notes Payable	$152,000	_____	$201,000	_____
Total Current Liabilities	_____	_____	_____	_____
Long-Term Liabilities				
Mortgages Payable	$16,000	_____	$20,000	_____
Long-Term Notes Payable	$42,000	_____	$58,000	_____
Total Long-Term Liabilities	_____	_____	_____	_____
Total Liabilities	_____	_____	_____	_____
Owners' Equity	$106,000	_____	$118,000	_____
Total Liabilities And Owners' Equity	_____	_____	_____	_____

CHAPTER 15 Financial Statements and Ratios

2. YACHTS Complete the following horizontal analysis for a portion of the balance sheet for Pleasure Yacht, Inc. Note that figures are shown in thousands of dollars. Round to the nearest tenth of a percent. (See Example 1.)

Pleasure Yacht, Inc. (in thousands)

	Last Year	This Year	Increase or (Decrease) Amount	Percent
Assets:				
Current Assets				
Cash	$42,000	$52,000		
Notes Receivable	$6,000	$8,000		
Accounts Receivable	$120,000	$148,000		
Inventory	$120,000	$153,000		
Total Current Assets	$288,000	$361,000		
Plant Assets				
Land	$8,000	$10,000		
Buildings	$11,000	$14,000		
Fixtures	$13,000	$15,000		
Total Plant Assets	$32,000	$39,000		
Total Assets	$320,000	$400,000		

In Exercises 3–6, find (a) the current ratio and (b) the acid-test ratio. Round each ratio to the nearest hundredth. (c) Do the ratios suggest that the company is financially healthy, according to the guidelines given in the text? Dollar figures are in thousands of dollars. (See Examples 3 and 4.)

3. Pleasure Yacht, Inc., has the balance sheet given above and current liabilities of $204,000.

(a) _____
(b) _____
(c) _____

4. **JUICE COMPANY** Jungle Juice has current liabilities of $356,800, cash of $32,800, notes and accounts receivable of $248,500, and an inventory valued at $82,400.

(a) _____
(b) _____
(c) _____

5. **CADILLAC DEALER** Wagner Cadillac has current assets of $2,210,350, current liabilities of $1,232,500, total cash of $480,500, notes and accounts receivable of $279,050, and an inventory of $1,450,800.

(a) _____
(b) _____
(c) _____

6. **OXYGEN SUPPLY** BlueTex Oxygen Supply has current assets of $2,234,000, current liabilities of $840,000, total cash of $339,000, notes and accounts receivable of $1,215,000, and an inventory of $680,000.

(a) _____
(b) _____
(c) _____

A portion of a comparative balance sheet is shown next. First complete the chart, and then find the current ratio and the acid-test ratio for the indicated year. Round each ratio to the nearest hundredth.

Western Auto Supply (in thousands)

	Amount Last Year	Percent Last Year	Amount This Year	Percent This Year
Current Assets				
Cash	$15,000	_____	$12,000	_____
Notes Receivable	$6,000	_____	$4,000	_____
Accounts Receivable	$18,000	_____	$22,000	_____
Inventory	$24,000	_____	$26,000	_____
Total Current Assets	$63,000	84%	$64,000	80%
Total Plant Assets	$12,000	_____	$16,000	_____
Total Assets	_____	100.0%		100.0%
Total Current Liabilities	$25,000	_____	$30,000	_____

7. This year _____

8. Last year _____

Find the ratio of net income after taxes to average owners' equity for the following. (See Example 5.) Round to the nearest tenth of a percent.

9. AUTOMOTIVE REPAIR Jesse's Auto Repair had a stockholders' equity of $845,000 at the beginning of the year and $928,500 at the end of the year. Net income after taxes for the year was $54,400.

9. _____

10. TRANSMISSION REPAIR Owners' equity at Amgen Transmission was $372,600 at the beginning of the year and $402,100 at the end of the year. Net income after taxes for the year was $55,003.

10. _____

Calculate the current ratio and the acid-test ratio for the following companies. Are the companies healthy, based on the guidelines given in the text?

11. Akron Books: Current assets: $268,700
Current liabilities: $294,200
Liquid assets: $109,900

11. _____

12. Ashe Hair Salon: Current assets: $54,750
Current liabilities: $17,200
Liquid assets: $24,650

12. _____

13. Explain why the acid-test ratio is a better measure of the financial health of a firm than the current ratio. (See Objective 3.)

14. Explain why increased risk requires a higher return on investment. (See Objective 3.)

CHAPTER 15 QUICK CHECK ANSWERS

1. 2016, 21.3%; 2017, 27.6%; large increase in accounts payable
2. accounts payable, 32.1% increase; total assets, 1.8% increase
3. .54
4. 2.19
5. 11.2%

Chapter 15 Quick Review

Chapter Terms Review the following terms to test your understanding of the chapter. For each term you do not know, refer to the page number found next to that term.

accountant [p. 611]
acid test [p. 629]
acid-test ratio [p. 629]
assets [p. 623]
average owners' equity [p. 629]
balance sheet [p. 623]
banker's ratio [p. 628]
comparative balance sheet [p. 627]
comparative income statement [p. 616]
cost of goods sold [p. 611]
current assets [p. 623]

current liabilities [p. 623]
current ratio [p. 628]
direct competitor [p. 630]
ending inventory [p. 612]
financial ratios [p. 628]
fixed assets [p. 623]
gross profit [p. 611]
gross profit on sales [p. 611]
gross sales [p. 611]
horizontal analysis [p. 618]
income statement [p. 611]
initial inventory [p. 612]
liabilities [p. 623]

liquid assets [p. 629]
liquidity [p. 628]
long-term liabilities [p. 623]
market share [p. 618]
NASDAQ [p. 619]
net income [p. 611]
net income after taxes [p. 611]
net income before taxes [p. 611]
net sales [p. 611]
net worth [p. 623]
operating expenses [p. 611]
overhead [p. 611]
owners' equity [p. 623]

plant and equipment [p. 623]
plant assets [p. 623]
publicly held corporation [p. 611]
quick ratio [p. 629]
ratio of net income after taxes to average owners' equity [p. 629]
returns [p. 611]
short-term liabilities [p. 623]
stockholders' equity [p. 623]
total revenue [p. 611]
vertical analysis [p. 616]

CONCEPTS

15.1 Finding the gross profit and net income

1. Find the net sales.
2. Determine the cost of goods sold.
3. Find gross profit from the formula.

    Gross profit = Net sales − Cost of goods sold

4. Find the operating expenses.
5. Find the net income from the formula.

    Net income = Gross profit − Operating expenses

15.2 Finding the percent of net sales of individual items

1. Determine net sales using the formula.

    Net sales = Gross sales − Returns

2. Use the formula

    Percent of net sales = $\dfrac{\text{Particular item}}{\text{Net sales}}$

    for each item.

EXAMPLES

Candy-You-Love had a cost of goods sold of $218,509, operating expenses of $103,217, gross sales of $432,119, and no returns. Find the gross profit and net income before taxes.

Gross profit = Gross sales − Returns − Cost of goods sold
            = $432,119 − $0 − $218,509
            = **$213,610**

Net income before taxes = Gross profit − Operating expenses
            = $213,610 − $103,217
            = **$110,393**

Bill's Appliances lists the following information for a month.

Gross sales = $340,000   Salaries and wages = $19,000
Returns = $15,000   Rent = $8000
Cost of goods sold = $210,000   Advertising = $12,000

Express each item as a percent of net sales. Round to the nearest tenth of a percent.

Net sales = $340,000 − $15,000 = **$325,000**
Gross profit = $325,000 − $210,000 = **$115,000**
Total expenses = $19,000 + $8000 + $12,000 = **$39,000**
Net income = $115,000 − $39,000 = **$76,000**

Find all the desired percents to the nearest tenth of a percent by dividing each item by net sales.

Percent gross sales = $\dfrac{\$340,000}{\$325,000}$ = **104.6%**

Percent returns = $\dfrac{\$15,000}{\$325,000}$ = **4.6%**

Percent cost of goods sold = $\dfrac{\$210,000}{\$325,000}$ = **64.6%**

CHAPTER 15 Financial Statements and Ratios

CONCEPTS	EXAMPLES
	Percent gross profit $= \dfrac{\$115{,}000}{\$325{,}000} = 35.4\%$
	Percent expenses $= \dfrac{\$39{,}000}{\$325{,}000} = 12\%$
	Percent salaries and wages $= \dfrac{\$19{,}000}{\$325{,}000} = 5.8\%$
	Percent net income $= \dfrac{\$76{,}000}{\$325{,}000} = 23.4\%$
	Percent rent $= \dfrac{\$8{,}000}{\$325{,}000} = 2.5\%$
	Percent advertising $= \dfrac{\$12{,}000}{\$325{,}000} = 3.7\%$

15.2 Comparing income statements with published charts

In one chart, list the percent of items from a published chart and from the particular company.

From the preceding example, prepare a vertical analysis of Bill's Appliances.

The table shows that this store has very low expenses and a slightly higher than average gross profit.

Bill's Appliances

	Cost of Goods	Gross Profit	Total Expenses	Net Income	Wages	Rent	Advertising
Bill's Appliances	64.6%	35.4%	12%	23.4%	5.8%	2.5%	3.7%
Appliances	66.9%	33.1%	26%	7.2%	11.9%	2.4%	2.5%

15.2 Preparing a horizontal analysis chart

1. List last year's and this year's values for each item.
2. Calculate the amount of the increase or decrease of each item.
3. Calculate the percent increase or decrease by dividing the change by last year's amount.

The results of a horizontal analysis of the portion of a business is given. Calculate the percent increase or decrease in each item.

Salmon Farm

	This Year	Last Year	Increase or (Decrease) Amount	Percent
Gross Sales	$735,000	$700,000	$35,000	5.0%
Returns	$5,000	$10,000	($5,000)	(50.0%)
Net Sales	$730,000	$690,000	$40,000	5.8%
Cost of Goods Sold	$530,000	$540,000	($10,000)	(1.9%)
Gross Profit	$200,000	$150,000	$50,000	33.3%

15.3 Constructing a balance sheet

List all of the current assets, other assets, current liabilities, and other liabilities on one page. Subtract total liabilities from total assets to find stockholders' equity.

A cell phone kiosk in a local mall has cash of $28,300, accounts receivable of $49,250, and inventories of $4900. Other assets consist of equipment and a truck with a total value of $24,300. Accounts payable are $9300, and loans total $12,200. There is no long-term debt. Other liabilities amount to $12,400.

Balance Sheet

Current Assets:		Current Liabilities:	
Cash and Equivalents	$28,300	Accounts Payable	$9,300
Accounts Receivable	$49,250	Loans and Notes Payable	$12,200
Inventories	$4,900	Total Current Liabilities	$21,500
Total Current Assets	$82,450	Other Liabilities:	$12,400
Other Assets:		Total Liabilities	$33,900
Equipment and Truck	$24,300		
Total Assets	$106,750	Stockholders' Equity	$72,850
		Total Liabilities and Equity	$106,750

CONCEPTS	EXAMPLES
15.4 Determining the value of the current ratio 1. Determine the current assets. 2. Find the current liabilities. 3. Divide current assets by current liabilities.	Caribbean Tours has $250,000 in current assets and $110,000 in current liabilities. Find the current ratio. $$\text{Current ratio} = \frac{\text{Current assets}}{\text{Current liabilities}}$$ $$= \frac{\$250,000}{\$110,000} = 2.27 \text{ (rounded)}$$
15.4 Finding the value of the acid-test ratio 1. Determine the liquid assets. 2. Find the current liabilities. 3. Divide liquid assets by liabilities.	If Carribean Tours has $125,000 in liquid assets, find the acid-test ratio. $$\text{Acid-test ratio} = \frac{\text{Liquid assets}}{\text{Current liabilities}}$$ $$= \frac{\$125,000}{\$110,000} = 1.14 \text{ (rounded)}$$
15.4 Determining the ratio of net income after taxes to the average owners' equity 1. Find the net income after taxes. 2. Determine the average owners' equity for the year using the formula. $$\text{Average owners' equity} = \frac{\text{Owners' equity at beginning} + \text{Owners' equity at end}}{2}$$ 3. Divide the net income by the average owners' equity.	At the beginning of the year, Caribbean Tours had owners' equity of $140,000. At the end of the year, owners' equity was $180,000. The net income after taxes for the agency was $25,000. Find the ratio of net income after taxes to average owners' equity. Net income after taxes = $25,000 Average owners' equity $$= \frac{\$140,000 + \$180,000}{2} = \$160,000$$ Ratio of net income after taxes to average owners' equity $$= \frac{\$25,000}{\$160,000} = 15.6\% \text{ (rounded)}$$

Case Study

BICYCLE SHOP

Last year, John Smarter started a business selling and repairing bicycles. He is somewhat successful, but he is tired of working 60 hours a week and dealing with part-time employees. With borrowed funds, Smarter believes he will be able to expand to a larger store and hire one or two full-time people to help. Help Smarter complete his financial statements to give to the banker.

1. Prepare an income statement and find net income after taxes

Gross Sales	$238,300	Salaries and Wages	$78,690
Returns	$3,600	Rent & Utilities	$11,800
Inventory on Jan. 1	$53,400	Advertising	$3,200
Cost of Goods Purchased	$98,500	Miscellaneous Expenses	$4,800
Freight	$2,300	Income taxes	$3,600
Inventory on Dec. 31	$48,900		

```
Gross Sales                              _____
Returns                                  _____
Net Sales                                _____
    Inventory, Jan. 1            _____
    Cost of Goods Purchased      _____
    Freight                      _____
    Total Cost of Goods Purchased       _____
    Total of Goods Available for Sale   _____
    Inventory, Dec. 31                  _____
Cost of Goods Sold                       _____
Gross Profit                             _____
    Expenses
        Salaries and Wages       _____
        Rent & Utilities         _____
        Advertising              _____
        Miscellaneous Expenses   _____
        Total Expenses                   _____
Net Income before Taxes                  _____
    Income taxes                         _____
NET INCOME AFTER TAXES
```

2. Year-end figures for the bike shop are: $62,000 in cash, $10,700 in accounts receivable, $48,900 in inventory, land and building worth $0, improvements worth $41,500. It also has $4,300 in notes payable, and $14,800 in accounts payable. Long-term notes payable are $12,300, no mortgages. Prepare a balance sheet and find the owners' equity.

ASSETS		
Current Assets		
Cash	_____	
Accounts Receivable	_____	
Inventory	_____	
Total Current Assets		_____
Plant Assets		
Land and Buildings	_____	
Improvements	_____	
Total Plant Assets		_____
Total Assets		_____
LIABILITIES		
Current Liabilities		
Notes Payable	_____	
Accounts Payable	_____	
Total Current Liabilities		_____
Long-Term Liabilities		
Long-Term Notes Payable	_____	
Mortgages Payable	_____	
Total Long-Term Liabilities		_____
Total Liabilities		_____
OWNERS' EQUITY		
Owners' Equity		_____
TOTAL LIABILITIES AND OWNERS' EQUITY		_____

[1]Owners' equity = Total Assets − Total Liabilities
= $163,100 − $31,400
= $131,700

INVESTIGATE

Publicly held companies must make their financial statements available to anyone who wishes to look at them. Choose your favorite publicly held company and find recent financial statements using the Internet—a librarian can help if needed. Calculate the current ratio, acid-test ratio, and ratio of net income after taxes to average owners' equity.

Case in Point Summary Exercise

APPLE, INC.

www.apple.com

Facts:

- 1976: Founded
- 1984: Introduced Macintosh computer
- 2001: Introduced iPod
- 2007: Introduced iPhone
- 2010: Introduced iPad
- 2015: Introduced Apple watch

Apple, Inc. was founded by Steve Jobs and Steve Wozniak on April 1, 1976 in Cupertino, California. It was originally named Apple Computer, Inc. based on its original focus on building computers. Management changed the name to Apple, Inc. in 2007 to reflect the fact that the firm was far more focused on a variety of consumer electronics. The firm has revolutionized several industries with its creative and artful products including the iPod (music), iPhone (smart phone), iPad (books, periodicals, movies, music, games, applications, and web content), iTunes (playing and downloading digital music and video files), and Apple watch (computer on your wrist).

Apple has about 500 retail stores in 20 countries, a worldwide online presence, more than 100,000 employees, and annual revenue over $200 billion. It manages a global supply chain, with much of its production in Asia, but with high sales volumes in the United States, Europe, and Asia.

1. Assume the following figures are from a store that sells products that compete with Apple. Prepare an income statement and find net income after taxes for the quarter.

Gross Sales	$834,200	Salaries and Wages	$193,200
Returns	$4,500	Rent & Utilities	$68,900
Inventory on Jan. 1	$84,200	Advertising	$13,900
Cost of Goods Purchased	$346,500	Insurance & Payroll Taxes	$19,400
Freight	$9,100	Miscellaneous Expenses	$18,700
Inventory on Dec. 31	$96,200	Income taxes	$37,800

```
Gross Sales                                     _____
Returns                                         _____
Net Sales                                                  _____
    Inventory, Jan. 1                _____
    Cost of Goods Purchased          _____
    Freight                          _____
    Total Cost of Goods Purchased               _____
    Total of Goods Available for Sale           _____
    Inventory, Dec. 31                          _____
Cost of Goods Sold                                         _____
Gross Profit                                               _____
    Expenses
        Salaries and Wages                      _____
        Rent & Utilities                        _____
        Advertising                             _____
        Insurance and Payroll Taxes             _____
        Miscellaneous Expenses                  _____
            Total Expenses                                 _____
Net Income before Taxes                                    _____
    Income Taxes                                           _____
NET INCOME AFTER TAXES                                     _____
```

2. At the end of the year, the store had the following: $84,500 in cash, $2100 in accounts receivable, land and building with a fair market value of $186,500, improvements worth $82,100. It has a note payable on store shelves with a remaining balance of $4900 and $58,400 in accounts payable for supplies. The only long-term debt is the mortgage of $134,200 on the building. Prepare a balance sheet for the firm at the end of the year and find the owners' equity.

ASSETS		
Current Assets		
Cash	_____	
Accounts Receivable	_____	
Inventory	_____	
Total Current Assets		_____
Plant Assets		
Land and Buildings	_____	
Improvements	_____	
Total Plant Assets		_____
Total Assets		_____
LIABILITIES		
Current Liabilities		
Notes Payable	_____	
Accounts Payable	_____	
Total Current Liabilities		_____
Long-Term Liabilities		
Long-Term Notes Payable	_____	
Mortgages Payable	_____	
Total Long-Term Liabilities		_____
Total Liabilities		_____
OWNERS' EQUITY		
Owners' Equity		_____
TOTAL LIABILITIES AND OWNERS' EQUITY		_____

^{1}Owners' equity = Total assets − Total liabilities
                 = $451,400 − $197,500
                 = $253,900

Discussion Question: Explain the purpose of the income statement and balance sheet. Can an individual or family also build an income statement and balance sheet? Explain.

Chapter 15 Test

To help you review, the numbers in brackets show the section in which the topic was discussed.

1. Benni's Fish Co. had gross sales of $756,300 with returns of $285. The inventory on January 1 was $92,370, and the cost of goods purchased during the year was $465,920. Freight costs during the year were $1205. Total inventory on December 31 was $82,350. Salaries and wages totaled $84,900, advertising was $2800, rent was $42,500, utilities were $18,950, taxes on inventory and payroll were $4500, and miscellaneous expenses totaled $18,400. Income taxes were $25,450. Complete the following income statement. **[15.1]**

Benni's Fish Co.
Income Statement
Year Ending December 31

Gross Sales	_____
Returns	_____
Net Sales	_____
Inventory, January 1	_____
Cost of Goods Purchased	_____
Freight	_____
Total Cost of Goods Purchased	_____
Total of Goods Available for Sale	_____
Inventory, December 31	_____
Cost of Goods Sold	_____
Gross Profit	_____
Expenses	
Salaries and Wages	_____
Rent	_____
Advertising	_____
Utilities	_____
Taxes on Inventory and Payroll	_____
Miscellaneous Expenses	_____
Total Expenses	_____
Net Income before Taxes	_____
Income Taxes	_____
Net Income after Taxes	_____

2. Complete a horizontal analysis for the following portion of an income statement. Round to the nearest tenth of a percent. **[15.2]**

China Imports, Inc.
Comparative Income Statement
(in thousands)

			Increase or (Decrease)	
	Last Year	This Year	Amount	Percent
Net Sales	$60,000	$95,000	_____	_____
Cost of Goods Sold	$40,000	$63,000	_____	_____
Gross Profit	$12,000	$16,000	_____	_____

3. Complete the following chart for Alberta Heights Service Station. Express each item as a percent of net sales, and then write in the appropriate average percent from the chart on page 617. Round to the nearest tenth of a percent. **[15.2]**

Alberta Heights Service Station

	Amount (in thousands)	Percent	Average Percent
Net Sales	$1200	100%	100%
Cost of Goods Sold	_____	_____	_____
Gross Profit	$325	_____	_____
Net Income	$112	_____	_____
Wages	$129	_____	_____
Rent	$72	_____	_____
Total Expenses	$213	_____	_____

Find (a) the current ratio and (b) the acid-test ratio for each firm. (Round to the nearest hundredth.) **[15.4]**

4. J & L Construction:
   - Current assets: $2,482,500
   - Current liabilities: $1,800,200
   - Cash: $850,000
   - Notes and accounts receivable: $680,100
   - Inventory: $952,400

   (a) _____
   (b) _____

5. Ben Franklin Tobacco:
   - Current assets: $154,000
   - Current liabilities: $146,500
   - Cash: $22,000
   - Notes and accounts receivable: $32,500
   - Inventory: $99,500

   (a) _____
   (b) _____

Find the ratio of net income after taxes to average owners' equity for each of the following firms. (Round to the nearest tenth of a percent.) **[15.4]**

6. Talisman Imports:
   - Net income after taxes: $148,200
   - Owners' equity
     - beginning of year: $472,600
     - end of year: $514,980

   6. _____

7. Burns Commercial Properties:
   - Net income after taxes: $8,465,000
   - Owners' equity
     - beginning of year: $28,346,000
     - end of year: $36,450,000

   7. _____

Budgeting and Business Statistics 16

CHAPTER CONTENTS

16.1 Planning and Budgeting

16.2 Frequency Distributions and Graphs

16.3 Mean, Median, and Mode

CASE IN POINT

BEV JOHNSON has always been interested in food and restaurants. She worked in a restaurant as a teenager and then later worked in a bistro when an exchange student in France. She received her college degree in Restaurant Management and eventually opened her own business, naming it after herself: Bev's Deli.

Bev uses budgeting to help her control and track expenses on a monthly basis. She uses statistics to help her better understand trends and related data.

The first topic in this chapter is planning and budgeting. Business managers do this all the time, but it is a very worthwhile endeavor for individuals and families too. Look at these clippings to see a family and a business owner that did not plan and budget well enough.

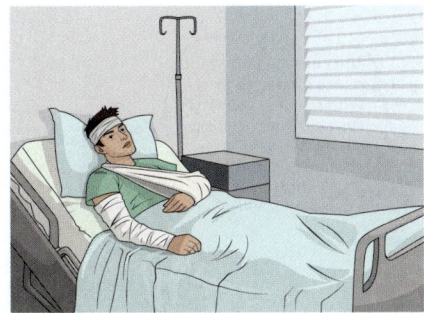

Family Needs Help!

Tony Chavez is a diesel mechanic and his wife is a school teacher. Their incomes allowed them to borrow to buy a house and two new cars. But they have a lot of debt and no savings, and the family was in a car accident. Everyone is okay, except for Tony who will not be able to work for several months. Bills have piled up this month, and it is becoming clear that they are likely to lose one automobile and perhaps their home, which would force them to move back into an apartment. The family is desperate!

Area Business Declares Bankruptcy!

After working in a wedding shop for several years, Janice Walker opened her own business. It grew nicely and was profitable by the second year. However, the largest employer in town has announced a big layoff due to a recession. Sales at her shop are falling and Janice doesn't have any savings. She cannot pay the rent for her business space this month. She doesn't know where to turn and can't sleep.

Read on to find out how to reduce the chances of something like the above happening to you.

16.1 Planning and Budgeting

OBJECTIVES

1. Understanding expected and unexpected expenses.
2. Controlling expenses.
3. Controlling your financial future using planning and budgeting.

OBJECTIVE 1 Understanding expected and unexpected expenses. Your financial life begins as you choose a field to study in college, borrow money, and then find work in a career. It is best to choose a field and a career that interest you. You are likely to work in that field for a long time, although people sometimes change careers later in life. The first and a very important step is to work hard to get the right education for entry and success in your chosen field.

The tables in Section 6.1 (page 209) and Section 6.2 (page 219) may help you think about education and careers. The first shows that average income tends to increase sharply with level of education; the second shows average earnings for people in several different careers. Use the tables to stimulate your thinking—do not limit yourself to just the information in the tables, since they are incomplete. There are many other great careers and not everyone makes the average income in any one career. It is also a good idea to talk to people who have actually worked in a career you are considering. They can tell you what it is really like and give you valuable suggestions. Be prepared with a list of really good questions.

There are many factors to consider in choosing a career. What college degree will you need? How long will you have to go to college and what will the costs be? How will you pay for college? Where will you find a job? Will you have to move to a different city? How stable is employment likely to be in that career? What is a typical workday like? Will you enjoy working in that field?

646 CHAPTER 16 Budgeting and Business Statistics

Although these are career-oriented questions, there are many other reasons to pursue an education. It will help you be more aware of what is going on around you. It will give you the knowledge and critical thinking skills to better understand what is happening in the world and help you throughout your life. Employers want to hire people who can think critically, communicate well, and get along with others. No matter what your field of study, a college education will serve you well.

It is also worthwhile to try to keep risk at a manageable level, for example, by making sure you work for a stable employer that pays benefits and by not taking on too much debt. As the table in Section 13.5 (page 550) shows, you can reduce risk of serious illness and all the associated costs and suffering if you do a few things: quit smoking, exercise frequently, do not drive while intoxicated, and stick to a diet rich in vegetables. Financial risk to you and your family can also be reduced through planning and budgeting.

Some young people say it is impossible to save these days, and for those without stable income and/or very high bills it probably is. For one thing, student loans make it more difficult for many new graduates to save at first. Others can only save maybe $50 a month and say, "Why bother? We'd rather just eat out and go to a movie tonight." Yet it is important to consider saving in the larger scheme of things. Think about this section in terms of your life after graduating from college and once you have a good job.

Everyone has regular, ongoing expenses, which we can call **expected expenses**. But many **unexpected expenses** also occur with little or no warning. In fact, we can start a list—the blank rows at the bottom are for you to make your own additions.

> **Quick TIP**
> Can you think of other things to add to each list?

Expected Expenses	Unexpected Expenses
Rent	Vehicle breaks down
Groceries	Loss of job
Car payment	Divorce
Child care	Medical bills
Car insurance	Spouse decides to go back to school
Entertainment	Brother loses job and needs help
Clothes and school supplies	Unexpected pregnancy
Smart phone, cable TV	Cannot work due to an injury
Health insurance	Someone breaks in and steals everything
Medications	Roof starts leaking

It is easy to determine your total income after taxes if you have a steady job and do not have investments. Many people compare their monthly income to expected expenses and don't think too much about, or plan for, unexpected expenses. However, unexpected expenses are a fact of life, aren't they? Have you ever had a sudden expense that created a problem for you? We all have and so have all businesses. The odds are higher that you will have unexpected expenses if you are married or living with a significant other, and especially after having a child or two. You never know when the bills will come, or how high they will be.

What is the best way to deal with unexpected expenses? The answer lies in trying to get a rough estimate on what unexpected expenses might reasonably be and then setting aside some money where it WILL NOT BE SPENT for day-to-day things such as eating out. Since you never know in advance what unexpected expenses will hit you, at first you just have to make your best guess of how much you will need. In time, you will develop a better sense of how much to set aside.

> **Quick TIP**
> Saving a little every month reduces the chances of a financial shock to your family from an unexpected expense such as your car dying unexpectedly resulting in you having to buy another when you do not have the down payment.

The way to avoid becoming one of the people in the clippings at the beginning of this section is to plan (estimate or even guess how much you need to save) and then do it. It is worth doing even if you can save only $50 a month. It won't look like much at first: $50 saved at the end of the first month, $100 saved at the end of the second month, $150 saved at the end of the third month, etc. However, be patient! By saving, you are moving some of the unexpected expenses you will face at some point into the "expected expenses" category. The table above now looks like the following.

Expected Expenses	Unexpected Expenses
Rent	Hopefully there aren't any big ones!
Groceries	
Car payment	
Child care	
Insurance	
Entertainment	
Clothes and school supplies	
Smart phone, cable TV	
Health insurance	
Medications	
Saving for unexpected expenses	

Here is another way to think about it.

| Save some each month | → | Now and then something unexpected generates bills | → | Pay the bills with savings resulting in less or no debt |

Quick TIP

It is important NOT to draw money out of savings for day-to-day things such as entertainment or eating out. That will destroy your savings and defeat the purpose of saving.

When you avoid (or minimize) debt, you avoid (reduce) interest payments—which is yet another expense, as we saw in Chapter 12 on business and consumer loans. So that saves you money. And you may earn a little interest on your savings, which also helps a little (see Chapters 10 and 11 on compound interest and annuities). At this point, you are starting to control your financial life rather than it controlling you.

We have not yet talked about saving for larger purchases or expenses that you expect to have at some point in the future. For example, what will it cost to replace your automobile in two or three years when needed? What down payment must you have to buy a house in, say, 6 years? How much should you save to try to help your kids go to college in 15 years? Finally, what about retirement? The process is the same for these goals as for the unexpected expenses listed above. Estimate the amount needed for each goal, and try to also set that amount aside in a different account since you really do not want to touch these funds for years. You might have to wait until you graduate from college and have a good job before you can plan and save at this level, but a professional job and careful planning and budgeting will allow you to do so in time.

According to the Economic Policy Institute, the median savings of families nearing retirement is about $17,000. This means that one-half of all families nearing retirement have savings of less than $17,000 and one-half of families nearing retirement have savings of more than $17,000. Think about it: How long can a family of two adults in their 60s live on $17,000? If they have worked they will also get Social Security payments, but the average Social Security check is about $1300 per month, and one spouse often receives less than the other due to different work histories. So these families may have an income of around $2000 per month from Social Security and $17,000, and that is ALL they have to live on for the remainder of their lives. They are likely to have problems replacing a car, paying unexpected medical bills, and helping their kids with any financial problems. They will also probably have to limit things like going out to eat or taking costly vacations.

As we saw in Chapter 11, saving early makes a huge difference over time. You don't have to save a lot each month—but start early, and then just keep saving and let compound interest work for you!

OBJECTIVE 2 Controlling expenses. Life is full of expenses, many of which are hard to control! Some people never have enough money to pay all their bills and therefore depend a lot on borrowing. Believe it or not, this is also true for many people making more than $100,000 a year. Are you impressed by a family with a new, large home with a manicured lawn and two new cars in the garage? Don't be—they may be up to their ears in debt and might have to file for bankruptcy in a few years. Let's look at an example that shows how to start working on a budget.

Examining Monthly Expenses — EXAMPLE 1

After the birth of their first child, Sarah and James Livingston realize they have too much debt, so they decide to work on creating a budget. They quickly see that some of the payments on their debt are made weekly, others monthly. They decide not to include their payments on credit cards, since those were based on past spending and borrowing and the couple is determined to change those habits. They draw up the table below, but have no idea what to put down for "Other expenses."

Item	Expense	Frequency	Annual Cost
Rent	$700	monthly	$8,400 = $700 × 12 months in a year
Utilities	$95	monthly	$1,140
Groceries	$195	weekly	$10,140 = $195 × 52 weeks in a year
Car payment	$190	monthly	$2,280
Car insurance	$130	monthly	$1,560
Phone/Internet/TV	$140	monthly	$1,680
Entertainment	$130	weekly	$6,760
Health insurance	$175	monthly	$2,100
Other expenses	?	?	?
		Total	$34,060

This amounts to about $2838 a month of spending, not including payments on credit cards or other expenses. However, they realize that they need to get a sense of those other expenses, and include it in the table to understand what is really going on.

> **QUICK CHECK 1**
>
> Prepare a similar table for yourself, your family, or your parents' family. Try to think of every category of ongoing, regular expenses.

Quick TIP
Controlling costs is important for every family and business.

OBJECTIVE 3 Controlling your financial future using planning and budgeting. Budgeting requires that you first have a good understanding of all your income and expected expenses and a reasonable estimate of your unexpected expenses. Once you have these, you can plan for the future. Controlling costs is always part of the process. Business managers do detailed budgeting and planning for their firms at least once a year. If they can do it for their firms, you can do it for your family.

You do not need exact numbers, but they should be reasonably close. The budgeting and planning process will make you more aware of income and expenses, which in turn will help you become more aware of your finances.

Estimating Other Expenses — EXAMPLE 2

The Livingstons decided to track every dollar spent for three months and found that their average "Other expenses" came to $482 a month. They were surprised it was that high. The costs included: clothes, diapers and the cost of a bed for the baby, shoes for dad, replacing a broken window, copays for visits to doctors and a dentist, an unexpected repair on the car, and other things.

They decide to work to reduce their "Other expenses" to about $330 a month by James repairing the car himself, Sarah getting a friend to cut her hair rather than using a hairdresser, and the couple shopping more carefully. They also decide to cut entertainment costs, drop cable TV, and maybe save a little on utilities. On the basis of what they have learned about managing finances and unexpected costs, they decide to put $80 a month into savings.

Quick TIP
The Livingston's know it will be tough to cut expenses, but they are committed.

Item	Expense	Frequency	Annual	Monthly Budget
Rent	$700	monthly	$8,400	$700 = $8,400 ÷ 12 months
Utilities	$85*	monthly	$1,020	$85
Groceries	$195	weekly	$10,140	$845
Car payment	$190	monthly	$2,280	$190
Car insurance	$130	monthly	$1,560	$130
Phone/Internet	$95*	monthly	$1,140	$95
Entertainment	$70*	weekly	$3,640	$303 (rounded)
Health insurance	$175	monthly	$2,100	$175
Other expenses	$330	monthly	$3,960	$330
Saving for future unexpected expenses	$80	monthly	$960	$80
		Totals	$35,200	$2,933

*These costs were reduced by the Livingston's.

The totals do not include payments on credit-card debt that the Livingstons must still pay, nor do they include savings for big things like replacing a car, but it is a start. Although they must live on this tight budget for about eight months until the credit-card debt is paid down and the car loan is paid off, they plan to stick to it. They understand they are making short-term sacrifices for themselves and their baby. Even more importantly, they have resolved to more carefully manage their finances in the future.

QUICK CHECK 2

If possible, go back three months and determine your average "Other expenses" cost, which includes all expenses not included in your list in Example 1. This may be difficult due to many small, scattered purchases, some of which were made using cash. If you cannot get the historical data, take your best guess at average "Other expenses" and estimate what you need to put into Savings for future unexpected expenses. Prepare a table similar to the one in Example 2. You can track your actual expenses for the next three months to find your average and then use that more accurate figure.

The next step is to actually observe what happens from month to month, which is likely to result in your making changes to the budget and adjusting your lifestyle to reduce expenses. Tracking actual expenses against your budget will increase your awareness and help you budget and plan in the future.

Comparing Actual Expenses to Budget — **EXAMPLE 3** // The Livingstons work very hard the first month to cut costs after completing their budget and plan. They then prepare the following table to compare actual to budgeted expenses for the first month.

Item	Expense	Frequency	Annual	Monthly Budget*	Actual Expense for Month
Rent	$700	monthly	$8,400	$700	$700
Utilities	$85	monthly	$1,020	$85	$84
Groceries	$195	weekly	$10,140	$845	$867
Car payment	$190	monthly	$2,280	$190	$190
Car insurance	$130	monthly	$1,560	$130	$130
Phone/Internet/TV	$95	monthly	$1,140	$95	$95
Entertainment	$70	weekly	$3,640	$303	$360
Health insurance	$175	monthly	$2,100	$175	$175
Other expenses	$330	monthly	$3,960	$330	$385
Saving for future unexpected expenses	$80	monthly	$960	$80	$80
Totals			$35,200	$2,933	$3,066

Comparing the last two columns, the Livingstons spent $133 more than their budget for the month and they have saved $80 for future unexpected expenses. They also made large enough payments on their credit cards to reduce the debt a little and avoided new charges. Entertainment costs were higher than expected because Sarah's sister and family visited them for several days. But the Livingstons are excited about their progress and intend to continue. They will have to be very careful because we all know how easy it is to be lulled into buying something (eating out, new clothes for the baby, Starbucks drinks, movies with friends, etc.).

If I do not budget and take control of my expenses, who will? What would be the result if I do not control my expenses?

QUICK CHECK 3

If you can, get your actual expenses for last month and compare to the budget you created for yourself in Example 2. If not, estimate the numbers as best you can. In the future, you can track actual expenses against your budget and see how you are doing. If you aren't getting the results you want, you need to find a way to increase income, decrease expenses, or both.

Budgeting and Planning in a Business

EXAMPLE 4 — CASE IN POINT

Bev Johnson went through a budgeting and planning process before opening Bev's Deli. She knows how important it is to control expenses and quickly get income above expenses. She has saved enough to pay the costs to fix up the space and buy all the furniture and supplies she will need on opening day. But she is worried about the coming year and needs the business to make a profit soon. She plans to work long hours and receive no wages for at least a few months. So her remaining savings are really going to be stretched. The deli will undoubtedly lose money at first and she will also have to pull funds out of her savings to live.

Bev will begin with two employees working part-time. She knows that many costs, including food, beverages, wages and supplies, are proportional to sales—the higher the sales, the higher these expenses. However, the table below includes her best estimates for the first few months based on her expectations and research. Although she has worked in restaurants, she has never started one of her own, and is not confident she has good estimates for all costs. So she includes a high number for "Other expenses" according to the level of her uncertainty. Actual expenses for the first month are also shown in the table.

Item	Expense	Frequency	Monthly Budget	Actual Expenses for First Month
Rent	$1,000	monthly	$1,000	$1,000
Utilities	$280	monthly	$280	$340
Insurance	$900	semiannually	$150	$150
Wages and payroll taxes	$800*	weekly	$3,467	$4,012
Food and beverages	$1,700	weekly	$7,367	$8,491
Supplies	$280	weekly	$1,213	$1,140
Advertising	$200	monthly	$200	$160
Other expenses	$1,400	monthly	$1,400	$1,210
		Totals	**$15,077**	**$16,503**

*Bev takes no salary.

Projected Sales: Bev projects that sales will start at about $2800 per week and hopefully grow to $8000 per week by the end of her first year. Actual sales for the first month of over $15,000 were a little better than she expected. But sales will have to be much higher in order for her to take a salary out of the deli. To grow the business, Bev works hard to provide great food, a nice environment for customers, and friendly service. Part of her marketing strategy is to please customers to the point where they tell their friends about the new deli in town.

Sales of $15,077 during the first month exceeded her expectations but fell short of actual expenses for the month of $16,503, or by about $1400. Part of the reason actual expenses were higher than expected was that sales and thus wages, food and beverage costs, and supplies were higher than expected. Of course, higher sales a good thing, but higher sales results in slightly higher costs.

Bev was surprised at how much food was wasted in the kitchen during the first month and believes she can reduce the waste. She also notes that workers are more efficient now than in the first week, so that will help control costs as sales increase. Of course, she will have to hire more people if sales continue to increase. She considers increasing the price of the meals, but she is afraid customers might then go to competing restaurants in the area. So she does not change prices.

QUICK CHECK 4

In Example 4, sales were 10% more than Bev had forecast. Increase the monthly costs of wages and payroll taxes, food and beverages, and supplies by 10% for the extra sales and find the new total monthly budget. Then find the difference between the new monthly budget and actual expenses for the first month. Given that sales were higher than expected requiring an adjustment to higher costs, was she far off of her budget?

16.1 Exercises // MyLab Math

Answer the following. Round all numbers to the nearest dollar.

1. The Patton family has the following annual income: his after-tax salary—$24,600, her after-tax salary for part-time work—$8200, and childcare payments from her former spouse—$2400 a year. Without a lot of thought, they quickly jot down these expected expenses.

Rent and utilities	$695 a month
Phone/Internet	$120 a month
Groceries and pharmacy	$200 a week
Insurance	$300 a month
Child care	$150 a week
Automobile	$200 a month
Medical	$500 a year

   Determine their monthly after-tax income, average monthly expense based on the numbers above, and the difference between the two. What have they left out that should be considered as they work toward a budget? On the basis of your own personal experience, are they likely to be able to handle their financial affairs without borrowing more?

2. Marsha Patterson is a college student. She works part-time as a bartender and earns about $240 a week after taxes, including tips. She has a scholarship that pays $2000 per semester and she attends two semesters a year. She hopes to work full-time in the summer and should make $1400 a month after taxes for 3 months, not including the $240 a month she will also earn from bartending through the summer. She uses her dad's car at no cost other than gasoline and her own car insurance. She lives at home for free and quickly jots down these ongoing expenses.

Smart phone	$80 per month
Clothes and personal items	$200 per week
Insurance	$125 per month
Entertainment	$50 per week
Tuition	$6800 per semester
Textbooks and school supplies	$350 per semester

   To make it easy, assume the income she receives in the summer as well as the scholarship funds are spread out over 12 months in equal amounts. Find her average monthly after-tax income, average monthly expense based on the numbers above, and the difference between the two. What expenses has she not considered that she should have? Based on the numbers above, how much does she need to borrow per month on average? Assuming all numbers stay the same and ignoring anything she might have failed to list, find the debt she will accumulate in an interest-free Stafford loan if she goes to school for 4 years (assume 48 months).

3. Belinda McCabe has just divorced and is now a single working woman with a toddler. She earns $27,200 a year after income taxes from her job as an office manager and will receive $500 a month in child care from her ex. She received some stock in the divorce that pays a dividend of $325 per quarter. She lists "Other expenses" at $800 per month, as she isn't yet quite sure what they will be, given that her son has health issues.

Rent and utilities	$650 a month
Phone/Internet	$120 a month
Groceries and pharmacy	$140 a week
Insurance	$210 a month
Child care	$170 a week
Automobile	$200 a month
Medical and dental	$800 a year
Other expenses	$800 a month

Find her average monthly after-tax income, average monthly expense based on the numbers above, and the difference between the two. Based on your experience, is her budget reasonable? Does she have enough income?

4. Roberta and Ben Rodriquez hope to retire when they turn 65 and 66, respectively, so they are working on a budget for their retirement years. They will get about $1480 and $1140 a month, respectively, from Social Security. Their home is paid off, although they do have to pay taxes, upkeep, and insurance on the house, which they estimate to be $700 per month. Roberta will receive a pension of about $720 a month, and a financial planner tells them they should be able to withdraw $1000 a month from their retirement plans for the remainder of their lives. They expect income taxes to be $200 to $300 a month and won't have any payroll taxes unless one or the other has to work part-time.

They want to put $520 a month in an account, which will allow them to replace a car every 4 or 5 years. Due to Ben's many fishing trips and the fact that they like plays, dining out, and short vacations, they include a fairly significant cost for "Other expenses."

House expenses	$700 a month	
Utilities	$180 a month	
Phone/Internet/cable TV	$150 a month	
Groceries and pharmacy	$160 a week	
Car insurance	$110 a month	
Medical insurance	$650 a month	Medicare and a plan to supplement Medicare
Automobile repairs and gasoline	$200 a month	
Medical and dental	$2500 a year	unsure of costs during retirement
Annual vacation and trips	$4000 a year	approximate
Gifts to kids/grandkids	$1000 a year	
Income taxes	$300 a month	use the higher figure to be conservative
Car replacement	$520 a month	
Other expenses	$800 a month	

Find their average monthly after-tax income, average monthly expense based on the numbers above, and the difference between the two. Are they likely to be able to meet their budget on an ongoing basis? If not, which expenses do you think they can reasonably reduce and how can they increase their income?

5. Describe the steps needed to create a realistic budget. (See Objectives 1 and 2.)

6. What does *planning* refer to in this section? (See Objectives 2 and 3.)

7. What are the advantages and disadvantages of budgeting and planning?

8. Use the Internet to find an estimate of average or median income of newly graduated college students in your state (or in the country). Use the chapters on income and payroll taxes to estimate their taxes. Then prepare a budget based on what you think reasonable living expenses would be for a student just starting out in the area in which you live, using your best estimate of local costs. You can use either a single person or a couple.

9. Ask people you know who are over 35 years of age if they budget and plan. This might be your parents, brothers or sisters, aunts and uncles, or family friends. (You might need to explain the terms "planning" and "budgeting" as used in this section to them, but that does not mean they are not doing some of the same things discussed here.)

QUICK CHECK ANSWERS

Answers to the first three Quick Check exercises will vary, since they are based on the spending patterns of students and their families.

4. Monthly budget—$16,282; the difference between the new monthly budget and actual expenses for the first month—$221. No, her budget is very close to actual.

16.2 Frequency Distributions and Graphs

OBJECTIVES
1. Construct and analyze a frequency distribution.
2. Make a bar graph.
3. Make a line graph.
4. Draw a circle graph.

The word **statistics** refers to data, whether from business, economics, education, or any other field. However, it also refers to techniques used to analyze data to help us better understand the world around us, including what is happening in a firm. In the **information society** we live in today, many employees are involved in analyzing data to find relevance and meaning. Becoming good at this will improve your career prospects. Charts and graphs are an excellent and very common way to present data. You will see them everywhere you look.

CASE IN POINT Bev Johnson uses only fresh, high-quality ingredients at her deli, and she is quickly building a reputation for quality. However, the deli has been open only a few months, so she must continue to carefully watch sales and costs. Most restaurants fail in the first year.

OBJECTIVE 1 Construct and analyze a frequency distribution. It can be difficult to interpret or find patterns in a large group of numbers. One way of analyzing the numbers is to organize them in a table that shows the frequency of occurrence of the various numbers. This type of table is called a **frequency distribution**.

Constructing a Frequency Distribution

EXAMPLE 1 Bev Johnson is analyzing sales during her first 16 weeks in business. The weekly sales data are in thousands of dollars. Read down the columns, beginning with the left column, for successive weeks of the year.

$2.3 $5.9 $6.1 $6.4
$4.9 $4.3 $6.5 $7.2
$3.9 $4.9 $5.2 $6.1
$4.6 $5.2 $6.1 $7.2

Construct a table that shows each weekly sales amount. Then go through the data and place a **tally mark** (|) next to each corresponding value to create a frequency distribution table.

SOLUTION

Sales (thousands)	Tally	Frequency	Sales (thousands)	Tally	Frequency
$2.3	|	1	$5.9	|	1
$3.9	|	1	$6.1	|||	3
$4.3	|	1	$6.4	|	1
$4.6	|	1	$6.5	|	1
$4.9	||	2	$7.2	||	2
$5.2	||	2			

The frequency distribution shows that the most common weekly sales amount was $6100, although there were two weeks with each of the following in sales: $4900, $5200, and $7200.

QUICK CHECK 1

Weekly sales during the spring at a local nursery (in thousands) were: $21, $23, $21, $25, $23, and $21. Construct a frequency distribution.

It is difficult to get meaning from the frequency distribution above, as it gives too much information. We need something that better summarizes things. This can be done by combining weekly sales into somewhere between five and ten groups. We arbitrarily chose the sales groupings in the left column of the following table and then counted the number of weeks that sales fell into each grouping.

Weekly Sales History at Bev's Deli (Grouped Data)

Sales (thousands)	Frequency (number of weeks)
$2.1–$3.0	1
$3.1–$4.0	1
$4.1–$5.0	4
$5.1–$6.0	3
$6.1–$7.0	5
$7.1–$8.0	2

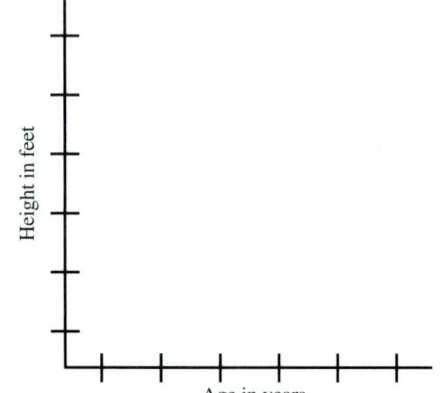

Analyzing a Frequency Distribution

EXAMPLE 2 When sales are $5000 or less per week, Johnson can take no salary, and the business loses money. She can take a small to modest salary out of the company when sales are between $5000 and $10,000 per week, but she can take a good salary when sales are $10,000 or more per week.

SOLUTION

(a) During how many weeks did Johnson take no salary?   6 weeks
(b) During how many weeks did Johnson take a small salary?   10 weeks
(c) During how many weeks did Johnson take a good salary?   0 weeks

Business owners often take little or no salary for the first several months to a year after starting a business. As a result, it is very important that someone starting a business has enough financial resources to get through the first year.

Careful planning, and budgeting ahead of time help an entrepreneur determine the funds needed to start a business. Lack of good planning and budgeting is one reason so many new businesses fail.

QUICK CHECK 2

Bev's Deli needs sales of more than $5000 per week in order to break even so that all expenses can be paid from sales. Use the grouped data table to find the number of weeks during which the restaurant broke even.

OBJECTIVE 2 Make a bar graph. The next step in analyzing these data is to use them to make a **graph**. A graph is a visual presentation of numerical data. One of the most common graphs is a **bar graph**, where the height of a bar represents the frequency of a particular value.

The bar graph that follows shows the same information as in the table in the Solution to Example 1. Using the bar graph, you can clearly see that weekly sales of $6.1 (thousand) was most common. It is more intuitive for most of us to look at a graph than a table full of numbers.

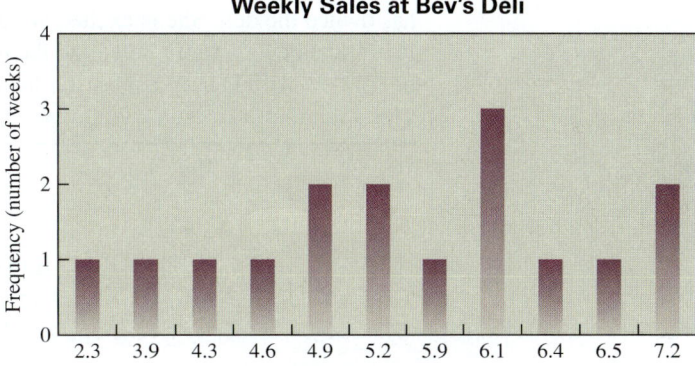

The prior bar graph contains a lot of information. We can simplify and perhaps better understand if we use the grouped data in the table above Example 2 (Bev's Deli Grouped Sales History) to create the following bar graph. It is also important to title a chart or graph fully and accurately. Doing so will help you to remember what you have created and to communicate the information to others.

Quick TIP

Take the time to create a good title and x- and y-axis labels when constructing a chart, graph, or table.

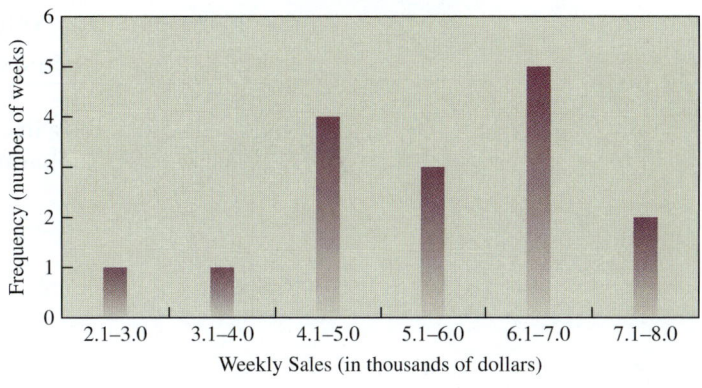

The grouped data shows that sales were commonly between $4100 and $7000 per week, which is close to Johnson's break-even level of sales of $5000 or more.

However, one very important factor is missing from the graphs above: time. Bev desperately needs to understand the trend of sales over time. Only then will she have a good idea about whether her business is growing, and growing fast enough to provide her an income before she runs out of savings.

OBJECTIVE 3 Make a line graph. Bar graphs show which numbers occurred and how many times, but do not necessarily show the order in which the numbers occurred. To discover any trends that may have developed, draw a **line graph** with time in weeks on the *x*-axis.

Drawing a Line Graph EXAMPLE 3 Use the raw data from Example 1 to show the growth in sales by week in a line graph.

SOLUTION

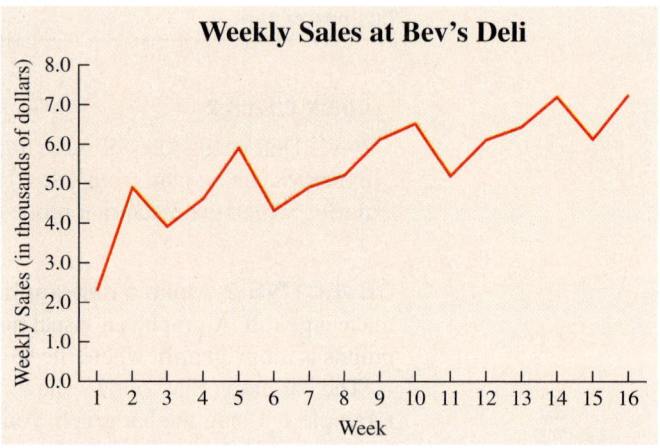

Quick TIP
Does the line graph of sales versus time help you understand what is happening? Is it easier to see the trend using the graph compared to the raw data in Example 1?

It is apparent from the line graph that sales have continued to grow in the 16 weeks that Bev has owned the deli. She is excited about the trend and determined to keep it going since her livelihood depends completely on the restaurant. Similarly to many small business owners, she has put everything she has into her business. She knows that most restaurants fail in the first year, and failure would be tough on her, both emotionally and financially.

QUICK CHECK 3
After opening the deli, Johnson figured out that her expenses were higher than anticipated. Her break-even point is actually about $5800 in sales per week. Use the graph above to estimate the number of weeks during which Bev's deli lost money.

One advantage of line graphs is that two or more sets of data can be shown on the same graph. For example, suppose the manager of a local convenience store wants to compare total sales, profits, and overhead using the following historical data with an estimate for 2018.

Year	Total Sales	Overhead	Profit
2015	$740,000	$205,000	$83,000
2016	$860,000	$251,000	$102,000
2017	$810,000	$247,000	$21,000
2018 (est.)	$1,040,000	$302,000	$146,000

Separate lines can be made on a line graph for each category, so that it is easier to compare. This type of graph is called a **comparative line graph**.

16.2 Frequency Distributions and Graphs 657

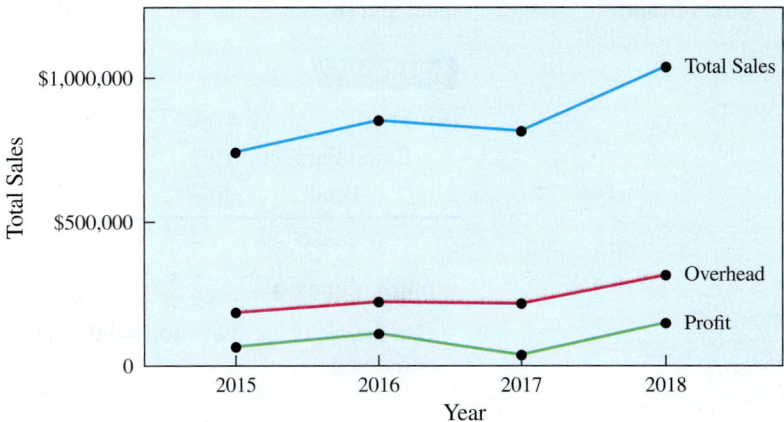

Total Sales, Overhead, and Profit

OBJECTIVE 4 Draw a circle graph. A sales manager for Afkin Medical Supplies records the expenses to keep a sales force on the road. First, she converts each expense to a percent of the total expenses. Then she generates the data needed to create a **pie chart**, also called **circle graph**. According to a convention that began over 4000 years ago, a circle or pie chart has 360 **degrees** (360°). To find the number of degrees of the circle graph associated with each expense, multiply the percent by 360° as shown.

Afkin Medical Supplies – Midwest Region

Item	Expense	Percent of Total	Degrees
Travel	$246,000	30%	30% × 360° = 108°
Lodging	$180,400	22%	22% × 360° = 79° (rounded)
Food	$82,000	10%	10% × 360° = 36°
Entertainment	$82,000	10%	10% × 360° = 36°
Sales Meetings	$65,600	8%	8% × 360° = 29° (rounded)
Other	$164,000	20%	20% × 360° = 72°
Total	$820,000	100%	Total 360°

Think of 108° as $\frac{108°}{360°}$ of a circle which reduces to $\frac{3}{10}$, and so on, for the other values in the last column. The Afkin sales manager used the above data in a spreadsheet and created the following pie chart, which she then presented to her bosses.

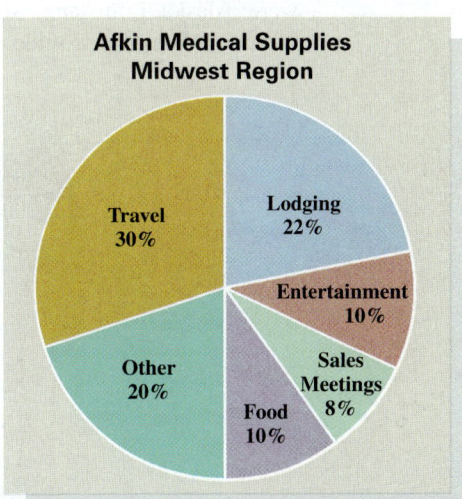

Quick TIP
Do you notice anything missing in the chart? For example, what time period is represented by the data?

Once you have the pie chart, you may no longer need the raw data. For example, we can use the pie chart to see that travel and lodging together accounted for 52% of the expenses by adding 30% + 22%.

Pie charts are excellent ways to view certain data, and they can be used to show comparisons when one item is very small compared to another. For example, an item representing 1% of the total could be drawn as a small, but noticeable, slice of a circle graph.

658 CHAPTER 16 Budgeting and Business Statistics

Interpreting a Circle Graph **EXAMPLE 4** Use the pie chart above and find the percent of total expenses spent on (a) food and entertainment and (b) travel and sales meetings.

SOLUTION

(a) Food   10%
    Entertainment 10%
    Total   20%

(b) Travel   30%
    Sales Meetings 8%
    Total   38%

QUICK CHECK 4

Use the pie chart above to find the total percent spent on sales meetings and entertainment combined.

Graphs and charts are used to help visualize, understand, and communicate data. For example, the following line graph shows that the percent of the U.S. population living in middle-income households has fallen from 61% of the population in 1971 to about 49% in 2018. The data show* that income inequality has increased in the United States, but it gives no hint as to why this has happened. Many believe that the growing income inequality has become a major social problem and they are working to find a solution.

Quick TIP

Graphs and charts may be very helpful for many, but what about blind or seriously visually impaired students? Actually there is software to help these students.

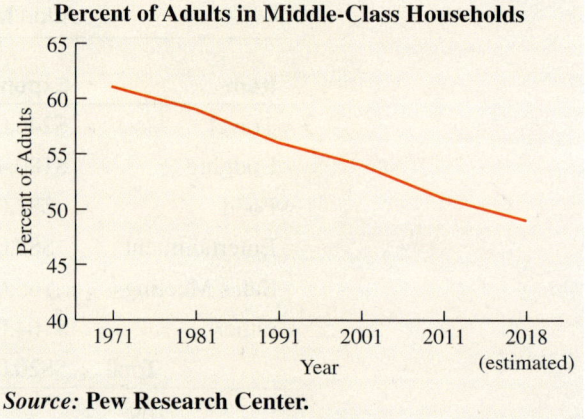

Source: Pew Research Center.

There are a wide variety of types of graphs and charts, including some very creative ones. For example, the following graph shows unemployment by state. Notice that there are large differences in unemployment between states. The states with the highest unemployment in 2017 were: Alaska, Louisiana, New Mexico and West Virginia. This graph shows data at one point in time; it does not provide unemployment data for a different date such as September 2018.

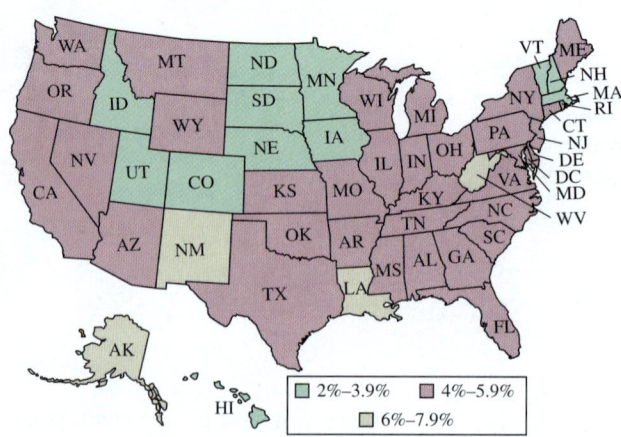

Source: U.S. Bureau of Labor Statistics.

*The word *data* is plural, so we say "The data show …" and not "The data shows …."

Beware of poorly designed graphs and charts that mislead. For example, does the bar chart below refer to a state, a region, or a country? The years decrease as you go from left to right along the *x*-axis. So are unemployment rates going up or down with time? It is confusing. What is the source of the data? There is a HUGE amount of information on the Internet today, but much of it is NOT valid. It is extremely important to use only reliable data sources—anyone can make up junk and put it on the Internet.

Quick TIP

Be careful to fully label any graph you make and check it for accuracy. Also be careful looking at others graphs since data may be poorly labeled or manipulated thereby leading you to the wrong conclusion.

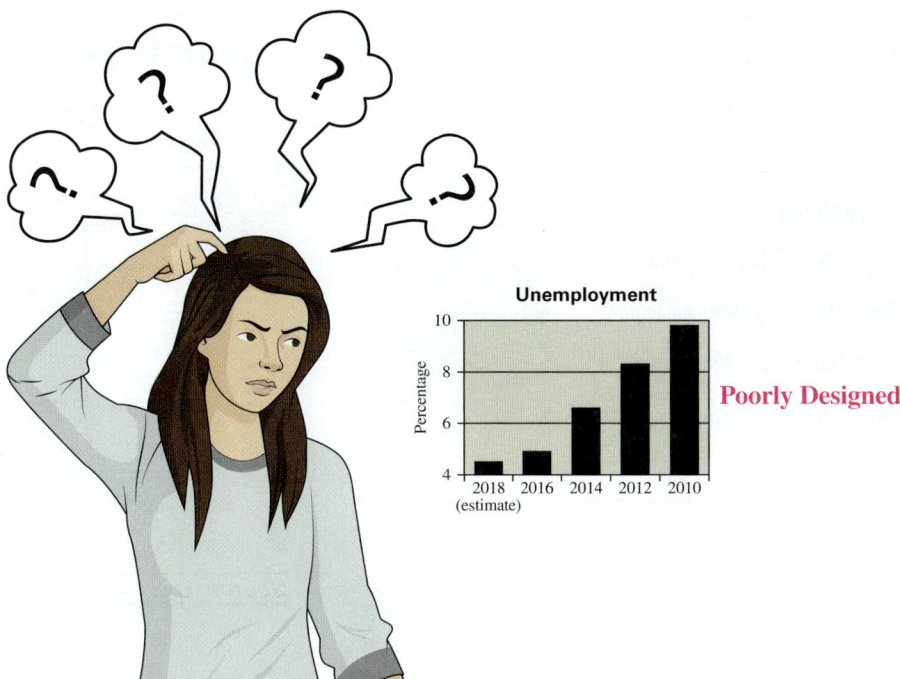

Poorly Designed

Actually, the data contained in the graph above is the unemployment rate in the United States taken from the Bureau of Labor Statistics, which we would expect to be reliable. The bar graph below uses the same data, but the graph is organized very differently. In it, you can see the dramatic story of unemployment rates in the United States, which fell from a high of 9.8% of workers unemployed near the end of the Great Recession in 2010 to an estimated 4.3% unemployed today. If we make the simplifying assumption that there were 125 million people wanting to work during those years, then there were

12.3 million unemployed workers in 2010 versus
5.6 million unemployed workers estimated for 2018

The difference between the two numbers is 6.7 million people. Imagine the financial and emotional difficulties of the 6.7 million additional unemployed in 2010 who could not find work, not to mention those able to find only low-paying jobs. So, we can use the graph to build a very powerful story if the data are accurate (from a reliable source), we understand the data, and we think about it carefully.

Quick TIP

The data are from a reliable source and indicate that the unemployment rate fell drastically between 2010 and 2018.

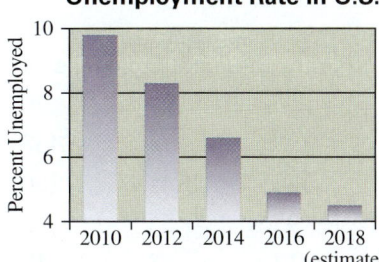

Source: **U.S. Bureau of Labor Statistics.**

16.2 Exercises // MyLab Math

The shaded sections below contain solutions to help you get a QUICK START on the various types of exercises.

Answer Exercises 1–3 using the bar graph on population growth. Answer Exercises 4–6 using the line graph on people who live together.

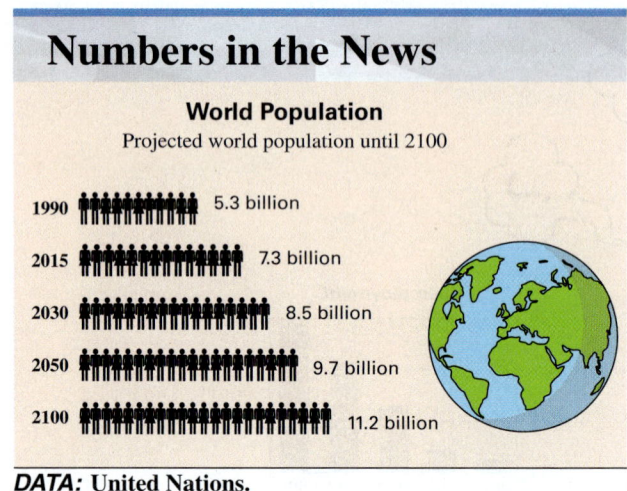

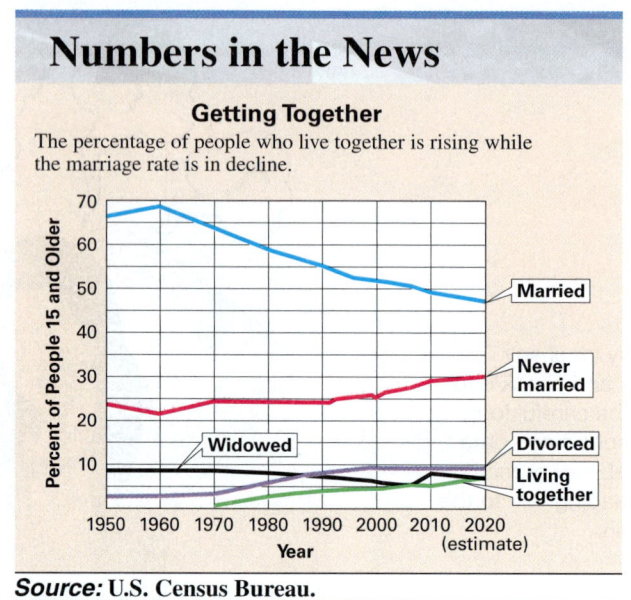

1. Estimate world population in 2020.
   a little more than 7.5 billion so about 7.7 billion

   1. 7.7 billion

2. Find the percent increase in world population from 2015 to 2030.

   2. _____

3. In 2050, assume that 56% of the world's population lives in Asia and that 5.2% of the world's population lives in the U.S. Estimate the population of Asia and of the U.S.

   3. _____

4. Approximately what percent of the U.S. population was married in 1950 and in 2020.

   4. _____

5. What has happened to the percent of people divorced during the past 60 years?

   5. _____

6. Estimate the percent of people living together in 2020.

   6. _____

The following list gives the numbers of college credits completed by 30 employees of the Franklin Bank.

College Credits
Franklin Bank Employees

74	133	4	127	20	30
103	27	139	118	138	121
149	132	64	141	130	76
42	50	95	56	65	104
4	140	12	88	119	64

Use these numbers to complete the following table with number of credits grouped. (See Examples 1 and 2.)

	Number of Credits	Number of Employees
7.	0–24	4
8.	25–49	_____
9.	50–74	_____
10.	75–99	_____
11.	100–124	_____
12.	125–149	_____

13. Make a line graph using the frequencies that you found.

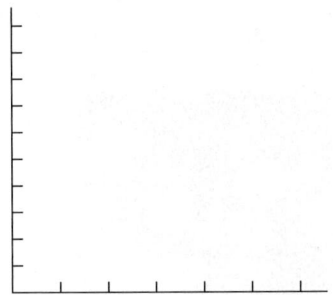

14. How many employees completed fewer than 25 credits?   14. 4

15. How many employees completed 50 or more credits?   15. _____

16. How many employees completed from 50 to 124 credits?   16. _____

17. How many employees completed from 0 to 49 credits?   17. _____

Weekly sales in thousands of dollars are given for one Home Depot store. The numbers are in chronological order going down the columns. For example, sales for the first and fourth weeks, respectively, are $302,000 and $304,000.

302	304	318	301	330	337	335	348	339
265	275	279	283	322	349	330	325	334
315	288	299	326	325	342	328	347	

Use the numbers to complete the following table with sales grouped. (See Example 1.)

	Sales (in thousands)	Frequency		Sales (in thousands)	Frequency
18.	260–269	1	19.	270–279	2
20.	280–289	_____	21.	290–299	_____
22.	300–309	_____	23.	310–319	_____
24.	320–329	_____	25.	330–339	_____
26.	340–349	_____			

662 CHAPTER 16 Budgeting and Business Statistics

27. Make a bar graph using your answers to Exercises 18–26.

28. Make a line graph using the original numbers.

29. How many weeks did sales equal or exceed $300,000? 29. _____

30. How many weeks did sales fall below $270,000? 30. _____

The following numbers show the scores for 80 students on a marketing test.

79	60	74	59	55	98	61	67	83	71
71	46	63	66	69	42	75	62	71	77
78	65	87	57	78	91	82	73	94	48
87	65	62	81	63	66	65	49	45	51
69	56	84	93	63	60	68	51	73	54
50	88	76	93	48	70	39	76	95	57
63	94	82	54	89	64	77	94	72	69
51	56	67	88	81	70	81	54	66	87

Use these numbers to complete the following table. (See Example 1.)

	Score	Frequency		Score	Frequency
31.	30–39	1	**32.**	40–49	6
33.	50–59	_____	**34.**	60–69	_____
35.	70–79	_____	**36.**	80–89	_____
37.	90–99	_____			

38. Make a bar graph showing your answers to Exercises 31–37.

39. How many students passed the test (passing is 70)? 39. _____

40. If a grade of B is given for a score of 80 or higher, how many students received a B or better? 40. _____

41. How many students failed the test (scored below 70)? 41. _____

42. How many students scored from 60 to 79? 42. _____

During one recent period Angela Rueben, a student, had $1400 in expenses, as shown in the following table. Find all numbers missing from the table in Exercises 43–48. (See Objective 4.)

Item	Dollar Amount	Percent of Total	Degrees of a Circle	Item	Dollar Amount	Percent of Total	Degrees of a Circle
43. Rent $\frac{72°}{360°} = .20 = 20\%$	$280	**20%**	72°	44. Clothing $\frac{\$210}{\$1400} = .15 = 15\%; .15 \times 360° = 54°$	$210	15%	**54°**
45. Books	$140	10%	___	46. Entertainment	$210	___	54°
47. Savings	$70	___	___	48. Other	___	___	126°

49. Draw a circle graph using Rueben's information. (See Objective 4.)

50. What percent did Rueben spend on rent, clothing, and books?

50. _____

51. What percent did Rueben spend on savings and entertainment?

51. _____

Solve the following application problems.

52. **FURNITURE** Annual sales at Antique Furnishings are divided into five categories as follows.

Item	Annual Sales
Finishing materials	$25,000
Desks	$80,000
Bedroom sets	$120,000
Dining room sets	$100,000
Other	$75,000

Make a circle graph showing this distribution. (See Example 4.)

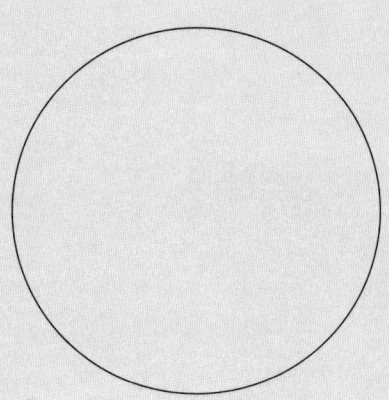

53. BOOK PUBLISHING Armstrong Publishing had 25% of its sales in mysteries, 10% in biographies, 15% in cookbooks, 15% in romance novels, 20% in science, and the rest in business books. Draw a circle graph with this information. (See Example 4.)

54. WAR DEATHS The bar chart shows an estimate of the number of Americans killed and wounded in various wars. Show this data in a circle graph. Round degrees to the nearest degree and percents to the nearest tenth of a percent. Which two wars had the highest number of killed and wounded?

55. **TOP TAX RATES** The graphic shows the highest tax rate on high-income individuals in the United States. Show the data using a bar chart. What conclusion(s) can you draw from the figure?

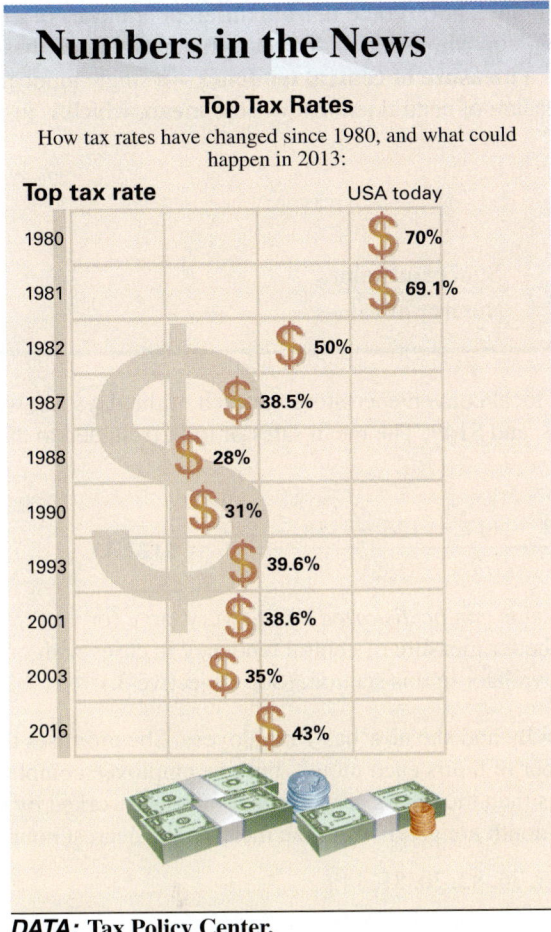

DATA: Tax Policy Center.

56. List the advantages of using a graph over a table when looking for trends.

57. Cut out three graphs from newspapers or magazines and tape them to your homework assignment. Be sure to explain the data in each case.

QUICK CHECK ANSWERS

1. $21, 3; $23, 2; $25, 1
2. 10 weeks
3. 8 weeks
4. 18%

16.3 Mean, Median, and Mode

OBJECTIVES

1. Find the mean of a list of numbers.
2. Find the weighted mean.
3. Find the median.
4. Find the mode.

OBJECTIVE 1 Find the mean of a list of numbers. Managers are often faced with the problem of analyzing raw data. For example, reports come in from different branches of a company or sales forecasts vary depending on who made the forecast. In analyzing all these data, one of the first things to look for is a **measure of central tendency**—a single number that shows the "middle" of the data. One measure of central tendency is the **mean**, which is just the **average** of a set of numbers.

> **Finding Mean**
>
> $$\text{Mean} = \frac{\text{Sum of all values}}{\text{Number of values}}$$

For example, suppose milk sales at a local convenience store for each of the days last week were $86, $103, $118, $117, $126, $158, and $149. The mean sales of milk (rounded to the nearest cent) follows.

$$\text{Mean} = \frac{86 + 103 + 118 + 117 + 126 + 158 + 149}{7} = \$122.43$$

One criticism of the mean is that its value *can be distorted* by one very large (or very small) value, as shown in the next example. A better measure of central tendency in cases with one abnormally large (or small) value is shown later in this section. (See Objective 3.)

Finding the Mean **EXAMPLE 1** Bev Johnson's sales at her deli grew rapidly and she now has 7 employees. She promises them that each can work about the same number of hours each month, but one employee complains that she worked considerably more hours than the others. The number of hours worked by each of the seven employees during the past month are given. Find the mean to the nearest hour.

75, 63, 76, 82, 70, 81, 149

SOLUTION

Add the numbers and divide by 7, since there are 7 numbers. Check that the sum of the numbers is **596**.

→ distorted by the 1 large number of 149

$$\text{Mean} = \frac{596}{7} = 85.14$$

The mean of 85 seems too high, since one employee clearly worked many more hours than the other six employees. The mean without this value of 149 is the sum of the remaining 6 numbers divided by 6.

$$\text{Mean} = \frac{447}{6} = 74.5$$

This value seems more in line with the average number of hours worked. Perhaps there was an unusual reason the one employee worked 149 hours (e.g., someone else was sick and could not work.)

> **QUICK CHECK 1**
>
> First find the mean of the following numbers: 24, 65, 25, 24, and 28. Then eliminate the largest value and find the mean of the remaining numbers.

Averages are commonly used in graphs and elsewhere as well. For example, the line graph on the next page to the left shows the average age at which people marry in the United States. Apparently, on average people are waiting longer to get married. The bar chart on the next page to the right shows average annual income by level of education. It shows that the average income goes up with each additional level of education. On average, a person with a master's degree makes more than $2\frac{1}{2}$ times as much as someone who did not get a high school diploma. However, by definition an average does not give the specific number for every person in a group. For example, not everyone with an associate's degree will make $41,496 a year. Some will make more, others less.

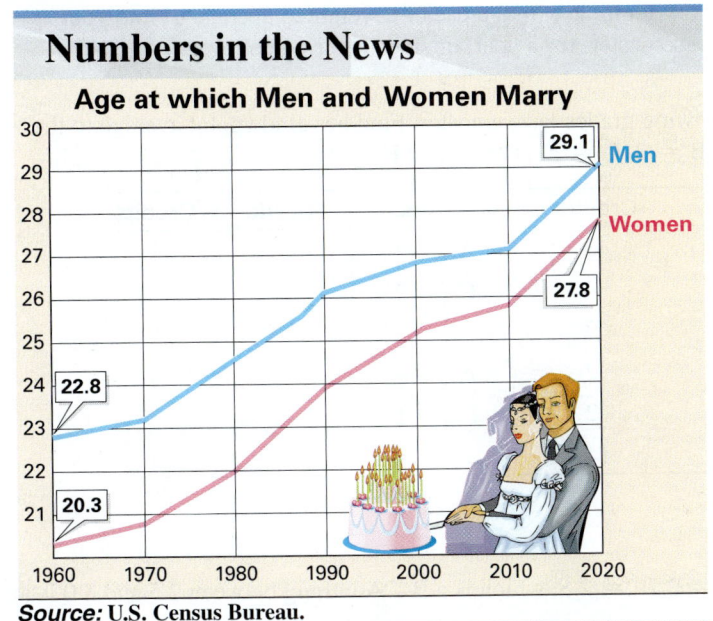

Numbers in the News

Age at which Men and Women Marry

Source: U.S. Census Bureau.

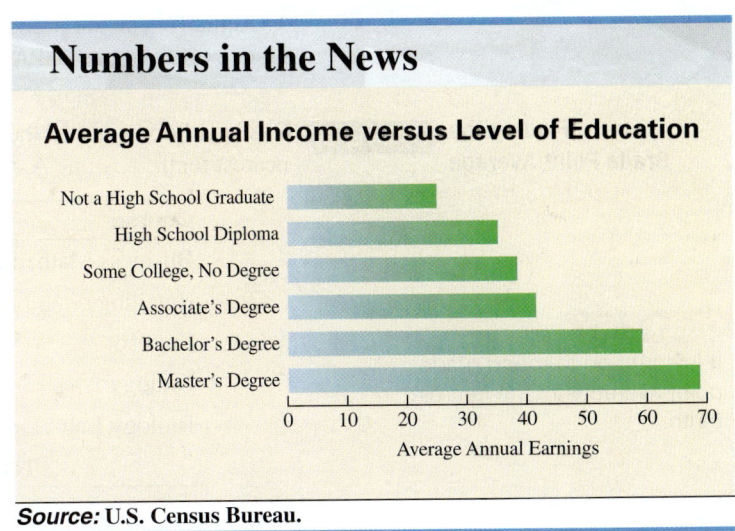

Numbers in the News

Average Annual Income versus Level of Education

Source: U.S. Census Bureau.

OBJECTIVE 2 Find the weighted mean. Some of the items in a list might appear more than once. In this case, it is necessary to find the **weighted mean**, or **weighted average**, where each value is the product of the number itself and the number of times it occurs.

Finding the Weighted Mean

EXAMPLE 2 Find the weighted mean of the numbers given in the following table.

Value	Frequency
3	4
5	2
7	1
8	5
9	3
10	2
12	1
13	2

SOLUTION

According to this table, the value 5 occurred 2 times, 8 occurred 5 times, 12 occurred 1 time, and so on. To find the mean, multiply each value by the frequency for that value. Then add the products. Add the values in the Frequency column to find the total number of values and then divide.

Value	Frequency	Product
3	4	12
5	2	10
7	1	7
8	5	40
9	3	27
10	2	20
12	1	12
13	2	26
Totals	20	154

$$\text{Weighted mean} = \frac{154}{20} = 7.7$$

QUICK CHECK 2

Find the weighted mean of the following sample of average family income in one neighborhood: 12 families, $27,300 per year; 15 families, $33,700 per year; and 5 families, $42,100 per year.

668 CHAPTER 16 Budgeting and Business Statistics

Actually, you are probably very familiar with the use of a weighted average to find your grade point average (GPA) for a semester, for a year, or since entering college.

Finding the Grade Point Average

EXAMPLE 3 Svetlana Popov earned the following grades last semester. Find her grade point average to the nearest tenth. Assume A = 4, B = 3, C = 2, D = 1, and F = 0.

Course	Credits	Grade	Grade × Credits
Business Mathematics	3	A (= 4)	3 × 4 = 12
Retailing	4	C (= 2)	4 × 2 = 8
English	3	B (= 3)	3 × 3 = 9
Biology	2	A (= 4)	2 × 4 = 8
Biology Lab	2	D (= 1)	2 × 1 = 2
Totals	14		39

Quick TIP
It is common to round grade point averages to the nearest tenth.

SOLUTION

Her grade point average is $\frac{39}{14} = 2.79 = 2.8$, which is a B–. Anything between 2.5 and 3.0 is a B– at her school.

This problem is solved using a scientific calculator as follows.

Note: Refer to Appendix B for calculator basics.

QUICK CHECK 3

Becky Johnson received an A in English, a C in History, and a C in Biology. All three classes are three-credit classes. An A is 4 points, a B is 3 points, and a C is 2 points. Find her grade point average to the nearest tenth.

OBJECTIVE 3 Find the median. As we saw in Example 1, the mean is a poor indicator of central tendency when there is one very large or one very small number. This effect can be avoided by using another **measure of central tendency**—method to estimate the middle of a set of data—called the **median**. The median divides a group of numbers in half—half the numbers lie at or above the median, and half lie at or below the median. However, the exact process to find the median depends on whether the list has an odd or even number of values.

Finding the Median
1. List the numbers from smallest to largest as an **ordered array**.
2. Find the median.
    (a) If there are an odd number of numbers, then the median is the number in the middle.
    (b) If there are an even number of numbers, then the median is the average of the two numbers in the middle.

Finding the Median

EXAMPLE 4 Find the median of the following weights (in pounds).
(a) 30 lb, 25 lb, 28 lb, 23 lb, 24 lb
(b) 14 lb, 18 lb, 17 lb, 10 lb, 15 lb, 19 lb, 18 lb, 20 lb

SOLUTION

(a) First list the numbers from smallest to largest.

23, 24, 25, 28, 30   ordered array

There are 5 numbers, so divide 5 by 2 to get 2.5. The next larger whole number is 3, so the median is the third number, **or 25**. The numbers 23 and 24 are less than 25, and the numbers 28 and 30 are greater than 25.

(b) First list the numbers from smallest to largest.

$$10, 14, 15, 17, 18, 18, 19, 20 \quad \text{ordered array}$$

There are 8 numbers, so divide 8 by 2 to get 4. The median is the mean of the numbers in the fourth and fifth positions.

$$\text{Median} = \frac{17 + 18}{2} = 17.5$$

> **QUICK CHECK 4**
>
> The heights of the students in one club at an elementary school are: 48, 41, 49, 47, and 51 inches. Find the median height.

Looking at the graph below and to the left, you can figure out that the median household income for Asians is a little over twice the median household income for blacks. Part of the reason is because Asians in the United States are far more likely to have advanced college degrees such as a master's or doctorate. In fact, over 21% of Asians in the U.S. have an advanced degree, in contrast to only 8.2% of blacks, and, as we have seen, higher education often leads to higher income.

According to the graph below and to the right, the median earnings of men is about 27% higher than that of women. This is NOT because more men have bachelor's or advanced degrees than women, since similar numbers of men and women attain those degrees. Part of the difference may be explained by the fact that women tend to get degrees in fields with lower median incomes, such as education, whereas men may be more oriented toward higher-paying fields such as engineering. However, a number of published research papers strongly suggest that there is still discrimination against women in terms of pay. It is important to note that there are many managers who work to give equal pay for men and women with equivalent backgrounds, job requirements, and responsibilities.

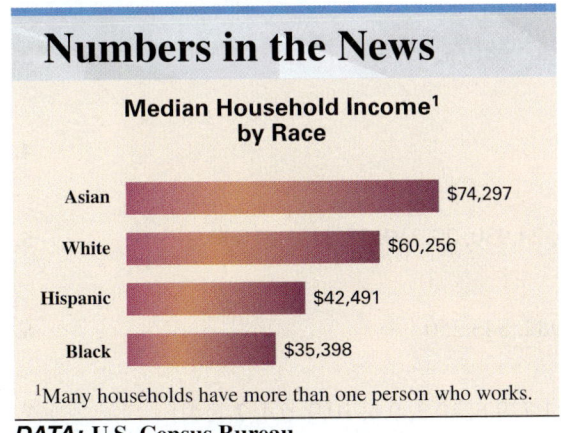

OBJECTIVE 4 Find the mode. The last important statistical measure discussed in this section is called the **mode**. The mode is the number that occurs most often. For example, if 10 students earned scores of

$$74, 81, 38, 74, 82, 80, 100, 92, 74, 85$$

on a business law examination, then the mode is 74. This is because more students obtained this score than any other score.

670 CHAPTER 16 Budgeting and Business Statistics

A data set in which every number occurs the same number of times is said to have *no mode*. A data set in which two different numbers occur the same number of times, with each occurring more often than any other number in the data set, is said to be bimodal. The prefix "bi" means two and the root "modal" refers to modes, so **bimodal** means *two modes*.

Finding the Mode **EXAMPLE 5** Professor Miller gave the same test to both his day and evening sections of Business Math at American River College. Find the mode of the tests given in each class. Which class has the lower mode?

(a) Day Class: 85, 92, 81, 73, 78, 80, 83, 80, 74, 69, 80, 65, 71, 65, 80, 93, 54, 78, 80, 45, 70, 76, 73, 80, 71, 68

(b) Evening Class: 68, 73, 59, 76, 79, 73, 85, 90, 73, 69, 73, 75, 93, 73, 76, 70, 73, 68, 82, 84, 77

Quick TIP
It is not necessary to place the numbers in numerical order when looking for the mode, but it helps with a large set of numbers.

SOLUTION

(a) The number **80** is the mode for the day class because it occurs more often than any other number. It occurs 6 times.

(b) The number **73** is the mode for the evening class because it occurs more often than any other number. It occurs 6 times as well.

The evening class has the lower mode.

QUICK CHECK 5

Five economists were asked to forecast the growth rate of the gross domestic product (GDP) for next year. Find the mode of their forecasts: 3.7%, 3.4%, 4.1%, 3.5%, and 3.4%.

16.3 Exercises // MyLab Math

The shaded sections below contain solutions to help you get a QUICK START on the various types of exercises.

Find the mean for the following lists of numbers. Round to the nearest tenth. (See Example 1.)

1. Gallons of spoiled milk: 3.5, 1.1, 2.8, .8, 4.1
   $\frac{3.5 + 1.1 + 2.8 + .8 + 4.1}{5} = 2.5$

   1. **2.5**

2. Weeks premature: 2, 3, 1, 4, 5, 2
   $\frac{2 + 3 + 1 + 4 + 5 + 2}{6} = 2.8$

   2. **2.8**

3. Guests at open school board meetings: 40, 51, 59, 62, 68, 73, 49, 80

   3. _____

4. Algebra quiz scores: 32, 26, 30, 19, 51, 46, 38, 39

   4. _____

5. Number attending football games: 21,900, 22,850, 24,930, 29,710, 28,340, 40,000

   5. _____

6. Annual salaries: $38,500, $39,720, $42,183, $21,982, $43,250

   6. _____

7. Average number of defects per 1000: 10.6, 12.5, 11.7, 9.6, 10.3, 9.6, 10.9, 6.4, 2.3, 4.1

   7. _____

8. Weight of dogs (pounds): 30.1, 42.8, 91.6, 51.2, 88.3, 21.9, 43.7, 51.2

   8. _____

9. When is it better to use the median, rather than the mean, for a measure of central tendency? Give an example. (See Objective 3.)

10. List some situations in which the mode is the best measure of central tendency to use to describe the data. (See Objective 4.)

Find the weighted mean for the following. Round to the nearest tenth. (See Example 2.)

11. | Value | Frequency |
|---|---|
| 9 | 3 |
| 12 | 4 |
| 18 | 2 |

$9 \times 3 = 27$
$12 \times 4 = 48$
$18 \times 2 = \underline{36}$
$9 \quad 111$
$\frac{111}{9} = 12.3$

_____12.3_____

12. | Value | Frequency |
|---|---|
| 9 | 3 |
| 12 | 5 |
| 15 | 1 |
| 18 | 1 |

$9 \times 3 = 27$
$12 \times 5 = 60$
$15 \times 1 = 15$
$18 \times 1 = \underline{18}$
$10 \quad 120$
$\frac{120}{10} = 12$

_____12_____

13. | Value | Frequency |
|---|---|
| 12 | 4 |
| 13 | 2 |
| 15 | 5 |
| 19 | 3 |
| 22 | 1 |
| 23 | 5 |

_____

14. | Value | Frequency |
|---|---|
| 25 | 1 |
| 26 | 2 |
| 29 | 5 |
| 30 | 4 |
| 32 | 3 |
| 33 | 5 |

_____

15. | Value | Frequency |
|---|---|
| 104 | 6 |
| 112 | 14 |
| 115 | 21 |
| 119 | 13 |
| 123 | 22 |
| 127 | 6 |
| 132 | 9 |

_____

16. | Value | Frequency |
|---|---|
| 243 | 1 |
| 247 | 3 |
| 251 | 5 |
| 255 | 7 |
| 263 | 4 |
| 271 | 2 |
| 279 | 2 |

_____

672 CHAPTER 16 Budgeting and Business Statistics

Find the grade point averages for the following students. Assume that
$A = 4, B = 3, C = 2, D = 1,$ *and* $F = 0$. *Round to the nearest tenth. (See Example 3.)*

17. Credits	Grade
4	B
2	A
5	C
1	F
3	B

18. Credits	Grade
3	A
3	B
4	B
2	C
4	D

Find the median for the following lists of numbers. (See Example 4.)

19. Number of hacking attacks on a network per month: 140, 85, 122, 114, 98

85, 98, **114**, 122, 140
↑
Median

19. 114

20. Cost of new laptops: $569, $478, $620, $515, $598

20. _____

21. Number of books loaned: 125, 100, 114, 150, 135, 172

21. _____

22. Calories in menu items: 346, 521, 412, 515, 501, 528, 298, 621

22. _____

23. Number of students taking business math: 37, 63, 92, 26, 44, 32, 75, 50, 41

23. _____

24. Number of orders: 1072, 1068, 1093, 1042, 1056, 1005, 1009

24. _____

Find the mode or modes for each of the following lists of numbers. (See Example 5.)

25. Porosity of soil samples: 21%, 18%, 21%, 28%, 22%, 21%, 25%
If the data are listed according to how many times each number appears,
18%, 21%, 22%, 25%, 28%
     21%
     21%
     ↑——— 21% is the mode since 21% is listed more than any other number.

25. 21%

26. Number of students graduating with honors: 85, 69, 72, 69, 103, 81, 98

26. _____

27. Age of retirees: 80, 72, 64, 64, 72, 53, 64

27. _____

28. Number of pages read: 86, 84, 83, 84, 83, 86, 86

28. _____

29. Number of fifth-grade students: 32, 38, 32, 36, 38, 34, 35, 30, 39

29. _____

30. Number of people on flights from Chicago to Denver: 178, 104, 178, 150, 165, 165, 82

30. _____

A quality-control inspector in a plant that manufactures electric motors measured the following shaft diameters (in thousandths of an inch).

$$35, \ 33, \ 32, \ 34, \ 35, \ 34, \ 35, \ 35, \ 34$$

Using these numbers, find each of the following. (Round to the nearest hundredth.)

31. The mean _____   **32.** The median _____

The quality-control inspector subsequently determined that he had made a mistake when he wrote 32 thousandths of an inch in the list above. Eliminate this number from the list and find each of the following.

33. The mean _____   **34.** The median _____

35. If you want to avoid a single extreme value having a large effect on the average, would you use the mean or the median? Explain your answer.

36. Does an employer look at the mean, median, or mode grade on a college transcript when considering hiring a new employee? Which do you think the employer should look at? Explain.

QUICK CHECK ANSWERS

1. 33.2; 25.25
2. $32,612.50
3. 2.7
4. 48 inches
5. 3.4%

Chapter 16 Quick Review

Chapter Terms Review the following terms to test your understanding of the chapter. For each term you do not know, refer to the page number found next to that term.

average [p. 666]
bar graph [p. 655]
bimodal [p. 670]
circle graph [p. 657]
comparative line graph [p. 656]
degrees [p. 657]
expected expenses [p. 646]
frequency distribution [p. 654]
graph [p. 655]
information society [p. 645]
line graph [p. 656]
mean [p. 666]
measure of central tendency [p. 666]
median [p. 668]
mode [p. 669]
ordered array [p. 668]
pie chart [p. 657]
statistics [p. 654]
tally mark [p. 654]
unexpected expenses [p. 646]
weighted average [p. 667]
weighted mean [p. 667]

CONCEPTS

16.1 Planning and budgeting

Managers and families have to budget and carefully plan. Budgets are created by looking at past expenses and estimating what expenses will be for the next few months. It is important to consider expected expenses but to also make allowances for unexpected expenses (Other expenses).

EXAMPLES

Here is a partial budget for a local coffee and breakfast shop along with actual expenses for one month.

	Budget	Actual
Rent	$1,200	$1,200
Utilities	$320	$302
Wages	$5,100	$5,830
Supplies	$2,400	$2,641
Other expenses	$1,500	$1,680
Totals	$10,520	$11,653

Actual expenses exceeded the budget by $1133, so the manager needs to take a second look at both the budget and what caused expenses to be so high for the month.

16.2 Constructing a frequency distribution from raw data

1. Construct a table listing each value and the number of times this value occurs.
2. Combine the pieces of data into groups.

Construct a frequency distribution for weekly sales, in thousands, at a concrete plant.

$22, $20, $22, $25, $18, $19, $22, $24, $24, $29, $19

Data	Tally	Frequency
$18	|	1
$19	||	2
$20	|	1
$22	|||	3
$24	||	2
$25	|	1
$29	|	1

Class	Frequency
$18–$20	4
$21–$23	3
$24–$26	3
$27–$29	1

CONCEPTS

16.2 Constructing a bar graph from a frequency distribution

Draw a bar for each class using the frequency of the class as the height of the bar.

16.2 Constructing a line graph

1. Plot each year on the horizontal axis.
2. For each year, find the value of sales, and plot a point at that value.
3. Connect all points with straight lines.

16.2 Constructing a circle graph

1. Determine the percent of the total for each item.
2. Find the number of degrees of a circle that each percent represents.
3. Draw the circle.

EXAMPLES

Construct a bar graph from the frequency distribution of the preceding example.

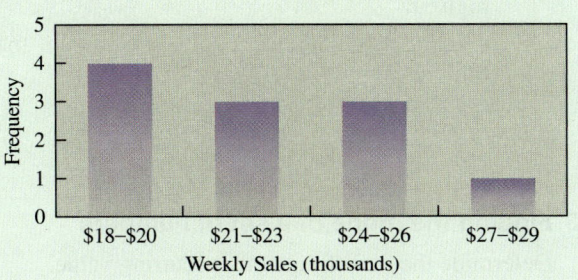

Construct a line graph for the following sales data.

Year	Total Sales
2015	$850,000
2016	$920,000
2017	$695,000
2018	$975,000

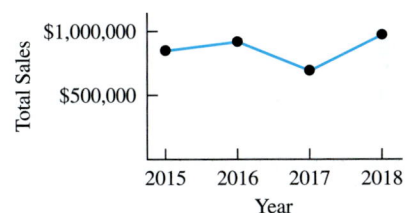

Construct a circle graph for the following table, which lists expenses for a short business trip.

Item	Amount
Car	$200
Lodging	$300
Food	$250
Entertainment	$150
Other	$100
	$1000

Item	Amount	Percent of Total
Car	$200	$\frac{$200}{$1000} = \frac{1}{5} = 20\%$; $360° \times 20\% = 360 \times .20 = $ **72°**
Lodging	$300	$\frac{$300}{$1000} = \frac{3}{10} = 30\%$; $360° \times 30\% = 360 \times .30 = $ **108°**
Food	$250	$\frac{$250}{$1000} = \frac{1}{4} = 25\%$; $360° \times 25\% = 360 \times .25 = $ **90°**
Entertainment	$150	$\frac{$150}{$1000} = \frac{3}{20} = 15\%$; $360° \times 15\% = 360 \times .15 = $ **54°**
Other	$100	$\frac{$100}{$1000} = \frac{1}{10} = 10\%$; $360° \times 10\% = 360 \times .10 = $ **36°**

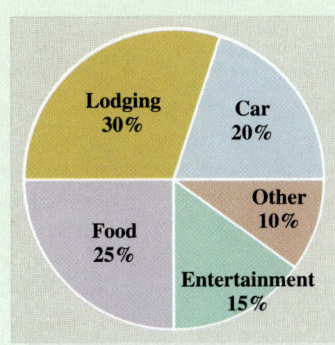

16.3 Finding the mean of a set of numbers

1. Add all numbers to obtain the total.
2. Divide the total by the number of pieces of data.

The quiz scores for Pat Phelan in her business math course were as follows:

85 79 93 91 78 82 87 85

Find Pat's quiz average.

$$\text{Mean} = \frac{85 + 79 + 93 + 91 + 78 + 82 + 87 + 85}{8} = \mathbf{85}$$

CONCEPTS	EXAMPLES
16.3 Finding the median of a set of numbers 1. Arrange the data in numerical order from lowest to highest. 2. Select the middle value or the average of the two middle values.	Find the median for Pat Phelan's grades from the preceding example. The data arranged from the lowest to highest are 78 79 82 85 85 87 91 93 The middle two values are 85 and 85. The average of these two values is $$\frac{85 + 85}{2} = 85$$
16.3 Finding the mode of a set of numbers Determine the most frequently occurring value.	Find the mode for Pat Phelan's grades in the preceding example. The most frequently occurring score is 85 (it occurs twice), so the mode is 85.

Case Study

WATCHING A SMALL BUSINESS GROW

Jeremy Clark decided to call his business The Soft Touch. He opened his first kiosk selling candles and other gifts in a nearby mall on February 1. In June, he opened a second kiosk selling the same gifts. Sales at the two kiosks (stores) are shown in thousands of dollars.

	Feb.	Mar.	Apr.	May	June	July	Aug.	Sept.	Oct.
Store 1	6.5	6.8	7.0	6.9	7.5	7.8	8.0	7.6	8.2
Store 2	—	—	—	—	8.2	6.2	8.2	8.7	9.6

1. Find the median, mean, and mode sales for each store to the nearest tenth.

2. Plot sales for both stores on the same line graph, with month on the horizontal axis and sales on the vertical axis.

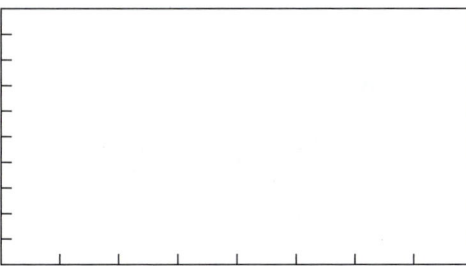

3. What trends are apparent from the preceding line graph?

INVESTIGATE

Find three graphs and charts using newspapers, magazines, and/or the Internet. Then explain each of them in writing. Graphs and charts can be great tools for communicating with customers, fellow workers, or even your boss. How can you make sure that a graph or chart clearly communicates the message that you wish it to convey?

Case in Point Summary Exercise

BOBBY FLAY

www.bobbyflay.com

Facts:

- 1991: Opened Mesa Grill in New York
- 1996: On Food Network for first time
- 2010: On "Boy Meets Grill" weekly on TV
- 2017: Has published 12 books

Bobby Flay opened Mesa Grill in New York City in 1991 with "a new twist on southwestern cuisine." He owns restaurants in New York, Las Vegas, and the Bahamas in addition to Bobby's Burger Palace with 19 locations across 11 states. Flay prides himself on creating a working environment in which his employees are able to express their creativity and grow professionally. His organization currently offers a scholarship every year to one high school senior who plans to attend the French Culinary Institute.

Bobby Flay has been one of Bev Johnson's favorite TV personalities for several years. She watches his shows when she has time and brings recipes from the show into her deli from time to time. Her business has continued to grow, and sales during the recent holidays are shown in thousands of dollars.

	October	November	December
Week 1	$7.4	$8.0	$8.9
Week 2	$8.0	$8.2	$8.9
Week 3	$7.4	$9.8	$9.2
Week 4	$8.3	$10.6	$7.1

1. Find the mean and median sales for each month.

2. Make a line graph of sales versus weeks. Why do you think sales were high in the last two weeks of November and then low in the last week in December?

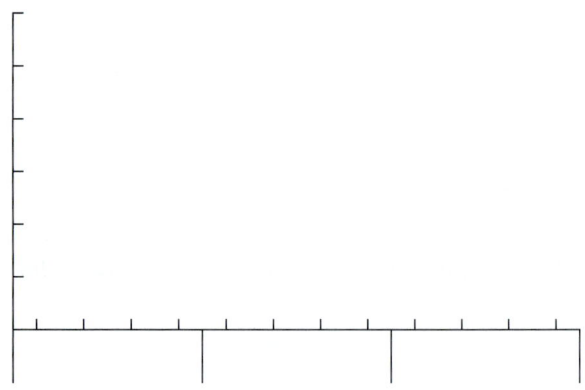

3. Make a pie chart using monthly sales data for October, November, and December.

Discussion Question: *Is it meaningful to collect data when managing a business? It is meaningful to use the data to build graphs? Explain.*

Chapter 16 Test

To help you review, the numbers in brackets show the section in which the topic was discussed.

1. The following are the numbers of cases of motor oil sold per week, by a regional distributor on the East Coast, for each of the past 20 weeks.

12,450	11,300	12,800	10,850	14,100
14,900	12,300	11,600	12,400	12,900
13,300	12,500	13,390	12,800	12,500
15,100	13,700	12,200	11,800	12,600

    Use these numbers to complete the following table. [16.2]

Cases of Motor Oil	Number of Weeks
10,000–10,999	____
11,000–11,999	____
12,000–12,999	____
13,000–13,999	____
14,000–14,999	____
15,000–15,999	____

2. How many weeks had sales of 13,000 cases or more? [16.2]

    2. _____

3. Use the numbers in the table above to draw a bar graph. Be sure to put a title and labels on the graph. [16.2]

4. During a 1-year period, the campus newspaper at Midwest Community College had the following expenses. Find all numbers missing from the table. [16.2]

Item	Dollar Amount	Percent of Total	Degrees of a Circle
Newsprint	$12,000	20%	____
Part-time wages	$6,000	____	36°
Wire service	$18,000	30%	____
Salaries	$18,000	30%	____
Other	$6,000	10%	____

5. Draw a circle graph using the information in test question 4. [16.2]

6. What percents of the expenses were for newsprint, part-time wages, and wire service? [16.2]

    6. _____

680 CHAPTER 16 Budgeting and Business Statistics

Find the mean for each of the following. Round to the nearest tenth if necessary. **[16.3]**

7. Number attending classical music recitals: 220, 275, 198, 212, 233, 246

    7. _____

8. Length of boards (centimeters): 12, 18, 14, 17, 19, 22, 23, 25

    8. _____

9. Weekly commission ($): 458, 432, 496, 491, 500, 508, 512, 396, 492, 504

    9. _____

Find the weighted average for each of the following to the nearest tenth. **[16.2]**

10.
Volume (quarts)	Frequency
6	7
10	3
11	4
14	2
19	3
24	1
	_____

11.
Sales ($)	Frequency
150	15
160	17
170	21
180	28
190	19
200	7
	_____

Find the median for each of the following lists of numbers. Do not round. **[16.3]**

12. Numbers of actors trying out for a part: 22, 18, 15, 25, 20, 19, 7

    12. _____

13. Number of text messages received per hour: 41, 39, 45, 47, 38, 42, 51, 38

    13. _____

14. Hours worked per day: 7.6, 9.3, 11.8, 10.4, 4.2, 5.3, 7.1, 9.0, 8.3

    14. _____

15. Trees planted: 58, 76, 91, 83, 29, 34, 51, 92, 38, 41

    15. _____

Find the mode or modes for the following lists of numbers. **[16.2]**

16. Contestants' ages: 51, 47, 48, 32, 47, 71, 82, 47

    16. _____

17. Customers served: 32, 51, 74, 19, 25, 43, 75, 82, 98, 100

    17. _____

18. Defectives per batch: 96, 104, 103, 104, 103, 104, 91, 74, 103

    18. _____

Solve the following application problems.

19. Stein's Fine Furnishings had the following sales (in thousand of dollars).

Year	Sales
2014	$754
2015	$782
2016	$853
2017	$592
2018	$680

    The geographic area in which Stein's operates had a serious business recession (economic slowdown) in 2017. Do you think the recession may have affected the business? Support your view by drawing a line graph.

    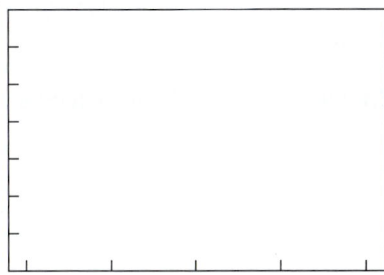

20. Ted Smith is an investment advisor at Merrill Lynch and receives a base salary plus a commission based on performance. His wife developed a serious illness at the beginning of 2017 and her condition slowly improved through the balance of the year. Is there evidence to suggest that his wife's illness affected his work?

	Quarterly Commissions			
Year	Quarter 1	Quarter 2	Quarter 3	Quarter 4
2015	$14,250	$12,375	$15,750	$13,682
2016	$13,435	$14,230	$11,540	$15,782
2017	$8,207	$7,350	$10,366	$11,470

    Support your view by drawing a line graph. Be sure to label the quarter in which Mrs. Smith became ill.

    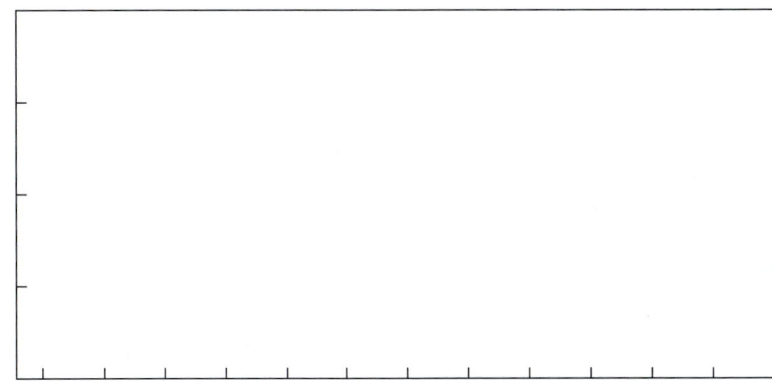

21. Sarah Smith decided to take over her father's bowling alley after he had some health issues. She quickly noted that her father's bookkeeping and budgeting processes were sloppy, so she resolved to do better. She created the following estimate of expenses and tracked one month's actual expenses to compare to her budget.

Expenses		Monthly Budget	Actual Expenses for the Month
Rent	$3,500 per month		$3,500
Utilities	$4,200 per year		$309
Wages and Payroll Taxes	$8,700 per month		$9,204
Insurance	$3,204 per year		$267
Supplies	$4,350 per quarter		$1,807
Subcontract maintenance	$1,300 per month		$1,390
Reserve to replace equipment	$7,056 per year		$583
Other expenses	$1,800 per month		$2,294
Totals			

Complete the Monthly Budget column including Total and compare against actual expenses.

Appendix A

The Metric System

The metric system is a system of measurements, weights, and temperatures that is used nearly everywhere in the world other than the United States. It is a system of measurements and weights that is based on decimal numbers. Even some firms in the U.S. use the metric system, particularly those that build products that are sold in Europe or Asia. Notice that measures of time are the same in both the English and metric systems.

Length		Weight	
1 foot	= 12 inches (in.)	1 pound (lb)	= 16 ounces (oz)
1 yard (yd)	= 3 feet (ft)	1 ton (T)	= 2000 pounds (lb)
1 mile (mi)	= 5280 feet (ft)		

Capacity		Time	
1 cup (c)	= 8 fluid ounces	1 week (wk)	= 7 days
1 pint (pt)	= 2 cups	1 day	= 24 hours (hr)
1 quart (qt)	= 2 pints (pt)	1 hour (hr)	= 60 minutes (min)
1 gallon (gal)	= 4 quarts (qt)	1 minute (min)	= 60 seconds (sec)

1 meter
1 meter is 39.37 inches.

1 yard
1 yard is 36 inches.

Prefixes in the Metric System

deca- = 10 times
kilo- = 1000 times
deci- = $\frac{1}{10}$ times
centi- = $\frac{1}{100}$ times
milli- = $\frac{1}{1000}$ times

OBJECTIVES

1. Learn the metric system.
2. Learn how to convert from one system to the other.

OBJECTIVE 1 Learn the metric system. The basic unit of length in the metric system is the **meter**. A meter is a little longer than a yard. For shorter lengths, the units **centimeter** and **millimeter** are commonly used. The prefix "centi" means hundredth, so 1 centimeter is one-hundredth of a meter. Thus

$$100 \text{ centimeters} = 1 \text{ meter} \quad \left(\text{or } 1 \text{ centimeter} = \tfrac{1}{100} \text{ meter}\right)$$

The prefix "milli" means thousandth, so 1 millimeter means one-thousandth of a meter. Thus

$$1000 \text{ millimeters} = 1 \text{ meter} \quad \left(\text{or } 1 \text{ millimeter} = \tfrac{1}{1000} \text{ meter}\right)$$

Meter is abbreviated m, centimeter is cm, and millimeter is mm.
Convert from centimeters to meters by moving the decimal point, as shown in the following example.

Converting Length Measurements **EXAMPLE 1** Convert the following measurements:

(a) 6.4 m to cm
(b) .98 m to mm
(c) 34 cm to m

SOLUTION

(a) A centimeter is a small unit of measure (a centimeter is about $\frac{1}{2}$ the diameter of a penny) and a meter is a large unit (a little over 3 feet), so many centimeters make a meter. Therefore, *multiply* by **100** to convert meters to centimeters.

$$6.4 \text{ m} = 6.4 \times 100 = 640 \text{ cm}$$

(b) Multiply by **1000** to convert meters to millimeters.

$$.98 \text{ m} = .98 \times 1000 = 980 \text{ mm}$$

A-1

(c) A meter is a large unit of measure that is a little longer than a yard. There are 100 centimeters in one meter. To convert 34 centimeters to meters, divide by 100:

$$34 \text{ cm} = \frac{34}{100} = .34 \text{ m}$$

QUICK CHECK 1

Convert 6 meters to millimeters and 3 centimeters to meters.

Long distances are measured in **kilometer** (km) units. The prefix "kilo" means one thousand. Thus,

$$1 \text{ kilometer} = 1000 \text{ meters} \quad \left(\text{or } 1 \text{ meter} = \tfrac{1}{1000} \text{ kilometer}\right)$$

Since a meter is slightly longer than a yard, 1000 meters is longer than 1000 yards, which is 3000 feet. Actually, there are about 3281 feet in a kilometer. A kilometer is about .62 miles (3281 ÷ 5280 feet in a mile).

The basic unit of volume in the metric system is the **liter** (L), which is a little more than a quart. You may have noticed that Coca Cola is sometimes sold in 2-liter plastic bottles. Again the prefixes "milli" and "centi" are used. Thus,

$$1 \text{ liter} = 100 \text{ centiliters}$$
$$1 \text{ liter} = 1000 \text{ milliliters}$$

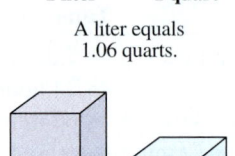

1 liter   1 quart
A liter equals 1.06 quarts.

1 kilogram   1 pound
A kilogram equals 2.2 pounds.

Milliliter (mL) and **centiliter** (cL) are such small volumes that they find their main uses in science. In particular, drug doses are often expressed in milliliters.

Weight is measured in **grams** (g). A nickel weighs about 5 grams. **Milligrams** (mg; one-thousandth of a gram) and **centigrams** (cg; one-hundredth of a gram) are so small that they are used mainly in science. A more common measure is the **kilogram** (kg), which is 1000 grams. A kilogram is equivalent to about 2.2 pounds.

$$1000 \text{ grams} = 1 \text{ kilogram}$$

Converting Weight Measurements   **EXAMPLE 2**   Convert the following measurements.

(a) 650 g to kg
(b) 9.4 L to cL
(c) 4350 mg to g

SOLUTION

(a) A gram is a small unit, and a kilogram is a larger unit. Thus, *divide* by **1000** to convert grams to kilograms.

$$650 \text{ g} = \frac{650}{1000} = .65 \text{ kg}$$

(b) *Multiply* by **100** to convert liters to centiliters.

$$9.4 \text{ L} = 9.4 \times 100 = 940 \text{ cL}$$

(c) *Divide* by **1000** to convert milligrams to grams.

$$4350 \text{ mg} = \frac{4350}{1000} = 4.35 \text{ g}$$

QUICK CHECK 2

Convert 2 kilograms to grams and 700 centiliters to liters.

OBJECTIVE 2 Learn how to convert from one system to the other. It is not as easy for Americans to think in the metric system as it is for Europeans or Asians who already use the system daily. It helps to have a table such as the following when making conversions.

APPENDIX A The Metric System A-3

English-Metric Conversion Table

FROM METRIC	TO ENGLISH	MULTIPLY BY	FROM ENGLISH	TO METRIC	MULTIPLY BY
Meters	Yards	1.09	Yards	Meters	.914
Meters	Feet	3.28	Feet	Meters	.305
Meters	Inches	39.37	Inches	Meters	.0254
Kilometers	Miles	.62	Miles	Kilometers	1.609
Grams	Pounds	.00220	Pounds	Grams	454
Kilograms	Pounds	2.20	Pounds	Kilograms	.454
Liters	Quarts	1.06	Quarts	Liters	.946
Liters	Gallons	.264	Gallons	Liters	3.785

Quick TIP
Either the English or metric system can be used for measurements. Some countries use one scale and other countries use the other.

Converting Metric to English

EXAMPLE 3 Use data in the table above to make the following conversions.
(a) 15 meters to yards
(b) 39 yards to meters
(c) 47 meters to inches
(d) 87 kilometers to miles
(e) 598 miles to kilometers
(f) 12 quarts to liters

SOLUTION

(a) Look at the table for converting meters to yards and find the number **1.09**. Multiply 15 meters by **1.09**.

$$15 \times 1.09 = 16.35 \text{ yards}$$

(b) Read the yards-to-meters row of the table. The number **.914** appears. Multiply 39 yards by **.914**.

$$39 \times .914 = 35.646 \text{ meters}$$

(c) 47 meters = $47 \times 39.37 = 1850.39$ inches
(d) 87 kilometers = $87 \times .62 = 53.94$ miles
(e) 598 miles = $598 \times 1.609 = 962.182$ kilometers
(f) 12 quarts = $12 \times .946 = 11.352$ liters

QUICK CHECK 3

Convert a speed limit of 70 miles per hour to kilometers per hour, and convert 3 pounds to grams.

Temperature in the metric system is measured in degrees **Celsius** or **centigrade** (abbreviated C). In the Celsius scale, water freezes at 0°C and boils at 100°C. This is more sensible than degrees **Fahrenheit** (abbreviated F) in use now, in which a mixture of salt and water freezes at 0°F, and 100°F represents the temperature inside the mouth of an individual who lived in the 1700s: Gabriel Fahrenheit.

Converting from Fahrenheit to Celsius

Step 1 Subtract 32.
Step 2 Multiply by 5.
Step 3 Divide by 9.

These steps can be expressed by the following formula.

$$C = \frac{5(F - 32)}{9} = \frac{5}{9}(F - 32)$$

A-4 APPENDIX A The Metric System

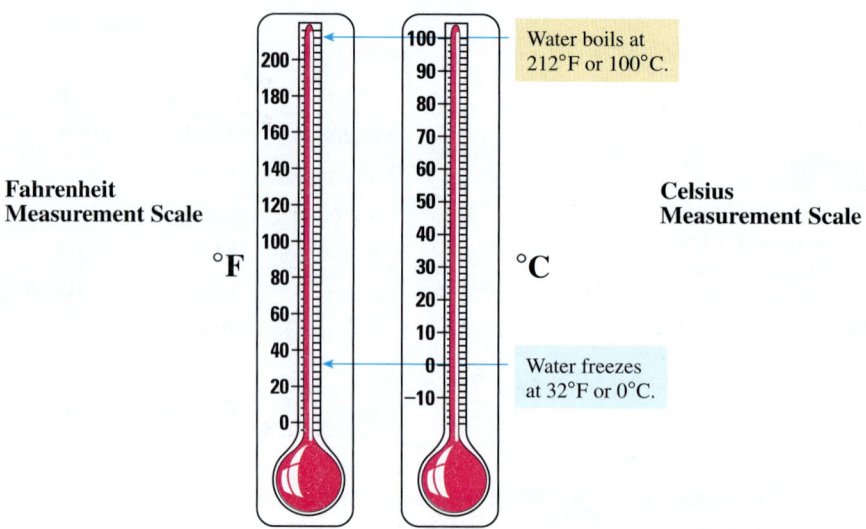

Converting Fahrenheit to Celsius

EXAMPLE 4 Convert 68°F to Celsius.

SOLUTION

Use the steps on the previous page.

Step 1 Subtract 32: $68 - 32 = 36$
Step 2 Multiply by 5: $36 \times 5 = 180$
Step 3 Divide by 9: $\dfrac{180}{9} = 20$

Thus, $68°F = 20°C$.

QUICK CHECK 4

Convert 85°F to Celsius.

Converting from Celsius to Fahrenheit

Step 1 Multiply by 9.
Step 2 Divide by 5.
Step 3 Add 32.

These steps can be expressed by the following formula.

$$F = \dfrac{9 \times C}{5} + 32$$

Converting Celsius to Fahrenheit

EXAMPLE 5 Convert 11°C to Fahrenheit.

SOLUTION

Use the previous steps.

Step 1 Multiply by 9: $9 \times 11 = 99$
Step 2 Divide by 5: $99 \div 5 = 19.8$
Step 3 Add 32: $19.8 + 32 = 51.8°F$

Thus, $11°C = 51.8°F$.

QUICK CHECK 5

Convert 35°C to Fahrenheit.

The next example shows that −40°C is equal to −40°F. Note that this temperature is below 0° on both scales and it is very cold. In fact, it is much colder than the inside of the freezer of your refrigerator which is probably in the neighborhood of 0°F.

Converting Celsius to Fahrenheit

EXAMPLE 6 Convert −40°C to degrees Fahrenheit.

SOLUTION

$$F = \frac{9 \times C}{5} + 32$$

$$F = \frac{9 \times (-40)}{5} + 32$$

$$F = -72 + 32$$

$$F = -40°$$

So, −40°C is equal to −40°F.

QUICK CHECK 6

Convert 30°C to Fahrenheit.

An interesting and easy way to convert from Celsius to Fahrenheit and from Fahrenheit to Celsius is shown next. It may be easier for your to remember than the equations given above, but you are effectively doing exactly the same thing. Example 7 shows how it is used.

Converting from Celsius to Fahrenheit or from Fahrenheit to Celsius:

Step 1   Add 40 to the temperature given to you.
Step 2   Multiply the sum by 9/5 if converting from Celsius to Fahrenheit or by 5/9 if converting from Fahrenheit to Celsius.
Step 3   Subtract 40 from the number found in Step 2.

Converting between Celsius and Fahrenheit

EXAMPLE 7 Use the short cut given above to convert:

(a) 15°C to Fahrenheit and
(b) 78°F to Celsius.

SOLUTION

(a) Step 1 $15 + 40 = 55$ Add 40.

    Step 2 $55 \times \dfrac{9}{5} = 99$ Multiply by $\dfrac{9}{5}$.

    Step 3 $99 - 40 = 59°F$ Subtract 40.

(b) Step 1 $78 + 40 = 118$ Add 40.

    Step 2 $118 \times \dfrac{5}{9} = 65.6$ (rounded)   Multiply by $\dfrac{5}{9}$.

    Step 3 $65.6 - 40 = 25.6°C$ Subtract 40.

QUICK CHECK 7

Convert (a) 10°C to Fahrenheit and (b) 65°F to Celsius.

The relationship between degrees Celsius and degrees Fahrenheit is that of a straight line. You can see it on the next graph where we have plotted a few points found above and connected them with a straight line.

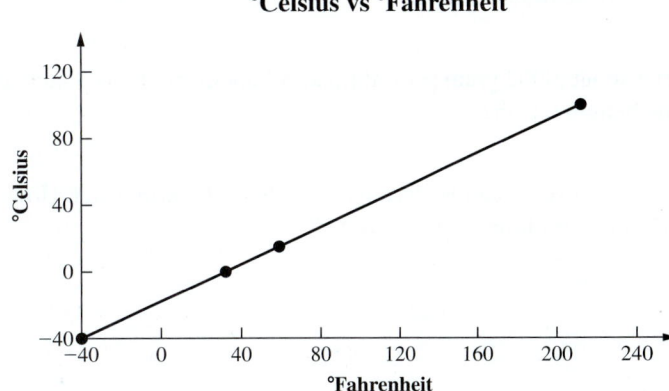

°Celsius vs °Fahrenheit

A-6 APPENDIX A The Metric System

Appendix A Exercises — MyLab Math

Convert the following measurements.

1. 68 cm to m _____
2. 934 mm to m _____
3. 4.7 m to mm _____
4. 7.43 m to cm _____
5. 8.9 kg to g _____
6. 4.32 kg to g _____
7. 39 cL to L _____
8. 469 cL to L _____
9. 46,000 g to kg _____
10. 35,800 g to kg _____
11. .976 kg to g _____
12. .137 kg to g _____

Convert the following measurements. Round to the nearest tenth.

13. 36 m to yards _____
14. 76.2 m to yards _____
15. 55 yards to m _____
16. 89.3 yards to m _____
17. 4.7 m to feet _____
18. 1.92 m to feet _____
19. 3.6 feet to m _____
20. 12.8 feet to m _____
21. 496 km to miles _____
22. 138 km to miles _____
23. 768 miles to km _____
24. 1042 miles to km _____
25. 683 g to pounds _____
26. 1792 g to pounds _____
27. 4.1 pounds to g _____
28. 12.9 pounds to g _____
29. 38.9 kg to pounds _____
30. 40.3 kg to pounds _____

31. One nickel weighs 5 grams. How many nickels are in 1 kilogram of nickels? 31. _____

32. Seawater contains about 3.5 grams of salt per 1000 milliliters of water. How many grams of salt do 5 liters of seawater contain? 32. _____

33. Helium weighs about .0002 gram per milliliter. A balloon contains 3 liters of helium. How much does the helium weigh? 33. _____

34. About 1500 grams of sugar can be dissolved in 1 liter of warm water. How much sugar can be dissolved in 1 milliliter of warm water? 34. _____

35. Find your height in centimeters.

35. _____

36. Find your height in meters.

36. _____

Convert the following Fahrenheit temperatures to Celsius. Round to the nearest degree.

37. 104°F _____

38. 86°F _____

39. 536°F _____

40. 464°F _____

41. 98°F _____

42. 114°F _____

Convert each of the following Celsius temperatures to Fahrenheit.

43. 35°C _____

44. 100°C _____

45. 10°C _____

46. 25°C _____

47. 135°C _____

48. 215°C _____

In most cases today, medical measurements are given in the metric system. In each of the following problems, a doctor's prescription is given. Decide whether the dosage is, or is not, reasonable.

49. 1940 grams of Kaopectate after each meal

49. _____

50. 76.8 centiliters of cough syrup every 2 hours

50. _____

51. 943 milliliters of antibiotic every 6 hours

51. _____

52. 1.4 kilograms of vitamins every 3 hours

52. _____

QUICK CHECK ANSWERS

1. 6000 mm; .03 m
2. 2000 g; 7 L
3. 112.63 km per hour; 1362 g
4. 29.4°C
5. 95°C
6. 86°F
7. (a) 50°F
   (b) 18.3°C (rounded)

Appendix B

Basic Calculators

OBJECTIVES
1. Learn the basic calculator keys.
2. Understand the $\boxed{C}$, $\boxed{CE}$, and $\boxed{ON/C}$ keys.
3. Understand the floating decimal point.
4. Use the $\boxed{\%}$ and $\boxed{1/x}$ keys.
5. Use the $\boxed{x^2}$, $\boxed{y^x}$, and $\boxed{\sqrt{}}$ keys.
6. Use the $\boxed{a^{b/c}}$ key.
7. Solve problems with negative numbers.
8. Use the calculator memory function.
9. Solve chain calculations using order of operations.
10. Use the parentheses keys.
11. Use the calculator to solve problems.

The first all-transistor **calculator** was introduced in 1966 and weighed 55 pounds. Today's calculators are much faster and can do a lot more than the early models—some are even more powerful than the earliest computers. Calculators used to be something to hold in your hand; today there are apps for your smart phone that will do everything calculators will do.

Many types of calculators are available, from the inexpensive basic calculator to the more complex **scientific**, **financial**, and **graphing** calculators. In this appendix, we discuss the basic calculator, which has a percent key, reciprocal key, exponent keys, square-root key, memory function, order of operations, and parentheses keys. In Appendix C, the financial calculator with its associated financial keys is discussed.

> **Quick TIP**
> The various calculator models differ significantly. *Use the instruction booklet that came with your calculator* for specifics about that calculator.

OBJECTIVE 1 Learn the basic calculator keys. Most calculators use **algebraic logic**. Some problems can be solved by entering number and function keys in the same order as you would solve a problem by hand. Other problems require a knowledge of the order of operations when entering the problem.

Using the Basic Keys

EXAMPLE 1 Perform the following operations.
(a) $12 + 25$ (b) $456 \div 24$

SOLUTION

(a) The problem $12 + 25$ is entered as

$$\boxed{1}\boxed{2}\boxed{+}\boxed{2}\boxed{5}\boxed{=}$$

and 37 appears as the answer.

(b) Enter $456 \div 24$ as

$$\boxed{4}\boxed{5}\boxed{6}\boxed{\div}\boxed{2}\boxed{4}\boxed{=}$$

and 19 appears as the answer.

QUICK CHECK 1

Find $279.04 \div 12.8$.

OBJECTIVE 2 Understand the $\boxed{C}$, $\boxed{CE}$, and $\boxed{ON/C}$ keys. Most calculators have a $\boxed{C}$ key. Pressing this key erases everything in most calculators and prepares them for a new problem. Some calculators have a $\boxed{CE}$ key. Pressing this key erases only the number displayed, thus allowing for the correction of a mistake without having to start the problem over. Many calculators combine the $\boxed{C}$ key and $\boxed{CE}$ key and have an $\boxed{ON/C}$ key. This key turns the calculator on and is also used to erase the calculator display. If the $\boxed{ON/C}$ is pressed after the $\boxed{=}$ key, or after one of the operation keys ($\boxed{+}$, $\boxed{-}$, $\boxed{\times}$, $\boxed{\div}$), everything in the calculator is erased. If the wrong operation key is pressed, simply press the correct key and the error is corrected. For example, in 7 $\boxed{+}$ $\boxed{-}$ 3 $\boxed{=}$ 4, pressing the $\boxed{-}$ key cancels out the previous $\boxed{+}$ key entry.

OBJECTIVE 3 Understand the floating decimal point. Most calculators have a **floating decimal** that locates the decimal point in the final result.

Calculating with Decimal Numbers

EXAMPLE 2 Jennifer Videtto purchased 55.75 square yards of vinyl floor covering at $18.99 per square yard. Find her total cost.

SOLUTION

Proceed as follows.

$$55.75 \; \boxed{\times} \; 18.99 \; \boxed{=} \; 1058.6925$$

The decimal point is automatically placed in the answer. Since money answers are usually rounded to the nearest cent, the answer is $1058.69.

QUICK CHECK 2

Find the cost of 7 boxes of paper costing $24.87 each.

To use a machine with a floating decimal, enter the decimal point as needed. For example, enter $47 as

$$\boxed{4}\boxed{7}$$

with no decimal point, but enter $.95 as follows.

$$\boxed{.}\boxed{9}\boxed{5}$$

One problem in utilizing a floating decimal is shown by the following example.

Placing the Decimal Point in Money Answers

EXAMPLE 3 Add $21.38 and $1.22.

SOLUTION

$$21.38 \; \boxed{+} \; 1.22 \; \boxed{=} \; 22.6$$

The final 0 is left off. Remember that the problem deals with dollars and cents, and write the answer as $22.60.

QUICK CHECK 3

Add 95.07 and 100.33.

OBJECTIVE 4 Use the $\boxed{\%}$ and $\boxed{1/x}$ keys. The $\boxed{\%}$ key moves the decimal point two places to the left when used following multiplication or division.

Using the $\boxed{\%}$ Key

EXAMPLE 4 Find 8% of $4205.

SOLUTION

$$4205 \; \boxed{\times} \; 8 \; \boxed{\%} \; \boxed{=} \; 336.4 = \$336.40$$

QUICK CHECK 4

Find the real estate commission of 6% on a home sold for $235,000.

APPENDIX B Basic Calculators B-3

The $\boxed{1/x}$ key replaces a number with the reciprocal of that number.

Using the $\boxed{1/x}$ Key EXAMPLE 5 Find the inverse or reciprocal of 40.

SOLUTION

$$40 \;\boxed{1/x}\; .025$$

QUICK CHECK 5

Find the inverse of 80.

OBJECTIVE 5 Use the $\boxed{x^2}$, $\boxed{y^x}$, and $\boxed{\sqrt{}}$ keys. The product of 3 × 3 can be written as follows.

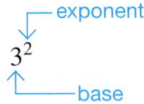

$$3^2 \quad \begin{array}{l}\leftarrow \text{exponent} \\ \leftarrow \text{base}\end{array}$$

The exponent (2 in this case) shows how many times the base is multiplied by itself (multiply 3 by itself or 3 × 3). The $\boxed{x^2}$ key can be used to quickly find the square of a number.

Using the $\boxed{x^2}$ Key EXAMPLE 6 Find 5^2 and 8.5^2.

SOLUTION

$$5 \;\boxed{x^2}\; 25 \quad \text{and} \quad 8.5 \;\boxed{x^2}\; 72.25 \quad \left\{\begin{array}{l}\text{Pushing } \boxed{=} \text{ is usually not necessary} \\ \text{when using the } \boxed{x^2} \text{ key.}\end{array}\right.$$

QUICK CHECK 6

Find 25^2.

The $\boxed{y^x}$ key raises any base number y to a power x. Use as follows.

1. Enter the base number first.
2. $\boxed{y^x}$
3. Enter the exponent.
4. $\boxed{=}$

Using the $\boxed{y^x}$ Key EXAMPLE 7 Find 5^3.

SOLUTION

$$5 \;\boxed{y^x}\; 3 \;\boxed{=}\; 125$$

QUICK CHECK 7

Find 4.8^3.

Since $3^2 = 9$, the number 3 is called the **square root** of 9. Square roots of numbers are written with the symbol $\sqrt{}$.

$$\sqrt{9} = 3$$

B-4   APPENDIX B   Basic Calculators

Using the $\sqrt{}$ Key   **EXAMPLE 8**   Find each square root.
(a) $\sqrt{144}$ (b) $\sqrt{20}$

SOLUTION

(a) Enter

$$144\;\sqrt{}$$

and 12 appears in the display. The square root of 144 is 12.

(b) The square root of 20 is

$$20\;\sqrt{}\quad 4.472136$$

which may be rounded to the desired position.

QUICK CHECK 8
Find the square root of 729.

OBJECTIVE 6   Use the $\boxed{a^{b/c}}$ key.   Many calculators have an $\boxed{a^{b/c}}$ key that can be used for problems containing fractions and mixed numbers. A mixed number is a number with both a whole number and a fraction, such as $7\frac{3}{4}$, which equals $7 + \frac{3}{4}$. The rules for adding, subtracting, multiplying, and dividing both fractions and mixed numbers are given in Chapter 2. Here, we simply show how these operations are done on a calculator.

Using the $\boxed{a^{b/c}}$ Key with Fractions   **EXAMPLE 9**   Solve the following.
(a) $\dfrac{6}{11} + \dfrac{3}{4}$ (b) $\dfrac{3}{8} \div \dfrac{5}{6}$

SOLUTION

Quick TIP
The calculator automatically shows fractions in lowest terms and as mixed numbers when possible.

(a) $6\;\boxed{a^{b/c}}\;11\;\boxed{+}\;3\;\boxed{a^{b/c}}\;4\;\boxed{=}\;1\dfrac{13}{44}$

(b) $3\;\boxed{a^{b/c}}\;8\;\boxed{\div}\;5\;\boxed{a^{b/c}}\;6\;\boxed{=}\;\dfrac{9}{20}$

QUICK CHECK 9
Solve $\frac{7}{8} + \frac{9}{16}$.

Using the $\boxed{a^{b/c}}$ Key   **EXAMPLE 10**   Solve the following.
(a) $4\dfrac{7}{8} \div 3\dfrac{4}{7}$ (b) $\dfrac{5}{3} \div 27.5$ (c) $65.3 \times 6\dfrac{3}{4}$

SOLUTION

(a) $4\;\boxed{a^{b/c}}\;7\;\boxed{a^{b/c}}\;8\;\boxed{\div}\;3\;\boxed{a^{b/c}}\;4\;\boxed{a^{b/c}}\;7\;\boxed{=}\;1\dfrac{73}{200}$

(b) $5\;\boxed{a^{b/c}}\;3\;\boxed{\div}\;27.5\;\boxed{=}\;0.060606061$

(c) $65.3\;\boxed{\times}\;6\;\boxed{a^{b/c}}\;3\;\boxed{a^{b/c}}\;4\;\boxed{=}\;440.775$

QUICK CHECK 10
Solve $5\frac{3}{4} \div 4\frac{1}{2}$.

OBJECTIVE 7   Solve problems with negative numbers.   There are several calculations in business that result in a **negative number**, or **deficit amount**.

APPENDIX B Basic Calculators B-5

Working with Negative Numbers

EXAMPLE 11 The amount budgeted for advertising last month was $4800, while $5200 was actually spent. Find the balance remaining in the advertising account.

SOLUTION

Enter the numbers in the calculator.

$$4800 \; [-] \; 5200 \; [=] \; -400$$

The minus sign in front of the 400 indicates that there is a deficit or negative amount. This value can be written as −$400 or sometimes as ($400), which indicates a negative amount. Some calculators place the minus sign after the number, as 400−.

QUICK CHECK 11

The following transactions were made to a checking account with a balance of $1482.05: a deposit of $185, an electronic bill payment of $122.17, and a withdrawal of $100 at an ATM machine. Find the new balance.

Negative numbers may be entered into the calculator by using the $[-]$ key before entering the number. For example, if $3000 is now added to the advertising account in Example 11, the new balance is calculated as follows.

$$[-] \; 400 \; [+] \; 3000 \; [=] \; 2600$$

The new account balance is $2600.

The $[+/-]$ key can be used to change the sign of a number that has already been entered. For example, 520 $[+/-]$ changes +520 to −520.

OBJECTIVE 8 Use the calculator memory function. Many calculators feature memory keys, which are a sort of electronic scratch paper. These **memory keys** are used to store intermediate steps in a calculation. On some calculators, a key labeled $[M]$ or $[STO]$ is used to store the numbers in the display, with $[MR]$ or $[RCL]$ used to recall the numbers from memory.

Other calculators have $[M+]$ and $[M-]$ keys. The $[M+]$ key adds the number displayed to the number already in memory. For example, if the memory contains the number 0 at the beginning of a problem, and the calculator display contains the number 29.4, then pushing $[M+]$ will cause 29.4 to be stored in the memory (the result of adding 0 and 29.4). If 57.8 is then entered into the display, pushing $[M+]$ will cause

$$29.4 + 57.8 = 87.2$$

to be stored. If 11.9 is then entered into the display, with $[M-]$ pushed, the memory will contain

$$87.2 - 11.9 = 75.3$$

The $[MR]$ key is used to recall the number in memory as needed, with $[MC]$ used to clear the memory.

Scientific calculators typically have one or more **memory registers** in which to store numbers. These memory keys are usually labeled $[STO]$ for store and $[RCL]$ for recall. For example, 32.5 can be stored in memory register 1 by

$$32.5 \; [STO] \; 1$$

or it can be stored in memory register 2 by 32.5 $[STO]$ 2, and so forth. Values are retrieved from a particular memory register by using the $[RCL]$ key followed by the number of the register. For example, $[RCL]$ 2 recalls the contents of memory register 2.

With a scientific calculator, a number stays in memory until it is replaced by another number or until the memory is cleared. The contents of the memory are saved even when the calculator is turned off.

Quick TIP
Always clear the memory before starting a problem; not doing so is a common error.

Using the Memory Registers

EXAMPLE 12 An elevator technician counted the number of people entering an elevator and also measured the weight of each group of people. Find the average weight per person.

Number of People	Total Weight
6	839 pounds
8	1184 pounds
4	640 pounds

SOLUTION

First find the total weight of all three groups and store in memory register 1.

839 [+] 1184 [+] 640 [=] 2663 [STO] 1

Then find the total number of people.

6 [+] 8 [+] 4 [=] 18

Finally, divide the contents of memory register 1 by the 18 people.

[RCL] 1 [÷] 18 [=] 147.9444444 pounds

This value can be rounded as needed.

QUICK CHECK 12

The following animals were brought to the market: 7 calves with a total weight of 2989 pounds; 10 calves with a total weight of 4490 pounds; and 17 calves with a total weight of 6953 pounds. Find the average weight of a calf to the nearest tenth of a pound.

OBJECTIVE 9 Solve chain calculations using order of operations. Long calculations involving several operations (adding, subtracting, multiplying, and dividing) must be done in a specific sequence called the **order of operations** and are called **chain calculations**. The logic of the following order of operations is built into most scientific calculators and can help us work problems without having to store a lot of intermediate values.

Order of Operations

Step 1	Do all operations inside parentheses first.
Step 2	Simplify any expressions with exponents and find any square roots.
Step 3	Multiply and divide from left to right.
Step 4	Add and subtract from left to right.

Quick TIP The order of operations is discussed in detail in Appendix D.

Using the Order of Operations

EXAMPLE 13 Solve the following.

(a) $3 + 7 \times 9\dfrac{3}{4}$ (b) $42.1 \times 5 - 90 \div 4$

SOLUTION

Most calculators automatically keep track of the order of operations for us.

(a) 3 [+] 7 [×] 9 [a^b/c] 3 [a^b/c] 4 [=] $71\dfrac{1}{4}$

(b) 42.1 [×] 5 [−] 90 [÷] 4 [=] 188

QUICK CHECK 13

Solve $9^2 + 2.7 \cdot 14$.

OBJECTIVE 10 Use the parentheses keys. The parentheses keys can be used to help establish the order of operations in a more complex chain calculation. For example, $\frac{4}{5+7}$ can be written as $\frac{4}{(5+7)}$, which can be solved as follows.

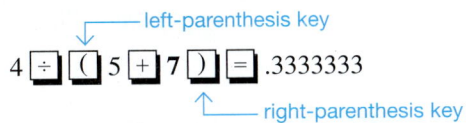

Using Parentheses

EXAMPLE 14 Solve the following problem.

$$\frac{16 \div 2.5}{39.2 - 29.8 \times .6}$$

SOLUTION

Think of this problem as follows:

$$\frac{(16 \div 2.5)}{(39.2 - 29.8 \times .6)}$$

Using parentheses to set off the numerator and denominator will help you minimize errors.

$$\boxed{(}\ 16\ \boxed{\div}\ 2.5\ \boxed{)}\ \boxed{\div}\ \boxed{(}\ 39.2\ \boxed{-}\ 29.8\ \boxed{\times}\ .6\ \boxed{)}\ \boxed{=}\ 0.3001876$$

QUICK CHECK 14

Solve $\dfrac{5^3 - 16}{253 - 4^2 \cdot 9}$

OBJECTIVE 11 Use the calculator to solve problems. Scientific calculators are great tools to help you solve problems.

Finding Sale Price

EXAMPLE 15 A digital camera with an original price of $560 is on sale at 10% off. Find the sale price.

SOLUTION

If the discount from the original price is 10%, then the sale price is 100% − 10% of the original price.

$$560\ \boxed{\times}\ \boxed{(}\ 100\ \boxed{-}\ 10\ \boxed{)}\ \boxed{\%}\ \boxed{=}\ 504$$

On some calculators the following keystrokes will also work.

$$560\ \boxed{-}\ 10\ \boxed{\%}\ \boxed{=}\ 504$$

The sale price is $504.

QUICK CHECK 15

Find the price of a new home that was originally priced at $239,000 if the price was reduced by 15%.

Applying Calculator Use to Problem Solving

EXAMPLE 16 A home buyer borrows $125,000 at 6% on a house, for 30 years. The monthly payment on the loan is $5.996 per $1000 borrowed. Annual taxes are $3800, and fire insurance is $620 a year. Find the total monthly payment including taxes and insurance.

SOLUTION

The monthly payment is the *sum* of the monthly payment on the loan *plus* monthly taxes *plus* monthly fire insurance costs. The monthly payment on the loan is the number of thousands in the loan times the monthly payment per $1000 borrowed.

To the nearest cent, this amount rounds to $1117.83.

QUICK CHECK 16

An investor borrows $1,200,000 on a piece of commercial property. The monthly payment on the loan is $8.43 per $1000 of debt. Annual taxes and insurance are $28,200 and $6240, respectively. Find the monthly payment.

Applying Calculator Use to Problem Solving

EXAMPLE 17 A Japanese company produces robotic dogs for the retail market. The dogs talk, cock their heads, bark, and walk. The company sells the dogs to distributors for $256.80 each.

(a) Find the revenue to the manufacturer if the company sells 26,340 robotic dogs.
(b) Find the final price to the customer if the robotic dogs are marked up an average of 87% on cost.

SOLUTION

(a) Revenue = Number sold × Price of each

26340 ⊠ 256.80 ⊟ 6764112   or $6,764,112

(b) Sales price = Cost × (1 + Markup percent)

256.80 ⊠ 1.87 ⊟ 480.216   or $480.22

QUICK CHECK 17

It cost a manufacturer an average of $38.20 per pair for each of the 180,400 pair of shoes produced in one month. The markup on the shoes is 35% based on cost. Find the total revenue to the company from the sale of the month's production.

Appendix B Exercises — MyLab Math

Solve the following problems on a calculator. Round each answer to the nearest hundredth.

1. 384.92
    407.61
    351.14
    + 27.93

2. 85.76
    21.94
    + 39.89

3. 6850
    321
    + 4207

4. 781.42
    304.59
    + 261.35

5. 4270.41
    − 365.09

6. 3000.07
    − 48.12

7. 384.96
    − 129.72

8. 36.84 − 12.17

9. 365
    × 43

10. 27.51
    × 1.18

11. 3.7 × 8.4

12. 62.5 × 81

13. $\dfrac{375.4}{10.6}$

14. $\dfrac{9625}{400}$

15. 96.7 ÷ 3.5

16. 103.7 ÷ .35

Solve the following chain calculations. Round each answer to the nearest hundredth.

17. $\dfrac{9 \times 9}{2 \times 5}$

18. $\dfrac{15 \times 8 \times 3}{11 \times 7 \times 4}$

19. $\dfrac{87 \times 24 \times 47.2}{13.6 \times 12.8}$

20. $\dfrac{2 \times (3 + 4)}{6 + 10}$

21. $\dfrac{2 \times 3 + 4}{6 + 10}$ _____

22. $\dfrac{4200 \times .12 \times 90}{365}$ _____

23. $\dfrac{640 - .6 \times 12}{17.5 + 3.2}$ _____

24. $\dfrac{16 \times 18 \div .42}{95.4 \times 3 - .8}$ _____

25. $\dfrac{14^2 - 3.6 \times 6}{95.2 \div .5}$ _____

26. $\dfrac{9^2 + 3.8 \div 2}{14 + 7.5}$ _____

Solve the following problems. Reduce any fractions to lowest terms or round to the nearest hundredth.

27. $7\dfrac{5}{8} \div \left(1 + \dfrac{3}{8}\right)$ _____

28. $\left(5\dfrac{1}{4}\right)^2 \times 3.65$ _____

29. $\left(\dfrac{3}{4} \div \dfrac{5}{8}\right)^3 \div 3\dfrac{1}{2}$ _____

30. $\sqrt{6} \times \dfrac{3^2 + 2\frac{1}{2}}{7 \times \frac{5}{6}}$ _____

31. Describe in your own words the order of operations to be used when solving chain calculations. (See Objective 9.)

32. Explain how the parentheses keys are used when solving chain calculations. (See Objective 10.)

Solve the following application problems on a calculator. Round each answer to the nearest cent.

33. A college bookstore bought 397 used copies of a computer science book at a net cost of $46.40 each; 125 used copies of an accounting book at $38.40 each; and 740 used copies of a real estate text at $28.30 each. Find the total paid by the bookstore.

33. _____

34. Judy Martinez needs to file her expense account claims. She spent 5 nights at the Macon Holiday Inn at $104.19 per night and 4 nights at the Charlotte Super 8 motel at $86.80 per night. She then rented a car for 8 days at $36.40 per day. She drove the car 916 miles with a charge of $.28 per mile. Find her total expenses.

34. _____

35. In Virginia City, the sales tax is 6.5%. Find the tax on each of the following items: **(a)** a new car costing $17,908.43 and **(b)** a digital tv costing $1463.58.

(a) _____
(b) _____

36. Marja Strutz bought a five-year-old commercial fishing boat equipped for sardine fishing at a cost of $78,250. Additional safety equipment was needed at a cost of $4820, and sales tax of $7\frac{1}{4}$% was due on the boat and safety equipment. In addition she was charged a licensing fee of $1135 and a Coast Guard registration fee of $428. Strutz will pay $\frac{1}{3}$ of the total cost as a down payment and will borrow the balance. How much must she borrow?

36. _____

B-10 APPENDIX B Basic Calculators

37. Becky Agnosti and her husband bought a small townhouse for $155,000. They paid $8000 down and agreed to make payments of $1002.80 per month for 30 years. By how much does the down payment and the sum of the monthly payments exceed the purchase price?

    37. _____

38. Linda Smelt purchased a 16-unit apartment house for $620,000. She made a down payment of $150,000, which she had inherited from her parents, and agreed to make monthly payments of $5050 for 15 years. By how much does the sum of her down payment and all monthly payments exceed the original purchase price?

    38. _____

39. Ben Hurd wishes to open a small repair shop but has only $32,400 in cash. He estimates that he will need $15,000 for equipment, $2800 for the first month's rent on a building, and about $28,000 operating expenses until the business is profitable. How much additional funding does he need?

    39. _____

40. Koplan Kitchens wishes to expand their retail store. In order to do so, they must first purchase the $52,000 lot next door to them. They then anticipate $240,000 in construction costs plus an additional $57,000 for additional inventory. They have $100,000 in cash and must borrow the balance from a bank. How much must they borrow?

    40. _____

41. A college bookstore buys a used textbook for $24.50 at the end of a semester and sells it at the beginning of the next semester for $60. Find the percent of markup on selling price to the nearest percent.

    41. _____

42. A homebuilder spent the following when building a small cottage: $37,800 for a cleared parcel of land with utility hookups, $59,600 for materials, and $24,300 for labor and other expenses. He then sold the home for $136,500. Find the percent markup over cost to the nearest percent.

    42. _____

QUICK CHECK ANSWERS

1. 21.8
2. $174.09
3. 195.4
4. $14,100
5. .0125
6. 625
7. 110.592
8. 27
9. $1\frac{7}{16}$
10. $1\frac{5}{18}$
11. $1444.88
12. 424.5 pounds (rounded)
13. 118.8
14. 1
15. $203,150
16. $12,986
17. $9,303,228

Appendix C

Financial Calculators

OBJECTIVES

1. Learn the basic conventions used with cash flows.
2. Learn the basic financial keys.
3. Understand which keys to use for a particular problem.
4. Use the calculator to solve financial problems.

The use of calculators is so common in business that special calculators called *financial calculators* have been designed for managers and others.

OBJECTIVE 1 Learn the basic conventions used with cash flows. Some financial calculators require that inflows (cash received) and outflows (cash spent) be separated. As a result, we will use the following convention in this appendix.

1. Inflows of cash (cash received) are **positive**.
2. Outflows of cash (cash paid out) are **negative**.

For example, assume that you are making regular investments into an account. Your payments are *outflows* of cash to you and should be considered *negative* numbers. The future value of your savings will eventually be returned to you as an inflow of cash, thereby as a positive number.

OBJECTIVE 2 Learn the basic financial keys. Financial calculators have special functions that allow the user to solve financial problems involving time, interest rates, and money. Many of the compound interest problems presented in this text can be solved using a financial calculator. Most financial calculators have financial keys similar to those shown below.

These keys represent the following functions (see Chapter 10 for a full definition of each term).

[n]—The number of compounding periods

[I]—The interest rate *per compounding period*

[PV]—Present value, the value in *today's* dollars

[PMT]—The amount of a level payment (e.g., $625 per month); this is used for annuity type problems.

[FV]—Future value, the value at *some future date*

Note: Different financial calculators sometimes give slightly different answers to the same problems due to rounding.

OBJECTIVE 3 Understand which keys to use for a particular problem. Most simple financial problems require only four of the five financial keys described earlier. Both the number of compounding periods [n] and the interest rate per compounding period [i] *are needed for each financial problem*—these two keys will always be used. Which two of the remaining three financial keys ([PV], [PMT], and [FV]) are used depends on the particular problem. Using the convention described under Objective 1, one of these values will be negative and one will be positive. The process of solving a financial problem is to enter values for the three variables that are known, *then press the key for the unknown*, fourth variable.

> **Quick TIP**
>
> Different financial calculators look and work somewhat differently from one another. You *must refer to the instruction book* that came with your calculator to determine how the keys are used with that particular calculator.

C-1

For example, if you wish to know the future value of a series of known, equal payments, enter the specific values for [n], [i], and [PMT]. Then press [FV] for the result. Or, if you wish to know how long it will take for an investment to grow to some specific value at a given interest rate, enter values for [PV], [i], and [FV]. Then press [n] to find the required number of compounding periods.

Many financial calculators require that you enter a cash inflow as a positive number and a cash outflow as a negative number. Also, be sure to clear all values from the memory of your calculator before working a problem.

OBJECTIVE 4 Use the calculator to solve financial problems.

Finding FV, given n, i, and PV

EXAMPLE 1 Barbara and Ivan Cushing invest $2500 that they received from the sale of the old family car in a stock mutual fund that has recently grown at 8% compounded quarterly. Find the future value in 5 years if the fund continues to do as well.

SOLUTION

The present value of $2500 (a cash outflow is entered as a negative number) is compounded at 2% per quarter (8% ÷ 4 = 2%) for 20 quarters (5 × 4 = 20). Enter values for [PV], [i], and [n].

$$-2500 \; [PV] \; 2 \; [i] \; 20 \; [n]$$

Then press [FV] to find the compound amount at the end of 5 years.

[FV] $3714.87, which is the future value

QUICK CHECK 1

A lump sum of $28,000 is deposited in a retirement account earning 10% compounded quarterly. Find the future value in 10 years.

Finding FV, given n, i, and PMT

EXAMPLE 2 Joan Jones plans to invest $100 at the end of each month in a mutual fund that she believes will grow at 9% per year compounded monthly. Find the future value at her retirement in 20 years.

SOLUTION

Two hundred forty payments (12 × 20 = 240) of $100 each (cash outflows entered as a negative number) are made into an account earning .75% per month (9% ÷ 12 = .75%). Enter values for [n], [PMT], and [i].

$$240 \; [n] \; - \; 100 \; [PMT] \; .75 \; [i]$$

Press [FV] for the result.

[FV] $66,788.69, which is the future value

Quick TIP

The order in which data are entered into the calculator does not matter—just remember to press the financial key for the unknown value last.

QUICK CHECK 2

Benjamin Delton deposits $2500 at the end of each year into a retirement account earning 8% per year. Find the future value in 45 years.

Any one of the four values used to solve a particular financial problem *can be unknown*. Look at the next three examples in which the number of compounding periods [n], the payment amount [PMT], and the interest rate per compounding period [i], respectively, are unknown.

Finding *n*, given *i*, *PMT*, and *FV* **EXAMPLE 3** Mr. Trebor needs $140,000 for a new farm tractor. He can invest $8000 at the end of each month in an account paying 6% per year compounded monthly. How many monthly payments are needed?

SOLUTION

The $8000 monthly payment (cash outflow) will grow at .5% per compounding period ($6\% \div 12 = .5\%$) until a future value of $140,000 (cash inflow at a future date) is accumulated. Enter values for [PMT], [i], and [FV].

$$-8000 \; [PMT] \; .5 \; [i] \; 140000 \; [FV]$$

Press [n] to determine the number of payments.

$$[n] \; 17 \; \text{monthly payments of } \$8000 \; \text{each are needed}$$

Actually, 17 payments of $8000 each into an account earning .5% per month will grow to slightly more than $140,000.

$$-8000 \; [PMT] \; .5 \; [i] \; 17 \; n$$

Press [FV] to determine the future value.

$$[FV] \; \$141{,}578.41, \text{ which is the future value}$$

The 17th payment would need to be only

$$\$8000 - (\$141{,}578.41 - \$140{,}000) = \$6421.59$$

to accumulate exactly $140,000.

> **QUICK CHECK 3**
>
> Appalachian Coal, Inc., needs $1,280,000 for a new dredge. The firm can invest $21,200 at the end of each quarter in a fund earning 2% per quarter. Find the number of quarters needed to save up the required funds.

Finding *PMT*, given *n*, *i*, and *FV* **EXAMPLE 4** Jane Abel wishes to have $1,000,000 at her retirement in 40 years. Find the payment she must make at the end of each quarter into an account earning 8% compounded quarterly to attain her goal.

SOLUTION

One hundred sixty payments ($40 \times 4 = 160$) are made into an account earning 2% per quarter ($8\% \div 4 = 2\%$) until a future value of $1,000,000 (cash inflow at a future date) is accumulated. Enter values for [n], [i], and [FV].

$$160 \; [n] \; 2 \; [i] \; 1000000 \; [FV]$$

Press [PMT] for the quarterly payment.

$$[PMT] \; -\$878.35, \text{ which is the required quarterly payment of cash}$$

One hundred sixty payments of $878.35 at the end of each quarter into an account earning 8% compounded quarterly will grow to $1,000,000.

> **QUICK CHECK 4**
>
> Bill Watson has only 13 years until he retires and he wants to have $1 million at that time. He decides to make payments into a mutual fund at the end of each year, one that he hopes will earn 9% per year. Find the yearly payment needed.

Finding *i*, given *n*, *PV*, and *FV* **EXAMPLE 5** Tom Fernandez bought 200 shares of stock in an oil company at $33.50 per share. Exactly three years later, he sold the stock at $41.25 per share. Find the annual rate, rounded to the nearest tenth of a percent, that Mr. Fernandez earned on this investment.

SOLUTION

In three years, the per-share price increased from a present value of $33.50 to a future value of $41.25. The purchase of the stock is a cash outflow, and the eventual sale of the stock is a cash inflow. It is not necessary to multiply the stock price by the number of shares—the interest rate associated with the return on the investment is the same whether 1 share or 200 shares are used. Enter values for [n], [PV], and [FV].

$$3\ \boxed{n} \quad -33.50\ \boxed{PV} \quad 41.25\ \boxed{FV}$$

Press [i] for the annual interest rate.

$$\boxed{i}\ 7.18\%, \text{ or about } 7.2\% \text{ per year}$$

Mr. Fernandez's return on his original investment grew at 7.2% per year.

> **QUICK CHECK 5**
>
> Fang Hzu bought a commercial lot for $148,400 and sold it six years later for $245,000. Find the annual interest rate to the nearest tenth of a percent.

Interest rates can have a great influence on both individuals and businesses. Individuals borrow for homes, cars, and other personal items, whereas firms borrow to buy real estate, expand operations, or cover operating expenses. A small difference in interest rates can make *a large difference* in costs over time, as shown in the next example.

Comparing Monthly House Payments **EXAMPLE 6** Hank and Francesca Wilson want to borrow $165,000 on a 30-year home mortgage. Find the monthly payment at interest rates of **(a)** 6% and **(b)** 8%. Find **(c)** the monthly savings at the lower rate and **(d)** the total savings if they make payments for the entire 30-year life of the mortgage.

SOLUTION

(a) Enter a present value of $165,000 (cash inflow) with 360 compounding periods ($30 \times 12 = 360$) and a rate of .5% per month ($6\% \div 12 = .5\%$) and press [PMT] to find the monthly payment.

$$165000\ \boxed{PV} \quad 360\ \boxed{n} \quad .5\ \boxed{i}$$

[PMT] **−$989.26** is the monthly payment.

(b) Enter the values again using a rate of .66667% ($8\% \div 12 = .66667\%$ rounded)

$$165000\ \boxed{PV} \quad 360\ \boxed{n} \quad .66667\ \boxed{i}$$

[PMT] **$1210.72** is the monthly payment.

(c) Difference between payments = **$1210.72 − $989.26**
= $221.46 per month

(d) Find the total difference in payments during the 30-year mortgage as follows.

$$\$221.46 \times 360 \text{ payments} = \$79{,}725.60$$

The lower rate will save a total of $79,725.60 in payments over the 30 years. Interest rates *do* make a difference!

> **QUICK CHECK 6**
>
> Bernie Slotsky borrowed $168,000 for 30 years. Find the monthly payment at interest rates of **(a)** 6.5% and **(b)** 7.5%. **(c)** Then find the difference in the monthly payments.

Planning for Retirement **EXAMPLE 7** Courtney and Nathan Wright plan to retire in 25 years and will need $3500 per month for 20 years.

(a) Find the amount needed at retirement to fund the monthly retirement payments, assuming the funds earn 9% compounded monthly while payments are being made.

(b) Find the amount of the quarterly payment they must make for the next 25 years to accumulate the necessary funds, assuming earnings of 12% compounded quarterly during the accumulation period.

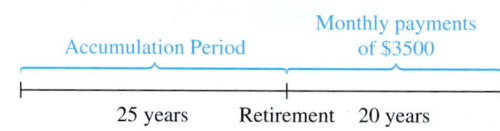

SOLUTION

(a) The accumulated funds at the end of 25 years are, at their retirement, a present value that must generate a cash inflow to the Wrights of $3500 per month for 240 months $(20 \times 12 = 240)$ assuming earnings of .75% per month $(9\% \div 12 = .75\%)$. Enter values for [n], [i], and [PMT].

$$240 \; [n] \; .75 \; [i] \; 3500 \; [PMT]$$

Press [PV] to find the amount needed at the end of 25 years.

[PV] **$389,007.34** is the amount they must accumulate

(b) The Wrights have 25 years of quarterly payments (100 payments that are cash outflows) in an account earning 3% per quarter $(12\% \div 4 = 3\%)$ to accumulate a future value of $389,007.34. The question is what quarterly payment is required. Enter values for [n], [i], and [FV].

$$100 \; [n] \; 3 \; [i] \; 389007.34 \; [FV]$$

Press [PMT] to find the quarterly payment needed.

[PMT] **$640.57** is the required quarterly payment

Thus, the Wrights must make 100 end-of-quarter deposits of $640.57 each into an account earning 3% per quarter in order to subsequently receive 20 years of payments of $3500 per month, assuming 9% per year during the time that payments are made.

QUICK CHECK 7

Tom and Jane Blackstone plan to retire in 20 years. At that time, they believe they will need $60,000 per year for 25 years, not including income from Social Security. **(a)** If funds earn 7% per year during retirement, find the amount needed to fund their retirement when they retire, to the nearest dollar. **(b)** Find the end-of-the-year payment they must make for the next 20 years to have the needed amount at retirement, assuming funds earn 8% per year.

Appendix C Exercises // MyLab Math

Using a financial calculator, solve the following problems for the missing quantity. Round dollar answers to the nearest cent, interest rates to the nearest hundredth of a percent, and number of compounding periods to the nearest whole number. Assume that any payments are made at the end of the period.

	n	i	PV	PMT	FV
1.	20	10%	$5800	—	_____
2.	7	8%	$8900	—	_____
3.	10	3%	_____	—	$12,000
4.	16	4%	_____	—	$8200
5.	7	8%	—	$300	_____
6.	25	2%	—	$1000	_____
7.	30	_____	—	$319.67	$12,000
8.	50	_____	—	$4718.99	$285,000
9.	360	1%	$83,500	_____	—

APPENDIX C   Financial Calculators

	n	i	PV	PMT	FV
10.	180	.5%	$125,000	_____	—
11.	_____	4%	$85,383	$5600	—
12.	_____	2%	$3822	$100	—

Solve each of the following application problems.

13. Juanipa Manglimont inherited $23,500 from her father. She placed the money in an investment earning 6% compounded quarterly. Find the future value at the end of 5 years.

13. _____

14. At the end of each month, Tina Ramirez has $50 taken out of her paycheck and invested in an account paying .5% per month. Find the future value at the end of 14 years.

14. _____

15. After a down payment, Mr. and Mrs. Thrash borrowed $140,000 on a 30-year loan on a home at 6% per year. Find the monthly payment.

15. _____

16. Terrance Walker wishes to have $20,000 in 10 years when his son begins college. What payment must he make at the end of each quarter into an investment he hopes will earn 10% compounded quarterly?

16. _____

17. The *Daily Gazette* needs $340,000 for a new printing press. The *Gazette* can invest $12,000 per month in an investment paying .8% per month. Find the number of payments that must be paid before reaching the goal. Round to the nearest whole number.

17. _____

18. Cathy Cockrell anticipates that she will need $70,000 when her son Sam enters college. She can save $500 per month and earn 7% per year compounded monthly. How long will it take her to save the needed funds?

18. _____

19. Mr. and Mrs. Peters wish to build a home and must borrow $110,000 on a 30-year mortgage to do so. Find the highest acceptable annual interest rate, to the nearest tenth of a percent, if they cannot afford a monthly payment above $845.

19. _____

20. Jim Blalock needs to borrow $28,000 for a new work truck but cannot afford a payment of more than $700 per month. If a bank will finance the truck for 4 years, find the maximum interest rate Blalock can afford.

20. _____

QUICK CHECK ANSWERS

1. $75,181.79
2. $966,264.04
3. 40 quarters
4. $43,566.56
5. 8.7%
6. (a) $1061.87
   (b) $1174.68
   (c) $112.81
7. (a) $699,215
   (b) $15,279.39

Appendix D

Exponents and the Order of Operations

OBJECTIVES
1. Understand the basics of exponents.
2. Multiply and divide using exponents.
3. Apply the order of operations.
4. Evaluating expressions containing exponents.
5. Solving application problems.

OBJECTIVE 1 Understand the basics of exponents. Exponents show repeated multiplication of a quantity by itself. The quantity being multiplied by itself is called the **base**. The number of times it is be multiplied by itself is the **exponent**. For example, $6 \cdot 6$ can be written as follows.

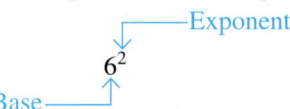

Read 6^2 any of the following ways: "6 to the second power," "6 raised to the second power," or "6 squared." Note that the base is the number 6 and the exponent is the number 2, which is written in a smaller type to the upper right of the number 6. Here are several more examples.

	Read as	Base	Exponent	Equal to
5^4	"5 to the 4th power"	5	4	$5 \cdot 5 \cdot 5 \cdot 5$
x^2	"x to the 2nd power, or x squared"	x	2	$x \cdot x$
y^3	"y to the 3rd power, or y cubed"	y	3	$y \cdot y \cdot y$
e^4	"e to the 4th power"	e	4	$e \cdot e \cdot e \cdot e$

If both the base and the exponent are known, we can evaluate the term by simply multiplying.

$$6^2 = 6 \cdot 6 = 36 \quad \text{or} \quad 5^4 = 5 \cdot 5 \cdot 5 \cdot 5 = 625$$

A few important special situations involving exponents are shown in the following table.

Explanation	Examples
Any nonzero value to the power of 0 is defined to be equal to 1.	$4^0 = 1$
	$x^0 = 1$ as long as x is not equal to zero
0 to the 0th power is not defined.	0^0 has no meaning
In other words, it has no meaning and should not be used.	y^0 is undefined when $y = 0$
Any value to the power of 1 is the value itself.	$5^1 = 5$
	$z^1 = z$
	$0^1 = 0$

Note that the base can be 0 as long as the exponent is not zero. For example,
$$0^3 = 0 \cdot 0 \cdot 0, \quad \text{which equals 0.}$$

Writing and Evaluating Exponents

EXAMPLE 1 (a) Write $t \cdot t \cdot t$ and $4 \cdot 4$ using exponents.
(b) Evaluate 9^3 and $(25.1)^2$.
(c) Simplify 54^0 and x^1.

SOLUTION

(a) $t \cdot t \cdot t = t^3$ $\qquad$ $4 \cdot 4 = 4^2$
(b) $9^3 = 9 \cdot 9 \cdot 9 = 729$ $\qquad$ $(25.1)^2 = 25.1 \cdot 25.1 = 630.01$
(c) $54^0 = 1$ $\qquad$ $x^1 = x$

D-2 APPENDIX D Exponents and the Order of Operations

> **QUICK CHECK 1**
> (a) Write $s \cdot s$ and $9.7 \cdot 9.7 \cdot 9.7$ using exponents.
> (b) Evaluate 8^2 and 2^4.
> (c) Simplify r^0 when $r \neq 0$ and then simplify p^1.

OBJECTIVE 2 Multiply and divide using exponents. There are several convenient rules that make it easier to work with exponents. We will show one rule at a time and give examples using both integers and variables. Then the concept behind each rule is explained using an example.

RULE 1 $X^m \cdot X^n = X^{(m+n)}$ When multiplying with the same base, add the exponents.

Examples: $7^2 \cdot 7^3 = 7^{(2+3)} = 7^5$ Add the exponents.
$y^4 \cdot y^2 = y^{(4+2)} = y^6$ Add the exponents.

Concept: $7^2 \cdot 7^3 = (7 \cdot 7) \cdot (7 \cdot 7 \cdot 7) = 7^5$ The number of 7s being multiplied is found by adding the two exponents.

RULE 2 $\dfrac{X^m}{X^n} = X^{(m-n)}$ When dividing with the same base, subtract the exponents.

Examples: $\dfrac{9^5}{9^2} = 9^{(5-2)} = 9^3$ Subtract the exponents.

$\dfrac{Z^4}{Z^2} = Z^{(4-2)} = Z^2$ Subtract the exponents.

Concept: $\dfrac{9^5}{9^2} = \dfrac{9 \cdot 9 \cdot 9 \cdot 9 \cdot 9}{9 \cdot 9} = 9^3$ The number of 9s remaining after canceling is found by subtracting the exponents.

RULE 3 $(X \cdot Y)^n = X^n \cdot Y^n$ Apply the exponent to both values in the parentheses.

Examples: $(8 \cdot 9)^2 = 8^2 \cdot 9^2$ Apply the exponent to both numbers.
$(S \cdot Z)^n = S^n \cdot Z^n$ Apply the exponent to both variables.

Concept: $(8 \cdot 9)^2 = (8 \cdot 9) \cdot (8 \cdot 9)$ Multiply $(8 \cdot 9)$ by itself.
$= (8 \cdot 8) \cdot (9 \cdot 9) = 8^2 \cdot 9^2$ The exponent applies to both numbers.

RULE 4 $\left(\dfrac{X}{Y}\right)^n = \dfrac{X^n}{Y^n}$ Apply the exponent to both the numerator and denominator.

Examples: $\left(\dfrac{7}{9}\right)^2 = \dfrac{7^2}{9^2}$ Apply the exponent to both numbers.

$\left(\dfrac{Z}{T}\right)^3 = \dfrac{Z^3}{T^3}$ Apply the exponent to both variables.

Concept: $\left(\dfrac{7}{9}\right)^2 = \dfrac{7}{9} \cdot \dfrac{7}{9} = \dfrac{7 \cdot 7}{9 \cdot 9} = \dfrac{7^2}{9^2}$ Apply the exponent to the numerator and to the denominator.

RULE 5 $(X^m)^n = X^{m \times n}$ Multiply the two exponents.

Examples: $(5^2)^3 = 5^{2 \cdot 3} = 5^6$ Multiply the two exponents.
$(R^4)^2 = R^{4 \cdot 2} = R^8$ Multiply the two exponents.

Concept: $(5^2)^3 = (5^2) \cdot (5^2) \cdot (5^2)$
$= 5^{(2+2+2)} = 5^{3 \cdot 2} = 5^6$

APPENDIX D Exponents and the Order of Operations D-3

Here are all the rules for working with exponents in one place.

Rules for Exponents—Quick Directions	
1. $X^m \cdot X^n = X^{(m+n)}$	Add the exponents.
2. $\dfrac{X^m}{X^n} = X^{(m-n)}$	Subtract the exponents.
3. $(X \cdot Y)^n = X^n \cdot Y^n$	Apply the exponent n to each variables in the parentheses.
4. $\left(\dfrac{X}{Y}\right)^n = \dfrac{X^n}{Y^n}$	Apply the exponent n to both the numerator and denominator of the fraction.
5. $(X^m)^n = X^{m \times n}$	Multiply the two exponents.

Note In Rule 3, notice that $(X \cdot Y)^n$ and $X^n \cdot Y^n$ can be written without the multiplication dot ($\cdot$) as $(XY)^n$ and $X^n Y^n$, respectively.

In Rule 1, it is important to realize that you cannot combine two values with different bases. For example, you cannot combine the following into one variable raised to a power because the bases are not the same.

$$X^2 \cdot Y^3$$

It has already been written as simply as possible unless the values for X and Y are known.

Nor can you simplify $\dfrac{y^3}{t^2}$ because these bases also differ.

Working With Exponents

EXAMPLE 2 Simplify each of these using the preceding rules.

(a) $\dfrac{y^4}{y^2}$

(b) $(9^2)^3$

(c) $z^2 \cdot z^3$

(d) $(5 \cdot 6)^2$

(e) $\left(\dfrac{r}{S}\right)^3$

(f) T^1

(g) $x^3 \div x^3$

(h) 0^0

SOLUTION

(a) $\dfrac{y^4}{y^2} = y^{(4-2)} = y^2$

(b) $(9^2)^3 = 9^{2 \cdot 3} = 9^6$

(c) $z^2 \cdot z^3 = z^{(2+3)} = z^5$

(d) $(5 \cdot 6)^2 = 5^2 \cdot 6^2$

(e) $\left(\dfrac{r}{S}\right)^3 = \dfrac{r^3}{S^3}$

(f) $T^1 = T$

(g) $x^3 \div x^3 = x^{(3-3)} = x^0 = 1$

(h) 0^0 has no meaning; it is undefined.

QUICK CHECK 2

Simplify each using the rules given above.

(a) $\dfrac{x^5}{x^3}$

(b) $(5^3)^2$

(c) $x^4 \cdot x^2$

(d) $(x \cdot y)^2$

(e) $\left(\dfrac{x}{t}\right)^3$

(f) 7^1

(g) $y^2 \div y^2$

(h) 0^0

OBJECTIVE 3 Apply the order of operations. Every expression must be evaluated in the sequence given by the **order of operations** shown in the following table. Be careful to always do the operations in an expression in the order listed in the table.[1]

Order of Operations–Calculate from the Top Down

1. Do all operations inside parentheses.
2. Evaluate square roots and exponents.
3. Multiply and divide from left to right.
4. Add and subtract from left to right.

---

[1] More practice is given using the order of operations in the appendix on basic calculators.

APPENDIX D   Exponents and the Order of Operations

To see why it is important to agree on the order of operations, look at the following. How should we evaluate it?

$$7 \cdot 8 - 3$$

Should we subtract first, or should we multiply first? A different answer is found depending on which mathematical operation is done first. There are only two math operations in the expression, namely, multiplication and subtraction. Since multiplication (item 3 in the preceding table) is listed above subtraction (item 4 in the preceding table), multiply first. Then subtract.

$$7 \cdot 8 - 3$$
$$56 - 3 \quad \text{First multiply 7 by 8.}$$
$$53 \quad \text{Then subtract.}$$

Quick TIP
It is very important to use the rules for order of operations when working with expressions or equations.

It is very important that you always use the order of operations when evaluating expressions. In complicated expressions, it is best to *do only one mathematical operation at a time* to prevent mistakes. The solution in the next example does that—only one operation is done at a time.

Evaluating Arithmetic Expressions   **EXAMPLE 3**   Use the order of operations to evaluate each.
(a) $7^2 - (5 \times 2)$
(b) $6 \cdot 5^3 + 3$

SOLUTION

Notice that operations are done in the sequence listed in the order of operations.

(a) $7^2 - (5 \times 2)$

$7^2 - 10$ First, evaluate the contents in the parentheses.
$49 - 10$ Next, evaluate the term with the exponent.
39 Finally, subtract.

(b) $6 \cdot 5^3 + 3$

$6 \cdot 125 + 3$ First, evaluate 5^3, which is $5 \cdot 5 \cdot 5 = 125$.
$750 + 3$ Next, multiply.
753 Finally, add.

QUICK CHECK 3

Evaluate using the order of operations.
(a) $1 + 6 \cdot 5^2$
(b) $(5 - 1)^2 \cdot 2 + 9$

OBJECTIVE 4   Evaluating expressions containing exponents. To evaluate an expression with a variable, first substitute the value of the variable into the equation. Then do the operations in the sequence listed in the order of operations to evaluate.

Evaluating an Expression with an Exponent   **EXAMPLE 4**   Evaluate each of the following.
(a) $X^3 - 7.8 \cdot 3$ when $X = 8$
(b) $(R - .7)^2 \cdot (9.2 - 1)$ when $R = 19$

SOLUTION

(a) $X^3 - 7.8 \cdot 3$

$8^3 - 7.8 \cdot 3$ Substitute 8 for the variable X.
$512 - 7.8 \cdot 3$ First evaluate 8^3, which is $8 \cdot 8 \cdot 8 = 512$.
$512 - 23.4$ Next, multiply.
488.6 Finally, subtract.

(b) $(R - .7)^2 \cdot (9.2 - 1)$

$\quad(19 - .7)^2 \cdot (9.2 - 1)$ Substitute 19 for the variable R.

$\quad\quad\quad(18.3)^2 \cdot (8.2)$ Do the operations inside both parentheses.

$\quad\quad\quad\quad 18.3^2 \cdot 8.2$ Remove parentheses since only one number is inside each.

$\quad\quad\quad\quad 334.89 \cdot 8.2$ Evaluate the exponent.

$\quad\quad\quad\quad 2746.098$ Finally, multiply.

QUICK CHECK 4

Evaluate using the order of operations.
(a) $(y + x)^2 - 12.5$ when $y = 10$ and $x = 4$
(b) $15 + (r - 1)^2$ when $r = 4$

OBJECTIVE 5 Solving application problems. The following example shows an application that uses exponents.

Solving an Application Problem **EXAMPLE 5** The manager of Benthos Diving has perfected and patented a new valve for deep-sea scuba diving. An economist friend tells him his profit can be approximated as follows.

$$P = \$.15N^2 + \$168N - \$300{,}000$$

where P = annual profit and N = number of valves sold in a year. Find the expected profit if 1100 valves are sold in the first year.

SOLUTION

$\quad\quad P = \$.15N^2 + \$168N - \$300{,}000$

$\quad\quad\quad = .15 \cdot 1100^2 + 168 \cdot 1100 - 300{,}000$ Substitute 1100 for N.

$\quad\quad\quad = .15 \cdot 1{,}210{,}000 + 168 \cdot 1100 - 300{,}000$ Evaluate the exponent.

$\quad\quad\quad = 181{,}500 + 184{,}800 - 300{,}000$ Multiply before adding.

$\quad\quad\quad = 366{,}300 - 300{,}000$ Add.

$\quad\quad P = \$66{,}300$

The expected profit is only $66,300. The balance of the revenue is "eaten up" by costs, which include labor, materials, rent, interest on bank loans, etc.

QUICK CHECK 5

Profit at a startup pizzaria can be approximated by: $P = 0.00135n^2 + 12n - \$122{,}500$ where n is the number of pizzas sold in a year. Estimate profit if 8000 pizzas are sold.

Appendix D Exercises // MyLab Math

Write each using exponents.

1. $p \cdot p$
2. $3 \cdot 3$
3. $r \cdot r \cdot r$
4. $7 \cdot 7 \cdot 7$
5. $x \cdot x \cdot x \cdot x$
6. $5 \times 5 \times 5$

Evaluate each of the following.

7. 7^2
8. $(.75)^2$
9. X^0 when $X = 7$
10. 5^1
11. 19^0
12. 12^3

APPENDIX D Exponents and the Order of Operations

Simplify each of the following, leaving exponents in each answer.

13. $(t \cdot g)^4$
14. $(S^3)^2$
15. $9^2 \cdot 9^2$
16. $\dfrac{6^5}{6^3}$
17. $\left(\dfrac{3}{4}\right)^2$
18. $\dfrac{7^m}{7^n}$
19. $(x \cdot y)^2$
20. $\left(\dfrac{y}{R}\right)^3$

Evaluate the following expressions.

21. $17 - 3 \cdot 4$
22. $9 \cdot 8 - 7$
23. $5 \cdot 4^2 + 3$
24. $(9.1 - 1) \cdot 13$
25. $191 - 5^3$
26. $(14 - 7)^2 - 3 \cdot 8$
27. $\dfrac{2^5}{2^3} \cdot 5$
28. $1 - \left(\dfrac{3}{4}\right)^2$
29. $\dfrac{12^3}{12^3} \cdot 75^2$
30. $17.2^3 + (5 - 2^2)$
31. $(16 - 2 \cdot 7)^0$
32. $(4^1 + 2^3 \div 2)^1$

Substitute the value(s) for the variable(s) and then evaluate.

33. $x^2 - 4 \cdot 2; x = 13$
34. $(9 - y)^2 + 5y; y = 3$
35. $7r \div 3^2; r = 27$
36. $\left(\dfrac{s}{x}\right) - 4; s = 42, x = 6$
37. $(y^2 - 7.8) \cdot 3t; y = 10, t = 2$
38. $\left(\dfrac{12}{x}\right)^2 \cdot r^2; x = 3, r = 9$
39. $(S - 7)^n \cdot 9.2 - 1; S = 13, n = 1$
40. $\dfrac{Gr^2}{2} \cdot 7 - 2^2; G = 21, r = 6$

Solve the following application exercises.

41. The daily cost to produce a new memory chip for a smartphone is given by $C = \$.17N^2 + \$12N + \$18{,}900$, where C = daily cost and N = average number produced in a day. Find the daily cost if $N = 420$.

42. The daily profit from selling a new Barbie doll is given by $P = \$.027N^2 + \$4.50N - \$62{,}700$, where P = daily profit and N = average number sold per day. Find the daily profit if $N = 1860$.

43. The future value of an investment is given by $M = P(1 + i)^t$, where M = maturity value, P = amount initially invested, i = interest rate per compounding period written as a decimal, and t = number of compounding periods. Find the future value of a $4800 investment expected to earn 5% per year for 4 years.

44. The future value of an investment is given by $M = P(1 + i)^t$, where M = maturity value, P = amount initially invested, i = interest rate per compounding period written as a decimal, and t = number of compounding periods. Find the future value of a $15,000 investment expected to earn 7% per year in a mutual fund for 40 years.

QUICK CHECK ANSWERS

1. (a) $s^2; 9.7^3$ (b) 64; 16 (c) $1; p$
2. (a) x^2 (b) 5^6 (c) x^6 (d) x^2y^2 (e) $\dfrac{x^3}{t^3}$ (f) 7 (g) 1 (h) undefined
3. (a) 151 (b) 41
4. (a) 183.5 (b) 24
5. $59,900

Appendix E

Graphing Equations

OBJECTIVES

1. Understand dependent and independent variables.
2. Choose the axes and labels carefully.
3. Graph linear equations.
4. Graph nonlinear equations.

OBJECTIVE 1 Understand dependent and independent variables. A **dependent variable** is one whose value *depends on* the value of a different variable. On the other hand, an **independent variable** is one whose value *does not depend on* the value of another variable. Think about the amount of rain in the past 24 hours and amount of water flowing down a local creek. Does the amount of rain in the past 24 hours depend on the amount of water flowing down the creek? No! However, the amount of water flowing down the creek probably does depend to some degree on the amount of rain in the past 24 hours. So, the dependent variable is the amount of water flowing down the creek and the independent variable is the amount of rain in the past 24 hours. Here is an example from business.

Managers at Andiamo's Pizza have studied monthly sales and advertising data and have found that the relationship can be estimated as follows:

$$\text{Sales} = 37.5 \cdot \text{advertising} + \$12{,}200$$
$$S = 37.5A + 12{,}200$$

Since the volume of sales should depend on the amount of advertising dollars spent, the variable S (*sales* dollars) is the dependent variable and the variable A (*advertising* dollars) is the independent variable. Another way to think about it is to realize that the amount of advertising does not depend on sales. Rather, the amount of advertising is chosen by management. More advertising should call more attention to the restaurant and result in higher sales. So, sales depend on advertising.

Equations with variables have one dependent variable and at least one independent variable. Here are some common examples—the key is to decide which variable depends on the other.

Dependent Variable		Independent Variable
Real estate commission	depends on	sale price of house
Sales tax	depends on	amount of purchase
Interest earned	depends on	amount invested

Actually, there can be more than one independent variable in an equation. For example, the amount of *interest earned* in an account at a bank depends on *amount invested, interest rate,* and *length of time* funds are on deposit. In this *case, interest earned* is the dependent variable and the other three variables (*amount invested, interest rate,* and *length of time* funds are on deposit) are all independent variables.

Identifying Independent and Dependent Variables  Identify the independent and dependent variables for each of the following by asking yourself which variable depends on another variable or variables.

(a) Income, income tax to be paid

(b) Amount of debt on VISA card, monthly interest charge

(c) Property tax, assessed value of a house

(d) Interest rate earned, years to retirement, amount at retirement, annual investment

SOLUTION

	Dependent Variable		Independent Variable(s)
(a)	Income tax to be paid	depends on	income
(b)	Monthly interest charge	depends on	amount of debt on VISA card
(c)	Property tax	depends on	assessed value of a house
(d)	Amount at retirement	depends on	annual investment, years to retirement, and interest rate earned

QUICK CHECK 1

Identify the independent and dependent variables for each.
(a) FICA tax, wages
(b) cost of new car, sales tax

OBJECTIVE 2 Choose the axes and labels carefully. Data is often shown using a **line graph**, which is a graph with a line or curve drawn on it. A graph does not have the precision of an algebraic equation in that you cannot use it to find an exact number. However, line graphs are commonly used since they help convey ideas and trends so well.

Before constructing a graph, think about each axis. An **axis** is a horizontal or vertical line along the edge of a graph that usually has a **scale** with units of measurement. An axis is required in order to construct the graph, and it is an important aide to help understand the information and data in the graph. The horizontal axis is commonly used for the *independent variable,* and it is often called the **x-axis.** It should have a label below it unless the value on the axis is obvious to anyone reading the graph. Here is one choice for the x-axis for the equation introduced above: $S = 37.5A + 12{,}200$.

Advertising
Independent variable

The purpose of a graph is to *clearly communicate* data and ideas. The x-axis above can be improved by adding a scale showing different amounts of advertising. Commonly, the amount of Advertising at the leftmost point, which is called the **origin**, is $0. However, the **origin** can be a number other than 0. Here we have added the origin at the left.

Origin
Advertising

The next questions to ask are: What scale should be used for the x-axis, and how should it be labeled? A scale divides the length of the axis into sections that show amounts for the independent variable. In our example, it should be based on the amounts of advertising that managers would consider. If they would consider up to $2000 per month in advertising, then the scale should divide $2000 into 4 to 8 intervals. A reasonable label is "Advertising (in dollars)."

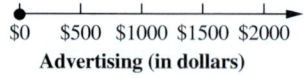

$0 $500 $1000 $1500 $2000
Advertising (in dollars)

However, the following choice of a scale may also be appropriate.

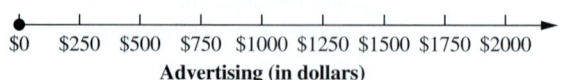

$0 $250 $500 $750 $1000 $1250 $1500 $1750 $2000
Advertising (in dollars)

However, this last axis seems a little "too busy" with too many numbers on it, so we will use the axis just above the last one.

In a similar fashion, the **y-axis** is chosen to represent the reasonable values for the outcome of *Sales* based on the values of *Advertising* shown on the x-axis. The y-axis is commonly vertical. To determine the scale on the y-axis, first find the minimum and maximum amount of sales that are possible using the equation from above: $S = 37.5A + 12{,}200$.

Minimum sales occur if Advertising (A) is $0: $S = 37.5 \cdot \$0 + \$12,200 = \$12,200$

Maximum sales occur at the maximum advertising of $2000: $S = 37.5 \cdot \$2000 + \$12,200 = \$87,200$

Therefore, the scale on the *y*-axis should go from below $12,200 to a little over $87,200. A reasonable scale is shown here.

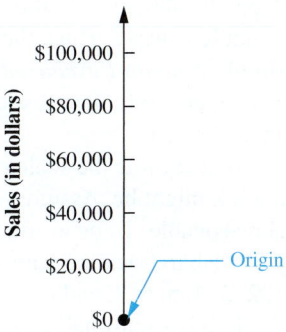

Combining our chosen *x*- and *y*-axes at the origin results in the following.

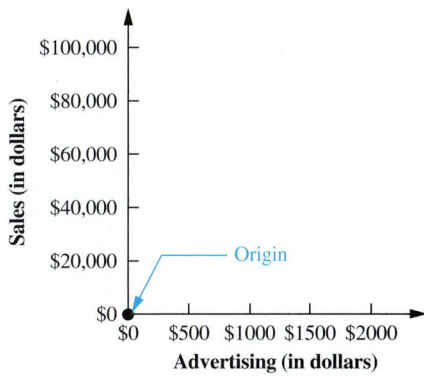

Remember, the purpose of a graph is to communicate CLEARLY. We have not yet graphed the equation, but is there something we can do to improve the quality of the communication in the basic graph itself? Yes, we can add a title to it. The title is very important, and it should clearly indicate the main theme of the information or data in the graph. In our example, the title should include some or all of the following words: Andiamo's Pizza, sales, advertising, and monthly data. The graph should be labeled so well that you can put it aside and still understand the graph a month or two later when you look at it again.

Quick TIP

Always try to use a title, a label for the axis, and a scale that will be clearly understood.

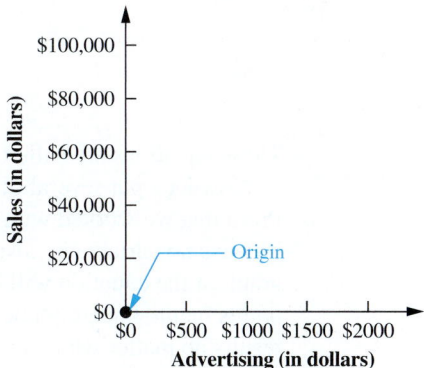

Date Graph Developed: November 2018

Shortly we will see how to graph the relationship between advertising and sales for Andiamo's Pizza.

Setting Up a Reasonable Graph

EXAMPLE 2 At the request of a financial planner, the Hernandez family is setting up a personal budget. They wish to graph the amount spent per week on groceries for their family of three. Help them set up an appropriate graph.

SOLUTION

First identify the dependent and independent variables. Since the amount of food purchased depends on the week, the dependent variable is *amount of food purchased* and the independent variable is *week*. Thus, the y-axis label should be *Amount of Food Purchased (in dollars)* or simply *Amount Purchased (in dollars)*. The x-axis label should be *Week* (either week number 1, 2, 3, etc., or a date associated with the first day of successive weeks such as Jan. 9, Jan. 16, Jan. 23, etc.).

To determine the scale on the y-axis, decide what the maximum amount of food purchased in a week might be. Assuming a maximum purchase of $300, a scale of $0, $100, $200, and $300 is reasonable. To determine the scale on the x-axis, decide on the number of weeks for which you wish to track the data. Assuming you want to track the data for 8 weeks, the scale would be 1, 2, 3, 4, 5, 6, 7, and 8.

Here is a reasonable graph.

> **Quick TIP**
> The first step in building a graph is to identify the independent and dependent variables.

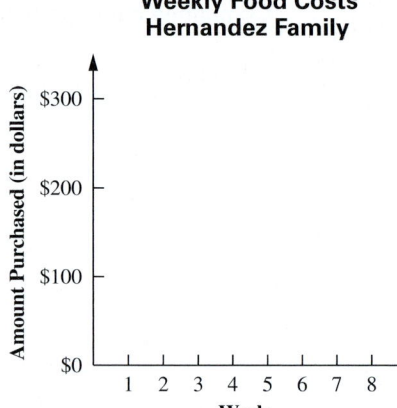

QUICK CHECK 2

Help an economist set up a graph showing the number of homes under construction in one region versus the 30-year fixed mortgage interest rate. The number of homes under construction varies from 112,000 to 289,000 and the interest rate varies from 3% to 10%.

OBJECTIVE 3 Graph linear equations. A **linear equation** is an equation that does not have any variables raised to a power (such as x^2) or multiplied by another variable (such $x \cdot y$). Here are some examples of linear equations:

$$S = 37.5A + 12{,}200$$
$$y = 35x + 94$$
$$P = 1700 - 47.3s$$

These equations are called linear equations, since the graph of each is a straight line.

Now let's put several of these ideas together and graph the equation of sales at Andiamo's Pizza that we worked with above: $S = 37.5A + \$12{,}200$. First, note that it is a linear equation because no variable is raised to a power, nor are any variables multiplied by one another. So, the graph of the equation will be a straight line. For a straight line, we need to plot only two data points on the line graph before connecting them with a ruler. The exact same straight line results no matter which two points we use, and the same straight line would occur if we used three or more points.

> **Graphing a Straight Line**
>
> Step 1 Pick two different values of the independent variable.
> Step 2 For each value chosen in Step 1, calculate the corresponding value of the dependent variable.
> Step 3 Draw a straight line through the two points.

It is best to chose values of the independent variable that are not too close to one another. Reasonable *Advertising* expenses range from $0 to $2000, so we arbitrarily choose advertising values of $500 and $2000. Next, substitute these two values for A on the right side of the equation to find the predicted sales.

Amount of *Advertising* (A)	Corresponding *Sales* (S) amount
$500	$S = 37.5 \cdot \$500 + \$12{,}200 = \$30{,}950$
$2000	$S = 37.5 \cdot \$2000 + \$12{,}200 = \$87{,}200$

The next step is to graph the following data points on the graph developed above. Often the points are written in this format: (value of independent variable, value of dependent variable), or as shown in the right column below. We sometimes refer to this notation as (x, y) where x is the value for the independent variable and y is the value for the dependent variable.

Advertising	Sales	Data Point
$500	$30,950	($500, $30,950)
$2000	$87,200	($2000, $87,200)

y-variable (dependent)
x-variable (independent)

To plot the data point ($500, $30,950), place a point directly above $500 on the *x*-axis and aligned horizontally to $30,950 on the *y*-axis. You cannot see the difference between $30,950 versus, say, $31,200 on the graph, so do the best you can. As stated earlier, graphs do not have the precision of an algebraic equation, but they often show the overall relationship more clearly than an algebraic equation. Next, plot ($2000, $87,200) by placing a dot aligned with $2000 on the *x*-axis and with $87,200 on the *y*-axis.

Monthly Sales & Advertising
Andiamo's Pizza

Date Graph Developed: November 2018

Finally, connect the lines with a ruler.

**Monthly Sales & Advertising
Andiamo's Pizza**

Date Graph Developed: November 2018

The formula $S = 37.5A + \$12{,}200$ is the algebraic form of the relationship. However, the graph is a visual representation of the exact same thing. These are two very different ways of showing the *same information*. Graphs are extremely common in business, since they allow us to use our visual ability to quickly understand a relationship or pattern. Is it easier for you to work with a graph or an equation?

The **y-intercept** is the point at which the line intersects the y-axis. The y-axis for this graph corresponds to $0 in advertising. You can either substitute $0 into the equation in place of A and solve to find that the y-intercept is $12,200 or estimate the y-intercept by looking at the graph. Monthly sales would be $12,200 even with no advertising.

Constructing a Graph — **EXAMPLE 3** — The real estate commission in Albuquerque, New Mexico, is 6% of the sales price of the home. The median price of homes sold in one part of Albuquerque last month was $223,000, but prices of homes sold ranged from $100,000 to $480,000 during the month. Write an equation for commission versus sales price and graph it in a well-labeled graph.

SOLUTION

The algebraic relationship is based on the fact that *Real Estate Commission* equals 6% of *Sales Price*, so commission depends on price as follows.

$$C = .06 \times P$$

where C = Real Estate Commission
P = Selling Price of Home

Since the independent variable *Selling Price of Home* (P) ranged between $100,000 to $480,000, a reasonable scale for the x-axis would be $0, $100,000, $200,000, $300,000, $400,000, and $500,000. Using thousands, the scale would be $0, $100, $200, $300, $400, and $500. Although we can start the x-axis at $100 (in thousands) on the left since that is the lowest value of the independent variable, we choose to use $0 for the origin.

The dependent variable is *Real Estate Commission*, which can be as high as 6% of $480,000, or $28,800. So, a reasonable scale for the y-axis would be $0, $10,000, $20,000, and $30,000. If the label on the y-axis indicates that the commissions are in thousands of dollars, then the numbers shown along the axis would be $0, $10, $20, and $30.

Again, the equation is linear, so we need to plot only two data points. Remember to choose values for the independent variables that are spread out a little—we arbitrarily choose $100,000 and $300,000.

Independent Variable	Dependent Variable	Data Point (thousands of dollars)
$100,000	$C = .06 \times \$100{,}000 = \$6000$	($100, $6)
$300,000	$C = .06 \times \$300{,}000 = \$18{,}000$	($300, $18)

Now construct the chart with a good title, appropriate axes, and a reasonable scale.

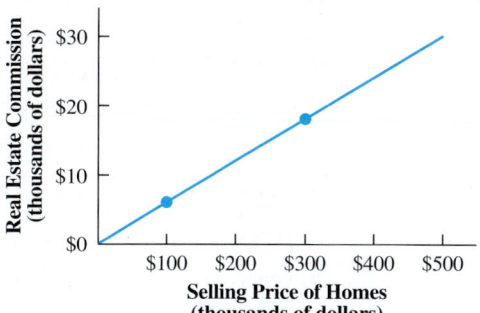

Quick TIP
Software such as Microsoft Excel can be used to create professional-looking graphs.

> **QUICK CHECK 3**
>
> Graph the following equation: Sales tax $= 0.085 \cdot S$, where S is sales not including taxes. Sales for one month range from $0 to $100,000.

OBJECTIVE 4 Graph nonlinear equations. Up to this point, we have worked only with linear equations, which graph as straight lines. However, many situations in business require **nonlinear equations**, which have graphs that are not straight lines. Compound interest calculations used daily by banks and others require the use of the following nonlinear equation. It shows how to find the maturity value of an investment of P dollars at an interest rate of i for n years.

$$M = P(1 + i)^n$$

where $M =$ total amount accumulated at the end of the investment period

$P =$ principal amount invested

$i =$ interest rate per year written as a decimal number (e.g., write 6% as 0.06)

$n =$ number of years the investment is held

The dependent variable M (maturity value) depends on the three independent variables: P (original investment), i (interest rate), and n (number of years). The equation is nonlinear since it includes an exponent and also since the variable P is being multiplied by $(1 + i)$, which has a variable itself. Thus, it will not graph as a straight line. As a result, it is *necessary to plot several data points* before connecting them with a smooth line—two data points are not enough for a nonlinear graph.

Graphing an Equation that Involves Exponents

EXAMPLE 4 Trisha Patterson inherits $8000 from her grandmother and decides to put the money into a mutual fund that she hopes will earn 5% per year. She hopes the accumulated funds will help her 2-year-old pay for college. Graph the equation $M = P(1 + i)^n$ using the data given.

Quick TIP
Before graphing an equation with exponents or variables multiplied together, plot several points.

SOLUTION

The original investment is $P = \$8000$ and the interest rate per year is $i = 5\%$, or .05. Putting the values for P and i into the equation simplifies it.

$$M = \$8000 \cdot (1 + .05)^n$$

Or, simpler still, as

$$M = \$8000 \cdot (1.05)^n$$

A reasonable scale for the x-axis is 0, 5, 10, 15, and 20 years. Next, find the value of M that corresponds to each of these values of the independent variable. Notice below that the largest value on the y-axis is $21,226, so label the y-axes as Maturity Value (in thousands), and use this scale: $0, $5, $10, $15, $20, $25.

Dependent Variable	Maturity Value (M), Rounded to the Nearest Dollar
0 years	$M = \$8000 \cdot (1.05)^n = \$8000 \cdot 1.05^0 = \$8000 \cdot 1 = \$8000$
5 years	$M = \$8000 \cdot (1.05)^n = \$8000 \cdot 1.05^5 = \$8000 \cdot 1.27628 = \$10{,}210$
10 years	$M = \$8000 \cdot (1.05)^n = \$8000 \cdot 1.05^{10} = \$8000 \cdot 1.62889 = \$13{,}031$
15 years	$M = \$8000 \cdot (1.05)^n = \$8000 \cdot 1.05^{15} = \$8000 \cdot 2.07893 = \$16{,}631$
20 years	$M = \$8000 \cdot (1.05)^n = \$8000 \cdot 1.05^{20} = \$8000 \cdot 2.65330 = \$21{,}226$

Finally, plot the points (0, $8), (5, $10), (10, $13), (15, $17) and (20, $21) and connect the points with a smooth line.

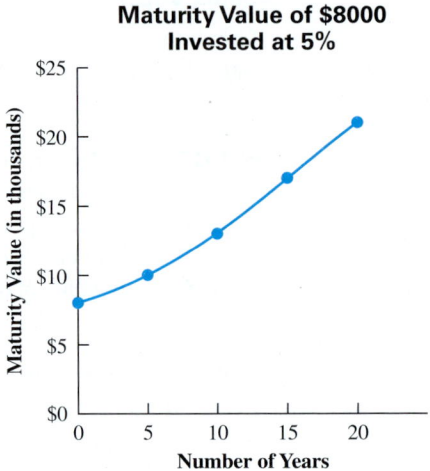

Notice that the line is not a straight line. Rather, it curves upward slightly, showing that the maturity value grows more rapidly with time. In actuality, in the later years the investment is earning interest on larger and larger amounts since interest is being paid on interest already paid to the account. Nonlinear relationships such as these are both very common and very important in business. As we will see in the chapter on annuities, the increasing amounts of interest earned over time is a key to building up investments.

QUICK CHECK 4

The chief financial officer at Siena College puts $1,000,000 in an investment expected to earn 4.5% per year for the next 5 years. Use $M = P(1 + i)^n$ to graph the maturity value for the next five years. The variable i = interest rate as a decimal number, n = number of years of the investment, and P = original investment.

A common application of exponents uses a number represented by the symbol e, which is approximately equal to 2.7182818. This number is so commonly used in business, engineering, and scientific calculations that it can be found on most calculators. Is it on yours? Maturity value M depends on the following three things:

- The original investment of P dollars
- Invested for t years
- At an interest rate of r per year.

The formula for the maturity value assuming continuous compounding is given here.

$$M = Pe^{rt}$$

The value for e is a constant that is found on your calculator, so you will not have to key it in each time. Since e is raised to the power of r (interest rate) multiplied by t (time), the equation is nonlinear. The next example shows the huge difference both time and interest rate make in accumulating money for a future event, such as saving for an automobile or a down payment on a home.

Graphing an Equation that Uses the Number e

EXAMPLE 5 Sara Walker won a scratch-off lottery. After buying a Toyota Camry, she has $7000 to invest. A financial advisor shows her the difference that interest rate and length of investment can make using the following two investment options.

(a) A conservative investment expected to earn 4% per year for 10 years, or

(b) A mutual fund expected to earn 7% for 40 years.

Graph maturity value against year for both of these investment options on the same graph.

SOLUTION

It is necessary to make a separate calculation for option (a) and (b) at each of the years to be used on the x-axis. Option (a) is only for 10 years, so we choose to calculate the maturity values for this option for years 0, 5, and 10. However, option (b) is for 40 years, so we will calculate the maturity values for it for 0, 5, 10, 20, 30, and 40 years. Here are the calculations for maturity value for both options for 5 years, with maturity value rounded to the nearest dollar.

(a) $M = Pe^{rt} = \$7000 \cdot e^{.04 \times 5} = \$7000 \cdot e^{.2} = \$7000 \cdot 1.22140 = \$8550$

(b) $M = Pe^{rt} = \$7000 \cdot e^{.07 \times 5} = \$7000 \cdot e^{.35} = \$7000 \cdot 1.41907 = \$9933$

The remaining maturity values are calculated in the same way using the appropriate interest rate and number of years in place of the variables r and t, respectively.

The initial investment is $7000 under both option (a) and (b) above, so $7000 is the amount at time 0 for both options. Check the maturity values in the table that follows to make sure they are correct. Notice that no calculation is needed under option (a) beyond 10 years.

	Maturity Values	
Year	Option (a)	Option (b)
0	$7,000	$7,000
5	$8,550	$9,933
10	$10,443	$14,096
20	*	$28,386
30	*	$57,163
40	*	$115,113

Initial investment.

Sara is amazed that a $7000 investment can grow to about $115,000 in 40 years. The amount of interest that the investment will earn in 40 years is found by subtracting the original investment from the maturity value as follows.

$$\$115,113 - \$7000 = \$108,113$$

She likes the thought of her money working for her over the years. Option (b) clearly results in far more money than option (a), but of course it takes much longer to accumulate.

The left column of the table above, labeled *Year*, shows the values for the x-axis. The y-axis of the graph should go higher than $115,113, so we choose to label it as *Maturity Value* (in thousands) with a scale of $0, $25, $50, $75, $100, $125.

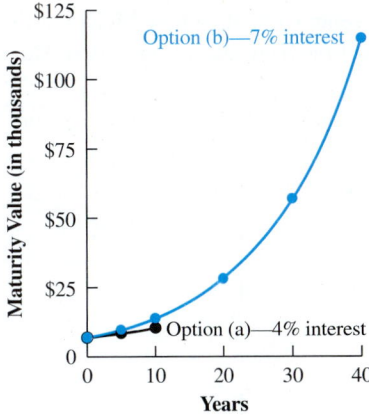

Maturity Value of a $7000 Investment at 4% and 7%

The difference in the effects of the two different interest rates of 4% and 7% is shown by the difference in the two curves up to year 10. Option (b) has the higher interest rate and results in larger maturity values. So, both interest rate and time are very important to building investments. The chapter on annuities and sinking funds (Chapter 11) will show you practical ways to accumulate a large sum of money over time.

E-10 APPENDIX E Graphing Equations

> **QUICK CHECK 5**
>
> Use $M = Pe^{rt}$ to graph the maturity value of a $10,000 investment ($P$), when $r = 3.5\%$ and $t = 3$ years.

Appendix E Exercises // MyLab Math

Identify the dependent and independent variable(s) based on your internal sense of which variable depends on the other(s).

		Dependent	Independent
1.	Interest earned, total investment		
2.	Calories eaten, weight gain		
3.	Profit, sales		
4.	Number of employees, payroll		
5.	Weight of calf, age of calf		
6.	Total sales, number of smart phones sold		
7.	Sales, profit, total expenses		
8.	Size of roof, material used, cost of new roof		

Construct a reasonable graph for each of the following.

9. **FLOODING** Due to severe flooding in the past, administrators in Pakistan carefully watch the flow of water in the Indus River which flows out of China and India and into Pakistan. Water flows are estimated each day based on the height of the river in meters taken at the upstream monitoring station. One of the worst floods on record occurred when the height of the water at the monitoring station peaked at 32.1 meters. It resulted in 1600 deaths and affected 14 million people.

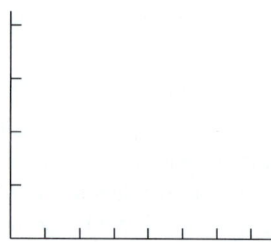

10. **MCDONALDS** An entrepreneur sets up a new McDonald's franchise and plans to carefully watch weekly sales for the first 10 weeks of operation. He hopes weekly sales will approach $16,000.

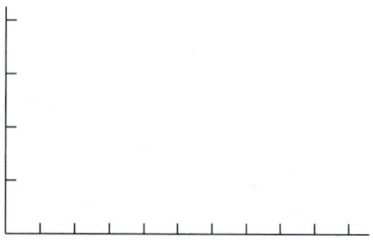

11. **APPLE, INC.** An investor loves her iPad and iPhone and is debating whether to buy stock in the company that makes them, Apple Inc. She wants to look at a graph of the firm's after-tax quarterly profit for each quarter of the past year. Quarterly profits range up to nearly $8 billion.

12. **QUALITY** A quality-control inspector is concerned with the number of bottles of beer that are not being capped appropriately at a bottling machine on a production line. Although the production line runs very rapidly, he has a machine that scans every bottle for quality problems. He doesn't expect more than 30 defective bottles in an hour, and he wants a graph that will show results for a 24-hour day.

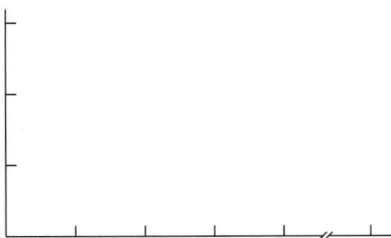

Construct the graphs indicated.

13. **MONTHLY SALES** Monthly sales for one product line are based on advertising and can be approximated by $S = 40.2A + \$17,500$. Graph this relationship if advertising does not exceed $2000 per month.

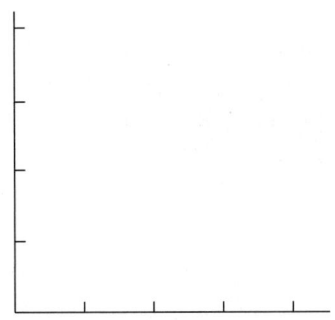

E-12 APPENDIX E Graphing Equations

14. **STOCK PRICE** Recently, the share price of LemonKing Inc. has increased with the firm's earnings per share of stock according to $S = 24.1E + \$12.3$, where $S =$ share price and $E =$ quarterly earnings per share of stock. Graph this relationship, assuming that earnings per share do not exceed $.90.

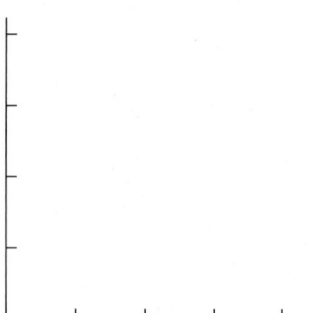

15. **ENTREPRENEUR** Bill Thomas is thinking about working part-time tutoring gifted children. He charges $25 for each tutoring sessions with a student and will rent a small office with a total cost of $275 per month. First, write an equation for his monthly profit based on the number of students he tutors in a month. (*Hint:* monthly profit = monthly revenue − monthly expense.) Then, graph the equation assuming that he will tutor between 15 and 80 students in any given month.

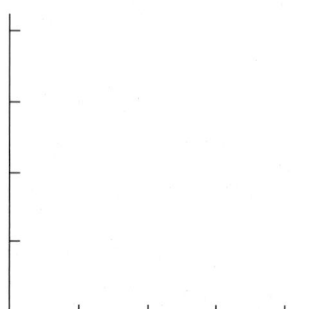

16. **CELL PHONE** Verizon advertises a low-cost beginner special on cell phone use to attract new users. The monthly cost is $14.95 a month plus $.07 per minute for each minute of time (excluding taxes) used for calls made in the United States. Write an equation for monthly cost versus cell phone use and graph it, assuming users do not use over 400 minutes per month.

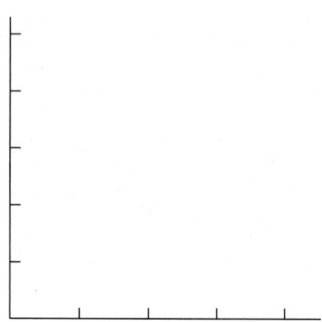

17. **LAWN MOWING** Calvin Tigard is only 15, but he has decided to start a lawn-mowing business. He plans to charge an average of $40 to mow and trim each yard. He thinks that the cost of buying and maintaining the equipment needed plus operational expenses such as gasoline will be about $7 per yard mowed. Write an equation for weekly profit and graph it, assuming he does not mow more than 8 yards per week.

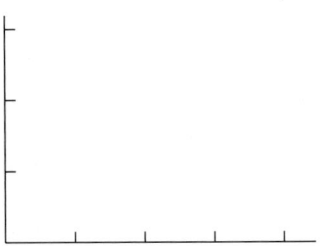

18. **VACATION** After college, Tina Duncan and her husband decide to drive the 3000 miles across the United States to visit her grandparents. They can average 60 miles per hour on the highways, but Tina is not sure how many hours per day they will drive. Write an equation to show miles driven per day based on the number of hours driven and graph it, assuming they are not willing to drive more than 10 hours in any one day.

19. **COMPUTER SALES** Computer Shoppe is a high-volume seller of computers on the Internet. The total cost of selling computers can be approximated by $C = \$95{,}000 + \$680N + \$.21N^2$, where C = total annual cost and N = number of computers sold in a month. Graph this nonlinear equation, assuming fewer than 1000 computers are sold. (*Hint:* Since it is a nonlinear equation, you must plot several different values of the independent variable before connecting points with a smooth curve.)

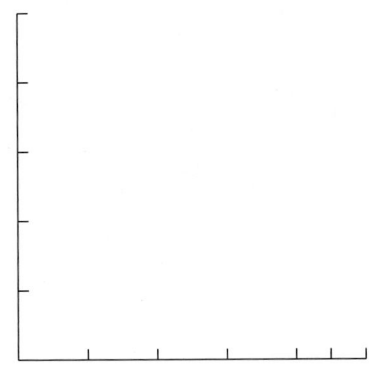

E-14 APPENDIX E Graphing Equations

20. **SMARTPHONES** Smartphones.com sells smartphones cheaply, both through their store location and on the Internet. The total weekly profit is approximated by $P = -\$85{,}000 + \$645N - \$.3N^2$, where P = total weekly profit and N = number of smartphones sold during the week. Use the equation to create a graph, assuming that smartphone sales range between 200 and 800 per week. (*Hint:* Since this is a nonlinear equation, you must plot several different values of the independent variable before connecting points with a smooth curve.)

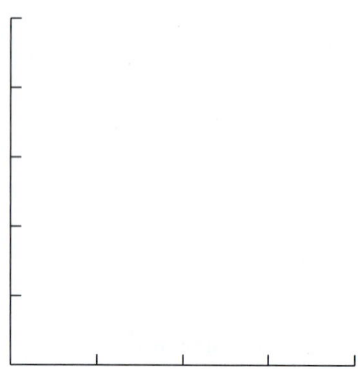

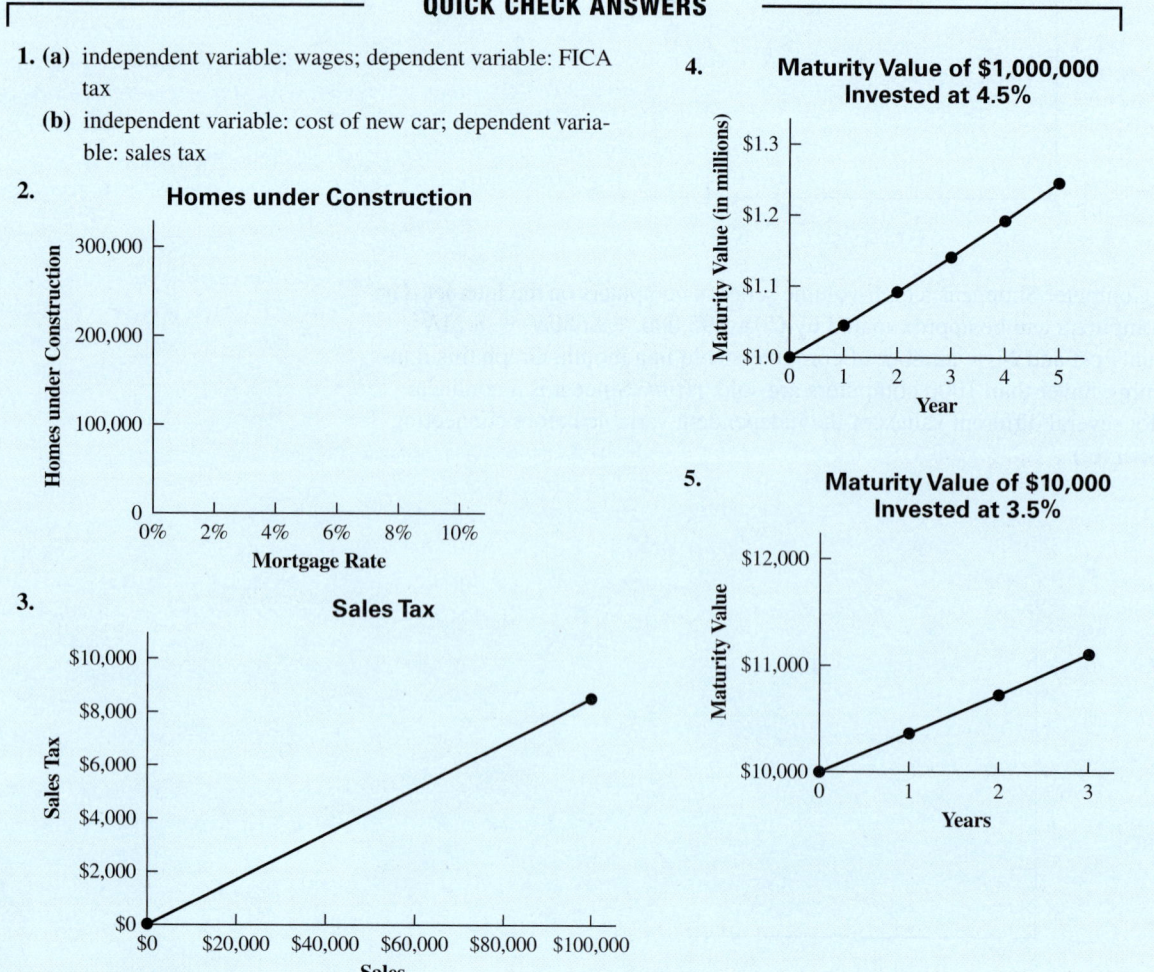

QUICK CHECK ANSWERS

1. **(a)** independent variable: wages; dependent variable: FICA tax
   (b) independent variable: cost of new car; dependent variable: sales tax

Answers to Selected Exercises

CHAPTER 1

Section 1.1 Exercises (Page 8)

1. seven thousand, forty **3.** thirty-seven thousand, nine hundred one **5.** four million, six hundred fifty thousand, fifteen **7.** 2070; 2100; 2000 **9.** 46,230; 46,200; 46,000 **11.** 106,050; 106,100; 106,000 **15.** 210 **17.** 2186 **19.** 1396 **21.** 983,493 **23.** 668 **25.** 2877 **27.** 21,546 **29.** 6,088,899 **31.** Totals horizontally: $293,267; $387,795; $426,869; $373,100; $1,481,031; Totals vertically: $269,761; $267,502; $206,932; $246,587; $244,616; $245,633; $1,481,031 **33.** 9374 **35.** 117,552 **37.** 1,696,876 **39.** 8,107,899 **41.** Estimate: 12,760; Exact: 12,605 **43.** Estimate: 600; Exact: 545 **45.** Estimate: 30,000; Exact: 29,986 **47.** $37 \times 18 = 666$; 66,600 **49.** $376 \times 6 = 2256$; 22,560,000 **51.** 1241 R1 **53.** 458 R21 **57.** 2385 R5 **59.** 58 R4 **61.** two million, four hundred forty-three thousand **63.** three million, two hundred thousand **65.** 854,795 boxes **67.** 56,312,700 **69.** 200,000 chips **71.** 500 items per hour **73.** $4503 **75.** 128,197,000 **77.** 32,293,000 **79.** 12,500 stores **81.** Amazon; fewer than 100 stores **83.** Costco; $10.3 billion

Learning Catalytics (Page 2)

1. two hundred sixty-one million, forty thousand, nine hundred **2.** 14

Section 1.2 Exercises (Page 17)

1. 5208 sandwiches **3.** 2 trillion additional miles per year **5.** 1,189,100,000 metric tons **7.** 8589 pounds **9.** $64 **11.** 6,098,400 square feet **13.** $1050 **15.** $20,961 **17.** $375 **19.** 20 seats

Learning Catalytics (Page 14)

1. 102 **2.** 400

Section 1.3 Exercises (Page 21)

1. thirty-eight hundredths **3.** five and sixty-one hundredths **5.** seven and four hundred eight thousandths **7.** thirty-seven and five hundred ninety-three thousandths **9.** four and sixty-two ten-thousandths **13.** 438.4 **15.** 97.62 **17.** 1.0573 **19.** 3.5827 **21.** $6.00 **23.** $.58 **25.** $3.99 **27.** 3.5; 3.52; 3.522 **29.** 2.5; 2.55; 2.548 **31.** 27.3; 27.32; 27.325 **33.** 36.5; 36.47; 36.472 **35.** .1; .06; .056 **37.** $5.06 **39.** $32.49 **41.** $382.01 **43.** $42.14 **45.** $.00 **47.** $1.50 **49.** $2.00 **51.** $752.80 **53.** $26 **55.** $0 **57.** $12,836 **59.** $395 **61.** $4700 **63.** $379 **65.** $722

Learning Catalytics (Page 20)

1. thirty-two and sixty-eight thousandths **2.** 17.1

Section 1.4 Exercises (Page 25)

1. $40 + 20 + 9 = 69$; 68.46 **3.** $6 + 4 + 5 + 7 + 2 = 24$; 23.82 **5.** $2000 + 5 + 3 + 7 = 2015$; 2171.414 **7.** $6000 + 500 + 20 + 8 = 6528$; 6666.061 **9.** $2000 + 70 + 500 + 600 + 400 = 3570$; 3451.446 **11.** 173.273 **13.** 59.3268 **17.** $15,138.19 **19.** $11.10 per pound **21.** $20 - 7 = 13$; 13.16 **23.** $50 - 20 = 30$; 31.507 **25.** $300 - 90 = 210$; 240.034 **27.** $8 - 3 = 5$; 4.848 **29.** $5 - 2 = 3$; 3.0198 **31.** $43,815.81

Learning Catalytics (Page 24)

1. 26.99 **2.** 2.8

Section 1.5 Exercises (Page 31)

1. $100 \times 4 = 400$; 406.56 **3.** $30 \times 7 = 210$; 231.88 **5.** $40 \times 2 = 80$; 89.352 **7.** 1.9152 **9.** 9.3527 **11.** .002448 **13.** $268.25 **15.** $418.10 **17.** 8.075 **19.** 27.442 **21.** 57.977 (rounded) **25.** $14,790 **27.** 50.9 mpg **29.** 11 months **31. (a)** .43 inch **(b)** 4.3 inches **33.** $129.25 **35. (a)** $70.05 **(b)** $25.80

Learning Catalytics (Page 28)

1. 31 **2.** 21

Case Study (Page 37)

1. $20,496 **2.** $4372 **3.** 178 guests; $24 **4.** $46.30 **5.** 22 months

Case in Point Summary Exercise (Page 38)

1. $2190.77 **2.** 19.5 hours; $168.68 **3.** $200.56; 213 customers **4.** $673.28; $14,280.19

Chapter 1 Test (Page 39)

1. 840 **2.** 22,000 **3.** 672,000 **4.** 50,000 **5.** 900,000 **6.** $606 **7.** $8399 **8.** $21.06 **9.** $364.35 **10.** $7246 **11.** 181.535 **12.** 498.795 **13.** 133.6 **14.** 3.7947 **15.** 15.8256 **16.** 8.0882 **17.** 11.56 **18.** 350 **19.** 23.8 **20.** $219.54 **21.** $3942.90 **22.** 14.454 gallons saved **23.** $17.31 **24.** $.79 per pound **25.** 253 seedlings **26.** 9 milliliters **27.** $47.80

CHAPTER 2

Section 2.1 Exercises (Page 46)

1. $\frac{29}{8}$ **3.** $\frac{17}{4}$ **5.** $\frac{38}{3}$ **7.** $\frac{183}{8}$ **9.** $\frac{55}{7}$ **11.** $\frac{364}{23}$ **13.** $3\frac{1}{4}$ **15.** $2\frac{2}{3}$ **17.** $3\frac{4}{5}$ **19.** $3\frac{7}{11}$ **21.** $1\frac{62}{63}$ **23.** $7\frac{8}{25}$ **27.** $\frac{1}{2}$ **29.** $\frac{5}{8}$ **31.** $\frac{3}{5}$ **33.** $\frac{11}{12}$ **35.** 1 **37.** $\frac{7}{12}$ **41.** ✓ ✓ x x ✓ x x **43.** ✓ ✓ ✓ ✓ x x ✓ **45.** ✓ ✓ x ✓ ✓ x ✓ ✓ **47.** ✓ x ✓ x x x x x

Learning Catalytics (Page 42)

1. $\frac{2}{5}$ **2.** $\frac{19}{5}$

Section 2.2 Exercises (Page 51)

1. 16 **3.** 36 **5.** 48 **7.** 42 **9.** 24 **11.** 180 **13.** 480 **15.** 2100 **19.** $\frac{3}{5}$ **21.** $\frac{8}{10}$ **23.** $\frac{1}{2}$ **25.** $\frac{17}{48}$ **27.** $1\frac{23}{36}$ **29.** $\frac{13}{14}$ **31.** $2\frac{1}{15}$ **33.** $2\frac{11}{36}$ **35.** $1\frac{13}{30}$ **37.** $\frac{9}{20}$ **39.** $\frac{7}{24}$ **43.** $\frac{23}{24}$ cubic yard **45.** $\frac{47}{60}$ inch **47.** $\frac{7}{24}$ of the contents **49.** $\frac{3}{4}$ cups **51.** $\frac{3}{16}$ inch **53.** $\frac{7}{24}$ **55.** $\frac{1}{2}$ in work and travel, and class time **57.** $\frac{1}{2}$ inch **59.** $\frac{1}{12}$ mile

Learning Catalytics (Page 48)

1. 12 **2.** $\frac{7}{12}$

Section 2.3 Exercises (Page 57)

1. $97\frac{4}{5}$ **3.** $80\frac{3}{4}$ **5.** $97\frac{7}{40}$ **7.** $53\frac{17}{24}$ **9.** $4\frac{3}{8}$ **11.** $3\frac{11}{24}$ **13.** $6\frac{1}{4}$ **15.** $3\frac{11}{12}$ **19.** $22\frac{7}{8}$ hours **21.** $116\frac{1}{2}$ inches **23.** 130 feet **25.** $1\frac{5}{8}$ cubic yards

Learning Catalytics (Page 56)

1. $7\frac{1}{3}$ **2.** $6\frac{1}{4}$

Section 2.4 Exercises (Page 64)

1. $\frac{3}{10}$ **3.** $\frac{99}{160}$ **5.** $\frac{9}{32}$ **7.** $4\frac{3}{8}$ **9.** $9\frac{1}{3}$ **11.** $\frac{1}{3}$ **13.** $4\frac{7}{12}$ **15.** $1\frac{7}{9}$ **17.** $\frac{3}{5}$ **19.** $1\frac{1}{2}$ **21.** $\frac{2}{3}$ **23.** $2\frac{2}{3}$ **25.** $\frac{3}{20}$ **27.** $8\frac{2}{5}$ **31.** $12 **33.** $18.75 **37.** 9¢ **39.** 36 yards **41.** 12 homes **43.** $2632\frac{1}{2}$ inches **45.** 32 strawberry cheesecakes **47.** 471 gallons **49.** 88 dispensers **51.** 5 trips

Learning Catalytics (Page 60)

1. $\frac{8}{35}$ **2.** $1\frac{1}{4}$

AN-1

Section 2.5 Exercises (Page 70)

1. $\frac{3}{4}$ **3.** $\frac{6}{25}$ **5.** $\frac{73}{100}$ **7.** $\frac{17}{20}$ **9.** $\frac{17}{50}$ **11.** $\frac{111}{250}$ **13.** $\frac{5}{8}$ **15.** $\frac{161}{200}$ **17.** $\frac{12}{125}$ **19.** $\frac{3}{80}$ **21.** $\frac{3}{16}$
23. $\frac{1}{625}$ **27.** .25 **29.** .375 **31.** .667 (rounded) **33.** .778 (rounded)
35. .636 (rounded) **37.** .88 **39.** .883 (rounded)

Learning Catalytics (Page 68)

1. $\frac{4}{5}$ **2.** .375

Case Study (Page 74)

1. $288,000 **2.** $\frac{5}{12}$; $\frac{1}{4}$; $\frac{1}{12}$; $\frac{1}{16}$; $\frac{1}{16}$; $\frac{1}{8}$ **3.** Miscellaneous: $\frac{1}{8}$; Insurance: $\frac{1}{16}$; Advertising: $\frac{1}{16}$; Utilities: $\frac{1}{12}$; Rent: $\frac{1}{4}$; Salaries: $\frac{5}{12}$ **4.** 150°; 90°; 30°; 22.5°; 22.5°; 45°

Case in Point Summary Exercise (Page 75)

1. $193\frac{1}{2}$ inches **2.** $467\frac{5}{8}$ square inches **3.** $2805\frac{3}{4}$ square inches
4. 4 side panels

Chapter 2 Test (Page 76)

1. $\frac{5}{6}$ **2.** $\frac{7}{8}$ **3.** $\frac{7}{11}$ **4.** $8\frac{1}{8}$ **5.** $4\frac{2}{3}$ **6.** $2\frac{2}{3}$ **7.** $\frac{31}{4}$ **8.** $\frac{94}{5}$ **9.** $\frac{147}{8}$ **10.** 30 **11.** 120 **12.** 72
13. $\frac{7}{8}$ **14.** $15\frac{1}{16}$ **15.** $36\frac{5}{16}$ **16.** 36 **17.** $1\frac{1}{2}$ **18.** $24\frac{1}{8}$ pounds **19.** $340
20. $35\frac{7}{8}$ gallons **21.** 117 pizzas; $5\frac{1}{2}$ ounces **22.** $\frac{5}{8}$ **23.** $\frac{41}{50}$ **24.** .25 inch
25. .875 inch

CHAPTER 3

Section 3.1 Exercises (Page 84)

1. 25% **3.** 72% **5.** 203.4% **7.** 362.5% **9.** 87.5% **11.** .05% **13.** 345%
15. 3.08% **17.** .625 **19.** .65 **21.** .125 **23.** .125 **25.** .0025 **27.** .8475
29. 1.75 **31.** .5; 50% **33.** $\frac{7}{8}$; 87.5% **35.** $\frac{1}{125}$; .008 **37.** 10.5; 1050%
39. $\frac{13}{20}$; 65% **41.** $\frac{1}{200}$; .5% **43.** .$\overline{3333}$; $33\frac{1}{3}$% **45.** $2\frac{1}{2}$; 250%
47. $\frac{17}{400}$; .0425 **49.** 2.5; 250% **51.** $10\frac{3}{8}$; 10.375 **53.** $\frac{1}{400}$; .25%
55. $\frac{3}{8}$; .375

Learning Catalytics (Page 80)

1. 60% **2.** .654

Section 3.2 Exercises (Page 90)

1. 62 homes **3.** $604 **5.** 4.8 feet **7.** 10,185 miles **9.** 182 cell phones
11. 148.44 yards **13.** $5366.65 **17.** 238 people **19.** $268.30
21. 64 females **23.** 8.95 ounces **25.** 4853 accidents **27. (a)** 35%
(b) 455,247 **29.** $239.25 **31.** 2156 products **33.** 3,602,866 vehicles
35. $51,844.20 **37.** $6296.40 **39.** $199.89 **41.** $87.58

Learning Catalytics (Page 87)

1. 3.384

Section 3.3 Exercises (Page 96)

1. 2120 **3.** 325 **5.** 2000 **7.** 4800 **9.** 44,000 **11.** 20,000 **13.** $90,320
15. 1080 **17.** 312,500 **19.** 65,400 **21.** 40,000 **25.** 125.0 million
27. 7761 students **29.** $4500 **31.** $3333.33 **33.** 1749 owners
35. $185,500

Learning Catalytics (Page 94)

1. 50

Supplementary Exercises (Page 98)

1. 16 ounces **3.** $288,150 **5.** 478,175 Mustangs **7.** 162 calories
9. $39,000 **11.** 230 companies **13.** 55 companies

Section 3.4 Exercises (Page 102)

1. 10 **3.** 50 **5.** 28.3 **7.** 76 **9.** 4.1 **11.** 5.9 **13.** 1.3 **15.** 250 **17.** 27.8
21. 6.2% **23.** 18% **25.** 8.7% **27.** 58.0% **29.** 69.1%

Learning Catalytics (Page 100)

1. 25%

Supplementary Exercises (Page 104)

1. 97 people **3.** 40% **5.** 1100 boaters **7.** $370.14 **9.** $568.80
11. 4% **13.** 4 million riders **15.** 62.1% **17.** China—2.8%;
India—17.4%; U.S.—10.6%; Indonesia—14.3%; Brazil—10.1%;
Pakistan—29.6%; Nigeria—44.5%. Nigeria is expected to grow
most rapidly followed by Pakistan.
19. (1376 + 1311 + 258 + 189)/7500 = 41.8%. Many Asian
countries are not listed in the table (e.g., Korea, Vietnam, Japan,
etc.), so Asians actually made up far more than 41.8% of world
population in 2015. **21.** 5760 items **23.** $5886 **25.** $6528
27. 5742 deaths **29. (a)** 36% **(b)** 64%

Section 3.5 Exercises (Page 111)

1. $375 **3.** $27.91 **5.** $25 **7.** $854.50 **11.** $195,500 **13. (a)** $950
(b) $76 **15.** 2,619,048 restaurants **17.** $66.12 **19.** 28 million
21. 39.5 million **23.** $3864 **25.** $145.24 million **27.** 51.2 million
29. 30,000 students **31.** 695 deaths **33.** 219

Learning Catalytics (Page 108)

1. 115.5

Case Study (Page 117)

Apple—29.5%; Exxon Mobil $85.29; McDonald's—$123.79;
Coca-Cola Co.—$41.88; Wal-Mart Stores—.9%

Case in Point Summary Exercise (Page 118)

1. 3.2% **2.** $1,181,640 **3.** $13,488 **4.** $82,976.25

Chapter 3 Test (Page 119)

1. 37.5% **2.** $\frac{7}{20}$ **3.** .024 **4.** $\frac{7}{50}$ **5.** 587.5% **6.** 1.75 **7.** 300 home sales
8. $3.15 **9.** $\frac{6}{25}$ **10.** 1920 purchase orders **11.** $\frac{7}{8}$ **12.** 137.5% **13.** 43
14. 1391 **15.** $26,563.95 **16.** 322.1 million **17. (a)** 11% **(b)** $4488
per year **18.** 75% **19.** $186.25 **20.** 1200 backpacks **21.** 6.2%
22. $1.47 billion

CHAPTER 4

Section 4.1 Exercises (Page 128)

1. 3 **3.** 31 **5.** 3 **7.** 7 **9.** 2 **11.** 30 **13.** 7.5 **15.** 10 **17.** 4 **19.** 1.5 **21.** 7
23. 1

Learning Catalytics (Page 122)

1. $x = 5$ **2.** $y = 3$

Section 4.2 Exercises (Page 135)

1. $27 + x$ **3.** $22 + x$ **5.** $x - 4$ **7.** $x - 3\frac{1}{2}$ **9.** $3x$ **11.** $\frac{3}{5}x$ **13.** $\frac{9}{x}$ **15.** $\frac{16}{x}$
17. $2.1(4 + x)$ **19.** $7(x - 3)$ **21.** $15y$ **23.** $472 - x$ **25.** $73 - x$
27. $\frac{\$20,210}{x}$ **29.** $21 - x$ **31.** 13 **33.** 1.5 **35.** 1 **37.** $1\frac{3}{7}$ **39.** 59 systems
41. 207 employees **43.** $349 **45.** $30,000 office; $105,000 retail
47. 3 new; 19 experienced **49.** 81 Altimas; 39 Sentras

Learning Catalytics (Page 130)

1. $3x - 12$

Section 4.3 Exercises (Page 142)

1. $586.50 **3.** $10,080 **5.** $16.50 **7.** 14 **9.** 2250 **11.** 151.2
13. $749.86 **15.** 7.5 **17.** 7 **19.** 24,000 **21.** $L = \frac{A}{W}$ **23.** $V = \frac{nRT}{P}$
25. $P = \frac{M}{(1 + i)^n}$ **27.** $i = \frac{A - P}{P}$ **29.** $D = \frac{M - P}{MT}$ **31.** $h = \frac{2A}{b + B}$
33. $2.40 **35.** $93.80 **37. (a)** $427 **(b)** $502.50 **39.** $236 million
41. $3.71 **43.** $644,400 **45.** $390 **47.** .13, or 13% **49.** 4 years
51. $4000 **53.** $4500

Learning Catalytics (Page 139)

1. 60

Section 4.4 Exercises (Page 152)

1. $\frac{9}{32}$ **3.** $\frac{27}{1}$ **5.** $\frac{4}{3}$ **7.** $\frac{3750}{1}$ **9.** $\frac{225}{1}$ **11.** $\frac{8}{5}$ **13.** $\frac{1}{6}$ **15.** $\frac{4}{15}$ **17.** $\frac{9}{2}$ **19.** T **21.** F **23.** F
25. F **27.** F **29.** F **31.** T **33.** T **35.** F **37.** T **39.** 7 **41.** 72 **43.** 105
45. $3\frac{1}{2}$ **47.** 24 **49.** 8 **53.** 1575 tickets **55.** $744,000 **57.** 28.6 minutes
59. 2.4°F **61.** 1020 miles **63.** $713,211.20 **65.** $128,000
67. 350 miles **69.** 3,500,000 cubic meters **71.** $177

Learning Catalytics (Page 148)

1. $\frac{3}{5}$ **2.** 50 pounds

Case Study (Page 159)

1. $10,460; $16,400 **2.** $3000 **3.** Answers will vary.

Case in Point Summary Exercise (Page 160)

1. Answers will vary but here are some points. Sales increased to 2000 then fell and fell sharply by 2010 (due to the financial crisis). Sales picked back up by 2017. General Motors has a lot of competition but still has the largest share of the market.
2. 60,720 vehicles **3.** 128,490 vehicles **4.** 22,756 vehicles

Chapter 4 Test (Page 162)

1. 51 **2.** 50.7 **3.** 16.3 **4.** $5\frac{1}{4}$ **5.** 258 **6.** 136 **7.** 56 **8.** $13\frac{1}{2}$ **9.** 7 **10.** $1\frac{2}{3}$
11. 21 **12.** 3.6 **13.** $I = 504$ **14.** $S = 324$ **15.** $G = 3183.6$
16. $M = 493.50$ **17.** $R = .05$ **18.** $A = 12,000$ **19.** $D = 200$
20. $I = 75$ **21.** $d = .84$ **22.** $I = 120$ **23.** $\frac{A}{L}$ **24.** $\frac{d}{t}$ **25.** $\frac{I}{PR}$ **26.** $\frac{P-1}{T}$
27. $\frac{A-P}{PR}$ **28.** $\frac{D}{1-DT}$ **29.** $\frac{1}{5}$ **30.** $\frac{9}{22}$ **31.** $\frac{8}{5}$ **32.** $\frac{1}{6}$ **33.** $\frac{7}{12}$ **34.** $\frac{1}{6}$ **35.** true
36. false **37.** false **38.** true **39.** true **40.** true **41.** 175 **42.** 39 **43.** 5
44. 27 **45.** 8 **46.** 32 **47.** $\frac{7}{2}$ or $3\frac{1}{2}$ **48.** $\frac{15}{2}$ or $7\frac{1}{2}$ **49.** 12 **50.** 47 **51.** 60
52. 8 **53.** 15 **54.** 2 **55.** 45; 46 **56.** 119; 121 **57.** $1.74 **58.** 19 men; 28 women **59.** $2056 **60.** $430,000 **61.** $12 **62.** $114.25 **63.** $20
64. $1200

Cumulative Review: Chapters 1–4 (Page 166)

1. 65,500 **3.** 78.4 **5.** 3609 **7.** 24,092 **9.** 85 **11.** 35.174 **13.** 12.218
15. $98 **17.** $31,658.27 **19.** $\frac{8}{9}$ **21.** $7\frac{2}{15}$ **23.** $13\frac{13}{24}$ **25.** 3 **27.** $12\frac{1}{2}$ square feet **29.** 130 feet **31.** $\frac{13}{20}$ **33.** 87.5% **35.** 2170 home loans **37.** 25%
39. 1.03% **41.** 97,757 copies **43.** 22,500% **45. (a)** 52% **(b)** .555 million metric tons **47.** .075 million metric tons **49.** 6 **51.** $7\frac{1}{5}$
53. $l = 337.5 **55.** $R = .07$ **57.** $T = \frac{I}{PR}$ **59.** $P = \frac{M}{(1+i)^n}$ **61.** $\frac{20}{29}$
63. $\frac{7}{20}$ **65.** 46 **67.** $52.98 **69.** $10,924

CHAPTER 5

Section 5.1 Exercises (Page 178)

1. $14.20 **3.** $20.00 **5.** $17.10 **7.** $21.90

9.
Date	March 8, 2018
Amount	$1835.42
To	Wayne Plumbing
For	Septic
Total	$6215 13
Deposits	+1594 12
This Check	−1835 42
Other (+ or −)	−100 00
Balance	$5873 83

11.
Date	October 7, 2018
Amount	$138.94
To	PetSmart
For	Pet Food
Total	$2105 09
Deposits	+832 47
This Check	−138 94
Other (+ or −)	−280 65
Balance	$2517 97

17.
Date	October 10, 2018
Amount	$39.12
To	Post Office
For	Postage
Total	$5972 89
Deposits	+775 50
This Check	−39 12
Other (+ or −)	−650 07
Balance	$6059 20

19. 9412.64; 8838.86; 8726.71; 9479.99; 10,955.68; 10,529.13; 9891.20; 9825.58; 9577.41; 9913.26; 9462.76 **21.** 574.86; 384.36; 462.65; 620.07; 581.31; 405.43; 784.71; 587.51; 562.41; 487.41; 1209.76

Learning Catalytics (Page 172)

1. A check is used to pay a bill. **2.** 14.2

Section 5.2 Exercises (Page 185)

1. $1595.36 **3.** $1387.67 **5.** $1352.98 **7.** $203.86 **9.** $66.48
11. $1064.72 **13.** $991.89 **15.** $972.05 **17.** $60.21 **19.** $29.62

Learning Catalytics (Page 182)

1. checking account, overdraft, ATM cash withdrawal, returned deposit, cashier's check, stop-payment order, etc.

Section 5.3 Exercises (Page 192)

1. $4870.24 **3.** $7690.62 **5.** $18,314.72 **11.** 421 $371.52; 424 $429.07; 427 $883.69; 429 $35.62; $1719.90; $6875.09; $701.56; $421.78; $689.35; $8687.78; $1719.90; $6967.88; $6965.92; $8.75; $6957.17; $10.71; $6967.88 **13.** 765 $63.24; 768 $135.76; $199.00; $5636.51; $220.16; $5856.67; $199.00; $5657.67; $5866.85; $209.30; $5657.55; $.12; $5657.67

Learning Catalytics (Page 188)

1. c

Case Study (Page 199)

1. $8178.46 **2.** $204.46 **3.** $9810.36 total of checks outstanding
4. $4882.58 deposits not recorded **5.** $7274.56

Case in Point Summary Exercise (Page 200)

Checks Outstanding 763—$723.35, 764—$3290.41; Total $4013.76; $13,272.17, 0, $13,272.17, −$4013.76, $9258.41; $9266.79, −$8.95, $9257.84, $.57, $9258.41; Wiffy's check register balances this month.

Chapter 5 Test (Page 202)

1. $19.90 **2.** $9.40 **3.** $17.40

4.
Date	August 6, 2018
Amount	$6892.12
To	WBC Broadcasting
For	Air time
Total	$16,409 82
Deposits	
This Check	−6,892 12
Other (+ or −)	
Balance	$9517 70

Answers to Selected Exercises

5.
Date	August 8, 2018
Amount	$1258.36
To	Lakeland Weekly
For	Ad
Total	$9517 70
Deposits	+1572 00
This Check	−1258 36
Other (+ or −)	
Balance	$9831 34

6.
Date	August 14, 2018
Amount	$416.14
To	W. Wilson
For	Freelance Art
Total	$9831 34
Deposits	10,000 00
This Check	−416 14
Other (+ or −)	
Balance	$19,415 20

7. $1709.55 **8.** $81.99 **9.** $1627.56 **10.** $56.96 **11.** $1570.60
12. 3221 $82.74; 3229 $69.08; 3230 $124.73; 3232 $51.20; $327.75; $4721.30; $758.06; $32.51; $298.06; $5809.93; $327.75; $5482.18; $5474.60; $2.00; $5472.60; $9.58; $5482.18

CHAPTER 6

Section 6.1 Exercises (Page 210)

1. 40; 0; $16.20 **3.** 38.75; 0; $28.05 **5.** 40; 5.25; $17.22 **7.** $448; $126; $574 **9.** $380; $160.31; $540.31 **11.** $16.95; $339.00; $0; $339.00 **13.** $21.60; $576.00; $97.20; $673.20 **15.** $13.77; $367.20; $58.52; $425.72 **17.** 50.5; 10.5; $4.75; $479.75; $49.88; $529.63 **19.** 53.5; 13.5; $6.25; $668.75; $84.38; $753.13 **21.** 35; 6; $14.10; $329.00; $84.60; $413.60 **23.** 39.5; 3.75; $16.20; $426.60; $60.75; $487.35 **25.** 39.75; 3.5; $32.25; $854.63; $112.88; $967.51 **29.** $443.08; $886.15; $1920; $23,040
31. $1200; $2400; $2600; $62,400 **33.** $1660; $1798.33; $3596.67; $43,160 **35.** $427.86 **37.** $467.25 **39.** $832
41. $509.60 **43.** $550.40 **45.** $891.80 **47. (a)** $825 weekly **(b)** $1650 biweekly **(c)** $1787.50 semimonthly **(d)** $3575 monthly

Learning Catalytics (Page 205)

1. (d) **2.** (b)

Section 6.2 Exercises (Page 219)

1. $93.12 **3.** $153.60 **5.** $99.60 **7.** $105.36 **11.** $407.85
13. $371.90 **15.** $420.28 **17.** $471.50 **19.** $608.10 **21.** $1766.10
23. $1405 **25.** $688.40 **27.** $845.82 **29.** $628.61

Learning Catalytics (Page 214)

1. (a) **2.** nurse, firefighter, social worker, bank teller

Section 6.3 Exercises (Page 225)

1. $26.04; $6.09 **3.** $28.72; $6.72 **5.** $52.99; $12.39 **7.** $189.39 **9.** $308.99 **11.** $41.56 **13.** $368.80; $76.07; $444.87; $27.58; $6.45; $4.45 **15.** $568; $106.50; $674.50; $41.82; $9.78; $6.75
17. $467.20; $122.64; $589.84; $36.57; $8.55; $5.90
19. (a) $28.62 **(b)** $6.69 **21. (a)** $95.67 **(b)** $22.38 **(c)** $15.43
23. $7221.60; $1688.92 **25.** $3609; $844.04 **27.** $3328.61; $778.46

Learning Catalytics (Page 222)

1. (d)

Section 6.4 Exercises (Page 235)

1. $106 **3.** $34 **5.** $111 **7.** $26 **9.** $42 **11.** $27 **13.** $19.32
15. $28.00 **17.** $69.11 **19.** $10.07; $522.12 **21.** $642.12; $3516.54
23. $173.51; $1929.15 **25.** $201.69; $2542.08 **27.** $637.10; $2828.96 **29.** $223.31; $1426.47 **35.** $8748.05 **37.** $53,332.52
39. $44,323.10 **41.** $693.45 **43.** $817.85 **45.** $3670.86

Learning Catalytics (Page 228)

1. 1189.5

Case Study (Page 242)

1. $818 **2.** $368.10 **3.** $1186.10 **4.** $73.54 **5.** $17.20 **6.** $165.48
7. $11.86 **8.** $52.19 **9.** $628.83 **10.** 53.0%

Case in Point Summary Exercise (Page 243)

1. Chavez, S., $345; $21.39; $5.00; $24.67; $10.59; $3.45; $279.90
Parton: $581.90 $36.08; $8.44; $26.21; $17.86; $5.82; $462.49
Dickens: $280.50; $17.39; $4.07; $26.68; $8.61; $2.81; $210.94
2. $450; $497.40; $947.40; $58.74; $13.74; $76.29; $29.09; $9.47; $695.07 **3.** FICA—$267.20; Medicare—$62.50; Federal Tax—$153.85; Total—$483.55 **4.** State income tax—$66.15; State disability insurance—$21.55; Total—$87.70

Chapter 6 Test (Page 244)

1. 40; 6.5; $537.30 **2.** 40; 7.5; $639.60 **3. (a)** $655 weekly
(b) $1310 biweekly **(c)** $1419.17 semimonthly
(d) $2838.33 monthly **4.** $425 **5.** $3532.50 **6. (a)** $685.10
(b) $160.23 **7. (a)** $523.90 **(b)** $160.23 **8.** $23 **9.** $32 **10.** $102
11. $52 **12.** $0 **13.** $1494.20 **14.** $741.87 **15.** $766.61
16. (a) $71.68 **(b)** $16.76 **(c)** $11.56 **17. (a)** $162.71 **(b)** $38.42
18. (a) $4552.55 **(b)** $1064.71 **19. (a)** $5255.20 **(b)** $1229.04
20. $3283.14

CHAPTER 7

Section 7.1 Exercises (Page 252)

1. $226.80 **3.** $126.36 **5.** $3610.36 **7.** $4935.88 **9.** $57.00
11. $28.40 **13.** $501.20 **15.** foot **17.** pair **19.** kilogram **21.** case
23. drum **25.** liter **27.** gallon **29.** cash on delivery
33. .9 × .8 = .72 **35.** .9 × .9 × .9 = .729
37. .75 × .95 = .7125 **39.** .6 × .7 × .8 = .336
41. .95 × .9 × .85 = .72675 **43.** $267.52 **45.** $14.02
47. $722.93 **49.** $218.88 **51.** $16.83 **53.** $714.42 **55.** $972
57. $640 **63.** $525.69 **65. (a)** 20/15 **(b)** $4.08 **67. (a)** 15/20
(b) $.55 **69.** $1280.45 **71.** $91,188.60 **73.** $326.40 undercharged

Learning Catalytics (Page 247)

1. 61.47 **2.** (a)

Section 7.2 Exercises (Page 258)

1. .72; 28% **3.** .68; 32% **5.** .504; 49.6% **7.** .5184; 48.16%
11. $720 **13.** $2280 **15.** $388 **17.** $5933.81 **19.** $1740
21. $207.09 **23. (a)** $25.89 wholesale **(b)** $28.76 retailer's price **(c)** $2.87

Learning Catalytics (Page 257)

1. 92.5%

Section 7.3 Exercises (Page 264)

1. May 14; June 3 **3.** July 25; Sept. 8 **5.** Oct. 1; Oct. 11 **7.** $1.70; $92.20 **9.** $0; $81.25 **11.** $21.60; $1120.55 **15.** $4542.69
17. $2,303,107.80 **19. (a)** Jan. 28; Feb. 7; Feb. 17 **(b)** Mar. 9
21. (a) Apr. 25 **(b)** May 5

Learning Catalytics (Page 261)

1. (d)

Section 7.4 Exercises (Page 270)

1. Mar. 10; Mar. 30 **3.** Sept. 18; Oct. 8 **5.** Dec. 22; Jan. 11
7. $20.47; $661.81 **9.** $10.69; $345.51 **11.** $117.42; $2817.98
13. $106.20; $3433.80 **17. (a)** Dec. 13 **(b)** $2334.93
19. $6586.09 **21.** $1495.58 **23. (a)** $1467.39 **(b)** $549.51
25. (a) June 10 **(b)** June 30 **27.** $1509.75 due **29. (a)** $3350.52
(b) $1052.06

Learning Catalytics (Page 267)

1. (b)

Case Study (Page 276)

1. $712.80 **2.** $78.56 **3.** 17.7%

Case in Point Summary Exercise (Page 277)

1. $17,750.66 **2.** October 15 **3.** November 4 **4.** $17,966.52
5. $10,309.28; $8189.76

Chapter 7 Test (Page 278)

1. $225.65 **2.** $784.80 **3. (a)** .63 **(b)** 37% **4. (a)** .576 **(b)** 42.4%
5. Mar. 15 **6.** May 30 **7.** Jan. 15 **8.** Dec. 19 **9. (a)** $788.80
(b) $773.02 **(c)** $811.77; $560; $52; $101.20; $75.60; $788.80;
$15.78; $773.02; $811.77 **10.** $91,300.78 **11. (a)** July 20
(b) $3041.28 **12.** $437.48 **13. (a)** Builders Supply **(b)** $1.91
14. (a) $1882.59 **(b)** $1900.39 **15.** $2527.08 **16. (a)** $1717.53
(b) $1198.47

CHAPTER 8

Section 8.1 Exercises (Page 285)

1. 140%; $4.96; $17.36 **3.** 100%; 20%; $27.17; $5.43
5. 100%; 130%; $168.00; $218.40 **7.** $2.70; $11.70
9. 60%; $19.20 **11.** $61.44; 40% **13.** $33.80; 25%
17. $148.64 markup **19.** $118.94 **21.** $221.40 selling price
23. (a) $95.96 **(b)** 25% **(c)** 125% **25. (a)** 126% **(b)** $567 selling
price **(c)** $117 markup

Learning Catalytics (Page 281)

1. (c)

Section 8.2 Exercises (Page 294)

1. 75%; $21.00; $28.00 **3.** 58%; 42%; $105.00 **5.** 50%; 100%;
$2025; $4050 **7.** $1920; $2400.00 **9.** $8.46; $22.26; 61.3%
11. $750; $1050; 28.6% **13.** 50% **15.** 15.3% **19. (a)** $1250 selling
price **(b)** $812.50 cost **(c)** 65% **21. (a)** $1507 **(b)** $577 **(c)** 38.3%
(d) 62.0% **23.** $1.13

Learning Catalytics (Page 288)

1. 31.0%

Supplementary Exercises (Page 296)

1. $559.52 **3.** 19.1% **5.** $6.98 **7.** $119 **9. (a)** 60% **(b)** $297.67
(c) $119.07 **11. (a)** $32.40 **(b)** 25.9% **13. (a)** $24.90 **(b)** 12.5%
(c) 14.2% **15.** $35.00

Section 8.3 Exercises (Page 301)

1. 25%; $645 **3.** 30%; $18.48 **5.** 20%; $5.20 **7.** $120; $20; none
9. $16; $22; $6 **11.** $385; $250; $60 **15.** 41% **17.** $30.01 operating loss **19. (a)** $77.15 operating loss **(b)** $18.77 absolute loss
21. (a) $88.40 **(b)** $58.40 **(c)** 33.9% **(d)** Absolute loss of $9.60

Learning Catalytics (Page 298)

1. 6.5%

Section 8.4 Exercises (Page 309)

1. $22,673 **3.** $60,568 **5.** 13.10; 13.96 **7.** 8.42; 7.60 **9.** 4.69; 4.66
11. $182; $195; $170 **13.** $2352; $2385; $2312.50 **17.** 5.43 turnover at cost **19. (a)** $1012.50 **(b)** $1125 **(c)** $1040 **21. (a)**
$1251.20 **(b)** $1430 **(c)** $1040 **23.** $30,660

Learning Catalytics (Page 304)

1. 1,352,920

Case Study (Page 317)

1. $125 **2.** $2062.50 **3.** $375 **4.** none

Case in Point Summary Exercise (Page 318)

1. Steep Alpine $293; Cliff Hoppers $327 **2.** Steep Alpine
$404.34; Cliff Hoppers $451.26 **3.** $9109 **4.** .54 **5.** Steep Alpine
$242.60; Cliff Hoppers $270.76 **6.** $1804.94 operating loss

Chapter 8 Test (Page 319)

1. (a) 20 **(b)** 120 **(c)** 76.80 **2. (a)** 138 **(b)** 365.50 **(c)** 138.89
3. (a) 80 **(b)** 20 **(c)** 33.60 **4. (a)** 75 **(b)** 25 **(c)** 18.45 **5.** 20%
6. 50% **7.** $200; $14; none **8.** $72; $99; $27 **9.** 5.76; 5.73
10. $165.90 **11. (a)** $6.54 **(b)** 48.6% **(c)** 32.7% **12.** $25 **13.** $4200
14. 40% **15. (a)** $37.99 **(b)** 19.0% **(c)** 23.5% **16.** 28%
17. (a) $131.10 operating loss **(b)** $45.60 absolute loss
18. $130,278 average inventory **19.** $12,978 **20. (a)** $11,775
(b) $13,350

Cumulative Review: Chapters 5–8 (Page 321)

1. $1958.20 **3.** $1749.75 **5.** $1727.88 **7.** $4359.38 **9.** $240.47
11. 62.2% **13.** Nov. 15; Dec. 5 **15.** $400; $280; $32 **17.** $93.11
19. 12.88 **21.** $6489

CHAPTER 9

Section 9.1 Exercises (Page 331)

1. $209; $4009 **3.** $440; $5940 **5.** 68 **7.** 99 **9. (a)** $2493.15
(b) $2527.78 **(c)** $34.63 **11. (a)** $2547.95 **(b)** $2583.33 **(c)** $35.38
13. Helen Spence **15.** Donna Sharp **17.** 90 days **19.** Jan. 25, 2019
21. Oct. 18; $5064 **23.** May 9; $6591.38 **25. (a)** $138,750
(b) $2,138,750 **27.** $13,125 **29. (a)** Oct. 3 **(b)** $14,016.81
31. (a) Sept. 6 **(b)** $84,200 **33.** $86.17 **35. (a)** Sept. 30
(b) $136,106.67

Learning Catalytics (Page 324)

1. Interest is earnings on money left in a loan or investment for a
period of time.
2. (d)

Section 9.2 Exercises (Page 339)

1. $14,000 **3.** $5040 **5.** $156,000 **7.** 11.8% **9.** 9.5% **11.** 7.5%
13. 120 days **15.** 120 days **17.** 5 months **19.** $10,200 **21.** 4.8%
23. 10.8% **25. (a)** $10,800 **(b)** $11,250 **27.** 76 days **29.** 8.6%
31. (a) $12,000 **(b)** $600 **33.** 208 days **35. (a)** 6.8% **(b)** 7.1%

Learning Catalytics (Page 335)

1. (a)
2. .08

Section 9.3 Exercises (Page 348)

1. $234; $7566 **3.** $950; $18,050 **5.** $408.33; $21,991.67
7. Jun. 20; $62,480 **9.** Dec. 9; $9572.92 **11.** Feb. 8; $23,600
13. (a) $160 **(b)** $5840 **15.** 200 days **17.** 10% **19.** $66,232.39
21. (a) $3780 **(b)** 13.3% **23.** 105 days **25. (a)** 166,107,382.60 yen

AN-6 Answers to Selected Exercises

(b) 8.1% **27.** (a) $11,262.50 (b) 8.9% **29.** (a) $24,625,000 (b) $25,000,000 (c) $375,000 (d) 6.09%

Learning Catalytics (Page 343)

1. (b)
2. .8 years

Section 9.4 Exercises (Page 356)

1. 107 days **3.** 53 days **5.** $10,179 **7.** $24,812.50 **9.** $6362.75; 37 days; $78.47; $6284.28 **11.** $2044; 49 days; $33.39; $2010.61 **13.** $17,355; 42 days; $228.43; $17,571.57 **15.** $30,829.37; 83 days; $814.09; $31,285.91 **17.** (a) $4968 (b) $367,632 **19.** (a) $238,750 (b) 93 days (c) $5166.67 (d) $244,833.33 **21.** (a) $311,250 (b) $304,713.75 **23.** (a) $24,150 (b) $538.46 (c) $24,461.54 (d) 6.71%

Learning Catalytics (Page 352)

1. 4063.5

Supplementary Exercises (Page 360)

1. (a) $660 (b) $18,660 **3.** $96,000 **5.** 200 days **7.** (a) $750 (b) $20,750 **9.** 12.1% **11.** Apr. 12; $9,475,000 **13.** $15,000 **15.** $18,208.12 **17.** (a) $1711.11 (b) $29,711.11 (c) 130 days (d) $1180.19 (e) $28,530.92 **19.** (a) $3843.89 (b) $72,191.09 (c) $4191.09 (d) $347.20

Case Study (Page 368)

3. (a) $1,037,500 (b) $78,292,500 (c) 2.65%

Case in Point Summary Exercise (Page 369)

1. $92,727.27; $7727.27 **2.** $7437.50; $92,437.50 **3.** Simple interest loan from Union Bank; $289.77 **4.** Bank One—10.91%; Union Bank—10.50%

Chapter 9 Test (Page 370)

1. $1020.83 **2.** $508.75 **3.** $832.29 **4.** $148.06 **5.** $13,013.01 **6.** $102,354.88 **7.** $11.19 **8.** 9.2% **9.** 333 days **10.** $43,000 **11.** $26,595.74 **12.** $359.33; $9440.67 **13.** $162.29; $10,087.71 **14.** (a) $14,550 (b) 9.3% **15.** $28,626.58 **16.** $9034.40 **17.** (a) $19,812.50 (b) $20,000 (c) $187.50 (d) 3.79% **18.** 44 days; $144.07; $9285.93 **19.** (a) $452.81 (b) $8997.19 (c) loses $2.81

CHAPTER 10

Section 10.1 Exercises (Page 380)

1. $16,325.87; $4325.87 **3.** $30,906.76; $2906.76 **5.** $40,841.23; $8491.23 **7.** $13,855.83; $1555.83 **9.** $60,476.40; $15,476.40 **11.** $1296; $1417.39; $121.39 **13.** $1440; $2606.60; $1166.60 **15.** (a) $11,431.57 (b) $2931.57 **17.** (a) $5707.08 (b) $1207.08 **19.** (a) 31,669.25 yuan (b) 6669.25 yuan **21.** $966; 10%; 6.14% **23.** (a) $4,283,208 (b) $483,208 **25.** (a) $28,137.75 (b) $45,152.75 (c) $17,015 **27.** (a) $8787.45 (b) $10,079.40 (c) second is larger

Learning Catalytics (Page 373)

1. Yes, individuals can earn interest and compound interest can be used over many years to build a very large estate.
2. 1.061208

Section 10.2 Exercises (Page 390)

1. $39.75 **3.** $50.48 **5.** $101.88 **7.** $3978.78 **9.** $17,412.96 **13.** (a) $3284.75 (b) $24.75 **15.** (a) $11,638.49 (b) $118.49 **17.** $4163.24; $4247.34 **19.** (a) $849,467.14 (b) $49,467.14

21. $2350 **23.** Loss of $583.20 **25.** (a) $183,636.14 (b) $185,481.59 (c) $1845.45

Learning Catalytics (Page 384)

1. (d)
2. Inflation is the increase in prices that occurs over time. Inflation will affect your life: a car costing $25,000 today may cost around $50,000 in 25 years.

Section 10.3 Exercises (Page 397)

1. $10,327.33; $1972.67 **3.** $7674.01; $1675.99 **5.** $9792.06; $9060.94 **7.** (a) $30,315.20 (b) $9684.80 **9.** $3503.40 **11.** (a) $793,054 (b) $592,618 **13.** (a) $26,620 (b) $20,990

Learning Catalytics (Page 394)

1. 48 payments
2. (d)

Case Study (Page 401)

1. $4,053,386; $3,016,084 **2.** $2,539,384; $1,889,530 **3.** $2,666,330; $1,984,000 **4.** $2,539,384; $1,889,530; $2,666,330; $1,984,000; $4,053,386; $3,016,084

Case in Point Summary Exercise (Page 402)

1. 2007—$4,235,680; 2008— –$7,891,680; 2009— –$8,550,370; 2010—$480,000; 2011–$512,960 **2.** $5,030,500; not enough to offset 2008 loss **3.** (a) $16,442,050 (b) $7,708,691

Chapter 10 Test (Page 403)

1. $11,906.56; $3206.56 **2.** $16,127.04; $4127.04 **3.** $13,170.42; $3370.42 **4.** $28,535.40; $6035.40 **5.** $50.52 **6.** $530.60 **7.** $286.28 **8.** $7509.25 **9.** $10,380.83 **10.** $38,291.99 **11.** $16,077.25 **12.** $4120.01 **13.** $4647.31 **14.** $36,428.34 **15.** Loss of $676 **16.** Loss of $975 **17.** (a) $4408.99 (b) $3801.87 **18.** (a) $15,173.40 (b) $13,481.41 **19.** (a) $283,233.48 (b) $206,955.87 **20.** (a) $59 million (b) $93 million

Cumulative Review: Chapters 9–10 (Page 405)

1. $272 **3.** $8500.11 **5.** 9.5% **7.** 80 days **9.** $270; $8730 **11.** $4875 **13.** $1216.65 **15.** $86.07; $12,686.07 **17.** $583.49; $416.51 **19.** $1250; $23,750 **21.** $10,871.40 **23.** $20.01

CHAPTER 11

Section 11.1 Exercises (Page 414)

1. $25,319.14; $9119.14 **3.** $201,527.78; $51,527.78 **5.** $139,509.30; $41,509.30 **7.** $7603.12; $1603.12 **9.** $207,485.32; $36,485.32 **11.** $51,985.25; $6385.25 **15.** (a) $423,452.16 (b) $290,452.16 **17.** (a) $11,277.89 (b) $3277.89 **19.** $44,502

Learning Catalytics (Page 408)

1. (b)

Section 11.2 Exercises (Page 422)

1. $14,762.54 **3.** $30,493.92 **5.** $18,991.59 **9.** $810,043.65 **11.** (a) $70,016.48 (b) $9983.52 **13.** (a) $48,879.76 (b) No **15.** (a) First offer (b) $2769.99 **17.** $194,361.41

Learning Catalytics (Page 416)

1. car, house, boat, refrigerator, smart phone, etc.

Answers to Selected Exercises   AN-7

Section 11.3 Exercises (Page 427)

1. $2784.12 **3.** $715.29 **5.** $2271 **7.** $6847.56 **11. (a)** $144,026
(b) $55,844 **13. (a)** $5800 **(b)** $797,000 **15.** $26,792
17. (a) $360,065 **(b)** $263,281 **19. (a)** $1200 **(b)** $6511.80
(c)

Payment Number	Amount of Deposit	Interest Earned	Total in Account
1	$6511.80	$0	$6511.80
2	$6511.80	$260.47	$13,284.07
3	$6511.80	$531.36	$20,327.23
4	$6511.80	$813.09	$27,652.12
5	$6511.80	$1106.08	$35,270.00
6	$6511.80	$1410.80	$43,192.60
7	$6511.80	$1727.70	$51,432.10
8	$6510.62	$2057.28	$60,000.00

21.

Payment Number	Amount of Deposit	Interest Earned	Total in Account
1	$713,625	$0	$713,625.00
2	$713,625	$10,704.38	$1,437,954.38
3	$713,625	$21,569.32	$2,173,148.70
4	$713,625	$32,597.23	$2,919,370.93

23. $1707 **25.** $6819.92

Learning Catalytics (Page 424)

1. (d)

Supplementary Exercises (Page 431)

1. (a) $15,210.93 **(b)** $3210.93 **3. (a)** $77,985.46 **(b)** $37,985.46
5. $1,472,513.85 **7.** $106,216.95 **9.** $267,986.10

Payment Number	Amount of Deposit	Interest Earned	Total in Account
1	$267,986.10	$0	$267,986.10
2	$267,986.10	$21,438.89	$557,411.09
3	$267,996.02	$44,592.89	$870,000.00

11. (a) $286,748 **(b)** $6265.44

Section 11.4 Exercises (Page 439)

1. $6.59 **3.** $.15 **5.** 937,800 shares **7.** 11 **9.** $.67 **11.** 1.6% **13.** $2.40
15. up $1.23 **17.** $14.01 **19.** $53.112; $544 **21.** $5645; $105.60
23. $39,858; $1535.80 **27.** 3.4% **29.** 3.1% **31.** 1.4% **33.** 25
35. 23 **37.** 23 **39.** .85; 2.5% **41.** $18,133 **43. (a)** $13,211.40
(b) $17,866.85 **(c)** $4655.45

Learning Catalytics (Page 433)

1. You are a shareholder and own part of Apple, Inc.
2. (a)

Section 11.5 Exercises (Page 447)

1. $1054.09 **3.** Feb. 1, 2046 **5.** 4.559% **7.** $19,482.60 **9.** $90,651
11. $29,467.50 **15. (a)** $3,872,440 **(b)** $190,000 **(c)** 4.9%
17. (a) $24,263.25 **(b)** $475 **(c)** 2.0% **19. (a)** $3600
(b) $97,151.40

Learning Catalytics (Page 443)

1. borrow, from profits, issue stock, issue bonds
2. (c)

Case Study (Page 452)

1. $620,703.67 **2.** $97,090.40 **3.** $67,090.40 **4.** $658,703.61
5. Almost, about $38,000 short. **6.** Answers will vary.

Case in Point Summary Exercise (Page 453)

1. $6985 **2.** $132,555.25 **3.** $4,801,500,000 **4. (a)** $367,314,750
(b) $288,090,000 **5. (a)** $5,456,904,750 **(b)** five billion, four
hundred fifty-six million, nine hundred four thousand, seven
hundred fifty dollars **6.** $1,684,944 **7.** $2,869,596

Chapter 11 Test (Page 454)

1. $9897.47 **2.** $96,355.40 **3.** $930,909 **4.** $86,210.18
5. $67,880.28 **6.** $47,575.42 **7.** $6801.69 **8.** $39,884.63
9. $25,752.96 **10.** $264,795.02 **11.** $22,696.72 **12.** $4562.80
13. $8702 **14.** $8395 **15.** $5835.60 **16.** $17,976 **17.** $22,492
18. $2322 **19. (a)** $17,164 **(b)** $456 **20. (a)** $23,800 **(b)** $1050
(c) 4.4% **21.** Highest and lowest prices for the year were $65.90
and $20.10. Weekly volume was 7,451,500 shares. Dividend
yield is .8% of current price. PE ratio is 27. Stock closed at $63.44.
The stock price was up $1.23 for the week. Earnings were $5.47
last year and are expected to be $2.53 this year and $2.89 next
year. The last quarterly dividend was $.12 per share. **22.** Annual
interest is 7.950% of $1000 = $79.50 per bond. Bond matures
on June 15, 2032. Bond closed at $1173.89. Yield to maturity is
5.369%. Estimated volume for the week is $209,183,000.

CHAPTER 12

Section 12.1 Exercises (Page 463)

1. $51.40 **3.** $15.02 **5.** October, $6.12; $419.17; November,
$419.17, $5.87, $541.68; December, $541.68, $7.58, $529.22;
January, $529.22, $7.41, $211.71 **9.** $20.42 **11.** $4.87 **13.** $17.30
15. (a) $104 **(b)** $65 **(c)** $39 **17. (a)** $2641.56 **(b)** $39.62
(c) $2885.84 **19. (a)** $312.91 **(b)** $4.69 **(c)** $285.94
21. (a) $139.71 **(b)** $2.10 **(c)** $74.32

Learning Catalytics (Page 457)

1. three or four times higher
2. (b)

Section 12.2 Exercises (Page 471)

1. $2050; $250 **3.** $1800; $300 **5.** $3543; $643 **7.** 8.1% **9.** 11.6%
11. 7.6% **13.** 12.25% **15.** 10.25% **17.** 11.25% **21. (a)** $59,840
(b) $83,966.60 **(c)** $9166.60 **(d)** 9.9% **23. (a)** 13.3% **(b)** 13%
25. 10.00% **27. (a)** $18,800 **(b)** $25,178.50 **(c)** $4378.50
(d) 10.25% **29. (a)** 732,212.32 pesos **(b)** 13.75%
31. (a) $97,689.60 **(b)** 10.75%

Learning Catalytics (Page 467)

1. 16,590.88

Section 12.3 Exercises (Page 478)

1. $8307.31; $307.31 **3.** $9198.18; $698.18 **5.** $9862.15; $507.15
7. $231 **9.** $103.18 **11.** $28.38 **15. (a)** $294.19 **(b)** $9194.19
17. (a) $69,936.02 **(b)** $2286.02 **19. (a)** $41,176.11 **(b)** $1876.11
21. (a) $17.31 **(b)** $457.69 **23. (a)** $1700 **(b)** $170 **(c)** $5980
25. (a) $87,716.70 **(b)** $10,916.70 **(c)** $4930.12 **(d)** $43,547.68

Learning Catalytics (Page 475)

1. 286

Section 12.4 Exercises (Page 486)

1. $1703.13 **3.** $404.73 **5.** $4185.60 **7.** $161.93 **9.** $147.61;
$899.62 **11.** $360.11; $3085.28 **13.** $255.45; $3577
17. (a) $1900.70 **(b)** $13,628

19.

Payment Number	Amount of Payment	Interest for Period	Portion to Principal	Principal at End of Period
0	—	—	—	$4000.00
1	$1207.68	$320.00	$887.68	$3112.32
2	$1207.68	$248.99	$958.69	$2153.63
3	$1207.68	$172.29	$1035.39	$1118.24
4	$1207.70	$89.46	$1118.24	$0

21.

Payment Number	Amount of Payment	Interest for Period	Portion to Principal	Principal at End of Period
0	—	—	—	$14,500.00
1	$374.83	$132.92	$241.91	$14,258.09
2	$374.83	$130.70	$244.13	$14,013.96
3	$374.83	$128.46	$246.37	$13,767.59
4	$374.83	$126.20	$248.63	$13,518.96
5	$374.83	$123.92	$250.91	$13,268.05

23.

Payment Number	Amount of Payment	Interest for Period	Portion to Principal	Principal at End of Period
0	—	—	—	$35,000.00
1	$1196.30	$408.33	$787.97	$34,212.03
2	$1196.30	$399.14	$797.16	$33,414.87
3	$1196.30	$389.84	$806.46	$32,608.41
4	$1196.30	$380.43	$815.87	$31,792.54
5	$1196.30	$370.91	$825.39	$30,967.15

Learning Catalytics (Page 482)

1. (c)

Section 12.5 Exercises (Page 495)

1. $2601.00 **3.** $889.68 **5.** $1240.14 **9.** $1495.92 **11.** $2050.96
13. $1319.95 **15.** Yes, qualified **17.** Monthly
payment = $122.5 \times \$9.28 = \$1136.80$

Payment Number	Total Payment	Interest Payment	Principal Payment	Balance of Principal
0	—	—	—	$122,500.00
1	$1136.80	$765.63	$371.17	$122,128.83
2	$1136.80	$763.31	$373.49	$121,755.34

Learning Catalytics (Page 490)

1. (d)
2. Companies such as Equifax maintain a credit history on you and summarize it in a number called a FICO score. A high score means you have good credit and can borrow cheaply and vice versa.

Case Study (Page 501)

1. $495.00; $419.99; $1104.43; $149.45; $2168.87 **2.** $2168.87; $215.00; $290.00; $279.17; $2953.04 **3.** $392.32; $285.61; $654.33; $149.45; $187.00; $290.00; $279.17; $2237.88
4. $715.16

Case in Point Summary Exercise (Page 503)

1. $1715 **2.** $229,875; $76,625 **3.** $100,000 **4.** $1530.45
5. $430.45

Chapter 12 Test (Page 505)

1. $20,900 **2.** $932.56 **3.** 11.75% **4.** 12.25% **5.** 11% **6.** $4326.13
7. (a) $235.51 (b) $2502.17 **8.** $4811.72 **9.** $1924.50 **10.** $1805.40
11. $1209.69 **12.** $518.29 **13.** (a) $6365.56 (b) $1,281,800.80
14. (a) $1398.25 (b) $360,580

Cumulative Review: Chapters 11–12 (Page 507)

1. $9214.23; $1214.23 **3.** $17,855.03; $2855.03 **5.** $13,367.29
7. $403.45 **9.** $11,957.94 **11.** (a) $2345 (b) 15 (c) 1.5%
13. (a) $3098.48 (b) $298.48 (c) $2300 (d) 12.00%
15. (a) $2118.14 (b) $3017.68 **17.** (a) $14,570 (b) $14,952.46
19. $133,270.55

CHAPTER 13

Section 13.1 Exercises (Page 515)

1. $34,000 **3.** $71,150 **5.** $325,125 **7.** .27% **9.** .80% **11.** .15%
13. (a) $4.84 (b) $48.40 (c) 48.4 **15.** (a) 7.08% (b) $7.08
(c) 70.8 **19.** $5861.60 **21.** $5384.40 **23.** $5978.70 **25.** $7836
27. $460,525 **29.** $9660 **31.** (a) The second parish (b) $75.24

Learning Catalytics (Page 511)

1. (a)

Section 13.2 Exercises (Page 529)

1. $23,131 **3.** $103,820 **5.** $44,533 **7.** $21,790 **9.** $26,490;
$3509.75 **11.** $51,750; $6835 **13.** $73,030; $10,027 **15.** $69,062;
$13,036.75 **17.** $65,250; $10,610 **19.** $2795.75 tax refund
21. $684.31 tax refund **23.** $5173.50 tax due **27.** $8596
29. $8247.50 **31.** $6150.65 **33.** $5303.05

Learning Catalytics (Page 518)

1. (c)
2. (d)

Section 13.3 Exercises (Page 537)

1. $1700 **3.** $2194.50 **5.** $9298.72 **7.** $70,344.83 **9.** $19,850
11. $36,500 **13.** $60,000; $20,000 **15.** $292,500; $260,000;
$97,500 **17.** $18,804 **19.** $702.65 **23.** (a) $136,986.30
(b) $43,013.70 **25.** (a) $30,681.82 (b) $14,318.18
27. A: $274,000 B: $182,666.67 C: $91,333.33 **29.** 1: $125,000
2: $41,666.67 3: $83,333.33 **31.** (a) $218,750 (b) A: $87,500;
B: $131,250

Learning Catalytics (Page 532)

1. (d)

Section 13.4 Exercises (Page 546)

1. $1031 **3.** $1173 **5.** $2601.60 **9.** $1126 **11.** $1570.80
13. (a) $25,000; (b) $11,500 **15.** (a) $4300; (b) $850
17. (a) $1778; (b) $6936; (c) $100,000; (d) $15,100

Learning Catalytics (Page 541)

1. $12,300
2. younger drivers have more accidents

Section 13.5 Exercises (Page 554)

1. $256; $130.56; $66.56; $23.24 **3.** $849.10; $433.04; $220.77;
$77.10 **5.** $516.80; $263.57; $134.37; $46.93 **7.** $319.50;
$162.95; $83.07; $29.01 **9.** $2973.10; $1516.28; $773.01;
$269.96 **13.** $364 **15.** (a) $100.50 (b) $384 **17.** $566 per year;
$16,980 **19.** (a) $444.72 (b) $226.72 (c) $79.18

Learning Catalytics (Page 549)

1. 3.3 years

Case Study (Page 559)

1. $11,790.75 **2.** $39,940.40 **3.** $439.88 **4.** $52,171.03 **5.** $1328.97

Case in Point Summary Exercise (Page 560)

1. $5952 **2.** $2214.65 **3.** $6063 **4.** $1006 **5.** $945 **6.** $13,966

Chapter 13 Test (Page 562)

1. $5.76; $57.60 **2.** 9.35%; $9.35 **3.** $53,895; $9245
4. $14,687; $1468.70 **5.** $2290.74 **6.** $8892.40 **7.** $4135.25
8. $8449.14 **9.** $42,613.64 **10. A:** $36,000 **B:** $21,600 **C:** $14,400
11. $2213.70 **12.** $872 **13.** $158.48; $80.82; $41.20; $14.39
14. $1940.80; $989.81; $504.61; $176.22 **15. (a)** $853 **(b)** $417
16. $6400 to repair his truck

CHAPTER 14

Section 14.1 Exercises (Page 569)

1. 20% **3.** 12.5% **5.** 5% **7.** $6\frac{2}{3}$% **9.** 1.25% **11.** 2% **13.** $450
15. $800 **17.** $9360 **19.** $2850 **21.** $5000 **23.** $2775 **25.** $73,000

27.

Year	Computation	Amount of Depreciation	Accumulated Depreciation	Book Value
0	—	—	—	$12,000
1	$(33\frac{1}{3}\% \times \$9000)$	$3000	$3000	$9000
2	$(33\frac{1}{3}\% \times \$9000)$	$3000	$6000	$6000
3	$(33\frac{1}{3}\% \times \$9000)$	$3000	$9000	$3000

29.

Year	Computation	Amount of Depreciation	Accumulated Depreciation	Book Value
0	—	—	—	$51,200
1	$(16\frac{2}{3}\% \times \$37,200)$	$6,200	$6,200	$45,000
2	$(16\frac{2}{3}\% \times \$37,200)$	$6,200	$12,400	$38,800
3	$(16\frac{2}{3}\% \times \$37,200)$	$6,200	$18,600	$32,600
4	$(16\frac{2}{3}\% \times \$37,200)$	$6,200	$24,800	$26,400
5	$(16\frac{2}{3}\% \times \$37,200)$	$6,200	$31,000	$20,200
6	$(16\frac{2}{3}\% \times \$37,200)$	$6,200	$37,200	$14,000

33. (a) $55,000 depreciation **(b)** $1,025,000 book value
35. (a) 12.5% **(b)** $9000 **(c)** $79,000

Learning Catalytics (Page 565)

1. (d)

Section 14.2 Exercises (Page 576)

1. 40% **3.** 25% **5.** $13\frac{1}{3}$% **7.** 20% **9.** $33\frac{1}{3}$% **11.** 4% **13.** $3000
15. $9000 **17.** $1900 **19.** $3360 **21.** $1215 **23.** $6834 **25.** $750

27.

Year	Computation	Amount of Depreciation	Accumulated Depreciation	Book Value
0	—	—	—	$14,400
1	$(50\% \times \$14,400)$	$7200	$7,200	$7,200
2	$(50\% \times \$7,200)$	$3600	$10,800	$3,600
3	$(50\% \times \$3,600)$	$1800	$12,600	$1,800
4		$1800*	$14,400	$0

*To depreciate to $0 scrap value.

29.

Year	Computation	Amount of Depreciation	Accumulated Depreciation	Book Value
0	—	—	—	$14,000
1	$(40\% \times \$14,000)$	$5600	$5,600	$8,400
2	$(40\% \times \$8,400)$	$3360	$8,960	$5,040
3	$(40\% \times \$5,040)$	$2016	$10,976	$3,024
4		$524*	$11,500	$2,500
5		$0	$11,500	$2,500

*To depreciate to $2500 scrap value.

33. $1153 **35.** $8476 **37. (a)** 25% **(b)** $1450 **(c)** $4425 **(d)** $1375

Learning Catalytics (Page 573)

1. States make their own rules.

Section 14.3 Exercises (Page 583)

1. $\frac{4}{10}$ **3.** $\frac{6}{21}$ **5.** $\frac{7}{28}$ **7.** $\frac{10}{55}$ **9.** $1640 **11.** $10,000 **13.** $7500 **15.** $7700
17. $2700 **19.** $890 **21.** $2400

23.

Year	Computation	Amount of Depreciation	Accumulated Depreciation	Book Value
0	—	—	—	$3900
1	$(\frac{3}{6} \times \$3420)$	$1710	$1710	$2190
2	$(\frac{2}{6} \times \$3420)$	$1140	$2850	$1050
3	$(\frac{1}{6} \times \$3420)$	$570	$3420	$480

25.

Year	Computation	Amount of Depreciation	Accumulated Depreciation	Book Value
0	—	—	—	$10,800
1	$(\frac{6}{21} \times \$8400)$	$2400	$2400	$8,400
2	$(\frac{5}{21} \times \$8400)$	$2000	$4400	$6,400
3	$(\frac{4}{21} \times \$8400)$	$1600	$6000	$4,800
4	$(\frac{3}{21} \times \$8400)$	$1200	$7200	$3,600
5	$(\frac{2}{21} \times \$8400)$	$800	$8000	$2,800
6	$(\frac{1}{21} \times \$8400)$	$400	$8400	$2,400

29. $248,571 **31.** $4887 **33. (a)** $\frac{8}{36}$ **(b)** $2360 **(c)** $10,620
(d) $4750

Learning Catalytics (Page 580)

1. 55

Supplementary Exercises (Page 587)

1. $2,050,000 **3.** $7400 **5.** $1242 **7.** $3760 **9.** $63,180
11. $13,680; $10,260; $6840; $3420 **13.** $37,575 **15.** $10,080
17. (a) $5580 **(b)** $5220

Section 14.4 Exercises (Page 593)

1. $.75 **3.** $.029 **5.** $.24 **7.** $30 **9.** $25,300 **11.** $17,280 **13.** $2775
15. $3108

19.

Year	Computation	Amount of Depreciation	Accumulated Depreciation	Book Value
0	—	—	—	$6800
1	(1350 × $1.26)	$1701	$1701	$5099
2	(1820 × $1.26)	$2293	$3994	$2806
3	(730 × $1.26)	$920	$4914	$1886
4	(1100 × $1.26)	$1386	$6300	$500

Learning Catalytics (Page 591)

1. (d)

Section 14.5 Exercises (Page 600)

1. 19.2% **3.** 6.56% **5.** 20.00% **7.** 3.636% **9.** 5.76% **11.** 2.564% **13.** $1751 **15.** $43,050 **17.** $4800 **19.** $6254 **21.** $16,920 **23.** $2756 **25.** $122,084

27.

Year	Computation	Amount of Depreciation	Accumulated Depreciation	Book Value
0	—	—	—	$10,980
1	(33.33% × $10,980)	$3660	$3,660	$7,320
2	(44.45% × $10,980)	$4881	$8,541	$2,439
3	(14.81% × $10,980)	$1626	$10,167	$813
4	(7.41% × $10,980)	$813*	$10,980	$0

*due to rounding in prior years

29.

Year	Computation	Amount of Depreciation	Accumulated Depreciation	Book Value
0	—	—	—	$122,700
1	(10% × $122,700)	$12,270	$12,270	$110,430
2	(18% × $122,700)	$22,086	$34,356	$88,344
3	(14.4% × $122,700)	$17,669	$52,025	$70,675
4	(11.52% × $122,700)	$14,135	$66,160	$56,540
5	(9.22% × $122,700)	$11,313	$77,473	$45,227
6	(7.37% × $122,700)	$9,043	$86,516	$36,184
7	(6.55% × $122,700)	$8,037	$94,553	$28,147
8	(6.55% × $122,700)	$8,037	$102,590	$20,110
9	(6.56% × $122,700)	$8,049	$110,639	$12,061
10	(6.55% × $122,700)	$8,037	$118,676	$4,024
11	(3.28% × $122,700)	$4,024*	$122,700	$0

*due to rounding in prior years

33. $4855 **35.** $490 **37. Year 1:** $21,165; **Years 2–5:** $22,050

Learning Catalytics (Page 596)

1. An office building lasts a lot longer than an automobile.

Case Study (Page 606)

1. $114,000 **2.** $61,560 **3.** $228,000 **4.** $57,000 straight-line; $24,624 double-declining-balance; $38,000 sum-of-the-years'-digits

Case in Point Summary Exercise (Page 607)

1. $13,300 **2.** $5724 **3.** MACRS; 5-year property **4.** $23,670; $37,872; $22,723 **5.** $34,085 **6.** 71.20% **7.** No additional depreciation is allowed

Chapter 14 Test (Page 608)

1. 25%; 50%; $\frac{4}{10}$ **2.** 20%; 40%; $\frac{5}{15}$ **3.** $12\frac{1}{2}$%; 25%; $\frac{8}{36}$ **4.** 5%; 10%; $\frac{20}{210}$ **5.** $7900 **6.** $21,375 **7. Year 1:** $2700; **Year 2:** $2025; **Year 3:** $1350; **Year 4:** $675 **8.** $4788 **9.** $43,000 **10.** $5702 **11. (a)** Year 1: $4836; Year 2: $2666; Year 3: $3007; Year 4: $4712 **(b)** Year 1: $15,264; Year 2: $12,598; Year 3: $9591; Year 4: $4879 **12.** $2,443,924

CHAPTER 15

Section 15.1 Exercises (Page 614)

1. (a) $316,350 **(b)** $88,050 **(c)** $65,350 **3.** Gross Sales, $852,300; Returns, $42,800; Net Sales, $809,500; Inventory, January 1, $174,690; Cost of Goods Purchased, $345,790; Freight, $18,107; Total Cost of Goods Purchased, $363,897; Total of Goods Available for Sale, $538,587; Inventory, December 31, $158,200; Cost of Goods Sold, $380,387; Gross Profit, $429,113; Salaries and Wages, $168,240; Rent, $48,200; Advertising, $24,300; Utilities, $11,600; Taxes on Inventory, Payroll, $13,880; Miscellaneous Expenses, $21,900; Total Expenses, $288,120; Net Income before Taxes, $140,993; Income Taxes, $34,800; Net Income, $106,193

Learning Catalytics (Page 611)

1. (d)

Section 15.2 Exercises (Page 620)

1. 45.2%; 32.6% **3.** 50.7%; 29.2% **5.** Cost of Goods Sold, 67.1%, 67.9%; Gross Profit, 32.9%, 32.1%; Wages, 12.5%, 12.3%; Rent, 2.5%, 2.4%; Advertising, 1.5%, 1.4%; Total Expenses, 24.6%, 23.5%; Net Income before Taxes, 8.3%, 8.6% **9.** Gross Sales, 100.3%, 100.7%; Returns, .3%, .7%; Net Sales, $1,850,000, $1,680,000; Cost of Goods Sold, 59.6%, 56.5%; Gross Profit, 40.4%, 43.5%; Wages, 13.6%, 14.8%; Rent, 4.4%, 4.6%; Advertising, 6.0%, 7.3%; Utilities, 1.7%, 1.0%; Taxes on Inv., Payroll, .9%, 1.1%; Miscellaneous Expenses, 3.4%, 3.5%; Total Expenses, 30.1%, 32.2%; Net Income before Taxes, $192,000, 10.4%, $189,000, 11.3% **11.** 64.8%, 35.2%, 23.4%, 11.7%, 7.9%, 4.9%, 1.8%

Learning Catalytics (Page 616)

1. 24.3%

Section 15.3 Exercises (Page 625)

1. Cash, $273; Notes Receivable, $312; Accounts Receivable, $264; Inventory, $180; Total Current Assets, $1029; Land, $466; Buildings, $290; Fixtures, $28; Total Plant Assets, $784; Total Assets, $1813; Notes Payable, $312; Accounts Payable, $63; Total Current Liabilities, $375; Mortgages Payable, $212; Long-Term Notes Payable, $55; Total Long-Term Liabilities, $267; Total Liabilities, $642; Owners' Equity, $1171; Total Liabilities and Owners' Equity, $1813

Learning Catalytics (Page 623)

1. (b)
2. (c)

Section 15.4 Exercises (Page 631)

1. Cash, 13.1%, 13%; Notes Receivable, 1.9%, 2%; Accounts Receivable, 37.5%, 37%; Inventory, 37.5%, 38.3%; Total Current Assets, $288,000, 90%, $361,000, 90.3%; Land, 2.5%, 2.5%; Buildings, 3.4%, 3.5%; Fixtures, 4.1%, 3.8%; Total Plant Assets, $32,000, 10%, $39,000, 9.8%; Total Assets, $320,000, $400,000; Accounts Payable, 1.3%, .8%; Notes Payable, 47.5%, 50.3%;

Total Current Liabilities, $156,000, 48.8%, $204,000, 51%; Mortgages Payable, 5%, 5%; Long-Term Notes Payable, 13.1%, 14.5%; Total Long-Term Liabilities, $58,000, 18.1%, $78,000, 19.5%; Total Liabilities, $214,000, 66.9%, $282,000, 70.5%; Owners' Equity, 33.1%, 29.5%; Total Liabilities and Owners' Equity, $320,000, 100%, $400,000, 100% **3. (a)** 1.77 **(b)** 1.02 **(c)** Ratios are fine. **5. (a)** 1.79 **(b)** .62 **(c)** No, acid-test ratio is low **7.** 2.13; 1.27 **9.** 6.1% **11.** .91; .37; not healthy; very low liquidity

Learning Catalytics (Page 627)

1. 3.1%

Case Study (Page 638)

1. Gross Sales, $238,300; Returns, $3,600; Net Sales, $234,700; Inventory, Jan. 1, $53,400; Cost of Goods Purchased, $98,500; Freight, $2,300; Total Cost of Goods Purchased, $100,800; Total of Goods Available for Sale, $154,200; Inventory, Dec. 31, $48,900; Cost of Goods Sold, $105,300; Gross Profit, $129,400; Salaries and Wages, $78,690; Rent & Utilities, $11,800; Advertising, $3,200; Miscellaneous Expenses, $4,800; Total Expenses; $98,490, Net Income Before Taxes, $30,910; Income taxes, $3,600; Net Income After Taxes, $27,310 **2.** Cash, $62,000; Accounts Receivable, $10,700; Inventory, $48,900; Total Current Assets, $121,600; Land and Buildings, $0; Improvements, $41,500; Total Plant Assets, $41,500; Total Assets, $163,100; Notes Payable, $4,300; Accounts Payable, $14,800; Total Current Liabilities, $19,100; Long-term Notes Payable, $12,300; Mortgages, $0; Total Long-Term Liabilities, $12,300; Total Liabilities, $31,400; Owners' equity, $131,700; Total Liabilities and Owners' Equity, $163,100

Case in Point Summary Exercise (Page 640)

1. Gross Sales, $834,200; Returns, $4,500; Net Sales, $829,700; Inventory, Jan. 1, $84,200; Cost of Goods Purchased, $346,500; Freight, $9,100; Total Cost of Goods Purchased, $355,600; Total of Goods Available for Sale, $439,800; Inventory, Dec. 31, $96,200; Cost of Goods Sold, $343,600; Gross Profit, $486,100; Salaries and Wages, $193,200; Rent & Utilities, $68,900; Advertising, $13,900; Insurance and Payroll taxes, $19,400; Miscellaneous Expenses, $18,700; Total Expenses, $314,100; Net Income Before Taxes $172,000; Income taxes, $37,800; Net Income After Taxes, $134,200 **2.** Cash, $84,500; Accounts Receivable, $2,100; Inventory, $96,200; Total Current Assets, $182,800; Land and Buildings, $186,500; Improvements, $82,100; Total Plant Assets, $268,600; Total Assets, $451,400; Notes Payable, $4,900; Accounts Payable, $58,400; Total Current Liabilities, $63,300; Long-term Notes Payable, $0; Mortgages, $134,200; Total Long-Term Liabilities, $134,200; Total Liabilities, $197,500; Owners' equity, $253,900; Total Liabilities and Owners' Equity $451,400

Chapter 15 Test (Page 642)

1. Gross Sales, $756,300; Returns, $285; Net Sales, $756,015; Inventory, January 1, $92,370; Cost of Goods Purchased, $465,920; Freight, $1,205; Total Cost of Goods Purchased, $467,125; Total of Goods Available for Sale, $559,495; Inventory, December 31, $82,350; Cost of Goods Sold, $477,145; Gross Profit, $278,870; Salaries and Wages, $84,900; Rent, $42,500; Advertising, $2,800; Utilities, $18,950; Taxes on Inventory and Payroll, $4,500; Miscellaneous Expenses, $18,400; Total Expenses, $172,050; Net Income Before Taxes, $106,820; Income Taxes, $25,450; Net Income After Taxes, $81,370 **2.** Net Sales, $35,000, 58.3%; Cost of Goods Sold, $23,000, 57.5%; Gross Profit, $4,000, 33.3% **3.** Cost of Goods Sold, $875, 72.9%, 76.8%; Gross Profit, 27.1%, 23.2%; Net Income, 9.3%, 6.3%; Wages, 10.8%, 8.5%; Rent, 6%, 2.3%; Total Expenses, 17.8%, 16.9% **4. (a)** 1.38 **(b)** .85 **5. (a)** 1.05 **(b)** .37 **6.** 30.0% **7.** 26.1%

CHAPTER 16

Section 16.1 Exercises (Page 651)

1. Monthly income—$2933; Monthly expenses from table: $2873; Difference between the two—$60 a month. They have not considered entertainment and Other expenses, which should include an estimate for unexpected expenses. They may also have forgotten to consider credit-card payments. No, the $60 difference between monthly income and expenses is too low given what they are not considering. **3.** Monthly income—$2875; Monthly expenses—$3390. Expenses exceed income by $515 per month. Her budget seems reasonable, but she clearly does not have enough income for the expenses she projects.

Learning Catalytics (Page 645)

1. not enough money to pay the bills, nothing left to save for future bills, no retirement savings

2. (d)

Section 16.2 Exercises (Page 660)

1. 7.7 billion **3.** 5.4 billion Asians and .5 billion Americans **5.** It has increased. **7.** 4 **9.** 6 **11.** 5

13.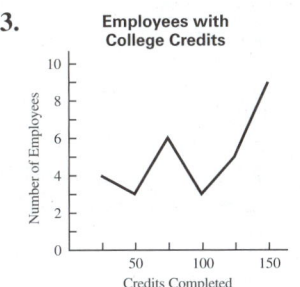

15. 23 **17.** 7 **19.** 2 **21.** 1 **23.** 2 **25.** 6

27.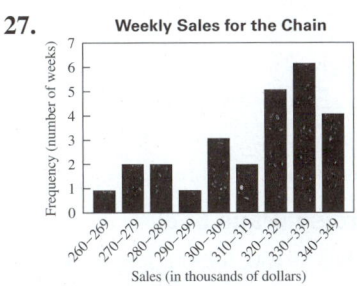

29. 20 **31.** 1 **33.** 13 **35.** 17 **37.** 8 **39.** 38 **41.** 42 **43.** 20% **45.** 36° **47.** 5%, 18°

49.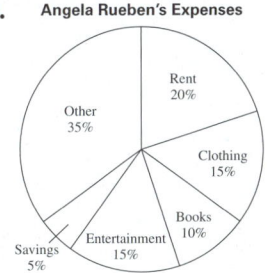

51. 20%

53. Revenues for Armstrong Publishing

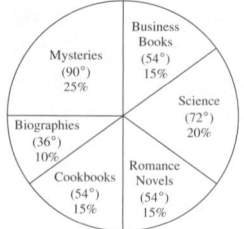

55.

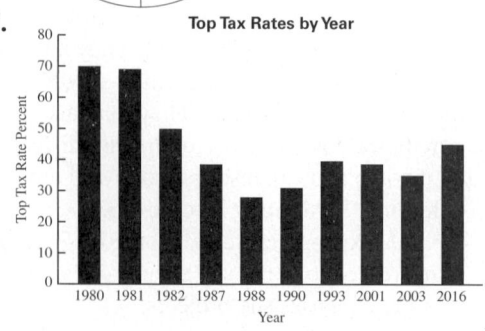

Learning Catalytics (Page 654)

1. answers will vary

Section 16.3 Exercises (Page 670)

1. 2.5 **3.** 60.3 **5.** 27,955 **7.** 8.8 **11.** 12.3 **13.** 17.2 **15.** 118.8 **17.** 2.6 **19.** 114 **21.** 130 **23.** 44 **25.** 21% **27.** 64 **29.** Bimodal with modes 32 and 38 **31.** 34.11 **33.** 34.38

Learning Catalytics (Page 666)

1. 26
2. 61

Case Study (Page 677)

1. Store 1 mean = $7.4; Store 2 mean = $8.2; Store 1 median = $7.5; Store 2 median = $8.2; Store 1 has no mode; Store 2 mode = $8.2

2.

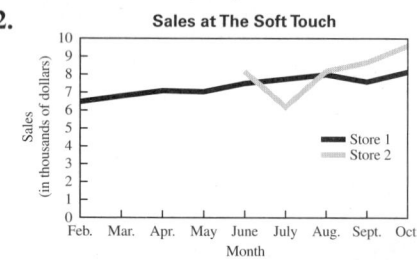

3. Sales at Store 2 are growing faster than at Store 1.

Case in Point Summary Exercise (Page 678)

1. October $7775, $7700; November $9150, $9000; December $8525, $8900 **2.** In the United States, sales at many restaurants are typically high during mid-to late November associated with shopping for Christmas and then low during the week of Christmas.

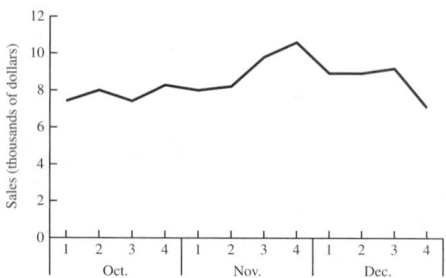

3. Sales at Bev's Deli

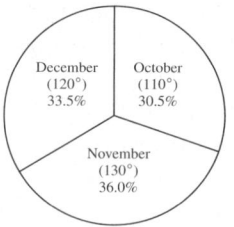

Chapter 16 Test (Page 679)

1. 1, 3, 10, 3, 2, 1 **2.** 6

3.

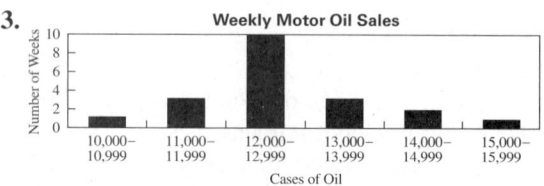

4. Newsprint, 72°; Part-time wages, 10%; Wire Service, 108°; Salaries, 108°; Other, 36°

5. Expenses at Midwest Community College

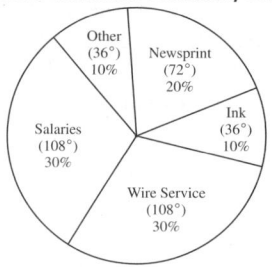

6. 60% **7.** 230.7 **8.** 18.8 centimeters **9.** $478.90 **10.** 11.3 **11.** 173.7 **12.** 19 **13.** 41.5 **14.** 8.3 **15.** 54.5 **16.** 47 **17.** no mode **18.** bimodal, 103 and 104

19.

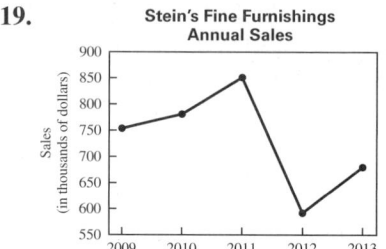

It appears that the recession had an effect on the business.

20.

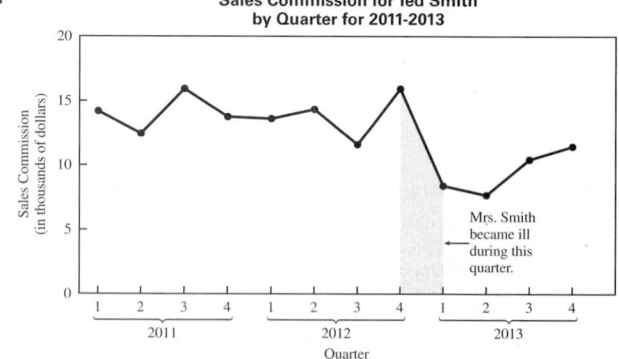

It appears that Mr. Smith's work was affected.

21. Rent—$3,500; Utilities—$350; Wages and payroll taxes—$8,700; Insurance—$267; Supplies—$1,450; Subcontract maintenance—$1,300; Reserve to replace equipment—$588; Other expenses—$1,800; Totals—$17,955, $19,354.
Actual Expenses for the Month exceed the Monthly Budget by $1,399 with Wages and payroll taxes, Supplies, and Other expenses in particular coming in higher than expected. Sarah needs to figure out why and what to do about the higher costs.

APPENDICES

Appendix A Exercises (Page A-6)

1. .68 m **3.** 4700 mm **5.** 8900 g **7.** .39 L **9.** 46 kg **11.** 976 g
13. 39.2 yards **15.** 50.3 m **17.** 15.4 feet **19.** 1.1 m **21.** 307.5 miles
23. 1235.7 km **25.** 1.5 pounds **27.** 1861.4 g **29.** 85.6 pounds
31. 200 nickels **33.** .6 g **37.** 40°C **39.** 280°C **41.** 37°C **43.** 95°F
45. 50°F **47.** 275°F **49.** Not rea-sonable **51.** Not reasonable

Learning Catalytics (Page A-1)

1. 39 inches

2. 16 cups

Appendix B Exercises (Page B-8)

1. 1171.60 **3.** 11,378 **5.** 3905.32 **7.** 255.24 **9.** 15,695 **11.** 31.08
13. 35.42 **15.** 27.63 **17.** 8.1 **19.** 566.1397059 = 566.14 (rounded)
21. .625 = .63 **23.** 30.57004831 = 30.57 **25.** 0.915966387 = .92
27. $5\frac{6}{11}$ **29.** .493714286 = .49 **33.** $44,162.80 **35.** (a) $1164.05
(b) $95.13 **37.** $214,008 **39.** $13,400 **41.** 59%

Learning Catalytics (Page B-1)

1. $202.96

Appendix C Exercises (Page C-5)

1. $39,019.50 **3.** $8929.13 **5.** $2676.84 **7.** $1.5% **9.** $858.89
11. 24 **13.** FV = $31.651.09 **15.** PMT = $839.37 **17.** $n = 26$
19. $i = 8.5\%$

Learning Catalytics (Page C-1)

1. $540.44

Appendix D Exercises (Page D-5)

1. p^2 **3.** r^3 **5.** x^4 **7.** 49 **9.** 1 **11.** 1 **13.** t^4g^4 **15.** 9^4 **17.** $\dfrac{3^2}{4^2}$ **19.** x^2y^2
21. 5 **23.** 83 **25.** 66 **27.** 20 **29.** 5625 **31.** 1 **33.** 161 **35.** 21 **37.** 553.2
39. 54.2 **41.** $53,928 **43.** $5834.43

Learning Catalytics (Page D-1)

1. 20.5

Appendix E Exercises (Page E-10)

1. Interest earned; Total Investment **3.** Profit; Sales **5.** Weight of Calf; Age of Calf **7.** Profit; Sales. Total Expenses

9.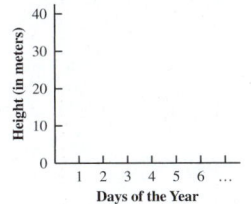

11.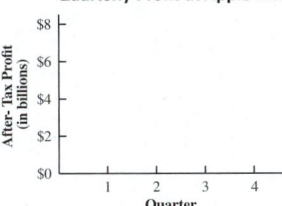

13.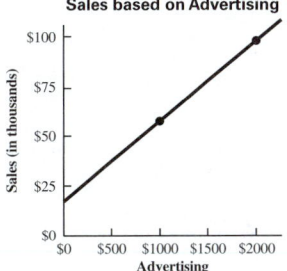

15.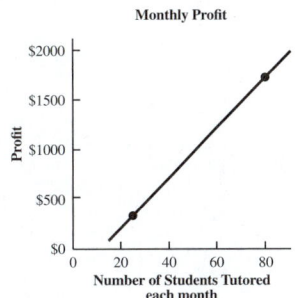

17.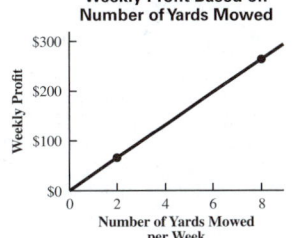

19.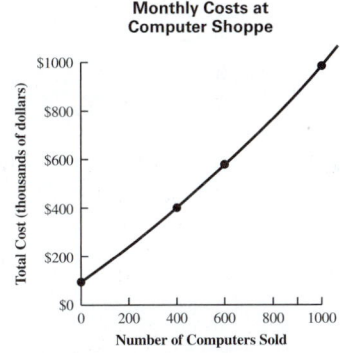

Learning Catalytics (Page E-1)

1.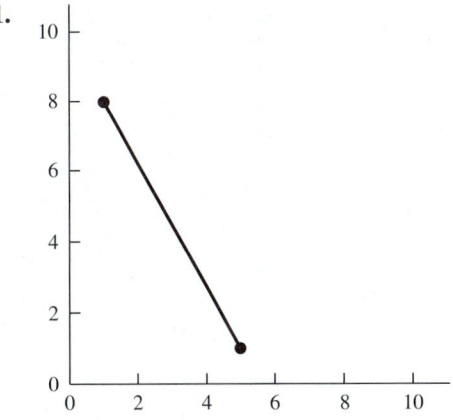

Glossary

401 (k): A retirement plan for individuals working for private-sector companies.

403 (b): A retirement plan for employees of public schools and certain tax-exempt organizations.

A

Absolute, or gross, loss: The loss resulting when the selling price is less than the cost.

Accelerated depreciation: A technique to increase the depreciation taken during the early years of an asset's useful life.

Accelerated mortgages: Mortgages with payoffs of less than 30 years, such as 15, 20, or 25 years.

Accountant: A person who maintains financial data for a firm or individual and then prepares the income tax return.

Accounts payable: A business debt that must be paid.

Accumulated depreciation: A running balance or total of the depreciation to date on an asset.

Acid-test ratio: The sum of cash, notes receivable, and accounts receivable, divided by current liabilities.

ACRS (Accelerated cost recovery system): A depreciation method introduced as part of the Economic Recovery Tax Act of 1981.

Actual physical counts: A method consisting of actually taking a physical count of inventory. Common in the past, it is now used only periodically to confirm inventories listed on computers.

Actual rate of interest: The true annual percentage rate that can be used to compare loans.

Actuary: A person who determines insurance premiums.

Addends: The numbers added in an addition problem.

Addition rule: The same number may be added or subtracted on both sides of an equation.

Adjustable rate mortgage: A home loan where the interest rate is adjusted up or down depending on a benchmark interest rate.

Adjusted bank balance: The actual current balance of a checking account after reconciliation.

Adjusted gross income: An individual's or family's income for a year, including all sources of income, and after subtracting certain expenses, such as moving expenses.

Algebraic logic: Rules used by most calculators for entering and evaluating arithmetic expressions.

Allowances: The number of allowances chosen by the taxpayer determines the amount withheld from gross income during the year.

American Express: A widely accepted credit card that requires an annual fee.

Amortization table: A table showing breaking a regular payment into interest and that which reduces debt. It also shows remaining debt.

Amortize: The process of paying off a loan with a sequence of periodic payments over a period of time.

Amount of an annuity: The future value of the annuity.

Amount of depreciation: The dollar amount of depreciation taken. This is usually an annual figure.

Annual meeting: Corporations have annual meetings for stockholders.

Annual percentage rate (APR): The true annual percentage rate which can be used to compare loans. The federal Truth-in-Lending Act requires lenders to state the APR.

Annual percentage rate table: A table used to find the annual percentage rate (APR) on a loan or installment purchase.

Annual rate of depreciation: The percent or fraction of the depreciable amount or declining balance to be depreciated each individual year of an asset's useful life.

Annuity: Periodic payments of a given, fixed amount of money.

Annuity due: An annuity whose payments are made at the beginning of a time period.

APR (Annual percentage rate): The true annual percentage rate that can be used to compare loans. It is required by the federal Truth-in-Lending Act.

"AS OF": A later date that appears on an invoice. The given sales terms may start at this time.

Assessed value: The value of a piece of property. Set by the county assessor, assessed value is used in figuring property taxes.

Assessment rate: The assessed valuation of a property is found by multiplying the fair market value by the assessment rate.

Asset: An item of value owned by a firm.

ATM (Automated teller machine): A machine that allows bank customers to make deposits, withdrawals, and fund transfers.

Automatic savings transfer account: A bank account that automatically transfers funds from one account to another.

Average: *See* mean.

Average cost method: An inventory valuation method whereby the cost of all purchases during a time period is divided by the number of units purchased.

Average daily balance method: A method used to calculate interest on open-end credit accounts.

Average inventory: The sum of all inventories taken divided by the number of times inventory was taken.

Average owner's equity: Sum of owner's equity at the beginning and end of the year divided by 2.

B

Bad check: A check that is not honored because there are insufficient funds in the checking account.

Balance brought forward (Current balance): The amount left in a checking account after previous checks written have been subtracted.

Balanced: In agreement. When the bank statement amount and the depositor's checkbook balance agree, they are balanced.

Balance sheet: A summary of the financial condition of a firm at one point in time.

Bank discount: A bank fee charged on a note. It is subtracted from the face value to find the proceeds loaned.

Banker's interest: A method used to calculate interest by dividing exact number of days by 360.

Banker's ratio: *See* current ratio.

Bank statement: A monthly statement prepared by a bank that lists all charges and deposits to a checking account. Historically banks mailed the statements, but many people now look at their monthly statements on the web.

Bankrupt: A company or individual whose liabilities exceed assets can declare bankruptcy, which is a legal process of working with debtors to pay off debts.

Bar code: A code placed on the side of a product that allows it to be scanned and then recognized and priced by a computer system.

Bar graph: A graph using bars to compare various numbers.

Base: The starting point or reference point or that to which something is being compared.

bbl.: Abbreviation for *barrel*.

Bimodal: A set of data with two modes.

Blank endorsement: A signature on the back of a check by the person to whom the check is made.

Board of directors: A group of people who represent the stockholders of a corporation.

Bodily injury insurance: A type of automobile insurance that protects a driver in case he or she injures someone with a car.

Bond: A contractual promise by a corporation, government entity, or church to repay borrowed money at a specified rate and time.

Book value: The cost of an asset minus depreciation to date.

Bouncing a check: Writing a check without sufficient funds in the account.

Break-even point: The cost of an item plus the operating expenses associated with the item. Above this amount, a profit is made; below it, a loss is incurred.

Broker: A person who sells stocks, bonds, and other investments owned by others.

Budget deficits: Annual expenses that exceed annual revenue, which commonly happens with the U.S. government.

Business account: The type of checking account used by businesses.

bx.: Abbreviation for *box*.

C

C: Roman numeral for 100.

Canceled check: A check is canceled after the amount of the check has been transferred from the payer's bank account into the account of the receiver of the check.

Cancellation: A process used to simplify multiplication and division of fractions.

Capital: The amount of money originally invested in a firm. The difference between the total of all the assets and the total of all the liabilities is called the capital or net worth.

Capital gains: Profits made on investments such as stocks or real estate.

cart.: Abbreviation for *carton*.

Cash discount: A discount offered by the seller allowing the buyer to take a discount if payment is made within a specified period of time.

Cashier's check: A check written by a financial institution, such as a bank, that is guaranteed by the institution.

Cash value: Money that has built up in an ordinary life insurance policy.

Casualty or theft loss: Loss due to a casualty (e.g. fire or theft) that is deductible on a personal income tax return.

Catastrophic event: A major (harmful) event that occurs rarely. For example, a financial crisis is a catastrophic event that can devastate the finances of families.

Centi-: A prefix used in the metric system meaning hundredth. (For example, a centiliter is one one-hundredth of a liter.)

Centimeter: One one-hundredth of a meter. There are 2.54 centimeters to an inch.

Central tendency: The middle of a set of data.

Certificate of deposit (CD): A savings account in which a minimum amount of money must be deposited and left for a minimum period of time.

Chain calculations: Long calculations done on a calculator.

Chain discount: Two or more discounts that are combined into a single discount.

Check 21 Act: A federal law that took effect in October 2004 and allows banks to take electronic photos of all cancelled checks and exchange checks electronically, so the banks no longer have to mail cancelled checks.

Check register: A table usually found in a checkbook that is used by the check writer to list all checks written, deposits and withdrawals made, and ATM transactions.

Checks outstanding: Checks written that have not reached and cleared the bank as of the statement date.

Check stub: A stub attached to a check and retained as a record when a check is written.

Circle graph: A circle divided into parts that are labeled and often colored or shaded to show data.

Claims: Either paperwork or an electronic filing with an insurance company requesting payment for an insured expense.

COD: A method of shipping goods that requires cash on delivery of goods.

Coinsurance: The portion of a loss that must be paid by the insured.

Collateral: Assets foreclosed on by a lender should the borrower default on payments.

Collision insurance: A form of automobile insurance that pays for car repairs in case of an accident.

Commission: A fee charged by a broker for buying and selling either stocks or bonds.

Commissions: Payments to an employee that represent a certain percent of the total sales produced by the employee's efforts.

Common denominator: Two or more fractions with the same denominator are said to have common denominators.

Common stock: Ownership of a corporation, held in portions called shares.

Comparative balance sheet: An analysis for two or more periods that compares asset categories such as cash.

Comparative income statement: A vertical analysis for two or more years that compares incomes or balance sheet items for each year analyzed.

Comparison graph (Comparative line graph): A graph showing how two or more things change with respect to one another.

Compensatory time (Comp time): Time off given to an employee to compensate for previously worked overtime.

Compound amount: The future value of an investment.

Compounding period: The interval of time at which interest is added to the account. For example, interest compounded quarterly results in interest being added to the account every quarter.

Compound interest: Interest charged or received on both principal and interest.

Comprehensive insurance: A form of automobile insurance that pays for damage to a car caused by fire, theft, vandalism, and weather.

Consolidated statement: A financial statement showing the combined results of all subsidiaries of a firm.

Consumer installment loan: A loan made to individuals on personal property, such as a car, that requires regular payments such as every month.

Consumer price index (CPI): A measure of the cost of living calculated by the government and used to estimate inflation.

Contents: The items in a home or business such as furniture, wine, jewelry, computers, etc.

Continuous inventory: Inventory systems that continuously monitor inventory levels.

Conventional loan: A loan made by a bank, savings and loan, or other lending agency that is not guaranteed or insured by the federal government.

Corporation: A form of business that gives the owners limited liability.

Cosign: Signing a loan with someone else. Parents sometimes cosign at the bank on a car loan applied for by their son or daughter. The cosigner must pay back the loan if the primary borrower does not.

Cost: The total cost of an item, including shipping, insurance, and other charges. Most often, the cost is the basis for calculating depreciation of an asset.

Cost (Cost price): The price paid to the manufacturer or supplier after trade and cash discounts have been taken. This price includes transportation and insurance charges.

Cost of goods sold: The amount paid by a firm for the goods it sold during the time period covered by an income statement.

Cost of living index: A measure of the cost of living calculated by the government and used to estimate inflation. It is the same as the consumer price index or CPI.

Country club billing method: A billing method that provides copies of original charge receipts to the customer.

Coupon rate: The rate of interest paid on a bond.

cpm.: Abbreviation for *cost per thousand*.

Credit card (transactions): The purchase or sale of goods or services using a credit card in place of cash or a check.

Credit history: The history built up by an individual when purchasing things and making payments. A lender checks a person's credit history before making a loan.

Credit score: Three national credit-reporting agencies (Equifax, TransUnion, and Experian) keep financial records on U.S. citizens and calculate a credit score on each indicating the creditworthiness of that individual. Also called a FICO score.

Credit union: A financial institution similar to a bank, except that it is owned by its member customers.

Credit union share draft account: A credit union account that may be used as a checking account.

Cross multiply: Multiply the numerator of the number on the left by the denominator of the number on the right. Then multiply the denominator of the number on the left by the numerator of the number on the right.

Cross-products: The equal products obtained when each numerator of a proportion is multiplied by the opposite denominator.

Crowdsourcing: Obtaining information or money from a large group of people to get needed services or funding.

cs.: Abbreviation for *case*.

ct.: Abbreviation for *crate*.

ctn.: Abbreviation for *carton*.

Current assets: Cash or items that can be converted into cash within a given period of time, such as a year.

Current liability: Debts that must be paid by a firm within a given period of time, such as a year.

Current ratio: The quotient of current assets and current liabilities.

Current yield: The annual dividend per share of stock divided by the current price per share.

cwt.: Abbreviation for *per hundredweight* or *per one hundred pounds*.

D

Daily interest charge: The amount of interest charged per day on a loan.

Daily overtime: The amount of overtime worked in a day.

Debit card: A card that results in a debit to a bank account when the card is used for a purchase.

Decimal: A number written with a decimal point, such as 4.3 or 7.22.

Decimal equivalent: A decimal that has the same value as a fraction.

Decimal point: The starting point in the decimal system (.).

Decimal system: The numbering system based on powers of 10 and using the 10 one-place numbers 0, 1, 2, 3, 4, 5, 6, 7, 8, and 9, which are called *digits*.

Declining-balance depreciation: An accelerated depreciation method.

(200%) Declining-balance method: An accelerated method of depreciation using twice, or 200% of, the straight-line rate.

Decrease problem (Difference problem): A percentage problem in which something is taken away from the base. It may require you to find the base.

Decreasing term insurance: A form of life insurance in which the insured pays a fixed premium until age 60 or 65, with the amount of life insurance decreasing periodically.

Deductible: An amount paid by the insured, with the balance of the loss paid by the insurance company.

Deductions: Amounts that are subtracted from the gross earnings of an employee to arrive at the amount of money the employee actually receives.

Defaulting on debt: Failure to pay back a debt.

Deflation: Occurs when prices of goods and services fall over time.

Degrees: A measurement of temperature; a reference to a college degree; or a measure of angles (i.e., there are 360° in a circle).

Denominator: The number below the line in a fraction. For example, in the fraction $\frac{7}{9}$, 9 is the denominator.

Dependents: An extra deduction is allowed on income taxes for each dependent.

Deposits in transit: Deposits that have been made but have not yet been recorded by a bank.

Deposit slip: A slip for listing all currency and checks that are part of a deposit into a bank account.

Depreciable amount: The amount of an asset's value that can be depreciated.

Depreciation: The decrease in value of an asset caused by normal use, aging, or obsolescence.

Depreciation schedule: A schedule or table showing the depreciation rate, amount of depreciation, book value, and accumulated depreciation for each year of an asset's life.

Difference (Remainder): The answer in a subtraction problem.

Differential piece rate: A rate paid per item that depends on the number of items produced.

Digits: One-place numbers in the decimal system. They are 0, 1, 2, 3, 4, 5, 6, 7, 8, and 9.

Direct competitor: A firm that produces products or offers services very similar to another company.

Direct deposit: Allows deposits (such as payroll) to be made directly into your account.

Direct payment: Allows you to authorize electronic payments from your account.

Disability coverage: Insurance coverage in the event of a disability.

Discount: (1) To reduce the price of an item. (2) The amount subtracted from the face value of a note to find the proceeds loaned.

Discount broker: A stockbroker who charges a reduced fee to customers (and, generally, reduced services).

Discount date: The last date on which a cash discount may be taken.

Discounting a note: Cashing or selling a note before the note is due from the maker.

Discount method of interest: A method of calculating interest on a loan by subtracting the interest from the amount of the loan. The borrower receives the amount borrowed less the discounted interest.

Discount period: The discount period is the period from the time of sale of a note to the note's due date.

Discount rate: The discount rate is a percent that is multiplied by the face value and time to find bank discount.

Discover: A credit card that sometimes pays the cardholder back a percentage of the amount charged.

Distributive property: The property that states the product of the sum of two numbers equals the sum of the individual products; that is $a(b + c) = ab + ac$.

Diversify: To own different types of assets such as stocks and real estate, or stocks in several different types of businesses along with some bonds. Diversification reduces the risk of large losses.

Dividend: (1) The number being divided by another number in a division problem. (2) A return on an investment; money paid by a company to the holders of stock.

Divisor: The number being divided into the dividend.

Double-declining balance: A method of accelerated depreciation that doubles depreciation in the early years compared to straight-line depreciation.

Double time: Twice the regular hourly rate. A premium often paid for working holidays and Sunday.

Dow Jones Industrial Average: A frequently quoted average price of the stocks of 30 large industrial companies.

doz.: Abbreviation for *dozen*.

Drawing account: An account from which a salesperson can receive payment against future commissions.

drm.: Abbreviation for *drum*.

E

ea.: Abbreviation for *each*.

Effective rate: The true rate of interest.

Effective rate of interest: The true annual percentage rate that can be used to compare loans. It is required by the federal Truth-in-Lending Act.

Electronic banking: Banking activities that take place over a network, such as the World Wide Web.

Electronic bill pay: Allows bills to be paid using the Internet.

Electronic commerce: Purchases that take place over a network, such as the World Wide Web.

Electronic funds transfer: Moving money electronically over a network, such as the World Wide Web.

Electronic payment: Bills that are paid using the Internet.

Electronic product code: A tag on a product that includes information about that product. It may succeed the bar code.

End-of-month dating (EOM): A system of cash discounts in which the time period begins at the end of the month the invoice is dated. *Proximo* and *prox.* have the same meaning.

Endorse: Sign the back of a check so that it can be deposited.

Endowment policy: A life insurance policy guaranteeing the payment of a fixed amount of money to a given individual whether or not the insured person lives.

Equation: Two algebraic expressions that are equal to one another.

Escrow account into which monies are paid: *See* impound account.

Exact interest: A method of calculating interest based on 365 days per year.

Exchange traded funds (ETFs): These funds are similar to mutual funds except they are not as actively managed. A particular

ETF tries to match the performance of an index such as the Dow Jones Averages or perhaps a market sector such as energy.

Executive officers: The top few officers in a corporation.

Expected expenses: Expenses incurred by a family or business that are regular, expected, and generally ongoing such as rent, car payment, salaries, etc.

Expenses: The costs a firm must pay to operate and sell its goods or services.

Expression: A mathematical phrase containing numbers and variables and various mathematical operations such as $(5x - 16) \times 9^2$.

Extension total: The number of items purchased times the price per unit.

Extra dating (ex., x): Extra time allowed in determining the net payment date of a cash discount.

F

Face value: The amount shown on the face of a note.

Face value of a bond (Par value of a bond): The amount the company has promised to repay.

Face value of a policy: The amount of insurance provided by the insurance company.

Factoring: The process of selling accounts receivable for cash.

Factors: Companies that buy accounts receivable.

Fair Labor Standards Act: A federal law that sets the minimum wage and also a 40-hour workweek.

Fair market value: The price for which a piece of property could reasonably be expected to be sold in the market.

FAS (Free alongside ship): A method of shipping goods in which the seller pays for transportation of the goods to the port from which they will be shipped. The buyer of the goods must pay for all costs in moving the goods from the shipping port to his facility.

Federal Insurance Contributions Act (FICA): An emergency measure passed by Congress in the 1930s that established the so-called social security tax. *See* FICA tax.

Federal Reserve Bank: Today, all banks are part of the Federal Reserve system. The Federal Reserve is our national bank.

Federal Truth-in-Lending Act: An act passed in 1969 that requires all interest rates to be given as comparable percents.

Federal Unemployment Tax Act (FUTA): An Act that requires employers to pay an unemployment insurance tax. The tax is paid to the federal government.

FHA loan: A real estate loan that is insured by the Federal Housing Administration, an agency of the federal government.

FICA tax (Social Security tax): The amount of money deducted from the paychecks of almost all employees, used by the federal government to pay pensions to retired people, survivors' benefits, and disability.

FICO: Three national credit-reporting agencies (Equifax, Trans Union, and Experian) keep financial records on U.S. citizens and calculate a credit score on each indicating the credit-worthiness of that individual. Also called a credit score.

FIFO: A method of inventory accounting in which the first items received are considered to be the first ones shipped.

Finance charge: The difference between the cost of something paid for in installments and the cash price.

Financial ratio: A number found using financial data that is used to compare different companies within the same industry.

Fixed assets: Assets owned by a firm that will not be converted to cash within a year.

Fixed liabilities: Items that will not be paid off within a year.

Fixed-rate loan: A loan made at a fixed, stated rate of interest.

Flat-fee checking account: A checking account in which the bank supplies check printing, a bank charge card, and other services for a fixed charge per month.

Float: The time between an actual deposit to an account and the moment those funds are available for use.

Floating decimal: A feature on most calculators that positions the decimal point where it should be in the final answer.

FOB (Free on board): A notation sometimes used on an invoice. "Free on board shipping point" means the buyer pays for shipping. "Free on board destination" means the seller pays for shipping.

Foreclose (or Foreclosure): The process by which a lender takes back the property when payments are not made.

Form 941: The Employer's Quarterly Federal Tax Return form that must be filed by the employer with the Internal Revenue Service.

Form 1040A: The form used by most federal income taxpayers.

Form 1040EZ: A simplified version of the 1040A federal income tax form.

Fraction: An indication of a part of a whole. (For example, $\frac{3}{4}$ means that the whole is divided into 4 parts, of which 3 are being considered.)

Frequency distribution table: A table showing the number of times one or more events occur.

Fringe benefits: Benefits offered by an employer, not including salary, that can include medical, dental, life insurance, and day care for employee's children.

Front-end rounding: Rounding so that all digits are changed to zero except the first digit.

Future amount: The value of an investment at a future date. It is the same as future value.

Future value: The value, at some future date, of an investment.

G

GI (VA) loan: A loan guaranteed by the Veterans Administration and available only to qualified veterans.

Grace period: The period between the due date of a payment and the time the lending institution assesses a penalty for the payment being late, usually a few days after the payment is due.

Gram: The unit of weight in the metric system. (A nickel weighs about 5 grams.)

Graph: A visual presentation of numerical data.

Great Depression: A period of time in the 1930s when global economic activity was greatly reduced and unemployment levels soared to beyond 25% in many countries.

Gr gro. (Great gross): Abbreviation for 12 gross ($144 \times 12 = 1728$).

Gro.: Abbreviation for *gross*.

gross: A dozen dozen, or 144 items.

Gross earnings: The total amount of money earned by an employee before any deductions are taken.

Gross loss: *See* absolute loss.

Gross profit: The difference between the amount received from customers for goods and what the firm paid for the goods.

Gross profit on sales: *See* gross profit.

Gross sales: The total amount of money received from customers for the goods or services sold by the firm.

H

Half-year convention: Method of depreciation used for the first year the property is placed in service.

Head of household: An unmarried person who has dependents can use the head of household category when filing income taxes.

Health insurance: Insurance for services related to illness or accident including doctors, hospitals, prescriptions, medical specialists, etc.

High: The highest price reached by a stock during the day.

Homeowner's policy: An insurance policy that covers a home against fire, theft, and liability.

Horizontal analysis: An analysis that shows the amount of any change from last year to the current year, both in dollars and as a percent.

I

Identity theft: When someone gathers enough information about you to fraudulently establish credit cards or borrow money using your name and personal information.

Impound account (Escrow account): An account at a lending institution into which taxes and insurance are paid on a monthly basis by a borrower on real estate. The lender then pays the tax and insurance bills from this account when they become due.

Improper fraction: A fraction with a numerator larger than the denominator. (For example, $\frac{7}{5}$ is an improper fraction; $\frac{1}{9}$ is not.)

Incentive rate: A payment system based on the amount of work completed.

Income statement: A summary of all the income and expenses involved in running a business for a given period of time.

Income tax: The tax based on income that both individuals and corporations are required to pay to the federal government and sometimes to a state.

Income tax withholding: Federal income tax that the employer withholds from gross earnings.

Income-to-monthly-payment ratio: A ratio used to determine from an income standpoint whether a prospective borrower meets the lender's qualifications.

Increase problem (Amount problem): A percentage problem in which something has been added to the base. Usually the base must be found.

Index fund: A mutual fund that holds the stocks that are in a particular market index such as the Dow Jones Industrial Average.

Indicator words: Key words that help indicate whether to add, subtract, multiply, or divide.

Individual retirement account (IRA): An account designed to help people prepare for future retirement.

Inflation: Inflation results in a continuing rise in the cost of goods and services. *See* consumer price index (CPI).

Information society: A society in which the creation, manipulation, and use of information is significant economically and politically. Advanced societies today are typically information societies.

Installment loan: A loan that is paid off with a sequence of periodic payments.

Insurance: Individuals and firms purchase insurance from insurance companies to protect them in the event of an unexpected loss.

Insured: A person or business that has purchased insurance.

Insurer: The insurance company.

Intangible assets: Assets such as patents, copyrights, or customer lists that have a value that cannot be immediately converted to cash, unlike jewelry or stocks.

Interest: A charge paid for borrowing money or a fee received for lending money.

Interest-bearing checking account: A checking account that earns interest.

Interest-in-advance notes: *See* simple discount note.

Interest rate per compounding period: Interest rates are usually given on an annual basis; however the interest rate per compounding period is needed when making many financial calculations.

Interest rate spread: The difference between the interest rate charged to borrowers and the interest rate paid to depositors.

Internal Revenue Service: The branch of the U.S. federal government responsible for collecting taxes.

Internet banking: Banking done over the Internet.

Internet service providers: An organization that provides services to access the Internet.

Inventory: The value of the merchandise that a firm has for sale on the date of balance sheet.

Inventory-to-net-working-capital ratio: Inventory divided by working capital, where working capital is current assets minus current liabilities.

Inventory turnover: The number of times during a certain time period that the average inventory is sold.

Inventory turns: The number of times a year that a firm turns over its average inventory.

Invoice: A printed record of a purchase and sales transaction.

Invoice amount: List price minus trade discounts.

Invoice date: The date an invoice is printed.

Issuer: The firm that issues a bond.

Itemized billing method: A billing method that provides an itemization of the customer's charge purchases but not copies of the original charge receipts.

Itemized deductions: Tax deductions, such as interest, taxes, and medical expenses, that are listed individually on a tax return in order to affect the total amount of taxes payable at the end of the year.

J

Joint return: An income tax return filed by both husband and wife.

K

Kilo-: A prefix used in the metric system to represent 1000.

Kilogram: A unit of weight in the metric system meaning 1000 grams. One kilogram is about 2.2 pounds.

Kilometer: One thousand meters. A kilometer is about .6 mile.

L

Late fees: Fees required because payments were made after a specific due date.

Least common denominator: The smallest whole number that all the denominators of two or more fractions evenly divide into. (For example, the least common denominator of $\frac{3}{4}$ and $\frac{5}{6}$ is 12.)

Left side: The left-hand side of an equation.

Length of a loan: The length of time until a loan is repaid.

Level premium: A level premium insurance policy requires the same payment throughout its life.

Liability: An expense that must be paid by a firm.

LIFO: A method of inventory accounting in which the most recent items received are considered to be the first ones shipped.

Like fractions: Fractions with the same denominator.

Like terms: Two terms in an algebraic expression with the same variable can be added together (combined). For example, $2x + 5x = 7x$.

Limited liability: A form of protection that shields a company and its shareholders from having to pay large sums of money in the event that the company loses a lawsuit.

Limited-pay life insurance: Life insurance for which premiums are paid for only a fixed number of years.

Line graph: A graph that uses lines to compare numbers.

Liquid assets: Cash or items that can be converted to cash quickly.

Liquidity: The ability of a firm or individual to raise cash quickly without being forced to sell assets at a loss.

List price: The suggested retail price or final consumer price given by the manufacturer or supplier.

Liter: A measure of volume in the metric system. One liter is a little more than one quart.

Loan amount: The amount of a loan.

Loan reduction schedule: *See* repayment schedule.

Long-term care coverage: Insurance that pays for long-term care such as nursing home expenses.

Long-term liabilities: Money owed by a firm that is not expected to be paid off within a year.

Long-term notes payable: The total of all debts of a firm, other than mortgages, that will not be paid within a year.

Low: The lowest price reached by a stock during the day.

Lowest terms: The form of a fraction if no number except the number 1 divides evenly into both the numerator and denominator.

M

MACRS (Modified accelerated cost recovery system): A depreciation method introduced as part of the Tax Reform Act of 1986.

Maintenance charge per month: The charge to maintain a checking account (usually determined by the minimum balance in the account).

Maker of a note: A person borrowing money from another person.

Manufacturers: Businesses that buy raw materials and component parts and assemble them into products that can be sold.

Margin: The difference between cost and selling price.

Marital status: An individual can claim married, single, or head of household when filing income taxes.

Markdown: A reduction from the original selling price. It may be expressed as a dollar amount or as a percent of the original selling price.

Marketing channels: The path of products and services beginning with the manufacturer and ending with the consumer.

Market share: Percent of the total market for a specific product or service that is controlled by one company or entity.

Markup (Margin, Gross profit): The difference between the cost and the selling price.

Markup on cost: Markup that is calculated as a percent of cost.

Markup on selling price: Markup that is calculated as a percent of selling price.

Markup with spoilage: The calculation of markup including deduction for spoiled or unsaleable merchandise.

MasterCard: A credit-card plan (formerly known as Master Charge).

Maturity value: The amount that a borrower must repay on the maturity date of a note.

Mean: The sum of all the numbers divided by the number of numbers.

Median: A measure of central tendency in which one-half of the observations are above this number and one-half are below it.

Medical insurance: Insurance on an individual that covers medical expense in the event of illness or accident.

Medicare tax: The amount of money deducted from the paychecks of almost all employees, used by the federal government to pay for Medicare.

Memory function: A feature on some calculators that stores results internally in the machine for retrieval and future use.

Merchant batch header ticket: The bank form used by businesses to deposit credit-card transactions.

Meter: A unit of length in the metric system that is slightly longer than one yard.

Metric system: A system of weights and measures based on decimals, used throughout most of the world.

Milli-: A prefix used in the metric system meaning thousandth. (For example, a milligram is one one-thousandth of a gram.)

Millimeter: One one-thousandth of a meter. There are 25.4 millimeters to an inch.

Mills: A way of expressing a real estate tax rate that is based on thousandths of a dollar.

Minuend: The number from which another number (the subtrahend) is subtracted.

Mixed number: A number written as a whole number and a fraction. (For example, $1\frac{3}{4}$ and $2\frac{5}{9}$ are mixed numbers.)

Mobile banking: Banking from a smart phone.

Mobile ticketing: Purchasing tickets to a movie or play on a smart phone.

Mode: The number that occurs most often in a group of numbers.

Modified accelerated cost recovery system: See MACRS.

Money market account: An interest-bearing account offered by many banks, savings and loans, and brokerage firms. These accounts pay interest but allow the user to withdraw funds without penalty.

Money order: A document that looks similar to a check and is issued by a bank, other financial institution, or a retail store. It is often used in place of cash.

Mortgage: A loan on a home.

Mortgages payable: The balance due on all mortgages owed by a firm.

Multiple carrier insurance: The sharing of risk by several insurance companies.

Multiplicand: A number being multiplied.

Multiplication rule: The same nonzero number may be multiplied or divided on both sides of an equation.

Multiplier: A number doing the multiplying.

Mutual fund: A mutual fund accepts money from many different investors and uses it to purchase stocks or bonds of numerous companies.

N

NASDAQ composite index: A commonly quoted stock index composed of the stock prices of several technology companies.

Negative numbers: Numbers that are the opposite of positive numbers.

Net cost: The cost or price after allowable discounts have been taken. See net price.

Net cost equivalent: The decimal number derived from the complement of the single trade discount. This number multiplied by the list price gives the net cost.

Net earnings: The difference between gross margin and expenses. After the cost of goods and operating expenses are subtracted from total sales, the remainder is net profit.

Net income: The difference between gross margin and expenses.

Net pay: The amount of money actually received by an employee after deductions are taken from gross pay.

Net payment date: The date by which an invoice must be paid.

Net price: The list price less any discounts. See net cost.

Net proceeds: The amount received from the bank for a discounted note.

Net profit: See net earnings.

Net sales: The value of goods purchased by customers after the value of goods returned is subtracted.

Net worth (Capital, Stockholder's equity, Owner's equity): The difference between assets and liabilities.

No-fault insurance: A guarantee of reimbursement (provided by the insured's own insurance company) for medical expenses and costs associated with an accident no matter who is at fault.

Nominal rate: The interest rate stated in connection with a loan. It may differ from the annual percentage rate.

Nonsufficient funds (NSF): When a check is written on an account for which there is an insufficient balance, the check is returned to the depositor for nonsufficient funds.

No scrap value: The value of an item is assumed to be zero at the end of its useful life.

Notes payable: The value of all notes owed by a firm.

Notes receivable: The value of all notes owed to a firm.

NOW account (Negotiable order or withdrawal): Technically a savings account with special withdrawal privileges. It looks the same and is used the same as a checking account.

Number of compounding periods: The number of times interest is paid out or (theoretically) added to an investment. It is a calculation and may or may not involve an actual credit to an investment at each compounding period.

Numerator: The number above the line in a fraction. (For example, in the fraction $\frac{5}{8}$, 5 is the numerator.)

O

Odd lot: Fewer than 100 shares of stock.

Open-end credit: Credit with no fixed number of payments. The consumer continues making payments until no outstanding balance is owed. Most credit cards offer open-end credit

Operating expenses (Overhead): Expenses of operating a business. Wages, salaries, rent, utilities, and advertising are examples.

Operating loss: The loss resulting when the selling price is less than the break-even point.

Ordered array: A list of numbers arranged from smallest to largest.

Order of operations: The rules that are used when evaluating long arithmetic expressions.

Ordinary annuity: An annuity whose payments are made at the end of a given period of time.

Ordinary dating: A method for calculating the discount date and the net payment date. Days are counted from the date of the invoice.

Ordinary interest: A method of calculating interest, assuming 360 days per year. See banker's interest.

Ordinary life insurance (Whole life insurance, Straight life insurance): A form of life insurance whereby the insured pays a constant premium until death or retirement, whichever occurs first. Upon retirement, monthly payments are made by the company to the insured until the death of the insured.

Other expenses: Certain expenses that are deductible on a personal income tax return.

Overdraft: An event that results when there is not enough money in a bank account to cover a check that is written from that account.

Overdraft fee: A fee charged when a bank account is in overdraft.

Overhead: Expenses involved in running a firm. See operating expenses.

Over-the-limit fees: Fees charged when the balance on a credit-card account exceeds the account's credit limit.

Overtime: The number of hours worked by an employee in excess of 40 hours per week or 8 hours per day.

Owner's equity: See net worth.

P

Part: The result of multiplying the base times the rate.

Partial payment: A payment made on an invoice that is less than the full amount of the invoice.

Partial product: Part of the process of getting the answer in a multiplication problem.

Par value of a bond: *See* face value of a bond.

Payee: The person who lends money and will receive repayment on a note.

Payer of a note: A person borrowing money from another person. *See* maker of a note.

Payroll: A record of the hours each employee of a firm worked and the amount of money due each employee for a given pay period.

Payroll card: A card maintained by employers showing the name of employee, dates of pay period, days, times, and hours worked.

Payroll ledger: A chart showing all payroll information.

Percent (Rate): Some parts of a whole: hundredths, or parts of a hundred. (For example, a percent is one one-hundredth. Two percent means two parts of a hundred, or $\frac{2}{100}$.)

Percentage method: A method of calculating income tax withholding that is based on percentages.

Per debit charge: A charge per debit made on an account.

Periodic inventory: A physical inventory taken at regular intervals.

Permanent life insurance: Life insurance that can be continued until death, regardless of the age of the insured.

Perpetual inventory: A continuous inventory system normally involving a computer.

Personal account: The type of checking account used by individuals.

Personal computer banking: Banking using a personal computer and the Internet.

Personal exemption (Exemption): A deduction allowed each taxpayer for each dependent and the taxpayer himself or herself.

Personal identification number (PIN): A lettered or numbered code that allows a person with a credit or debit card to gain access to credit or cash.

Personal property: Property such as a boat, a car, or a stereo.

Piecework: A method of pay by which an employee receives so much money per item produced or completed.

Plant assets: *See* fixed assets.

Point-of-sale terminal: A machine that allows a customer to make purchases using a credit or debit card.

Policy: A contract outlining the insurance agreement between an insured and an insurance company.

Policy limits: The maximum amount that an insurance company will pay as defined in the policy.

Postdating: Dating in the future; on an invoice, "AS OF" dating.

pr.: Abbreviation for *pair*.

Preferred stock: A type of stock that offers investors certain rights over holders of common stock.

Premium: The amount of money charged for insurance policy coverage.

Premium factor: A factor used to adjust an annual insurance premium to semiannually, quarterly, or monthly.

Prepaid card: A card on which a user has stored monetary value which allows her to make purchases.

Present value: The amount that must be deposited today to generate a specific amount at a specific date in the future.

Price-earnings (PE) ratio: The price per share divided by the annual net income per share of stock.

Prime interest rate: The interest rate banks charge their largest and most financially secure borrowers.

Prime number: A number that can be divided without remainder by exactly two distinct numbers: itself and 1.

Principal: The amount of money either borrowed or deposited.

Privately held corporation: A corporation that has relatively few owners, or perhaps a single owner. Its stock is not traded on a large exchange such as the New York Stock Exchange.

Proceeds: The amount of money a borrower receives after subtracting the discount from the face value of a note.

Product: The answer in a multiplication problem.

Promissory note: A business document in which one person agrees to repay money to another person within a specified amount of time and at a specified rate of interest.

Proper fraction: A fraction in which the numerator is smaller than the denominator. (For example, $\frac{2}{3}$ is a proper fraction; $\frac{9}{5}$ is not.)

Property damage insurance: A type of automobile insurance that pays for damages that the insured causes to the property of others.

Proportion: A mathematical statement that two ratios are equal.

Proximo (Prox.): *See* end-of-month dating.

Publicly held corporations: Corporations that are owned by the public and have stock that trades freely.

Purchase invoice: A list of items purchased, prices charged for the items, and payment terms.

Q

Qualifying for a loan: A person applying for a loan is said to qualify if his or her credit history, income, and financial statement satisfy the requirements of the lending institution.

Quick ratio: The quotient of liquid assets and current liabilities.

Quota: An expected level of production. A premium may be paid for surpassing quota.

Quotient: The answer in a division problem.

R

Radio frequency identification (RFID): A chip that responds to an electronic scan with information about the product.

Rate: Parts of a hundred. *See* percent.

Rate of interest: The percent of interest charged on a loan for a certain time period.

Ratio: A comparison of two numbers often shown as a fraction.

Real estate: Real property such as a home or a parcel of land.

Receipt-of-goods dating (ROG): A method of determining cash discounts in which time is counted from the date that goods are received.

Reciprocal: A fraction formed from a given fraction by interchanging the numerator and denominator.

Reconciliation: The process of checking a bank statement against the depositor's own personal records.

Recourse: Should the maker of a note not pay, the bank may have recourse to collect from the seller of the note.

Recovery classes: Classes used to determine depreciation under the modified accelerated cost recovery system.

Recovery period: The life of property depreciated under the accelerated cost recovery system.

Recovery year: The year of life of an asset when using the MACRS method of depreciation.

Reduced net profit: The situation that occurs when a markdown decreases the selling price to a point that is still above the break-even point.

Refinance: Borrowers can go to a lender and refinance their existing loan with a different interest rate, period, and payment.

Regulation DD: A Federal Reserve System document that specifies how interest paid to savers is to be calculated.

Regulation Z: A Federal Reserve System document that implements the Truth-in-Lending Act.

Renter's coverage: Insurance that covers only the possessions of a renter and not the house or apartment in which the possessions are kept.

Repayment schedule: A schedule showing the amount of payment going toward interest and principal and the remaining debt after each payment has been made.

Repeating decimals: Decimal numbers that do not terminate but that contain a series of digits or numbers that repeat.

Replacement cost: The cost of replacing property that is completely destroyed.

Repossess: The taking back of property by a lender when payments have not been made to the lender.

Residual value: *See* scrap value.

Restricted endorsement: A signature or imprint on the back of a check that limits the ability to cash the check.

Retailer: A business that buys from wholesalers and sells to consumers.

Retail method: A method used to estimate inventory value at cost that utilizes both cost and retail amounts.

Returned check: A check that was deposited and then returned due to lack of funds in the payer's account.

Return on average total assets: Net income divided by average total assets.

Returns: The total value of all goods returned by customers.

Revolving charge account: A charge account that never has to be paid off.

Right side: The right-hand side of an equation.

Risk: The potential of gaining or losing something that is valuable such as money or health.

Roth IRA: Contributions to a Roth Individual Retirement Account (Roth IRA) are not deductible when made. However, funds in the account grow tax free and withdrawals are not taxed once the account holder reaches a certain age.

Rounding: Writing a number with fewer digits of accuracy.

Rounding whole numbers: Reduction of the number of nonzero digits in a whole number.

Round lot: A multiple of 100 shares of stock.

Rule of 78: A method of calculating a partial refund of interest that has already been added to the amount of a loan. This calculation is done when the loan is paid off early.

S

Salary: A fixed amount of money per pay period.

Salary plus commission: Earnings based on a fixed salary plus a percent of all sales.

Sale price: The price of an item after markdown.

Sales invoice: *See* purchase invoice.

Sales tax: A tax placed on sales to the final consumer. The tax is collected by the state, county, or local government.

Salvage value: *See* scrap value.

Savings account: An interest-paying account that allows day-to-day savings and withdrawals.

Schedule 1: The part of the 1040A federal tax form that is used to list all interest and dividends.

Scrap value (Salvage value): The value of an asset at the end of its useful life. For depreciation purposes, this is often an estimate.

SDI deduction: State disability insurance pays the employee in the event of disability and is paid for by the employee.

Self-employed people: People who work for themselves instead of for the government or for a private company.

Series discount: *See* chain discount.

Service fees: Fees charged by banks for bouncing a check, monthly service charge, etc.

Shift differential: A premium paid for working a less desirable shift, such as the swing shift or the graveyard shift.

Simple discount note: A note in which the interest is deducted from the face value in advance.

Simple interest: Interest received on only the principal.

Simple interest note: A note in which interest = principal × interest rate × time in years.

Single discount equivalent: A series, or chain, discount expressed as a single discount.

Single return: An income tax return filed by a single person.

Sinking fund: A fund set up to receive periodic payments in order to pay off a debt at some time in the future.

sk.: Abbreviation for *sack*.

Sliding scale: Commissions that are paid at increasing levels as sales increase.

Smart card: A card with a microchip that receives electronic signals, makes calculations, and sends signals. It is sometimes used for identification.

Social Security tax: *See* FICA tax.

Solution: The number that makes an equation true when it is substituted in place of the variable.

Special endorsement: A signature on the back of a check that passes the ownership of the check to someone else.

Specific identification method: An inventory valuation method that identifies the cost of each item.

Split-shift premium: A premium paid for working a split shift, for example, for an employee who is on 4 hours, off 4 hours, and then on 4 hours.

Square root: The square root of 49 is 7, since $7 \times 7 = 49$.

Stafford loan: A loan taken out by college students to help pay tuition.

Standard and Poor's 500: An index made up from the stock prices of 500 leading U.S. companies.

Standard deduction: A tax preparer may use the higher of the itemized deductions or the standard deduction established by the government.

Stated rate: The interest rate stated in connection with a loan. It may differ from the annual percentage rate.

State income taxes: An income tax that is paid to a state government on income earned in that state.

Statement: Usually sent out monthly by the bank, a list of all charges and deposits made against and to a checking account.

Statement fee: A fee charged by some banks for a paper copy of a monthly statement.

Statistics: Refers both to data and to the techniques used in analyzing data.

Stock: A form of ownership in a corporation that is measured in units called *shares*.

Stockbroker: A person who buys and sells stock at the stock exchange.

Stock exchange: An institution where stock shares are bought and sold.

Stockholders: Individuals who own stock in a particular company.

Stockholder's equity: *See* net worth.

Stock ratios: Ratios calculated from the financial statements of a company—used to determine the financial health of the firm.

Stock turnover: *See* inventory turnover.

Stop payment: A request from a depositor that the bank not honor a check that the depositor has written.

Straight commission: A salary that is a fixed percent of sales.

Straight life insurance: *See* ordinary life insurance.

Straight-line depreciation: A depreciation method in which depreciation is spread evenly over the life of the asset.

Strict monthly budget: A list of monthly expenses that a family is expected not to exceed.

Substitution: Method for checking the solution to an equation.

Subtrahend: The number being subtracted or taken away in a subtraction problem.

Suggested retail price: The price at which a manufacturer recommends its product be sold to a customer.

Sum: The total amount; the answer in addition.

Sum-of-the-years'-digits method: An accelerated depreciation method that results in larger amounts of depreciation taken in earlier years of an asset's life.

Supply chain: The chain of businesses and processes involved with moving raw materials, components, and final goods to the end user.

T

Tangible assets: Assets such as a car, machinery, or computers.

Taxable income: Adjusted income subject to taxation.

Tax assessor: A government official who is in charge of the office that determines the amount of tax due on real estate.

Tax deduction: Any expense that the Internal Revenue Service allows taxpayers to subtract from adjusted gross income.

Taxes: Individuals and corporations must pay taxes to government entities such as schools, cities, counties, states, and the federal government. Here are a few of the many types of taxes: sales taxes, property taxes, gasoline taxes, income taxes, and estate taxes.

Tax preparation: The preparation of an income tax return that is then sent to the Internal Revenue Service.

T-bill: A short-term note issued by the federal government that pays interest to the holder of the note. Issuing T-bills allows the federal government to raise cash without having to borrow money from a bank and pay interest.

Telephone transfer account: An interest-bearing checking account into which funds may be transferred by the customer over the telephone.

Term: A number, variable, or product or division of a number and a variable, such as $5x$ or $7 + b$.

Term insurance: A form of life insurance providing protection for a fixed length of time.

Term of an annuity: The length of time that an annuity is in effect.

Term of a note: The length of time between the date a note is written and the date the note is due.

Terms: The area of an invoice where cash discounts are indicated if any are offered. The words "terms discount" are often used in place of "cash discount."

Territorial ratings: Ratings used by insurance companies that describe the quality of fire protection in a specific area.

Texting: A method of communicating to a cell phone(s). Some banks communicate with account holders by texting.

Time-and-a-half rate: One and one-half times the normal rate of pay for any hours worked in excess of 40 hours per week or 8 hours per day.

Time card: A card filled out by an employee that shows the number of hours worked by that employee.

Time deposit account: A savings account in which the depositor agrees to leave money on deposit for a certain period of time.

Time rate: Earnings based on hours worked, not for work accomplished.

Total installment cost: Includes the down payment plus the sum of all payments.

Total revenue: The total of all revenue from all sources.

Trade discount: A discount offered to businesses. This discount is expressed either as a single discount (like 25%) or a series discount (like 20/10) and is subtracted from the list price.

Transaction: An action done by an individual through a bank, such as making a deposit.

Transaction register: Shows the checks written and deposits made on a checking account.

True rate of interest: *See* effective rate of interest.

Turnover at cost: A measure of how much inventory has turned over, or sold, based on the cost of each item.

Turnover at retail: A measure of how much inventory has turned over, or sold, based on the selling price of each item.

U

Underinsured: An individual or firm that does not have enough insurance.

Underinsured motorist: A motorist who does not carry enough insurance to cover the costs of an accident.

Underinsured motorist coverage: Insurance coverage that insures against an uninsured or underinsured motorist who causes damage to your vehicle or people riding in your vehicle.

Underwater: The situation where homeowners owe more on their home than it is worth.

Underwriters: Term applied to any insurer. Usually associated with an insurance company.

Unearned interest: Interest that a company has received but has not yet earned so that it is not shown in revenues.

Unexpected expenses: Expenses incurred by a family or business that are not regular and not expected, such as the transmission going out on a car that is only three years old.

Uniform product code (UPC): A code printed on retail product packages that helps identify the particular item. Scanners pick up the UPC, which allows a computer to post the price for the item and decrease the inventory of that item by 1 in the database.

United States Rule: The rule by which a loan payment is first applied to the interest owed, with the balance used to reduce the principal amount of the loan.

Unit price: The cost of one item.

Units-of-production: A depreciation method by which the number of units produced determines the depreciation allowance.

Universal life policy: A policy whose premiums flow into a general account from which the insurance company makes investments.

Unlike fractions: Fractions with different denominators.

Unlike terms: Two terms in an algebraic expression that do not have the same variables, so they cannot be added together. For example, $2x + 5y$ are unlike terms and cannot be combined into one term.

Unpaid-balance method: A method used to calculate interest on open-end credit accounts.

Unreimbursed employee expenses: Certain expenses that are deductible on a personal income tax return.

Useful life: The estimated life of an asset. The Internal Revenue Service gives guidelines of useful life for depreciation purposes.

V

Valuation of inventory: Determining the value of merchandise in stock. Four common methods are specific-identification, average cost, FIFO, and LIFO.

Variable: A letter that stands for a number.

Variable commission: A commission whose rate depends on the total amount of the sales.

Variable interest rate loan: A loan on which the interest rate can go up or down.

Variable life policy: A life insurance policy that allows the owner to invest the funds within the policy in different types of investments.

Vertical analysis: The listing of each important item on an income statement as a percent of total net sales or each item on a balance sheet as a percent of total assets.

Visa: A credit-card plan (formerly known as Bank Americard).

W

Wage: A rate of pay expressed as a certain amount of dollars per hour.

Wage bracket method: A method of calculating income tax withholding that is based on tables that list income ranges.

Weighted average cost: A method for calculating the arithmetic mean for data where each value is weighted (or multiplied) according to its importance.

Whole life insurance: *See* ordinary life insurance.

Whole number: A number with no fraction or decimal part.

Wholesaler: A business that buys directly from the manufacturer or other wholesalers and sells to the retailer.

Withholding allowance: An allowance for the employee, spouse, and dependents that determines the amount of withholding tax taken from gross earnings.

Withholding tax: The funds withheld from an employee's paycheck and sent to the federal or state government on behalf of that individual.

With recourse: An understanding that the seller of a note is responsible for payment of the note if the original maker of the note does not make payment. The note is sold with recourse.

Workers' compensation: Insurance purchased by companies to cover employees against work-related injuries.

W-2 form: The wage and tax statement given to the employee each year by the employer.

W-4 form: A form usually completed at the time of employment, on which an employee states the number of withholding allowances being claimed.

Y

Yield to maturity: The interest rate yield to the maturity of a bond or other financial instrument. It includes any interest payments, time to maturity, and payoff at maturity.

Youthful operator: A driver of a motor vehicle who is under a certain age, usually 25.

Subject Index

A

Absolute loss, 300–301
Accelerated cost recovery system (ACRS), 565, 596
Accelerated depreciation, 573
Accelerated mortgages, 491
Accountants, 611
Accounts payable, 623
Accounts receivable, 623
Accumulated depreciation, 565, 566
Acid test, 629
Acid-test ratio, 629, 630
ACRS. *See* Accelerated cost recovery system
Actual selling price, 298
Actuaries, 541
Addends, 3
Addition
    of decimal numbers, 24
    of fractions, 48, 50
    of mixed numbers, 56
    verbal expressions involving, 130–131
    of whole numbers, 3–4
Addition rule, 122
Adjustable rate mortgages (ARMs), 493
Adjusted bank balance, 190
Adjusted gross income (AGI), 519
ADP. *See* Automatic Data Processing
Adult operators, 544
AGI. *See* Adjusted gross income
Alcorn's Boutique, 159
Algebraic logic, B-1
Allowances, 217
Alphabet, 610
Amortization, 467, 482
Amortization schedule, 483, 491–492
Amortization tables, 484, 490
Amount
    in addition, 3
    on invoices, 247
Annual meetings, 433
Annual percentage rate (APR)
    in average daily balance method, 461
    formula for, 468–469
    legislation on, 467
    simple discount notes, 346, 347
    tables in calculating, 469–470
Annuities
    case studies, 407, 452–453
    defined, 408
    due, 411
    finding amount of, 409–410
    ordinary, 408
    present value of ordinary, 416–421
    reasons to buy, 408
    sinking funds and, 424–427
    terminology, 408–409
Apple, Inc., 323, 369, 433, 434, 610, 618–619, 624, 640–641, 654
Application problems
    equations in solving, 132–134
    estimating answers in, 14–15
    indicator words in, 14
    solving, D-5
    solving for rate in, 101
    steps in solving, 14–17, 132, 136
Approximate annual percentage rate (APR), 468
ARMs. *See* Adjustable rate mortgages
AS OF dating, 263
Assessed value, 512

Assets, 623
Automated Teller Machines (ATMs), 172, 182
Automatic Data Processing (ADP), 205
Automobiles. *See* Motor vehicles
Average (statistical), 666
Average cost method, 306
Average daily balance method, 460–461
Average inventory, 304
Average owners' equity, 629
Axis, defined, E-2

B

Baker's Pottery, 559
Balance brought forward, 176
Balanced bank accounts, 190
Balance sheets, 623–624, 627–630
Bank accounts, interest-bearing, 384
Bank discount, 343, 352. *See also* Discounts
Banker's interest, 328
Banker's ratio, 628
Banking
    account reconciliation, 188–191
    accounts with interest, 173, 384
    case studies, 171, 199–201
    checking account types, 173
    check registers, 177
    check stubs, 176
    credit-card sales in, 182–184
    deposit slips in, 174–176
    parts of a check, 174
    service charges, 174, 190
    services available, 182
Bank of America, 368, 372, 402
Bankruptcy, 324, 443
Bank statements, 188–191
Bank withdrawals, 372
Barcodes, 305
Bar graphs, 655–656
Barron's, 434
Base. *See also* Percent(s)
    defined, 87, D-1
    exponents, 142
    finding, 94–95
    markup on selling price as, 288–289
    terminology associated with, 88
Bed Bath & Beyond, 246, 277
Bev's Deli, 644
Bimodal data sets, 670
Blank endorsements, 175
Boards of directors, 433
Bodily injury insurance, 541
Bonds
    commission charge and cost of, 446
    defined, 443
    monthly income from, 446–447
    tables, 443–445
Bond tables, 443–445
Book value, 565, 566–567. *See also* Depreciation
Borrowing, subtraction, 5, 57
Bounced checks, 188
Break-even point, 299
Brynski, Sarah, 204
Budget deficits, 444
Budgeting, 645–650
    case studies, 644, 650
    controlling expenses, 647–648
    expected and unexpected expenses, 645–647
    future planning, 648–650

Building classifications, 533
Buildings, on balance sheet, 623
Business checking accounts, 173
Business formulas. *See* Formulas
Business planning and budgeting, 650
Buying. *See also* Cash discounts
    case studies, 246, 276–277
    creating invoices, 247, 248
    invoice abbreviations, 248
    series discounts, 249–252, 257–258
    shipping terminology, 247–248
    supply chain, 247

C

Calculating discounts separately, 249–250
Calculators, B-1–B-8, C-1–C-5
Canceled checks, 176
Cancellation, 60
Capital (financial), 433
Capital Curb and Concrete, 564, 607
Carrying (addition), 56
Cars. *See* Motor vehicles
Cash, on balance sheet, 623
Cash discounts. *See also* Buying
    dating methods, 261–264, 267–270
    earned vs. not earned, 262–263
    finding net cost after, 261
    on postdated invoices, 263–264
Cashier's check, 182
Cash on delivery (COD), 247
Cash value, 551
Casualty or theft losses, 523
Catastrophic events, 532
CDs. *See* Certificates of deposit
Celsius (C) temperature, A-3–A-5
Centigrade (C) temperature, A-3–A-5
Centigrams (cg), A-2
Centiliters (cL), A-2
Centimeters, A-1–A-2
Century 21, 79–80, 118
Certificates of deposit (CDs), 182, 378
Chain calculations, B-6
Chain discounts, 249
Charity donations, 523
Checking accounts, 173, 384. *See also* Banking
Check processing, 176
Checks, 174, 188
    outstanding, 188–190
Check stubs, 176
Chevron, 437
Circle graphs, 657–659
Citigroup, Inc., 456, 503–504
COD. *See* Cash on delivery
Coinsurance clauses, 534
Collateral, 330
Collision insurance, 543–544
Commas, defined, 2
Commission charge, 446
Commission rate, 216–219
Common denominators, 48, 50–51
Common fractions, 42, 61. *See also* Fractions
Common stock, 434
Comparative balance sheet, 627
Comparative income statement, 616–617
Comparative line graphs, 656–657
Compensatory (comp) time, 208–209
Complement of a discount, 250–251
Compound amount, 373, 408
Compounding periods, 375–376

I-1

I-2 Subject Index

Compound interest. *See also* Interest
   on bank accounts, 384–387
   calculations of, 373
   case studies, 372, 401–402
   compounding periods in, 375–376
   defined, 374
   formula for, 376–377
   future value and, 394–397
   inflation and, 387–389
   present value and, 394–397
   simple interest formula in, 374–375
   simple interest vs., 324
   students learning about, 377
   tables in calculating, 377–379
Comprehensive insurance, 543–544
Comp time. *See* Compensatory time
Consolidating your loans, 462
Consumer installment loans, 324
Consumer price index (CPI), 388
Continuous inventory systems, 305
Contributions (retirement account), 519. *See also* Gifts to charity
Conversion formulas, 292
Corporations, 433
Cosigners, 325
Cost (bonds), 446
Cost (depreciation), 565
Cost (selling), 281
Cost of goods sold, 611, 612
Cost of living index, 388
Countrywide, 372
Countrywide Financial, 402
Coupon rate, 443
Coverage (insurance), 532
CPI. *See* Consumer price index
Credit cards. *See also* Debt
   consumer finance statistics, 458
   service fees for, 184
   students using, 182, 457
Credit-card sales, 182–184
Credit history, 173, 467
Credit limits, 457
Credit scores, 494
Cross products, 149, 150
Crowdsourcing, 172
Current assets, 623
Current balance, 190
Current liabilities, 623
Current ratio, 628, 630
Current yield, 435–436

D

Daily compounding, 375, 384–387. *See also* Compound interest
Daily overtime, 208–209
Data analysis. *See* Statistics
Days
   in finance charge calculations, 460–461
   in simple interest loans, 326–328
   worked to pay taxes, 511
Death, reducing risk of, 550
Debit cards, 172, 182, 184, 457. *See also* Credit cards
Debt. *See also* Credit cards; Loans
   budgeting to control, 647–648
   case studies, 456, 501–504
   cost of having, 443
   credit score impacting, 494
   defaulting on, 462
   government problem with, 443, 444
   loan consolidation, 461–462
   open-end credit, 457
   revolving charge accounts, 457–458
   tips for avoiding, 461
Decimal equivalents, 69

Decimal numbers
   addition of, 24
   case studies, 2, 37, 38
   converting fractions to, 69
   converting percents to, 82
   creating fractions from, 68
   creating percents from, 80–81
   defined, 20
   in depreciation, 568
   dividing, 29–31
   estimating, 24–25
   fractional percents as, 83–84
   multiplying, 28–29
   reading, writing, and rounding, 20–21
   subtraction of, 25
Decimal parts, 20
Decimal points, 2, B-2
Decimal system, 2
Declining-balance depreciation, 565, 573–575, 582, 596
Decrease problems, 108–110, 258
Decreasing term insurance, 550
Deductible (insurance), 543
Deductions, 205. *See also* Payroll; Tax deductions
Defaulting on your debt, 462
Deficit amounts, B-4–B-5
Deflation, 389
Degrees, in circle graphs, 657
Denominators, 42
Dental expenses, deducting, 523
Dependents, 521
Dependent variables, E-1–E-2
Deposits, 172, 385–386
Deposits in transit (DIT), 189
Deposit slips, 175–176
Deposit ticket, 175
Depreciable amount, 566
Depreciation
   case studies, 564, 606–607
   choosing method of, 575
   declining-balance, 565, 573–575
   formulas, 566, 573, 581, 591
   modified accelerated cost recovery system, 596–599
   straight-line, 565–568
   sum-of-the-years'-digits, 580–582
   taxes and, 565, 575, 582, 596, 597
   terminology, 565
   units-of-production, 591–593
Depreciation fractions, 580–581
Depreciation per unit, 591
Depreciation schedules
   double-declining balance method, 574–575
   modified accelerated cost recovery system, 599
   straight-line method, 567–568
   sum-of-the-years' digits method, 581–582
   units-of-production method, 593
Depression. *See* Great Depression
Diabetes, 550
Difference (subtraction), 5
Differential piece rate, 215
Digits, 2
Direct deposits, 172
Direct payments, 172
Disability insurance, 224–225, 537
Discount brokers, 434
Discount period, 352, 354, 355
Discount rate, 343, 352
Discounts. *See also* Cash discounts; Simple discount notes; Trade discounts
   bank, 343, 352
   complements of, 250–251
   on notes, 354, 355
   series, 249–252, 257–258
   on shipping and insurance, 247

Discover card, 457
Distance (metric system), A-2, A-3
Distribution chain, 247
Distributive property, 126
DIT. *See* Deposits in transit
Dividends, 7, 433
Divisibility rules, 45
Division
   of decimals, 29–31
   of fractions, 62, 63
   of mixed numbers, 62
   rules of divisibility, 45
   using exponents, D-2–D-3
   verbal expressions involving, 131
   of whole numbers, 3, 7–8
Divisor, 7, 62
Dollars per $100, property tax as, 513
Dollars per $1000, property tax as, 513
The Doll House, 510, 560–561
Donations, 523
Double-declining-balance method, 573–575
Double-declining-balance rate, 573
Double time, 208
Dow Jones Industrial Average, 437
Draws, 218
Due annuities, 411
Due date
   calculating, 262–264, 267–268, 355
   of promissory note, 330
Dugally, Tom, 79

E

Earnings, 205–207, 207–208. *See also* Payroll
EC. *See* Electronic commerce
Economic Recovery Tax Act of 1981, 596
Education, pay relating to, 209, 666–667. *See also* Students
Effective rate of interest, 346
EFT. *See* Electronic funds transfer
Einstein, Albert, 373
Electronic banking. *See also* Banking
Electronic (online) banking, 172
Electronic commerce (EC), 172
Electronic product code (EPC) devices, 305
Employee's Withholding Allowance Certificate (W-4 form), 228–229
Employee theft, 188
Employer's Quarterly Federal Tax Return (Form 941), 235
Ending inventory, 612
End-of-month (EOM) dating, 267–268
Endorsements, on checks, 175, 176
Endowment policies, 551
English-metric conversion table, A-3
English system, A-1–A-3
EOM. *See* End-of-month dating
EPC. *See* Electronic product code devices
Equations. *See also* Formulas
   in application problem solutions, 132–134
   case studies, 159–161
   combining like terms in, 125–126
   defined, 122
   distributive property in, 126–127
   as mathematical expressions, 130–131
   multiple operations in solving, 124–125
   rules for solving, 122–124
   steps in solving, 127
   terminology of, 122
   from words, 131–132
Equipment, on balance sheet, 623
Equivalent cash price, 420–421
Escrow accounts, 493
Estimation, 14–15, 24–25
Exact interest, 328
Exchange-traded funds (ETFs), 434, 437

Executive officers, 433
Expected expenses, 645–647
Expenses
   controlling, 647–648
   expected and unexpected, 645–647
Exponents, 142, 376, D-1–D-5
   division using, D-2–D-3
   evaluating expressions, D-4–D-5
   multiplication using, D-2–D-3
Expressions, 122
Extension totals, 247
Extra (ex., x) dating, 268–269
Exxon Mobil, 437, 610

F

Face value, 329, 343–345, 443, 533
Factoring, 355
Factors, 355
Fahrenheit, Gabriel, A-3
Fahrenheit (F) temperature, A-3–A-5
Fair Isaac Corporation (FICO), 494
Fair Labor Standards Act, 206
Fair market value, 512
FAS. *See* Free alongside ship
Federal Deposit Insurance Corporation (FDIC), 384
Federal Insurance Contributions Act (FICA), 222. *See also* Social Security
Federal Reserve Bank, 328, 469
Federal Reserve System, 469
Federal Unemployment Tax Act (FUTA), 235
Federal withholding, 229–233, 234
FICA. *See* Federal Insurance Contributions Act
FICO. *See* Fair Isaac Corporation
Fidelity, 412, 434
FIFO method. *See* First-in, first-out method
Finance charges, 458, 468
Financial calculators, B-1, C-1–C-5
Financial ratios, 628–630. *See also* Formulas
Financial statements
   balance sheet, 623–624, 627–630
   case studies, 610, 638–641
   income statements, 611–613, 616–617
First-in, first-out (FIFO) method, 307–308
Fixed assets, 623
Fixed-rate loans, 493
Flay, Bobby, 678
Floating decimal, B-2
Float time, 176
FOB destination. *See* Free on board destination
FOB shipping point. *See* Free on board shipping point
Food Network, 678
Ford Motor Company, 607
Forecasting, 121, 160–161. *See also* Formulas
Foreclosures, 330, 493–494
Foreman, George E., 276
Form 941, 235
Form 1099, 519, 520
Form 1040A, 525–528
Form 1040EZ, 525
Formula for compound interest, 376–377
Formula for markdown, 298
Formulas. *See also* Equations
   amortization, 482–483
   annual percentage rate, 468–469
   balance sheet accounts, 623, 629–630
   for base, 94
   case studies, 159–161
   coinsurance, 534
   conversion, 292
   defined, 139
   depreciation, 566, 573, 581, 591
   evaluating, 139–140, 142
   income statement accounts, 611, 612, 616, 618
   inventory turnover, 304–305
   for percent, 87
   present value, 394, 416
   property taxes, 513
   for rate, 100
   ratios and proportion, 148–151
   simple interest, 374–375
   solving for specific variables, 140–141
   in solving word problems, 141
   stock ratios, 435–437
Form W-2, 519, 520
Form W-4, 228–229
Fortune magazine, 280
401(k) plans, 412
403(b) plans, 412
Fractions. *See also* Mixed numbers
   adding and subtracting, 48, 50
   case studies, 41, 74, 75
   common denominator, 50–51
   converting decimals to, 68
   converting mixed numbers to, 43
   converting percents to, 82–83
   creating decimals from, 69
   creating mixed numbers from, 43–44
   creating percents from, 81–82
   defined, 42
   in depreciation, 580–581
   dividing, 62, 63
   least common denominator, 48–50
   lowest terms, 44–45
   multiplying, 60, 63
   of percents, 83–84
   rules of divisibility, 45
   types of, 42–43
Free alongside ship (FAS), 247
Free on board (FOB) destination, 247
Free on board (FOB) shipping point, 247
Frequency distributions, 654–655
Fringe benefits, 235
Front-end rounding, 4–5
FUTA. *See* Federal Unemployment Tax Act
Future amount, 373
Future planning, 648–650
Future value
   of annuity, 408
   in business valuation, 396–397
   calculating, 375, 394–396
   of compound amount, 374
   defined, 373
   present value vs., 394

G

Gender, household income by, 669
General Motors (GM), 121, 160–161, 433
George Foreman grills, 276
Gifts to charity, 523
GM. *See* General Motors
Google, 640
Government. *See also* Federal Reserve Bank; Legislation
   debt problems, 443, 444
   income and expense statistics, 518
   inflation affected by, 389
   interest rates affected by, 343
   issuing T-bills, 347–348, 355–356
Grace period, 458
Grade point averages, 668
Grams (g), A-2, A-3
Graphing calculators, B-1
Graphing equations, E-1–E-9
   linear, E-4–E-7
   nonlinear, E-7–E-9
Graphs, 655–659
Great Depression, 372, 437
Great Recession of 2007, 437
Great Recession of 2008-2012, 493
Gross earnings. *See also* Payroll
   calculating, 205–207
   defined, 205
   education relating to, 209
   finding net earnings from, 234
   insurance as percent of, 224
Gross loss, 300
Gross profit, 281, 611
Gross profit on sales, 611
Gross sales, 611

H

Half-year convention, 597
Head of household, 521
Health Care Act of 2010, 222
Hernandez, Jessica, 323
The Hershey Company, 654
Holmes, Oliver Wendell, Jr., 511
Home Depot, 41, 75
Homeowners, insurance spending by, 537
Homeowner's insurance policies, 532, 537
Horizontal analysis, 618–620, 628
Hourly wage, 205, 215–216
Household budgeting, 648–649
Housing, interest rate affecting, 324
Hundredths, 80. *See also* Percent(s)

I

Identity theft, 173
Impound accounts, 493
Improper fractions, 42. *See also* Fractions
Incentive rates, 214
Income
   inequality in, 658
   linked to education level, 209, 666–667
   median, 669
Income statements, 611–613, 616–617
Income taxes. *See also* Internal Revenue Service (IRS); Taxes
   finding balance due or refund on, 524–525
   finding liability for, 519
   finding taxable income, 522–523
   forms, 520, 525–528
   on life insurance proceeds, 551
   recordkeeping guidelines, 519
   reporting, 519
   standard deduction, 521
   on W-2 forms, 520
Income tax liability, 519
Income tax method, 596
Income tax withholdings, 228–235. *See also* Taxes
Increase problems, 108–110
Independent variables, E-1–E-2
Index funds, 437–438
Indicator words, 14
Individual retirement accounts (IRAs), 412–413, 519
Industry average charts, 617
Inflation, 387–389
Information society, 654
Initial inventory, 612
Inspection, 48
Installment cost, 468
Installment loans, 324, 467–470
Insurance. *See also* Medicare; Social Security
   case studies, 510, 559–561
   coinsurance formula, 534
   discounts on, 247
   finding annual premium for, 533–534
   life, 549–553
   motor vehicle, 541–546
   multiple-carrier, 535–536
   other types of, 537
   people needing more, 549

I-4 Subject Index

Insurance (*continued*)
   as percent of earnings, 224
   rising cost of, 533
   role of, 511, 532
   terminology, 532–533
   unemployment insurance tax, 235
Insurance policies, 532–533
Intangible assets, 565
Interest. *See also* Compound interest; Simple interest
   banker's, 328
   charged by banks, 172
   defined, 173, 443
   effective rate of, 346
   ordinary, 328
   in simple discount notes, 343
   tax-deductible, 523
   tax reporting, 519, 520
   unearned, 477–478
   variable, 493
Interest-bearing bank accounts, 182, 384–387. *See also* Banking; Compound interest
Interest-bearing checking accounts, 384
Interest-in-advance notes, 343
Interest paid (checking accounts), 182, 384
Interest payments, 95
Interest rate per compounding period, 375
Interest rates, 95. *See also* Rates
Interest rate spreads, 427
Internal Revenue Service (IRS). *See also* Income taxes; Taxes
   on depreciation, 565
   Publication 946, 597
   quarterly payment to, 234–235
   on retirement plans, 413
   role of, 519
Inventory. *See also* Selling
   average, 304
   on balance sheet, 623
   initial and ending, 612
   tracking methods, 305
   turnover, 304–305
   valuation methods, 306–307
Inventory shrinkage, 304
Inventory turns, 304
Investments, finding amount of, 95. *See also specific investments*
Invoices, 247, 248. *See also* Buying
Invoice totals, 247
IRAs. *See* Individual retirement accounts
Itemized billing, 458, 459
Itemized deductions, 523–524

J
J. B. Sherr Co., 247, 248
Jackson & Perkins Wholesale, Inc., 171, 200–201
James, Ben, 121
Jobs, Steve, 323
Johnson, Darnell, 407, 453

K
Kilograms (kg), A-2
Kilometers (km), A-2
Knuckle method, 328
Kohl's, 457

L
Land, on balance sheet, 623
Last-in, first-out (LIFO) method, 308
Late fees, 458
Layoffs, preparing for, 205
Least common denominator (LCD), 48–50
Left side (equations), 122

Legislation
   on banking, 384
   on loans, 346, 467
   on payroll, 204, 206, 222–224, 235
   on taxes, 596
Length (metric system), A-1–A-2, A-3
Level-premium term policy, 550
Liabilities, 433, 623
Liability insurance, 537, 541
Life expectancies, 549
Life insurance, 549–553
LIFO method. *See* Last-in, first-out method
Like fractions, 48
Like terms, 125–126
Limited liability, 433
Limited-payment life insurance, 551
Linear equations, graphing, E-4–E-7
Line graphs, 656–657, E-2
Liquid assets, 629
Liquidity, 628–629
List price, 247, 257–258
Loan amount, 343
Loan consolidation, 461–462, 501–502
Loan payoff table, 485
Loans. *See also* Debt; Real estate loans
   amount of, 343
   consumer installment, 324
   early payoff of, 475–478
   installment, 467–470
   personal property, 482–485
Long-term care insurance, 537
Long-term liabilities, 623
Long-term notes payable, 623
Losses, 299–301, 523
Lowest terms, 44–45

M
MACRS. *See* Modified accelerated cost recovery system
Maintenance charge per month, 174
Makers (promissory notes), 329
Manufacturing, in supply chain, 247
Margin, defined, 281
Marital status, 228
Markdown, 298–301
Markup, 281, 289
Markup formula, 281–285, 289–291
Markup on cost, 282, 291–292
Markup on selling price, 288–294
Markup percent, 283–285, 292–293
Marriage, average age for, 666, 667
MasterCard, 184, 457, 459
Mathematical models, 132
Mattel Inc., 560
Maturity (bonds), 443
Maturity date, 329, 443–444
Maturity value
   defined, 329
   in finding proceeds, 355
   formula for, 325–326, 353, 354
   in simple discount notes, 343
Mayo Clinic, 407, 453
McDonald's, 401, 433
Mean, 666–668
Measure of central tendency, 666
Median, 668–669
Medical expenses, 523
Medical insurance, 537, 542
Medicare. *See also* Insurance
   finding amount of, 223–224
   history of, 204, 222–223, 243
   on W-2 forms, 520
   withholding for, 222
Memory keys, B-5–B-6
Merrill Lynch, 347, 372, 402, 444

Mesa Grill, 678
Meters, A-1–A-2
Method of cross products, 149
Method of prime numbers, 48–49
Metric system, A-1–A-5
Microsoft, 433, 640
Milligrams (mg), A-2
Milliliters (mL), A-2
Millimeters, A-1–A-2
Mills, property tax as, 513–514
Minuend, 5
Miscellaneous deductions, 523
Mixed numbers. *See also* Fractions
   addition of, 56
   converting fractions to, 44
   converting to fractions, 43–44
   defined, 42
   dividing, 62
   multiplying, 60–62
   subtracting, 57
Mobile banking, 172
Mobile payments, 182
Mobile ticketing, 172
Mode, 669–670
Modified accelerated cost recovery system (MACRS), 565, 596–599
Money market accounts, 182, 384
Money order, 182
Monthly compounding, 375. *See also* Compound interest
Months, length of each, 262, 326–328
Mortgage insurance, 550
Mortgage rates, 490
Mortgages, 490. *See also* Real estate loans
Mortgages payable, 623
Motor vehicles
   cost of financing, 467
   insurance, 541–546
Multiple-carrier insurance, 535–536
Multiplicand, 6
Multiplication
   of decimals, 28–29
   of fractions, 60, 63
   of mixed numbers, 60–62
   by omitting zeros, 6–7
   using exponents, D-2–D-3
   verbal expressions involving, 131
   of whole numbers, 3, 5–6
Multiplication rule, 122
Multiplier, 6
Mutual funds, 437–438, 446–447

N
NASDAQ Composite Index, 437, 619
Negative cash flow, C-1
Negative numbers, B-4–B-5
Net cost, 249, 257–258, 261
Net cost equivalent, 250–252
Net earnings, 281
Net income, 611
   after taxes, 611, 629
   before taxes, 611
Net pay, 205, 234. *See also* Payroll
Net price, 249
Net profit, 281, 299
Net sales, 611
Net worth, 623
New York Stock Exchange (NYSE), 433
Nike, 433, 434
No-fault insurance, 544
Nominal rates, 346, 467
Nonlinear equations, graphing, E-7–E-9
Nonsufficient funds (NSF), 182. *See also* Returned checks
No salvage value, 565, 567

Notary service, 182
Notes. *See* Simple discount notes
Notes payable, 623
Notes receivable, 623
NSF. *See* Nonsufficient funds
Numerator, 42
NYSE. *See* New York Stock Exchange

O

Olympic Sports, 317
Omitting zeros, multiplication by, 6–7
Online (electronic) banking, 172
Open-end credit, 457
Operating expenses, 281, 611
Operating loss, 299–301
Operations, mathematical, 3, 124–125. *See also specific types*
Ordered arrays, 668
Order of operations, B-6, D-3–D-4
Ordinary annuities, 408. *See also* Annuities
Ordinary dating method, 261–262
Ordinary interest, 328
Ordinary life insurance, 551
Origin, defined, E-2
Original numbers, 8
Overdraft protection, 182
Overdrafts, 182
Overhead, 281, 611
Over-the-limit fees, 458
Overtime. *See also* Payroll
    daily, 208–209
    legislation on, 206
    for piecework, 214–216
    salaried employees, 210
    weekly, 206–208
Overtime premium method, 207
Owners' equity, 623, 629

P

Parentheses keys, B-7
Part (percent problems), 87, 88
Partial payments, 269–270
Partial products, 6
Par value, 443
Passbook accounts, 384
Payees (promissory notes), 329
Payers (promissory notes), 329
Payment period, 408
Payments. *See also* Amortization schedule
    in bank statement reconciliation, 199
    direct, 172
    electronic, 172
    methods, 183
    mobile, 182
    monthly loan, 485
Pay periods, 209, 231–233
Payroll. *See also* Overtime; Salary; Wages
    case studies, 204, 242–243
    commission, 216–219
    common pay periods for, 209
    employer filing responsibilities, 235
    federal withholding tax, 229–233
    gross earnings calculations, 205–207
    net pay after deductions, 234
    piecework, 214–216
    quarterly IRS payments, 234–235
    state withholding tax, 233–234
    tax calculations, 222–224
Payroll ledger, 205
PE ratio. *See* Price-earnings ratio
Percentage method, 231–233
Percent formula, 87
Percent(s)
    basic formula, 87
    case studies, 79–80, 117, 118
    creating decimals from, 82
    creating fractions from, 82–83
    defined, 80
    finding base in, 94–95
    finding rate in, 100–101
    fractional, 83–84
    increase and decrease problems, 108–110
    of markdown, 298–299
    of markup, 283–285, 292–293
    problem components, 87
    problem format, 89–90
    property tax as, 512
    sales tax calculation, 88–89
    solving for part, 87–88
    terminology, 88
    writing decimals as, 80–81
    writing fractions as, 81–82
Per-check charge (per-debit charge), 173
Periodic inventory systems, 305
Perishables, selling price for, 293–294
Permanent life insurance, 551
Perpetual inventory systems, 305
Personal checking accounts, 173
Personal exemptions, 521
Personal finance, students studying, 377
Personal identification number (PIN), 172
Personal income tax, 228
Personal property, vs. other property types, 482
Personal property loans, 482–485
Physical inventory, 305
Piecework rates, 214–216
Pie charts, 657
PIN. *See* Personal identification number
Planning, for future, 648–650
Plant and equipment, 623
Plant assets, 623
Point of sale (POS), 172, 173
Policies (insurance), 532
Policy limits, 542
POS. *See* Point of sale
Positive cash flow, C-1
Postdating, 263–264
Preferred stock, 434
Premium (insurance), 533, 541, 551–552
Premium factor, 552
Prepaid card, 172
Prepayment of loans, 475–478
Present value, 373
    of an annuity, 416–421
    in business valuation, 396–397
    formula, 394, 416
    future value vs., 394
    tables in calculating, 394–396
Present value tables, 418–419
Price-earnings (PE) ratio, 435, 436–437
Prime numbers, 48–49
Prime rate, 324
Principal, 324, 329, 335–336
Privately held corporations, 433
Proceeds, 343, 352–354
Processing checks, 176
Processing information (checks), 176
Products (multiplication), 6
Promissory notes, 329–330
Proper fractions, 42. *See also* Fractions
Property damage coverage, 542
Property tax, 511–514, 559. *See also* Taxes
    formula, 513
Property-tax rate, 513
Proportions, 149–151. *See also* Equations; Formulas
Proximo (prox.) dating, 267–268
Publication 946, 597
Publicly held corporations, 433
Purchase invoice, 247

Q

Quarterly compounding, 375. *See also* Compound interest
Quarterly interest, 375, 386–387
QuickBooks, 205
Quick ratio, 629
Quotas, 215
Quotient, 7, 43, 131

R

Race, household income by, 669
Radio frequency identification (RFID) chips, 305
Rate of commission, 217
Rates. *See also* Compound interest; Percent(s); Simple interest
    affecting future value, 373
    government power over, 343
    in percent problems, 87
    in simple interest formula, 324
    solving for, 101, 336–337
Ratings, for fire insurance, 534
Ratio of net income after taxes to average owners' equity, 629
Ratios, 148–151, 618. *See also* Equations; Formulas
Raw materials, in supply chain, 247
Real estate, 482, 523
Real estate amortization table, 490
Real estate loans
    accelerated, 491
    escrow accounts on, 493
    finding monthly payments on, 490–491
    fixed vs. variable rate, 493
    repayment schedules for, 491–492
Real property, 482
Reasonable graphs, setting, E-4
Receipt-of-goods (ROG) dating, 267–268
Recession of 2009, 324, 372. *See also* Great Recession of 2008-2012
Reciprocals, 124
Reconciliation (bank statements), 188–191
Recourse, 352
Recovery classes, 596–597
Recovery periods, 596–597
Recreational Equipment Inc. (REI), 280, 318
Reduced net profit, 299
Reduced price, 298–299
Refinancing, 490
Refund (loan prepayment), 477
Registers, 176
Regular IRAs, 412
Regulation DD, 384
Regulation Z, 467
REI. *See* Recreational Equipment Inc.
Renter's insurance, 537
Repayment schedules, 491–492
Replacement cost, 534
Repossession, 482
Residual value, 565
Restricted endorsement, 175, 176
Retail, 247, 289. *See also* Buying; Selling
Retail method of estimating inventory, 308–309
Retirement accounts. *See* Annuities; Individual retirement accounts (IRAs)
Retirement plans, self-employed, 117
Returned checks, 188. *See also* Nonsufficient funds (NSF)
Returned-deposit item, 182
Returns (merchandise), 217, 611
Revolving charge accounts, 457–458
RFID chips. *See* Radio frequency identification chips
Rhyme method, 328
Right side (equations), 122

I-6 Subject Index

Risk
    and discounting a note, 352
    to shareholders, 437
ROG dating. *See* Receipt-of-goods dating
Rose Gardens, 171, 200–201
Roth IRAs, 412
Rounding, 2–4, 20–21, 251
Rule of 78, 476–478
Rules for solving equations, 122–124
Rules of divisibility, 45

S

Salary, 209–210. *See also* Payroll
Salary plus commission, 218–219
Sale price, 298
Sales, earnings based on, 216–219
Sales invoice, 247
Sales tax, 88–89, 94–95
Salvage value, 565, 574
Savings
    in layoff preparations, 205
    planning and budgeting, 646–647
Savings accounts, 384. *See also* Banking
Scale, defined, E-2
Schedule 1 (Interest and Ordinary Dividends for Form 1040A Filers), 528
School district funding, 512. *See also* Students
Schwab, 412, 434
Scientific calculators, B-1
Scrap value, 565
SDI deductions. *See* State disability insurance deductions
Self-employment, 117, 224
Selling. *See also* Inventory
    case studies, 280, 317–318
    markdown, 298–301
    markup, 281–285, 288–294
    price of perishables, 293–294
    terminology, 281
Selling price, 281, 288–294
Semiannual compounding, 375. *See also* Compound interest
Series discounts, 249–252, 257–258
Service charges (banking), 174, 190
Service fees, 172, 184
Shift differentials, 208
Shipping, 247
Short-term liabilities, 623
Simple discount notes. *See also* Simple interest
    discounting, 352–354
    finding bank discount and proceeds, 343
    finding effective interest rate, 346–347
    finding face value, 344–346
    simple interest notes vs., 343
    terms used with, 343
    U.S. Treasury bills, 347–348
Simple interest. *See also* Interest; Simple discount notes
    affecting costs, 324
    calculations of, 373
    case studies, 323, 368–369
    compound interest vs., 324
    exact and ordinary, 328–329
    finding number of days, 326–328
    finding principal, rate, and time, 335–338
    formula for, 139, 374–375
    maturity value, 325–326
    note due dates, 330
    note terminology, 329–330
    solving for, 324–325
Simple interest notes, 329
Single discount equivalents, 257
Single discounts, 249
Sinking funds, 424–427
Sinking funds table, 425–426
Sliding scale commissions, 217–218
Sliding scale discounts, 264
Smart card, 172
Smelter, James, 280
Smoking, statistics by age, 82
Social Security. *See also* Insurance
    finding amount of, 223–224
    history of, 222–223, 243
    multi-country comparison, 224
    in retirement planning, 413
    on W-2 forms, 520
    withholding for, 222
Social Security Act, 204, 243
The Soft Touch, 677
Solutions (equations), 122
Special endorsement, 175, 176
Specific identification method, 306
Split-shift premiums, 208
Stafford loans, 343, 468
Standard deduction, 521
Starbucks, 204, 243
State disability insurance (SDI) deductions, 224–225
Stated rates, 346, 467
State income tax, 233–234, 523
Statistics
    case studies, 644, 677–678
    defined, 654
    frequency distributions, 654–655
    graphs, 655–659
    mean, 666–668
    median, 668–669
    mode, 669–670
Stock
    capital from sale of, 433
    finding current yield on, 435–436
    finding daily prices of, 434–435
    indexes, 437
    in mutual funds, 437–438
    price-earnings ratio on, 436–437
    types of, 433–434
Stockbrokers, 434
Stock certificates, 433
Stock exchanges, 433
Stockholders, 433
Stockholders' equity, 623
Stock ratios, 435–436
Stock tables, 434–435
Stocks, capital from sale of, 433
Stock turnover, 304. *See also* Inventory
Stop-payment order, 182
Straight commission, 216
Straight life insurance, 551
Straight-line depreciation
    other methods vs., 574–575, 582
    overview, 565–568
    tax regulations, 596
Strict monthly budget, 462
Students
    credit card use by, 182, 457
    grade point average calculations, 668
    pay linked to education level, 209, 666–667
    personal finance education for, 377
    school funding for, 512
Subsidized Stafford loans, 468
Substitutions (equations), 122
Subtraction
    of decimals, 25
    of fractions, 48, 50
    of mixed numbers, 57
    verbal expressions involving, 130–131
    of whole numbers, 3, 5
Subtrahend, 5
SUBWAY, 1, 38
Suggested retail price, 247
Sum (addition), 3
Sum-of-the-years'-digits depreciation, 565, 580–582, 596
Suppliers, 247
Supply chain, 247

T

Table of days, 327
Tally marks, 654
Tangible assets, 565
Taxable income, 521
Tax deductions, 521, 523–524
Tax Equity and Fiscal Responsibility Act of 1982, 596
Taxes. *See also* Income taxes; Internal Revenue Service (IRS)
    calculating federal, 229–233, 234
    calculating state, 233–234
    case studies, 510, 559–561
    days worked to pay, 511
    deductible, 523
    depreciation and, 565, 575, 582, 596, 597
    home loan interest reducing, 490
    payroll, 222–224
    property, 511–514
    on retirement accounts, 412
    sales, 88–89, 94–95
Tax preparation, 523
Tax rate schedules, 522
Tax Reform Act of 1984, 596
Tax Reform Act of 1986, 596
T-bills, 347–348, 355–356
Temperature (metric system), A-3–A-4
1099 Interest Income form, 519, 520
Tennis rackets, cost of, 307
Term (of financial products)
    annuities, 408
    defined, 443
    promissory note, 329
Term insurance, 550
Terms (equations), 122
Territorial ratings, 533
Theft
    employee, 188
    identity, 173
    losses from, 523
Time. *See also* Compound interest; Simple interest
    affecting future value, 373
    in metric system, A-1
    in simple discount notes, 343
    in simple interest formula, 324
    solving for, 337–338
Time-and-a-half rate, 206
Time cards, 205
Time deposits, 378
Time rates, 214
Total (in addition), 3
Total balance, 176
Total revenue, 611
Toyota, 433
Trade discount, 249
Trader Joe's, 606
Traditional banking. *See also* Banking
Trailing zeros, 24
Transaction costs, 172, 176
Transaction register, 177
True rate, 346
Truth in Lending Act, 346, 467
Truth in Savings Act of 1991, 384
Turnover at cost, 304
Turnover at retail, 304
200% method, 573

U

Underinsured motorist insurance, 544
Underwater (mortgages), 493, 503–504
Underwriters, 533
Unearned interest, 477–478
Unemployment insurance tax, 235
Unemployment rates, 209, 658–659
Unexpected expenses, 645–647
Uniform Product Codes (UPC), 173
Uninsured motorist insurance, 544
United States Rule, 475–476
Unit price, 247
Units of production, 591
Units-of-production depreciation, 565, 591–593
Universal life insurance, 551–552
Universal product code (UPC), 305
Unlike fractions, 48, 50
Unlike terms, 125
Unpaid balance method, 458–459
Unreimbursed job expenses, 523
Unsubsidized Stafford loans, 468
UPC. *See* Uniform Product Codes; Universal product code
U.S. Treasury bills, 347–348, 355–356
Useful life, 565, 568

V

Vanguard, 412
Variable commission, 217–218
Variable interest rates, 493
Variable life insurance, 551
Variable rate loans, 493
Variables (equations), 122
Variables, choosing dependent and independent, E-1–E-2
Vertical analysis, 616–617, 627
Visa, 184, 457, 458
Volume (metric system), A-1, A-2, A-3

W

Wage bracket method, 229–231
Wages, 218, 519, 520. *See also* Payroll
Wal-Mart, 437
Weekly overtime, 206–208
Weight (metric system), A-1, A-2, A-3
Weighted average, 667–668
Weighted-average method, 306–307
Weighted mean, 667–668
W-4 form, 228–229
Whole life insurance, 551
Whole numbers
  adding, 3–4
  application problems, 14–17
  case studies, 2, 37, 38
  defined, 2
  dividing, 3, 7–8
  dividing decimals by, 29–30
  fractions in operations with, 63
  multiplying, 3, 5–6
  rounding, 2–4
  subtracting, 3, 5
Wholesale, in supply chain, 247
Wiffy, Barbara, 171
Withholding allowance, 228–229
Woodline Moldings and Trim, 74
Word problems, 130–131, 141
Worker's compensation, 537
Wozniak, Steven, 323
W-2 Wage and Tax Statement, 519, 520

X

x-axis, E-2
X (extra) dating, 268–269

Y

y-axis, E-2
Yield to maturity, 444
y-intercept, E-6
Youthful operators, 544–545

Credits

FM
Page xiii: (t) Courtesy of Gary Clendenen; Page xiii: (b) Courtesy of Stanley Salzman; Page xiii: (tl) Valentine Tchoukhonine; Page xiv: Art Vandalay/Photodisc/Getty Images.

Chapter 1
Page 1: Michael Neelon/Alamy Stock Photo; Page 2: Denis Rozhnovsky/Shutterstock; Page 5: Beth Anderson/Pearson Education, Inc; Page 6: Courtesy of Stanley Salzman; Page 7: ZUMA Press Inc/Alamy Stock Photo; Page 10: 97/iStock/Getty Images; Page 11: Michelle Albers/Fotolia; Page 12: (b) Evron.info/Fotolia; Page 12: (t) Mahathir Mohd Yasin/Shutterstock; Page 15: David Sacks/DigitalVision/Getty Images; Page 16: Ajit Solanki/AP Images; Page 17: Beth Anderson/Pearson Education, Inc; Page 18: (c) Beth Anderson/Pearson Education, Inc; Page 18: (b) Miklav/Fotolia; Page 18: (t) Martazmata/Shutterstock; Page 20: Beth Anderson/Pearson Education, Inc; Page 22: (t) Celso Pupo/Fotolia; Page 22: (b) Mohamed Kasim Navfal/123RF; Page 25: Kameleon007/E+/Getty Images; Page 26: Wavebreak Media Ltd/123RF; Page 28: Beth Anderson/Pearson Education, Inc; Page 38: Beth Anderson/Pearson Education, Inc.

Chapter 2
Page 41: (t) Don Smetzer/PhotoEdit; Page 41: (b) BanksPhotos/iStock/Getty Images; Page 43: Elnur/Fotolia; Page 47: Elnur/Fotolia; Page 54: (t) Berc/Fotolia; Page 54: (b) Doug Menuez/Forrester Images/Photodisc/Getty Images; Page 61: Vlad Ivantcov/Fotolia; Page 65: Defun/Fotolia; Page 66: Klotz/123RF; Page 67: Silvano Rebai/Fotolia; Page 75: Photodisc/Getty Images.

Chapter 3
Page 79: (t) Beth Anderson/Pearson Education, Inc; Page 79: (b) Ken Hurst/Fotolia; Page 80: Mr.Markin/Fotolia; Page 81: Goodluz/Fotolia; Page 83: Reuters/Alamy Stock Photo; Page 87: Iriana Shiyan/Fotolia; Page 88: Dennis MacDonald/Alamy Stock Photo; Page 89: Dean Bertoncelj/Shutterstock; Page 91: (c) Arieliona/Shutterstock; Page 91: (t) Chris Selby/Alamy Stock Photo; Page 91: (b) Chris Ryan/Ojo Images/Getty Images; Page 92: Rob Van Petten/Photodisc/Getty Images; Page 93: (t) Adam Gregor/Shutterstock; Page 93: (b) Clarence Holmes Photography/Alamy Stock Photo; Page 95: Aurinko/123RF; Page 97: Courtesy of Gary Clendenen; Page 98: (b) Dimis/Fotolia; Page 98: (t) Paul Hebditch/Alamy Stock Photo; Page 98: (c) Pbpgalleries/Alamy Stock Photo; Page 101: Jon Lord/Alamy Stock Photo; Page 102: LCpl Michael Dye/U.S. Marine Corps; Page 103: (t) Dmitri Maruta/Fotolia; Page 103: (b) Sielan/Fotolia; Page 104: (b) Richard Ransier/Corbis/VCG/Getty Images; Page 104: (c) Alexander/Fotolia; Page 105: Terry Vine/Photodisc/Getty Images; Page 106: Courtesy of Gary Clendenen; Page 109: Cosmin4000/iStock/Getty Images; Page 110: Love Silhouette/Shutterstock; Page 112: (b) Maksymowicz/Fotolia; Page 112: (t) 3d Brained/Shutterstock; Page 113: (t) David J. Green/Lifestyle 2/Alamy Stock Photo; Page 113: (b) Nikita Tv/Shutterstock; Page 114: Hypervision Creative/Shutterstock; Page 118: Beth Anderson/Pearson Education, Inc; Page 119: Craig Lassig/EFE News Agency/Alamy Stock Photo.

Chapter 4
Page 121: Ed Aldridge/Shutterstock; Page 132: Raymond Boyd/Michael Ochs Archives/Getty Images; Page 134: Tetiana Butovchenko/123RF; Page 137: Dieter76/Fotolia; Page 141: Maxal Tamor/Fotolia; Page 144: (l) Marcovarro/Fotolia; Page 144: (r) Courtesy of Gary Clendenen; Page 145: (t) Courtesy of Gary Clendenen; Page 145: (b) Milkos/123RF; Page 148: Jonathan Weiss/Shutterstock; Page 150: Ian O'Leary/Dorling Kindersley, Ltd.; Page 151: Fdsmsoft/Fotolia; Page 153: Vladimir Galkin/123RF; Page 154: (b) Courtesy of Gary Clendenen; Page 154: (t) Erikreis/iStock/Getty Images; Page 155: (b) Bluesun/Fotolia; Page 155: (t) Anibal Trejo/Shutterstock; Page 156: Courtesy of Gary Clendenen; Page 159: SW Productions/Photodisc/Getty Images; Page 160: Kristoffer Tripplaar/Alamy Stock Photo.

Chapter 5
Page 171: Hannu Viitanen/123RF; Page 173: Sergey Yechikov/Shutterstock; Page 180: Yuhorakushin/123RF; Page 183: Comstock Images/Stockbyte/Getty Images; Page 184: Yuri Arcurs/E+/Getty images; Page 185: (t) PJPhoto69/Fotolia; Page 185: (b) Julietphotography/Shutterstock; Page 186: (t) LHB Photo/Alamy Stock Photo; Page 186: (b) Kali9/E+/Getty Images; Page 199: Gemenacom/Fotolia; Page 200: Jupiterimages/Photos.com/Getty Images.

Chapter 6
Page 204: Beth Anderson/Pearson Education, Inc; Page 206: (t) DrRave/E+/Getty Images; Page 206: (b) Marty Bicek/Modesto Bee/ZUMA Press Inc/Alamy Stock Photo; Page 212: Adam Gault/Ojo Images/Getty Images; Page 215: Jupiterimages/Stockbyte/Getty Images; Page 217: (b) Klikk/iStock/Getty Images; Page 217: (t) Iriana88w/123RF; Page 221: Christophe Testi/Shutterstock; Page 222: Sueddeutsche Zeitung Photo/Scherl/Alamy Stock Photo; Page 226: Pio3/Shutterstock; Page 228: Arvind Balaraman/Fotolia; Page 234: Michael Jung/Shutterstock; Page 235: Internal Revenue Service; Page 238: VILevi/Fotolia; Page 243: Silver Burdett Ginn/Pearson Education, Inc.

Chapter 7
Page 246: RosaIreneBetancourt 6/Alamy Stock Photo; Page 249: (b) Alex Hinds/Shutterstock; Page 249: (t) Dionisio Dias Filho/123RF; Page 250: Jeremy O'Donnell/Getty Images; Page 254: (b) Iofoto/Shutterstock; Page 254: (t) Wavebreakmedia/Shutterstock; Page 255: (t) Olga_Sweet/Yay Media As/Alamy Stock Photo; Page 255: (b) Serezniy/123RF; Page 257: Hxdbzxy/Shutterstock; Page 258: Bartkowski/Fotolia; Page 259: Bfk/Shutterstock; Page 262: Dean Bertoncelj/Shutterstock; Page 264: (b) Leszek Kobusinski/Shutterstock; Page 264: (t) Robyn Mackenzie/Shutterstock; Page 265: Meunierd/Shutterstock; Page 266: CaroleGomez/iStock/Getty Images; Page 267: Joshua Resnick/123RF; Page 268: JPstock/Shutterstock; Page 269: Maria Dryfhout/Shutterstock; Page 270: PaulCowan/iStock/Getty Images; Page 272: (b) Brand X Pictures/Stockbyte/Getty Images; Page 272: (t) Scanrail/123RF; Page 272: (c) Muzsy/Shutterstock; Page 273: German S/Shutterstock; Page 276: Anthony Harvey/Getty Images; Page 277: Sean Pavone/Shutterstock.

Chapter 8

Page 280: Chris Cross/Caiaimage/Getty Images; Page 281: Gorillaimages/Shutterstock; Page 282: (b) Ken Gillespie Photography/Alamy Stock Photo; Page 282: (t) Sportpoint/iStock Editorial/Getty Images; Page 283: (t) StockphotoVideo/Shutterstock; Page 283: (b) George Dolgikh/Shutterstock; Page 284: (t) Photofollies/Shutterstock; Page 284: (b) Sander Van Der Werf/Shutterstock; Page 286: (b) Parcan/Shutterstock; Page 286: (t) Paul Maguire/123RF; Page 287: (b) Bhartia/iStock/Getty Images; Page 287: (t) Grzegorz Czapski/Shutterstock; Page 288: Radu Razvan/Shutterstock; Page 289: Mercury Green/Shutterstock; Page 290: Elliot Burlingham/123RF; Page 291: (t) Nerthuz/123RF; Page 291: (b) Maksim Shebeko/123RF; Page 293: Steven Cukrov/123RF; Page 295: (t) Kaband/Shutterstock; Page 295: (b) Vima/Shutterstock; Page 296: (t) Kirstypargeter/iStock/Getty Images; Page 296: (c) Ronen/123RF; Page 296: (b) Africa Studio/Shutterstock; Page 297: (b) Comstock/Stockbyte/Getty Images; Page 297: (t) Tspider/123RF; Page 298: Db238/Fotolia; Page 299: (t) Stockbyte/Getty Images; Page 299: (b) Ruslan Kudrin/Shutterstock; Page 300: Palto/Shutterstock; Page 302: (t) Jupiterimages/Creatas/Getty Images; Page 302: (b) Kraig Scarbinsky/Photodisc/Getty Images; Page 303: Geptays/Shutterstock; Page 305: (t) Joshhh/Fotolia; Page 305: (c) Sufi/Shutterstock; Page 305: (b) Huseyinbas/Shutterstock; Page 306: Maygutyak/Fotolia; Page 308: Scanrail/123RF; Page 311: Frances L Fruit/Shutterstock; Page 312: Sergey Peterman/123RF; Page 317: (l) Ultimathule/Shutterstock; Page 317: (r) 2Happy/Shutterstock; Page 318: Tomas Marek/123RF; Page 322: Gradt/Fotolia.

Chapter 9

Page 323: (b) DMstudio House/Shutterstock; Page 323: (t) Jupiter Images/Stockbyte/Getty Images; Page 325: Nick White/Cultura/Getty Images; Page 329: Courtesy of Stanley Salzman; Page 332: Portadown/Shutterstock; Page 333: (b) Adisa/Fotolia; Page 333: (t) Johan Larson/123RF; Page 334: Comaniciu Dan/Shutterstock; Page 335: Lev/Fotolia; Page 337: WDG Photo/Shutterstock; Page 338: Berc/Fotolia; Page 340: Inhauscreative/E+/Getty Images; Page 341: (b) Nikolai Sorokin/Fotolia; Page 341: (t) Keneva Photography/Shutterstock; Page 342: Pyzata/Fotolia; Page 344: Leonard Zhukovsky/Shutterstock; Page 345: Core Imagery/Alamy Stock Photo; Page 349: Margrit Hirsch/Shutterstock; Page 350: Shirley/Fotolia; Page 352: Dalibor Sevaljevic/123RF; Page 354: Kar Tr/Shutterstock; Page 355: Bankerwin/Shutterstock; Page 358: Anirut Rassameesritrakool/123RF; Page 360: Chris Driscoll/Shutterstock; Page 361: Lorado/iStock/Getty Images; Page 362: (b) Jarp5/123RF; Page 362: (t) Purino/Shutterstock; Page 368: Jonathan Weiss/Shutterstock; Page 369: Neil Fraser/Alamy Stock Photo.

Chapter 10

Page 372: (l) Everett Historical/Shutterstock; Page 372: (r) BravoKiloVideo/Shutterstock; Page 373: Harris & Ewing Collection/Library of Congress Prints and Photographs Division [LC-DIG-hec-31012]; Page 374: Hongqi Zhang/123RF; Page 375: Ed Aldridge/Shutterstock; Page 379: Philip Date/Hemera/Getty Images; Page 381: Tupungato/Shutterstock; Page 385: Rob Marmion/123RF; Page 391: Ian Pilbeam/Alamy Stock Photo; Page 392: (b) Andres Rodriguez/Fotolia; Page 392: (t) Laurentiu Lordache/123RF; Page 394: Cathy Yeulet/123RF; Page 396: Qingqing/Shutterstock; Page 397: Robh/E+/Getty Images; Page 401: Kevin Britland/Alamy Stock Photo ; Page 402: TA Images/Alamy Stock Photo.

Chapter 11

Page 407: MBI/Alamy Stock Photo; Page 408: Art Konovalov/Shuterstock; Page 409: UpperCut Images/Getty Images; Page 410: Wavebreakmedia/Shutterstock; Page 415: Sasacvetkovic33/iStock/Getty Images; Page 418: Fotoinfot/Shutterstock; Page 421: (b) Nano/E+/Getty Images; Page 421: (t) Kurhan/Shutterstock; Page 422: Deshakalyan Chowdhury/AFP/Getty Images; Page 424: (t) Guillermo Lobo/iStockphoto360/Getty Images; Page 424: (b) Snehit/123RF; Page 426: Comstock Images/Stockbyte/Getty Images; Page 428: (c) Jason Poston/Shutterstock; Page 428: (t) Anna G/Shutterstock; Page 428: (b) Ivonne Wierink/123RF; Page 431: (t) Kenneth Summers/Shutterstock; Page 431: (b) Pack-Shot/Shutterstock; Page 432: Nigel Hicks/Alamy Stock Photo; Page 436: Peter Etchells/Shutterstock; Page 440: AlenKadr/Shutterstock; Page 444: Monkey Business Images/Shutterstock; Page 447: Itonggg/Shutterstock; Page 452: Imtmphoto/Shutterstock; Page 453: Ken Wolter/Shutterstock; Page 455: RosaIreneBetancourt 11/Alamy Stock Photo.

Chapter 12

Page 456: Maxfx/Fotolia; Page 459: Cobalt/123RF; Page 461: Otomobil/Shutterstock; Page 462: Wavebreak Media Ltd/123RF; Page 464: Ipixguy/Shutterstock; Page 465: Jupiterimages/Stockbyte/Getty Images; Page 466: Zina Seletskaya/Shutterstock; Page 467: Ed Aldridge/Shutterstock; Page 468: Africa Studio/Shutterstock; Page 470: Dick Luria/Photodisc/Getty Images; Page 472: Bomshtein/Shutterstock; Page 473: Msujan/Shutterstock; Page 474: Dimasobko/Fotolia; Page 475: Keith Brofsky/Photodisc/Getty Images; Page 476: Branex/Fotolia; Page 478: Stephen Coburn/Shutterstock; Page 479: Manfredxy/Shutterstock; Page 480: (b) Tad Denson/Shutterstock; Page 480: (t) Loza-Koza/Shutterstock; Page 481: Yuri Bizgajmer/123RF; Page 483: Sascha Burkard/Fotolia; Page 485: Bartlomiej Magierowski/Alamy Stock Photo; Page 487: Skaljac/Fotolia; Page 489: Pietro Scozzari/AGE Fotostock/Alamy Stock Photo; Page 492: Lindasj22/Shutterstock; Page 493: Valentine Tchoukhonine; Page 501: Andres Rodriguez/Fotolia; Page 503: Tupungato/Shutterstock.

Chapter 13

Page 510: David Coll Blanco/Alamy Stock Photo; Page 511: Everett Collection Inc/Alamy Stock Photo; Page 513: Ariel Skelley/Digital Vision/Getty Images; Page 514: Ermejoncqc/Shutterstock; Page 516: (t) Matthew Lees Dixon/iStock/Getty Images; Page 516: (c) Kevin Foy/Alamy Stock Photo; Page 516: (b) T.M.O.Buildings/Alamy Stock Photo; Page 523: 2xSamara.com/Shutterstock; Page 524: Pavel L Photo and Video/Shutterstock; Page 531: (b) Farina3000/Fotolia; Page 531: (t) Video1/E+/Getty Images; Page 533: Dustie/Shutterstock; Page 536: Ian Marlow/Alamy Stock Photo; Page 537: (r) Lisa S/Shutterstock; Page 537: (l) Monkey Business Images/Shutterstock; Page 541: Tom Wang/Shutterstock; Page 543: Scott Leman/Fotolia; Page 547: Courtesy of Stanley Salzman; Page 548: AnKudi/Fotolia; Page 553: H. Tuller/Shutterstock; Page 555: EchOlav/Shutterstock; Page 559: George Doyle/Stockbyte/Getty Images; Page 560: David Coll Blanco/Alamy Stock Photo; Page 563: Kunpisit/Shutterstock.

Chapter 14

Page 564: Courtesy of Stanley Salzman; Page 566: Araya Pacharabandit/123RF; Page 568: (t) Art_zzz/Fotolia; Page 568: (b) Petr Nad/Shutterstock; Page 570: Courtesy of Stanley Salzman; Page 572: Eyeidea/Shutterstock;

Page 577: Bernard Weil/ZUMA Press/Newscom; Page 578: Antoniodiaz/Shutterstock; Page 581: Andre Babiak/Alamy Stock Photo; Page 585: Irabel8/Shutterstock; Page 588: (t) Vladimir Wrangel/Shutterstock; Page 588: (b) RosaIreneBetancourt 12/Alamy Stock Photo; Page 589: Photoiron/Fotolia; Page 591: Zuchero/Fotolia; Page 595: Erwin Purnomo Sidi/123RF; Page 602: Ribeiroantonio/Shutterstock; Page 603: Courtesy of Stanley Salzman; Page 607: Jim West/Alamy Stock Photo.

Chapter 15

Page 610: DMstudio House/Shutterstock; Page 613: Autumn's Memories/Shutterstock; Page 614: BCFC/Shutterstock; Page 620: (b) Vetkit/Fotolia; Page 620: (t) Lisa F. Young/Shutterstock; Page 620: (c) Rich Carey/Shutterstock; Page 621: Ermess/Shutterstock; Page 623: Alessio Ponti/Fotolia; Page 631: Alexey Stiop/Shutterstock; Page 633: Gwright/Alamy Stock Photo; Page 638: Michael Blann/DigitalVision/Getty Images; Page 640: Anton Ivanov/Shutterstock; Page 642: Bernard Van Dierendonck/Look Die Bildagentur der Fotografen GmbH/Alamy Stock Photo; Page 643: Deligodin Evgeny/Shutterstock.

Chapter 16

Page 644: Monkeybusinessimages/iStock/Getty Images; Page 655: Monkey Business Images/Shutterstock; Page 662: Kzenon/Shutterstock; Page 666: Jupiterimages/Photos.com/Getty Images; Page 670: Courtesy of Gary Clendenen; Page 677: James Brunker/Alamy Stock Photo; Page 678: Matt Sayles/Invision for Breeders' Cup/AP Images; Page 681: Andersen Ross/Photodisc/Getty Images.

Appendix B

Page B8: Tom Strattman/Getty Images.